SYNTHETIC LUBRICANTS AND HIGH-PERFORMANCE FUNCTIONAL FLUIDS

edited by
Ronald L. Shubkin
Ethyl Corporation
Baton Rouge, Louisiana

Marcel Dekker, Inc. New York • Basel • Hong Kong

Library of Congress Cataloging-in-Publication Data

Synthetic lubricants and high-performance functional fluids / edited by
Ronald L. Shubkin.
p. cm. -- (Chemical industries ; 48)
Includes bibliographical references and index.
ISBN 0-8247-8715-3
1. Synthetic lubricants. 2. Hydraulic fluids. I. Shubkin,
Ronald L. II. Series.
TJ1077.S96 1992
621.8'9--dc20 92-25564
CIP

This book is printed on acid-free paper.

Marcel Dekker, Inc.
270 Madison Avenue, New York, New York 10016

Current printing (last digit):
10 9 8 7 6 5 4 3 2

PRINTED IN THE UNITED STATES OF AMERICA

SYNTHETIC LUBRICANTS AND HIGH-PERFORMANCE FUNCTIONAL FLUIDS

CHEMICAL INDUSTRIES

A Series of Reference Books and Textbooks

Consulting Editor

HEINZ HEINEMANN
Berkeley, California

Preface

While in many applications the primary function of a synthetic fluid is lubrication, it is by no means the only function. Heat transfer, power transmission, electrical insulation, and corrosion inhibition are only a few of the other tasks that fluids may be expected to perform. In many instances the requirements of the tasks exceed the performance capabilities of natural fluids, and the synthesis of new products to meet the extraordinary demands is necessary. In other cases, synthetic fluids have proven to be more cost-effective than natural products in meeting the requirements of a given application. In still other cases, environmental or toxicological considerations have mandated the use of synthetic functional fluids, which are relatively nontoxic in comparison to many naturally derived fluids. Synthetic fluids, designed to provide performance characteristics above and beyond those available from fluids derived from most natural sources, are the subject of this book.

The intention of this book is to provide a one-volume overview of the various types and uses of synthetic functional fluids currently available. The first section of the book is devoted to the different fluids. Each chapter is focused on a particular class of compounds and is authored by a recognized expert in the field. The fluids covered include all the well-established commercial products as well as some experimental fluids that have the potential of becoming commercialized in the near future. To the extent possible, each chapter includes a discussion of:

Historical development
Chemistry
Property and performance characteristics
Manufacture, marketing, and economics
Outlook

The second section of the book deals with applications. These chapters focus on specific application areas for which synthetic fluids are either currently used or being seriously considered for use. Again, the authors are recognized experts in the application area. The chapters cover:

Historical development
Synthetic fluids appropriate to application
Comparative performance data
Current commercial practice

The third and last section of the book deals with long-range trends in the use of synthetic fluids in major industries or areas of interest. The authors of these chapters have been chosen because of their ability to take a long-term view. They have been asked to consider not only where their field of interest is today, but where it is headed in the future. They have in effect been asked to become prognosticators of future developments in synthetic functional fluids. Some chapters in this section of the book deal with industrial markets, whereas other chapters deal with economic and environmental trends. The chapters on industrial markets deal with:

Current equipment and fluids
Developmental equipment and fluids
Long-range trends

I am particularly pleased to be able to conclude this book with a chapter on the environmental impact associated with the use of synthetics. There are two critical areas of concern. The first involves the conservation of our planet's limited supply of natural resources. Synthetics extend these assets not only by simple replacement but, in many applications, by dramatically increasing service lifetimes. The second area of environmental concern involves toxicity and biodegradability. In both areas, synthetics are becoming recognized as environmentally responsible solutions to a sensitive issue.

I wish to extend my gratitude to all of my colleagues and friends who encouraged and supported my efforts to bring this work together. Thanks also to Ethyl Corporation, the management of Ethyl's Research and Development Division, and the Ethyl Industrial Chemicals Division for their support in this undertaking. I especially wish to thank those people in Ethyl's Research and Developmental Department with whom I have had the good fortune to work on the PAO project. Special thanks also goes to Ms. Norma Delaune, who helped me in handling the correspondence and in preparation of the tables and graphs. Finally, a word of thanks to my wife, Swee—without her encouragement and understanding I could never have finished the project.

Ronald L. Shubkin

Introduction

In the most fundamental sense, a lubricant may be defined as a substance that has the ability to reduce friction between two solid surfaces rubbed against each other. The use of natural products to achieve this effect dates to antiquity. Art decorations on the inner wall of the Egyptian tomb of Tehuti-Hetep (ca. 1650 B.C.) indicate that olive oil on wooden planks was used to facilitate the sliding of large stones, statues, and building materials. Egyptian chariots dating to 1400 B.C. have been uncovered that have small amounts of greasy materials, presumed to be either beef or mutton tallow, on the axles. In addition to animal fat and vegetable oil, petroleum products have been used for lubrication in a primitive fashion for more than a millennium. In fact, Herodotus (484–424 B.C.) described methods of producing bitumen and a lighter oil from petroleum.

The first synthetic hydrocarbon oils were produced by the prominent chemists Charles Friedel and James Mason Crafts in 1877. Standard Oil Company of Indiana commercialized a synthetic hydrocarbon oil in 1929, but it was unsuccessful because of a lack of demand. However, the onset of World War II and the subsequent shortages of petroleum feedstocks in Germany, France, and Japan revitalized interest in synthetic lubricants. Moreover, the German disaster at the Battle of Stalingrad in 1942 demonstrated the inadequacy of petroleum products in extremely cold weather. The lubricants used in tanks, aircraft, and other military vehicles gelled, and the engines used in the vehicles could not be started. An intense German research effort to find alternative lubricants ensued, which led to the first manufacture of synthetic products by olefin polymerization.

Interest in ester-based lubricants appears to date back to the Zurich Aviation Congress in 1937. Triesters were later synthesized from fatty acids and trimethanolethane. These materials performed so well that more than 3500 esters were prepared and evaluated in Germany between 1938 and 1944. Meanwhile, the first diester base stocks were developed in the United States at the Naval Research Laboratory between 1942 and 1945. The British began using esters in 1947 as lubricants for turboprop aircraft where conventional mineral oils failed to give satisfactory performance at high temperatures. With the introduction of jet engines the problem became even more complex because of the necessity of providing lubrication at very low temperatures to the front of the engine and at very high temperatures to the rear. The need for fire-resistant hydraulic fluids in aircraft further spurred the development of synthetics.

From the end of World War II until the mid-1970s, various synthetic fluids were developed to meet the ever-increasing demands of the newer and more efficient high-performance engines and machines being developed. The use of synthetics, however, was limited to those applications where the required performance characteristics could not be met by petroleum-based products.

The oil embargo of 1974 and the subsequent escalation of petroleum prices brought a new and urgent need for the conservation of oil reserves and the development of alternative raw materials. By early 1987, the pressures on the world petroleum supply had abated, but the lessons of the late 1970s had not been forgotten. The increased emphasis on cost-effectiveness had a twofold effect on the growth of synthetics. First, the fluids

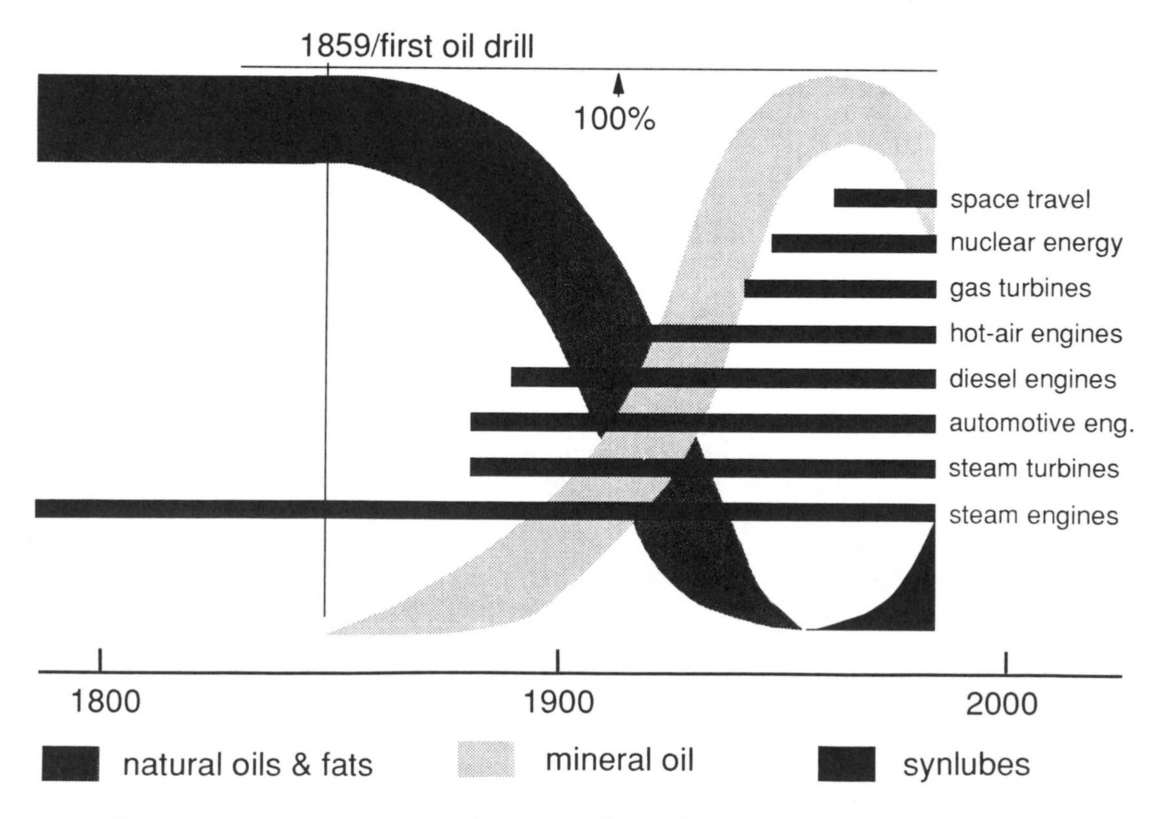

Figure 1 History of lubrication. (Courtesy of G. van der Waal, Unichema International.)

themselves were found to offer advantages of longer drain intervals, less down-time, and greater fuel efficiency. Second, a new generation of engines were being developed that required fluid performance characteristics that were becoming increasingly difficult to achieve with mineral oil-based products.

Finally, the 1990s brought a new urgency to the concept of environmental responsibility. This development has led to an increased need for functional fluids that are both biodegradable and low in toxicity.

Figure 1 is a graphical representation of the history of lubrication. Overlaying the chart showing the relative consumption of natural oils and fats, mineral oils, and synthetic lubricants are time lines showing the introduction and continued use of various types of equipment requiring lubrication. The total consumption of lubricants is normalized to 100% at any particular point in time. It may be seen that prior to 1859, lubrication functions were performed by natural oils and fats. Starting soon after the drilling of the first oil well in 1859, mineral oils began to displace the natural oils and fats. By the 1970s, mineral oils had essentially displaced natural oils and fats, but synthetics were starting to come into use. As we begin the 1990s, synthetics still represent a low percentage of the total, but the graph certainly has the same appearance as in the late 1800s. The idea that synthetic lubricants could displace mineral oils seems farfetched, and perhaps it is, but estimates for worldwide crude oil reserves indicate that we may run out of crude oil sometime in the next century.

Ronald L. Shubkin

Contents

Part III. TRENDS

Contributors

Wilfried J. Bartz *Technische Akademie Esslingen, Ostfildern, Germany*

Paul A. Bessette *William F. Nye Company, New Bedford, Massachusetts*

Mary Jo Bieberich *Naval Surface Warfare Center, Annapolis, Maryland*

William L. Brown *Union Carbide Chemicals and Plastics Company, Inc., Tarrytown, New York*

Edward W. Casserly *Pennzoil Products Company, The Woodlands, Texas*

Douglas R. Chrisope *Ethyl Petroleum Additives, Inc., St. Louis, Missouri*

Nye A. Clinton *Union Carbide Chemicals and Plastics Company, Inc., Tarrytown, New York*

John M. Collins *Ethyl Corporation, Southfield, Michigan*

Raymond B. Dawson *Ethyl Corporation, Baton Rouge, Louisiana*

Thomas W. Del Pesco *E. I. du Pont de Nemours and Co., Inc., Deepwater, New Jersey*

Donna H. Demby *GE Silicones, Waterford, New York*

Hans Dressler *Indspec Chemical Corporation, Pittsburgh, Pennsylvania*

Brad F. Droy* *Ethyl Corporation, Baton Rouge, Louisiana*

Giuseppe Fisicaro *AgipPetroli, Rome, Italy*

John D. Fotheringham *BP Chemicals, Ltd., Grangemouth, Stirlingshire, Scotland*

Giampaolo Gerbaz *AgipPetroli, Rome, Italy*

Allen Gross *GE Silicones, Overland Park, Kansas*

Lois J. Gschwender *Air Force Wright Laboratory, Wright-Patterson Air Force Base, Ohio*

Philip S. Korosec *Ethyl Corporation, Baton Rouge, Louisiana*

James F. Landry *Castrol Specialty Products, Irvine, California*

Michael P. Marino *FMC Corporation, Philadelphia, Pennsylvania*

Paul L. Matlock *Union Carbide Chemicals and Plastics Company, Inc., Tarrytown, New York*

J. William Miller *CPI Engineering Services, Inc., Midland, Michigan*

Alan J. Mills *Castrol International, Ltd., Pangbourne, Berkshire, England*

Gunner E. Nelson *Ethyl Corporation, Baton Rouge, Louisiana*

Andrew G. Papay *Ethyl Petroleum Additives, Inc., St. Louis, Missouri*

F. Alexander Pettigrew *Ethyl Corporation, Baton Rouge, Louisiana*

Charles Platteau *Ethyl S.A., Brussels, Belgium*

Steven James Randles *ICI Chemicals and Polymers, Ltd., Middlesbrough, Cleveland, England*

Dale A. Ruesch *Halocarbon Products Corporation, River Edge, New Jersey*

Glenn D. Short *CPI Engineering Services, Inc., Midland, Michigan*

Ronald L. Shubkin *Ethyl Corporation, Baton Rouge, Louisiana*

Robert E. Singler *Army Materials Technology Laboratory, Watertown, Massachusetts*

Carl E. Snyder, Jr. *Air Force Wright Laboratory, Wright-Patterson Air Force Base, Ohio*

Stanley J. Stoklosa *GE Silicones, Waterford, New York*

Clifford G. Venier *Pennzoil Products Company, The Woodlands, Texas*

E. Ian Williamson *The College of Petroleum and Energy Studies, Oxford, England*

---

*Current affiliation: *Woodward-Clyde Consultants, Baton Rouge, Louisiana*

SYNTHETIC LUBRICANTS AND HIGH-PERFORMANCE FUNCTIONAL FLUIDS

1
Polyalphaolefins

Ronald L. Shubkin
Ethyl Corporation
Baton Rouge, Louisiana

I. INTRODUCTION

Saturated olefin oligomers are a class of synthetic high-performance functional fluids that have been developed to meet the increasingly stringent demands being placed on today's working fluids. The term polyalphaolefin, or PAO, is commonly used to designate such fluids, and that designation will be used in this chapter. The term PAO was first used by Gulf Oil Company (later acquired by Chevron), but it has now become an accepted generic appellation for hydrocarbons manufactured by the catalytic oligomerization (polymerization to low-molecular-weight products) of linear α-olefins having six or more (usually 10) carbon atoms (1).

Technological advances are often accompanied by a variety of problems and complications not previously anticipated. Advances in the function and efficient operation of modern machines and engines have brought new challenges relating to the satisfactory use and performance of existing functional fluids. Among these challenges are:

Operation under increasingly severe conditions.
The need for more cost-effective and hence competitive operations.
The need to lessen dependence on the availability of crude oil stocks.
The specialized performance requirements of emerging end-use applications.
The necessity of accounting for the critically important, but long-ignored, toxicological and biodegradable characteristics of the fluids being used.

Today, mineral oil base stocks are being refined to give products that are certainly superior to those available only a few years ago. But the limits to which mineral oils can be economically refined are being strained. In order to satisfactorily address the challenge of solving the problems noted here, industry is turning to synthetic alternatives.

PAOs are gaining rapid acceptance as high-performance lubricants and functional fluids because they exhibit certain inherent, and highly desirable, characteristics (1). Among these favorable properties are:

A wide operational temperature range.
Good viscometrics (high viscosity index).
Thermal stability.
Oxidative stability.
Hydrolytic stability.
Shear stability.
Low corrosivity.
Compatibility with mineral oils.
Compatibility with various materials of construction.
Low toxicity.
Manufacturing flexibility that allows "tailoring" products to specific end-use application requirements.

II. HISTORICAL DEVELOPMENT

A. Technical

Synthetic oils consisting only of hydrocarbon molecules were first produced by the prominent chemists Charles Friedel and James Mason Crafts in 1877 (2). Standard Oil Company of Indiana attempted to commercialize a synthetic hydrocarbon oil in 1929 but was unsuccessful because of a lack of demand. In 1931, Standard Oil, in a paper by Sullivan et al., disclosed a process for the polymerization of olefins to form liquid products (3). These workers employed cationic polymerization catalysts such as $AlCl_3$ to polymerize olefin mixtures obtained from the thermal cracking of wax. At about the same time that the work at Standard Oil was being carried out, H. Zorn of I. G. Farben Industries independently discovered the same process (4).

The first use of a linear α-olefin to synthesize an oil was disclosed by Montgomery et al. in a patent issued to Gulf Oil Company in 1951 (5). $AlCl_3$ was used in these experiments, as it was in the earlier work with olefins from cracked wax.

The use of free-radical initiators as α-olefin oligomerization catalysts was first patented by Garwood of Socony-Mobil in 1960 (6). Coordination complex catalysts, such as the ethylaluminum sesquichloride/titanium tetrachloride system, were disclosed in a patent issued to Southern et al. at Shell Research in 1961 (7).

The fluids produced by the various catalyst systems described above contained oligomers with a wide range of molecular weights. The compositions and internal structures of these fluids resulted in viscosity/temperature characteristics that gave them no particular advantage over the readily available and significantly less expensive mineral oils of the day.

In 1968, Brennan at Mobil Oil patented a process for the oligomerization of α-olefins using a BF_3 catalyst system (8). Prior to that time, BF_3 catalysis had given irreproducible results. Brennan showed that the reaction could be controlled if two streams of olefins were mixed in the reactor. The first stream contained the olefin plus a BF_3 • ROH complex, where ROH is an alcohol. The second stream contained the olefin saturated with gaseous BF_3. Of particular interest was the fact that this catalyst system produced a product consisting of a mixture of oligomers that was markedly peaked at the trimer.

Shubkin of Ethyl Corporation showed that H_2O (9), as well as other protic cocatalysts such as alcohols and carboxylic acids (10), could be used in conjunction with BF_3 to produce oligomers of uniform quality. The experimental technique employed a molar excess of BF_3 in relation to the cocatalyst. The excess was achieved by sparging the reaction medium with BF_3 gas throughout the course of the reaction or by conducting the reaction under a slight pressure of BF_3. These studies showed that the oligomerization products exhibited pour points that were well below those anticipated for such compounds, even when dimeric products were allowed to remain in the final mixture. The molecular structure of the dimer was believed to consist of a straight carbon chain containing a single methyl group near the middle. Such branched structures were known to exhibit relatively high pour points. More pertinent to the current subject, these were the first patents to address the potential importance of PAOs derived from BF_3 • ROH catalyst systems as synthetic lubricants. Shubkin et al. later showed that the unique low-temperature properties could be attributed to a high degree of branching in the molecular structure (11).

B. Commercial

The commercial development of PAO fluids as lubricants and high-performance functional fluids began in the early 1970s, but significant growth in markets and in the variety of end-use applications did not begin until the latter part of the 1980s. During this time, a handful of companies played significant roles with both research and development (R&D) and market development efforts (12).

1. Mobil Oil Corporation

Mobil Oil Corporation was the first company to introduce a PAO-based synthetic lubricant. In 1973, Mobil began marketing a synthetic motor oil for use in automotive engines in overseas markets. Circulating oils and gear oils were added to the Mobil line in 1974. The first U.S. test marketing of Mobil 1 synthesized engine lubricant began in the autumn of 1974. The test was expanded to eight cities in September 1975, and to all Mobil marketing areas in April 1976. Mobil 1 was initially an SAE 5W–20 product, but it was later replaced by a 5W–30 fluid based on PAO and a neopentyl polyol ester. The polyol ester improved additive solubility and increased seal swell.

Mobil's product distribution was extended to Canada, Japan, and several European countries in 1977. Also in 1977, Mobil introduced Delvac 1, a PAO-based product aimed at the truck fleet market. Mobil also pioneered PAO-based industrial lubricants with its line of Mobil SHC products.

Mobil's PAO plant in the United States has an estimated annual capacity of 25,000 metric tons. A new plant at Notre Dame de Gravenchon, France, reportedly has an annual capacity of 13,000 metric tons. Mobil purchases 1-decene for its PAO production.

In addition to the low-viscosity PAOs, Mobil also produces two grades of high-viscosity PAO. The annual sales for these products is believed to be around 4,000 metric tons.

2. Gulf Oil Company

Gulf Oil Company appears to have had an interest in synthetic hydrocarbons in the 1940s. Developmental work at the Gulf laboratories in Harmarville, Pennsylvania, continued into the 1960s and 1970s. In 1974, Gulf built a semi-works plant with a capacity of 1,125 metric tons per year. The first commercial sale from this plant was in December 1974.

During the years 1976–1980, Gulf introduced an arctic super duty 5W–20 CD/SE crankcase lubricant plus an arctic universal oil/transmission oil. In Canada, Gulf began marketing PAO-based gear lubricants, synthetic greases, and a partial synthetic 5W–30 crankcase oil.

Gulf began commercial production in the PAO plant in Cedar Bayou, Texas, in December 1980. The initial production capacity was 15,400 metric tons per year, and the facility was strategically located next to Gulf's olefin plant. In 1981–1983, Gulf added several new PAO-based products to their line of synthetic fluids. These included Gulf Super Duty II, a full synthetic 0W–30 crankcase oil, Gulf SL-H, a hydraulic fluid for high- and low-temperature operation, and Gulf Syngear, a 75W–90 gear oil for long life and fuel economy. In addition to their fully formulated products, Gulf marketed PAO to the merchant market under the tradename Synfluid Synthetic Fluids. Gulf Oil Corporation was acquired by Chevron Corporation in 1984.

3. Chevron Corporation

Prior to 1984, Chevron marketed a single synlube-based product. That product was Chevron Sub Zero Fluid, a 7.5W–20 CD/SE crankcase oil for use in construction equipment and vehicles employed in the Alyeska pipeline project in Alaska. In June 1984, Chevron acquired Gulf Oil Company. In late 1985, the PAO manufacturing and marketing responsibilities were transferred to the Oronite Division of Chevron Chemical. Chevron continued to offer the PAO-based arctic oil plus Chevron Tegra PAO-based synthetic lubricants, which included the old Gulf Syngear and three grades of compressor oils. Unlike Mobil, which chose to market aggressively under its own name, Chevron decided to focus on the merchant market. The capacity of the Chevron plant has been increased to approximately 24,000 metric tons.

Chevron, like Ethyl Corporation, but unlike Mobil, is basic in the α-olefin raw material.

4. Amoco

Amoco, formerly Standard Oil Company (Indiana), was probably the first U.S. petroleum company to investigate synthetic hydrocarbon fluids. The pioneering work by F. W. Sullivan in the early 1930s has already been mentioned (3). Those efforts led to a patent that described the aluminum chloride-catalyzed polymerization of olefins derived from cracked wax (13). An attempt to commercialize a synthetic lubricating fluid in 1929 was abandoned because of lack of demand.

In 1982, Amoco Oil Company began test marketing a 100% PAO-based lubricant. This venture was followed in April 1984 with the introduction of Amoco's Ultmate line of crankcase oils for both gasoline and diesel oils. Amoco later expanded the product line to include gear oils and grease bases. All of the PAO for the Ultmate products are purchased.

5. Ethyl Corporation

In 1970, Ethyl began conducting research on a process for the polymerization of linear α-olefins to form low-viscosity functional fluids. The concept was attractive since Ethyl was (and continues to be) one of the world's largest manufacturers of linear α-olefins. The target application was a hydraulic fluid specification for military jet aircraft. As it turned out, the specifications were written around an experimental fluid from Mobil, and the independent research at Ethyl led to a BF_3-catalyzed process and decene-based product similar to that developed by Mobil.

Ethyl chose not to commercialize its findings because of the small potential market that existed at that time. Following the oil embargo of 1974 and the subsequent introduction of Mobil 1, Ethyl reinstituted a PAO research program. Ethyl entered the merchant market for PAO-based fluids in the late 1970s through a toll manufacturing arrangement with Bray Oil in California. In 1981, Ethyl decided to build a market development unit (MDU) to manufacture PAO in Baton Rouge, Louisiana. The 7,000 metric ton MDU came on stream in mid-1982, and Ethyl intended that this plant would operate until the market had grown to a size that would justify a world-scale plant. Marketing of the PAO was handled by Ethyl's Edwin Cooper Division, which was responsible for the manufacture and marketing of Ethyl's lube oil additives, and the division tradename HiTEC was used for the fluids. The division name was later changed to Ethyl Petroleum Additives Division (EPAD).

Slow growth in the PAO market prompted Ethyl to shut down the MDU in 1985 and return to a toll arrangement. In 1987, Ethyl entered into an agreement with Quantum Chemical whereby Quantum would manufacture PAO from Ethyl's 1-decene. Ethyl's PAO sales in Europe began to grow rapidly, and a decision was made to build a plant at Ethyl's manufacturing site at Feluy, Belgium, where a large new α-olefin plant was also being planned.

In early 1989, Ethyl transferred responsibility for the PAO project from the Ethyl Petroleum Additives Division to the Industrial Chemical Division. This decision reflected the philosophy that PAO is a base stock rather than a lube additive, and the action allowed Ethyl to expand the scope of the sales effort to include a broader potential market. In keeping with the philosophy of PAO being a base stock, the tradename for the bulk fluids was changed to ETHYLFLO PAO fluids, and an aggressive marketing campaign in North America was launched.

In 1989, Quantum sold its Emery Division to Henkel, but retained its PAO plant at Deer Park, Texas, leaving Quantum in the difficult situation of having neither its own source of 1-decene nor its own marketing organization. In 1990, Ethyl purchased Quantum's Deer Park plant, which is located only a few miles from Ethyl's large α-olefin plant at Pasadena. The Deer Park facility has two PAO production trains and an annual capacity of 77,000 metric tons. Ethyl's 36,000 metric ton Feluy plant came on stream in January 1991.

6. Exxon Corporation

Exxon introduced Esso Ultra Oil in Europe in mid-1986. This lubricant is a partial synthetic oil containing PAO. Exxon has produced small quantities of PAO in its alkylation facility at its chemical plant in Port Jerome, France. Plans to convert that plant to full-scale PAO operation appear to have been shelved.

7. Quantum Chemical Corporation

Quantum Chemical Corporation is the name adopted in 1988 by the former National Distillers and Chemical Corporation. National Distillers entered the synthetic lubricants business in 1978 with the purchase of Emery Industries, an important producer of ester-based synlubes.

In December 1980, National Distillers announced the construction of a 15,400 metric ton PAO plant at their manufacturing facility in Deer Park, Texas. The plant did not actually come on stream until late 1983. In 1987, National Distillers entered into a manufacturing and marketing agreement with Ethyl Corporation, as described above. The 1-decene feedstock was supplied by Ethyl.

By 1989, Quantum had debottlenecked the PAO plant and built a second, larger plant at the same location, bringing the total capacity to 77,000 metric tons. In 1990, Quantum sold their PAO business and manufacturing site to Ethyl Corporation.

8. Castrol Limited

Castrol, originally The Burmah Oil Public Limited Company, and then Burmah-Castrol, has historically been an innovator in automotive lubricant marketing. In 1981, Castrol purchased Bray Oil Company, a small manufacturer of synthetic lubricants based in California. Bray Oil at that time had been toll producing PAO for Ethyl Corporation. Although Castrol maintained a strong interest in marketing synthetic lubricants, it chose to close the PAO plant and purchase its PAO requirements.

Castrol was an early marketer of synthetic automotive lubricants in Europe. They have introduced a full line of synthetic and semisynthetic gear lubes and compressor oils as well as higher-performance jet turbine oils, military hydraulic fluids, and jet lube products. They introduced Syntron X—a 5W–50 PAO-based automotive synlube—into the United Kingdom in 1988, and a new line of PAO-based automotive products, under the tradenames Syntorque and Transmax, was introduced into the United States in 1991.

9. Uniroyal Chemical Company

Uniroyal has produced high-viscosity PAOs ($KV_{100°C}$ = 40 and 100 cSt) since 1980 in a small plant at Elmira, Ontario. Uniroyal and Mobil are the only two producers of these grades of PAO in the world. Total production capacity is about 1,100 metric tons per year.

10. Neste Chemical

Neste Chemical has a PAO plant under construction in Berigen, Belgium. The facility is expected to come on stream in 1991 and to have a capacity of 20,000 metric tons.

11. Texaco

Texaco has conducted research on PAOs and holds several patents but has no commercial production.

12. Shell Chemical

Shell Chemical has conducted extensive research on PAOs but has never begun commercial manufacturing. Shell, along with Chevron and Ethyl, is basic in the α-olefin raw material.

III. CHEMISTRY

PAOs are manufactured by a two-step reaction sequence from linear α-olefins, which are derived from ethylene. The first step is synthesis of a mixture of oligomers, which are polymers of relatively low molecular weight.

$$\alpha\text{-Olefin} \rightarrow \text{dimer} + \text{trimer} + \text{tetramer} + \text{pentamer, etc.}$$

For the production of low-viscosity (2–10 cSt) PAOs, the catalyst for the oligomerization reaction is usually boron trifluoride (PAOs are commonly classified according to their approximate kinematic viscosity at 100°C—this convention will be used throughout this chapter). The BF_3 catalyst is used in conjunction with a protic cocatalyst such as water, an alcohol, or a weak carboxylic acid. It is necessary that the BF_3, a gas, be maintained in a molar excess relative to the protic cocatalyst. Although this stoichiometry may be accomplished by sparging the reaction mixture with a stream of BF_3, it is more practical, on a commercial basis, to conduct the reaction under a slight BF_3 pressure (10–50 psig). For convenience, a general designation for the catalyst system is $BF_3 \cdot ROH$, where ROH represents any protic species such as those noted earlier, and the presence of excess BF_3 is understood.

The $BF_3 \cdot ROH$ catalyst system is unique for two reasons. First, this catalyst combination produces an oligomer distribution that is markedly peaked at trimer. Figure 1.1 shows a gas chromatography (GC) trace indicating the oligomer distribution of a typical reaction product derived from 1-decene using a $BF_3 \cdot n\text{-}C_4H_9OH$ catalyst combination at a reaction temperature of 30°C. The chromatogram indicates that only a relatively small amount of dimer is formed. The bulk of the product is the trimer, with only much smaller amounts of higher oligomers present.

The second unique feature of the $BF_3 \cdot ROH$ catalyst system is that it produces products that have exceptionally good low-temperature properties. The extremely low pour-point values were puzzling to the early workers in the field until it was shown that the resulting oligomers exhibit a greater degree of skeletal branching than would be predicted by a conventional cationic polymerization mechanism (11).

The reason that BF_3 catalysis causes excess skeletal branching during the oligomerization process is unclear. The first researcher that recognized the phenomenon proposed a mechanism involving a skeletal rearrangement of the dimer (11). A later paper proposed that the monomer undergoes rearrangement (14). A third paper proposed that the excess branching arose from positional isomerization of the double bond in the monomer prior to oligomerization (15). In fact, the large number of isomers formed cannot be explained by any single mechanism, and the role of $BF_3 \cdot ROH$ in promoting the necessary rearrangements remains unexplained.

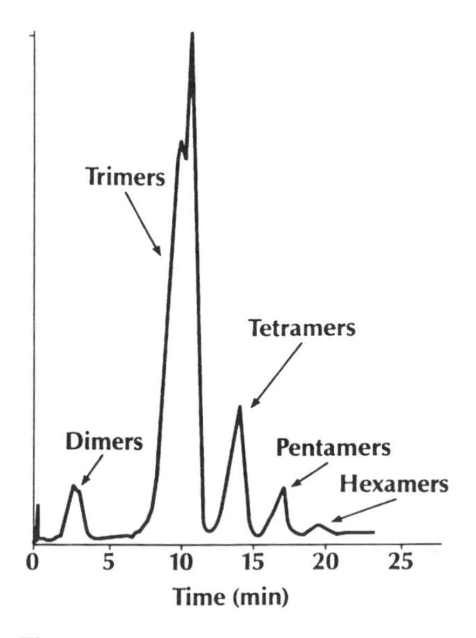

Figure 1.1 Gas chromatography of typical oligomer.

Even though the mechanism of the $BF_3 \cdot ROH$-catalyzed oligomerization remains to be fully elucidated, researchers have learned how to advantageously control the composition of the final PAO product so as to tailor the oligomer distribution to fit the requirements of specialized end-use applications (16). This customizing is done by manipulation of the reaction variables, which include:

Chain length of olefin raw material.
Temperature.
Time.
Pressure.
Catalyst concentration.
Cocatalyst type and concentration.
Cocatalyst feed rate.
Olefin feed rate.
Reaction quench and recovery procedures.
Hydrogenation catalyst and conditions.
Distillation.

In addition to controlling the relative distribution of the oligomers by manipulation of the reaction parameters, the PAO manufacturer also can make major alterations in the product properties by choice of starting olefin. Today, the commercial PAO market is dominated by decene-derived materials because these products have the broadest range of properties, but a knowledgeable producer has the option of choosing other starting olefins in order to better satisfy the requirements for a particular end-use application. More detail on the potential use of alternate olefin streams will be discussed later.

The crude reaction product is quenched with water or caustic, allowed to settle, and then washed again with more water to remove all traces of the BF_3 catalyst. Gaseous BF_3 can be recovered by concentrating the wash water and treating the solution with concentrated sulfuric acid.

A second step in the manufacturing process entails hydrogenation of the unsaturated oligomer. The hydrogenation may be carried out before or after distillation. Distillation is required to remove any unreacted monomer, to separate the dimer, which is marketed as a 2.0-cSt product, and in some cases to coproduce a lighter and a heavier grade of PAO.

The hydrogenation is typically performed over a supported metal catalyst such as nickel/kieselguhr or palladium/alumina. Hydrogenation is necessary to give the final product enhanced chemical inertness and added oxidative stability.

It is normally not possible to manufacture the higher viscosity (40 and 100 cSt) PAO products using the $BF_3 \cdot ROH$ technology. However, several other catalyst systems are known that can give the desired products. One class of catalysts employs alkylaluminum compounds in conjunction with $TiCl_4$ (7) or alkyl halides (17). The latter system is preferred by Uniroyal, which uses ethylaluminum sesquichloride with allyl chloride.

It has also been reported in a Mobil Oil European patent application that high-viscosity PAOs may be produced by dimerizing lower oligomers with peroxides (18). The patent describes the use of stoichiometric quantities of di-*tert*-butyl peroxide, which would probably not be economically feasible. On the other hand, a system that employs hydrogen peroxide directly or to regenerate an active intermediate might be commercially attractive. Mobil has also obtained a large number of patents describing the use of supported chromium catalysts (19). The system actually employed by Mobil for commercial manufacture has not been disclosed, but it is believed to employ an $AlCl_3$ catalyst.

IV. PROPERTIES

The physical and chemical properties of PAO fluids make them attractive for a variety of applications requiring a wider temperature operating range than can normally be achieved by petroleum-based products (mineral oils).

A. Physical Properties

1. Commercial PAOs

Table 1.1 lists the typical physical properties of the five grades of low-viscosity commercial PAOs available today. These products are all manufactured using 1-decene as the starting material, and the final properties are determined by control of the reaction parameters and (depending on the manufacturer) selective distillation of the light oligomers.

Table 1.1 shows that all commercial grades of low-viscosity PAOs have relatively high viscosity indices (VIs) of around 135. (Note: No VI is shown for PAO 2 because VI is undefined for fluids having a viscosity of less than 2.0 cSt at 100°C.) The viscosity of a high-VI fluid changes less dramatically with changes in temperature compared to the viscosity changes of a low-VI fluid. A practical consequence of this property is that PAOs do not require viscosity index improvers (VIIs) in many applications. The presence of a VII is often undesirable because many tend to be unstable toward shear. Once the VII begins to break down, the fully formulated fluid goes "out of grade" (i.e., fails to retain the original viscosity grade).

Several other important physical properties of commercial PAOs are shown in Table 1.1. All of the products have extremely low pour points, as well as low low-temperature viscosities. These properties make PAOs very attractive in the cold-climate applications for which they were first used. At the other end of the spectrum, all but the 2.0-cSt product have low volatilities as demonstrated by the low percent loss of material at 250°C in the standard NOACK volatility test. Low volatility is important in high-temperature operations to reduce the need for "topping up" and to prevent a fluid from losing its lighter components and thus becoming too viscous at low or ambient temperatures. Low volatility is also important as it relates to flash and fire points.

The typical physical properties of commercial high-viscosity PAO fluids are given in

Table 1.1. Physical Properties of Commercial Low-Viscosity PAOs

Parameter	PAO 2	PAO 4	PAO 6	PAO 8	PAO 10
KV at 100°C, cSt	1.80	3.84	5.98	7.74	9.87
KV at 40°C, cSt	5.54	16.68	30.89	46.30	64.50
KV at –40°C, cSt	306	2390	7830	18,200	34,600
Viscosity index	—	124	143	136	137
Pour point, °C	–63	–72	–64	–57	–53
Flash point, °C	165	213	235	258	270
NOACK,[a] % loss	99.5	11.8	6.1	3.1	1.8

[a]Volatility at 250°C.

Table 1.2. Physical Properties of Commercial High-Viscosity PAOs

Parameter	PAO 40	PAO 100
KV at 100°C, cSt	40–42	103–110
KV at 40°C, cSt	399–423	1260–1390
KV at –18°C, cSt	39,000–41,000	176,000–203,000
Pour point, °C	–36 to –45	–21 to –27
Flash point, °C	275–280	280–290
NOACK,[a] % loss	0.8–1.4	0.6–1.1

[a]Volatility at 250°C.

Table 1.2. The two grades available on the market today are the 40-cSt and 100-cSt fluids. As with the low-viscosity PAOs, these fluids have a very broad temperature operating range.

2. Comparison to Mineral Oils

The excellent physical properties of the commercial PAO fluids are most readily apparent when they are compared directly to those of petroleum-based mineral oils. The fairest comparison is to look at fluids with nearly identical kinematic viscosities at 100°C. The differences in both low- and high-temperature properties can then be examined.

Table 1.3 compares the physical properties of a commercial 4.0-cSt PAO with those of two 100N (neutral) mineral oils, a 100NLP (low pour) mineral oil, and a hydrotreated HVI (high viscosity index) mineral oil. The PAO shows markedly better properties at both high and low temperatures. At high temperatures, the PAO has lower volatility and a higher flash point. A relatively high flash point is, of course, often important for safety considerations. At the low end of the temperature scale the differences are equally dramatic. The pour point of the PAO is –72°C, while those of the three 100N mineral oils and the HVI oil are –15, –12, –15, and –27°C, respectively.

Table 1.4 compares a commercial 6.0-cS PAO with a 160HT (hydrotreated) mineral oil, a 240N oil, a 200SN (solvent neutral) mineral oil and a VHVI (very high viscosity

Table 1.3. 4.0-cSt Fluids

Parameter	PAO	100N	100N	100NLP	HVI
KV at 100°C, cSt	3.84	3.81	4.06	4.02	3.75
KV at 40°C, cSt	16.7	18.6	20.2	20.1	16.2
KV at –40°C, cSt	2390	Solid	Solid	Solid	Solid
Viscosity index	124	89	98	94	121
Pour point, °C	–72	–15	–12	–15	–27
Flash point, °C	213	200	212	197	206
NOACK,[a] % loss	11.8	37.2	30.0	29.5	22.2

[a]Volatility at 250°C.

Table 1.4. 6.0-cSt Fluids

Parameter	PAO	160HT	240N	200SN	VHVI
KV at 100°C, cSt	5.98	5.77	6.98	6.31	5.14
KV at 40°C, cSt	30.9	33.1	47.4	40.8	24.1
KV at –40°C, cSt	7830	Solid	Solid	Solid	Solid
Viscosity index	143	116	103	102	149
Pour point, °C	–64	–15	–12	–6	–15
Flash point, °C	235	220	235	212	230
NOACK,[a] % loss	6.1	16.6	10.3	18.8	8.8

[a]Volatility at 250°C.

index) fluid that is currently considered to be the best on the market. The broader temperature range of the PAO is again apparent. Table 1.5 makes similar comparisons for 8.0-cSt fluids.

The ability of PAO products to outperform petroleum-based products of similar viscosity at both ends of the temperature spectrum becomes easily understandable if one compares the gas chromatography traces. Figure 1.2 contains chromatograms run under identical conditions of a 4.0-cSt HVI oil and a 4.0-cSt PAO. The PAO product is essentially decene trimer with a small amount of tetramer present. The fine structure of the trimer peak is attributable to the presence of a variety of trimer isomers (same molecular weight, different structure). The HVI oil, on the other hand, has a broad spectrum of different molecular weight products. The oil contains low-molecular-weight materials that adversely affect the volatility and flash point characteristics. It also contains high-molecular-weight components that increase the low-temperature viscosity and linear paraffins that increase the pour point.

Figure 1.3 compares the GC traces of a very high quality 6.0-cSt VHVI fluid with a PAO of similar viscosity. The PAO has a well-defined chemical composition consisting of decene trimer, tetramer, pentamer, and a small amount of hexamer. The VHVI fluid, like the HVI fluid in the previous example, contains a wide range of components that degrade performance at both ends of the temperature scale.

Table 1.5. 8.0-cSt Fluids

Parameter	PAO	325SN	325N
KV at 100°C, cSt	7.74	8.30	8.20
KV at 40°C, cSt	46.3	63.7	58.0
KV at –40°C, cSt	18,200	Solid	Solid
Viscosity index	136	99	110
Pour point, °C	–57	–12	–12
Flash point, °C	258	236	250
NOACK,[a] % loss	3.1	7.2	5.1

[a]Volatility at 250°C.

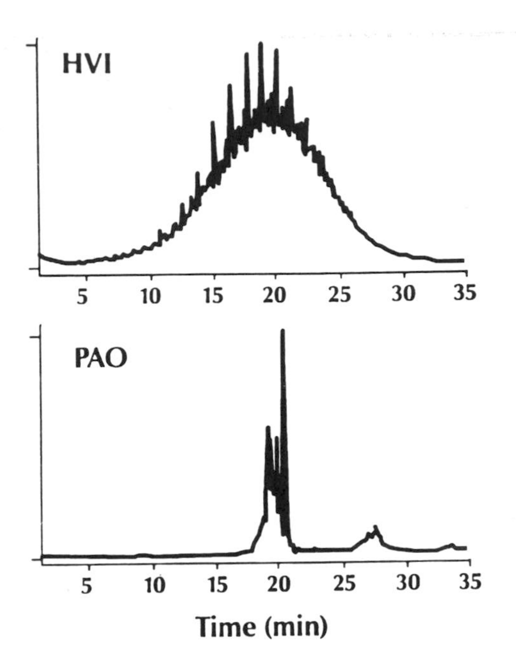

Figure 1.2 Gas chromatography traces of 4.0-cSt fluids.

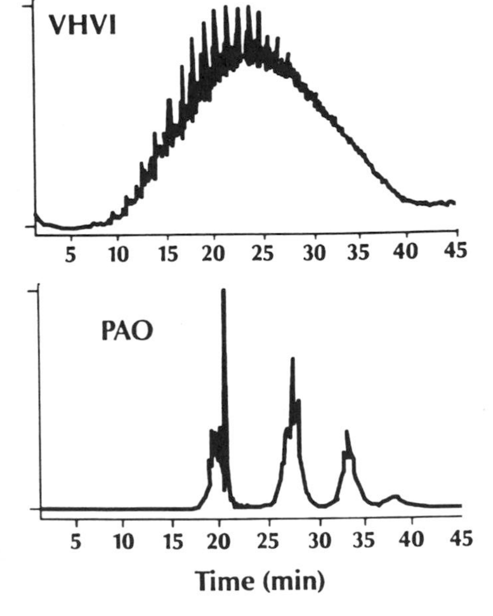

Figure 1.3 Gas chromatography traces of 6.0-cSt fluids.

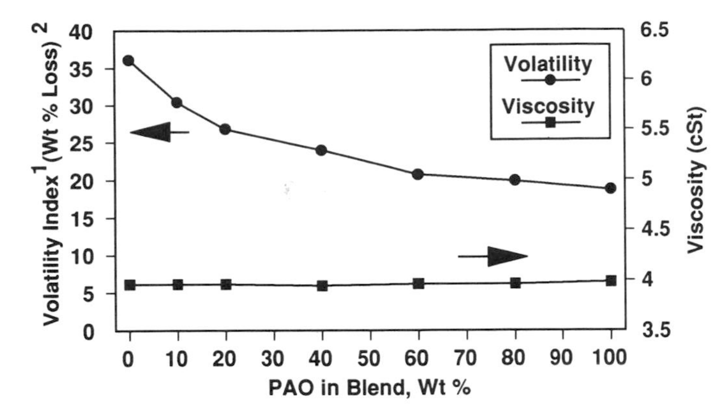

Figure 1.4 Effect of blending 4.0-cSt PAO with 100N mineral oil. (1) In-house test designed to give approximate correlation to D-972. (2) Weight percentage loss after 2.0 h at 204°C under flow of N_2.

3. *Properties of Blends*

The excellent combination of high- and low-temperature physical properties of PAOs, combined with their total miscibility with mineral oils, makes them attractive candidates for blending with certain base stocks in order to improve the base-stock quality and bring it into specification for a particular application. This practice has indeed become widespread (but little publicized) as refiners scramble to meet the newer and more stringent API classification requirements.

Figure 1.4 shows the effect on volatility and viscosity upon blending 4.0-cSt PAO with a light (100N) mineral oil (23). The "Volatility Index" depicted in this and the following figure is derived from an "in-house" test. A defined quantity of the test sample is placed in a small dish or "planchet," and the planchet is placed in an oven for 2.0 h at 204°C. A constant flow of nitrogen is maintained over the sample throughout the test. The values are not the same as obtained in the standard D-972 or NOACK tests, but they have been shown to correlate well on a relative basis. Small amounts of PAO have a dramatic effect in reducing the volatility of the mineral oil, while having essentially no effect on

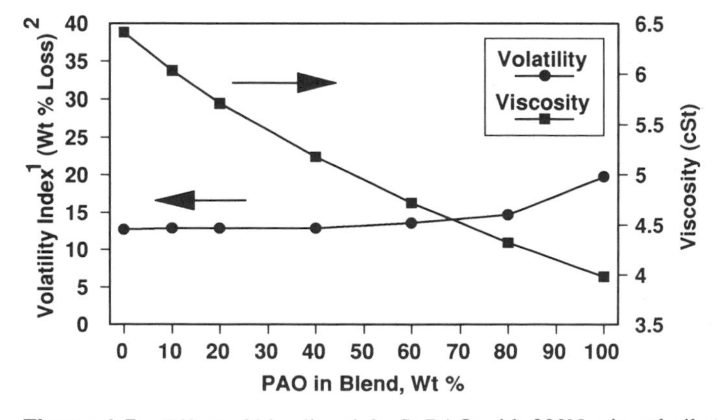

Figure 1.5 Effect of blending 4.0-cSt PAO with 200N mineral oil. (1) In-house test designed to give approximate correlation to D-972. (2) Weight percentage loss after 2 h at 204°C under flow of N_2.

viscosity. Figure 1.5 shows the effect of blending the same 4.0-cSt PAO with a heavy (200N) mineral oil. In this case, small amounts of the PAO have a large effect in reducing the viscosity of the mineral oil without increasing the volatility.

B. Chemical Properties

In addition to the physical properties, the chemical properties of a functional fluid must be considered. The most important chemical property requirements are that the fluid must be thermally stable and chemically inert. Under normal operating conditions a working fluid must not thermally degrade nor react with the atmosphere, the materials of construction, seals, paints, varnishes, performance-enhancing additives, other fluids with which it is intentionally contacted, or inadvertent contaminants.

1. Thermal Stability

Many of the operations for which a functional fluid is required are carried out at elevated temperatures. For this reason it is important that the fluid employed not be degraded under the operating conditions. The choice of an appropriate bench test, however, is often difficult. It is important that the test differentiate between thermal and oxidative degradation while simulating real-world operating conditions. It is also important that the test differentiate between thermal degradation and volatility. Some evaluations based on oven aging or thermogravimetric analysis (TGA) have led to erroneous conclusions because the loss in sample weight and/or increase in viscosity could be attributed to volatilization of the lighter components rather than to chemical degradation.

One test commonly employed that avoids the danger of misinterpreting volatility for thermal instability is the Panel Coker Thermal Stability Test. In this test, an aluminum panel heated to 310°C is alternately splashed by the test oil for 6 min and baked for 1.5 min. At the end of the test, the panels are rated for cleanliness. A completely clean panel has a rating of 10. Table 1.6 summarizes the results of one study that compared the performance characteristics of mineral oil and various synthetic base stocks for crankcase applications (20). Under these severe conditions, the mineral oil panel was covered with deposits, indicating a lack of thermal stability. An alkylated aromatic also performed poorly. By comparison, both a PAO of comparable viscosity and a dibasic ester per-

Table 1.6. Thermal Stability Panel Coker Test

Base fluid	Cleanliness
4.0-cSt Mineral oil	0
4.0-cSt PAO	8.0
5.0-cSt Alkylated aromatic	2.0
5.4-cSt Dibasic ester	8.0
4.0-cSt PAO/(polyol ester)	9.5

Test conditions	
Panel temp.	310°C
Sump temp.	121°C
Operation	6 min splash/1.5 min bake
Rating	10 = clean

formed well. The best performance was achieved using a mixture of PAO and a polyol ester. Dibasic and polyol esters are commonly used in conjunction with PAO in crankcase formulations.

The thermal stability of PAOs was also investigated regarding use in aviation lubricants (21). In this evaluation, thermal stability was determined by heating the fluid at 370°C under a nitrogen atmosphere for 6.0 h in a sealed autoclave. Thermal degradation was measured by the change in viscosity and by gas chromatographic analysis. The tests show that the thermostability of PAO products can be ranked as:

Dimer > trimer > tetramer

These findings are consistent with the molecular structures of the oligomers. The least thermally stable parts of the molecule are the tertiary carbon positions, that is, the points where there are branches in the carbon chains. The higher oligomers have more branches and are thus more subject to thermal degradation.

2. Hydrolytic Stability

For a functional fluid, the importance of inertness to reaction with water is important for a variety of reasons. Hydrolytic degradation of many substances leads to acidic products, which, in turn, promote corrosion. Hydrolysis may also materially change the physical and chemical properties of a base fluid, making it unsuitable for the intended use. Systems in which the working fluid may occasionally contact water or high levels of moisture are particularly at risk. Also at risk are systems that operate at low temperature or that cycle between high and low temperatures.

The excellent hydrolytic stability of PAO fluids was reported as a result of tests conducted to find a replacement for 2-ethylbutyl silicate ester as an aircraft coolant/dielectric fluid used by the U.S. military in aircraft radar systems (22). The test method required treating the fluids with 0.1% water (or 0.1% sea water) and maintaining the fluid at 170 or 250°F for up to 200 h. Samples were withdrawn at 20-h intervals, and the flash points were measured by the closed cup method. A decrease in flash point was interpreted as being indicative of hydrolytic breakdown to form lower-molecular-weight products. The PAO showed no decrease in flash point under any of the test conditions, while the 2-ethylbutyl silicate ester showed marked decreases. Figure 1.6 shows the results for tests at 250°F.

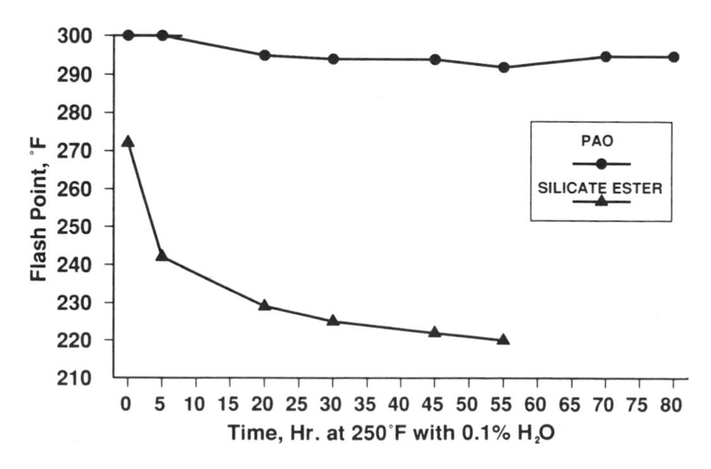

Figure 1.6 Hydrolytic stability.

3. Oxidative Stability

A high level of oxidative stability is essential to the performance of a functional fluid. In many applications the fluid is required to perform at elevated temperatures and in contact with air. The results from attempts toward evaluation of fluids for oxidative stability, however, are often confusing. The results are very dependent on the test methodology. Tests involving thin films tend to give different results than tests using bulk fluids. The presence or absence of metals that catalyze oxidation is not only very important, but different metals interact differently with different fluids. In addition, oxidative stability may be enhanced by the use of antioxidants, but different fluids respond differently to different antioxidants.

One set of experiments that attempted to differentiate between PAOs and mineral oils entailed using differential scanning calorimetry (DSC) (23). In this test, the fluid is heated in a pan at a controlled rate, and the temperature at which there is an onset of oxidation is determined by the accompanying exotherm. All of the commercial PAO products (with the exception of 2.0-cSt fluid) were tested. The onset temperatures for the six viscosity grades fell in the very narrow range of 187.3–191.6°C. Two 6.0-cSt mineral oils, on the other hand, gave values of 189.2 and 200.6°C, respectively. Quite a different result was reported for a laboratory oxidation test in which the fluid was heated at 163°C for 40 h in the presence of steel, aluminum, copper, and lead coupons (20). In this test a 4.0-cSt mineral oil exhibited a 560% viscosity increase and a light sludge appearance, while a 4.0-cSt PAO showed only a 211% viscosity increase and no sludge. These results seem to indicate better performance for the PAO, but the loss of weight by the lead coupon in the PAO was 2.8 times that of the coupon in the mineral oil. The same paper reports better performance for mineral oils in a rotary bomb test that measures the time for a specific pressure drop, but better performance for PAOs in beaker oxidation tests where the increase in viscosity is measured.

It has been reported that the failure of unstabilized PAO to outperform unstabilized mineral oil in oxidative stability tests may be attributed to the presence of natural antioxidants in the latter (24). The lack of inhibitors in the pure PAO is then given as the rationale for the greater responsiveness of the PAOs to the addition of small amounts of antioxidants. An interesting and somewhat similar rationale has been given for the unusually good responsiveness of PAOs to the addition of antiwear and other performance additives (25). These researchers from the All-Union Scientific Institute of Oil Refining in Moscow conclude that the efficiency of small concentrations of additives in PAO oils is related to the fast adsorption of the additives on the metal surfaces, which is promoted by weak inhibition of the process of additive diffusion in transportation from the bulk oil to the tribosurface, as the result of weak cohesive forces between the additive molecules and the PAO substrate.

The arguments noted above are supported by results obtained from oxidative stability testing of fully formulated part-synthetic engine oils (26). A thin-film oxygen uptake test (TFOUT) was used for these studies. This test is a modified rotary bomb oxidation procedure in which the bomb is charged with sample, a small amount of water, a fuel catalyst and a metal catalyst. The bomb is then pressurized with pure oxygen, placed in a bath at 160°C, and rotated axially at 100 rpm at a 30° angle from the horizontal. The time from the start of the test until a drop in pressure is noted is defined as the oxidation induction time of the oil. The test oils each contained 13.7% of a detergent-inhibitor package (DI) and 8.0% of a viscosity-index improver (VII). The base stock consisted of a 100SEN mineral oil blended with a 4.2-cSt PAO. Figure 1.7 shows that as the percentage of PAO in the sample was increased from 0 to 30%, the induction time for the onset of

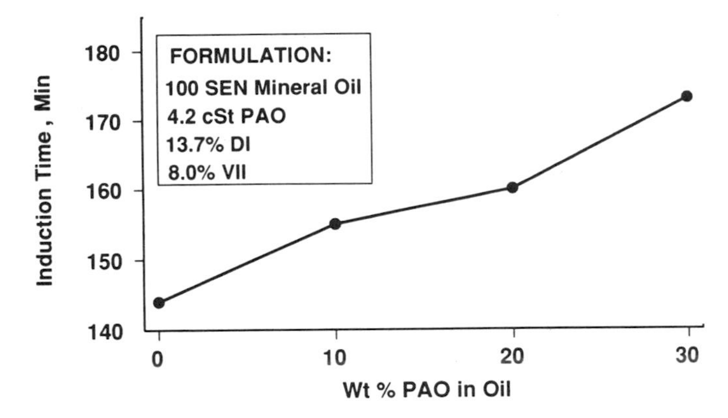

Figure 1.7 Thin-film oxygen uptake test fully formulated oils.

oxidation increased from 143 to 173 min. Two other papers of interest concerning the oxidative stability of PAOs are also referenced (27, 28).

V. APPLICATIONS AND PERFORMANCE CHARACTERISTICS

The use of PAO-based functional fluids is growing rapidly. Conventional applications, such as automotive crankcase, are being spurred on by tighter specifications and an increasing demand for higher performance. Nonconventional applications are also beginning to grow rapidly, especially where specific properties of PAO fluids give them particular advantages in performance, cost effectiveness, or environmental acceptability.

A. Overview of Application Areas

The following is a listing of both established and emerging application areas for PAOs. The list of applications has grown to such a degree in the last few years that a comprehensive review of the PAO performance attributes found advantageous in each and every application would require more space than is available here. Instead, where possible, a reference is cited so that the reader may refer to published information and data in the specific area of interest. Following this section, some performance data for areas of the broadest interest are presented. For detailed reviews of the most prominent areas of application, the reader is referred to the appropriate chapters in Part II of this book.

Engine crankcase (29, 30)
Hydraulic fluids (31)
Gear oils (32, 33)
Greases (4, 34–36)
Brake fluids
Shock absorbers
Automatic transmission fluids (37)
Metal Working Fluids

Compressor Oils (38–41)
Heat transfer media
Dielectric fluids (23, 42)
Gels for coating optical fibers
Off-shore drilling muds
Cosmetics and personal care products
Textiles
Polymers

B. Performance Testing for Automotive Applications

While physical properties are obviously important in choosing a fluid for a particular application, it is essential that the fluid be subjected to performance testing under conditions that simulate the limits to which the final product will be stressed. However, as indicated earlier, the list of applications for PAOs has grown to the point that it precludes a comprehensive discussion of performance testing for all applications. Because the requirements for the wide variety of automotive applications encompass much of the broader spectrum of applications, this section will focus on tests specifically designed and conducted by the automotive industry.

An excellent summary of the automotive testing conducted in the 1970s and early 1980s may be found in a collection of 26 papers published in one volume by the Society of Automotive Engineers (SAE) (43). In Appendix B of the SAE book, the editors summarize the "eight superior performance features of synthetic engine oils." Their conclusions are based on a compilation of data in the various papers. The eight features that they identify are:

1. Improved engine cleanliness. This is based on a test using four taxicabs employing a SAE 5W–20 PAO-based oil. Oil changes at 12,000 miles for 60,000 miles were followed by a 40,000 "no drain" period.
2. Improved fuel economy. The results of 10 different test programs involving a total of 182 vehicles showed a weighted average fuel savings of 4.2%.
3. Improved oil economy. In 10 different tests on oil consumption, the percent improvement in miles per quart ranged from 0% (for a military arctic lubricant) to 156%. The average improvement was 55.9%.
4. Excellent cold starting. Automobiles with 400 CID V-8 engines could be started at –39°F when the crankcase contained an SAE 5W–20 PAO-based synthetic oil. With a mineral oil of the same viscosity grade the lowest engine-starting temperature was –29°F.
5. Excellent low-temperature fluidity. For the two oils described in item 4, the PAO-based oil exhibited a pour point of –65°F, whereas the mineral oil had a pour point of –37°F.
6. Outstanding performance in extended oil drain field service. This conclusion was based on 100,000-mile tests using parkway police cruisers, which are normally operated at speeds ranging from 55 to 100 mph. The test vehicles used a PAO-based SAE 5W–20 "SE-CC" oil. Oil and filter changes were performed every 25,000 miles. The baseline consisted of a series of tests carried out in identical vehicles operated on SAE 10W–40 "SE" mineral oil with oil and filter changes every 5,000 miles.
7. High-temperature oxidation resistance. Viscosity increase was measured in a 2-liter Renault after 64 h of operation with an oil-sump temperature of 302°F. The synthetic oil showed a 10% increase in viscosity and the mineral oil showed a 135% increase. Both samples were SAE 10W–50 oils.
8. Outstanding single- and double-length SAE-ASTM-API "SE" performance tests. The results of all of these tests are presented in the reference. The PAO-based synthetic oils met or exceeded all of the requirements.

More recent data show that PAO-based fluids continue to provide superior performance for the increasingly sophisticated cars being built today. Today's automobiles tend to have smaller, more demanding engines. Increased emphasis on aerodynamics means less cooling under the hood, resulting in higher operating temperatures in both the engine and the transmission. In addition to the ability of PAO to meet this challenge with excellent thermal and oxidative stability, PAOs offer another advantage over mineral oils under these severe operating conditions. Both the thermal conductivity and the heat capacity of PAO fluids are about 10% higher than values for the comparable mineral oils. The net result is that PAO-lubricated equipment tends to run cooler.

The following subsections examine in somewhat greater detail the results of testing for all the major areas of automotive applications.

1. Crankcase

Tables 1.7–1.10 illustrate the results of tests related to the use of PAO in automotive crankcase applications (23). Table 1.7 contains data relating to the hot oil oxidation test (HOOT), which is designed to measure the thermal and oxidative stability of the fluid as it is splashed on hot metal surfaces inside an engine. A PAO and a mineral oil were compared employing identical additive packages at identical concentrations. The significantly superior performance of the PAO has two possible implications. First, the PAO-based fluid can be used for longer drain intervals, resulting in less down time and lower maintenance costs. Second, PAO can be used with lower levels of additives and other stabilizers, thus reducing the price differential between PAO and a comparable mineral oil.

Table 1.8 contains the results of the Petter W1 engine test after 108 h. The test measures both the increase in viscosity of the fluid and the amount of wear, as determined by bearing weight loss. In this test, the advantages of employing a part-synthetic oil mixture are shown. When PAO is used as only 25% of the base oil, the percent viscosity increase is halved.

The data in Table 1.9 were acquired from a Sequence IIIE engine test, which is commonly used in North America. Table 1.10 contains data relating to the VW Digiphant

Table 1.7. Hot Oil Oxidation Test (Automotive Crankcase)[a]

Fluid	Start ($KV_{40°C}$, cSt)	Finish ($KV_{40°C}$, cSt)	% Change
Mineral	95	146.3	54.0
PAO	94	96.8	3.0

[a]Conditions: same additive package at same concentration; temperature = 165°C; time = 5 days.

Table 1.8. Petter W1 Engine Test[a]

Oil	Grade	$KV_{40°C}$ % increase	Bearing weight loss, mg
Mineral	15W–40	108	14.1
PAO–25%	10W–40	54	9.7
PAO–50%	10W–40	45	11.5
PAO–100%	15W–40	20	14.5

[a]Conditions: same additive package at same concentration; time = 108 h.

Table 1.9. Sequence IIIE Engine Test (North America)

Oil	Grade	$KV_{40°C}$ % increase
Mineral	15W–40	167
PAO	5W–50	62

test, which is more widely used in Europe. In both tests a 5W–50 full-synthetic PAO-based oil is compared to a 15W–40 mineral oil. As indicated by the SAE classifications, the PAO-based oil is rated for operation at temperatures both lower and higher than the comparable mineral oil-based fluid. Nevertheless, the PAO lubricant still out-performed the mineral oil by a wide margin.

Another important feature that must be considered in automotive crankcase applications is low-temperature performance. Tables 1.11 and 1.12 compare the low-temperature characteristics of base fluid PAOs with HVI and VHVI mineral oils of comparable viscosity (23). The cold crank simulation test is of vital interest to any car owner who has ever lived in a cold climate. The advantage of a PAO-based formulation in the crankcase is immediate and obvious on a cold winter morning—it is the difference of being able to start the car or not.

Piston cleanliness is another important factor in choosing a crankcase oil. Table 1.13 presents the results of three different tests commonly used to rate piston cleanliness (23). The PAO formulations performed well compared to the mineral oils, even when used (as in the Fiat test) at only a 15% level in a part-synthetic formulation.

The results of a Caterpillar 1-G evaluation are given in Table 1.14 (23). Both a part-synthetic and a full-synthetic PAO-based oil outperformed an equivalent 10W–40 mineral oil.

Table 1.10. VW Digiphant Test (Europe)[a]

Oil	Grade	$KV_{40°C}$ % increase	$KV_{100°C}$ % increase
Mineral	15W–40	108	62
PAO	5W–50	25	9

[a]Time = 147 h.

Table 1.11. Low-Temperature Performance (Crankcase)

Oil	$KV_{100°C}$, cSt	Pour point, °C	Cold crank simulation, –25°C, mPa • s	Brookfield viscosity, –25°C, cP
PAO	3.90	–64	490	600
VHVI	3.79	–27	580	1160
HVI	4.50	–12	1350	Solid
100SN	3.79	–21	1280	Solid

Table 1.12. Low-Temperature Performance (Crankcase)

Oil	$KV_{100°C}$, cSt	Pour point, °C	Cold crank simulation, –25°C, mPa • s	Brookfield viscosity, –25°C, cP
PAO	5.86	–58	1300	1550
VHVI	5.38	–9	1530	Solid
HVI	5.84	–9	3250	Solid
HVI	5.79	–9	2740	Solid
150SN	5.17	–12	4600	Solid

Table 1.13. Piston Cleanliness

Test	Base oil	Grade	Piston merit
VW1431	Mineral	15W–40	63.70
	PAO	5W–50	72.60
Fiat TIPO	Mineral	15W–40	6.40
	PAO (15%)	15W–40	7.60
MWM "B"	Mineral	15W–40	73.00
	PAO (50%)	10W–40	82.80

Table 1.14. Caterpillar 1-G

Oil	Grade	Total groove fill (80% maximum)	Total weighted demerits (300 maximum)
Mineral	10W–40	76	294
Part synthetic	10W–40	67	243
Full synthetic	10W–40	53	103

Table 1.15. Hot Oil Oxidation Test[a] (Manual Transmission and Rear Axle Oils)

Time (h)	$KV_{100°C}$, cSt	
	PAO	Mineral oil
0	10.00	10.50
4	10.45	12.60
8	10.54	12.90
16	11.92	51.24
24	12.10	TVTM[b]

[a]Temperature of test = 200°C.
[b]TVTM = too viscous to measure.

2. Transmissions

Hot oil oxidation tests (HOOT) are used to screen oils for use in manual transmissions and rear axles. The test is conducted at a more severe temperature (200°C) than used in the evaluation of crankcase oils, and the kinematic viscosity at 100°C is measured at specified time intervals. A comparison of the performance of mineral and PAO-based fully formulated oils is shown on Table 1.15 (23). After 16 h, the viscosity of the PAO fluid increased only 19%, whereas the viscosity of the mineral oil fluid increased nearly 500%. After 24 h, the viscosity of the PAO fluid increased by only 21%, but that of the mineral oil product became too viscous to measure.

The HOOT is also used as an indicator of performance for automatic transmission fluids. A less viscous oil is used for automatic transmissions than for manual transmissions (7.5 cSt vs. 10.0 cSt), but the test is still conducted at 200°C. The results of the test are presented graphically in Fig. 1.8 (23). The PAO-based formulation showed only an 8.6% increase in 100°C viscosity after 24 h. The viscosity of the mineral oil formulation increased 550% in the same time period.

While the tests described above indicate that PAO-based transmission fluids show better durability and performance than mineral oils at a given temperature, another important phenomenon has been observed. Measurement of transmission lubricant temperatures under high-speed driving conditions shows that the synthetic-based oils run as much as 30°C cooler than their mineral oil counterparts (44).

3. Gears

The Mercedes Benz spur gear rig performance test is used to evaluate the performance of gear oils. In the test, the elapsed time to gear-tooth breakage is used as the indicator of performance. A SAE 75W–90 synthetic formulation showed a 60% improvement over an SAE 90 mineral oil (23). The data are presented in Table 1.16.

4. Seal Compatibility

Seal compatibility is an important factor for any functional fluid. Unlike mineral oils, PAO does not have a tendency to swell elastomeric materials. Early commercial PAO products were not formulated properly to allow for this difference in behavior. Consequently, early PAOs gained an undeserved reputation for leakage. Extensive tests have since shown that the addition of small quantities of an ester to the formulation easily alleviates this problem.

Recent work has indicated that the proper choice of other performance additives may eliminate the need to employ esters, but this approach is not yet in practice for crankcase applications. Table 1.17 shows the results obtained in the CCMC G5 seal compatibility test for base fluids (23). A 6.0-cS PAO was compared to a 150SN mineral oil. The four seal materials studied were acrylate, silicone, nitrile, and fluoroelastomer. The seals were evaluated at the end of the test for changes in tensile strength, elongation, volume (seal swell), and hardness. The PAO performance fell within the specification limits for all four elastomers. The mineral oil failed with silicone. Similar tests have been carried out with fully formulated part- and full-synthetic PAO oils. In all cases the fluids met the specifications.

Additional information on choosing the proper seal materials for use with PAO fluids may be found in references 45 and 46.

5. Economy

The performance benefits demonstrated by the various tests that have been described are meaningful to the automotive engineer or tribologist, but the average consumer is most interested in how much savings the use of a PAO-based product is going to generate.

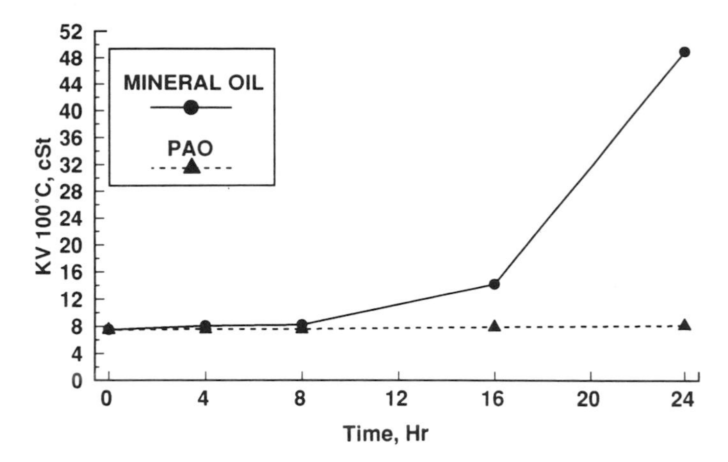

Figure 1.8 Hot oil oxidation test (automatic transmission, 200°C).

Table 1.18 describes the results of one study that considered both the increased fuel economy and the extended oil drain interval made possible with part- and full-synthetic PAO crankcase oils. The original calculations (47) have been updated to reflect current prices for gasoline and oil in North America. The calculations are based on 15,000 miles of driving and a "do-it-yourself" oil change regimen. A pump price of $1.20/gal for gasoline has been chosen, and the oil has been priced at $1.00, $2.00, and $4.00/qt for the mineral oil, the part-synthetic, and the full-synthetic, respectively. If the oil is changed every 5,000 miles, there is almost no cost differential for the three oils because of

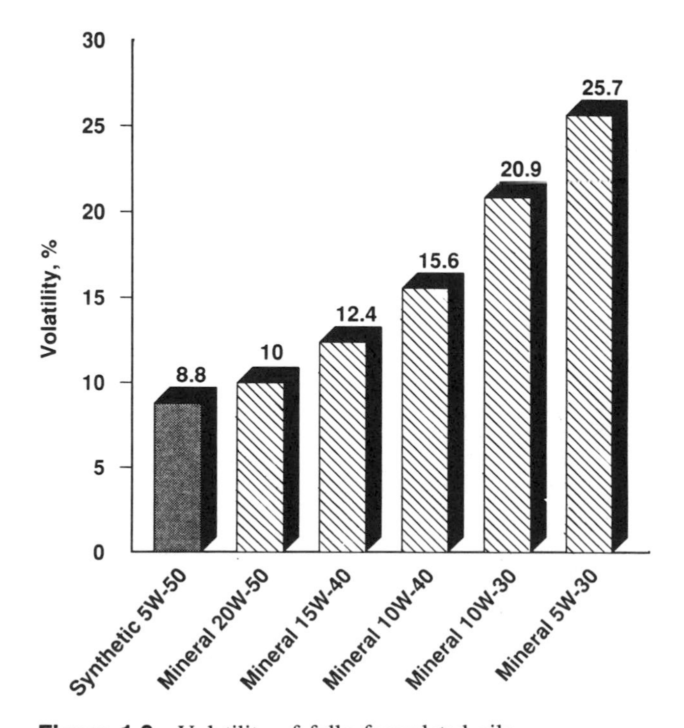

Figure 1.9 Volatility of fully formulated oils.

Table 1.16. Mercedes Benz Gear Rig Performance

Oil	Grade	Time to tooth breakage (h)
Mineral	90	85
PAO	75W–90	135

the improved fuel economy gained with the synthetics. For the 15,000 mile distance, the saving over the mineral oil formulation is $3.00 with the part-synthetic oil and a deficit of $3.00 is experienced with the full-synthetic. If, however, there is only one drain for the full-synthetic, the savings goes up to $11.00. In Europe, where gasoline is much more expensive and the differential in oil prices is less, the savings accrued by the use of synthetic crankcase oils will be much greater.

The use of lighter grades of crankcase oil is one answer to the need for increased fuel economy. The possible down-side to this strategy would be a concurrent increase in oil consumption and the loss of sufficient high-temperature viscosity for adequate engine protection. Studies show, however, that properly formulated PAO-based synthetic crankcase oils, with wide multigrade SAE performance classifications, can outperform mineral-oil based formulations in both fuel and oil consumption, while maintaining superior engine protection (43). Figures 1.9–1.12 from this study illustrate the point.

Figure 1.9 shows the relative volatility of a 5W–50 full-synthetic formulation com-

Table 1.17. Seal Compatibility (CCMC G5 Specification)

Elastomer	Tensile strength %	Elongation, %	Volume, %	Hardness, points
Fluoroelastomer				
150SN	+1.80	–12.0	+1.0	0
PAO	–1.20	–13.0	+0.6	0
Limits	–50/0	–60/0	0/+5	0/+5
Acrylate				
150SN	+5.40	–7.70	+3.40	0
PAO	–12.00	–30.00	–1.50	+4.0
Limits	–15/+10	–35/+10	–5/+5	–5/+5
Silicone				
150SN	–66.0	–60.0	+14.80	–16.0
PAO	–9.60	–15.0	+18.00	–13.0
Limits	–30/+10	–20/+10	0/+30	–25/0
Nitrile				
150SN	+11.0	–11.0	+2.40	+1.00
PAO	+13.0	–19.0	–1.90	+2.00
Limits	–20/0	–50/0	–5/+5	–5/+5

Table 1.18. Automotive Economy

Parameter	Mineral oil	Part synthetic	Full synthetic
Fuel			
Economy (miles/gal)	24	24.6(+2.5%)	25.2(+5%)
Use (gal/15,000 miles)	625	610	595
Cost ($1.20/gal)	750.00	732.00	714.00
Savings ($/15,000 miles)	—	18.00	36.00
Oil			
Cost ($/qt)	1.00	2.00	4.00
Cost of 3 × 5 qt changes plus 3 × $5 filters ($)	30.00	45.00	75.00
Additional cost ($)	—	15.00	39.00
Total savings			
Same oil drains ($)	—	3.00	–3.00
One drain for full-synthetic	—	3.00	11.00

pared to five different mineral oil fluids. For European driving, a limit of 15% maximum volatility is specified for CCMC (European Committee of Common Motor Manufacturers) G-3 performance, whereas Volkswagen specifies a more stringent 13%. It may be seen that a 15W–40 mineral oil-based formulation is required to meet this specification. The 5W–30 mineral oil formulation, which is used in North America for fuel economy and cold-starting reasons, does not come close to meeting the volatility standard.

Figure 1.10 compares the "high temperature/high shear" viscosity at 150°C of the full-synthetic 5W–50 formulation and the mineral oil formulations. The viscosity of the synthetic oil is even higher than the 20W–50 mineral oil. The outstanding performance of the synthetic oil is attributable to the naturally high VI of the PAO in combination with a shear-stable VI improver.

Figures 1.9 and 1.10 indicate that oil consumption should be under control with a full-synthetic formulation because of the superior volatility and viscosity performance. Figure 1.11 shows the results of a 12-car field test in which the oil consumption for the 5W–50 synthetic oil was compared with a 15W–50 mineral oil. The oil consumption for the synthetic oil was 25% less than for the mineral oil.

The data just presented for gasoline engines are equally valid for diesel engines. The CCMC D3 standard for super high-performance diesel (SHPD) engine oils can be met with a 5W–30 synthetic blend. Figure 1.12 shows that the full-synthetic SHPD oil gave approximately 2% increased fuel efficiency compared to the 15W–40 mineral oil SHPD across a range of typical driving modes.

C. Industrial Gear Oil Performance

The use of PAO-based gear oils in industrial settings can lead to important savings in energy consumption, as well decreased down-time and lower maintenance requirements. The wide range of operating temperatures allows the use of less viscous oils, which results in greater energy efficiency. The relatively low coefficient of friction for PAOs reduces the amount of internal friction created by the normal shearing of an oil film during operation.

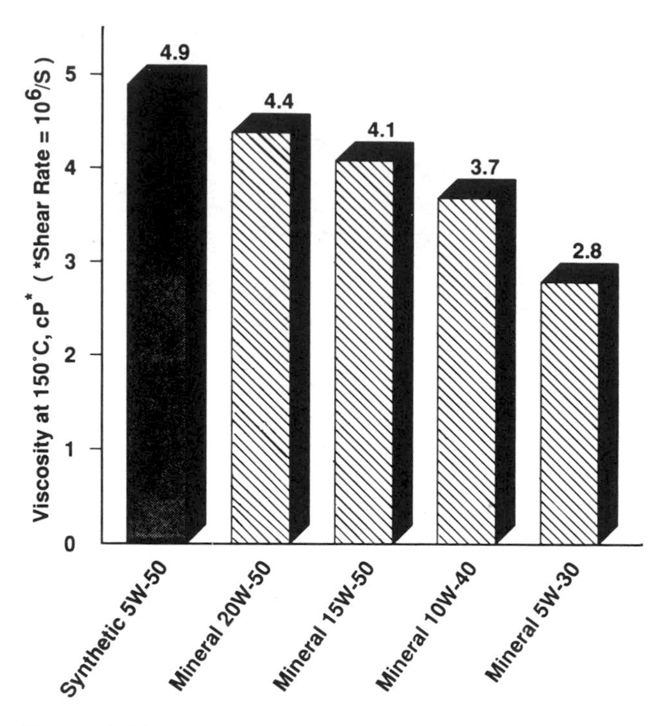

Figure 1.10 High-temperature/high-shear viscosity of fully formulated oils.

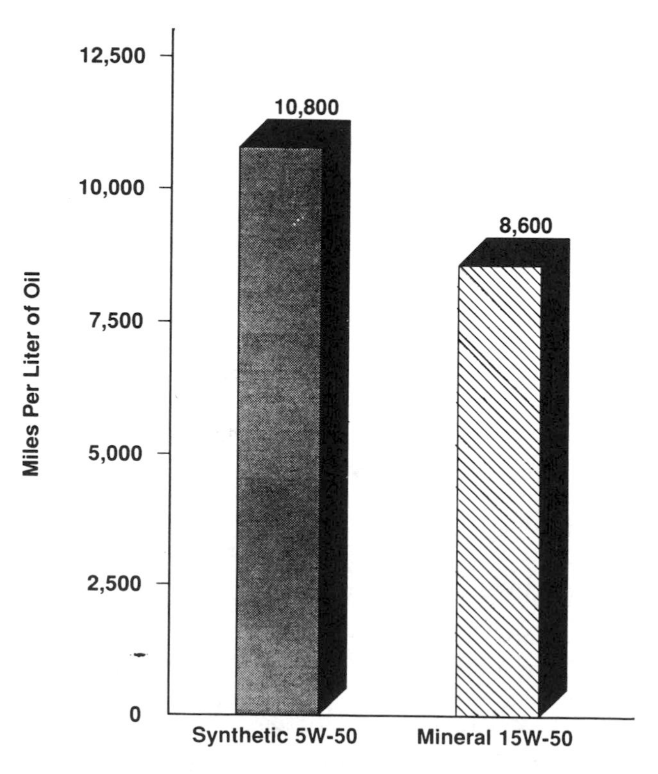

Figure 1.11 Average oil consumption for a range of modern cars.

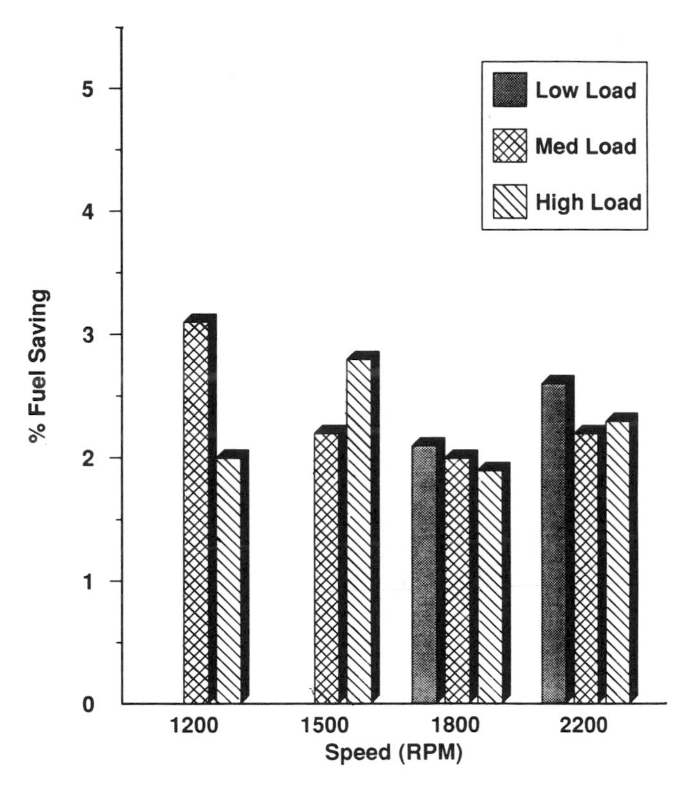

Figure 1.12 Fuel efficiency: super-high-performance diesel formulation. Percentage fuel saving for 5W–30 full synthetic versus 15W–40 mineral oil.

1. *Increased Efficiency*

Table 1.19 is a compilation of data from 10 reports relating to the benefits of increased efficiency found when industrial transmissions were switched from a mineral oil to PAO-based gear oils (48–57). The increases ranged from 2.2 to 8.8%. It is interesting to note that the efficiency increase observed in worm gears has a close positive correlation with the reduction ratio. This correlation exists despite the fact that the data were reported by different companies and were collected on different types of equipment.

2. *Cost Savings*

The open literature contains a number of reports of savings that have accrued to industrial concerns after they switched from a mineral oil to a PAO-based gear or bearing oil. Table 1.20 is a short tabulation of some of these reports (58). The table shows a diverse type of manufacturing for the companies included, and a diverse selection of applications for which the PAO-based lubricants were applied. The annual savings for these companies ranged from $12,000 to $98,000 per year. The largest reported savings on Table 1.20 was $98,000.00 per year when a PAO-based gear oil was used on the felt roll bearings in a paper mill. The high speed of the huge rolls in a paper mill is critical to their competitive operation, and the use of PAO-based fluids is becoming an important part of the overall strategy for cost-effective operation.

Table 1.19. Industrial Gear Oil Applications

Transmission type	Reduction ratio	Power (hp)	Load (% rated)	Efficiency increase (%)
Worm gear	10:1	3.0	100	2.2
Worm gear	15:1	1.55–2.0	100–130	3.8
Worm gear	25:1	6.5–8.1	100	4.4
Worm gear	30:1	3.2–3.9	96–117	5.6
Worm gear	50:1	3.0–6.0	100–200	7.7
Worm gear	50:1	0.5–1.0	50–100	8.8
Spur gear/chain	—	35	—	2
Spur gear	24:1	200	—	6
Series of nine worm gears	15:1	122–142	100–112	6
Series of five worm gears	39:1	75	—	5.8

D. Applications Sensitive to Health and Environmental Issues

1. Food Contact

The PAO base stocks are pure, saturated hydrocarbons. They contain no aromatics (except for small amounts in the 40- and 100-cSt fluids produced by Mobil) and no functional groups. As such, the toxicity is expected to be as low as or lower than the most highly refined white mineral oils.

The PAOs have Food and Drug Administration (FDA) approval for use in both "indirect" and "incidental" food-contact applications. They fall within the definition of a white mineral oil according to the Code of Federal Regulations, 21 CFR 178.3620, paragraph B. The applications for which FDA approval is required, and for which PAO is qualified, are listed in Table 1.21. In essence, PAO fluids may be used as a component of any material that contacts food or as a lubricant for any machinery that processes food. Direct food contact approval (i.e., as a component to be purposely ingested) has not yet been obtained, but probably could be obtained if there was an application that warranted the effort and expense of obtaining the approval.

Table 1.20. Savings with PAO-Based Gear Oils

Company type	Application	Annual savings
Soybean processing	Aeration blower	$2100/unit
Plastics	Bearing circulation system	$12,000
Copper wire	Line gears	$19,000
Paper mill	Felt roll bearings	$98,000
Steel mill	Fly ash blower shaft thrust bearings	$77,000
Pharmaceutical	Gear reducers	$70,000
Aluminum cans	Gear reducers	$35,000
Manufacturing	Various	$80,000

Table 1.21. FDA-Approved Applications for PAO

Section[a]	Application
175.105	Adhesives
176.200	Defoaming agents used in coatings
176.210	Defoaming agents used in the manufacture of paper and paperboard
177.2260	Production of resin-bonded filters
177.2600	Rubber articles (plasticizers) intended for repeated use
177.2800	Production of textiles and textile fibers
178.3570	Lubricants with incidental food contact
178.3910	Surface lubricants used in the manufacture of metallic articles (e.g., metallic foil)

[a]Food and Drug Administration, HHS-21 CFR Ch. 1 (4-1-88 edition).

2. Cosmetics and Toiletries

The PAO fluids are nontoxic when given orally to rats. The lethal dosage for 50% of the test subjects (LD_{50}) is greater than 5 g/kg of body weight. PAOs are also nonirritating to the eyes and skin of test animals, and they are not expected to induce sensitization reactions. They have low vapor pressures and therefore are not hazardous by inhalation. Subjectively, PAOs are said to have a better "feel" on human skin than white mineral oils. For all of these reasons, a small but growing market for PAO is developing in the cosmetics industry. A national brand of lipstick contains PAO as a major component.

3. Off-Shore Drilling

Regulations on the marine toxicity of fluids used to lubricate the drill head in off-shore drilling operations are becoming tighter, especially in the North Sea. The LC_{50} values for mineral oil-based drilling muds are in the range of 20,000–30,000 ppm. North Sea regulation is expected to require a minimum value of 50,000 ppm in the near future. The PAOs have been found to have an LC_{50} value of 200,000 ppm. Drilling muds formulated from experimental PAO fluids are currently being evaluated.

4. Miscellaneous

Other environmentally sensitive areas for which PAO fluids are being evaluated are: logging operations (chain saws), marine outboard engines, and hydraulic systems for large farm machinery.

In addition to low toxicity, it is important that fluids used in these applications exhibit biodegradation and low levels of bioaccumulation. Preliminary evaluations indicate that PAOs do not bioaccumulate and that their rate of biodegradation is faster than that of mineral oils of comparable viscosity. On the other hand, the rate of biodegradation is slower than for some ester-based drilling muds that are also undergoing evaluations for this application. For a full discussion of this very complex issue, the reader is referred to Chapter 25 of this book.

E. Military Applications

The earliest applications for PAO fluids were in the military. Mil-H-83282 is a specification for a hydraulic fluid for jet aircraft. The specification was built around an experimental 4.0-cSt decene-based PAO produced by Mobil in the late 1960s. The requirements included extreme low-temperature fluidity as well as high flash- and fire-point values. The latter requirement was to minimize the risk of loss due to fire in the event that a hydraulic line was severed by enemy gunfire. Mil-H-83282 remains an important military fluid today.

An interesting, if not publicized, example of superior performance for PAO came to light as a result of the war in the Persian Gulf in January 1991. Under harsh desert conditions, the U.S. weapons that were lubricated and cleaned with PAO-based oils performed better than similar Allied weapons using conventional fluids, resulting in some rush orders to the lubricant formulators from Allied commanders.

Table 1.22 contains a short summary of military specifications that either require or often use PAO fluids.

VI. MARKETS AND PRODUCTION CAPACITIES

By the end of 1990, world PAO demand had grown to 27.3 million (MM) gallons (85.6 M metric tons) (9). This volume represents a remarkable 14-fold increase since 1975, but still represents less than 0.05% of the total world lubricant base-stock market. During 1975–1980, demand for PAO grew at 33% per year. Synthetic engine oils were a novelty on the market during this period, and they were growing from a base near zero. Growth slowed during the 1980–1985 period to around 7%. Some early product entrants to the market were improperly formulated, and the resultant poor performance attached some stigma to the use of synthetics. The 1985–1990 time period saw a strong new interest in synthetic lubricants because of the enactment of stringent new specifications and governmental regulations that were difficult to meet with mineral oil base stocks. The growth rate for PAO during the last half-decade has been 19% per year.

A. Demand by Segment and Region

Strong growth for the PAO market is predicted to continue in the foreseeable future. Table 1.23 shows the expected rate of growth for PAO into the automotive, industrial, military, and emerging market segments. The total market is expected to grow from 185 MM lb in 1990 to 450 MM lb in 1995—an annual growth rate of about 20% per year (23, 58).

Although the size of the PAO markets in 1990 were approximately the same in Europe and North America, the breakdown by segments was considerably different. Figure 1.13 illustrates the fact that the European market was driven primarily by the automotive demand whereas the North American market was more balanced. In Europe, 78% of the PAO demand was for the automotive sector, with the rest going into industrial applications. In North America, the automotive and industrial markets each took about 38% of the PAO, while the military used 17%. The remainder went into "emerging" markets, which will be discussed in more detail later.

Table 1.24 is a breakdown of PAO market growth by both segment and region. It should be noted that the 1995 forecasts predict that the demand distribution by segment for PAO in Europe and North America will converge. North America will be catching up with Europe in the automotive applications area while Europe will be catching up with North America in industrial applications. Both continents are expected to undertake vigorous development of the "emerging" segments.

The development of markets and applications for PAO has been generally confined to

Table 1.22. Military Applications

Specification number	Applications	Lubricant Highlights
MIL-H-46170	Type I: Tank recoil and hydraulic systems Type II: Aerospace test stands	4-cSt PAO: ester: TCP PAO base stock specs Finished fluid specs
MIL-H-83282	Aircraft and missile hydraulic systems	4-cSt PAO: ester: TCP PAO base stock specs Finished fluid specs
MIL-H-(83282 low temperature)	Aircraft and missile hydraulic systems	Dimer/trimer ~3-cSt PAO: ester: TCP *No* pour point or VII additives Finished fluid specs only
MIL-G-10924	Multipurpose grease for all ground vehicles, artillery, and equipment	Typically 6-cSt PAO base stock Finished grease specs only Formulation and constituents Confidential and proprietary
MIL-L-63460	Small and large caliber weapons cleaner, lubricant, and preservative, –65° to 150°F	Mineral oil and/or synthetic based 2- and/or 3-cSt PAO Finished lube specs only
MIL-L-2104	I/C engine oil and power transmission fluids All types of military tactical/combat ground equipment	Mineral oil, synthetic, or combination base stock
MIL-L-2105	Gear oil for units, heavy-duty industrial type gear units, steering gear units, and universal joints	Mineral oil, synthetic, or combination base stock Finished lube specs by grade only
MIL-C-87252	Dielectric coolant for electronic applications Hydrolytically stable Replacing silicate ester coolant	PAO base stock specified 2-cSt Dimer ~99.5% PAO Oxidation/corrosion inhibitor Finished fluid specs only

Table 1.23. PAO Market Segment Growth

Segment	1990, MM lb	1995, MM lb	Growth rate, %/year
Automotive	110	230	18
Industrial	55	100	13
Military	15	20	6
Emerging	5	100	85
Total	185	450	20

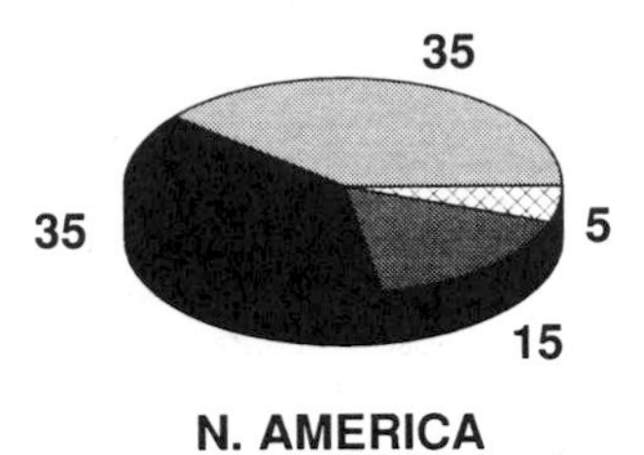

N. AMERICA

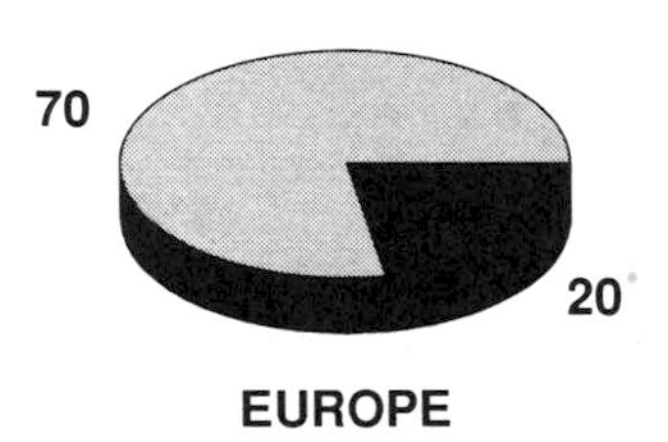

EUROPE

(PAO MARKET, MM LBS)

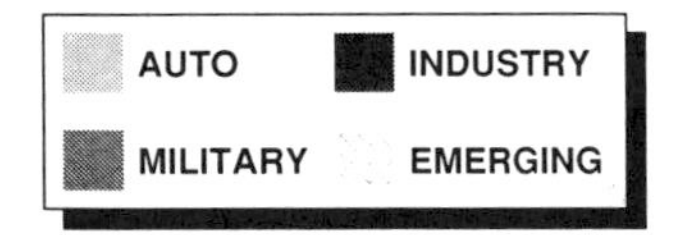

Figure 1.13 1990 PAO market.

Table 1.24. PAO Market Segment Growth by Region

Region and segment	1990, MM lb	1995, MM lb	Growth rate, %/year
North America			
Automotive	35	100	23
Industrial	35	60	11
Military	15	20	6
Emerging	5	40	51
Total	90	220	20
Europe			
Automotive	70	110	10
Industrial	20	40	15
Military	—	—	—
Emerging	—	60	N/A
Total	90	210	19
Far East			
Automotive	4	15	30
Industrial	1	5	38
Military	—	—	—
Emerging	—	—	—
Total	5	20	32

North America and Europe. In 1990, the Far East accounted for only 2.7% of the demand for PAO. Between 1990 and 1995, the consumption of PAO in the Far East is expected to grow at an annual rate of about 32%—reminiscent of the growth in the West during the 1975–1980 time frame. By 1995, the demand in the Far East will have grown to 4.4% of the total world-wide demand.

B. Emerging Markets

A substantial portion of the growth being forecast for PAO has been described as "emerging" markets. The term "emerging" is used to designate application areas where there is a high potential for PAO to capture a part of the market now being serviced by other types of fluids.

Table 1.25 lists eight areas where PAO fluids and formulations are being developed to fulfill specific requirements not being met by the fluids currently in use. The three driving forces for shifting from the current functional fluid to PAO are cost, performance, and toxicity. If the markets described in Table 1.25 develop as forecast, the total demand by 1995 could be as high as 165 mm lb.

C. PAO Production Capacity

At the end of 1990, the world-wide production capacity for PAO was 325 MM lb/year. Sales for 1990 were 57% of production capacity, which represented a major reversal of the demand/supply situation of the mid-1980s. Because of the shortage of PAO available at that time, formulators were forced to seek alternative (if sometimes less satisfactory) solutions for their performance requirements. There was a strong and understandable hesitancy among equipment manufacturers, formulators, and end users to place themselves in a precarious supply situation. As a result of the excellent supply situation that now exists, there is a new surge of activity in the development of new markets and applications for PAO fluids.

Table 1.26 is a summary of the PAO producers and their capacities in 1990, as well as a projection for 1995. In fact, much of the increased capacity for 1995 has already been

Table 1.25. Emerging Markets

Product line	Current volume, MM lb	Fluid type	1995 PAO potential, MM lb	Driving force
Polymer	200	WMO[a]	20	Toxicity/performance
Personal care	150	WMO/esters[a]	25	Toxicity
Refrigeration	45	Polyalkyl glycols	15	Performance
Textile	25	Silicones/WMO	5	Cost and performance
Dielectric fluids	90	Mineral/silicones/PCB	20	Cost/performance/toxicity
Drilling fluids	110	WMO	30	Toxicity and performance
Brake fluid	440	Polyethylene glycol/ silicones	30	Cost and performance
Shock absorbers	110	Mineral oil	20	Performance
Total	1170		165	

[a]White mineral oil.

Table 1.26. PAO Capacity (World-Wide)

Manufacturer	1990, MM lb/year	1995, MM lb/year	1995 Operating rate, %
Ethyl	170	250	90
Mobil	80	110	91
Chevron	55	55	100
Exxon	20	20	100
Neste	0	55	63
Other	—	25	55
Total	325	515	

realized in 1991. Ethyl Corporation brought their 80-MM lb/year plant in Feluy, Belgium, on-stream in January 1991. Neste will bring their 45-MM lb/year plant in Berigen, Belgium, on-stream by mid-1991. The operating rate for the combined PAO plants is expected to be around 87% by 1995.

D. Competitive Products

Highly refined mineral oils approach PAO in performance characteristics. These mineral oils fall into three categories.

1. Very High Viscosity Index (VHVI) Oils

The premier product derived from crude oil sources is Shell Oil's patented extra high viscosity index (XHVI) oil. It is produced in France and Australia from a special cut of refinery slack wax by a severe hydrocracking procedure. Shell's capacity is 150 MM lb/year, and it is thought to be limited by the availability of feedstock. The product exhibits very good performance characteristics, but it is deficient relative to PAO in both low-temperature properties and volatility (Table 1.4).

2. High Viscosity Index (HVI) Oils

The HVI base stocks are intermediate in properties between the VHVI fluids and conventional solvent-refined oils (Table 1.3). HVI oils are manufactured by a process that involves hydrotreating, redistilling, and solvent refining. HVI fluids were first produced by British Petroleum in 1976 and are now produced by BP at LaVera and Dunkerque in France. The HVIs are also produced by Modrica in Yugoslavia and DEA in Germany. The HVI fluids are less costly than either PAO or VHVI fluids, but 1.4–2.0 times more material is generally required to blend with an off-specification mineral oil to bring a formulation into 10W–30 specifications.

3. Polyinternalolefins

Polyinternalolefin (PIO) fluids are similar to PAO fluids in that they are both manufactured by the oligomerization of linear olefins. The olefins used for PIO manufacture, however, are derived from the cracking of paraffinic base stocks. The internal olefins are more difficult to oligomerize than the α-olefins derived from ethylene chain growth, and the products have VIs that are 10–20 units lower than comparable PAOs. Plans by Enichem for the start-up of a plant have reportedly been delayed, probably until 1992.

VII. CONCLUSION

A number of forces will drive the growth of high-performance functional fluids for the next decade and beyond. These forces derive from diverse societal needs, but they have a common goal rooted in the uniquely human belief that there must be a better way to do whatever it is that has to be done. Some of these forces and the consequences they imply for the growth of PAO fluids are discussed next.

A. Regulatory Forces

United States regulations for corporate average fuel economy (CAFE) are having a significant effect in both the design of new automobiles and the lubricant requirements and recommendations for them. In order to meet the fuel economy standards, which will now include a cold-start cycle specification, the original equipment manufacturers (OEMs) are being forced into recommending SAE 5W multigrade crankcase oils. In addition, increased emphasis on aerodynamics results in higher engine operating temperatures, which affects not only the crankcase lubricant but also the transmission fluid, the gear oils, and the greases. All of these factors will require fluids with lower low-temperature viscosity coupled with lower volatility, higher thermal and oxidative stability, higher heat capacity, and higher heat conductivity.

Consumer protection, worker safety, and environmental regulations are becoming increasingly stringent in their requirements for nontoxic, nonhazardous, environmentally friendly products. Regulatory agencies are beginning to recognize potential toxicological problems associated with white mineral oils. The PAOs are being put forward as high-performance, safe substitutes.

B. Performance and Cost-Effectiveness

The need for improved performance remains a critical factor in the drive toward increased usage of PAO-based lubricants and functional fluids. In many applications mineral oil-based products either cannot meet the more stringent requirements or are only marginally satisfactory. The use of PAOs for blending with marginal base stocks in order to bring them into specification is increasing.

Fleet operators, who are much more sensitive to cost-effectiveness than the general public, will continue to increase their usage of synthetics as they recognize the potential long-term savings.

Operators of large industrial machinery are beginning to recognize the increased cost-effectiveness of lubrication with PAO-based products. Machines operate at lower temperatures, are less subject to wear, require less maintenance and down-time, consume less oil, and operate longer between lubricant drain cycles. The value of PAO-based products for the lubrication of heavy-duty, off-road mobile equipment is also being recognized, especially in situations where routine maintenance is difficult.

C. Original Equipment Manufacturers

The diversity and regional availability of mineral oil base stocks make standardization based on mineral oils difficult. In those applications where performance requirements are exacting, there will be a shift by OEMs to require synthetic fluids in order to assure uniformity.

Industrial and automotive OEMs are under pressure from the consumer and from their competition to extend warranty periods. At the same time OEMs are under pressure to reduce the required amount of maintenance and down-time. Both avenues may be addressed by switching from mineral oil to PAO-based fluids. For the first time, General Motors will use a full-synthetic, PAO-based oil as the factory-fill and recommended crankcase fluid when it introduces the 1992 Chevrolet Corvette.

The latest trend to address consumer convenience as well as protection of equipment from inadvertent contamination of the working fluid is the "fill-for-life" concept. General Motors is studying a "fill-for-life" PAO-based automatic transmission fluid in its 1992 Cadillac Allante models.

D. Petroleum Companies and Blenders

Lubricant producers have historically had low profit margins. Base stock prices have been closely tied to crude oil prices, and the selling price for finished fluids has remained tied to base stock costs. Lubricant companies are beginning to recognize that high-performance, high-image products based on PAO afford the opportunity for higher selling prices and increased margins. European companies have been the leader in this regard, but North American companies are expected to catch up. While Mobil Oil has been the leader in the U.S. with Mobil 1 since the mid-1970s, most of the major lubricant producers have introduced, or plan to introduce, full-synthetic motor oils to the market in 1990s.

Another large factor in the forecasted growth for PAO is the recognition that there are insufficient high-quality base fluids to meet new product requirements. PAOs will be used to blend mineral oil stocks into specification. Some of these products will be sold and marketed as "part-synthetic" oils at a price between the top-tier mineral oils and the "full-synthetics." In other cases, the blender or formulator will use PAO in an "in again–out again" basis, depending on the availability of mineral oil base stocks of sufficiently high quality. In these cases, the consumers will never know that they have purchased a "part-synthetic."

E. Consumer

The role of the consumer will be important to the growth of PAO fluids. Manufacturers recommendations will have little effect if the consumer does not pay attention to them. Studies show that the traditional attitude in the United States has been that all oils are "pretty much the same," but that attitude is beginning to change. Consumers are becoming more aware of fuel economy, cleaner air, higher performance, lower maintenance, and longer vehicle life. All of these concerns, coupled with the increased availability of oils to meet the demand, will lead to a shift by a segment of the consumer population toward the premium synthetic oils.

F. New Technology

The final area that will provide an impetus to the increased use of PAO fluids will be the development of new technology. Two areas are clearly important. The first is the development of new additives and formulation packages specifically designed for use with PAO fluids. Formulation development is being actively pursued by PAO producers, additive manufacturers, formulators, lubrication specialty companies, and OEMs. Some

Table 1.27. Physical Properties: Effect of Olefin Chain Length

Property	Carbon number of initial olefin			
	8	10	12	14
KV at 100°C, cSt	2.77	4.10	5.70	7.59
KV at 40°C, cSt	11.2	18.7	27.8	41.3
KV at −18°C, cSt	195	409	703	1150
Viscosity index	82	121	152	154
Pour point, °C	<−65	<−65	−45	−18
Flash point, °C	190	228	256	272
NOACK,[a] % loss	55.7	11.5	3.5	2.3

[a]Volatility at 250°C.

of this work is in the form of joint efforts, and much of the information being developed is proprietary.

The second important area to be impacted by new technology is the development of new PAO fluids from new starting materials and/or with new catalyst systems. The objective in this research is to produce products with particular characteristics needed for specialty applications.

The use of alternative (other than 1-decene) olefin streams as the starting olefin for PAO manufacture offers the opportunity to "tailor-make" products for niche markets (16). Table 1.27 gives an indication of what happens when different linear α-olefins are reacted in an identical way. As mentioned earlier, decene was chosen as the raw material of choice by all of the PAO producers because it gives products with the broadest temperature operating range. For many applications, however, properties exhibited at one end of the temperature range may be more important than those at the other. For instance, a piece of industrial machinery that runs continuously at high temperature may have few, if any, low-temperature requirements but may require a very stringent volatility or flash-point specification. In such a case, a PAO based on 1-dodecene or 1-tetradecene may be more appropriate. Performance characteristics that can be enhanced by the appropriate choice of starting olefin and reaction conditions include volatility, pour point, viscosity index, low-temperature viscosity, flash and fire points, thermal and oxidative stability, and biodegradability (60).

The development of new catalyst systems for the production of olefin oligomers having specific isomer distributions also holds the potential for the development of new PAO products with enhanced characteristics (61).

REFERENCES

1. Shubkin, R. L. (1989). Synthetic lubricants, in *Alpha Olefins Applications Handbook,* G. R. Lappin and J. D. Sauer (Eds.), Marcel Dekker, New York, Chapter 13, pp. 353–373.
2. Gunderson, R. C. and A. W. Hart (1962). *Synthetic Lubricants,* Reinhold, New York.
3. Sullivan, F. W., Jr., V. Vorhees, A. W. Neeley, and R. V. Shankland (1931). *Ind. Eng. Chem.,* 23, 604.
4. Boylan, J. B. (1987). Synthetic basestocks for use in greases, *NGLI Spokesman,* 51(5), 188–195.

5. Montgomery, C. W., W. I. Gilbert, and R. E. Kline (1951). U.S. Patent 2,559,984, to Gulf Oil Co.
6. Garwood, W. E. (1960). U.S. Patent 2,937,129, to Socony-Mobil.
7. Southern, D., C. B. Milne, J. C. Moseley, K. I. Beynon, and T. G. Evans (1961). British Patent 873,064, to Shell Research.
8. Brennan, J. A. (1968). U.S. Patent 3,382,291, to Mobil Oil.
9. Shubkin, R. L. (1973). U.S. Patent 3,763,244, to Ethyl Corp.
10. Shubkin, R. L. (1973). U.S. Patent 3,780,128, to Ethyl Corp.
11. Shubkin, R. L., M. S. Baylerian, and A. R. Maler (1979). Olefin oligomers: Structure and mechanism of formation, presented at the *Symposium on Chemistry of Lubricants and Additives,* Division of Petroleum Chemistry, ACS, Washington, D.C., September 9–14. Also published in *Ind. Eng. Chem., Product Res. Dev.,* 19, 15 (1980).
12. Stewart, R. D., K. E. McCaleb, L. E. Rodgers, and T. Sasano (1988). *Synthetic Lubricants (Worldwide),* Report by Specialty Chemicals • SRI International, April.
13. Sullivan, F. W. (1934). U.S. Patent 1,955,260, April, to Standard Oil Company (Indiana).
14. Onopchenko, A., B. L. Cupples, and A. N. Kresge (1983). *Ind. Eng. Chem. Prod. Res. Dev.,* 22, 182–191.
15. Driscoll, G. L., and S. J. G. Linkletter (1985). *Synthesis of Synthetic Hydrocarbons via Alpha Olefins,* Air Force Wright Aeronautical Laboratories report designation AFWAL-TR-85-4066, May.
16. Shubkin, R. L., and M. E. Kerkemeyer (1990). *Tailor Making PAOs,* presented at the 7th International Colloquium on Automotive Lubrication, Technische Akademie Esslingen, Federal Republic of Germany, January 16–18. Also published in *J. Synth. Lubr.,* 8(1), (1991).
17. Loveless, F. C. (1977). U.S. 4,041,098, August 9, to Uniroyal, Inc.
18. Ashjian, H. (1989). EP Application 88312436.4, July 12, to Mobil Oil Corporation.
19. Wu, M. M. (1989). U.S. 4,827,073, May 2, to Mobil Oil Corporation.
20. Ripple, D. E., and J. F. Fuhrmann (1989). Synthetic basics—Performance comparisons of synthetic and mineral oil crankcase lubricant base stocks, *J. Synth. Lubr.,* 6(3), 209–232.
21. Koch, B. (1989). Thermal stability of synthetic oils in aviation applications, *J. Synth. Lubr.,* 6(4), 275–284.
22. Conte, Jr., A. Alfeo, M. Ruzansky, and R. Chen (1988). The hydrolytic stability of aircraft coolant/dielectric fluids, *J. Synth. Lubr.,* 5(1), 13–29.
23. Unpublished data, Ethyl Corporation.
24. van der Waal, G. (1987). Improving the performance of synthetic base fluids with additives, *J. Synth. Lubr.,* 4(4), 267–282.
25. Shkolnikov, V. M., O. N. Zvetkov, M. A. Chagina, and G. V. Kolessova (1990). Improvement of antioxidant and antiwear properties of polyalphaolefin oils, *J. Synth. Lubr.,* 7(30), 235–241 (1990).
26. Davis, J. E. (1987). Oxidation characteristics of some oil formulations containing petroleum and synthetic base stocks, *J. Am. Soc. Lubr. Eng.,* 43(3), 199–202.
27. Gunsel, S., E. E. Klaus, and J. L. Bailey (1986). *Evaluation of some Poly-Alpha-Olefins in a Pressurized Penn State Microoxidation Test,* presented at the 41st Annual ASLE Meeting, Toronto, May 12–15. Available as ASLE Preprint No. 86-AM-2C-1.
28. Gunsel, S., E. E. Klaus, and J. L. Duda (1987). *High Temperature Deposition Characteristics of Mineral Oil and Synthetic Lubricant Base Stocks,* presented at the STLE/ASME Tribology Conference. San Antonio, October 5–8. Available as STLE Preprint No. 87-TC-3B-3.
29. Neadle, D. J. (1986). Synthetic lubricants for turbocharged passenger cars, *J. Synth. Lubr.* 2(4), 311–327.
30. Papay, A. G., E. B. Rifkin, R. L. Shubkin, P. F. Jackisch, and R. B. Dawson (1979). *Advanced Fuel Economy Engine Oils,* SAE Paper 790947, SAE Fuels and Lubricants

Meeting, Houston, Tex., October 1–4. Also in *Synthetic Automotive Engine Oils,* Progress in Technology Series 22, Society of Automotive Engineers, Warrendale, Pa. (1981), pp. 237–248.
31. Law, D. A., J. R. Lohuis, J. Y. Breau, A. J. Harlow, and M. Rochette (1984). Development and performance advantages of industrial, automotive, and aviation synthetic lubricants, *J. Synth. Lubr.,* 1(1), 6–33.
32. Milligan, G. W., R. S. Robertson, and W. R. Murphy (1985). Synthetic gear oil performance, *Iron Steel Eng.* 63(6), 49–53.
33. Muraki, M., and Y. Kimura (1986). Traction characteristics of lubricating oils (5th report)—Traction characteristics of synthetic hydrocarbon oils, *J. JSLE, Int. Ed.,* 7, 119–124.
34. Baudouin, P., G. Chocha, and H. Raich (1985). Performance testing of a multiservice synthetic grease for industrial and automotive use, *J. Synth. Lubr.,* 2(3), 213–238.
35. Wunsch, F. (1991). *Synthetic Fluid Based Lubricating Greases,* presented at NLGI's 57th Annual Meeting, October, Denver, Colo. Also in *NLGI Spokesman,* 54(11), 454–464 (1991).
36. Tedrowm, L. E., and F. S. Sayles (1984). Field performance of synthesized hydrocarbon (polyalphaolefin) greases, *NLGI Spokesman,* 47(11), 395–398.
37. Willermet, P. A., C. C. Haakana, and A. W. Sever (1985). A laboratory evaluation of partial synthetic automatic transmission fluids, *J. Synth. Lubr.,* 2(1), 22–38.
38. Miller, J. W. (1984). Synthetic lubricants and their industrial applications, *J. Synth. Lubr.,* 1(2), 136–152.
39. Cohen, S. C. (1991). Development and testing of a screw compressor fluid based on two-stage hydrotreated base oils, and comparison with a synthetic fluid, *J. Synth. Lubr.,* 7(4), 267–279.
40. Daniel, G., M. J. Anderson, W. Schmid, and M. Tokumitsu (1982). Performance of selected synthetic lubricants in industrial heat pumps, *Heat Recovery Systems,* 2(4), 359–368.
41. Jayne, G. J. J., and A. P. Jones (1984). Progress in development of synthetic compressor oils, *Ind. Lubr. Tribol.,* 36(3), 90–98.
42. Robson, R. J. (1987). Polyalphaolefins as electrical insulating fluids, *Proceedings: Workshop on Substitute Insulation for Polychlorinated Biphenyls,* 1986, Report No. EL-5143-SR, Electric Power Research Institute, Palo Alto, Calif.
43. Patter, R. I., M. Campen, and H. V. Lowther. (Eds.) (1981). *Synthetic Automotive Engine Oils,* Progress in Technology Series 22, Society of Automotive Engineers, Warrendale, Pa.
44. Coffin, P. S., C. M. Lindsay, A. J. Mills, H. Lindenkamp, and J. Fuhrman (1990). The application of synthetic fluids to automotive lubricant development: Trends today and tomorrow, *J. Synth. Lubr.,* 7(2), 123–143.
45. Nagdi, K. (1989). Seal materials for synthetic lubricants and working fluids, *Proc. Conf. Synth. Lubr.,* pp. 304–32.
46. Nagdi, K. (1990). Polyalphaolefins and seal materials, *Eur. J. Fluid Power,* November, pp. 40–48.
47. Campen, M. (1981). A chemical introduction to "synthetic automotive engine oils"—Their sources, classes, advantageous properties, and fuel saving, cost/performance benefits, in *Synthetic Automotive Engine Oils,* Progress in Technology Series 22, Society of Automotive Engineers, Warrendale, Pa., pp. 1–10.
48. Anonymous (1983). Extending compressor valve cleaning periods with a synthetic compressor lubricant, *Fluid Lubr. Ideas,* 6(5), 24.
49. Anonymous (1983). Synthetic lubricants reduce downtime at Midwest power plant, *Fluid Lubr. Ideas,* 6(5), 20–22.
50. Anonymous (1982). Synthetic lubricant saves energy, increases oil-change interval, *Eng. Min. J.,* 183(10), 111.
51. Facciano, D. L., and R. L. Johnson (1985). Examination of synthetic and mineral based gear lubricants and their effect on energy efficiency, *NGLI Spokesman,* 48(11), 399–403.

52. Sinner, R. S. (1986). Synthetic lubricants—Why their extra cost can be justified, *Marine Eng. Rev.,* August, pp. 18, 20–21.
53. Black, P. A., and H. E. Knobel (1985). Synthetic lube oils improve performance, *The Motor Ship,* 66(782), 30–31.
54. Faufau, J., and T. C. Nick (1989). Synthetic lubricants can reduce downtime and extend bearing life, *Pulp & Paper,* 63(1), 127–128.
55. Fredel, W. (1984). Synthetic lubricants help Mosinee paper overcome temperature problems and save money, *Paper Trade J.,* 168(18), 56.
56. Schlenker, H. O. (1982). Synthetic lubricant upgrades worm gear capacities, *Power Transmission Design,* 24(7), 35–37.
57. Anonymous (1983). Improving industrial gear system performance with synthesized lubricants, *Fluid Lubr. Ideas,* 6(2), 9–10.
58. Edwards, D. J. (1983). Synthetic lubricants get tougher, *Plant Engineering,* August 18, 59–60.
59. Anonymous (1991). *Non-Conventional Lubricant Base Stocks 1990–2000 (A World Study),* report by Ozimek Data Corporation, March.
60. Kumar, G., and Shubkin, R. L. (1992). New polyalphaolefin fluids for specialty applications, paper presented at the *STLE Annual Meeting,* Philadelphia, PA, May 4–7.
61. Theriot, K. J., and R. L. Shubkin (1992). *A Polyalphaolefin with Exceptional Low Temperature Properties,* paper presented at the 8th International Colloquium on Tribology, Tribology 2000, sponsored by the Technische Akademie Esslingen, Federal Republic of German, January 14–16.

2
Esters

Steven James Randles
ICI Chemicals and Polymers, Ltd.
Middlesbrough, Cleveland, England

I. INTRODUCTION

Prior to the early nineteenth century, the main lubricants were natural esters contained in animal fats such as sperm oil and lard oil, or in vegetable oils such as rapeseed and castor oil. During World War II a range of synthetic oils was developed. Among these, esters of long-chain alcohols and acids proved to be excellent for low-temperature lubricants. Following World War II, the further development of esters was closely linked to that of the aviation gas turbine. In the early 1960s, neopolyol esters were used in this application because of their low volatilities, high flash points, and good thermal stabilities. Esters are now used in many applications including automotive and marine engine oils, compressor oils, hydraulic fluids, gear oils, and grease formulations. The low toxicity and excellent biodegradabilities of ester molecules now afford added benefits to those of performance.

II. MARKET POSITION

In 1992 the total consumption of lubricating oils in Western Europe is predicted to be over 6 million tonnes per year. Esters represent only a very small share (0.6%) of the total lubricant market. However, whereas the consumption of lubricating oils overall is stagnating, the synthetic oils are showing an average annual growth of 11% (1). Esters have a strong (8%), in some applications step change, growth rate. Figure 2.1 shows the growth

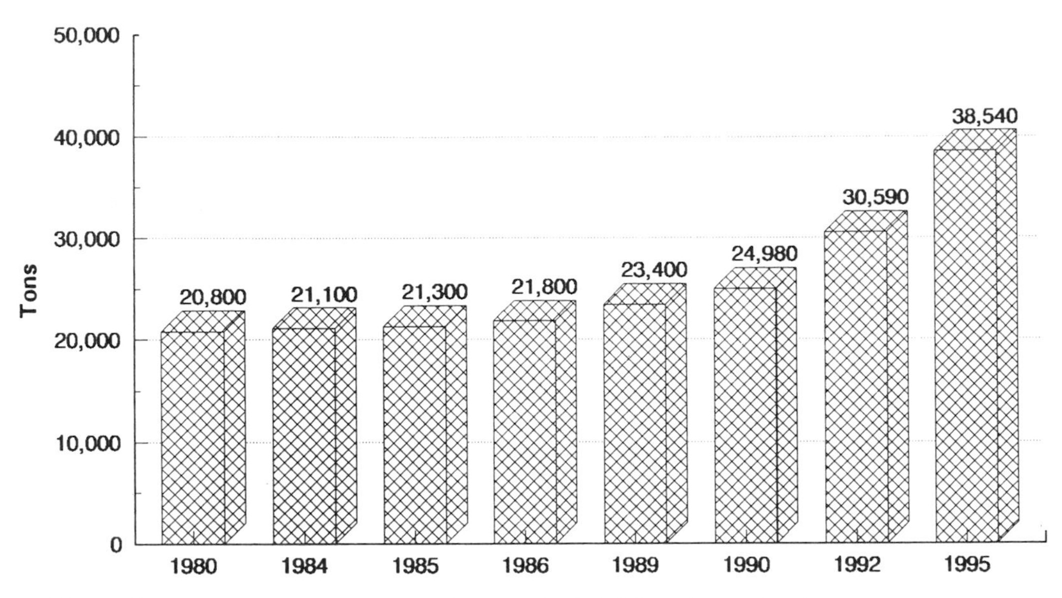

Figure 2.1 Sales of esters (Western Europe), 1980–1995 in thousands of tonnes.

in ester sales since 1980. Figure 2.2 shows projected sales (tonnes) of esters in 1992 split by application.

The volume of synthetics used in engine oils have substantially increased since 1988, with growth rates of esters being in the region of 15% pa. This is due to the increased use of synthetics in crankase oils. This move towards synthetics has been entirely performance driven. Growth in synthetic engine oils outside Western Europe is now being seen in North America, Southeast Asia, Japan, and South Africa.

The growth in aviation oils is expected to follow the growth in the aviation market. Step changes in growth could occur if and when new markets open up in Eastern Europe.

Step changes in the sales of esters into the two-stroke and hydraulic fluid market are expected to result from changes in environmental legislation in several European Economic Community nations, especially Germany.

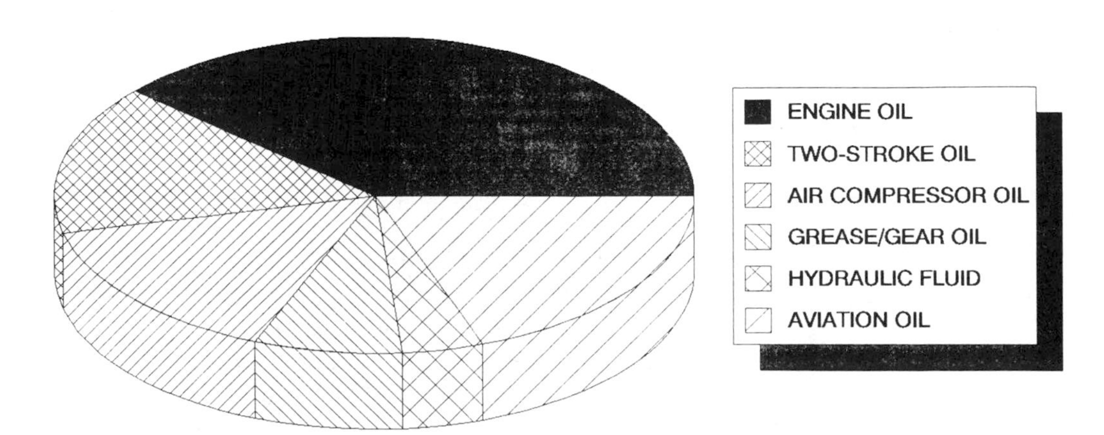

Figure 2.2 Sales of esters (Western Europe) in 1992 split by application.

The air compressor market is well established with only moderate market growth expected, unless legislation further restricts the use of mineral oil.

III. ESTER TYPES

The direct effect of the ester group on the physical properties of a lubricant is that it lowers the lubricants volatility and raises the flash point. This is due to the strong dipole moments, called the London forces, binding the lubricant together. The presence of the ester group also effects other properties, such as thermal stability, hydrolytic stability, solvency, lubricity, and biodegradability.

These properties are discussed at greater length later in the chapter. There are several major types of esters. These and their feedstocks are reviewed in Table 2.1. Table 2.2 summarizes the physical properties of these esters.

IV. MANUFACTURE OF ESTERS

The manufacturing process of esters consists of four distinct processes; esterification, neutralization, distillation, and filtration. This process is summarized in Fig. 2.3.

The fundamental reaction of acid+alcohol$\rightleftharpoons$ester+water is the basis for the produc-

Table 2.1 Types of Esters and Their Feedstocks

Ester type	Alternative name	Acid/ anhydride	Alcohol	Structure
Diester	Dioates	Diacids	Alcohol	Figure 2.12
Phthalate	1,2-Benzene dicarboxylate	Phthalic anhydride	Alcohol	Figure 2.14
Trimellitate	1,2,4-Benzene tricarboxylate	Trimellitic anhydride	Alcohol	Figure 2.16
Pyromellitate	1,2,4,5-Benzene tetracarboxylate	Pyromellitic anhydride	Alcohol	
Dimer acid ester	Dimerates	C_{36} Dimer acid	Alcohol	Figure 2.18
Polyols	Hindered esters	Mono acids	NPG TMP PE	Figure 2.20
Polyoleates		Oleic acid	NPG/TMP/PE	Figure 2.20

Commonly used feedstocks:

Alcohols: *n*-hexanol (C_6), *n*-heptanol (C_7), isoheptanol (C_7), *iso*-octanol (C_8), 2-ethyl hexanol (C_8), C_7–C_9 mixed cut, C_8 + C_{10} mixed cut, isononal (C_9), C_9–C_{11} mixed, *iso*-decanol (C_{10}), tridecanol (C_{13}).

Mono acids: valeric (C_5), heptanoic (C_7), pergalonic (C_9), C_8 + C_{10} cut, oleic acid (cut of C_{12}, C_{14}, C_{16}, and C_{18}).

Diacids: adipic, azelaic, sebacic, dodecanedioc.

Dimer acids: C_{36} dimer acids, hydrogenated C_{36} dimer acids.

Multifunctional alcohols: neopentylglycol (NPG), trimethylolpropane (TMP), pentaerythritol (PE).

Table 2.2 Summary of Ester Properties

Property	Diesters	Phthalates	Trimel-litates	C_{36} Dimerates	Polyols	Poly-oleates
Viscosity at 40°C, cSt	6–46	29–94	47–366	13–20	14–35	46–100
Viscosity at 100°C, cSt	2–8	4–9	7–22	90–185	3–6	10–15
Viscosity index	90–170	40–90	60–120	120–150	120–130	130–180
Pour point, °C	–70 to –40	–50 to –30	–55 to –25	–50 to –5	–60 to –9	–40 to +8
FLASHPOINT, °C	200–260	200–270	270–300	240–310	250–310	220–280
Thermal stability	Good	Very good	Very good	Very good	Excellent	Fair
Biodegradability	Excellent	Good	Poor	Fair	Excellent	Excellent
Cost (PAO = 1)	0.9–2.5	0.75–1.25	1.5–2.0	1–2	2–2.5	0.9–1.5

tion of esters. The reaction is reversible. For diesters, using excess alcohol and by removing excess water as it forms, the reaction is driven to completion. This usually takes several hours, the reaction being monitored by taking samples periodically for acid number determination. The use of an azeotroping agent such as 5% toluene (to aid water removal) is optional.

The acid and alcohol can be reacted either just thermally or more usually in the presence of a catalyst in an esterification reactor. Possible catalysts include sulfuric acid, *p*-toluene sulfonic acid, tetraalkyl titanate, anhydrous sodium acid sulfate, phosphorus oxides, and stannous octanoate.

After the ester has been formed, unreacted acid is neutralized using sodium carbonate or calcium hydroxide and is removed by filtration.

Typical reaction conditions for titanium catalysts are 230°C and 50–760 mm Hg vacuum. A significant amount of alcohol vaporizes along with the water, and must be recovered. This is accomplished by condensing the reactor vapors and decanting the resulting two-phase liquid mixture. The alcohol is then refluxed and returned to the reactor.

Polyol esters are made by reacting a polyhydric alcohol, such as neopentylglycol (NPG), trimethylolpropane (TMP), or pentaerythritol (PE), with a monobasic acid to give the desired ester. When making neopolyol esters, excess acid is used because the acid is more volatile than the neopolyol and is therefore easier to recover from the ester product.

V. PHYSICOCHEMICAL PROPERTIES OF ESTER LUBRICANTS

Mineral oil base stocks are derived from crude oil and consist of complex mixtures of naturally occurring hydrocarbons. Synthetic ester lubricants, on the other hand, are prepared from man-made raw materials having uniform molecular structures and therefore well-defined properties that can be tailored to specific applications.

Many lubricant requirements are translated into specific properties of an oil measurable by conventional laboratory tests such as viscosity, evaporation, or flash point. Other,

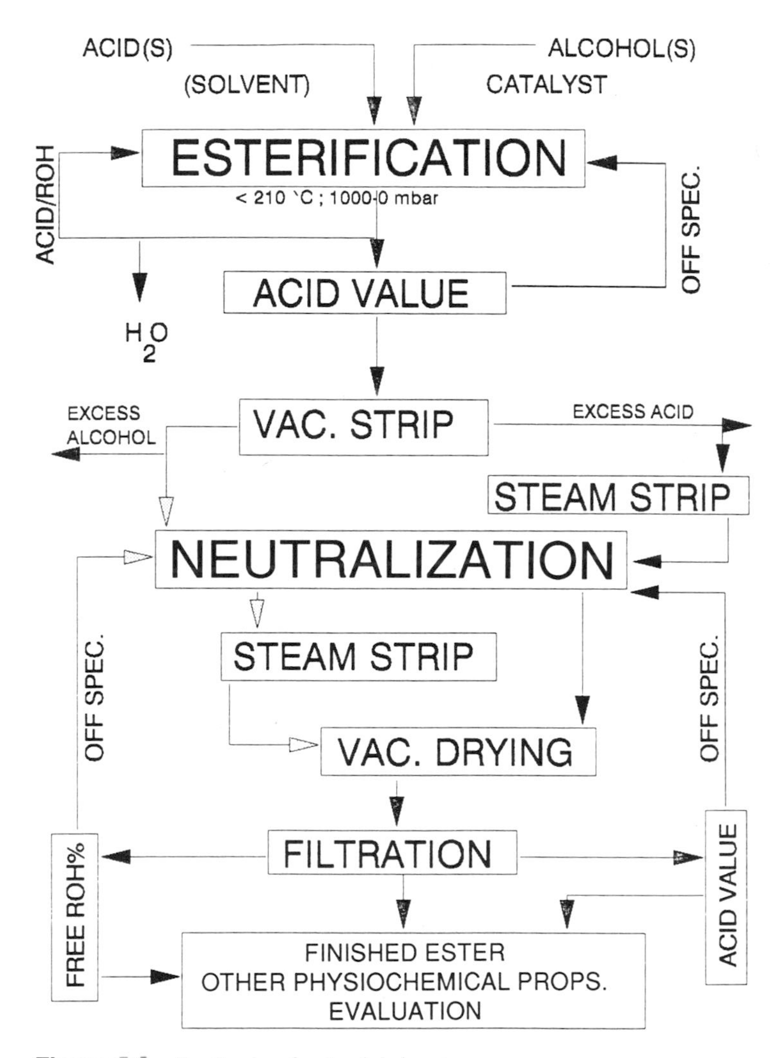

Figure 2.3 Synthesis of ester lubricants.

more critical requirements are related to the chemical properties of the lubricant, many of which can only be measured satisfactorily by elaborate and expensive special rigs developed to simulate performance.

A wide variety of raw materials can be used for the preparation of ester type base fluids. Various physical properties of the lubricant can be affected by which raw materials are chosen.

A. Viscosity

The viscosity of ester lubricants can be increased by:

Increasing the molecular weight of the molecule, by increasing the chain length of the acid, increasing the chain length of the alcohol, increasing the degree of polymerization, and increasing the functionality of the ester.

Increasing the size and/or the degree of branching.
Including cyclic groups in the molecular backbone.
Maximizing dipolar interactions.
Decreasing the flexibility of the molecule.

Figure 2.4 shows the change in viscosity at 40°C with acid chain length for a range of polyol esters. For very viscous molecules, branched aromatic esters or polymerized esters tend to be used. For low-viscosity esters short-chain NPG polyols or adipates are used.

B. Flow Properties

The viscosity index (VI) of a ester lubricant can be improved by:

Increasing the acid chain length.
Increasing the alcohol chain length.
Increasing the linearity of the molecule. Branching restricts the rotational freedom around the ester linkage and also increases the ratio of length to cross section. Both effects contribute to lowering the VI.
Not using cyclic groups in the backbone, which tends to lower VI even more than aliphatic branches.
Molecular configuration. Viscosity indices of polyol esters tend to be somewhat lower than their diester analogues, due to the more compact configuration of the polyol molecule.

The pour point of the lubricant can be improved by:

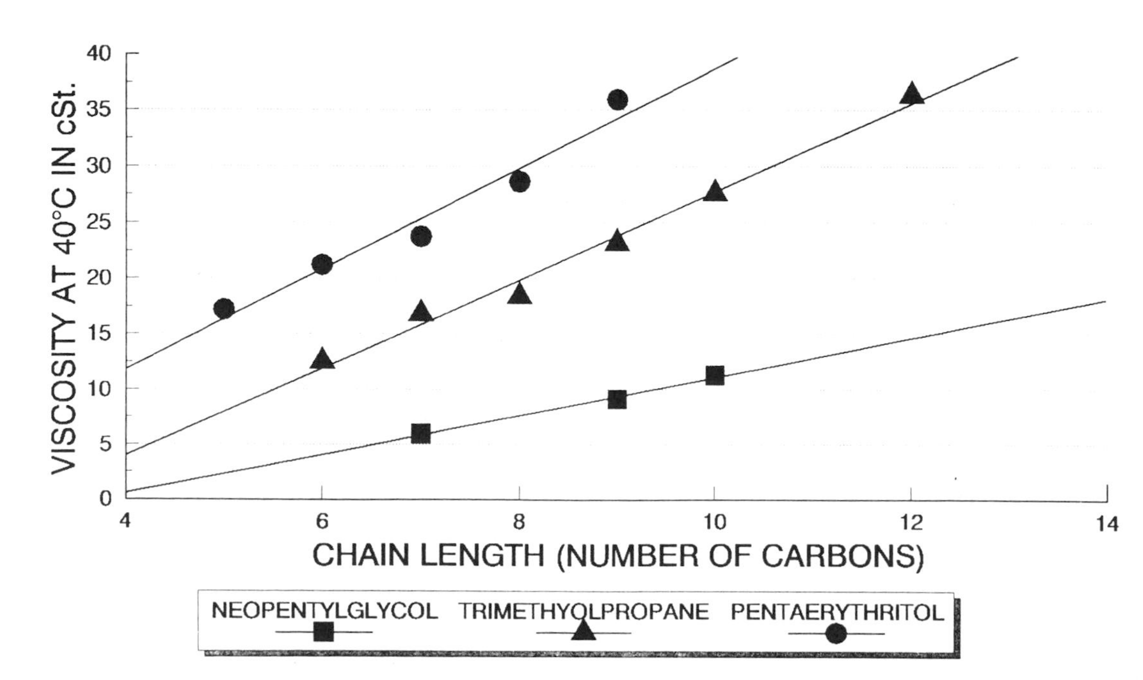

Figure 2.4 Variation of polyol viscosity at 40°C with linear acid chain length.

Increasing the amount of branching.
The positioning of the branch: branching in the center of the molecule gives better pour points than branches near the end.
Decreasing the acid chain length.
Decreasing the internal symmetry of the molecule. Esters made from mixtures of linear and branched acids have viscosity indices between those of the linear and branched acid esters, but have lower pour points than the esters used from either branched or linear acids.

Clearly, there is a trade-off between viscosity index and pour point: by increasing the linearity of the ester, the viscosity index improves but the pour-point properties decreases. Lubricants are required that have both a good pour point and a good VI.

C. Lubricity

Esters have a high degree of polarity due to the lone pair on the oxygen atom of the ester linkage. Polar molecules are very effective boundary lubricants as they tend to form physical bonds with metal surfaces. Esters are therefore more efficient lubricants than nonpolar mineral oils or very nonpolar polyalphaolefins (PAOs).

However, the lubricity of an ester in a fully formulated fluid is not always very easy to predict. As ester groups are polar they will effect the efficiency of antiwear additives. When a very polar base fluid is used, it will cover the metal surfaces instead of the antiwear additives. This can result in higher wear characteristics. Thus, although esters have superior lubricity properties than mineral oil, they are certainly less efficient than antiwear additives. Therefore, it is very important to choose the correct additive and to optimize its concentration to get the full lubricity benefit of using ester base stocks.

Esters can be classified in terms of their polarity, or nonpolarity, by using the formula (2):

$$\text{Nonpolarity index} = \frac{\text{total number of C atoms} \times \text{molecular weight}}{\text{number of carboxylate groups} \times 100}$$

Generally, the higher the nonpolarity index, the lower the affinity for the metal surface. Figure 2.5 depicts a graph that plots wear (as evaluated by the Shell four-ball test) versus nonpolarity index. Using the nonpolarity formula it can be seen that as a general rule increasing molecular weight improves overall lubricity. The scatter is large as the formula takes no account of the configurational aspects of the lubricant (i.e, whether it is branched or linear). Esters terminated by linear acids or alcohols have better lubricating properties than those made from branched acids/alcohols, and esters made from mixed acids/alcohols have intermediate lubricating properties between esters of linear acids/alcohols and esters of branched acids/alcohols.

Figure 2.6 shows a simplistic model of a diester on a metal surface. As the acid chain length increases, the carboxylate groups move further apart, reducing the overall polarity of the molecule and hence its affinity for the metal surface. As the chain length of the alcohol increases, the film thickness of the lubricant increases.

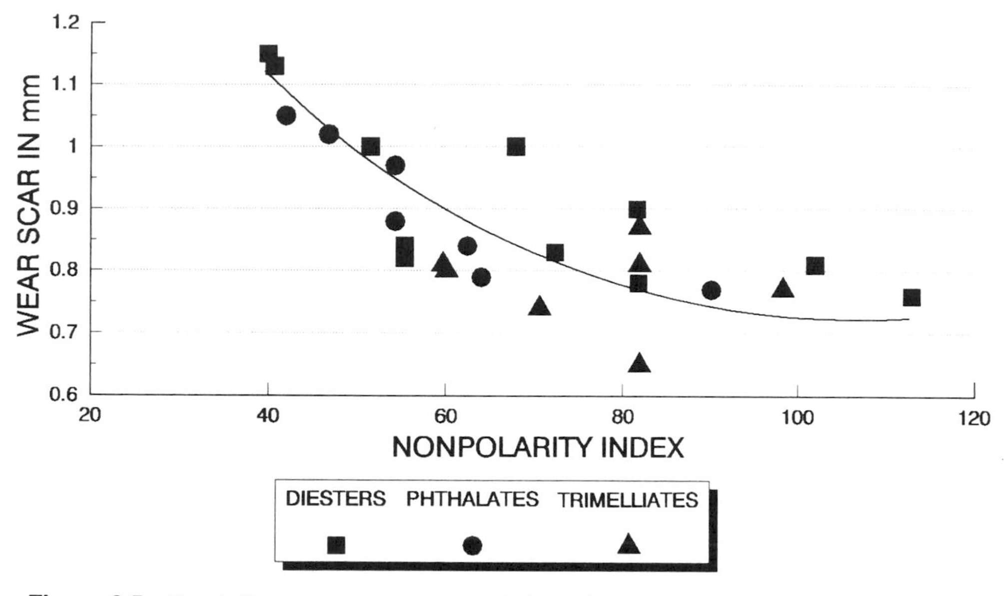

Figure 2.5 Four-ball wear scar versus nonpolarity index for a range of ester lubricants.

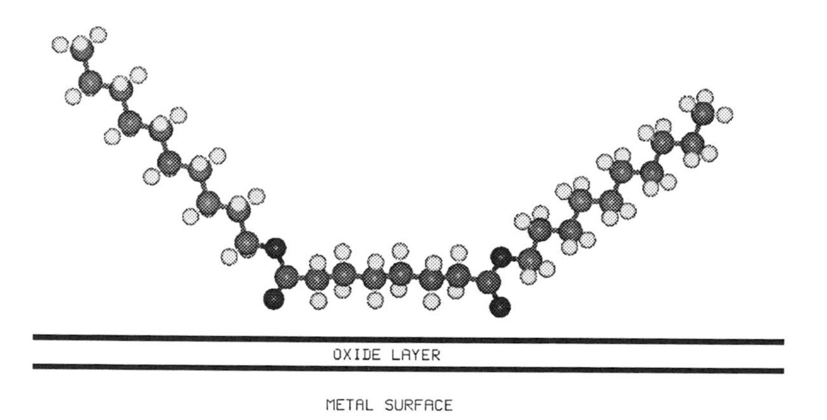

Figure 2.6 Lubricant–metal interface.

D. Thermal and Oxidative Stability

The ester linkage is an exceptionally stable one; bond energy determinations predict that the ester linkage is more thermally stable than the C–C bond.

Oxidative stability tests show that esters have an oxidative stability similar to that of mineral oil. This is because mineral oil contains natural antioxidants, whereas the ester components are pure chemicals. Usually, therefore, esters only show their superior oxidative performance over a mineral oil when both oils are blended with an antioxidant.

The olefinic (C=C) bond in the oleate chain of polyoleates makes such molecules unstable to oxidation, especially at high temperatures.

The advantage in thermal stability of polyol esters over diesters is well documented and has been investigated on a number of occasions. It has been found that the absence of hydrogen atoms on the beta-carbon atom of the alcohol portion of an ester leads to superior thermal stability. The presence of such hydrogen atoms enables a lower-energy decomposition mechanism to operate via a six-membered cyclic intermediate producing acids and l-alkenes (Fig. 2.7A). When beta-hydrogen atoms are replaced by alkyl groups this mechanism cannot operate and decomposition occurs by a free-radical pathway. This type of decomposition requires more energy and can only occur at higher temperature (Fig. 2.7B). Pentaerythritol-based polyols tend to be more thermally stable than polyols based on TMP, which in turn are more stable than those based on NPG.

Generally, short linear chains give better thermal stability than long branched chains. The desirable presence of side chains, which gives low pour point, incorporates tertiary carbon atoms, which are more susceptible to oxidative attack. Esters derived from geminal-dimethyl acids are the most robust of the branched species, as they are incapable of supporting the facile beta-hydrogen elimination degradation path.

Esters made from linear acids generally have higher flash points than those made from branched acid or linear and branched acids. Increasing molecular weight also increases flash points (Fig. 2.8). The volatility of esters depends on three main parameters:

1. Molecular weight.
2. Polarity.
3. Oxidative stability. Esters with low oxidative stability break down to form molecules of low molecular weight.

Volatility is usually evaluated by the NOACK test, which measures percent weight loss at 250°C. The reliance of volatility on molecular weight can be seen in Fig. 2.9.

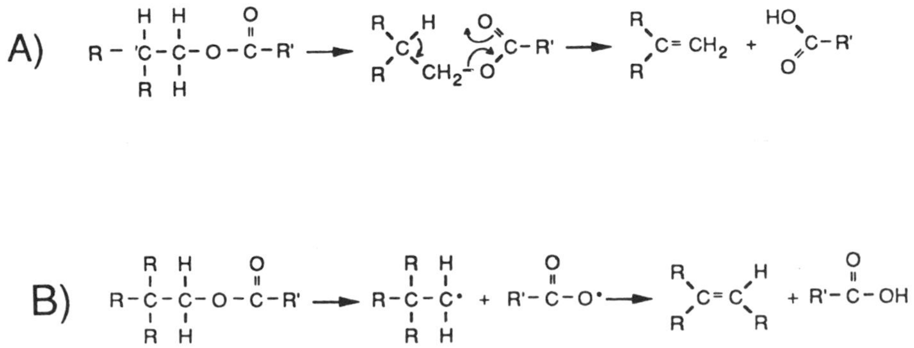

Figure 2.7 Thermal decomposition of esters.

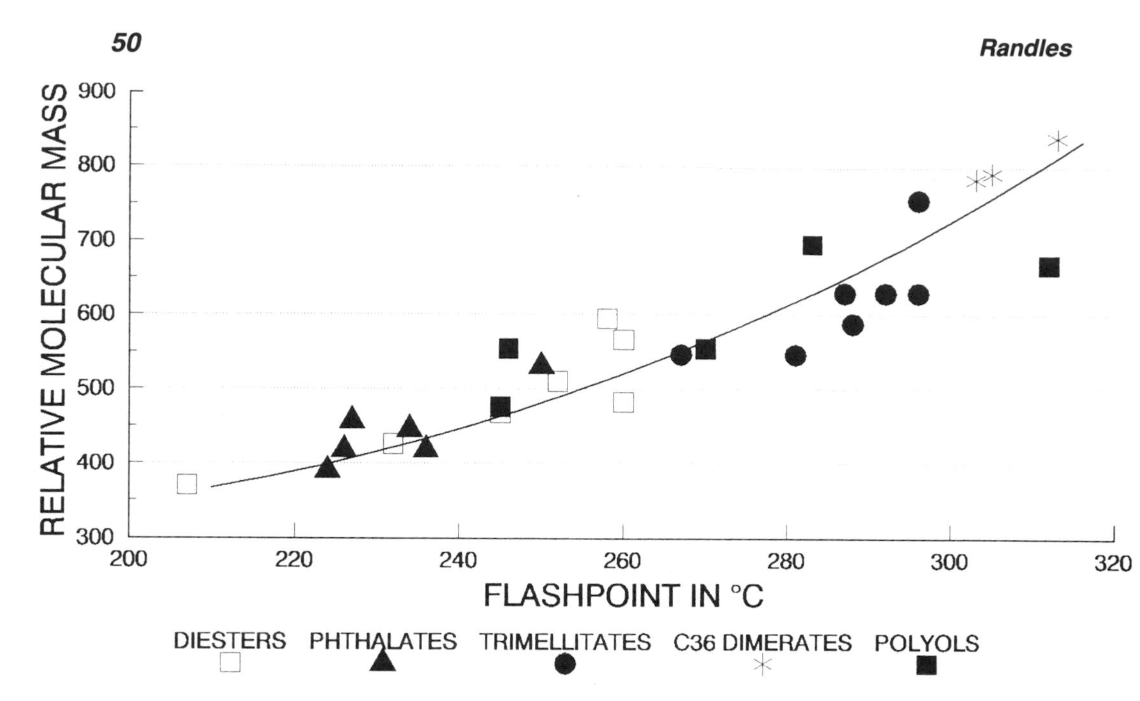

Figure 2.8 Plot of relative molecular mass versus flashpoint as evaluated by the COC (ASTM D-92) test.

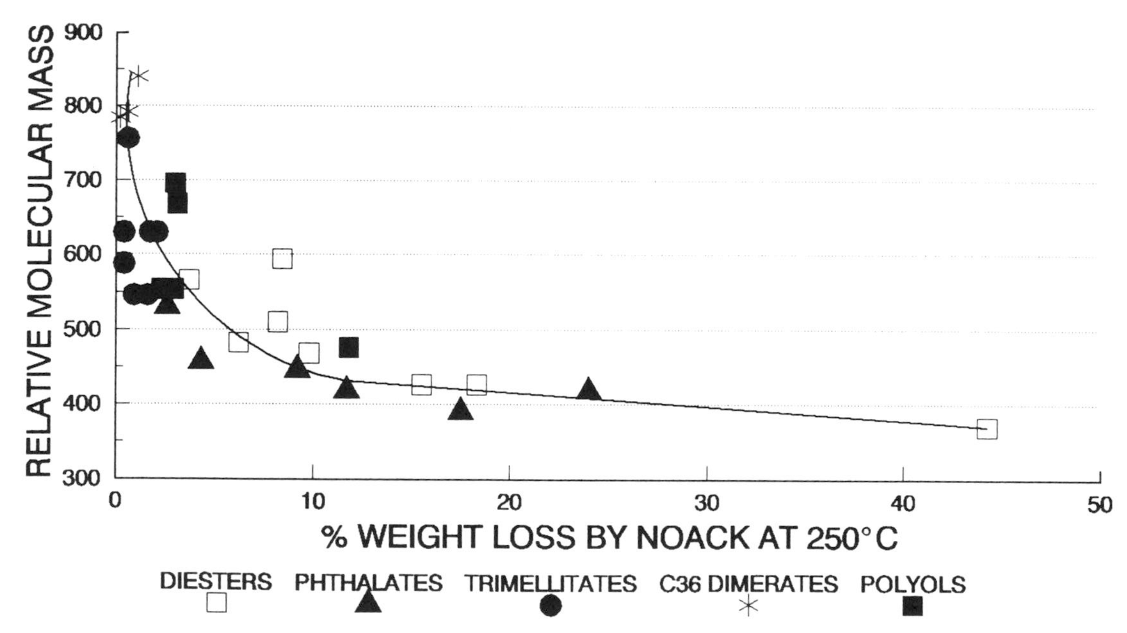

Figure 2.9 Plot of relative molecule mass versus percent weight loss of ester lubricant as evaluated by NOACK at 250°C (ASTM D-972).

E. Hydrolytic Stability

The hydrolysis of the ester, that is to say, their cleavage into an alcohol and an acid, has been the subject of many discussions in the past. However, this reaction has proved less disadvantageous in practice than had originally been feared. Ester lubricants must be hydrolytically stable because they are exposed to humid atmospheres during use and come into contact with appreciable quantities of water in many applications. The hydrolytic stability of esters depend on two main features: processing parameters and molecular geometry.

If the final processing parameters of the esters are not tightly controlled this can have a major effect on the esters hydrolytic stability. The following processing parameters can effect the hydrolytic stability of an ester:

Acid value.
Degree of esterification.
Catalyst used during esterification and the level remaining in the ester after processing.

Esters have to have a low acid value, a very high degree of esterification, and a low ash level before the effects of molecular geometry will begin to assert themselves.

Molecular geometry can effect hydrolytic stability in several ways. By sterically hindering the acid portion of the molecule hydrolysis can be slowed down, hindrance on the alcohol portion having relatively little effect. Geminal dibranched acids, such as neoheptanoic acids, have been used. However, there are penalties to be paid using these feedstocks, namely, very long reaction times to achieve complete esterification and poor pour points. The polar substituents on the acid portion of the molecule have also been shown to have have an effect on the hydrolytic stability of the ester. The hydrolytic stability of neopolyols can generally be regarded as good, and superior to that of dibasic esters.

F. Solvency

1. Compatibility with Additives and Other Lubricants

Generally, esters are fully compatible with mineral oils. This gives them several major advantages:

There are no contamination problems; esters can be used in machinery that previously used mineral oil.
Most additive technology is based on mineral oil and is therefore usually directly applicable to esters.
Esters can be blended with mineral oil (semisynthetics) to boost their performance.
Esters can be blended with other synthetics such as PAOs, which gives esters great flexibility, allowing them to be blended with other oils, giving unrivaled opportunities to balance the cost of a lubricant blend against its performance.

Polyalphaolefins, due to their low polarity, often result in solubility problems with various additives. This is especially true for viscosity index improvers (VII). In many applications esters are often blended with PAOs to overcome these solubility problems. As the following section shows, ester/PAO blends can have advantages in other areas as well.

2. *Elastomer Compatibility*

Elastomers that are brought into contact with liquid lubricants will undergo an interaction with the liquid that is diffusing through the polymer network. There are two possible kinds of interaction: chemical interactions and physical interactions.

Chemical interactions of elastomers with esters are rare. During a physical interaction of an ester lubricant and an elastomer, two different processes occur: an absorption of the lubricant by the elastomer causing swelling, and an extraction of soluble components out of the elastomer causing shrinkage.

The degree of swelling of elastomeric material can depend on:

The molecular size of the lubricant component: The larger the lubricant, the smaller the degree of swelling.
The closeness of the solubility parameters of the lubricant and the elastomer: Generally, the "like dissolves like" rule is obeyed.
The molecular dynamics of the lubricant: Linear molecules diffuse into elastomers quicker than branched or cyclic molecules.
The polarity of the lubricant: It is known that some elastomers are sensitive to polar ester lubricants. The nonpolarity index formulae can be used to model elastomeric seal swelling trends for specific ester types.

Several polar esters are well-known plasticizers used by industry. Nonpolar base stocks, such as PAOs, have a tendency to shrink and harden elastomers. By carefully balancing these compounds with esters, lubricants can be formulated so as to be neutral to elastomeric materials.

G. Environmental Aspects

Growing environmental awareness has turned the threat to our waters into one of the burning issues of our time. Many ways in which the environment can become polluted have followed in the wake of modern industrial development. Oils and oil-containing effluent have devastating consequences for fish stocks and other water fauna.

1. *Ecotoxicity*

In West Germany materials are classified according to their water-endangering potential or Wassergefährdungsklasse (WGK). Substances are given a ranking of between 0 and 3.

WGK	Definition of WGK ranking
0	Not water endangering
1	Slightly water endangering
2	Water endangering
3	Highly water endangering

Generally, esters have the following rankings:

Polyols, polyoleates, C_{36} dimerates, and diesters	0
Phthalates and trimellitates	1

This shows that esters have a low impact on the environment.

2. Biodegradability

The biochemistry of microbial attack on esters is well known in general outline and has been well reviewed. The main steps of ester hydrolysis (3), beta-oxidation of long-chain hydrocarbons (4), and oxygenase attack on aromatic nuclei (5) have been extensively investigated. The main features that slow or reduce microbial breakdown are:

The position and degree of branching (which reduces beta-oxidation).
The degree to which ester hydrolysis is inhibited.
The degree of saturation in the molecule.
The increase in molecular weight of the ester.

Figure 2.10 shows the biodegradabilities of a wide range of lubricants measured using the CEC-L-33-T-82 test. This test measures biodegradability after 21 days. Table 2.3 gives biodegradability data for a range of esters (6). As can be seen, diesters, polyols, and complex polyols are very biodegradable. A modified CEC test, where a flask was sacrificed every day, was used to compare the rate of biodegradation of a polyol ester against a mineral oil. The results can be seen in Fig. 2.11 (6).

H. Ester Types and Their Properties

1. Diesters

Diesters are made by reacting a diacid with a monofunctional alcohol (Fig. 2.12). Diesters have a very good VIs and pour points. The reason for this can be seen in Fig. 2.13. The linear diacid portion of the diesters gives the molecule very good VI, while the branched alcohol ends give the molecule a good pour point. As the branched portion of the diester is at the end of a linear portion, the free rotation around the ester linkage is still good. Therefore, there is an excellent trade-off between pour point and VI. One disadvantage of

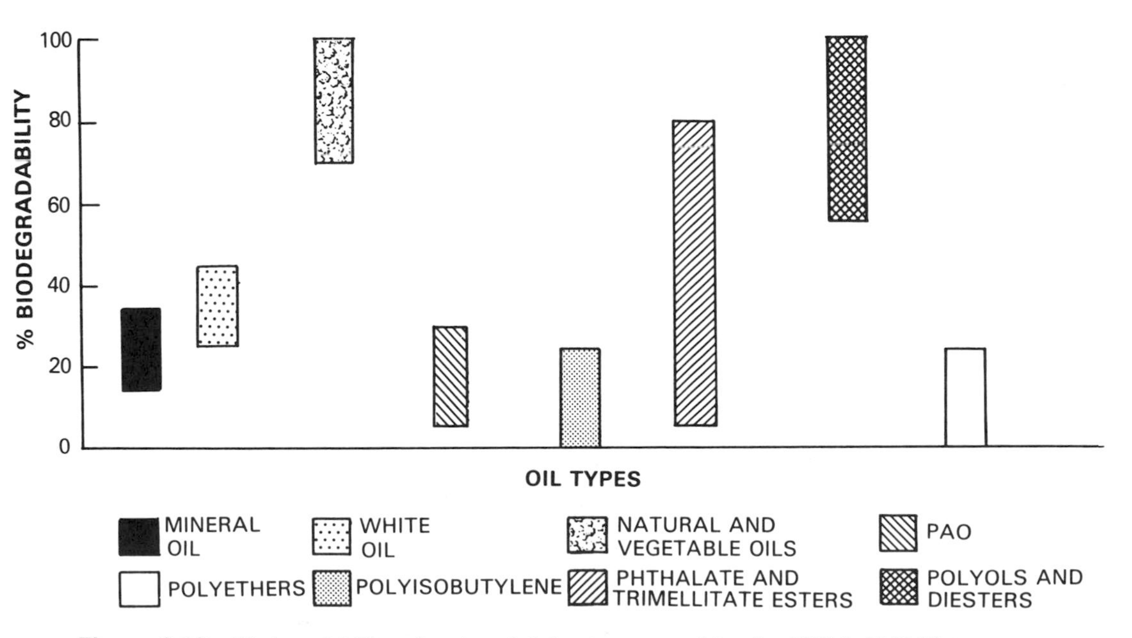

Figure 2.10 Biodegradability of various lubricants measured by the CEC-L-33-T-82 test.

Table 2.3 Biodegradability Ranges of Esters as Measured by the CEC-L-33-T-82 Test

Ester type	Percent biodegradability range (21-day test)
Diesters	75–100
Phthalates	45–90
Trimellitates	0–70
Pyromellitates	0–40
Dimerates	20–80
Polyols	70–100
Polyoleates	80–100
Complex polyols	70–100

diesters is that their low molecular weight results in low viscosities. This combined with their high polarities makes them quite aggressive to elastomeric seals. This can be reduced by using better elastomers or by blending with PAOs to nullify their swelling effects.

2. Phthalates

Phthalate esters are made by reacting phthalic anhydride with a monofunctional alcohol (Fig. 2.14). Phthalate esters are one of the most cost-effective esters, and as such they are often used in industrial applications, such as air compressors, to replace mineral oil. They can have low pour points or good VI but not both. The reason for this can be seen in Fig. 2.15. The bulky nature of the molecule results in a poor VI/pour point trade-off.

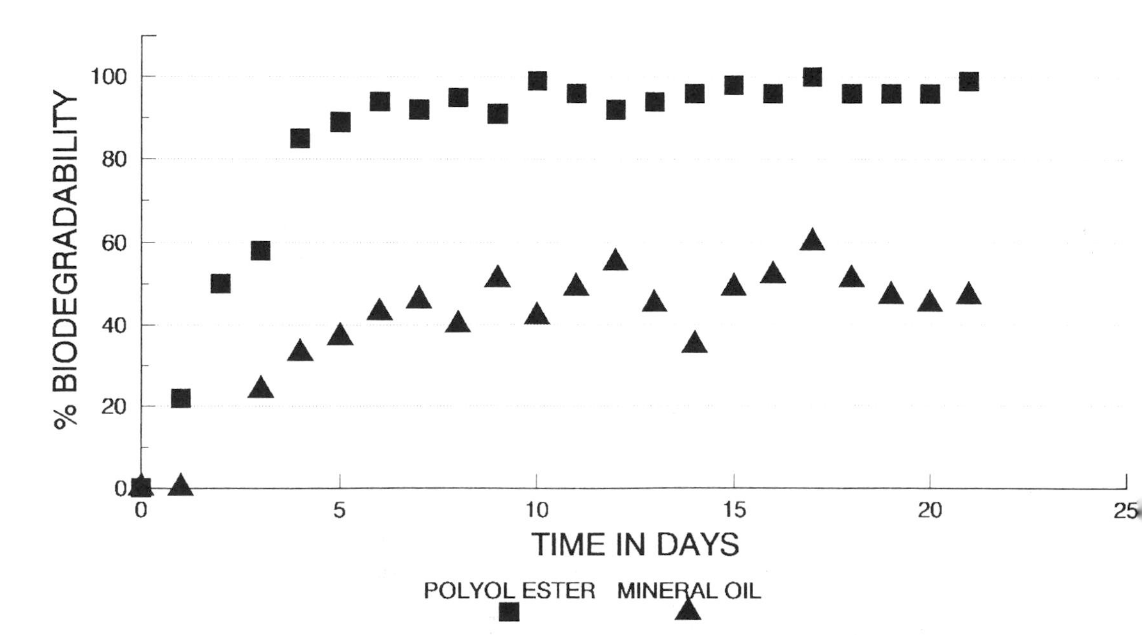

Figure 2.11 Biodegradability curve of a typical mineral oil and polyol ester as measured by the CEC-L-33-T-82 test.

$R^1OOC\text{-}(CH_2)_n\text{ - }COOR^{11}$ R^1, R^{11} = Linear, branched or mixed alkyl chain

n = 4 = Adipates
n = 7 = Azelates
n = 8 = Sebacates
n = 10 = Dodecanedioates

Advantages:
- High VI's
- Good pour points
- High biodegradabilities

Figure 2.12 Diesters (dioates).

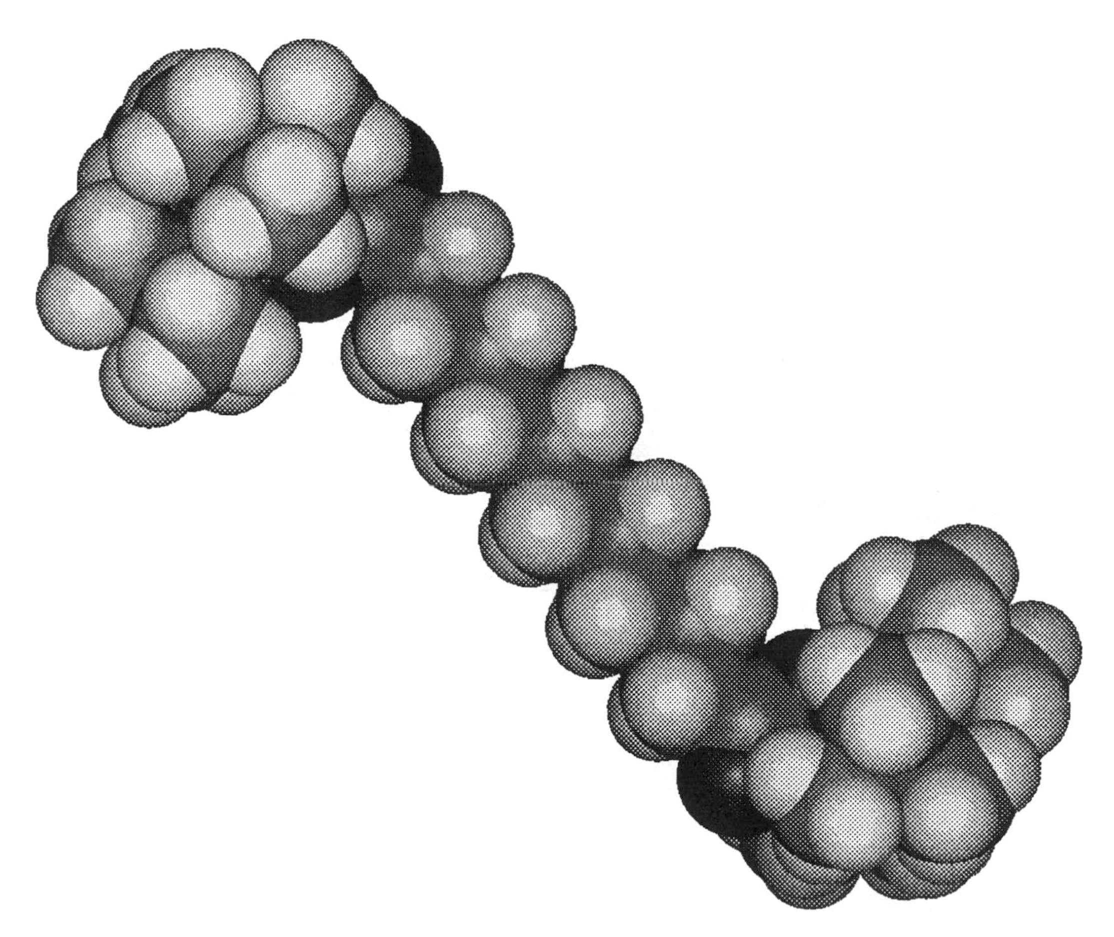

Figure 2.13 Diester structure.

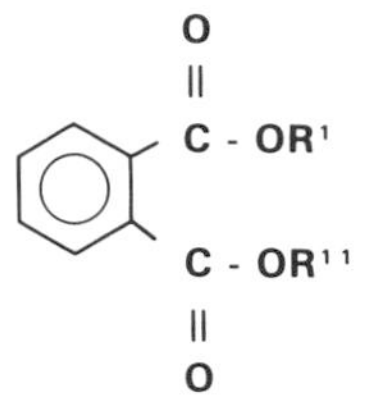

O
||
C - OR'
C - OR''
||
O

R',R'' = Linear, branched or mixed alkyl chain

Advantages: • A good balance of properties

Figure 2.14 Phthalate (1,2-benzene dicarboxylate).

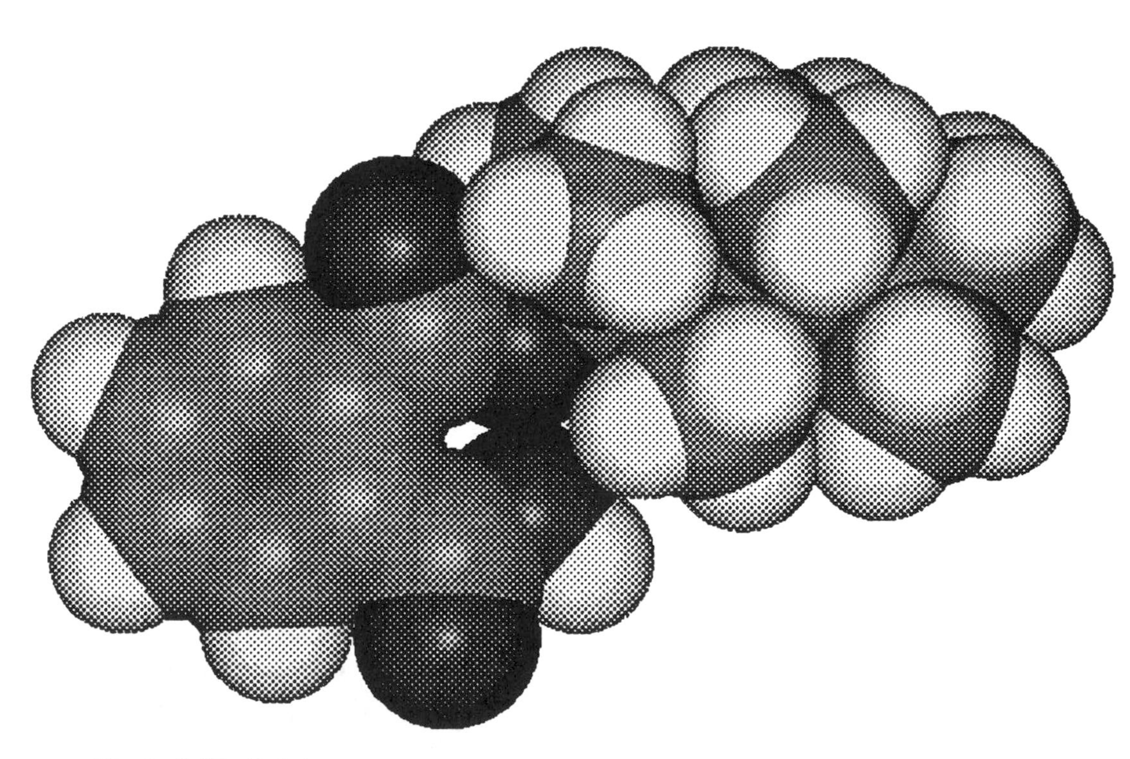

Figure 2.15 Phthalate structure.

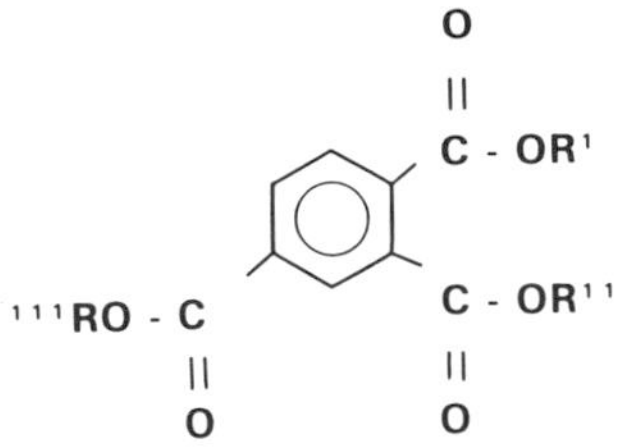

O
||
C - OR^{1}

^{111}RO - C C - OR^{11}
|| ||
O O

R^{1},R^{11},R^{111} = Linear, branched or mixed alkyl chain

Advantages:
- High viscosities
- Good thermal properties
- Low volatilities

Figure 2.16 Trimellitate (1,2,5-benzene tricarboxylate).

3. Trimellitates

Trimellitates are made by reacting trimellitic anhydride with a monofunctional alcohol (Fig. 2.16). They are often used instead of phthalates when a higher performance is required, such as in recipricating air compressors. Again, Fig. 2.17 shows that they are

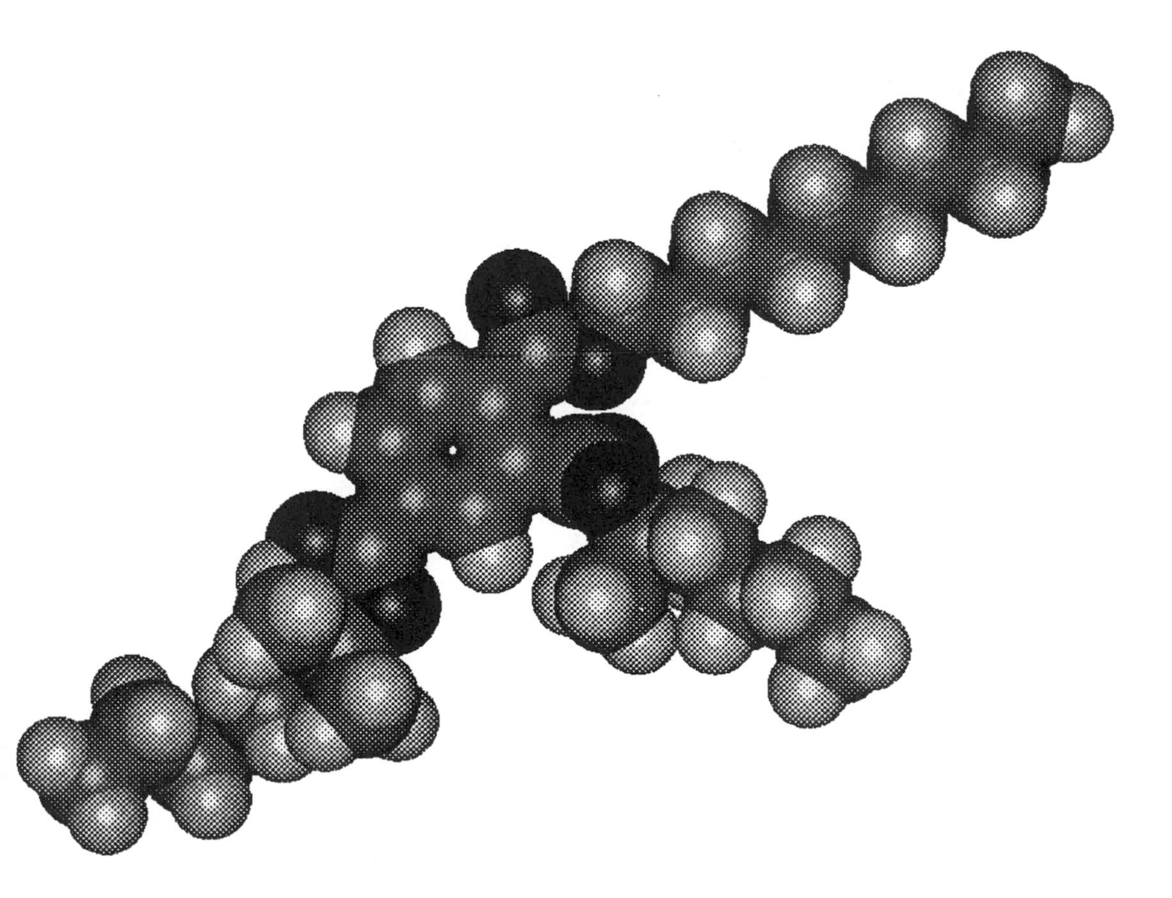

Figure 2.17 Trimellitate structure.

quite bulky and therefore have a poor VI/pour point trade-off. Trimellitates are very good lubricants, and because of their high molecular weights they have high flash points and low volatilities. They also have good thermal stability. When they do break down they form a soft carbonaceous varnish on metal surfaces, which in itself acts as a lubricant.

4. C_{36} Dimerates

Dimerates are made by reacting a C_{36} dimer acid (derived from tallow oil) and a monofunctional alcohol (Fig. 2.18). The molecule has several isomeric forms, one of which can be seen be seen in Fig. 2.19. The alcohol most used is 2-ethyl hexanol, making an ester that is usually referred to as 3608. This ester is often used in two-stroke oils and has some use in four-stroke oils. It has the advantage of being an excellent lubricant and having good thermal and oxidative stability. Its stability can be further improved by hydrodgenation; however, this halves the molecule's biodegradability and increases cost.

5. Polyols

Polyols are made by reacting a multifunctional alcohol with a monofunctional acid (Fig. 2.20). Their molecular configuration can be seen in Fig. 2.21. Polyols have the same disadvantages/advantages as diesters. They are, however, much more stable and tend to be used instead of diesters where temperature stability is important. A general rule of thumb is that a polyol is thought to be 40–50°C more thermally stable than a diester of the same viscosity. Esters give much lower coefficients of friction values than those of both PAO and mineral oil. In general, polyol esters based on TMP or PE give lower values than diesters. By adding 5–10% of an ester to a PAO or mineral oil the oil's coefficient of friction can be reduced markedly (7).

VI. APPLICATION AREAS

A. Engine Oils

In gasoline engines the application of electronic fuel injection, three-way exhaust catalysts, turbochargers, superchargers, and four-valve-per-cylinder technology is now common. Engine speeds have increased and gasoline consumption has been drastically

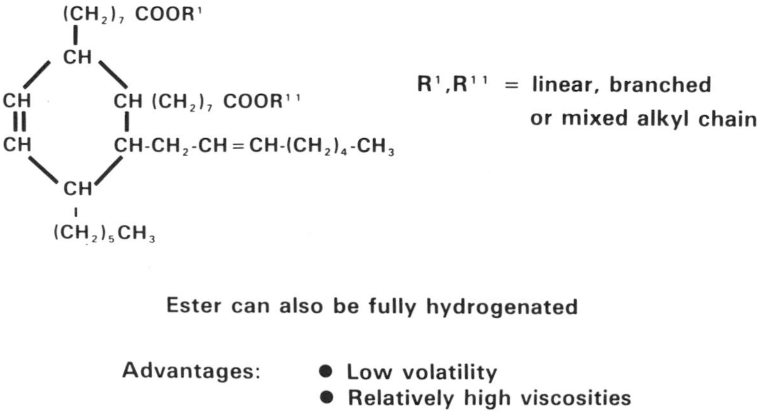

$(CH_2)_7\,COOR^{1}$

CH

CH — CH $(CH_2)_7\,COOR^{11}$

CH — $CH\text{-}CH_2\text{-}CH=CH\text{-}(CH_2)_4\text{-}CH_3$

CH

$(CH_2)_5CH_3$

R^1, R^{11} = linear, branched or mixed alkyl chain

Ester can also be fully hydrogenated

Advantages:
- Low volatility
- Relatively high viscosities

Figure 2.18 C_{36} Dimerate.

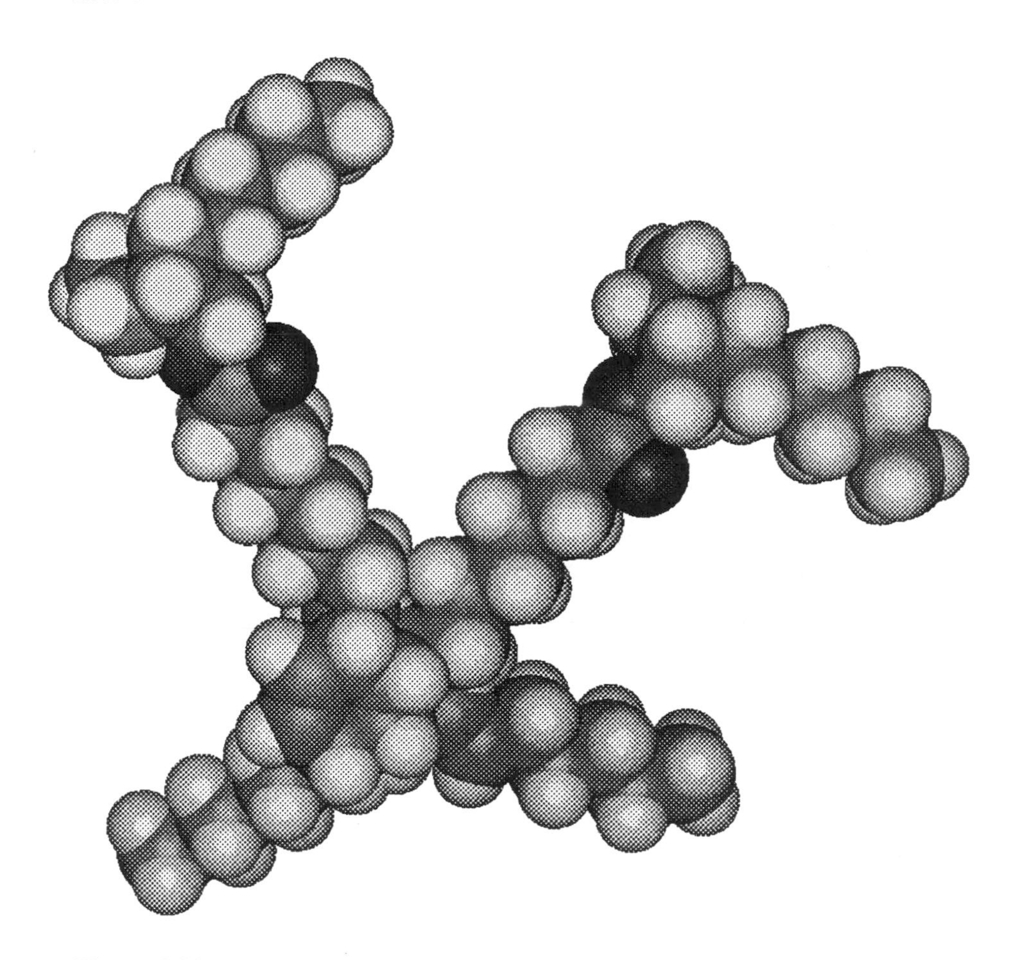

Figure 2.19 C_{36} Dimerate structure.

$C\ (CH_2CO_2R)_4$	Pentaerythritol esters
$CH_3CH_2C\ (CH_2CO_2R)_3$	Trimethyolpropane esters
$(CH_3)_2\ C\ (CH_2CO_2R)_2$	Neopentylglycol esters

R = Branched, linear or mixed alkyl chain

Advantages:
- Offer a range of viscosities
- Good thermal properties
- High biodegradabilities

Figure 2.20 Polyol (hindered ester).

reduced. The reduction in the lead content of gasoline and the addition of oxygenates to gasoline in some countries have certainly not made life easier on the engine or the lubricant (8).

With respect to diesel engines the most significant trends have been the wholesale conversion from naturally aspirated to turbocharged engines and the drive to fuel economy. Maximum engine torques have increased and compression ratios have been reduced. All this has immediate bearing on the performance required from the engine oil (8).

For both gasoline and diesel vehicles, electronic engine management systems have emerged and are part of all advanced engine designs. It is essential that the sensors used in these systems not be subject to deterioration caused by the lubricant (8).

It is now widely accepted that synthesized fluids, such as PAO/ester blends, offer a number of inherent performance advantages over conventional petroleum-based oils for the formulation of modern automotive engine oils. Practical benefits that may derive from their use include improved cold starting, better fuel and oil economy, plus improved engine cleanliness, wear protection, and viscosity retention during service. Fluid types used in the development of automotive crankcase oils, either commercialized or consid-

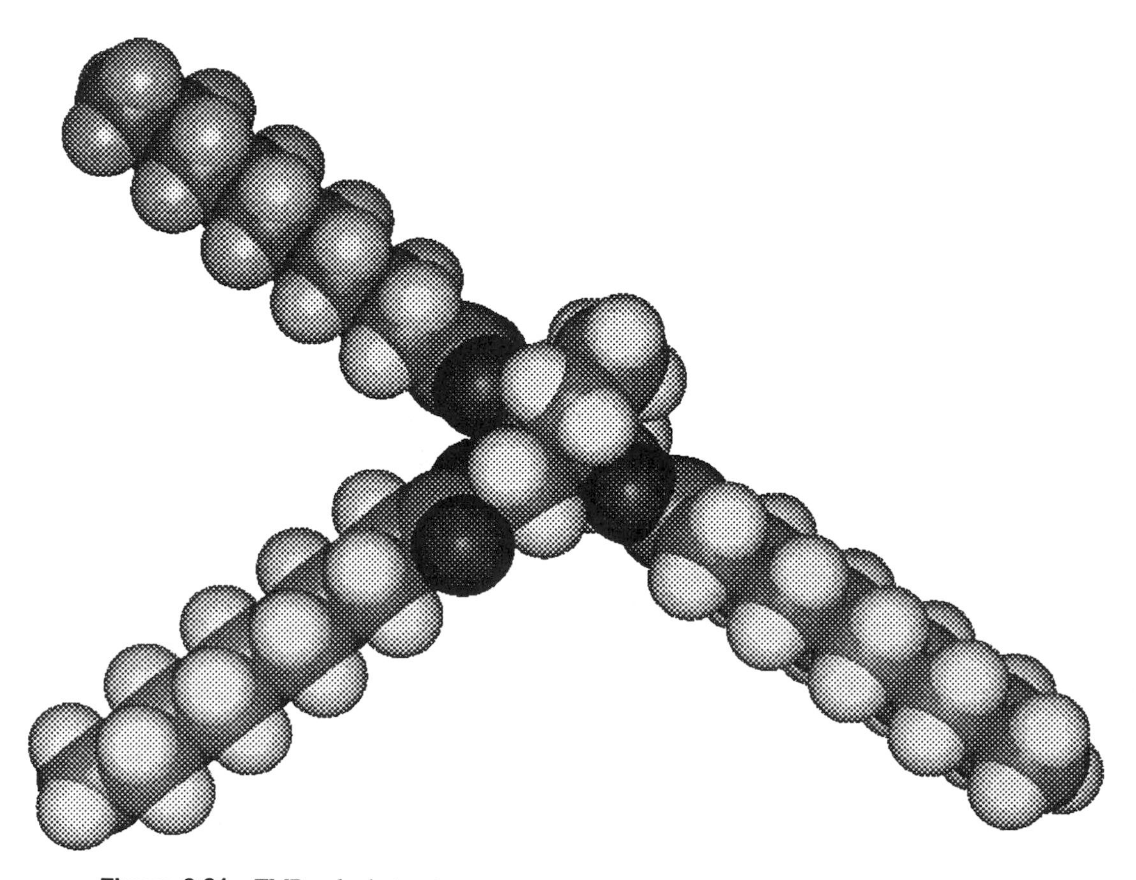

Figure 2.21 TMP polyol structure.

ered for commercialization, include PAOs (more correctly hydrogenated olefin oligomers), organic dibasic and polyol esters, alkylated aromatic hydrocarbons, and polyglycols. Experience from numerous laboratory engine bench and vehicle test programs conducted over the last 10 years has shown that a blend of PAO and an organic ester provides an excellent base fluid for the formulation of synthesized crankcase oils (9,10).

Low-temperature viscosity is perhaps the single most important technical feature of a modern crankcase lubricant. Cold starts are a prime cause of engine wear and can only be mitigated by immediately effective lubrication circulation. Low-temperature viscosity can also have the benefit of reducing start-up load and stresses, reducing battery current drain and making starting easier (8). Furthermore, motor vehicles are increasingly required to operate reliably in arctic conditions. Esters provide this essential low-temperature fluidity and, because of their low volatility, do so without any sacrifice of lubricant efficiency at high operating or ambient temperatures. Low volatility is especially important in the context of the modern trend toward smaller sump capacities and longer oil-change intervals.

B. Two-Stroke Oils

Ester lubricants (e.g., C_{36} dimer esters, trimellitates, and polyols) offer a number of advantages over mineral oils as the lubricant component of two-stroke engine mixtures. The clean-burn characteristics result in less engine fouling with much reduced ring stick and lower levels of dirt buildup on ring grooves, skirts, and undercrowns. Ignition performance and plug life are also enhanced. Owing to the presence of polar ester groups in the molecule, giving increased adhesiveness to metal surfaces, esters have much better lubricity than hydrocarbons, both conventional and synthetic. This removes the need to use brightstock and simultaneously permits the use of leaner burn ratios. In turn, this significantly reduces smoke levels. About 95% of particulates in the exhaust fumes have been found to be from unburnt lubricant (11). A 25% decrease in the amount of polyaromatic hydrocarbons (PAHs) in the exhaust emissions of a two-stroke engine has been found when a carboxylic ester has been used instead of a mineral oil (12). The PAHs have been found to be one of the major contributors to the carcinogenic nature of exhaust emissions. The excellent solubility of esters also allows them to be used without solvents (which are usually added to conventional two-stroke oil to help miscibility with the fuel).

The leaner burn ratios result in reduced oil emissions, which is a benefit in environmentally sensitive applications such as marine outboard engines and chainsaw motors. The high biodegradabilities of esters (e.g., polyols) and the low ecotoxicity and clean burn characteristics of ester formulations should allow them to be labeled as "environmentally considerate" in many countries, for example, "Blue Angel" in Germany.

Low-temperature performance is important in some applications, such as engines used to power snowmobile type vehicles. Therefore, esters with low pour points of down to –56°C are very suitable for fuels in these applications.

C. Compressor Oils

This sector of the market covers a wide range of compressor types, used for a number of different gases. The three basic functions of a compressor oil are to lubricate, reducing friction between moving parts; to seal at rings, vanes, or rotary screws; and to cool critical bearings and points of constant friction (13). Esters have the following advantages in compressor applications:

Compressors that use esters can be cheaper to run.
They have safety and ecological benefits.
They are excellent lubricants.

From the user's standpoint the main reason for choosing a synthetic over conventional hydrocarbon lubricants is to save money. This goal can be accomplished, in part, through extended drain intervals (up to 8- to 10-fold) and other associated cost savings, such as less frequent replacement of oil filters and air/oil separators. Esters can have better lubricity, which can mean reduced wear and fewer replacements of mechanical parts such as piston rings, seals, and bearings. Power cost savings is another factor, which can range from 1 to 7%. Resistance to breakdown means less sludge and varnish in the compressor and compressed gas system. There is also improved performance of newer compressors operating at higher speeds, temperatures, and pressures (14).

From a safety performance standpoint there are reduced fire and explosion hazards for compressors handling air and chemical gases. High outlet temperatures in modern air reciprocating compressor installations cause carbon buildup on compressor outlet/discharge valves. These carbon deposits can glow and cause fires and/or, in some cases, explosion. This problem is intensified in environments where air filtration is less than ideal and dust contributes to a buildup of dangerous deposits. Organic esters, especially the diesters and polyol esters, have improved oxidation resistance and low volatilities (3–10%) compared to those of mineral oils of similar viscosity. Esters therefore minimize buildup of deposits on the hot pistons and discharge valves. The high-temperature stability and solvency action of the organic esters therefore minimize the risks of fire and explosion.

The low volatility of esters results in reduced carryover. These properties, coupled with their higher flash and autoignition temperatures and low order of toxicity for vapor inhalation, ingestion, and skin irritation, make them considerably safer lubricants to use than mineral oil. Their low ecotoxicity and high biodegradabilities can also lessen their environmental impact.

The excellent lubricity, thermal stability, and conductivity of esters allow them to be used in high-performance compressor oils that cannot be formulated from traditional hydrocarbon lubricants. Diesters and phthalates have found their major application in air-compressor lubricants, but they are also used in compressors handling natural gas. Diesters generally have high viscosity indices, giving them a wide temperature range without the use of viscosity improvers, which can shear in this application. In reciprocating compressors, where oils of rather higher viscosity are preferred, trimellitate esters can be used. The solvency characteristics of esters are such that care has to be taken in the selection of elastomeric sealing materials; the use of Viton® is usually recommended. Diesters and polyol esters can also be blended with PAOs for use in the various compressor types. As discussed previously, the ratio of PAO to ester can be balanced to have a neutral effect on elastomers.

D. Refrigeration Lubricants

For the past 50 years, lubricants produced from naphthenic and paraffinic mineral oils have been used in refrigerator compressor systems. These oils were fully compatible with R12 and fully met the systems requirements. Due to the chemical differences between R12 and R134a, traditional mineral oils are not capable of meeting these requirements and their use would sacrifice both performance and reliability of the system. Ester lubricants based on simple and complex polyols have been developed that achieve the key characteristics of this application:

Good lubricity.
Good materials compatibility.
Good energy efficiency.
Good resistance to copper plating.
Good chemical, thermal, and hydrolytic stability.
Good solubility with 134a, mineral oil, and additives.

E. Aviation Lubricants

The bulk of aviation lubricant demand is for gas turbine lubricants for both military and civilian use. The requirements placed on the jet engine lubricant in terms of oxidation and aging stability cannot be met by hydrocarbon oils.

The first generation of oils (Type 1) were diesters, but these have slowly lost ground to the more expensive (Type 2) polyol esters over the last 25 years. Some diesters are still used in less demanding applications, such as for small private aircraft or turbo-prop engines. Type 2 aviation gas turbine lubricants are produced to a viscosity of 5 cSt (at 100°C). For some military applications, where operability at low temperatures is vital, the corresponding viscosity is reduced to 3 cSt.

F. Hydraulic Fluids

Ester-based hydraulic fluids have achieved significant success in Western Europe, although use in the United States has been slow to develop. The polyols base fluids are attractive because they are environmentally considerate, have a reasonable level of fire resistance, have good high- and low-temperature flow properties, have good thermal stabilities, and exhibit good wear performance.

The low ecotoxcity and high biodegradabilities of esters mean that they can obtain environmentally friendly "eco-label status" such as "Blue Angel" in Germany. This is a very important aspect, as it is estimated that 4% of hydraulic fluids leak into the environment, which in 1990 accounted for some 6,000 tonnes in West Germany alone (15).

Diesters have been used as fire-resistant hydraulic fluids (MIL-H-83288 spec). However, polyols now tend to be used, normally esters of oleic acid, most commonly the TMP ester of this acid. Polyol esters are classified as HFD U fluids. There are increasing concerns about the toxicity aspects around phosphate esters and the fact that their thermal decomposition products are highly noxious. Polyol esters are therefore beginning to replace phosphate esters in certain areas. Polyols have several advantages over phosphate esters:

They are cheaper.
They have better flow properties.
They are less aggressive to seals.
They are better lubricants.

Phosphate esters, however, are superior in their fire resistance.

Polyol ester fluids usually contain antioxidants, corrosion inhibitors, and antifoam additives. However, and very importantly, they do contain polymeric viscosity index improvers (usually based on styrene). High viscosity indicies have the effect of modifying the spray pattern of the fluid pumped through a nozzle at high pressure by increasing the droplet size. This produces a much narrower spray, which is much more difficult to ignite and enables the fluid to pass certain tests, in particular the well-known factory spray test.

VII. THE FUTURE FOR ESTERS

The need for lubricants that can operate at high temperature is causing a move from mineral oil to esters. Within the ester area itself there is a growing tendency to use polyols instead of diesters due to their better high temperature stability. Environmental pressures are also causing a move toward esters. To further improve the overall environmental impact of esters their chemistry is being modified so as to produce compounds that have high biodegradability, low toxicity, and clean engine emissions. Feedstocks from other chemical processes are being investigated so as to be able to produce esters from renewable resources or waste streams.

The degree of flexibility required by the lubricant is therefore growing. A new type of ester that can give this flexibility is complex esters. Their structures have the generalized formulas

Alcohol end-capped complex polyols: MA -$[(DA)_x$- $MFA]_n$-DA-MA

where MA is a monofunctional alcohol, DA a diacid, n a monomer unit 1 to 20, and MFA a multifunctional alcohol (x = 1, NPG, x = 2, TMP, or x = 3, PE) or hydroxyl-capped polyether.

Acid end-capped complex polyols: MAC-$[MFA\text{-}(DA)_x]_n$-MFA-$(MAC)_x$

where MAC is a monofunctional acid.

These structures form complex polymers, which can be polymerized to give the desired viscosity and polarity. They give lubricants with excellent biodegradabilities (80%+ by the CEC test) even at high viscosities. They also have excellent flow properties with VI well in excess of 150 without a corresponding poor pour point. Complex esters can also be used to blend with other synthetic lubricants to boost their properties. The great advantage of using complex esters for blending to viscosity targets is its inherently good load carrying properties. They can also be used as viscosity index improvers (VII). Unlike VII long-chain polymers, complex polyols do not exhibit the temporary loss of viscosity under forces exerted by gear testing. Because complex esters are shorter chain molecules they tend not to shear into smaller molecules.

Another way to improve ester performance involves incorporating additive type functional groups in the ester backbone, for example, secondary amines to act as antioxidants.

Overall, the great flexibility of ester chemistry will allow the use of esters well into the next century.

REFERENCES

1. Hanble, P. (1991). Esters, base oils for synthetic lubricants. *International Congress about Lubricants*, Lubricants for the Future and Environment, Brussels, September, p. 85.
2. Van der Waal, G. (1985). The relationship between chemical structure of ester base fluids and their influence on elastomer seals and wear characteristics. *J. Synth. Lub.*, 1(4), 281.

3. Macrae, R. A., and R. C. Hammond (1982). *Biotechnol. Gene. Eng. Rev.*, 3, 1093, 217.
4. Wyatt, J. M. (1982). Ph.D. thesis, University of Kent.
5. Cerniglia, C. E. (19). In *Petroleum Microbiology*, ed. R. M. Atlas, Macmillan, New York, p. 99.
6. Randles, S. J., and M. Wright (1991). Environmentally friendly ester lubricants for the automotive and engineering industries. Esters, base oils for synthetic lubricants, *International Congress about Lubricants*, Lubricants for the Future and Environment, Brussels, September, p. 197.
7. Van Der Waal, G. (1991). Esters: Synthetic base for lubricants, *International Congress about Lubricants*, Lubricants for the Future and Environment, Brussels, September, p. 251.
8. Coffin, P. S., C. M. Lindsay, A. J. Mills, H. Lindencamp, and J. Fuhrmann (1979). The application of synthetic fluid to automotive lubricant development trends today and tomorrow. *J. Synth. Lub.*, 7(2), 123.
9. O'Connor, B. M., and A. R. Ross (1989). Synthetic fluids for automotive gear oil applications: A survey of potential performance, *J. Synth. Lub.*, 6(1), 31.
10. Krulish, J. A. C, H. V. Lowther, and V. J. Miller (1977). An update of synthesized engine oil technology. SAE Paper No. 770634, Presented at *SAE Fuels and Lubricants Meeting*, Tulsa, OK, June.
11. Sugiura, K., and M. Kagaya (1977). A study of visible smoke reduction from a small two-stroke engine using various engine lubricants. SAE paper 770623, June.
12. Cosmachi, E., D. Cottia, L. Pozzoli, and R. Leoni (1988). PAH emissions of synthetic organic esters used as lubricants in two-stroke engines. *J. Synth. Lub.*, 3, 251.
13. Van Ormer, H. P. (1987). Trim compressed-air cost with synthetic lubricants. *Power*, February, 43.
14. Wits, J. J. (1989). Diester compressor lubricants in petroleum and chemical plant service. *J. Synth. Lub.*, 5(4), 321.
15. Kuhrt, W., and T. H. Mang (1990). Environmentally acceptable lubricants and legislation. Deutshes Scierstoff-Forum, Frankfurt, November.

3
Phosphate Esters

Michael P. Marino
FMC Corporation
Philadelphia, Pennsylvania

I. INTRODUCTION

Beginning with their development as antiwear additives before World War II, phosphate esters have been significant factors in the lubricants industry. Although some shortcomings, especially susceptibility to hydrolysis, poor viscosity index, and neurotoxicity, were identified during the early years of development, industry has learned to formulate and use these versatile fluids to great advantage. The phosphate esters that have survived today as commercially important products have proven to be outstandingly useful, stable, and nontoxic fluids and lubricants.

Phosphate esters are the most fire resistant of the reasonably priced synthetic base stocks. Their high fire points, flash points, and autoignition temperatures make them very difficult to ignite. Their low heats of combustion make them self-extinguishing fluids. Years of use have shown them to be excellent lubricating agents with exceptional thermal, hydrolytic, and oxidative stability.

In this chapter, we describe the chemistry and production of this class of phosphorus compounds that are esters of phosphoric acid. We outline the physical and chemical properties that make the phosphate esters practical, useful, and important industrial chemicals. We review the formulation of these compounds into lubricants and hydraulic fluids, placing emphasis on commercial applications. Lastly, we outline practical methods of employing phosphate esters to achieve optimum working life and performance in the industrial environment.

A. Historical Development

Although thousands of organophosphorus compounds have been synthesized, only those classified as phosphate esters, or organic salts of orthophosphoric acid, $O{=}P\,(OH)_3$, have become commercially significant in the field of synthetic lubricants. While mono-, di-, and trisubstituted esters have found commercial use, only the tertiary esters are used as synthetic lubricants or fluids. Of the tertiary esters, the triaryl esters are the most important.

Tertiary orthophosphate esters have been known for over 140 years, the trialkyls being synthesized about 1849 (1) and the triaryls about 1854 (2). The first significant use for phosphate esters developed after World War I when nitrocellulose laquers plasticized with tricresyl phosphate (TCP) became widely used as industrial and automotive coatings (3). In addition to plasticizing the coatings, the TCP reduced the flammability of the lacquer, a property that led to the investigation of phosphate esters for less flammable hydraulic fluids and lubricants. During the 1930s a number of investigators (4–7) defined the lubricating properties of phosphate esters, especially their usefulness as antiwear agents.

During World War II and the years immediately following, the development of increasingly sophisticated military and commercial aircraft using hydraulic rather than mechanical control systems created a need for the safest possible, that is, least flammable, hydraulic fluids. J. D. Morgan, working at the Cities Service Oil Company, was awarded patents in 1944 and 1946 (8, 9) on lubricant and hydraulic fluid compositions having wide operating temperature ranges (–40 to 200°F) based on tributyl and other trialkyl phosphates. Also in 1946, W. F. Hamilton and co-workers at the Lockheed Aircraft Corporation (10) were awarded a patent on what can be considered the forerunner of today's commercial aircraft hydraulic fluids.

The use of both trialkyl and triaryl phosphates as synthetic hydraulic fluids and lubricants was more thoroughly defined and developed in a major program jointly sponsored by the U.S. Navy and Air Force at the Shell Development Company between 1949 and 1953 (11). F. J. Watson at Shell was awarded several patents for fluid compositions based on tributyl and tricresyl phosphates (12a, 12b, 12c).

At about the same time, the Douglas Aircraft Company and Monsanto Chemical Company helped pioneer the use of phosphate esters in commercial jet aircraft. By the late 1950s such planes as the Douglas DC-8, Boeing 707, and Convair 880 were flying on Monsanto's Skydrol 500A fluid, which was based on a mixed alkyl aryl phosphate ester.

The use of phosphate esters in industrial processes developed concurrently with the use in aircraft. Tricresyl phosphate and other esters became widely used during the 1950s as deposit modifiers in leaded gasoline. By the late 1950s Monsanto, Celanese Corporation of America, and the E. F. Houghton Company had complete lines of phosphate ester-based industrial hydraulic fluids, which were readily adopted by the steel, aluminum, foundry, and casting industries. Each company also recommended phosphate ester products for use as fire-resistant compressor lubricants. The U.S. Navy in 1961 adopted Specification MIL-H-19457, a fire-resistant fluid based on trixylenyl phosphate, for the elevators of aircraft carriers.

Some early industrial fluids developed in both the United States and Europe were based on alkyl aryl phosphates. A number of these contained chlorinated aromatic hydrocarbons (chlorinated biphenyls), but these types of fluids have entirely disappeared from commercial use. The industrial phosphate ester fluids used today (aside from the aircraft fluids) are all based on triaryl phosphates and do not contain any kind of halogenated components.

II. CHEMISTRY

A. Structure

Various organic phosphorus compounds including phosphites, phosphonates, and phosphates have found application as additives in a wide variety of lubricant formulations as stabilizers, antiwear additives, antioxidants, metal passivators, and extreme pressure additives. The most widely used and best known of these compounds are the zinc dialkyl dithiophosphates found in almost all automotive engine lubricants. However, only one rather small group of phosphates has found significant use as synthetic base stocks. This group is the trisubstituted, or tertiary, esters of orthophosphoric acid, that is, those with the general structure:

```
        OR'
        |
  O  =  P - OR"
        |
        OR'''
```

Compounds in which one or two of the R groups are hydrogen are made commercially. These have found use as lubricant additives in the form of metallic or amide salts. However, all of the significant commercial synthetic lubricant base stocks are compounds in which all three R groups are organic moieties containing four or more carbon atoms. Thus, the important phosphate ester base stocks fall into three broad classes: triaryl, trialkyl, and aryl alkyl phosphates.

None of the commercially important base stocks contain nitrogen, sulfur, chlorine, or other elements substituted on the ring or alkyl chain. Therefore, the remainder of this chapter emphasizes those compounds in which the organic group contains only carbon and hydrogen.

The triaryl phosphates are the most significant commercial products. In this group of compounds all three organic groups can be the same, such as in tricresyl or trixylenyl phosphate, or they may be different, as in isopropylphenyl diphenyl phosphate or cresyl diphenyl phosphate. In commercial fluid base stocks, the location of substituents on the phenyl ring varies among the *ortho*, *meta*, and *para* positions.

Of the trialkyl phosphate esters, tributyl phosphate is by far the most important of the synthetic base stocks. Most of this finds use in the aircraft hydraulic fluid market.

Dibutylphenyl phosphate, also used in aircraft fluids, is the most common alkyl aryl ester. Although a significant patent estate developed on alkyl aryl esters and several of them were used widely at one time as industrial fluid base stocks, there is little significant commercial use of these esters today, other than the dibutyl phenyl ester.

B. Production

Although the tertiary phosphate esters can be considered as organic salts of orthophosphoric acid, H_3PO_4, they cannot be easily made by reaction with the acid. Because of the differences in the thermodynamics of the synthesis reactions, distinct commercial routes have developed to produce triaryl, trialkyl, and alkyl aryl esters.

1. Triaryl Phosphates

The simplest laboratory preparation, and most important commercial route, to triaryl esters is the phosphorylation of an aromatic alcohol—that is, a phenolic compound—with phosphorus oxychloride.

$$3ROH + POCl_3 \rightarrow (RO)_3P{=}O + 3HCl \quad (1)$$

Reaction conditions include elevated temperatures (up to 400°F), a metallic chloride catalyst such as magnesium or aluminum chloride, and an excess of the phenolic compound (13). These conditions are in part chosen to avoid the presence of the intermediate chloridates, $(RO)Cl_2P{=}O$ and $(RO)_2ClP{=}O$, which would reduce yields and produce acidic partial esters during subsequent processing.

Prior to the early 1970s, cresylic acids were the raw materials used for this preparation. The more common cresylic acids (cresol, xylenol, and mesitol) have one, two, and three methyl groups on the ring, respectively, in any of the *ortho, meta,* and *para* positions.

Trixylenyl phosphate became the most commonly used industrial fluid base stock of the triaryl esters. The range of viscosities required for the emerging industrial uses could be achieved by carefully selecting the xylenol isomers used as the starting material. An isomer mixture consisting of mostly 3,5-xylenol gives a higher viscosity than the a 2,6 mixture, for example.

Alternately, a feed to the reaction containing different ratios of phenol and/or cresylic acids could also serve as a means of producing variations in physical properties (see later sections of this chapter). Thus, tricresyl phosphate, mixed cresyl-xylenyl phosphates, and cresyl diphenyl phosphate became commonly produced phosphate esters.

Work by the Albright & Wilson Company (14) and the Ciba Geigy Corporation (15) in the United Kingdom in the late 1960s resulted in the development of a more easily controlled and less expensive route to the range of products desired. Work at both companies revolved around the catalytic alkylation of phenol with propylene or butylene and the subsequent reaction of this "alkylate" with phosphorus oxychloride as in reaction (1) above. This alkylate has become known in the industry as "synthetic" alkylate. (The cresylic acids are referred to as "natural" due to their derivation as by-products of coking and petroleum refining operations.) The esters made from the isopropyl and butyl phenols became known as "synthetic" esters.

In the synthetic process, the viscosity of the final product can be controlled in either or both of two ways: (a) by the degree of alkyklation of the phenol, that is, by the number of alkyl groups on the phenol ring, and (b) by using a variable, mixed feed of phenol and alkylphenol to the phosphorylation reaction (16). As the degree of alkylation increases, or the proportion of unalkylated phenol decreases, the viscosity of the product increases. In this manner a family of products can be produced tailored to the needs of the end user. Proper selection of the feed materials can produce synthetic esters with chemical and physical properties quite identical to the natural esters.

For both the natural and synthetic esters, various refining steps are used to produce the final product. The by-product hydrogen chloride (HCl) can be removed from the reaction by a variety of methods: heating, partial vacuum, sweeping with an inert gas, or reaction with an organic base such as pyridine. The most common method used combines heating and vacuum followed by scrubbing through water to recover the HCl as salable hydrochloric acid solution.

Following the HCl extraction, the crude product is refined. In one method, a series of distillation steps first removes the unreacted phenols and alkyl phenols for recycle and isolates the refined product, leaving the catalyst and high-boiling by-products in the still residue (17, 18). Alternately, the crude product can be distilled to remove unreacted raw materials, washed with aqueous alkali and water, and dried under vacuum (13).

The foregoing discussion of the raw materials and routes used to make the important phosphate esters indicates that most commercial triaryl esters are not symmetrical products. Indeed, the only truly symmetrical, pure triaryl phosphate of commercial importance is triphenyl phosphate, which is a solid at ambient temperatures and not useful as a fluid base stock.

Conversely, the asymmetry in the phosphate ester molecule is a significant determinant of its physical properties. Symmetrical products are crystalline or waxy solids. Controlling the degree of asymmetry produces liquids with varying physical properties. This asymmetry, then, can be tailored to a variety of application conditions.

As noted above, the assymetry can be introduced into the phosphate ester molecule by using a mixed feed to the reaction. As long as the reactivity of the phenolic compounds is reasonably close, this method is acceptable. This is the case in production of cresyl diphenyl, isopropylphenyl, and *t*-butylphenyl phosphates commercially important as base stocks. The same is true for tricresyl or trixylenyl phosphates, in which the similar reactivity of the *ortho, meta,* and *para* isomers makes the preparative reaction quite straightforward.

The products of a reaction in which two different aryl raw materials are used would be as shown by the following equation:

$$\begin{aligned} xROH + yR'OH + POCl_3 \rightarrow\ & (RO)_3P{=}O + (RO)_2(R'O)P{=}O \\ & + (RO)(R'O)_2P{=}O + (R'O)_3P{=}O \end{aligned} \quad (2)$$

A practical, commercial application of this reaction is the case in which R might be phenol and R' might be *tert*-butyl phenol. Furthermore, in practice, R' might be itself a mixture of mono-, di-, and tri-*t*-butyl phenols with the butyl groups possibly in the *ortho, meta,* and *para* positions. The composition, then, of the refined phosphate ester is a complex mixture of symmetrical and unsymmetrical trisubstituted esters whose composition is determined by the rates of reaction of phenol and the various alkyl phenols in the feed stock with phosphorus oxychloride.

Assymetry can also be introduced into the triaryl phosphate molecule by stepwise reaction of the sodium salt of a phenolic alcohol with an intermediate chloridate (13). The following reaction scheme is one of several possible alternatives:

$$2ROH + POCl_3 \rightarrow (RO)_2(Cl)P{=}O + 2HCl \quad (3)$$

$$R'ONa + (RO)_2(Cl)P{=}O \rightarrow (RO)_2(R'O)P{=}O + NaCl \quad (4)$$

The properties of the mixtures prepared by these reactions can be quite similar to those of symmetrical triaryl phosphates with a similar alkyl aryl content.

2. Trialkyl Phosphates

Trialkyl phosphates can be prepared using reactions similar to those used for triaryl compounds. However, because trialkyls are generally less stable than triaryls, the reaction [equation (1), phosphorylation of an alcohol] is usually carried out at more moderate temperatures. In order to drive the reaction to completion, greater excesses of alcohol are needed and the by-product hydrogen chloride must be removed as rapidly as possible.

The higher-molecular-weight trialkyl phosphates can be purified by stripping the unreacted alcohol, alkaline washing, and distillation drying, in steps similar to those used for the triaryl processes. The lower-molecular-weight esters, below tripropyl phosphate, can only be isolated by dry techniques due to their solubility in water.

Because of the inefficiencies of the aliphatic alcohol phosphorylation process, trialkyl phosphates are also commercially produced by the reaction of the sodium alkoxide with phosphorus oxychloride:

$$3RONa + POCl_3 \rightarrow (RO)_3P{=}O + 3NaCl \tag{5}$$

In this process, commonly referred to as the alkoxide process, the chlorine values precipitate rapidly as sodium chloride, NaCl. The NaCl can be removed by water washing, with further purification of the phosphate accomplished by distillation under vacuum.

Mixed or unsymmetrical trialkyl phosphates can be produced in ways similar to the unsymmetrical triaryl esters, that is, by using mixed alcohol feeds or by stepwise reaction of the intermediate chloridate with an alkoxide.

3. *Alkyl Aryl Phosphates*

The alkyl aryl phosphates, either alkyl diaryl or dialkyl aryl esters, can be made by the reaction of the appropriate purified intermediate alkyl or aryl chloridate with the desired alcohol under reactions and purification techniques similar to those described above.

Today, only dialkyl aryl phosphate esters are commercially significant in the synthetic fluids industry. These are based on lower-molecular-weight alkyls and are apparently best prepared (13) by the preparation of the dialkyl phosphoryl chloride (dialkyl phosphorochloridate),

$$2ROH + POCl_3 \rightarrow (RO)_2(Cl)P{=}O + 2HCl \tag{6}$$

which is purified by distillation under reduced pressure. The chloridate is then reacted with the sodium arylate in water:

$$RO_2(Cl)P{=}O + R'ONa \rightarrow (RO)_2(R'O)P{=}O + NaCl \tag{7}$$

The dialkyl aryl esters can then be isolated and purified by the techniques already described.

Much of the process development in recent years on the production of the trisubstituted phosphate esters has involved improvements in process efficiency (16, 17, 19–21), including development of continuous process steps (22), which have replaced batchwise operations of some earlier processes. Recent patents have been granted on the production of tertiary phosphate esters from phosphites (23) and phosphoric acid (24), but these are not commercially significant at this time.

III. PROPERTIES AND PERFORMANCE CHARACTERISTICS

A. Chemical Properties

Chemical inertness is one of the primary attributes that any lubricant or fluid base stock must possess. The fluid should not react with the metals or other materials from which the mechanical system is constructed. Since additives are commonly used, the base stock should not be reactive with or attacked by other classes of chemical compounds.

Phosphate esters have proven their chemical stability over many years of practical industrial service over a relatively wide temperature range. They generally are not attacked by most organic compounds and in fact show excellent solvent properties for most commonly used lubricant additives. Other aspects of chemical stability for synthetic base stocks, their thermal, oxidative, and hydrolytic stability, are more significant. The following discussion emphasizes these latter properties.

As was noted above, most of the commercially important trisubstituted phosphate ester fluids are actually mixtures in which a range of useful properties are achieved through asymmetry. (One major exception to this is tributyl phosphate, TBP.) In order to provide the most practical applications data, emphasis will be placed on commercially important products in their commercial form. To facilitate the presentation, the following abbreviations will be used for commonly occurring fluid/lubricant products:

IPPP	Isopropylphenyl phenyl phosphates
TBPP	*t*-Butylphenyl phenyl phosphates
ADP	Alkyldiaryl phosphates
TCP	Tricresyl phosphate
TXP	Trixylenyl phosphate
TBP	Tributyl phosphate
TBEP	Tributoxyethyl phosphate
TOP	Trioctyl phosphate

Where appropriate, to further describe the product if it is used commercially as a base stock, the designation just given will be followed by the ISO grade number. For example, IPPP/46 will define an ISO 46 phosphate ester base stock derived from isopropyl phenol; TBPP/32 will define an ISO 32 base stock made from *t*-butyl phenol.

1. *Thermal Stability*

Thermal stability will generally define the upper practical operating temperatures at which a fluid can be used. While in practical operating systems oxygen is usually present, study of the thermal stability in the absence of oxygen gives a clearer picture of the effect of temperature only. For example, localized hot spots at the asperities of a lubricated surface will place thermal stress on the fluid and cause degradation in the absence of oxygen. In practical applications, thermal stability is typically evaluated in relation to both temperature and time: that is, the shorter the time of exposure at a given temperature, the higher the temperature that can be tolerated.

A study by Blake et al. in 1960 (25) evaluated the thermal decomposition of a number of organic compounds under vacuum. The decomposition temperature (defined as the point at which the rate of pressure increase in the system reached 0.014 mm Hg/s) of tri(*n*-octyl) phosphate was reported as 195°C and that of triphenyl phosphate as 423°C. In the same study, a refined mineral oil decomposition temperature was reported at about 355°C.

An earlier study by Raley (26) using different criteria reported a decomposition temperature of triphenyl phosphate at 485°C and that of several tricresyl phosphate isomers in the range of about 375–395°C.

In a study of the ADP esters, Gamrath, Hatton, and Weesner (27) measured the relative stability of a wide variety of phosphates by determining acidity and weight loss at 245°C over a 24-h period. Some representative data from this study are presented in Table 3.1.

A more recent study of thermal stability (28) employed thermal gravimetric analysis (TGA) and differential scanning calorimetry (DSC). ASTM methods D-3850 and D-3350

Table 3.1 Stability of Selected Phosphate Esters at 245°C, 24 Hours[a]

Phosphate ester	Weight loss (%)	Acidity(ml NaOH/mole)
Tri(*n*-octyl)	3.2	12
Tri(2-ethylhexyl)	2.8	82
Tricresyl	0.4	3.5
n-Octyldiphenyl	0.8	31
n-Octyldicresyl	0.4	10
*n*Butyldiphenyl	2.4	9

[a]Data from Ref. 27.

were used, respectively, except that the samples were tested under nitrogen to eliminate oxidation effects. The DSC method estimates the onset of decomposition as the temperature at which an endotherm occurs as the sample is heated at a constant rate of 10°C/min. The TGA method provides a measure of the decomposition by determining the weight loss as the sample is heated. The data are recorded as the temperature at which a given percent weight loss is reached.

It should be noted that, in both DSC and TGA techniques in these ASTM methods, the temperature recorded as the "onset of decomposition" might be influenced by evaporation of the most volatile component if the test fluid is a mixture. Evaporation is endothermic in DSC; evaporation will result in weight loss in TGA. This is especially true for the synthetic triaryl phosphates, which are mixtures of monomeric compounds. Analysis of the vapor phase would be required to define the extent to which evaporation influences the data recorded.

The DSC data in Table 3.2 show that commercially used triaryl phosphates begin to show an endotherm, whether decomposition or evaporation, over 240°C, well above common operating temperatures. TBP, typical of the trialkyls, begins to decompose at a lower temperature, about 147°C. The TGA data in Table 3.3 also show that significant loss begins to occur well above common system operating temperatures.

Since triphenyl phosphate (TPP) is a significant component and the most volatile

Table 3.2 Relative Thermal Stability of Phosphate Esters under Nitrogen Atmosphere (ASTM D-3350) by Differential Scanning Calorimetry[a]

Phosphate ester	Initiation of decomposition (endotherm) temperature (°C)
TPP	246
TBPP/46	279
IPPP/32	245
IPPP/46	247
TCP	249
TBP	147

[a]Data from Ref. 28.

Table 3.3 Relative Thermal Stability of Phosphate Esters under Nitrogen (ASTM D-3850) by Thermal Gravimetric Analysis[a]

Weight loss (%)	Temperature (°C)								
	TPP	IPPP/22	IPPP/32	IPPP/46	TBPP/46	TCP	TBP	TBEP	TOF
10	261	274	272	265	301	278	154	221	208
20	281	292	294	285	320	298	173	242	231
30	294	304	307	297	333	310	183	254	242
50	310	320	324	313	350	325	196	269	257
75	323	334	339	327	365	325	207	279	268

[a]Data from Ref. 28.

species present in the commercial IPPP and TBPP fluids, the similarity seen in the data in Tables 3.2 and 3.3 between these fluids and TPP is to be expected. In a similar DSC study using a sealed sample container, TPP did not show significant physical signs of decomposition (color change, for example) until 360°C. As noted above, the degree to which the volatility of TPP influences the endotherm data on IPPP and TBPP would require analysis of the vapor phase, a step not performed in the study.

However, the data support other data and practical experience developed over the years. It has become generally accepted that the triaryl esters are the most thermally stable of the phosphates, more stable than either the trialkyl and alkylaryl esters.

Several studies (27, 29, 30) of the pyrolysis of phosphate esters have shown that the decomposition products are unsaturated hydrocarbons and acidic phosphate esters. Results are similar with both the trialkyl and alkylaryl esters, indicating that the weakest link in the decomposition is the aliphatic carbon-oxygen bond. The triaryl esters are more stable than either the trialkyl or alkyl aryl esters. Alkylation of the ring in the triaryls tends to reduce the thermal stability, but this in turn can be affected by the length and branching of the alkyl chain. As will be seen below, however, all three classes of phosphate esters exhibit sufficient thermal stability for many commercial applications, although the triaryls have achieved the widest use.

2. *Oxidative Stability*

The oxidative stability of phosphate esters has proven to be quite high and has not been a significant deterrent to their commercial use. Cho and Klaus (31) investigated the oxidative degradation of trialkyl and triaryl phosphates using the apparatus known as the Penn State microoxidation tester. After oxidation at specified time and temperature conditions in the presence of a metal catalyst, the reaction products were analyzed by gel permeation chromatography. The results of this study (Table 3.4) showed that the triaryl esters are again more stable than the trialkyl esters and that both types, without additives, were more stable than formulated organic ester aircraft engine (MIL-L-7808) lubricants.

The study showed no evidence of oxidation of the P–O–C bond, but rather that the relative resistance to oxidation was dependent on the ease of oxidation of the hydrocarbon structure of the alcohol from which the phosphate ester was derived. The higher-molecular-weight products of the oxidation were shown to be condensation products of the oxidized hydrocarbons.

A series of commercially available phosphate ester fluids and base stocks were

Table 3.4 Oxidative Stability of Phosphate Esters[a]

Phosphate ester	Time (min)	Temperature (°C)	Percent of Original Product		
			Unoxidized	Oxidized	Evaporated
Tributyl	5	225	13	6	51
	10	225	19	8	73
Tricresyl	30	225	85	1	14
	60	225	67	2	31
Tricresyl	30	250	57	3	40
Trixylenyl	30	250	65	4	31
Tricresyl	15	270	60	5	35
Trixylenyl	15	270	55	6	39
t-Butylphenyl	360	250	77	<1	22
diphenyl phosphate	180	270	81	<1	15

[a]As determined in the Penn State Microoxidation test apparatus, see ref. 31.

compared using DSC (ASTM D-3350) and TGA (ASTM D-3850) analyses with the tests being run in a flowing oxygen atmosphere in aluminum containers (28). These data (Table 3.5) are consistent with earlier studies. In addition to confirming the superior stability of the triaryl esters, the DSC and TGA data confirm earlier indications that triphenyl phosphate does not oxidize readily. This should be expected since a phenyl ring is more stable than an alkyl chain and TPP contains no alkyl sites at which oxidation can readily occur. In the ASTM D-3350 test, TPP evaporates before oxidation occurs.

Alkylation of the ring tends to lower the oxidative stability. However, there appears to be little difference between adding one or two methyl groups, as in tricresyl and trixylenyl esters. The addition of varying amounts of isopropyl groups also appears to have little effect compared to methyl groups under these conditions. The butyl phenol-derived esters are more stable than than either the methyl or isopropyl phenol-derived esters. No comment can be made regarding the effect of additives on the data in Table 3.5. The IPPP and TBPP products in the table are commercial fluids, and no data are available on the amount or type of antioxidants which might be present. The mineral oil and other esters (tricresyl, trixylenyl, tributyl, etc.) listed in Table 3.5 contained no additives.

Based on the thermal stability data, vapor pressure data (see Section III,B, Physical Properties), and other practical operating experience, the triaryl phosphates can be used in systems operating at temperatures up to 135°C for extended periods and up to 230°C for short periods.

3. Hydrolytic Stability

Phosphate esters are the reaction products of an aliphatic and/or aromatic organic alcohol and the inorganic phosphoric acid. Like all esters, the preparative reaction can be reversed in the presence of water. That is, hydrolysis can occur under the proper conditions. In fact, hydrolysis is the most important consideration in commercial application of phosphate and other organic esters.

Table 3.5 Oxidative Stability of Phosphate Esters under ASTM D-3350 (DSC) and D-3850 (TGA) Methods[a]

Oxygen flow 75 ml/min
Temperature 30–400°C
Aluminum sample container

Phosphate ester	Oxidation onset (°C)	Temperature (°C) at 1% Weight loss	5% Weight loss	10% Weight loss
100" SPN mineral oil[b]	195	155	205	225
Tributyl	175	65	100	115
Trioctyl	160	60	169	191
Tributoxyethyl	155	35	145	170
Triphenyl	—	185	220	235
Tricresyl	215	170	230	260
Trixylenyl	210	225	260	280
IPPP/32[c]	207–235	180–215	235–255	255–275
IPPP/46[c]	206–240	200–207	230–245	245–260
IPPP/68[c]	199–210	215–235	250–265	265–285
TBPP/46[c]	270–325	210–215	255–260	275
TBPP/68	290	100	235	255
TBPP/100[c]	313–320	145–235	200–275	240–290

[a]Data from Ref. 28.
[b]100-Second solvent-refined paraffinic neutral oil.
[c]The data represents the range of values obtained on samples of fluids or base stocks of various suppliers.

As in the thermal and oxidative degradation processes, the hydrolysis reaction will produce acidic products since it is the –O–C– bond that is attacked:

$$(RO)_3P{=}O + H_2O \rightarrow (RO)_2(OH)P{=}O + ROH \tag{7}$$

This hydrolysis reaction proceeds stepwise, first yielding the starting alcohol, phenol, or alkyl phenol and the disubstituted phosphoric acid ester. If driven to completion, phosphoric acid and the starting alcohol/phenol will be generated. Hydrolysis reactions are acid catalyzed and thus the hydrolytic degradation can be considered autocatalytic; that is, the acid products of hydrolysis further catalyze the decomposition.

Gamrath, Hatton, and Weesner (27) studied the relationship of hydrolytic stability to the structure of phosphate esters, with emphasis on alkyl diaryl esters. The esters were heated for 24 h at reflux with freshly distilled water; the acidity of both water and fluid layers was determined and expressed as moles of monobasic acid equivalents per mole of phosphate ester. The results are summarized in Table 3.6.

The authors concluded that the trialkyl and triaryl esters were generally more stable than the alkyl diaryl esters, although structure had a considerable affect. The data showed that higher-molecular-weight alkyl chains are more stable than lower, branched chains are more stable than straight chains, and the alkyl dicresyl esters are more stable than the corresponding alkyl diphenyl esters. However, all three types of esters have been successfully used as fluid base stocks.

ASTM method D 2619 is widely used to evaluate the hydrolytic stability of fluids. In

Table 3.6 Hydrolytic Stability of Selected Phosphate Esters[a]

Phosphate ester	Acid released per mole of phosphate ester
Tributyl	4.0
Tri-(2-ethylhexyl)	0.2
Tricresyl	1.2
n-Butyl diphenyl	27.8
t-Butyl diphenyl	7.3
n-Octyl diphenyl	14.6
6-Methylheptyl diphenyl	7.6
2-Ethylhexyl diphenyl	3.6
n-Octyl dicresyl	4.4
2-Ethylhexyl dicresyl	1.7

[a]Data from Ref. 27.

this test, the fluid contained in a beverage bottle with a copper strip is rotated slowly at 93°C for 48 h. The acidity of both the fluid and water layer is determined and the weight loss of the copper strip is measured. Unstabilized phosphates (Table 3.7) come quite close to passing the typical criteria. The "natural" esters are quite good, with TXP passing the standard criteria and the TBPP products coming quite close.

Both isopropylphenyl- and *t*-butylphenyl-derived trisubstituted esters are commercially used as fluid base stocks. Fluids made from both pass the typical criteria when properly formulated according to the producers' literature (Table 3.8).

In the practical application and use of phosphate esters, management of hydrolysis becomes the most important control element. While thermal and oxidative degradation can and do occur, hydrolysis from moisture contamination, accelerated by heat and acidic by-products, becomes the dominant cause of deterioration of phosphate ester fluids. Conversely, if hydrolysis is kept under control the phosphate ester fluid or lubricant can remain useful in commercial equipment for years without replacement (see Section III,E).

Table 3.7 Hydrolytic Stability of Unstabilized Phosphate Esters (ASTM D-2619)

Phosphate ester	Increase in fluid acidity (mg KOH/g)	Increase in water acidity (mg KOH/g)	Copper weight loss (mg/cm^2)
Tricresyl	0.095	9.1	0.30
Trixylenyl	Nil	Nil	0.03
TBPP/46	0.045	6.17	0.18
TBPP/68	0.09	8.98	0.34
TBPP/100	0.20	11.20	0.45
Standard criteria	0.2	5.0	0.3

[a]Data from Ref. 28.

Table 3.8 Hydrolytic Stability of Formulated Phosphate Ester Hydraulic Fluids (ASTM D-2619)[a]

Phosphate ester	Increase in fluid acidity (mg KOH/g)	Increase in water acidity (mg KOH/g)	Copper weight loss (mg/cm^2)
IPPP/46	0.05	0.4	0.05
TBPP/46	0.03	2.7	0.05
Standard criteria	0.2	5.0	0.3

[a]Data from Refs. 37 and 38.

B. Physical Properties

Phosphate esters commonly used as fluid and lubricant base stocks are all monomeric molecules and all contain the phosphorus-oxygen-carbon (P–O–C) bond system as the basic skeletal structure. While this P–O–C bond structure influences the basic properties, the organic moiety will determine most of the variation in physical properties exhibited by this class of compounds. As noted earlier, the range of properties is achieved mainly through variation of the size and number of the alkyl chains involved, whether alone in the trialkyl esters or attached to the aryl ring in the triaryl esters.

In trialkyl esters, the molecular weight of the alkyl group has the greatest influence on most properties and in a manner normally expected. That is, as the molecular weight increases, viscosity, boiling point, and pour points or freezing points increase and melting point decreases. The triaryl esters present a somewhat different situation, in that the simplest molecule, triphenyl phosphate, is a solid at ambient temperature, melting at about 49°C. As alkyl side chains are introduced, often introducing assymmetry at the same time, the melting point at first declines, giving useful liquids at ambient conditions; for example, cresyl diphenyl, tricresylphenyl, trixylenylphenyl, and many other alkylphenyl phosphates are liquids. Within the series of liquids behavior generally becomes normal; that is, viscosity, pour point, and boiling point increase as molecular weight increases.

The following sections emphasize the properties of the important fluid and lubricant base stocks, with other compounds included to illustrate trends or changes. Because of its use as a base stock, some data on TBP are included here, but its uses and properties have been more thoroughly reported by Schulz and Navratil (32).

1. Vapor Pressure and Boiling Point

Vapor pressure and boiling point (or range) play a significant role in the useful operating temperature range of a fluid. Phosphate ester base stocks generally have very high boiling points compared to other organic liquids with similar viscosities, as shown in Table 3.9. The trialkyls have the lowest boiling points; tri-*n*-butyl phosphate, used in aircraft and low-temperature formulations, boils at about 289°C.

The triaryl esters most widely used in industrial fluids have boiling points above 400°C at atmospheric pressure and over 250°C at 10 mm Hg. All of the triaryl base stocks, except for the ISO 100 grades, have a significant amount of triphenyl phosphate in them, and all are mixtures of isomers. Commercial products are more correctly described as having boiling ranges that are generally between 220° and 270°C at 4 mm Hg.

Table 3.9 Boiling Point of Phosphate Esters[a]

Phosphate ester	Boiling point (°C)	
	At 760 mm Hg	At 10 mm Hg
Trialkyls		
Methyl	196	
Ethyl	215	
n-Propyl	252	
n-Butyl	289	
Isobutyl	264	
Triaryls		
IPPP/22	365	255
IPPP/32	385	258
IPPP/46	396	262
IPPP/68	407	267
IPPP/100		272
TBPP/32	402	
TBPP/46	416	
TBPP/68	424	
TBPP/100	435	
TCP	427	
TXP	402	

[a]Data from refs. 13, 37, and 38.

Vapor pressure of phosphate esters can be estimated from the equation (33)

$$\log P = A/T + C \tag{8}$$

where P is the vapor pressure in mm Hg, T is the temperature (K), and A and C are constants. Table 3.10 contains values for A and C for several representative phosphate esters along with the calculated vapor pressures. The commercial triaryl phosphate base stocks—that is, the IPPP and TBPP esters, ISO grades 22–100—all have vapor pressures between 1 and 2 mm Hg at 210°C.

All of these data indicate that the phosphate ester base stocks used commercially are less volatile than mineral oils of similar viscosity and, when properly chosen, will provide excellent performance at most ambient working conditions, as well as being useful under vacuum conditions.

2. Viscosity

Viscosity and its variation with temperature significantly influence the design of equipment and the choice and performance of a lubricant/fluid. So much has been written on this subject that additional discussion need not be presented here. Among numerous other books and articles, the *Handbook of Lubrication Theory and Practice of Tribology* (34) is an excellent source of further information.

The viscosities of the commercially important phosphate esters fall in a wide and useful practical range (Table 3.11). The triaryl esters are marketed in ISO grades ranging from ISO 22 to ISO 100. Pour points of the triaryl esters generally fall between –10 and 20°C. The viscosity index (VI) of the triaryl esters is relatively poor, generally well below 100 and often near zero.

Table 3.10 Calculated Vapor Pressure of Phosphate Esters [a,b]
log P = –A/T + C

Phosphate ester	A	C	Vapor pressure at 27°C (mm Hg)	Latent heat of vaporization (kJ/mol)
Tri(*n*-butyl)	3207	8.586	7.9×10^{-3}	61.4
Tri(*m*-cresyl)	5787	11.55	1.8×10^{-8}	110.8
Tri(*p*-cresyl)	4535	9.44	2.1×10^{-6}	86.6
Dibutylphenyl	1921	4.99	3.9×10^{-2}	36.8
Tri(2-ethylhexyl)	3471	8.16	3.9×10^{-4}	66.6
t-Butylphenyl diphenyl	4444	9.24	–5.57333	85.0
Bis(*t*-butylphenyl) phenyl	4896	9.59	–6.7300	93.8
Triphenyl	4253	9.07	–5.10666	81.5
Cresyldiphenyl	3448	7.29	–4.20333	66.2

[a]Where P is pressure (mm Hg), T is temperature (°K), and A and C are constants.
[b]References 28, 32, and 33.

Table 3.11 Viscometric Properties of Phosphate Esters[a]

Phosphate ester	Viscosity (cSt) at 38°C	Viscosity (cSt) at 99°C	Viscosity index	Pour point (°C)
Tributyl	2.68	1.09	118	<–70
Tributoxyethyl	7.12	2.13	109	<–70
Tri-2-ethylhexyl	7.98	2.23	94	–70
Tri-*n*-octyl	8.48	2.56	148	– 1
Dibutylphenyl				
Di-2-ethylhexylphenyl	8.66	2.25	67	
Cresyldiphenyl	17.5	3.25	28	–34
Tri(3-isopropylphenyl)	42.7	5.53	59	
Tri(4-isopropylphenyl)	53.6	6.14	50	
Tricresyl	38.3	4.48	–12	–28
Trixylenyl	46.0	5.1	–32	–11
IPPP/22	18.9	3.4	40	–30
IPPP/32	31.3	4.4	20	–27
IPPP/46	45.3	5.3	14	–18
IPPP/68	77.9	6.3	<0	–11
IPPP/100	115	7.0	<0	– 5
TBPP/32	35	4.8	25	–10
TBPP/46	47	5.8	25	0
TBPP/68	62	7.0	25	5
TBPP/100	105	8.7	25	20
Aircraft fluid	1353	10.8	260	<–62

[a]Data from Refs. 27, 28, 32, 35, 37, 38.

The trialkyl esters have lower viscosity and pour points, and superior VIs, compared to the triaryls. The VIs of some trialkyl esters are near 100, comparable to paraffinic mineral oils.

The viscosity/temperature profiles of commercial phosphate ester hydraulic fluids are presented in Fig. 3.1. The slight variation in the slope of the bands results from the slightly different effects of the *t*-butyl phenyl group compared to the isopropyl phenyl group in each family of products. Bands for each of the product groups (IPPP and TBPP) are presented for the purpose of simplication. The upper edge of each band represents the ISO 100 grade and the lower edge the ISO 22 grade. The intermediate grades fall within the band with the plot line parallel to the band edges.

The useful operating range of either the alkyl or aryl family of esters can be extended by blending chemically or physically. That is, alkyl aryl esters (chemical "blends") can be prepared, or a trialkyl and a triaryl ester can be physically mixed to give improved viscosities, pour points, and VIs over the individual components. Commercially available polymers also can be used to improve the VI of phosphates. The aircraft hydraulic fluid products (35) are a primary example of the practical application of these techniques. These products (Table 3.12), 100% phosphate ester based with a VI improver, have pour points typical of the trialkyl phosphates, with the high-temperature viscosity more typical of the triaryl esters.

The viscosity index of straight triaryl esters can also be improved with typical polymeric VI improvers; usually less than 10% of the VI improver is adequate. Acrylic, styrenic, and other polymers have been used successfully.

Phosphate esters possess a correlative advantage compared to their low viscosity index—that is, their the excellent viscosity stability. The products are monomeric rather than polymeric. They have only stable benzene rings and/or relatively short alkyl chains. Thus, there is little in the molecular structure susceptible to the shearing forces normally found in hydraulic systems. This stability has been demonstrated (Fig. 3.2) in a field trial (36) of an electrohydraulic control fluid in a 300-MW steam turbine. The viscosity remained virtually unchanged for more than 5 years and remained well within equipment manufacturer operating specifications. The original fluid remained in the system during the entire period plotted in Fig. 3.2 with only minor additions of make-up fluid.

Phosphate esters can also be blended with other base stocks, synthetic or petroleum, to provide still broader properties.

As equipment is designed to operate at higher pressures, the variation of viscosity with pressure can become important. Table 3.13 shows the variation of viscosity with pressure of an isopropylphenol based ISO 46 commercial fluid (37).

3. Other Properties

A variety of other physical properties must be considered in use of lubricants and design of equipment. Tables 3.14–3.16 summarize data from a variety of sources including suppliers' technical literature (37–39).

a. Density and Specific Gravity. The unit weight of a fluid influences flow characteristics, the weight of a system, and other factors of consideration in equipment design (Table 3.14). Triaryl phosphates are more dense than the trialkyl esters. Within both families, the density and specific gravity decrease as the molecular weight increases. The alkyl aryl esters have intermediate values.

Phosphate esters are more dense than water and mineral oils. In a wastewater treatment system, for example, oil, being less dense than water, will rise to the surface of

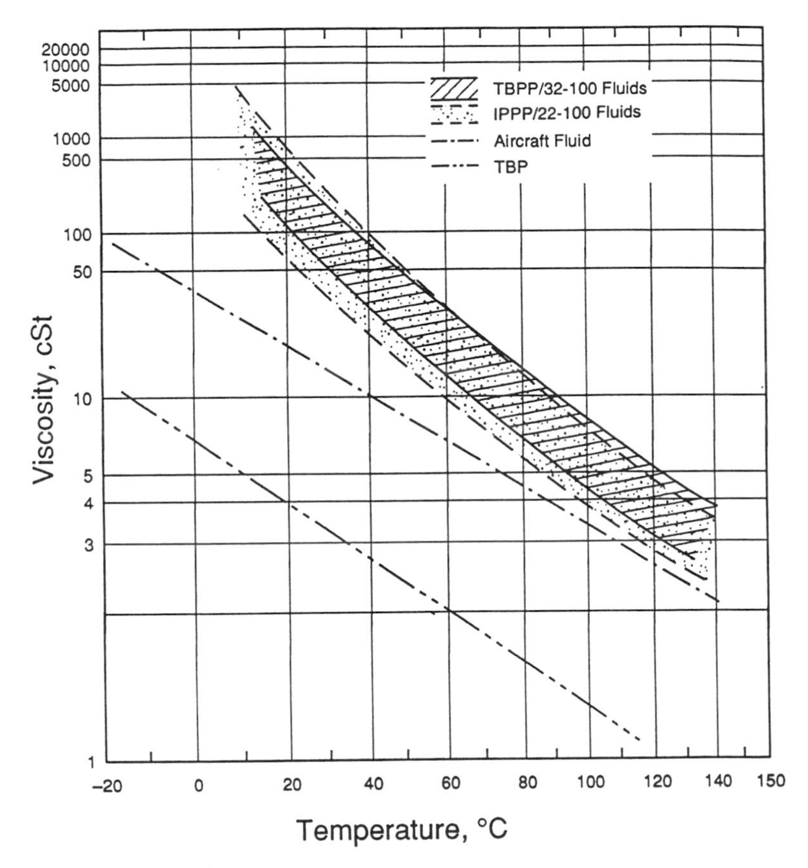

Figure 3.1 Viscosity-temperature relationship in industrial phosphate ester hydraulic fluids.

Table 3.12 Typical Physical Properties of Phosphate Ester Aircraft Hydraulic Fluid[a]

Viscosity (ASTM D-445)	
At −65°C	1353 cS
At 37.8°C	11 cS
At 98.9°C	4 cS
Viscisity index	260
Specific gravity 25/25°C	0.997
Pour point (ASTM D-97)	<−62°C
Neut. no. (ASTM D-974)	0.07
Flash point (ASTM D-92)	117°C
Fire point (ASTM D-92)	207°C
AIT (ASTM D-2155)	427°C

[a]Data from ref. 35.

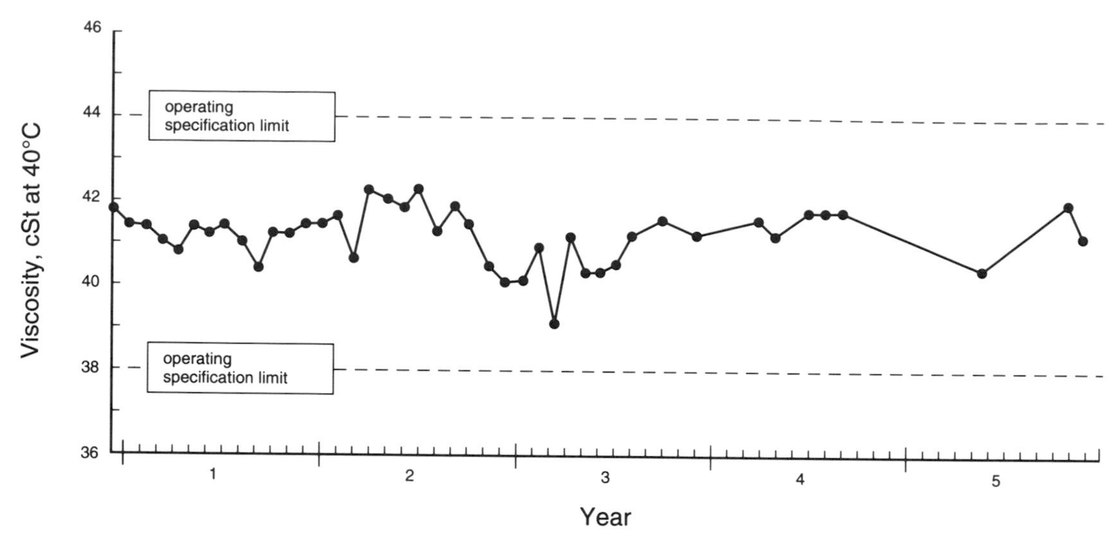

Figure 3.2 Viscosity stability of a TBPP/46 phosphate ester hydraulic fluid in an industrial application.

water and can be skimmed. Phosphate esters, being denser than water, will collect in low points of the system. Likewise, in a hydraulic system running on a triaryl phosphate ester fluid, significant levels of water from leakage or accidental contamination will collect on the surface of the fluid, usually in the sump.

b. Thermal Properties. Like density, specific heat (Cp) (28) is a characteristic property of a material and is significant in the design of heating or cooling capacity of a system. While the specific heat value of a phosphate ester varies with temperature, the values lie within a relatively narrow band over a broad temperature range, as shown in Table 3.15.

The thermal conductivity of typical triaryl phosphate fluids is approximately 3.04 ×

Table 3.13 Viscosity-Pressure Relationship of a phosphate Ester ISO 46 (IPPP/46) Hydraulic Fluid[a]

	Viscosity (cP) at		
Pressure	37.8°C	71°C	104°C
0	51.8	11.7	4.9
138	70.8	14.6	5.9
276	96.8	18.2	7.0
414	132	22.7	8.4
552[b]	185	29	10.5
689[b]	260	36	12
827[b]	350	45	15

[a]Data from ref. 37.
[b]Extrapolated data.

Table 3.14 Specific Gravity of Phosphate Esters[a]

Phosphate ester	Specific gravity	
	At 25/25°C	At 20/20°C
Trialkyls		
2-Ethylhexyl	0.923	
n-Octyl	0.915	
n-Butyl		0.980
Butoxyethyl		1.020
Triaryls		
Cresyldiphenyl	1.205	
Tricresyl[b]	1.161	1.167
Trixylenyl[b]	1.135	1.140
IPPP/32		1.164
IPPP/46		1.136
IPPP/68		1.090
TBPP/32	1.16	1.170
TBPP/46	1.13	1.138
TBPP/68	1.125	1.136
TBPP/100	1.120	1.124

[a]Data from refs. 13, 37–39.
[b]Typical commercial product, mixed isomers.

Table 3.15 Specific Heat of Phosphate Esters Between 60 and 150°C[a,b]

Phosphate ester	Specific heat, Cp (J/g/K)
Tributyl phosphate	1.58–2.04
Tributoxyethyl phosphate	2.17–2.20
Tricresyl phosphate	1.46–1.89
Trixylenyl phosphate	1.66–1.88
IPPP/22	1.93–2.33
IPPP/32	1.75–1.99
IPPP/46	1.70–1.97
IPPP/68	1.84–2.09
TBPP/22	1.78–2.10
TBPP/46	1.87–2.11
TBPP/100	1.94–2.09

[a]Determined by differential scanning calorimetry in an unpurged atmosphere under ASTM D3947-80.
[b]Data from ref. 28.

Table 3.16 Bulk Modulus and Compressability of an IPPP/46 Phosphate Ester Hydraulic Fluid at 37.8°C[a]

Pressure (bar)	Bulk modulus (bar$\times 10^4$)	Compressability (bar$^{-1}\times 10^{-5}$)
138	1.99	5.02
344	2.12	4.71
689	2.34	4.27
1034	2.53	3.95

[a]Data from ref. 37.

10^{-4} cal/s/cm/°C in the temperature range 35–95°C. The coefficient of thermal expansion between 25 and 50°C of these fluids is 6.9×10^{-4} cm^3/cm^3/°C. While some variation may exist between laboratories and among the various esters, the values above are representative of the entire class of commercial triaryl phosphate fluids. Phosphate esters generally are better heat-transfer agents than mineral oils.

c. Compressibility and Bulk Modulus. Compressibility of a fluid, or its reciprocal, bulk modulus, is an important factor in hydraulic system performance. A compressible fluid will result in sluggish operation, greater energy consumption, and heat buildup. Phosphate esters are less compressable than petroleum oils (Table 3.16) and thereby provide rapid hydraulic response and efficient operation.

C. Performance Properties

Fire resistance and lubrication characteristics represent the most important performance parameters that drive the use of phosphate esters in industrial systems. Although there is variation within the several families of phosphates, these two critical properties are inherent in the phosphate ester molecule: phosphate esters are excellent lubricants and the most fire resistant of the moderately priced synthetic fluids/lubricants.

1. Flammability

The resistance to burning, normally termed fire resistance in the fluids/lubricants field, can be measured by a variety of methods. Techniques have been developed to provide both laboratory screening methods and tests designed to simulate practical industrial performance in crisis situations. The most important tests that have been developed and accepted worldwide include:

Test	Method
Fire point	ASTM D-92
Flash point	ASTM D-92
Autoignition temperature	ASTM E-659[a]
Spray ignition	Factory Mutual Research Corp., Standard 6930
Hot surface ignition (hot channel)	Factory Mutual Research Corp., Standard 6930

Spray ignition	U.K. National Coal Board, Spec. 570/1970 (App. A)
Wick test	U.K. National Coal Board, Spec. 570/1970 (App. B)
Hot manifold	SAE Aeronautical Material Specification 3150C
Molten metal	Various, spray onto molten lead, aluminum, etc.
Spark ignition	DTD 5526
Compression ignition	U.S. Navy, MIL-H-19457c

[a]Data developed using earlier versions of ASTM D-286 and D-2155 continue to be quoted in trade and other literature. Both of these methods apparently gave 25–50°C higher temperatures than the current method E-659.

Representative fire, flash, and autoignition temperatures are presented in Table 3.17. Fire resistance data reported in suppliers technical literature are presented in Table 3.18. As the data in these tables show, phosphate esters readily pass all of the commonly used tests.

An important distinguishing characteristic of the phosphate esters, derived from the presence of phosphorus in the molecule as well as from the bond structure, is the heat of combustion. The triaryl phosphates have heats of combustion below 7800 kcal/kg (38), compared to mineral oils and other synthetic hydrocarbons with heats of combustion in the area of 9500 kcal/kg. The low heat release does not support combustion, and phosphate esters are self-extinguishing when the source of ignition is removed. This behavior is well demonstrated in the spray ignition tests listed above.

Table 3.17 Flash Point, Fire Point, and Autoignition Temperature of Phosphate Esters[a]

Phosphate ester	Flash point (°C) ASTM D-92	Fire point (°C) ASTM D-92	Autoignition temperature (°C) ASTM E-659
Tributyl	146–150	179	388[b]
Tri(2-ethylhexyl)	190	238	370[b]
Tributoxyethyl	224	252	260[b]
Triphenyl	224	310	635
Tricresyl	240–254[c]	338–346	600
Trixylenyl	255–343	320	535
IPPP/32	240–257[c]	330	543
IPPP/46	245–252[c]	327	500
IPPP/100	252–268[c]	335	515
TBPP/32	235	351	546
TBPP/46	252	332	535
TBPP/100	251	363	
Aircraft fluid	207	177	427

[a]Data from refs. 29, 32, 37–39.
[b]ASTM D-2155. This method usually gives higher values than the later method E-659.
[c]The range can represent differences resulting from testing in different laboratories as well as differences in product tested, such as isomer mixture.

Table 3.18 Performance of Commercial Phosphate Ester Hydraulic Fluids in Fire Resistance Tests[a]

Test	IPPP fluid	TBPP fluid	Test criteria
Fire point (°C) (ASTM D-92)	238–268	232–254	na
Flash point (°C) (ASTM D-92)	330–338	349–371	na
Autoignition temperature (°C)	515–585 (ASTM D-2155)	565 (ASTM E-659)	
Spray flammability (Factory Mutual)	Self-extinguishing in 4 s	Self-extinguishing in 5 s	10 s
Hot Manifold at 704°C (Factory Mutual)	No flashing or burning on tube	Self-extinguishes	
Factory Mutual Approval	Yes	Yes	
Hot channel	No ignition	No ignition	
Wick ignition	Self-extinguishing within 10 s	Self-extinguishing within 10 s	10 s
Molten metal at 815°C	No ignition		
Spark ignition	No ignition		

[a]Data from refs. 37–39.

Commercial industrial phosphate ester fluids generally are classified as HF-D type fluids and as Less Flammable Fluids, Group II, in the Factory Mutual Research Corporation rating system (40).

2. Lubricity

Phosphate esters have been proven as excellent lubricants in practical applications for many years. A number of early studies of phosphate esters (4–6, 41–43) led to their use as antiwear additives and defined the mechanism of their lubrication performance. It has been shown in these studies that phosphate esters react with ferrous metal surfaces to form iron phosphides and iron phosphates. These lower-melting compounds (at times called eutectics or alloys) are deformed by plastic flow and fill in low spots between asperities on the bearing surface, thus increasing the bearing area and reducing pressure and wear.

Laboratory data developed on commercial fluids, such as those presented in Tables 3.19 (four-ball wear) and Table 3.20 (vane pump test), corroborate the practical lubricant performance. The laboratory data show equal or lower rates of wear than those obtained with petroleum oils and other synthetic fluids.

3. Corrosion

Phosphate esters are generally not corrosive to common metals used in hydraulic and other systems. This is best illustrated by performance of a number of commercial fluids and lubricants in the corrosion/oxidation test FTMS (Federal Test Method) 5308.

Performance data of several phosphate ester products are presented in Tables 3.21 and 3.22. It should be kept in mind, however, that the decomposition products of thermal or hydrolytic reactions are acidic, which can be corrosive to metals. Copper and copper-bearing alloys will typically be the first metals to be attacked in such situations, and

Table 3.19 Four Ball Wear Tests of Phosphate Ester Fluids (ASTM D-2266)[a] at 40 kg Load, 1 Hour, and 1420 rpm

Fluid	Wear scar diameter (mm)
IPPP/22	0.68
IPPP/32	0.60
IPPP/46	0.62
IPPP/68	0.55
IPPP/100	0.55
TBPP/32	0.62
TBPP/46	0.54
TBPP/68	0.56
TBP/100	0.53

[a]Data from refs. 37 and 38.

Table 3.20 Performance of Phosphate Ester Fluids Vane Pump Wear Test (ASTM D-2882)[a]

Parameter	IPP/32	IPPP/46	IPPP/68	TBPP/46
Temperature (°C)	60	66	72	na
Duration (h)	100	100	250	100
Ring weight loss (mg)	2.5	3.7	3.7	14.0
Vane weight loss (mg)	34.8	7.4	2.0	55.3
Total weight loss (mg)	37.3	11.1	5.7	69.3

[a]Data from refs. 37 and 38.

nitrogen-containing metal passivating compounds, such as benzotriazole, can be included in the fluid formulation. Maintenance of acid number of the fluids at recommended levels (below 0.3 mg KOH/g) can avoid such potential problems.

4. Solvent Properties

Most phosphate esters are good solvents for other organic compounds, including hydrocarbons, alcohols, esters, ketones, and other relatively nonpolar materials. They are widely used as plasticizers for vinyl and cellulosic resins.

This strong solvency can be a benefit in that phosphates can be easily formulated with additives and other base stocks. It also can be a source of problems, as demonstrated by the poor compatibility with a number of common coatings, seals, and hose materials. Care is required in choosing materials of construction (see Section III,E). In systems where phosphate esters have been used to replace other fluids, this high solvency has also caused some problems because they clean dirt and other matter from internal equipment parts. Preflushing the system or the proper use of filters should be considered in such cases.

5. Additive Response

Being good solvents, phosphate esters are compatible with a wide variety of commonly used lubricant additives such as metal passivators, antioxidants, and VI improvers

Table 3.21 Performance of Phosphate Ester Aircraft Hydraulic Fluid in Standard Tests[a]

	Results	Industry standard
Thermal stability at 121°C, 168 h, 0.2–0.3% water		
Metal weight change (mg/cm^2)		
Steel	–0.02	+0.3 maximum
Magnesium	0.07	+5.0 maximum
Copper	0.00	+0.5 maximum
Cadmium-plated steel	0.02	+0.3 maximum
Aluminum	0.01	+0.2 maximum
Total acid no. change (mg KOH/g)	–0.07	+0.1 maximum
Viscosity change (cSt)		
At 38°C	0.87	+1.0 maximum
At 99°C	0.20	+0.3 maximum
Corrosion, hydrolysis, and oxidation stability at 82°C, 168 h, 0.8% water		
Metal weight change (mg/cm^2)		
Steel	0.02	+0.1 maximum
Magnesium	0.02	+0.2 maximum
Copper	0.00	+0.4 maximum
Cadmium-plated steel	0.00	+0.4 maximum
Aluminum	0.00	+0.1 maximum
Total acid no. change (mg KOH/g)	–0.06	+0.3 maximum
Viscosity change (cSt)		
At 38°C	0.33	+3.0 maximum
At 99°C	0.08	+1.0 maximum
Four ball wear test		
scar diameter (mm)		
At 4 kg load	0.12	0.45 maximum
At 10 kg load	0.37	0.50 maximum
At 40 kg load	0.69	0.55–0.75
Flammability tests		
Pipe wick (cycles)	34	25 minimum
Hot manifold at 794°C	No flame	No flame
Low-pressure spray	Self-existinguishing	Self-extinguishing

[a]Data from ref. 35.

(43–47). Additives are not always needed; at least one major supplier of hydraulic fluids claims excellent performance in typical industrial applications without any additives to the products. In some specialized applications, such as aircraft fluids (Table 3.21) and turbine lubricants (47, 48a, 48b), additives are used and the response of phosphate ester to them is quite good.

6. Chemical Stability

The hydrolytic and oxidative reactivity of the phosphate esters was described earlier. While it is important to understand the theory and mechanisms of these reactions, it is equally important that satisfactory performance in many applications has been well demonstrated. Figure 3.2 showed the excellent viscosity stability, and thus the lack

Table 3.22 Performance of Industrial Phosphate Ester Hydraulic Fluids in Corrosion/Oxidation Test,[a] Federal Test Method FTMS 5308.6

Properties	Limits	IPPP/46	TBPP/46
Total acid no.			
Before		0.040	0.021
After		0.056	0.105
Change	+2.0 maximum	0.016	0.084
Viscosity at 38°C (cSt)			
Before		46.50	43.85
After		47.21	45.07
Change	–5/+15	1.53	2.78
Metal weight change (mg/cm^2)			
Copper	+0.4	–0.022	–0.022
Steel	+0.2	Nil	0.007
Magnesium	+0.2	Nil	0.007
Aluminum	+0.2	Nil	0.015
Silver	+0.2	–0.007	Nil
Pass/Fail		Pass	Pass

[a]Data from ref. 28.

of chemical reactivity, of an ISO 46 electrohydraulic control fluid in field performance.

Since acidic products are formed from both the hydrolysis and oxidation of the esters, a rising acid number would indicate ongoing degradation of the fluid. Acid number data collected over a 5-year period in the previously mentioned system, shown in Fig. 3.3, clearly demonstrate that degradation had not occurred. Periodic additions of make-up fluid were the only additions to the system, and a fuller's earth by-pass filter system was employed. These data reinforce the effectiveness of a good maintenance program in achieving excellent operating performance from an industrial hydraulic fluid.

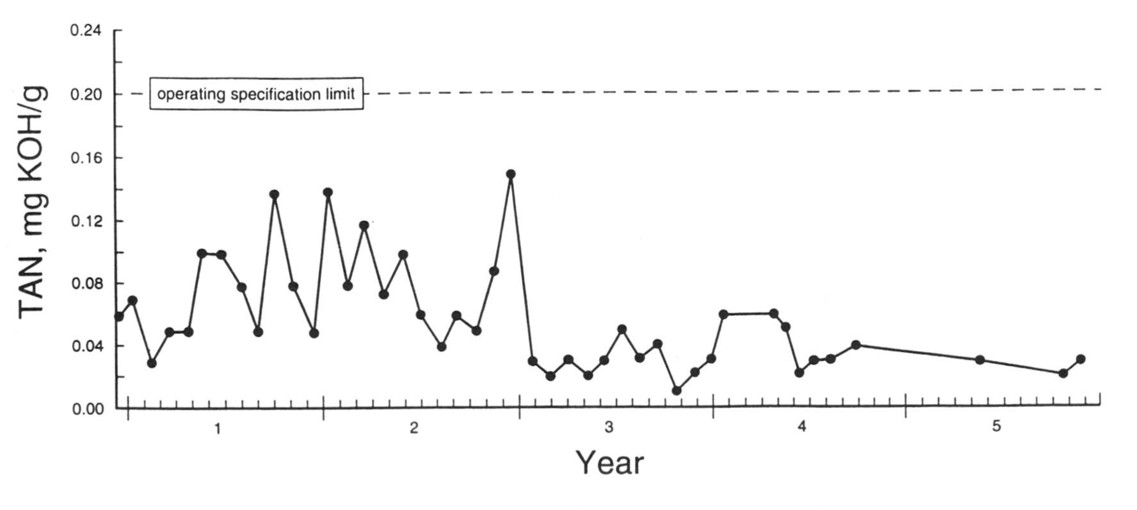

Figure 3.3 Acid number stability of a TBPP/46 phosphate ester hydraulic fluid in an industrial application.

7. *Foaming*

Foaming tendency and stability can be an issue in all hydraulic systems. Commercial phosphate ester fluids pass frequently specified tests, such as ASTM D-892, which determines the maximum foam height generated under standard air introduction rates as well as the rate of collapse of the foam.

Supplier's literature (37–39) indicates that there is little difference between the IPPP and TBPP base stocks over the ISO 22–100 grade range. All fluids generate less than 50 ml of foam, and complete collapse occurs within 20 s as specified by the test method. Some suppliers may use an antifoam additive to meet this specification. Silicones are effective, but other agents have been employed as well.

D. Toxicology

Much has been written and said about the toxicology of phosphate esters over the years. Unfortunately, considerable concern and confusion exist about this subject. In the following section we briefly outline the state of the situation as accurately as it is currently known. The user or potential user of phosphate esters should thoroughly investigate and evaluate the most recent data available so as to reach correct conclusions about the proper degree of concern to be held and the appropriate care to be taken in handling and use.

The use of recent data needs to be emphasized, because much of the early information about triaryl phosphates in particular was developed well before toxicological expertise and testing techniques reached today's level of sophistication. Often the precise identification of the compound or compounds involved in reported episodes or in laboratory tests was unknown or poorly defined. As will be seen below, one specific isomer has been identified as the most toxic of the triaryl phosphates, and its elimination results in orders-of-magnitude differences in the toxic responses observed. Assurance of, and confidence in, the definition of the compounds being evaluated is thus the most critical factor in determining the toxicity of the product.

It has been known since the early 1900s that exposure to organophosphate compounds can cause toxic effects in animals and humans. This phenomenon has been noted in a variety of compounds including phosphates, phosphites, and phosphonates. The effects can be visually noted in poor motor skills, which develop much after the exposure. Because of this, the effect has generally become known as organophosphate-induced delayed neurotoxicity (OPIDN). Neither carcinogenicity nor mutagenicity is involved in the observed effects.

Johnson (49) and Abou-Donia and Lapadula (50) have recently summarized and reviewed the history, symptoms, and mechanism of OPIDN. The major visible consequence of the delayed neurotoxicity is motor dysfunction, predominantly in the limbs. Tri-*ortho*-cresyl phosphate (TOCP) has been defined as the one of the more potent OPIDN neurotoxins in humans. The mechanism of attack generally begins with the inhibition of enzyme activity in the nervous system, with eventual damage to the system and impairment of neural transmissions. The degree of recovery, according to these authors, can often be related to the severity of the exposure. Mild exposures do not appear to result in permanent damage.

As evidence of OPIDN and the role of TOCP became better defined, the industry took extensive steps to reduce or eliminate this isomer from commercial products. For example, tricresyl phosphate is available today with little (<1%) *ortho* isomer content.

Current triaryl phosphate products, especially the synthetic esters made from isopropyl and *t*-butyl phenol, are widely used as fluid base stocks. These exhibit far less OPIDN than the TCP and TXP available through the 1960s. The trialkyl and alkylaryl esters commercially used today also show little to no neurotoxic effects (51, 52). Any effects seen are often at extremely high "limit" doses.

The synthetic triaryl phosphates are practically nontoxic by oral, dermal, and inhalation routes of exposure. Moreover, they are neither irritating to the skin or eyes nor mutagenic (53). Several trialkyl esters, specifically tributyl and trioctyl phosphates, however, are classed as a skin irritants. Detailed data are readily available from the Material Safety Data Sheets published by all of the various producers and other sources.

None of the triaryl phosphates that find significant use as base stocks are listed under the regulations of the International Agency for Research on Cancer (IARC), U.S. Occupational Health and Safety Administration (OSHA), American Conference of Governmental Industrial Hygienists (ACGIH), or the 1989 U.S. Superfund Amendment and Reauthorization Act Title III (SARA). TBP is not listed by IARC, OSHA, or SARA but does have a threshold limit value (TLV) of 2.5 mg/m^3 under ACGIH standards.

With normal and proper handling procedures and good industrial hygiene, phosphate esters can be safely used in industrial applications without undue concern for the well-being of workers and users.

E. Maintenance of Systems

Some of the performance data discussed above indicate that proper maintenance and control of the mechanical systems can assure excellent performance and result in exceptionally long fluid or lubricant life. The system maintenance program should involve three broad areas:

1. Design and preparation of the equipment and its components.
2. Equipment and maintenance programs to maintain the mechanical integrity of the system and prevent contamination of the fluid.
3. Operating procedures that include fluid conditioning and periodic chemical analysis of the fluid.

Design and preparation of the equipment is an important first step. Since phosphate esters are excellent solvents for many organic compounds, supplier recommendations regarding seals, hoses, filters, paints, and other accessory equipment should be followed carefully. Generally, seals and hoses should be made of fluoroelastomer or ethylene/propylene rubber. Epoxy coatings can be used, but painting of internal surfaces is not recommended. Most common metals can be used to fabricate parts and operating components. Suppliers of the phosphate ester fluids and lubricants have detailed recommendations on materials of construction; these suppliers should be consulted early in the design stage of new equipment or when conversion of older equipment to phosphate is being planned.

A number of reports have been prepared describing practical operating experience with phosphate ester fluids and lubricants (54–58). A recent report by the Electric Power Research Institute (59) presented an excellent review of worldwide operating experience with phosphates esters, including problems and successes, as well as procedures and control limits recommended for phosphate ester systems.

Hydrolysis presents the greatest risk to fluid stability, so maintenance of low water levels is important. Further, since hydrolysis forms acidic compounds, which can catalyze further hydrolysis, acid number can be used as an excellent control parameter to both determine the condition of the fluid and maintain its stability.

Maintenance of low water and acid levels is an operation known as fluid conditioning. This can be most readily accomplished by the use of bypass filters composed of earth, activated alumina, or ion-exchange resin media. Properly designed and maintained, this filtration conditioning will maintain moisture levels in the fluid below the recommended 0.1% by weight and the acid number below 0.2–0.3 mg KOH/g. Again, the best source of specific recommendations on products, procedures, and analytical methods should be the fluid suppliers.

IV. MANUFACTURE, MARKETING, AND ECONOMICS

A. Manufacturers

Over 85,000 metric tons per year (MTPY) of phosphate esters is produced worldwide (Table 3.23). Of this total, the triaryl and alkyl aryl esters represent 70,000–75,000 MTPY, with each of the major trialkyls (TBP, TBEP, and TOP) representing 3000–5000 MTPY. The largest use (Table 3.24), consuming 55,000–60,000 MTPY, is as a flame retardant plasticizer for a variety of plastic resins including cellulosics, polyvinyl chloride and polyphenylene oxide resins.

Lubricants and hydraulic fluids consume most of the remainder, 20,000–25,000 MTPY, mostly as a base stock for industrial and aircraft hydraulic fluids. Other base stock uses include specialty gas and steam turbine lubricants, and compressor lubricants.

Table 3.23 Estimated Worldwide Consumption (Metric Tons) of Phosphate Esters[a]

Region	Triaryls	Trialkyls	Total
North and South America	32–34	3–5	34–37
Europe and Africa	24–27	2–4	27–30
Asia	13–17	1–2	14–17
Total	71–74	8–10	85–88

[a]Author's estimates.

Table 3.24 Estimated World Markets (Metric Tons) for Phosphate Esters[a]

Use	Triaryls	Trialkyls	Total
Plasticizer	53–57	2–3	54–58
Lubricants/fluids	17–19	2–3	18–20
Lubricant additives	2–4	—	2–4
Other	2–4	3–5	6–8
Total	71–74	8–10	85–88

[a]Author's estimates.

Table 3.25 Major World Producers of Phosphate Esters

Producer	Location	Capacity (metric tons per year)	Products[a]
Ciba Geigy Corp.	Trafford Park, U.K.	40–50	TArP
FMC Corp.	Nitro, W.Va.	30–35	TArP, TAkP
Akzo Chemicals, Inc.	Gallipolis Ferry, W.Va.	25–30	AAP,TArP, TAkP
	Kashima-gun, Ibaraki, Japan	5–10	TArP
Bayer AG	Leverkusen, Germany	10–12	TArP, TAkP, AAP
Daihachi Chemical Industry Co.	Handa, Aichi, and Higashi, Osaka, Japan	5–10	TArP, TAkP
Ajinomoto Co., Inc.	Yokkaichi, Mie, Japan	5–10	TArP
Monsanto Company	Bridgeport, N.J.	4–5	AAP
	Newport, U.K.	3–4	AAP
Protex, SA.	Chateau Renault, France	—	TAkP
Croda Chemicals, Ltd.	Goole, U.K.	—	TAkP
BASF Aktiengellscaft	Ludwigschafen, Germany	—	TAkP
Showa Ether Co.	Aiko-gun, Kanagawa, Japan	—	TArP

[a]TArP, triaryl phosphates; TAkP, trialkyls; AAP, aklyl aryl phosphates.

Tricresyl phosphate and the isopropylphenyl, and *t*-butylphenyl phosphates also find wide use as antiwear additives in petroleum and synthetic base stock hydraulic fluids, tractor fluids, and aircraft turbine and piston engine lubricants. Virtually all of the TCP used in the lubricants industry, 1500–2000 MTPY, is used as an additive. Smaller volumes of phosphate esters are consumed in other industrial processes as antifoam agents (TBP), solvents in chemical manufacturing (TOP), uranium and other rare earth metal processing (TBP) (32), and as a leveling agent in floor finishes (TBEP).

The United States and Europe represent the major geographic consuming areas. Several producers exist in each of the major market areas (Table 3.23). Estimated world capacity totals over 115,000 MTPY (Table 3.25). Because most companies produce a variety of products at the same plant location and sometimes in common equipment, capacity figures are difficult to define, but it is estimated that capacity for the triaryl and aryl alkyl esters approaches 100,000 MTPY.

B. Suppliers

Phosphate esters are sold commercially by both basic chemical producers and fluid compounders under a variety of trade names to both the fluid/lubricant and lubricant additive markets. Table 3.26 lists the major suppliers, along with the more widely known trade names. The product list includes fluids that are pure phosphate ester base stocks, as well as those in which the phosphate is blended with either petroleum or other synthetic base stock.

Table 3.26 Suppliers of Phosphate Ester Hydraulic Fluids and Lubricants[a]

Company	Trade name[b]	Types of products
Ajinomoto Co., Inc.	—	Fluids
Akzo Chemicals	Fyrquel	Fluids, lubricants
	Syn-O-Ad	Additives
	Fyrtek	Fluids
Bayer AG	Disflammol	Additives
Chevron Corp.	HyJet	Aircraft fluid
Ciba Geigy Corp.	Reolube HYD	Fluids
	Reofos	Additives
	Reolube Turbo Fluid	Lubricants
Diahachi Chemical Industry Co.	—	Fluids, additives
D. A. Stuart Co.	Dasco	Fluids
E. F. Houghton Co.	Houghto-Safe	Fluids
FMC Corp.	Durad	Additives, fluids, lubricants
Metal Working Lubricants Co.	Metsafe	Fluids
Mobil Oil Corp.	Pyrogard	Fluids
Monsanto Co.	Skydrol	Aircraft fluid
Pennzoil Corp.	Pennsafe	Fluids
Pacer Lubricants, Inc.	Pyro-Safe PE	Fluids

[a]Data from ref. 40 and private communications.
[b]The names listed are trademarks or trade names of the respective companies. A number or letter designation often follows the trade name to further define the product. Some suppliers use the same trade name for fluids made from base stocks other than phosphate esters.

Except for the aircraft fluids in which TBP and/or an alkyl diaryl ester is the major component, all of the important fluids and lubricants are based on the synthetic triaryl phosphates produced from isopropylphenol or *t*-butylphenol.

C. Economics

Phosphate ester industrial fluids are more expensive to produce than petroleum and some of the other more common synthetic base stocks. Costs depend largely on petrochemical raw materials, such as butylene and phenol, and the relatively complex processing and purification procedures. Industrial fluids, such as those commonly used in primary metal refining and fabricating mills, generally sell for three to five times the price of petroleum oils, that is, in the (U.S.) $10–15 per gallon range. Other specialized fluids and lubricants have prices up to $20 per gallon; the highly refined products such as the aircraft and electrohydraulic control fluids reach over $30 per gallon in small quantities.

At these market prices, phosphate ester fluids are among the higher priced of the more common synthetic fluids such as organic acid esters and polyalphaolefins. They are not as expensive as others, such as silicone and fluorocarbon fluids.

The higher price for phosphate ester fluids is justified on performance characteristics, especially lubricating ability and resistance to burning. Phosphate esters are the safest in regard to fire resistance of the moderately priced less-hazardous fluids. Their high fire points and high flash points testify to the difficulty of igniting them, and their low heat of combustion makes them the only self-extinguishing fluids in their price class.

Therefore, when safety factors are of paramount importance, phosphate esters are the most cost-effective fluid base stocks yet developed.

V. OUTLOOK

Phosphate esters were among the earliest synthetic base stocks developed and are relatively mature products in their market segments. Engineering design improvements in materials of construction and operating pressures have enabled the use of lower-cost materials in many market segments pioneered by phosphate esters. In other market uses, such as in the aircraft and power generation (electrohydraulic control) markets where safety is critical and where high operating pressures and temperatures are common, phosphates have continued to grow and find wider use.

Being excellent lubricants as well as fire resistant, phosphate esters have always held promise in other industrial uses. Extensive laboratory and field development effort has been directed at the use of phosphates as turbine main bearing lubricants (57–59). Despite many years of effort, only small volumes have found continuing use in stationery gas turbines. However, recent developments in the Soviet Union (60) have resulted in the decision there to design main bearing systems that will use lubricants based on trixylenyl phosphate for all new large (over 800 MW) and small steam turbines.

Perhaps the most promising and most interesting use that may develop over the next decade is the use of phosphate esters as vapor-delivered lubricants for low heat rejection (adiabatic) engines. This work, pioneered by E. E. Klaus at Pennsylvania State University and supported by the U.S. Department of Energy and other agencies (61–63), continues to be the subject of active research efforts.

Adiabatic engines, while expected to achieve high fuel efficiencies and low emissions, will place exceptional strain on the thermal stability of the lubricants both in the combustion chamber and in the lubricant sump. Vapor delivery of the lubricant avoids the latter problem, because the bulk of the lubricant can be kept at moderate temperatures until it is delivered to the combustion chamber. Work to date has shown that tricresyl and tributyl phosphate vapors can be delivered directly to the walls of the combustion chamber in very low concentration, which, while providing adequate lubrication, will not result in excessive engine exhaust emissions or harmful deposit formation.

Phosphate esters may also play an important role as a co-base stock with other synthetics in the lubrication of low-heat rejection diesel engines being developed as intermediates or alternates to the truly adiabatic engine (64). This work is also being sponsored in part by the U.S. Department of Energy and several industrial companies.

ACKNOWLEDGMENT

The critical evaluation and editorial assistance of Mr. Douglas Placek and Mr. A. J. Raymond of the FMC Corporation were invaluable in the preparation of this chapter. Mr. Sundeep Shankwalkar, Ms. Deirdre LaMarche, and Ms. Victoria Sayers, all also of FMC Corporation, assisted greatly, the former in developing some of the thermal and fluid performance data, and the latter two in compiling the necessary literature searches. Their efforts, understanding, and patience are hereby acknowledged with sincere appreciation and thanks.

REFERENCES

1. Vogeli, F. (1849). *L.Annalen der Chemie*, 69: 190.
2. Williamson and Scrugham (1854). *L.Annalen der Chemie*, 92, 316.
3. Loeffler, A. T. (1956). *Armed Forces Chem. J.*, Sept-Oct., 6.
4. Evans, H. C., W. C. Davies, and W. J. Jones (1930). J. Chem. Soc., 1310.
5. Beech, O., J. W. Givens, and A. E. Smith (1940). *Proc. R. Soc.* A177, 90.
6. Beech, O., J. W. Givens, and A. E. Smith (1940). *Proc. R. Soc.* A177, 103.
7. Caprio, A. F. (to Celluloid Corp.) (1941). U.S. Patent 2,245,649.
8. Morgan, J. D. (to Cities Service Oil Co.) (1944). U.S. Patent 2,340,073.
9. Morgan, J. D. (to Cities Service Oil Co.) (1946). U.S. Patent 2,410,608.
10. Hamilton, W. F., M. F. George, Jr., and G. B. Weible (to Lockheed Corp.) (1946). U.S. Patent 2,392,530.
11. Vaughn, W. E. (1952). Organo Phosphorus Compounds, Shell Development Co., Final Technical Report, Sections A,B, and C, Report No. S-12401, to Office of Naval Research for Period June 1, 1949 to July 31, 1952.

12a. Watson, F. J. (to Shell Development Co.) (1951). U.S. Patent 2,549,270.

12b. Watson, F. J. (to Shell Development Co.) (1953). U.S. Patent 2,636,861.

12c. Watson, F. J. (to Shell Development Co.) (1953). U.S. Patent 2,636,862.

13. Gunderson, R. C., and A. W. Hart (1962). *Synthetic Lubricants* (cf. Chapter 4, Phosphate Esters, by R. E. Hatton), Reinhold, New York.
14. Electric Reduction Co. (assigned to Albright & Wilson Co.) (1965). British Patent 1.165,700.
15. Ciba Giegy Corp. (1966). British Patent 1,146,173.
16. Giolito, S. L. (to Stauffer Chemical Co.) (1982). U.S. Patent 4,351,780. (1982).
17. Giolito, S. L., and S. B. Marviss (to Stauffer Chemical Co.) (1985). U.S. Patent 4.559,184.
18. Aal, R. A., N. H. C. Chen, and J. K. Chapman, Jr. (to FMC Corp.) (1976). U.S. Patent 3,945,891.
19. Finley, J. H., & H. P. Liao (to FMC Corp.) (1984). U.S. Patent 4,482,506.
20. Finley, F. H., and H. P. Liao (to FMC Corp.) (1984). U.S. Patent 4,443,384.
21. Finley, J. H., and H. P. Liao (to FMC Corp.) (1984). U.S. Patent 4,438,048.
22. Smith, H. M., and R. D. Williams (to FMC Corp.) (1983). U.S. Patent 4,421,936.
23. DiBella, E. P. (to Borg Warner Chemical Corp.) (1984). U.S. Patent 4,469,644.
24. Segal, J., and L. M. Schorr (to Bromine Compounds, Ltd). (1989). British Patent 2,215,722.
25. Blake, E. S., W. C. Hammann, J. W. Edwards, T. E. Richards, and M. R. Oak (1960). American Chemical Society, Division of Petroleum Chemistry Symposium, Cleveland, April 5–14.
26. Raley, C. F., Jr. (1955). Wright Air Development Center, Technical Report 53-337.
27. Gamrath, H. R., R. E. Hatton, and W. E. Weesner (1954). *Ind. Eng. Chem.*, 46, 208.
28. Shankwalkar, S. G. and D. G. Placek (1992). Oxidation and weight loss of commercial phosphate esters. *Ind. & Eng. Chem. Res.* (in press).
29. Barrington, H. E., and R. A. Setterquist. (1957). *J. Am. Chem. Soc.* 79, 2605.
30. Noone, T. M. (1958). *Chem. Ind.*, 46, 1512.
31. Cho, L., and E. E. Klaus (1979). Oxidative degradation of phosphate esters, *ASLE Trans.*, 24, 1, 119–124.
32. Schulz, W. W., and J. D. Navratil (1984). *Science and Technology of Tributyl Phosphate*, Vols. I, IIa, and IIb, CRC Press, New York.
33. Dolby, A., and R. Keller (1957). Vapor pressure of some phosphate and phosphonate esters, *J. Phys. Chem.* 61, 1448.
34. Booser, E. R. (1984). *Handbook of Lubrication: Theory and Practice of Tribology*, Vol. II, CRC Press, New York.
35. Chevron International Oil Co. (1989). Technical Bulletin, Chevron HyJet®, A Phosphate Ester Aircraft Hydraulic Fluid. San Francisco, CA.

36. STLE Synthetic Lubricants Education Course (1991). Phosphate Esters Presentation, Montreal, May 1–2.
37. Ciba-Geigy Plastics and Additives Co. (1988). Technical Bulletin, Reolube® HYD Fire Resistant Fluids. Trafford Park, United Kingdom.
38. Azko Chemicals, Inc. (1986). Technical Bulletin 88-151, Fyrquel® Fire Resistant Hydraulic Fluids. Chicago, IL.
39. E. F. Houghton Co. (1990). Houghto-Safe® 1000 Series Phosphate Ester Fluids, Technical Bulletin No. 2-276-F 2M. Valley Forge, PA.
40. Factory Mutual Research Corp. (1991). *Factory Mutual System Approval Guide,* Equipment, Materials and Services for Property Conservation. Norwood, MA.
41. Klaus, E. E., and M. R. Fenske (1954). Fluids, Lubricants, Fuels and Related Materials, Wright Air Development Center, Technical Report 55-30, Part 3, November.
42. Bieber, H. E., E. E. Klaus, and E. J. Tweksbury (1968). A study of tricresyl phosphate as an additive for boundry lubrication. *ASLE Trans.,* 11, 155–161.
43. Godfrey, D. (1965). The lubrication mechanism of tricresyl phosphate on steel, *ASLE Trans.,* 8, 1–11.
44. Miles, P. (to Ciba-Geigy Corp.) (1988). U.S. Patent 4,919,833.
45. Mitsui Petrochemical Industries KK (1987). Japan Patent 870,328.
46. Wright, R. M. (to FMC Corp.) (1979). U.S. Patent 4,171, 272.
47. Dounchis, H. (to FMC Corp.) (1979). U.S. Patent 4,169,800.
48a. Stauffer Chemical Co. (1976). U.S. Patent 3,956,154.
48b. Stauffer Chemical Co. (1972). U.S. Patent 3,707,500 and 3,674,697.
49. Johnson, M. K. (1990). Organo phosphates and delayed neuropathy—Is NTE alive and well?, *Toxicol. Appl. Pharmacol.,* 102, 385–399.
50. Abou-Donia, M. B., and D. M. Lapadulla (1990). Mechanisms of organophosphorous ester-induced delayed neurotoxicity: Type I and Type II, *Annu. Rev. Pharmacol. Toxicol.,* 30, 405–410.
51. Carrington, C. D., D. M. Lapadulla, M. Othman, C. Farr, R. S. Nair, F. Johannsen, and M. B. Abou-Donia (1989). Assessment of the delayed neurotoxicity of tributyl phosphate, tributoxyethyl phosphate and dibutylphenyl phosphate, *Toxicol. Ind. Health,* 6, 415–423.
52. World Health Organization (1991). Environmental Health Criteria No. 112: Tributyl Phosphate, World Health Organization, Geneva.
53. FMC Corp. (1992). Technical Bulletin: Toxicology Profiles of Triaryl Phosphate Esters, Report No. CPG/S/89-018, revised.
54. Anzenberger, J. F. (1987). Evaluation of phosphate ester fluids to determine stability and suitability for continued use in gas turbines, *ASLE Lub. Eng.,* 43, 528–532.
55. Brown, K. J., and G. W. Staniewski (1989). Condition Monitoring and Maintenance of Steam Turbine Generator Fire Resistant Triaryl Phosphate Control Fluids, STLE Special Publication No. 27, pp. 91–96.
56. Shade, W. N. (1987). Field experience with degraded synthetic phosphate ester lubricants, *Lub. Eng.,* 43, 176–182.
57. Phillips, W. D., and J. Hartwig (1990). Fire Resistant Fluids for the Control and Lubrication of Steam and Gas Turbines, ASME Conference, Boston, Oct. 22–25.
58. Phillips, W. D., and G. D. Vilanskaya (1990). Recent operating experience in Europe and the Soviet Union with fire resistant turbine lubricants, *Proc. American Power Conference,* Chicago.
59. Electric Power Research Institute (1989). Evaluation of Fire Retardant Fluids for Turbine Bearing Lubricants, Final Report NP-6542, Project 2969-2, September.
60. Lysko, V. V. (1990). Fire Retardant Turbine Oils Based on Phosphoric Acid Esters, ASTM Symposium, San Francisco, June.
61. Klaus, E. E., G. S. Jeng, and J. L. Duda (1989). A study of tricresyl phosphate as a vapor delivered lubricant, *Lub. Eng.,* 45, 717–723 (1989).

62. Gunsel, S., E. E. Klaus, and R. W. Bruce (1989). Friction Characteristics of Vapor Deposited Lubricant Films, SAE Technical Paper No. 890148, SP 785, Worldwide Progress on Adiabatic Engines International Congress, Detroit, February 27–March 3.
63. Klaus, E. E., J. L. Duda, G. S. Jeng, N. S. Hakim, M. A. Groeneweg, and M. A. Belnaves (1987). Vapor phase tribology for advanced diesel engines, *U.S. Department of Energy Proc.—Coatings for Advanced Heat Engines Workshop,* Castine, Me.
64. Weber, K. E. (1990). Advanced low heat rejection diesel technology development, *U.S. Department of Energy, Proc.—Annual Automotive Technology Development Contractors Coordinating Meeting,* Dearborn, Mich., October 22–25.

4
Polyalkylene Glycols

Paul L. Matlock and Nye A. Clinton
Union Carbide Chemicals and Plastics Company, Inc.
Tarrytown, New York

I. INTRODUCTION

Polyalkylene glycols are one of the major classes of synthetic lubricants. A product of the second world war, they have found a variety of specialty applications as lubricants and functional fluids. As lubricants, they are used in applications where petroleum fails. As functional fluids, they are used where their chemical and physical properties provide unique advantages. They are the only major class of synthetic lubricants that are water soluble. This chapter presents information on polyalkylene glycols as used in lubrication.

II. CHEMISTRY OF SYNTHESIS

A. Nomenclature

Polyalkylene glycol is the common name for the homopolymers of ethylene oxide, propylene oxide, or the copolymers of ethylene oxide and propylene oxide. The ethylene oxide polymers are generally called poly(ethylene glycol)s or poly(ethylene oxide)s. The Chemical Abstracts nomenclature is oxirane polymer. The propylene oxide polymers are known as poly(propylene glycol)s or poly(propylene oxide)s with a Chemical Abstracts name of methyloxirane polymer. The copolymers are known as "oxirane, polymer with methyloxirane" or "oxirane, methyl polymer with oxirane," depending on which oxide was used in the greater amount. The Chemical Abstracts nomenclature does not distinguish between copolymers that are random or blocked (see later). The individual polymers and the copolymers all fall into the class of poly(alkylene oxide)s or, more commonly, poly(alkylene glycol)s. This latter name leads to the acronym PAGs. The acronym PAO has occasionally been used to indicate poly(alkylene oxide), but PAO is commonly used to designate polyalphaolefins.

B. Mechanism of Polymerization

Epoxide polymers are generally prepared from a starter that consists of an alcohol and a smaller amount of its metal alkoxide, usually the potassium or sodium salt. This solution is then reacted with epoxide. The epoxide reacts with the metal alkoxide form of one of the starter alcohol molecules to give an alkoxide derivative of a new alcohol. This new metal alkoxide is in equilibrium with all of the alcohols present, so that the next reaction of an epoxide can occur either with the molecule that has already reacted or with a different alcohol:

$$ROH + ROM + \overset{O}{\overbrace{CH_2-CHR'}} \longrightarrow ROH + ROCH_2CHR'OM \qquad (1)$$

$$ROH + ROCH_2CHR'OM \rightleftharpoons ROM + ROCH_2CHR'OH$$

If the epoxide is ethylene oxide, an ethyloxy group results:

$$ROH + \overset{O}{\overbrace{CH_2-CH_2}} \longrightarrow \underset{\text{ethyloxy}}{RO(CH_2CH_2O)H} \qquad (2)$$

If the epoxide is propylene oxide, a propyloxy group results:

$$ROH + \overset{O}{\overbrace{CH_2-\underset{\diagdown CH_3}{CH}}} \longrightarrow \underset{\text{propyloxy}}{RO(CH_2CH(CH_3)O)H} \qquad (3)$$

This equilibrium determines the molecular weight distribution of the resulting product. The epoxide monomers react with the metal salts of the alcohol at much faster rates than with the alcohols. Whichever alcohol is most acidic will tend to form the alkoxide salts and be the most reactive toward the epoxide. Once each starter alcohol has reacted with at least one epoxide, then all molecules in the system have approximately the same reactivity. Unless the parent alcohol is extremely unreactive, the fast exchange of metal salt between the growing polymer chains then results in what is nearly a Poisson distribution for molecular weight. The starter alcohols in commercial polymers use relatively reactive alcohols. The Poisson distribution is a much narrower distribution than the most probable or Gaussian distribution. In many applications the narrow distribution is critical, because it means that there is no significant fraction of low-molecular-weight, volatile, or low-boiling components. In addition, a narrow molecular weight distribution leads to a high viscosity index.

Polymerization of ethylene oxide produces a structure like the one shown below:

$ROCH_2CH_2\text{-}(OCH_2CH_2)_x\text{-}CH_2CH_2OH$

Ethylene oxide has two reactive sites, and the product is the same no matter which one reacts. The situation is different with propylene oxide. In this case the ring opening occurs predominantly to produce a secondary hydroxyl group:

```
          O                          H
         / \                        /
ROH  +  CH2-CH     ----> ROCH2C-OH     96%          (4)
              \                 \
               CH3               CH3
```

but a small fraction (about 4%) reacts to give a primary hydroxyl:

```
          O
         / \
ROH  +  CH2-CH  ---->        ROCH(CH3)CH2-OH    4%   (5)
              \
               CH3
```

This leads to polymer structure predominantly as follows:

```
ROCH2CH-(OCH2CH)xOCH2CH-OH
       \       \       \                             (6)
        CH3     CH3     CH3
```

Copolymers of ethylene oxide and propylene oxide have two types of structures, random and blocked. In the random polymer, the two epoxides are co-fed to the starter and will both be incorporated throughout the polymer. They both react to give a product that is itself reactive and is in the acid–base equilibrium with all the other alcohols and metal alkoxylates present. To a first approximation, the epoxides are incorporated in a random manner dependent on the relative mounts of each epoxide present and the molecular weight distribution is still approximated by the Poisson model. Polymers with this structure are identified as random copolymers. A portion of the structure of a random copolymer is:

```
R'-OCH2CH2-OCH2CH-OCH2CH-OCH2CH2-OCH2CH-OCH2CH2-OCH2CH2-OCH2CH-OR"
                 \      \               \                      \
                  CH3    CH3             CH3                    CH3
                                                                     (7)
```

In the block copolymer, an alternative structure is produced by reacting the starter first with one of the epoxides to produce a homopolymer. This can then be reacted with a different epoxide to produce a block copolymer. This name arises from the presence of a chain of one structure connected to a chain with a different structure. A schematic picture of a block copolymer is as follows:

$$H\text{-}(OCH_2CH_2)_Y\text{-}(OCH_2\underset{\displaystyle \backslash CH_3}{CH})_X\text{-}(OCH_2CH_2)_Y\text{-}OH \qquad (8)$$

block

The polyalkylene glycols that are used commercially as lubricants are of two main types:

1. Homopolymers of propylene oxide (polypropylene glycols), which are the water-insoluble type. These show limited solubility in oil.
2. Copolymers of ethylene oxide and propylene oxide, which are the water-soluble type.

All of the propylene oxide polymers that are commonly used in lubrication applications are started with butanol and have the following structure:

$$CH_3CH_2CH_2CH_2\text{-}(OCH_2\underset{\displaystyle \backslash CH_3}{CH})_X\text{-}OH \qquad (9)$$

The water-soluble polyalkylene glycols are typically started either from butanol or from a diol, like ethylene glycol. The resulting structures are:

$$CH_3CH_2CH_2CH_2\text{-}(OCH_2CH_2)_Y\text{-}(OCH_2\underset{\displaystyle \backslash CH_3}{CH})_X\text{-}OH$$

random

$$H\text{-}(O\underset{\displaystyle \backslash CH_3}{CH}CH_2)X\text{-}(OCH_2CH_2)_Y\text{-}OCH_2CH_2O\text{-}(CH_2CH_2O)_X\text{-}(OCH_2\underset{\displaystyle \backslash CH_3}{CH})_Y\text{-}OH \qquad (10)$$

random

Polymers consisting of all ethyloxy groups, the polyethylene glycols, are not often used as lubricants, since they tend to crystallize when their molecular weight reaches about 600. Nevertheless, solid polyethylene glycols are used in specialty lubrication applications where the use of a solid is an advantage.

The tendency of polyethylene glycol chains to crystallize affects the block polyalkylene glycols. If the blocks of ethyloxy groups are long enough in a block copolymer, pastes or waxes result. Block structures also tend to give the polymers surfactant properties in water. As a result, block polyalkylene glycols are often used as surfactants. However, surfactantlike properties are of little use for most lubrication applications.

The epoxide polymers formed by base-catalyzed reactions typically have molecular weights of less than 20,000. Traces of water in the monomer feed and minor side reactions limit the average molecular weight that can be achieved. The major side reaction for the base-catalyzed polymerization of propylene oxide is the rearrangement of propylene oxide to allyl alcohol. This was recognized as early as 1956 (1). The rearrangement involves deprotonation of the methyl group on the propylene oxide, followed by intramolecular ring opening:

$$RO^{-} + CH_3CH\overset{O}{\overbrace{-}}CH_2 \longrightarrow [RO{-}{-}H{-}{-}CH_2{-}CH\overset{O}{\overbrace{-}}CH_2]^{-} \longrightarrow ROH + CH_2{=}CH{-}CH_2O^{-} \quad (11)$$

This mechanism is supported by kinetic studies with 1,2-epoxypropane-3,3,3-D, which show a positive isotope effect (2). The allyl alcohol formed reacts with ethylene oxide and propylene oxide to form new polyalkylene glycol molecules, which are monoallyl ethers. This chain-transfer reaction limits the ultimate molecular weight that can be achieved with base catalysis.

An alternative technology for polymerization can produce much higher-molecular-weight ethylene oxide polymers. Using coordinate-initiated polymerization, it is possible to produce ethylene oxide polymers with molecular weights in excess of a million. A coordination catalyst suspended in a solvent that does not dissolve the polymer product is used. Ethylene oxide is added and the polymer, not being soluble in the medium, is produced as a granular solid. Polymerization is thought to take place by coordination of the epoxide to an electrophilic site on the catalysts. This coordination activates the epoxide for reaction with the growing chain.

C. Method of Synthesis

Ethylene oxide is a toxic material with a 1 ppm time-weighted exposure limit for 8 h of exposure and a short-term permissible limit of 5 ppm in a 15-min period according to OSHA. It is highly flammable, and has a wide flammable range in air of 3.0–100%. It can explosively decompose if exposed to an ignition source. The flammability is only heightened by a boiling point of 10.4°C, making it a gas at ordinary temperatures. It can be polymerized with acidic, basic, and coordination catalysts, a polymerization that is very exothermic. A very careful study of the hazards and procedures for safely handling ethylene oxide must be undertaken before use. Similar hazards exists with propylene oxide.

It is possible to prepare ethylene oxide, propylene oxide, and mixed ethylene oxide-propylene oxide polymers using glass equipment at atmospheric pressure. A nitrogen-flushed flask is charged with the starter solution and fitted with a dry-ice condenser. A small amount of the epoxide is fed to the heated flask (typically 100°C or more) and allowed to reflux from the dry ice condenser. The epoxide charge will be slowly consumed by the polymerization reaction and the reflux rate will decrease. More epoxide is added at a rate sufficient to keep the system at reflux (3). The rate can be increased by keeping the apparatus under a slight pressure from a dip tube emersed in an inert liquid. The higher pressure increases the concentration of monomer in the reaction solution. To make a random copolymer, the two oxides are co-fed, while a block copolymer requires sequential feeds of the two different epoxides. A similar system can be designed for coordinate-initiated polymerization.

The use of an autoclave for the polymerization will result in much faster rates because it is possible to operate at higher pressures, resulting in much higher liquid-phase concentrations of the monomers. The epoxide can be fed either by forcing it into the autoclave from a pressurized feed vessel with nitrogen pressure or by pumping it into the reactor. The reactor needs to be equipped with a cooling system and a control scheme to follow and control both pressure and temperature. The reactor is heated to the desired operating temperature and the epoxide is fed until the pressure has reached the desired level. As the reaction progresses the pressure will fall and more epoxide can be fed. Pure ethylene oxide vapor can explosively decompose when exposed to an ignition source. A sufficient amount of nitrogen present before the initiation of the epoxide feed will insure that the vapor phase does not reach the flammable limit at any time during the run. It is critical to keep the inventory of unreacted oxide in the reactor at a level such that the heat of polymerization (20 kcal/mol) can be removed by the cooling system. A critical factor in keeping the oxide concentration low is to keep the reactor at the desired temperature. If pressure is the control mechanism, then a low temperature in the reactor will allow the oxide to build to a potentially unsafe concentration. The reactor should have a safety relief device sized to handle a runaway reaction caused by loss of cooling. One of the authors has seen an autoclave and its high-pressure cell catastrophically destroyed with the autoclave top thrown many hundreds of feet. The cause was the inadvertent feeding of ethylene oxide at a low temperature allowing a large inventory of ethylene oxide to accumulate. This was followed by an uncontrolled polymerization that could not be contained.

To avoid exposure of personnel to unreacted ethylene oxide or propylene oxide, it is necessary to hold the reactor contents at temperature after the end of the feed until the concentration of unreacted epoxides has dropped to an acceptable level. This procedure is called a cook-out or digestion. A cook-out may be necessary during synthesis because the vessel will fill with liquid as the reaction proceeds and the polymer is produced. This will compress the nitrogen in the vessel and the partial pressure of the monomer will therefore decrease (the system is run by keeping pressure constant). The reaction rates will fall to unacceptably low levels. Venting of the excess nitrogen after a cook-out will allow feed to be resumed at faster rates. It may even be necessary to remove some of the reactor's liquid contents to allow room for further reaction. This is most likely to occur when synthesizing higher-molecular-weight products.

III. COMMERCIAL MANUFACTURE

A. Historical Development

Polyalkylene glycols are one of many important industrial chemicals developed during the Second World War. This work was performed by H. R. Fife, and to a lesser extent by R. F. Holden, as a joint development project between Union Carbide Chemicals and Plastics Company Inc. (then known as the Union Carbide and Carbon Corporation) and the Mellon Institute of Industrial Research in Pittsburgh, Pa. Union Carbide Chemicals and Plastics Company Inc. holds the original patents for the common lubricants (6–8).

The first preparations were similar to that used today. Sodium salts of alcohols were used as starters at reaction temperatures slightly above those used currently. Butanol was the starter alcohol of choice for monoethers. The products developed at this time are still manufactured today. The method used by Fife to neutralize his fluids was unusual. The crude fluids were diluted with water, acidified with carbon dioxide, extracted with hot water, and then stripped of water at high temperature. Decolorizing with Nuchar was the last step.

The first commercial scale syntheses were performed at Union Carbide Chemicals and Plastics Company Inc. production facility in Charleston, W.V.

B. Current Practice

The commercial preparation of poly(alkylene oxide)s is carried out in a manner analogous to that described for the laboratory autoclave. A semibatch stainless steel system with a recirculation loop and an agitator has been described. The reactions are carried out at 100–120°C at pressures of 60 psig. The oxide feed rate is controlled by pressure and feed times are on the order of 15 h or more (4).

Pressindustria Co. has reported a novel method for synthesizing polyalkylene glycols (5). The solution of growing polymer is sprayed through the head space of a horizontal reactor. The reaction with oxide monomer is reported to take place at the gas-liquid interface. Rapid reaction without large increases in pressure or temperature are reported. Cooling takes place with an external heat exchanger.

C. Current Manufacturers

A list of the major polyalkylene glycols for lubrication and functional fluid, their trademarks, and manufacturers is given in Table 4.1.

IV. CHEMICAL PROPERTIES

A. Oxidative and Thermal Stability

The bond strength of the carbon-carbon bond is 84 kcal/mole (ethane) and is slightly stronger than the 76 kcal/mole carbon-oxygen bond of an ether (dimethyl ether) (9). Other authors have reported that carbon-oxygen ether bonds are comparable to, or slightly stronger than, the usual carbon-carbon bonds (10). Thus, from a thermochemical standpoint, polyalkylene glycols are slightly less stable than typical hydrocarbons. In the absence of air, they can be used up to about 250°C.

The poly(alkylene oxide)s are all polyethers with an oxygen atom separated by two carbon atoms along the polymer backbone. As with all ethers, a secondary or tertiary carbon-hydrogen bond adjacent to the ether oxygen is susceptible to oxidative attack. The mechanism involves a free-radical abstraction of the hydrogen on the carbon alpha to the ether oxygen resulting in a carbon-based radical stabilized by the adjacent oxygen atom. This can then react with oxygen (air) to produce a peroxy radical. The chain process is continued with the peroxy radical abstracting a hydrogen atom to give another oxygen stabilized radical, as follows.

```
                                                  •
R• + R'O(CH2CHR")OR'   ---> RH + R'O(CH2CR")OR'

                                           O-O•
      •                                   /
R'O(CH2CR")OR' + O2 ---> R'O(CH2CR")OR'                          (12)

         OO•                                       OOH
        /                                         /                    •
R'O(CH2CR")OR' + R'O(CH2CHR")OR' ---> R'O(CH2CR")OR' + R'O(CH2CR")OR'
```

The process does not continue to build peroxide levels, because there are a number of mechanisms that lead to peroxide destruction. In the early stages of oxidation the peroxide will increase, but as the reaction proceeds the peroxide level reaches a steady state as carbonyl levels build. Further reaction will lead to the formation of acidic material. As the acidic oxidation products build the viscosity begins to drop. Apparently the oxidation builds peroxides that are converted to esters and other carbonyl compounds, often by chain cleavage. The carbonyls, particularly the aldehydes, are then further oxidized to give organic acids.

Table 4.1 Manufacturers of Polyalkylene Glycols

Type of molecule	Propyloxy groups (wt%)	Trade name	Manufacturer	Viscosity range, 40°C (cSt)
Monoalkyl ether	50	CAPROL MH	Asahi Denka Kogyo K.K.	20–1,000
		PLURACOL W	BASF Wyandotte Corporation	34–1,000
		BREOX 50-A	BP Chemicals	8–1,000
		BAYLUBE CL	Mobay Chemical Corporation	68–1,050
		NISSAN UNILUBE 50MB	Nippon Oil and Fats	8–1,000
		EMKAROX DA.11	ICI	32.5–1,050
		POLY-G WS	Olin Chemicals	20–1,000
		NEWPOL 50HB	Sanyo Chemical	8–130
		JEFFOX WL	Jefferson Chemical Co.	8–1,000
		UCON 50-HB	Union Carbide	8–1,015
Diol and triols	25% (typically)	CAPROL GH	Asahi Denkka Kogyo K.K.	280–17,850
		PLURACOL V	BASF Wyandotte	280–82,000
		BREOX 75-W	BP Chemical	272–50,800
		BAYLUBE HT	Mobay Chemical Corporation	450–18,200
		NISSAN UNILUBE 75-DE	Nippon Oil and Fats	19,400
		EMKAROX BB.31, FC.31	ICI	17,000–54,000
		NEWPOL 75H	Sanyo Chemical	17,000
		POLY G WT	Olin Chemical	91–17,000
		UCON 75-H	Union Carbide	51–54,000
Monoalkyl ether	100%	CAPROL M	Asahi Denka Kogyo K.K.	60–370
		BREFOX B	BP Chemicals	26–330
		UNILUBE MB	Nippon Oil and Fats	33–338
		EMKAROX 01	ICI	92–380
		NEWPOL LB	Sanyo Chemicals	
		POLY-G WI	Olin Chemicals	57–338
		JEFFOX OL	Jefferson Chemical	11–540
		UCON LB	Union Carbide	11–540

Oxidation of polyalkylene glycols produces esters, peroxides, ketones, and acids. The oxidation products can exist as part of a polymer chain, or, in cases of severe degradation, as low-molecular-weight species. These are polar organic materials. Polyalkylene glycols themselves are polar and will dissolve the oxidation products. In contrast, oils are nonpolar and their oxidation products are polar, consisting of peroxides and carbonyl species as well. Oils will not dissolve these polar species, and this contributes directly to their tendency to form sludge and varnish. The examination of the physical properties of the oxidized polyalkylene glycol will show a reduction in viscosity with time. The tendency to form insoluble material, sludge, or varnish is small, mostly as a result of having every third atom in an polymer chain being an oxygen atom. Consistent with this are Conradson carbon and Ramsbotton carbon (ASTM D-189 and D-524) values of less that 0.01%. When sludges do form from polyalkylene glycols, it is usually in oxygen-starved systems as a result of aldehyde condensation. Under conditions of exhaustive oxidation, the chain cleavage will have occurred to such an extent that the oxidation products evaporate. The volatilization of the fluid, together with the tendency not to form carbon or sludge, means that the polyalkylene glycol will be removed under high-temperature applications in a property known as clean burn-off. Clean burn-off is important in a number of the applications for these products. An example is as a carrier for graphite on chains being used in ovens or kilns. As with the pyrolysis products of any organic material, the vapors should be removed from the work place through good ventilation in an environmentally safe manner.

The oxidation of polyalkylene glycols could result in shorter-than-desired service life for some applications. Oxidation can effectively be controlled through the addition of antioxidants that interrupt the chain-transfer oxidation mechanism (11). Typical antioxidants that have been used include butylated hydroxyanisole, phenothiazine, hydroquinone monomethyl ether, butylated hydroxytoluene (12), and phenyl-alpha-naphthylamine (13, p. 109). The poly(alkylene oxide)s are dramatically stabilized toward oxidation by the addition of antioxidants. In many cases antioxidants at levels of a few hundred parts per million (ppm) are sufficient to stabilize against oxidative degradation under mild conditions, and higher levels will stabilize these systems under much more severe conditions. The uses of inhibited polyalkylene glycols as heat transfer fluids, quenchants, and calender lubricants are all examples of successful high-temperature applications. These applications show that if the system is protected against oxidative attack, either by the addition of an antioxidant or by removal of oxygen, then polyalkylene glycols will have very good high-temperature stability.

B. End-Group Chemistry

The polyalkylene glycols all have at least one hydroxyl group on the end of the molecule. If they have been produced from water or a multifunctional starter then they will have more than one hydroxyl group. The polyols used for urethane applications perform by virtue of the reaction of the hydroxyl groups reacting with isocyanate groups to give the urethane linkage. In this application, the primary hydroxyl group is more reactive than a secondary hydroxyl due to stearic factors. The urethane polyols are formed from propylene oxide. Since this results in a less reactive secondary hydroxyl end group it is necessary to end-cap the polyol with a small amount of ethylene oxide to increase the number of primary alcohol terminated molecules.

Other end-group reactions are used to functionalize polyalkylene glycols. These reactions use the known alcohol derivatization reactions. Esters are formed by reaction with either organic or inorganic acids. Ethers can be formed by conversion of the alcohol to an alkoxide followed by reaction with methyl chloride to give the methyl ether. Alternatively, it is possible to react the alcohol with a strong acid and an olefin to give an alkyl ether cap.

C. Coordination Chemistry

The presence of an ether oxygen atom at every third position of the polymer backbone leads to the rich coordination chemistry of these compounds. The use of these polymers as phase transfer agents has been reviewed (14). Complexations with phenols, phenolic resins, bromine, iodine, gelatin, sulfonic acids, mercuric salts, tannic acid, poly(acrylic acid), and urea all have been reported (15). The use of poly(ethylene oxide) polymers as flocculation agents is related to the absorption on colloidal silica, clay, and minerals.

The facile wetting of metal parts in lubrication applications is related to the ability of the polymer to associate with the metal surface. Like the complexation of other chemical species, the ability of the polyalkylene glycol to wet a metal surface is due to the presence of an ether oxygen atom at every third position of the polymer chain. This results in good extreme pressure and metal-working performance. The solution properties of these polymers in water are also directly related to the association of the water with the ether oxygens. This will be discussed in Section V,B.

V. PHYSICAL PROPERTIES

A. Properties of the Neat Base Fluids

The physical properties of poly(alkylene oxide)s are best understood by considering them as a series of homologous derivatives. Thus the polymers derived exclusively from ethylene oxide are considered as one class, differing primarily in molecular weight. The trends in properties can then be understood in relationship to structure.

The properties of the ethylene oxide polymers derived in principal from water (diols identified as PEGs) or from methanol (16) (identified as methoxy PEGs) are listed in Table 4.2. As can be seen from the data, the poly(alkylene oxide)s with a molecular weight above about 600 are crystalline solids. The alcohol-started products follow the same pattern. The structure in the solid state has been examined by infrared, x-ray diffraction, and Raman spectroscopy. It has been concluded that the molecules exist in a helical structure. The high symmetry of the structure leads to high crystal packing energies that favor crystallization.

The physical properties of the polymers of propylene oxide (17) are given in Table 4.3. These are derived from a butyl alcohol starter. In contrast to the polymers of poly(ethylene oxide), these polymers do not readily crystalize. Instead, they become too thick to flow at a temperature known as the pour point. The pour point for these polymers is very low. Even at temperatures below their pour point they do not crystalize but form a glasslike solid. The pendant methyl group on the backbone breaks up the crystal packing. These polymers have very high viscosity indices.

The random copolymers of ethylene oxide and propylene oxide have the properties (17) given in Table 4.4. This table has two families of structurally related compounds. The polymers of the first class are derived from a butyl starter and are formed from equal weights of ethylene oxide and propylene oxide. The polymers of the second class are

Table 4.2 Properties of Polyethylene Glycols

Molecular weight	Specific gravity, 20/20°C	Melting or freezing range (°C)	Solubility in water at 20°C (% by weight)	Viscosity at 99°C (mm^2/s, = cSt)
Polyethylene glycols				
200	1.127	−65[a]	Complete	4.3
300	1.127	−15 to −8	Complete	5.8
400	1.128	4–8	Complete	7.3
600	1.128	20–25	Complete	10.5
1,000	1.101[b]	37–40	70[c]	17.4
1,450	1.102[b]	43–46	70	25–32
3,350	1.072[d]	54–58	62	75–110
4,600	1.073[d]	57–61	50[c]	160–230
8,000	1.075[d]	60–63	50[c]	700–900
Polyethylene glycol monomethyl ethers				
350	1.097	−5–10	Complete	4.1
550	1.078[b]	15–25	Complete	7.5
750	1.084[b]	27–32	Complete	10.5
2,000	1.20	49–54	68[c]	63
5,000	1.20	57–63	58[c]	613

[a]Sets to a glass.
[b]At 55/20°C.
[c]Approximate.
[d]Density, g/ml at 80°C.

derived from water and are therefore diols. The oxide incorporated in these is 75 wt% ethylene oxide and 25 wt% propylene oxide.

The copolymers derived from equal amounts of ethylene oxide and propylene oxide have very low pour points. If we compare the pour points of these random copolymers with those of the polymers derived from propylene oxide, we find the values very similar.

Table 4.3 Physical Properties of Polypropylene Glycol Monobutyl Ethers

Molecular weight	Viscosity at 40°C (mm^2/s, = cSt)	ISO grade	Viscosity index	Pour point (°C)	Specific gravity, 20/20°C	Flash point, Penske-Martens closed cup (°C)
340	10.9	10	83	−56	0.960	152
640	25.6		161	−48	0.981	166
740	32.6	32	169	−46	0.983	143
1,020	56.65		184	−40	0.989	171
1,240	76.4		190	−37	0.994	171
1,420	104.0	100	196	−34	0.997	171
1,550	124.0		200	−32	1.000	171
2,080	227.0	220	214	−29	1.002	174
2,490	338.0	320	219	−23	1.002	177

Table 4.4 Physical Properties of Water-Soluble Polyalkylene Glycols

Molecular weight	Viscosity at 40°C (mm^2/s, = cSt)	ISO[a] grade	Viscosity index	Pour point (°C)	Specific gravity, 20/20°C	Flash point, Penske-Martens closed cup (°C)
50:50 Ethylene oxide:propylene oxide, monobutyl ethers						
270	8.33		97	−65	0.971	76[b]
520	19.1		165	−51	1.023	113
750	33.8	32	197	−43	1.031	174
970	52.0		212	−40	1.033	177
1,230	80.0		220	−38	1.049	177
1,590	132.0	160	230	−34	1.051	171
2,660	398.0		254	−32	1.062	177
3,380	700.0	680	269	−29	1.062	182
3,930	1,015.0	1000	281	−29	1.063	182
75:25 Ethylene oxide:propylene oxide, diols						
980	90.9	100	184	−15	1.097	188
2,470	282.0	320	207	4	1.096	219
6,950	1,800		282	4	1.094	177
15,000	17,850		414	4	1.097	191

[a]These are ISO viscosity grades for ambient temperature. Due to the high viscosity index, care should be exercised when considering other temperatures.
[b]Tag closed cup flash point

In fact, the copolymers seem to have slightly lower pour points at the higher molecular weights. This may be due to the slightly higher symmetry of the homopolymers of propylene oxide compared to the mixed copolymers. These copolymers have even higher viscosity indices than was the case with the polymers derived only from propylene oxide.

The random copolymers derived from block feeds of ethylene oxide and propylene oxide have physical properties that are dependent on the relative size of each block as well as the total molecular weight. Commercially available products are derived from poly-(propylene glycol), which is then reacted with ethylene oxide. Alternatively, a poly(ethylene glycol) molecule can be reacted with propylene oxide to produce what is referred to as a reverse blocked structure. These products are available as liquids, pastes, or flakeable solids.

B. Solution Properties

The aqueous solution properties of polyalkylene glycols are critical for many of their commercial applications. The solubilities of the lower-molecular-weight poly(ethylene oxide) polymers are shown in Table 4.1. These seem to be soluble at all temperatures. Careful examination at or above 100°C (sample under pressure) shows that the polymer comes out of solution as the temperature is raised. This loss of solubility in water with an increase in temperature is called a cloud point or separation temperature.

The copolymers of ethylene oxide and propylene oxide and the propylene oxide homopolymers have a more obvious temperature at which they come out of water solution. This temperature is a function of molecular weight and the proportion of ethylene oxide used in the synthesis. Increasing the molecular weight in an otherwise similar polyalkylene glycol series lowers the cloud point. Raising the amount of ethylene oxide relative to propylene oxide in the synthesis of polyalkylene glycols of otherwise similar structure raises the cloud point. Cloud point is very sensitive to salt concentration, which lowers the cloud point. The cloud point is generally below room temperature for the homopolymers of propylene oxide. They are generally not considered soluble in water, although the lower-molecular-weight examples are readily dissolved in cold water. The cloud points for common polyalkylene glycols (17) are shown in Fig. 4.1.

The cloud point is an important property in metal-working applications. Water-based metal-working fluids often contain 1–5% polyalkylene glycol. When exposed to the high temperature of the working surface, the polymer comes out of solution to provide the lubricity needed (see Section VI,D,7).

The hydration of polyalkylene glycols in aqueous solutions has been examined by differential scanning calorimetry. The more water-soluble poly(ethylene oxide) polymers were found to coordinate 2.8 moles of water per ether linkage, while the copolymers derived from equal weights of ethylene oxide and propylene oxide coordinated to 2.4 moles and the homopolymers from propylene oxide coordinated with only 1.6. This is consistent with the more soluble polymers requiring more energy (higher temperatures) to break the coordination sphere and become insoluble.

The hydration of the ether linkages affects the solution viscosities. Small amounts of

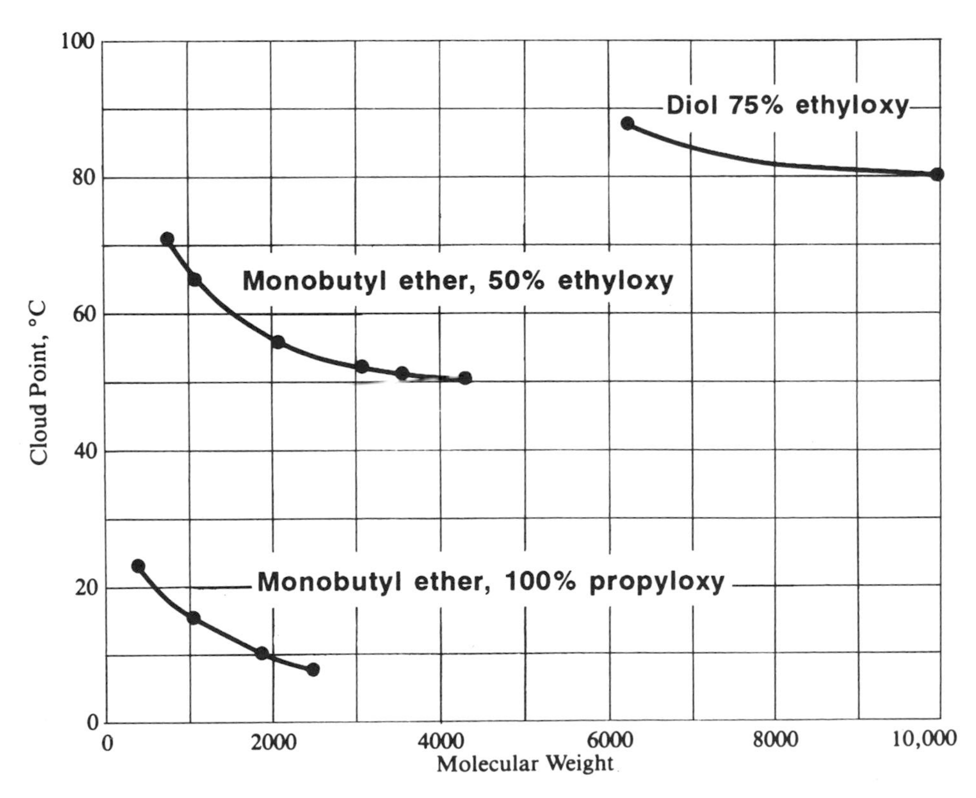

Figure 4.1 Cloud points of polyalkylene glycols. Cloud point is determined with 1% polymer in water.

water will actually raise the viscosity due to an increase of the effective molecular weight. This effect is decreased at elevated temperatures, the result of thermal energy breaking the water–ether association. The viscosity of water–polyalkylene glycol solutions (17) is shown in Fig. 4.2.

The viscosities of high-molecular-weight poly(ethylene oxide)s are given in Table 4.5 (15). Aqueous solutions of high-molecular-weight poly(ethylene oxide)s exhibit shear thinning. This effect is most pronounced at high molecular weight. As a result, the

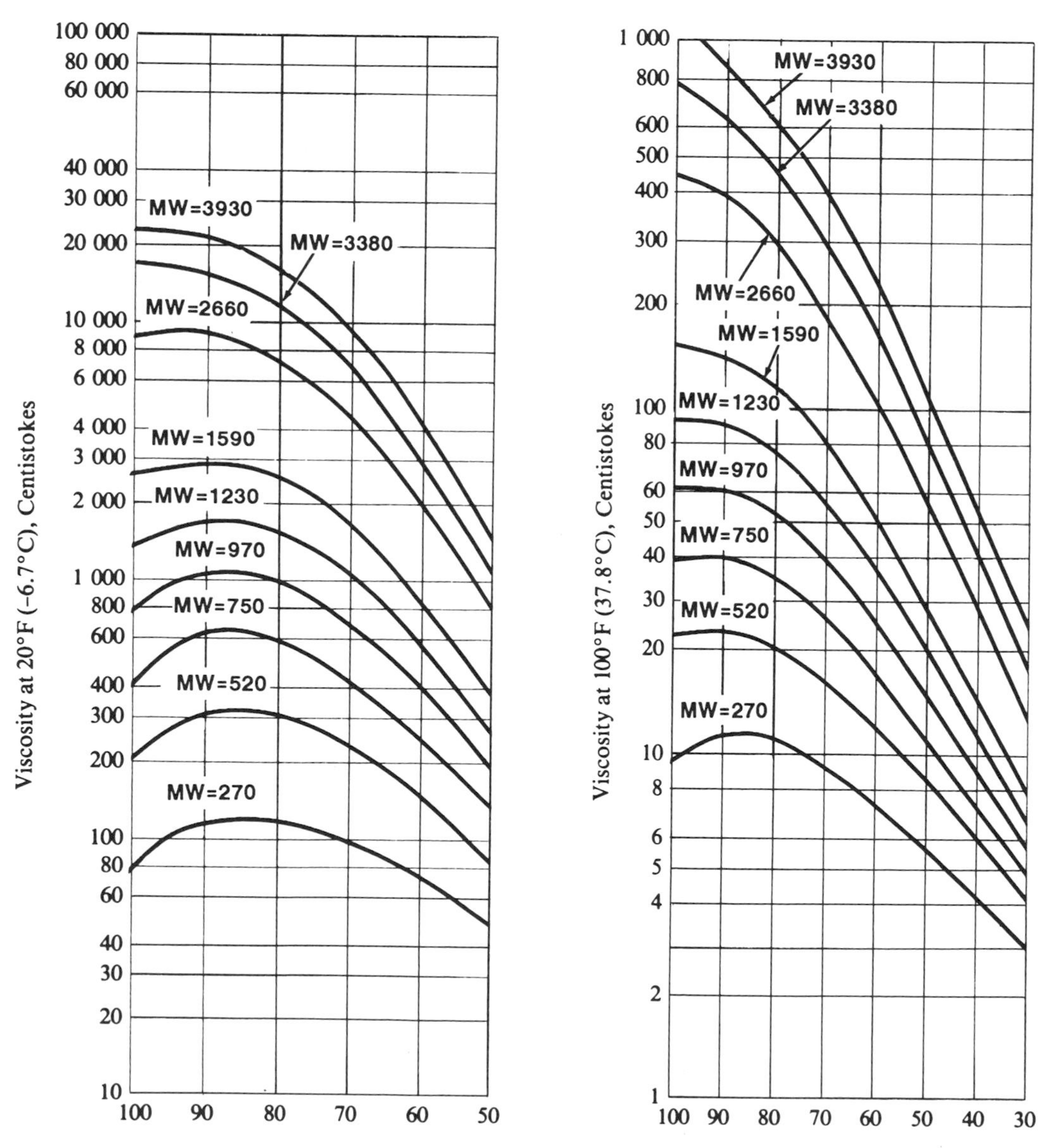

Figure 4.2 Aqueous viscosities of polyalkelene glycols. Monobutyl ether, 50% ethyloxy groups, volume percent in water.

Table 4.5 Solution Viscosities of High-Molecular-Weight Poly(ethylene oxide)s

Approximate molecular weight[a]	Viscosity range, 25°C (mPa, =cP)			
	5% Solution[b]	2% Solution[b]	1% Solution[b]	Brookfield spindle number/speed, rpm
100,000	12–50			1/50
200,000	65–115			1/50
300,000	600–1,200			1/10
400,000	2,250–4,450			1/2
400,000	2,250–3,350			1/2
600,000	4,500–8,800			2/2
900,000	8,800–17,000			2/2
1,000,000		400–800		1/10
2,000,000		2,000–4,000		3/10
4,000,000			1,650–5,000	1/2
5,000,000			>5,500	2/2

[a]Derived from rheological measurements.
[b]Percent by weight, slightly diluted with isopropanol used as a dispersing aid.

viscosity can vary over several orders of magnitude. At high shear rates, chain cleavage will occur, and permanent shear thinning will result.

C. Solubilities of Polyalkylene Glycols

Because polyalkylene glycols are polar molecules, they will dissolve in polar molecules. Those made from propylene oxide will dissolve in the less polar, hydrocarbon-type solvents. These principals are illustrated in Table 4.6 (17).

VI. APPLICATIONS OF POLYALKYLENE GLYCOLS

A. Advantages and Potential Problems

Polyalkylene glycols have found use as a petroleum lubricant replacement when the cost is justified by a performance advantage. Compared to petroleum lubricants, polyalkylene glycols show the following differences:

1. Lower pour point.
2. Higher viscosity index.
3. Lower tendency to form tar and sludge.
4. Increased solvency.
5. Wider range of solubilities, including water solubility.
6. Higher flash point.
7. Lower vapor pressure.
8. Lower ash and metals content.
9. Possess cloud points.

Comparison of polyalkylene glycols with petroleum should be done on a case-by-case basis, as petroleum lubricants can be formulated many different ways. Polyalkylene

Table 4.6 Solubilities of Polyalkylene Glycols

	1 Part PAG, 9 parts solvent			1 Part solvent, 9 parts PAG		
Solvent	Water-insoluble PAG[a]	Water-soluble PAG[b]	Water-soluble PAG[c]	Water-insoluble PAG[a]	Water-soluble PAB[b]	Water-soluble PAG[c]
Acetone	S[d]	S	S	S	S	S
Cyclohexane	S	I	I	S	S	S
Butyl ether	S	S	I	S	S	S
Dichloroethylene	S	S	S	S	S	S
Ethylene glycol	I	I	I	I	S	S
Glycerol	I	I	I	I	I	I
Heptane	S	I	I	S	S	I
Isopropanol	S	S	S	S	S	S
Methanol	S	S	S	S	S	S
2-Octanol	S	S	S	S	S	S
Toluene	S	S	S	S	S	S

[a]PAG, polypropylene glycol, monobutyl ether, molecular weight (MW)=1,550.
[b]PAG, 50:50 ethylene oxide:propylene oxide copolymer, monobutyl ether, MW=1,590.
[c]PAG, 75:25 ethylene oxide:propylene oxide copolymer, diol, MW=2,470.
[d]S, soluble; I, insoluble.

glycols owe their commercial existence to the ability to do what petroleum products cannot.

Changing from petroleum lubricants to polyalkylene glycol can present special problems. Machinery that has been used with petroleum often has wear grooves filled with carbonaceous material. Because of the good solvency characteristics of polyalkylene glycols, this carbonaceous material is often removed on changeover, revealing wear scars that were previously not visible. This wear is then incorrectly attributed to the polyalkylene glycol. Although polyalkylene glycols are compatible with most elastomers, this should be evaluated before change over. Polycarbonate and nylon machine parts are sometimes softened or embrittled. The good solvency properties tend to result in softened and lifted paint. Catalyzed epoxy, epoxy-phenolic, or modified phenolic coatings have performed well in contact with polyalkylene glycols.

B. First Commercial Uses

The first use of polyalkylene glycols was in water-based hydraulic fluids (18). These were first developed for the U.S. Navy (19) for use in military aircraft, and were being investigated as early as 1943. They were formulated from water, ethylene glycol, a polyalkylene glycol thickener, and an additive package. In military aircraft, it is important that fires not result if bullets or shrapnel severs hydraulic lines. The final test the Navy conducted was to fire a 50-caliber incendiary bullet, shredded by first passing through a steel baffle, through 1-gal cans of test fluid. This test was passed by water-based hydraulic fluids using polyalkylene glycol thickener (20).

More severe flammability requirements were to follow after the war. Hydraulic fluids to be used for missile ground handling equipment were developed that would not burn in a

100% gaseous oxygen atmosphere when the fluid was ejected at a pressure of 3,000 psi in the presence of a continuous electric discharge ignition source (21). Aqueous solutions of polyalkylene glycols could be formulated to pass this test.

When polyalkylene glycols were first developed, the high viscosity indices and low pour points were quickly identified (22). They were used in all-weather, heavy-duty brake fluids. Besides being fluid at temperatures that would cause petroleum products to freeze, they were also water tolerant. Small amounts of water contaminants would dissolve, not significantly changing the physical properties of the fluid nor crystallizing at low temperatures. This is a major use of polyalkylene glycols today.

Polyalkylene glycols were extensively used as aircraft engine lubricants in cold climates (23). Over 150,000 flying hours were accumulated, mostly in Alaska, using an inhibited polypropylene glycol monobutylether. The low pour point allowed aircraft engines to start at temperatures as low as –30°F without diluting the oil with fuel, a step that can be used to reduce oil viscosity. It was possible to hydraulically feather the propellers using the engine oil down to –60°F. Clean burn-off, an intrinsic property of polyalkylene glycols, resulted in low levels of carbon deposits and sludge, making engine cleanup easier during maintenance. Polyalkylene glycols were finally judged unsuitable for aircraft engine oils due to two factors: corrosion and deposits. Corrosion was due to the tendency of polyalkylene glycols to absorb water. Corrosion was principally a problem for engine parts exposed to moist air. Corrosion protection additives were not available at that time for polyalkylene glycols. The deposits formed were hard deposits consisting primarily of lead. The clean burn-off tendency of the fluid apparently was responsible for this. The lead deposits formed with petroleum as an engine lubricant are soft and have a lower lead content. It is believed that these unusual lead deposits resulted in valve sticking after about 300–400 h of operation (24), although no valve sticking was observed if valve clearances were adequate.

C. Early Commercial Applications of Polyalkylene Glycols

Lubrication engineers quickly developed new uses of polyalkylene glycols. The uses developed were for petroleum oil replacement in operations where petroleum oil was not entirely satisfactory and the higher cost of the polyalkylenc glycol could be justified. The desirable properties of the polyalkylene glycols include a low tendency to form carbon and sludge, clean burn-off, solvency, high viscosity index, tolerance for rubber and other elastomers, low pour point, and low flammability.

Polypropyleneglycol monobutylethers were tested extensively as lubricants for automobile engines (25). The fluids showed the expected low carbon, low sludge, clean engine parts, and satisfactory cranking at low temperature. Over 2 million miles of operation using these oil were experiences. This market was never developed.

Because polyalkylene glycols burn off cleanly, they are desirable to use in high-temperature applications where petroleum lubricants would form sludge. They have been used in glass factories to lubricate the turrets of hot-cut flare machines or to lubricate the bearings of rollers that smooth glass sheets. When mixed with graphite, polyalkylene glycols are very effective at lubricating bearings of carts being rolled into kilns. After the polyalkylene glycol has burned off, a soft, lubricating layer of graphite is left behind.

Polyalkylene glycols were found to have little or no solvent or swelling effects on most synthetic or natural rubbers. This gave rise to many uses where rubber parts needed to be lubricated, such as rubber shackles, joints, or O-rings, or in the manufacture of rubber parts where demolding lubricants were needed.

D. Current Applications

1. Fire-Resistant Hydraulic Fluids

Fire-resistant hydraulic fluids were the first use of polyalkylene glycols, and they continue to be used for this today. Polyalkylene glycols have been formulated into a range of hydraulic fluids for use where low flammability is important. These fluids are used in areas where fires cannot be tolerated, such as foundries, steel mills, and mines. In some cases the use of fire-resistant hydraulic fluid is mandated by law.

The strategy for formulating fire-resistant hydraulic fluids from polyalkylene glycols is as follows: Water in concentrations above 35% provides the fire resistance, ethylene glycol or diethylene glycol provides antifreeze protection, and the polyalkylene glycol provides viscosity. Amines are typically used to provide corrosion resistance, and anti-wear additives for ferrous and nonferrous metals are usually added.

Water–glycol-based polyalkylene glycol thickened hydraulic fluids are formulated in the viscosity region of low to medium viscosity oils. Their use temperature is limited from about –30 to 65°C. The upper temperature limit is a result of the high vapor pressure of the contained water. Higher temperatures can lead to cavitation and premature pump failure. These fluids have good seal compatibility. Water–glycol type fluids have been limited to pressures below that achievable with oils, typically less than 3,000 psi. Recent advances in formulating water–glycol fluids allow pressures up to 5,000 psi and higher to be used (26).

Some water-glycol fluids have been formulated from hydrophobe-capped polyalkylene glycols (27). The hydrophobe cap leads to associative thickening in aqueous solution, and less thickener is needed to achieve a target viscosity. Unfortunately, associative thickening is shear sensitive, and commercial hydraulic fluids based on this principle did not gain market acceptance.

Polyalkylene glycol-thickened aqueous fluids compete in the fire-resistant hydraulic fluid market with phosphate esters and oil in water emulsions. The oil-in-water emulsions are the most similar, in that they have a high water content to provide fire resistance. They are lower in viscosity, and as a result should be used in pumps with narrower tolerances, requiring the use of better filtration to prevent particles from damaging the moving parts. The lower viscosity also leads to higher wear. Because they have water contents as high as 80%, their upper temperature limit of use is 60°C due to the vapor pressure of the contained water. They are prone to microbial degradation, and use of biocides in the machinery should be considered.

The phosphate esters are the only widely used nonaqueous fluids employed as a fire-resistant hydraulic fluid. These are triaryl and trialkyl esters of phosphoric acid. Like all esters, they are subject to hydrolysis back to their parent acid. Hydrolysis rate depends on structure, with the longer alkyl chain esters being the most resistant. They are frequently used with a bleaching clay filter to remove the acidic hydrolysis and oxidation byproducts. Their seal compatibility is not as good as the water-based fluids, especially at elevated temperatures where fluorinated hydrocarbons are preferred. There are many tests to probe flammability, and at least in one of these, the spray ignition test with acetylene–oxygen flame (5. Lux. Rep. Part III), the triaryl phosphates fail (13, p. 326).

2. Brake Fluids

The use of polyalkylene glycols as the hydraulic fluid in braking systems for motor vehicles was one of the early uses of polyalkylene glycols. The properties that made them valuable were water miscibility, low pour point, high viscosity index, high boiling point, good elastomer compatibility, and low vapor pressure. Glycol ethers are commonly added, as are rust inhibitors and antioxidants.

The ability to absorb water is the property most useful in this application. At low temperatures, ice crystals would prevent the operation of the breaking system. At high temperatures, water could vaporize, causing "vapor lock." Even with the tendency to absorb water, the vapor lock temperature drops by about 80°C when the anhydrous system picks up 2% water (13, p 328). Water increase at low temperature increases the viscosity of the fluid. The lower the temperature, the greater the effect. This effect is due to hydrogen bonding, which is a weak interaction favored at reduced temperatures.

Silicone brake fluids are a significant alternative to the polyalkylene glycol fluids. They have been utilized for their greater temperature stability in the newer generation of cars that have higher under hood and brake operating temperatures. Their main disadvantage is that they are not as water tolerant as the polyalkylene glycol-based fluids. It is important not to mix the two types of brake fluids. They are insoluble in each other. In a mixture of the two types of fluids, the additives tend to partition between the two phases so that neither phase is properly protected against corrosion. Brake failure can result.

3. Textile Lubricants

Because many polyalkylene glycols are water soluble, they quickly became important in the textile industry. They are nonstaining and can be washed from the finished yarn or fabric with water (28). Polyalkylene glycols, when they are oxidized at moderate temperature with an adequate oxygen supply, do not form colored by-products. This is particularly advantageous in the textile industry, where color is a critical quality consideration.

The same reason that makes polyalkylene glycols useful as lubricants for fibers makes them important machine lubricants in the textile industry. If they come into contact with the textiles being worked, they are easily washed off. One disadvantage is that they tend to cause crazing of polycarbonate sight glasses.

4. Compressor Lubricants

The unique technical problems of lubricating compressors are those presented by the lubricant dissolving the process gas. If this occurs, the viscosity of the lubricating films will be reduced, resulting in inferior lubrication. In severe cases, the process gas can wash the lubricant away and the moving parts will be unlubricated (13, p. 295). Because the water-soluble polyalkylene glycols are not soluble in nonpolar media, they make excellent compressor lubricants for nonpolar gasses like ethylene, natural gas, landfill gas, helium, or nitrogen. Diesters are often used in these applications, although they do not have as favorable gas solubilities as the polyalkylene glycols.

For the compression of air, the largest compressor market, the tendency to not form a sludge or carbon deposits is particularly advantageous. Explosions in air compressors have been linked to the formation of sludge and carbon from petroleum lubricant. These carbonaceous deposits then catalyze the ignition of the petroleum lubricant (29). This is still mostly a petroleum market. The petroleum products for this market must pass very strict sludging and carbon formation tests. Phosphate esters and diesters have also been used.

5. Automotive Refrigeration Lubricants

Polyalkylene glycols used to play no role as refrigeration lubricants in automotive air-conditioning systems. These needs were satisfactorily met by petroleum. With the elimination of chlorofluorocarbons as refrigeration gasses, this situation is changing. The new refrigerant for automotive use is HFC-134a, which is 1,1,1,2-tetrafluoroethane. Petroleum is not soluble in HFC-134a and cannot be used. The likely new refrigerant lubricant will be a polyalkylene glycol (30). Various esters have also been tested. This market is still evolving.

6. Mill and Calender Lubricants

Petroleum meets the need of lubricating the large-diameter journal bearings, antifriction bearings, and gears that are present in the mills and calenders used by the rubber, textile, paper, and plastics industry up to temperatures of 350°F. Above this temperature, petroleum products tend to form sludges, resulting in increased maintenance. Polyalkylene glycols have been used in this application since they do not tend to form sludge or carbon deposits.

7. Metal-Working Lubricants

Polyalkylene glycols are often used as the lubricity base for water-based cutting and grinding fluids. In addition, they have been used in drawing, stamping, and rolling. They are often used in combination with extreme pressure additives like fatty acids and phosphate esters, where they give synergistic performance. The polyalkylene glycols have a high affinity for metals due to every third atom being an oxygen atom (13, p. 110). Of perhaps greater importance is the cloud point of the polyalkylene glycol. Under the high temperature generated by friction at the cutting surface, the polyalkylene glycol will cloud out, forming tiny droplets of concentrated lubricant. Studies have shown that the cloud point is an important phenomenon in metal cutting (31).

E. Market Size

Worldwide the polyalkylene glycol market for lubrication purposes is approximately 100 million pounds. The U.S. market represents about 50 million pounds. The market for polyalkylene glycols manufactured for other purposes, such as urethane foams or surfactants, are much larger than this. The polymers used for lubrication are used for other purposes, such as heat-transfer fluids, solder-assist fluids, and metal quenchants, that are not included in this estimate.

VII. ENVIRONMENTAL INFORMATION

A. Toxicology

Polyalkylene glycols enjoy a low degree of toxicity. As with any substance, material data sheets should be consulted for any specific fluid. The toxicity of the base fluid can be affected by additives.

Toxicity by ingestion is low. The toxicity is highest for the lower-molecular-weight products. Measured LD50 values range from a low of about 4 ml/kg to over 60 ml/kg when rats were used as the test animal. Long-term feeding studies have been done on dogs and rats with minimal effects.

Toxicity by skin contact is low. Toxicity by absorption is generally very low. The LD50 values by this route are generally over 10 ml/kg. Skin sensitization is a function of molecular weight, with the lower-molecular-weight fluids showing the greatest effect. In general, the effects, if any, are for transient redness of the skin.

Because of their low vapor pressure, toxicity by inhalation is very low for polyalkylene glycols. The exception is for the higher-molecular-weight copolymers of ethylene oxide propylene oxide that are monobutyl ethers. Here, mechanically generated mists were found to be toxic upon inhalation. The products of thermal degradation, like any organic product, are toxic. Mechanical ventilation should be reviewed where mists or thermal degradation products are likely to be present.

Eye injury is expected only for the lowest-molecular-weight polyalkylene glycols. The lower-molecular-weight fluids cause slight to moderate eye injury.

B. FDA Status

Because of the low degree of toxicity of polyalkylene glycols, they have been approved for a variety of uses where they might come into contact with food. These include the use in lubricants that are used to manufacture and otherwise process food (21 CFR 178.3570), and in a variety of foam control applications.

C. Aquatic Toxicity

Polyalkylene glycols have been tested for their effect on fathead minnows (32). Samples of mixed ethylene oxide, propylene oxide copolymers, both diols, and monobutyl ethers have very little effect. The most toxic fluid, which was the lowest molecular weight tested, had a LC50 of 8.1 g/liter. The polypropylene glycol monobutyl ether tested was more toxic. It was virtually insoluble in water, but showed a 10% kill rate at 10 mg/liter and a 20% kill rate at 50 mg/liter. This data is for base fluids. Formulated fluids could show higher degrees of aquatic toxicity.

D. Biodegradation of Polyalkylene Glycols

Polyalkylene glycols are biodegradable, but the speed and completeness of degradation depends on many factors. Foremost among these is the type of bacteria used. Using unacclimated biomass, as might be found in a municipal wastewater treatment facility, rates of biodegradation are slow, about 15% in 2 weeks (32) for water-soluble polymers. Bacteria can be chosen that will degrade polyethylene glycols of all molecular weights. By using a combination of *Flavobacterium* sp. and *Pseudomonas* sp., 99% of polyethylene glycol of 6,000 molecular weight can be degraded in 7 days (33). Each of these bacteria utilizes the other's metabolism products.

Polyalkylene glycols seem to metabolize from the ends. The hydroxyl end group is converted into a carboxylic acid. This is then followed by ether cleavage. Perhaps one reason that high-molecular-weight polyalkylene glycols have been reported to be nonbiodegradable (34) is the low concentration of hydroxyl end groups in high-molecular-weight polymers. As a result, the higher-molecular-weight polymers are degraded slowly. Another problem with many tests for biodegradability is that the polyalkylene glycol is used as the only carbon source (35). Biodegradation might be more rapid under actual sewage treatment conditions.

One technique that will increase the speed of biodegradation of polyalkylene glycols is to treat the waste stream with ozone (36, 37). This has the effect of breaking the chain into smaller pieces that can be more easily utilized by the bacteria.

Polyalkylene glycols do not inhibit bacteria, and therefore will not be expected to be detrimental to waste treatment facilities. The polyalkylene glycols tested were ethylene oxide and propylene oxide copolymers of intermediate molecular weight, and were either diols or monobutyl ethers. The IC50 value is over 1 g/liter, the highest level used in the test. When a 50:50 ethylene oxide-propylene oxide copolymer, monobutyl ether of molecular weight of 1,700 was used to examine the effect on bacteria, the respiration rate of the bacteria increased as the polymer concentration was increased to 10 g/liter, the highest level tested (38). Biomass acclimated to 1 g/liter of the polymer for 5 days was used. This is consistent with the bacteria using the polymer for food.

REFERENCES

1. St. Pierre, L. E., and C. C. Price (1956). *J. Am. Chem. Soc.*, 78, 3432.
2. Gee, G., W. C. E. Higginson, P. Levesley, and K. J. Taylor (1959). *J. Chem. Soc.*, 1338.
3. Furukawa, J., and T. Saegusa (1962). *Pure Appl. Chem.*, 4, 387.
4. Malkemus, J. D. (1956). *J. Am. Oil Chem. Soc.*, 33, 571.
5. Penati, A., C. Meffezzoni, and E. Moretti (1981). *J. Appl. Polym. Sci.*, 26, 1059.
6. Roberts, F. H., and H. R. Fife (1947). U.S. Patent 2,425,755, August 19.
7. Toussaint, W. J., and H. R. Fife (1947). U.S. Patent 2,425,845, August 19.
8. Fife, H. R., and F. H. Roberts (1948). U.S. Patent 2,448,845, September 7.
9. Mortimer, C. T. (1962). *Reaction Heats and Bond Strengths*, Pergamon Press, New York, pp. 129, 136.
10. Price, C. C. (1967). *The Chemistry of the Ether Bond*, S. Patai (ed.), Interscience Publishers, London, p. 500.
11. Lloyd, W. G. (1961). *J. Chem. Eng. Data*, 6, 541.
12. Emkarox Polyalkylene Glycols, ICI Corp., no date.
13. Klamann, D. (1984). *Lubricants and Related Properties*, Verlag Chemie, Weinheim.
14. Totten, G. E., and N. A. Clinton (1988). *Rev. Macromol. Chem. Phys.*, C28(2), 293.
15. Union Carbide (1981). *POLYOX Water Soluble Resins Are Unique*, Booklet 44029C, Union Carbide Chemicals and Plastics Company Inc., Danbury, Conn.
16. Union Carbide (1981). *CARBOWAX Polyethylene Glycols*, Booklet F-4772L-EODD, Union Carbide Chemicals and Plastics Company Inc., Danbury, Conn.
17. Union Carbide (1987). *UCON Fluids & Lubricants*, Booklet, P6-7640, Union Carbide Chemicals and Plastics Company Inc., Danbury, Conn.
18. Murphy, C. M., and W. A. Zisman (1949). *Lub. Eng.*, 5, 264.
19. Millett, W. H. (1948). *Iron Steel Eng.*, 41.
20. Union Carbide. UCON Hydrolubes, Union Carbide and Carbon Corporation, F-7380A.
21. Union Carbide. (1964). UCON Hydraulic Fluid M-1, Union Carbide Chemicals, F-40576A.
22. Union Carbide (1956). UCON Fluids and Lubricants, Union Carbide and Carbon Company, F-40134.
23. Rubin, B., and E. M. Glass (1950). *SAE Q. Trans.*, 4, 287.
24. Russ, J. M. (1947). *Symposium on Synthetic Lubricants*, Special Technical Publication No. 77, American Society for Testing Materials, Philadelphia.
25. Russ, J. M. (1946). *Lub. Eng.*, 151.

26. Lewis, W. E. F. (1989). U.S. Patent 4,855,070, August 8, assigned to Union Carbide Chemicals and Plastics Company Inc.
27. Schwartz, E. S., and C. A. Tincher (1985). U.S. Patent 4,493,780, January 15.
28. Trevor, J. S. (1950). *Textile Recorder*, 86.
29. Smith, T. G. (1964). Presentation at the Twentieth Annual Meeting of the National Conference on Fluid Power, October.
30. *Automotive Eng.* (1991). 99, 25.
31. Brown, W. L. (1987). *Lub. Eng.*, 44, 168.
32. Waggy, G. T., and J. R. Payne (1974). Union Carbide Chemicals and Plastics Company Inc., unpublished results.
33. Kawai, F. (1987). *Crit. Rev. Biotechnol.*, 6, 273.
34. Cox, D. P., and R. A. Conway (1976). In *Proc. Third Int. Biodegradation Symp.*, J. Miles and A. M. Kaplan (eds.), Applied Science Publishers, London, p. 835.
35. Watson, G. K., and N. Jones (1977). *Water Res.*, 11, 95.
36. Suzuki, J., N. Taumi, and S. Suzuki (1979). *J. Appl. Polym. Sci.*, 23, 3281.
37. Suzuki, J., H. Nakagawa, and H. Ito (1976). *J. Appl. Polym. Sci.*, 20, 2791.
38. Niehaus, W. R. (1975). Union Carbide Chemicals and Plastics Company Inc., Technical Bulletin TL-869A, April 15.

5
Alkylated Aromatics

Hans Dressler
Indspec Chemical Corporation
Pittsburgh, Pennsylvania

I. INTRODUCTION

Although alkylated aromatics were developed for functional fluid use as early as the 1928–1936 period (1), these products failed to gain commercial prominence, partly because of the higher cost of the synthetics versus the petroleum oils. Impetus for the further development of alkylaromatics, particularly in France, Germany, and Japan, was based on real or perceived shortages of petroleum products caused by supply interruptions due to war or other political manipulations, as well as special needs, such as good performance at low or high temperatures, which were not well served by the petroleum oils, no matter how refined.

One such instance was the German need for domestically produced lubricants of all kinds during World War II, which prompted the production of long-chain alkylaromatics. Indeed, Chemische Werke Rheinpreussen produced alkylnaphthalene-type lubricants from 1942 to 1945 on a 3,600 t/year scale. Expansion of production to 6000–12,000 t/year was planned, but the war ended before that was realized (2). These synthetics were made by the alkylation of excess naphthalene with chlorinated aliphatic hydrocarbons, obtained from paraffins (derived from the CO/H_2-based Fischer-Tropsch process) of boiling range 220–350°C. The alkylation catalyst was aluminum chloride. The crude alkylate was washed with caustic and water, bleached/clarified with clay, filtered, and fractionated. Several product cuts were obtained, including those used for transformer oils, bright stocks for motor oil blends, turbine oils, and steam cylinder oils. The viscosity indices of these products were reportedly mediocre at best.

Alkylnaphthalenes were also developed for use as pour-point depressants for mineral oils, and such products are still in use, as mentioned in some detail later on.

Another need-driven impetus for the development of alkylaromatics came with the research for petroleum in the arctic climates of Alaska and Canada, which required lubricants suitable for use at low as well as high temperatures. Military needs for functional fluids usable under these extreme conditions provided impetus as well. Use requirements increased with the building of the Alaskan pipeline in the 1960s.

In the 1970s, the OPEC Cartel shocked the world with dramatic price increases for petroleum and threats of curtailment of production. Fears of shortages of high-quality, petroleum-derived lubricant base stocks began to rise, as the U.S. oil reserves began to decline.

In this post-World-War II period, a relatively low-cost source of dialkylbenzene lubricant base stock existed in the thriving sodium alkylbenzenesulfonate segment of the synthetic detergent industry. Monoalkylbenzene was the feed for sulfonation, and the higher-boiling dialkylbenzenes were available for functional fluid use. In the 1950s, this detergent base was the branched-chain alkylbenzenes (BABs) made by alkylation of benzene with propylene oligomers (trimers to pentamers). However, the BAB sulfonates turned out to be poorly biodegradable, and foamy suds began to foul some streams and beaches. Thus, environmental concerns led to the development in the 1960s of linear, long-chain alkylbenzenesulfonates (LAS), which are more biodegradable. Several processes were developed to make the LAS at low cost. The dialkylbenzene byproducts changed as well and required reformulation of lubricants in a relatively short time, a costly undertaking.

In the 1980s, alkylaromatic lubricant manufacturers, as well as all industry, were involved in the turmoil of takeovers, mergers, joint ventures, acquisitions, and leveraged buyouts on a global scale. Abrupt changes in producers, products, and priorities of managements took place. On top of that, the cost and time of obtaining regulatory approval for new products began, in some cases, to exceed the ability and patience needed to take such risks. As of this writing, in early 1991, the Iraqi takeover of Kuwait, albeit brief, has again produced rising oil prices and related concerns.

Nevertheless, much research and development on new processes and new alkylaromatics as base stocks for functional fluids continued, influenced in good part by ever-rising health, safety, and environmental concerns. For example, the corrosive and/or toxic older Friedel-Crafts alkylation catalysts were found to be replaceable by a variety of solid catalysts, which can be easily removed and recycled.

As far as this review is concerned, an encyclopedic listing of developmental alkylaromatic products and processes is not intended or possible. Rather, the literature before 1970 is generally given by reference to prior reviews. The literature from 1970 on is presented with selected references, intended to show the trend and thrust of development in the last 20 years. Data on commercial products are given as available, but it should be kept in mind that the exact processes in actual use are not necessarily those described in the references.

II. CHEMISTRY AND DEVELOPMENTAL ACTIVITIES

A. General

The chemistry of the alkylation of aromatics with haloalkanes, alcohols or olefins in the presence of a Lewis or Bronsted acid catalyst is discussed in much detail in the literature

(3a,3b,4). This apparently mundane reaction holds many complexities as far as the development of an effective lubricant base stock is concerned, including the following:

1. The nature of the aromatic compound, ranging, for example, from benzene to condensed ring systems such as naphthalene, partially hydrogenated condensed ring systems such as indane and tetralin (1,2,3,4-tetrahydronaphthalene), and noncondensed polycyclic aromatics such as the terphenyls.
2. The nature of the alkylating agent—for example, olefins and chloroalkanes may give different products and byproducts.
3. The degree of alkylation, making mono-, di-, or polyalkylated aromatics.
4. The variations in the alkyl groups, including linear, branched or alicyclic features.
5. The position of the substituents on the arene ring, for example, substituents in *ortho, meta,* and *para* position to each other on the benzene ring. For condensed or noncondensed polycyclic aromatics, many more isomers are possible. As an example, for naphthalene, there are two possible monoalkyl naphthalenes (in the 1- or 2-position on the ring), 10 possible dialkylnaphthalenes with the same substituent, 14 for disubstitution/different substituents, and many more isomeric polyalkylnaphthalenes.
6. The position of the aryl group on the alkyl chain—generally mixtures are obtained.
7. The nature of the catalyst can influence the formation and amount of byproducts. For example, heating C_6 to C_{30} alpha-olefins with C_{10} to C_{24} alkylaromatic sulfonic acids at 200°C for 2 h was reported to give a product containing 92% internal olefins; similar results were obtained using an H-mordenite-type zeolite catalyst (5). As another example, there are many reports of olefin oligomerization by acid catalysts of all kinds; these oligomers have a slower rate of alkylation than the monomers, or if they remain in the product unreacted they are a cause of lowered oxidative stability.
8. The type of the catalyst—different catalysts may produce a different mix of products and byproducts.
9. Purity of raw materials. For example, impurities in the 78°C freezing point (f.p.) grade of coal-tar naphthalene can cause problems in alkylation.
10. The process variables, such as the mole ratio of reagents, the amount of catalyst, the reaction temperature and time, and the order and rate of addition of reagents may greatly influence the type of product made. For example, a high ratio of arene to alkylating agent favors monoalkylation but increases the problems of purification. As another example, a combination of high catalyst/high temperature may increase the rate of alkylation/dealkylation/realkylation and thus influence the product mix.

All of these factors impact not only on the development of a suitable functional fluid but on its cost of production as well.

B. Arenes Suitable for Alkylation

Both benzene and naphthalene of a suitable quality are available from petroleum or coal tar feed stocks in ample supply. Higher condensed-ring aromatics are of more limited interest and availability as a feed for high-performance fluids. Noncondensed polycyclic aromatics such as biphenyl are readily available. Partially hydrogenated aromatics such as indan or tetralin are available from petroleum or coal tar feeds or made by catalytic hydrogenation of indene or naphthalene. Methylnaphthalenes are also available from petroleum and coal tar processing.

A good, introductory source book on commercially available arenes was authored by Franck and Stadelhofer (6).

C. Alkylating Agents

For the benzene and naphthalene series, suitable alkylating agents are generally in the C_8 to C_{20} range.

The haloalkanes, for practical purposes usually the monohaloalkanes, are obtained by the chlorination of an excess of paraffin or paraffin fraction to minimize dichloroalkane byproducts.

The alcohols are obtained by the catalytic hydration of olefins or by the hydrogenation of fatty acid esters. Because of their higher cost compared to haloalkanes or olefins, the alcohols are of low commercial importance as alkylating agents in the C_8 to C_{20} range.

The C_8 to C_{20} olefins can be obtained by the dehydrohalogenation, often in situ, of haloalkanes, by thermal cracking of suitable feeds or by ethylene oligomerization. The last method is the commercially predominant one, particularly for making alpha-olefins, which have other large-scale uses, such as in ethylene copolymers and in the PAO synthetic lubricants.

D. Catalysts

A wide variety of Lewis and Bronsted acids and natural or synthetic solid catalysts are suitable for the alkylation of aromatic compounds with haloalkanes and olefins. Examples are sulfuric acid, hydrochloric acid, hydrofluoric acid, phosphoric acid, arylsulfonic acids, boron trifluoride (and its complexes with ethers, etc.), aluminum halides (without or with promoters such as hydrogen halides), ferric chloride, zinc chloride, antimony pentachloride, acid-washed clays such as montmorillonites, natural and synthetic zeolites and aluminosilicates, pillared clays, aluminophosphates, and others.

The important issues in the development/choice of catalyst are cost, activity, life, regenerability, and disposal.

E. Developmental Products and Processes

The older literature has been well covered in prior reviews (7a,7b; also 6, pp. 210–213). With few exceptions, this review presents selections from the patent literature from the mid-1970s to 1990. Quite often, patents cover a wide territory in their claims if not in the examples; this is particularly true for the aromatics claimed. Nonetheless, this section is organized into three predominant themes: alkylbenzenes, alkylnaphthalenes, and alkylated multiring aromatics.

1. *Alkylbenzenes*

A 1974 Phillips Petroleum Co. patent described the alkylation of aromatic hydrocarbons such as benzene with olefins and montmorillonite clay catalyst; olefin dimer and olefin oligomers were among the byproducts (8).

Di-C_8- to C_{28}-*t*-alkylbenzenes or -naphthalenes, prepared by the alkylation of the aromatics with an iso-olefin or a *t*-alkylhalide using $FeBr_3$ as the catalyst, were useful as oxidatively stable functional fluids (9).

Benzene or naphthalene was selectively tertiary-alkylated with a mixture of C_7 to C_{44} *t*-alkyl chlorides and iso-olefins in the presence of a Friedel-Crafts catalyst; the formation of secondary alkyl groups due to isomerization was reduced (10).

Fire-resistant hydraulic fluids with improved low-temperature characteristics were prepared by blending of a dialkylbenzene, a dialkylcarboxylate ester, and an orthosilicate (11).

Lubricants with VI in the 110–120 range and pour points as low as –70°F were prepared by dialkylating a monocyclic aromatic hydrocarbon and a minor amount of tetrahydronaphthalene with linear mono-C_6 to C_{18} olefins in the presence of aluminum chloride as the catalyst (12).

Metal [e.g., Al(III), In(III), Cr(III)] exchanged hectorite clays were found to be very effective catalysts for the alkylation of benzene with dodecene (13).

Aromatic hydrocarbons such as benzene were alkylated with an olefin such as 1-dodecene in the presence of a catalyst comprising a metal cation such as Al(III), In(III), or Cr(III) exchanged onto the surface of a synthetic nickeliferous hectorite (14).

Crystalline zeolites with high SiO_2/Al_2O_3 ratios were effective for the selective alkylation of benzene compounds with C_6 to C_{20} olefins to give alkylates enriched in the 2-phenylalkane isomers (15).

Benzene compounds were claimed to be alkylated with detergent-range olefins (C_8 to C_{22}) in the presence of a tungsten oxide on 70–90% silica; the catalyst gave clean alkylation without disproportionation of the olefin (16).

Didodecylbenzene was prepared by treating benzene with propylene tetramer in the presence of a complex $AlCl_3$ catalyst (17).

The use of a metal oxide substrate/tantalum(V) halide/oxide catalyst for the alkylation of benzene compounds with C_2 to C_{20} olefins was said to minimize equipment corrosion and isomerization of feed olefin (18).

The byproducts of cumene production, diisopropylbenzene and hexylbenzene, were alkylated with C_8 to C_{22} alpha-olefins in the presence of a metal oxide substrate/tantalum(V) halide/oxide catalyst to give lubricants with increased flash points and higher VI (19).

Electrical insulating oils for transformers, circuit breakers, condensers, and cables were made by blending refined mineral oils with alkylaromatics. These oils have good resistance to high temperature and corona discharge as well as good low-temperature performance (20).

Electrical insulating oils for use as condenser impregnants, transformer oils, and oil-filled cable oils were prepared by mixing a microbially treated paraffinic crude oil fraction of boiling point 275–390°C with an alkylaromatic mixture and an additive package (21).

A process was described that allowed for accurate and convenient control of the ratio of di- to monoalkylated aromatics by alkylating the arene with C_8 to C_{22} *n*-olefins in two stages, with separate amounts of olefin charged in each stage, and using HF as the catalyst (22).

Linear phenylalkanes were obtained by alkylating aromatics with olefins in the presence of amorphous, wide-pore silica/alumina catalysts; selectivity to monoalkylate was up to 89%, there was no skeletal isomerization of the alkyl chain to give nonlinear products, and oil stability was good since conventional Friedel-Crafts catalysts were not used (23).

The alkylation of C_2 to C_4 alkylbenzenes with C_{14} to C_{18} olefins in the presence of Friedel-Crafts catalysts gave lube oil base stocks with low (<–40°C) pour points and high (>95) VI; the oils may be hydrogenated to give specialty and lubricant base stocks and base stock additives (24).

A Japanese patent claimed that a mixture of a diorganopolysiloxane and a C_{10} to C_{15} alkylbenzene compound provided a hydraulic brake fluid with low moisture absorption (25).

Brominated alkylbenzenes were claimed as base stocks for hydraulic fluids of low flammability (26).

New catalysts for the alkylation of toluene and others with C_{10} to C_{15} olefin contained $Et_3OSiOAlEt_2$ and molybdenum or tungsten halides; these catalysts are less corrosive than conventional catalysts, and high olefin conversion is obtained (27).

The alkylation of aromatic compounds with C_{20} to C_{1300} olefinic oligomers was reported to give alkylaromatics with high VI, low pour points, and improved thermal stability (28).

A process for preparing long-chain alkylaromatic compounds by alkylation of an arene such as benzene or naphthalene with an olefin in the presence of a layered material containing titanate in the layers and oxide pillars separating the layers was claimed to give useful base stocks for lubricating oils (29).

2. *Alkylnaphthalenes*

A mixture of a mineral oil with a small amount of alkylated naphthalenes, acenaphthylene, phenanthrenes, and pyrenes as swelling agents was claimed to be suitble for hydraulic oils or lubricating oils having high oxidation stability and improved VI (30).

Naphthalene was alkylated with a mixture of long-chain alphaolefins in a 1 to 4–8 mole ratio using $AlCl_3$ as the catalyst to give a lubricant with average molecular weight. ~1500, VI ~120, viscosity of 50 cS at 100°C, pour point –30°C, and flash point >300°C (31).

A patent on the use of ZSM-4 zeolite catalyst reported that the alkylation of naphthalene with *n*-nonene gave only monoalkylate (32).

The use of a mixed methanesulfonic acid/pyro- or polyphosphoric acid catalyst for the alkylation of naphthalene with C_4 to C_{16} alphaolefins was said to inhibit dialkylation. The alkylates show utility as emulsion breakers in petroleum chemistry (33).

Long-chain (C_{12} to C_{16}) alkylnaphthalenes were reported to be good, nonmigratory additives for polyethylene as voltage stabilizers in high-voltage insulation compounds (34).

Multigrade blended petroleum oils, modified with VI improvers, show better low-temperature performance with the addition of a small amount of a combination of a condensation product of a chlorinated C_{10} to C_{50} wax with naphthalene and a C_{10} to C_{18} alkyl acrylate or methacrylate as pour-point depressants (35).

Benzylmethylnaphthalenes (1:1 mixture of alpha and beta isomers) were claimed to be highly radiation-resistant lubricants (36).

The alkylation of naphthalene with C_8 to C_{12} alkylchlorides in 1:2.5–5 mole ratio in the presence of $AlCl_3$ catalyst, followed by fractionation of the alkylate, gave a fraction of boiling point 220–250°C/ 0.2 mm Hg that was a good high-vacuum oil (37).

In the alkylation of naphthalene with 1-decene at 130°C in the presence of the sulfocationite KRS-40t, the selectivity toward alkylation versus oligomerization was ~50% (38).

Eicosylnaphthalene with 2,6-di-*t*-butyl-*p*-cresol added at 0.6 wt% was a lubricant with good oxidation resistance (39).

A Russian patent covered the use of a macroporous ion-exchange resin (sulfonated divinylbenzene–cross-linked polystyrene) for the alkylation of naphthalene with higher olefins (40).

Mono- and poly-C_{10} alkylnaphthalenes were said to be highly stable toward heat and thermal oxidation, making them suitable as heat-transfer agents (41).

The products of the alkylation of tetralin (1,2,,3,4-tetrahydronaphthalene) with styrene, using sulfuric acid as the catalyst, were, as such or perhydrogenated, useful as heat transfer oils, insulating oils and hydraulic oils (42).

Synthetic base oils for functional fluids and greases based on a wide range of mixtures of mono- and polyalkylated naphthalenes, with the alkyl groups in the C_{12} to C_{26} range, were disclosed. These oils ranged in viscosity from 10–18 cS/210°F (99°C), had good VI in the 105–136 range, flash points in the 505–560°F (262–293°C) range, pour points in the –40 to +5°F (–40 to –15°C) range, good lubricity, thermal stability, and low volatility. Several grease formulations with excellent performance at 305°F (152°C) were given (43).

The preparation of aralkyl aromatics by alkylation of aromatic hydrocarbons over H-type zeolites (e.g., H-Y zeolite), claimed to give high selectivity and olefin conversion, was patented (44).

Heteropolyacids (e.g., phosphotungstic acid on silica) were reported to be effective for the diaralkylation of naphthalene or tetraline with styrene or benzyl halides; the products were said to be useful starting materials for traction drive fluids, electrical insulating oils, rubber processing oils, and others (45). Similarly, aromatic hydrocarbons were dialkylated with styrenes using (activated) clay catalysts (46).

1-(1-Tetralyl)2-phenylpropane, structure **1**, was prepared by the alkylation of tetralin with alpha-methylstyrene in the presence of an alkali or alkaline earth metal catalyst; this product has a kinematic viscosity of 3.1 cS/100°C, VI=185, and pour point =25°C and has a high traction coefficient (47)

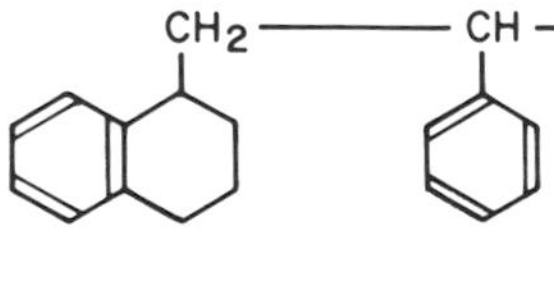

CH_2 — CH — CH_3

1

The alkylation of aromatic hydrocarbons with a mixture of C_{15} to C_{40} branched alkenes and C_{16} to C_{30} straight chain alkene was claimed to give an alkylate suitable as a lubricant; overbased calcium sulfonates prepared from it had less tendency to foam when used as a lubricating oil additive (48).

Mixed mono C_6 to C_{24} *sec*-alkylnaphthalenes with a ratio of alpha-to beta substitution on the naphthalene ring of at least 1.0 were shown in Japan to have very good oxidation stability and were useful as heat-transfer oils or as base stock for lubricants; the pour points of these oils were ≤–45°C; VI were not given (49). Similar products of high quality were prepared in high yields in the United States (50).

Mono- or dialkylnaphthalenes were prepared by alkylating naphthalene in the presence of an alicyclic hydrocarbon, such as decalin or bicyclohexyl, with alkylating agents over a dealuminized faujasite or Y-zeolite (51).

An improved process for making alkylnapththalene-based pour-point depressants for paraffinic lubricating oils was said to consist of the alkylation of naphthalene with chloroparaffins (14–18% Cl content, made from wax of freezing point 52–56°C) over a catalyst containing 1 wt% Al-alloy shavings and 0.75 wt% aluminum chloride (52).

Pour-point depressants of increased effectiveness were said to be made by alkylating naphthalene with C_{25} to C_{30} olefins (7–10 times the amount of naphthalene on a mole ratio basis) in the presence of 5–10% ammonium chloride (53).

Fluid pour-point depressants for petroleum lubricants or diesel fuels were made by the alkylation of naphthalene with chloroparaffin (containing $\leq$ 12% Cl) in the presence of 3 wt% aluminum chloride, at a chloroparaffin/naphthalene mole ratio of 9/1 (54).

3. *Alkylated Multiring Aromatics Other Than Naphthalenes*

Mono- or di-*sec*-butylbiphenyl (**2**) or -terphenyl (**3**) and mono- or di-*sec*-butylmethoxybiphenyl (**2**) or -methoxyterphenyl (**3**) were claimed to be nonspreading lubricants and fluids that are oxidatively stable at low and high temperatures (55).

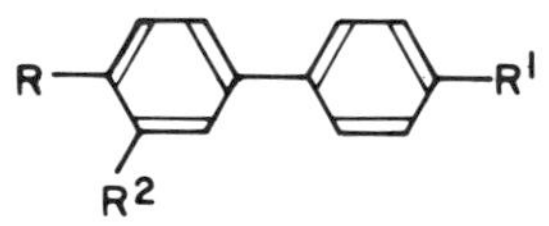

o-terphenyl and m-terphenyl

2 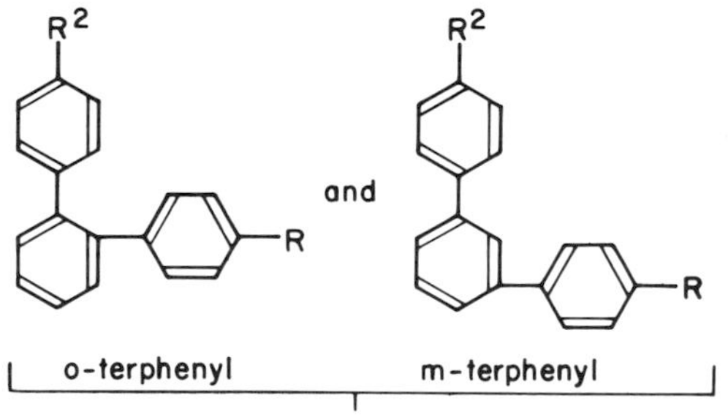 **3**

where R = sec-butyl; R^1 = sec-butyl, H or methoxy; R^2 = sec-butyl or H

Substituted indans, **4**, were said to be thermally and oxidatively stable oils and greases, or blending stocks with other lubricants for use as functional fluids (56).

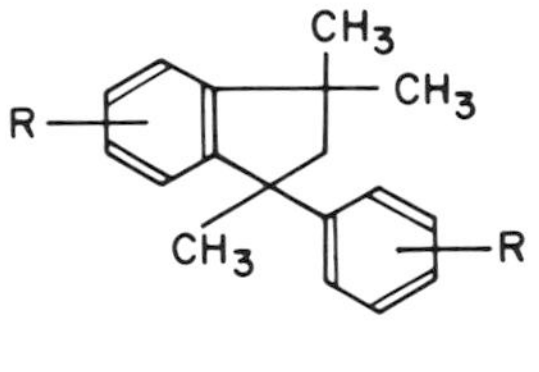

4

where R = H, alkyl, phenyl, carboxyalkyl, carboxyphenyl, phenoxy or thiophenoxy

Biphenyl, *o*, *m*-, or *p*-terphenyls, quaterphenyls, quinquiphenyls, hexaphenyls, and the like were alkylated with 1-olefins using catalysts such as BF_3 complexes, acid-treated clays, and mixtures thereof to give functional fluids (57).

Mixtures of alkylbiphenyls and alkylnaphthalenes with mono- or diolefins substituted with two condensed or noncondensed aromatic nuclei were reported to be useful as electrical insulating oils (58).

III. COMMERCIAL PRODUCTS AND DEMAND

A. Processes

No claim is made here of knowledge of the actual production processes for commercially available oils. Examples of likely, preferred routes follow.

Branched-chain type alkylbenzenes (BABs) as detergent precursers are made by the alkylation of excess benzene with propylene tetramer at 20–50°C in the presence of hydrofluoric acid or aluminum chloride as the catalyst. The monoalkylate is obtained in ~75% yield (59); the byproduct dialkylate is the potential lubricant base stock.

Linear chain-type alkylbenzenes as detergent precursers are made, for example, as follows (60). A C_{10} to C_{14} *n*-paraffin feed is obtained from the kerosene fraction of petroleum by a molecular sieve adsorption process or an urea adduct process. This paraffin fraction in at least 20% excess is chlorinated, for example, in the liquid phase at 200–300°F, at 98+% chlorine conversion and 10–30% paraffin conversion, with 80+% selectivity to monochloroparaffin [the dichloroparaffin byproduct leads to benzene alkylation byproducts such as diphenylalkanes (**5**) and tetralin (**6**) and indane derivatives (**7**)].

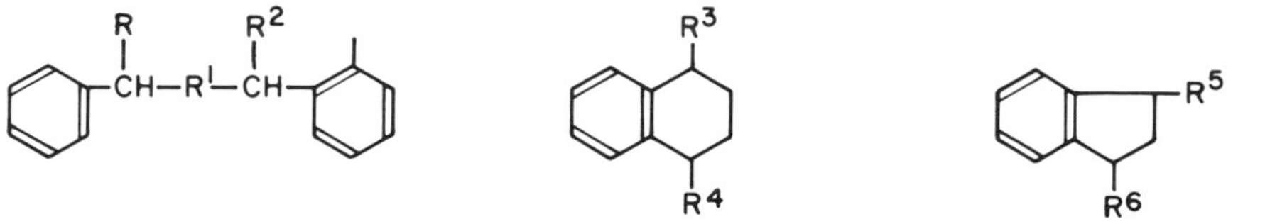

5 where $R + R^1 + R^2 = C_8 - C_{12}$

6 where $R^3 + R^4 = C_6 - C_{10}$

7 where $R^5 + R^6 = C_7 - C_{11}$

After removal of HCl and any unreacted chlorine, the mostly secondary chloroparaffin in excess paraffin is used directly to alkylate benzene (400–1400% excess) at 100–200°F and up to 50 psig with 3–10 mol% $AlCl_3$ catalyst (based on the alkylchloride) and at 100% chloroparaffin conversion. The alkylate is separated from the catalyst complex by decantation (the catalyst sludge is recycled with make-up fresh catalyst), caustic and water washed, and clarified/bleached by treatment with clay and filtration. The crude product is then distilled to remove unreacted paraffin and benzene, light ends (from degradation reactions of alkylbenzenes and/or lower alkylbenzenes made by alkylation with paraffin degradation products), the heart cut (monoalkylbenzene), and the heavy ends of interest as lube feed stocks. These heavy ends, obtained in an amount of 10–13 wt% of the heart cut, contain less than 10 wt% monoalkylbenzene; polyphenylalkanes, tetralins, or indanes can be removed or minimized by selective sulfonation or charcoal treatment. The dialkylbenzenes (DABs) that make up 80+% of the lube base stock have a structure represented by structure **8**.

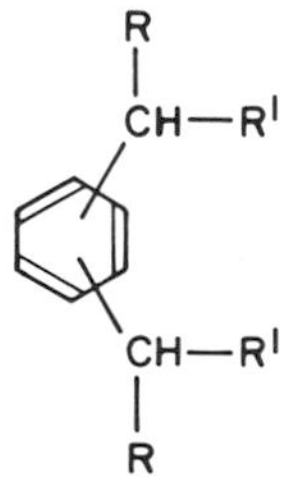

8

where R, R^1 = methyl or linear alkyl groups, $R + R^1 = C_7 - C_{19}$

These processes for mono-long-chain alkylbenzenes, where the DABs are obtained as a byproduct, can be tailored to give larger amounts of DABs, or to give DABs as the major product.

A process for linear alkylbenzenes by alkylation of benzene with linear olefins over a solid catalyst has been offered by Pla Espanola de Petroleas SA Madrid and UOP, Inc., Des Plaines, Ill. (61). This process, too, should give a dialkylbenzene fraction.

The dialkylbenzenes can be, and likely are being, used in blends with petroleum-based mineral oils, with naphthenic oils, and with other synthetic lubricants.

Finally, the alkylnaphthalene pour-point depressants are made by the condensation of an excess chloroparaffin and/or 1-olefin with naphthalene, optionally in a chlorinated solvent such as ethylene chloride, at between 60 and 180°C. References from the 1970s for the preparation of such materials were given earlier (52–54; old references are 64–67). Old trade names are Pourex and Paraflow. The structures of these pour-point depressants are not well described. From the literature, they appear to be poly(long-chain alkyl)naphthalenes, perhaps partially cross-linked (due to the presence of dichloroparaffins in the feed or due to the use of dichlorinated solvents). One possible structure of this type is **9**. The preferred range of average molecular weights for these pour-point depressants is 1,000–3,000.

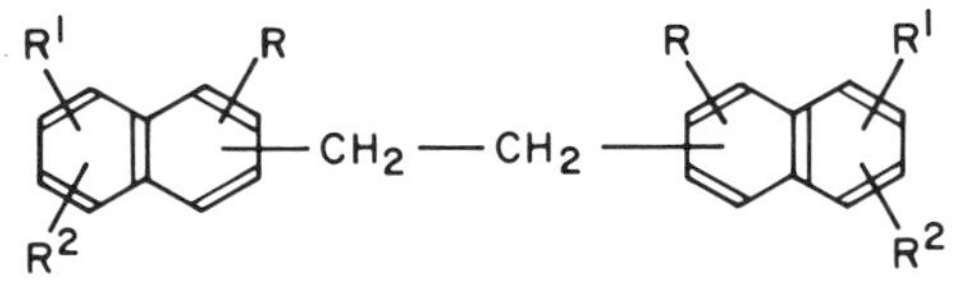

R^1 R R R^1

CH_2—CH_2

R^2 R^2

9

where R, R^1, R^2 = $C_{16}-C_{30}$ alkyl or H

B. Products and Markets

The good properties of the dialkylbenzene-type oils at low temperatures became quite important during the search for oil in Alaska and Canada in the 1960s and later during the construction of the Alaska pipeline in the 1970s. At the arctic temperature extremes of about –50 to +100°F, conventional lubricants for machinery performed poorly in the low-temperature region. In 1969, Conoco introduced formulated lubricants that were made from dialkylbenzene base stocks and that combined good viscosity indexes, low pour points, excellent additive solubility, and little elastomer (seal) swelling to meet the arctic performance needs.

One of Conoco's products was DN-600 Stock, which had the properties and approximate composition shown in Table 5.1. The mass spectral analysis of the DN-600 base oil also indicated the presence of minor amounts of indanes, tetralins, and naphthalenes. The data in Table 5.1 indicate that this base stock was probably made by the alkylation of benzene with chlorinated paraffins. Thus, this base oil was of the linear alkylbenzene type.

DN-600 synthetic motor oil, another Conoco product, was a multipurpose oil for year-round service as an engine crankcase oil, hydraulic oil, and torque converter fluid, formulated with antioxidant, rust inhibitor, and antiwear agents. The properties of this oil are given in Table 5.2. As a crankcase oil, it exceeded the requirements of Engine Service Classification SF-CC as defined by the API, SAE, and ASTM; it met the Military Specification MIL-L-46152B, Ford and General Motors specifications, and it was an approved Allison Type C-3 transmission fluid. It assured faster starting in cold weather as well as increased engine protection for vehicles operating in warm climates. It was compatible with engine seals and bearings.

Table 5.1 DN-600 Base Oil Properties and Composition[a]

Property	ASTM Method no. D-	Value
Specific gravity	2422	0.8628–0.8576
Gravity, API	287	32.5–33.5
Pounds per gallon, typical	—	7.16
Flash point COC, °C(°F), minimum	92	224(435)
Pour point, °C(°F), maximum	97	–54(–65)
Kinematic viscosity, (cS)		
40°C	445	26–27 typical
100°C		4.9 minimum
–40°C	2602	9,500 maximum
100°F	445	29–30 typical
210°F		5.0 minimum
Viscosity index	2270	100 minimum
Composition (wt% by GLC)		
Monoalkylbenzenes		1.5 maximum
Diphenylalkanes		12.5 maximum
Dialkylbenzenes		80.0 minimum
Postdialkylbenzenes		6.0 maximum
Chlorine		0.01 maximum

[a]Data from DN-600 specifications of Conoco, with permission.

Table 5.2 DN-600 Synthetic Motor Oil Product Typicals[a]

Property	Grade 5W–30	Test method
Gravity, API	29.9	ASTM D-287
Flash (°F)	400	ASTM D-92
Viscosity		
cS, –30°F[b]	10,000	ASTM D-445
cS, 40°C	53	ASTM D-445
cS, 100°C	9.38–10.35	ASTM D-445
SSU, 100°F	274	—
SSU, 210°F	58–61	—
Viscosity index	170	ASTM D-2270
Pour point	–50	ASTM D-97
Total base number	6.9	ASTM D-2896
Total acid number	3.1	ASTM D-664
Sulfated ash (wt%)	0.98	ASTM D-874
Foam sequences, I, II, III	10–0–0	ASTM D-892
Zinc (wt%)	0.13	—

[a]Data by permission of Conoco.
[b]Borderline pumping temperature.

Table 5.3 DN-600 Antiwear Hydraulic Oil Typical Specifications[a]

Property	Value	
Gravity °API	31	
Flash, °F, minimum	300	(Typical 325)
ASTM D-445 kinematic viscosity		
cS, –40°C maximum	3,000	(Typical 2,600)
cS, 100°C	3.3	
Viscosity index	160	
ASTM rust test A&B	Pass	
Foam test	Pass	
Vickers vane pump test	Pass	
Pour point (°F) maximum	–70	(Typical –80)

[a]Data by permission of Conoco.

As a hydraulic oil and torque converter oil, DN-600 synthetic motor oil protected hydraulic pumps as well as the best antiwear hydraulic oils. The excellent low-temperature fluidity prevented cavitation and gave quick hydraulic system response. This oil was rust and corrosion inhibited, with good oxidation stability in hot weather and resistance to foaming. Data for a specially formulated DN-600 antiwear hydraulic oil are given in Table 5.3.

DN-600 gear oil, by Conoco, is a multipurpose EP hypoid gear lubricant fully formulated to protect gears and bearings under a wide variety of load and temperature conditions. Its properties are shown in Table 5.4.

Conoco's Polar Start DN-600 SRI grease is a special, rust-inhibited, low-temperature grease with extreme pressure additives. It provides excellent protection for equipment

Table 5.4 DN-600 Gear Oil Typical Properties[a]

Property	SAE grade 75W	Test Method
Gravity, API	28.5	ASTM D-287
Flash (°F)	395 minimum	ASTM D-92
Pour point (°F)	–65 maximum	ASTM D-97
Viscosity		
Brookfield viscosity at –40°C (cP)	9,000	ASTM D-2983
cS, –34°C	5,900	ASTM D-445
cS, –18°C	1,032	ASTM D-445
cS, 40°C	28	ASTM D-445
cS, 100°C	5.1	ASTM D-445
SSU, 100°F	146	ASTM D-445
SSU, 210°F	43	ASTM D-445
Viscosity index	111	ASTM D-2270
Foam, sequence I, II, III	0–40–10	ASTM D-892

[a]Data by permission of Conoco.

subjected to salt water, muskeg (a bog of North America), mine water, and other corrosive environments. The properties of this grease are shown in Table 5.5.

Another description of formulated engine oils for gasoline and diesel engine hydraulic oils, gear oils, and greases using DN-600 base oil is given by Scott and McCloud (62).

In 1981, Conoco was acquired by DuPont Co., and these dialkylbenzene-type lubricants are no longer made by the originator. The alkylbenzene business was spun off to VISTA Chemical Co. In early 1991, VISTA was being acquired by RWE-DEA Co. of Germany (63).

VISTA has prepared a rough comparison of several synthetic base oils, including the dialkylbenzenes with conventional mineral oils, shown in Table 5.6.

VISTA Chemical Co. has pointed out the following additional advantageous qualities of the dialkylbenzenes: They have excellent compatibility with other synthetic and mineral oil base stocks formulation with conventional additives. They also can be formulated for the lowest-cost finished lubricant.

VISTA has provided some additional physical properties for the dialkylbenzene (DAB) base oil, listed in Table 5.7. The superiority of the DAB base oil in viscosity and pour point compared to paraffinic and naphthenic base is shown in Table 5.8. The DAB base oil has the desirable low viscosity at low temperatures and has a dramatically lower pour point of –70°F (–57°C).

Another type of alkylbenzene fluid is made by Shrieve Chemical Co., the Zerol 150 and 300 refrigeration fluids. This business was acquired by Shrieve Chemical from Chevron. Chevron's brochure on the Zerol fluids shows a branched, long-chain alkylated benzene structure, **10**. Thus the oil of structure **10** is probably prepared by the alkylation of benzene with a propylene oligomer. The Zerol fluids are recommended as lubricants for

Table 5.5 Polar Start DN-600 SRI Grease Typical Physical Properties[a]

Property	
NLGI no.	1
Thickener type	Lithium
Penetration	
60 Strokes	315
10,000 Strokes	325
Roll stability	
% change	3.2
Dropping point (°F)	345
Wheel bearing leakage (%)	4.5
Water washout (%, 100°F)	0.25
Rust properties	1
Four-ball EP wear (mm)	0.55
Weld (kg)	315
LWI	38.5
No seizure (kg)	50
Base oil properties	
Pour point (°F)	–70
Flash point (°F)	360

[a]Data by permission of Conoco.

Table 5.6 Comparison of Base Oil Performance[a] (1=Best, 5=Poorest)

Oil	Dialkylbenzene	Olefin oligomer[b]	Diester	Polyol-ester	Mineral oil
Volatility	2	3	2	1	5
Oxidation stability	2	5	4	1	2
Viscosity index (actual)	4(113)	2(137)	1(147)	3(130)	5(95)
Pour point (°F)	2(–70)	1(–90)	1(–96)	2(–70)	5(+5)
Additive solubility	1	5	2	1	1
Sludge-forming tendency	2	2	1	2	5
Seal compatibility	1	3	3	3	1
Cost[c] ($/gal)	5	~5	4–6	~7	<1

[a]Data by permission of VISTA Chemical Co.
[b]Hydrogenated oligomer.
[c]1=Least expensive, 7=most expensive.

Table 5.7 VISTA Chemical Co. DAB Base Oil[a]

Property	Typical value
Viscosity (cS)	
210°F	5.1
100°F	29.1
–40°F	8,600
Viscosity index	115
Pour point (°F)	–65
Flash point, COC (°F)	450
Evaporation loss, D-972	
6.5 h/400°F (wt%)	6.6
Specific gravity	0.865
Density (lb/gal)	7.2
Appearance	Clear and bright, pale yellow liquid
Distillation, D-1160 (°F)	
IBP	795
5%	819
50%	829
90%	845
EP	860
Lubricity, four-ball antiwear,	
10 s wear (mm)	
32 kg	0.73
40 kg	1.63

[a]Data by permission of VISTA Chemical Co.

refrigeration, air conditioning, and heat-pump compressors. Their advantages are said to include good viscosity at low temperature, very good low-temperature miscibility with refrigerant, good oil return and heat transfer, low flocculation point, low foaming, excellent compatibility with elastomers, metals, and other oils, and outstanding thermal stability. The good low-temperature viscosity gives less suction line pressure drop.

Table 5.8 Viscosity and Pour Point Comparison of Dialkylbenzenes with Paraffinic and Naphthenic Base Oils[a]

Property	DAB	Paraffinic	Naphthenic
Viscosity (cS)			
250°F	2.42	2.40	2.05
210°F	5.10	5.10	5.10
100°F	29.7	32.9	45.1
0°F	910	No flow	No flow
–40°F	9,540	No flow	No flow
Pour point (°F)	–70	0	–30

[a]Data by permission of VISTA Chemical Co.

Typical properties for the Zerol finds are given in Table 5.9. The viscosity indicates that the Zerol fluids may have a considerable content of dialkylbenzenes.

$$CH_3-CH(CH_3)-CH_2-C(CH_3)(C_6H_5)-[CH(CH_3)-CH_2]_x-CH(CH_3)-CH_3$$

10

where x = 1 - 3

The "floc point"—the temperature at which wax precipitates out of solution (and can clog capillary tubes, foul the evaporator wall, and reduce heat transfer, and can make the expansion valves stick)—is generally higher for petroleum-based mineral oils than for either Zerol 150 or 300, and this is an important advantage of the Zerols as refrigeration systems oils.

Recently, because of environmental concerns, a drive has begun to replace the Freons used in refrigeration and air-conditioning systems with products believed to be environmentally less harmful to the ozone layer in the atmosphere. The effect, if any, of this change to new refrigerants on the use of the branched-chain alkylbenzene lubricants has not yet been discussed in public.

Finally, the estimated consumption data and prevailing prices in 1987 for dialkylbenzenes (DABs) are given in Table 5.10. It is probable that the U.S. usage of DAB lubricants has diminished considerably since the completion of the Alaska pipeline and also because PAO lubricants have gained market share since then. The relatively high usage of DAB lubricants in Japan is due, in good part, to their use as electrical transformer insulating oils. In Western Europe, there is a sizeable use of dialkylbenzene oil in water-mixed metal-working oils. A modest amount is used in heat-transfer applications. There is also a significant use of DABs in electrical cable insulating oils.

The major participants in the DAB lubricant business are listed in Table 5.11.

Finally, the poly(long-chain alkyl)naphthalene pour-point depressants are likely to be manufactured by lubricant additive manufacturers and used neat or in blends with other, structurally different, pour-point depressants. Due to the commercial secrecy requirements, no commercial data have been published.

Table 5.9 ZEROL Refrigeration Fluid-Typical Tests

Property	Test method	ZEROL 150	ZEROL 300
Floc point (°C)	ASHRAE 86	<–73	<–55
Floc point (°F)		<–100	<–67
Viscosity			
Saybolt universal seconds	ASTM O-445, O-2161		
100°F		150	300
210°F		40	46
Kinematic centistokes	ASTM D-445		
40°C		28	57
100°C		4.1	5.8
Water content (ppm)	ASTM D-1533	25	25
Pour point (°C)	ASTM D-97	–40	–35
Pour point (°F)		–40	–31
Color	ASTM D-1500	<1	<1
Gravity:			
°API 15.6/15.6°C	ASTM D-287	30.5	31.0
Specific Gravity, 60/60°F	ASTM D-1298	0.87	0.87
Density (lbs/gal), 60/60°F	ASTM D-1298	7.26	7.27
Flash point (°C)	ASTM D-92	275	175
Flash point (°F)		347	347
Acid number (mg KOH/g)	ASTM D-664, D-974	0.01	0.01
Molecular weight	ASTM D-3592	330	375
Stability sealed tube (%R-22)	ASHRAE 97	0.07	0.15
Dielectric strength (kV), 2580	ASTM D-877	>35	>30
Specific heat (Btu/lb °F)	ASTM D-2766		
50°F (10°C)		0.428	0.428
100°F (38°C)		0.445	0.445
150°F (66°C)		0.462	0.462
Thermal conductivity (Btu/ft/h/ft^2/°F)	ASTM D-2717	0.062	0.062
Thermal expansion coefficient per °C, 0–100°C	ASTM D-1903	0.00074	0.00077

[a]Data from Chevron brochure on ZEROL fluids, reproduced with permission of Shrieve Chemical Co.

Table 5.10 Dialkylbenzene-Type Oil Consumption and Price Range, 1987[a]

Region	Tons	$/kg.
United States	2,600	0.80–1.20
Western Europe	8,200	0.80–1.20
Japan	4,400	1.30–2.75

[a]Source: Specialty Chemicals Update Program, Synthetic Lubricants—Worldwide, April 1988, SRI International.

Table 5.11 Producers of Dialkylbenzene-Type Oils

Company	Region
VISTA Chemical Co.[a]	United States
Shrieve Chemical Co.[a]	United States
Exxon Paramins[b]	France
Chevron Chemical SA[b]	France
Wintershall[b]	Germany
Mitsubishi Petrochemical Co. Ltd.[a]	Japan
Mihon Alkylate Co., Ltd.[a]	Japan
Nippon Petroleum Detergent Co., Ltd.[a]	Japan

[a]Source: Specialty Chemicals Update Program, Synthetic Lubricants—Worldwide, April 1988, SRI International.
[b]Source: E. I. Williamson, The College of Petroleum Studies, Oxford, U.K.

IV. CONCLUSION

The dialkylbenzenes (DABs) have been proven to be valuable base stocks for lubricants and functional fluids for a wide variety of applications. However, the DABs have been buffeted by large variations in demand, process changes, management changes, political manipulations, and high-quality competitive materials in the last 25 years. Development of alkylaromatics has continued in this period at a good pace, worldwide. New polyalkylated benzenes and a wide range of viscosity grade of alkylated naphthalenes have been developed, which have excellent properties for low- as well as high-temperature service. Today's high costs of development, including environmental agency approvals, are an impeding factor for commercialization. Nonetheless, with good feedstock availability and pricing, the newly designed alkylaromatics may well begin a rise to greater prominence as functional fluids than before.

REFERENCES

1. British Patent 323,100 (Dec. 3, 1928; to IG Farbenindustrie A.G.), Condensation Products of Polynuclear Compounds With Olefins; German Patent 565,249 (March 26, 1930; H. Zorn, M. Mueller-Cunradi, and W. Rosinski to IG Farbenindustrie A.G.), Lubricating Oil; U.S. Patent 1,815,072 (July 14, 1931; G. H. B. Davis to Standard Oil Development Co.), Lubricating Oil; U.S. Patent 2,030,832 (Feb. 11, 1936; F. H. McLaren to Standard Oil Co. of Indiana), Synthetic Lubricating Oils.
2. Koelbel, H. (1948). Synthesis of lubricants via the alkylation of naphthalene, *Erdoel Kohle,* 1, 308–318.

3a. Olah, G. A. (ed.) (1964). *Friedel-Crafts and Related Reactions,* vol. 2, part 1, chapters 14, 17, and 18, Interscience Publishers, New York.

3b. Roberts, R. M., and A. A. Khalaf (eds.) (1984). *Friedel-Crafts Alkylation Chemistry: A Century of Discovery,* Marcel Dekker, New York.

4. Asinger, F. (1968). *Mono-Olefins—Chemistry and Technology,* Pergamon Press, Oxford.
5. Jpn. Kokai Tokkyo Koho JP 59,219,240 [84,219,240] (Dec. 10, 1984; to Lion Corp.), Internal Olefins; PC Int. Appl. WO 90 03,354 (Apr. 5, 1990; N. Akiyama and M. Mori to Mitsubishi Monsanto Chemical Co.), Process for Producing Internal Olefins.
6. Franck, H. G., and J. W. Stadelhofer (1988). *Industrial Aromatic Chemistry,* Springer Verlag, New York.

7a. Klamann, D. (1984). *Lubricants and Related Products,* Verlag Chemie, Weinheim, pp. 105–106.

7b. Plaskunova, S. L., E. K. Ivanova, B. R. Serebryakov, and V. M. Shkolnikov (1983). Synthetic lubricating oils made with alkylbenzenes, *Soviet Chem. Ind.*, 15, 653–660, Eng. transl. from *Khim. Prom.*, 6, 328–331.
8. U.S. Patent 3,849,507 (Nov. 19, 1974; E A. Zuech to Phillips Petroleum Co.), Alkylation of Aromatic Hydrocarbons Using a Compacted Montmorillonite Clay Catalyst.
9. U.S. Patent 3,657, 370 (Apr. 18, 1972; W. C. Hammann and C. F. Hobbs to Monsanto Co.), Process for the Preparation of Di-tertiary-alkylaromatic Hydrocarbons.
10. U.S. Patent 3,678,123 (May 18, 1972; J. K. Boggs to Esso Research and Engineering Co.), Tertiary Alkylation Using an Admixture of Olefins and Tertiary Alkyl Chlorides.
11. U.S. Patent 3,793,207 (Feb. 19, 1974; M. L. Burrous to Chevron Research Co.), Fire-Resistant Hydraulic Fluid.
12. U.S. Patent 3,909,432 (Sept. 30, 1975; S. E. McGuire, J. R. Riddle, G. E. Nicks, O. C. Kerfoot, and C. D. Kennedy to Continental Oil Co.), Preparation of Synthetic Hydrocarbon Lubricants.
13. U.S. Patent 3,965,043 (June 22, 1976; G. E. Striddle to NL Industries), Catalyst for Alkylating Aromatic Hydrocarbons; U.S. Patent 3,992, 467 (Nov. 16, 1976; G. E. Striddle to NL Industries), Process for Alkylating Aromatic Hydrocarbons and Catalyst Therefor.
14. U.S. Patent 4,046,826 (Sept. 6, 1977; G. E. Striddle to NL Industries, Inc.), Process for Alkylating Aromatic Hydrocarbons With Synthetic Hectorite-type Clay Catalyst.
15. U.S. Patent 4,301,316 (Nov. 17, 1981; L. B. Young to Mobil Oil Corp.), Preparing Phenylalkanes.
16. U.S. Patent 4,358,628 (Nov. 9, 1982; L. H. Slaugh to Shell Oil Co.), Alkylation of Benzene Compounds With Detergent Range Olefins.
17. German Offen. DE 3,226,308 (May 19, 1982; M. Sauerbier and C. Schudok to Rhein-Chemie Rheinau G.M.b.H.), Catalyst and Its Use for Alkylation of Aromatic Compounds.
18. U.S. Patent 4,463,207 (July 31, 1984; T. H. Johnson to Shell Oil Co.), Arene Alkylation With Metal Oxide—Tantalum Halide/Oxide Catalysts; U.S. Patent 4,658,072 (Apr. 14, 1987; T. H. Johnson to Shell Oil Co.), Lubricant Composition.
19. U.S. Patent 4,658,072 (Aug. 22, 1984; T. H. Johnson to Shell Oil Co.), Alkylaromatic Lubricant.
20. Japan Patent 85,007,325 (Feb. 23, 1985; to Nippon Oil KK), Electrical Insulating Oils.
21. German (East) Patent DD 217,238 (Sept. 1, 1985; to VEB Petrochem Schwedt), Electrical Insulating Oils Production.
22. U.S. Patent 4,520,218 (May 28, 1985; R. C. Berg, T. P. Malloy, and B. V. Vora to UOP Inc.), Production of Dialkylated Aromatics.
23. European Patent Application 0160,145 (Jan. 22, 1986; H. A. Boucher to Exxon Research and Engineering Co.), Alkylation of Aromatic Molecules Using Wide Pore, Amorphous Silica-Alumina Catalyst.
24. European Patent Application 0168,534 (Jan. 22, 1986; H. A. Boucher to Exxon Research and Engineering Co.), Dialkylaromatic and Hydrogenated Dialkylaromatic Synthetic Lubricating and Specialty Oils.
25. Japan Patent 6 2192-494 (Aug. 24, 1987; to Ethylene Chemicals KK), Hydraulic Brake Fluid Composition for Automobiles.
26. German Offen. DE 3,526,873 (Jan. 29, 1987; H. Theunissen and R. Weber to Hydrocor-Forschungs-und Analytic G.m.b.H.), Preparation of Brominated Alkylbenzenes as Low Flammability Biodegradable Functional Fluids.
27. U.S. Patent 4,929,584 (May 29, 1990; L. H. Slaugh, T. H. Johnson, and R. J. Hoxmeier to Shell Oil Co.), Catalysts for Alkylation of Aromatic Compounds With Surfactant-range Olefins.
28. European Patent Application EP 377,305 (July 11, 1990; T. R. Forbus, S. C. H. Ho, B. P. Pelrine, and M. M. S. Wu to Mobil Oil Corp.), Alkylaromatic Lubricant Fluids.

29. U.S. Patent 4,912,277 (March 27, 1990; B. A. Aufdembrink, C. T. Kresge, Q. N. Le, J. Shim, and S. S. Wong to Mobil Oil Corp.), Process for Preparing Long Chain Alkyl Aromatic Compounds.
30. U.S. Patent 3,498,920 (March 3, 1970; C. W. Nichols, Jr. and M. I. Smith to Mobil Oil Corp.), Compositions of Liquid Paraffins Containing Mixtures of Alkyl-Substituted Polynuclear Aromatic Hydrocarbons as Swelling Agents.
31. Polish Patent 63,556 (Oct. 15, 1971; J. Ruta, M. Novak, L. Glus, S. Ligeza, and J. Mosurka to Przedsiebiorstwo Panstwowe Rafineria Nafty Jaslo), Synthetic Lubricant by Condensation of Olefins with Naphthalene.
32. U.S. Patent 3,716,596 (Feb. 13, 1973; E. Bowes to Mobil Oil Corp.), Alkylation and Dealkylation of Aromatics in the Presence of Crystalline Aluminosilicates.
33. U.S. Patent 3,959,399 (May 25, 1976; B. W. Bridwell and C. E. Johnson to Nalco Chemical Co.), Mono-alkylation of Naphthalene.
34. German Offen. 2,518,641 (Nov. 4, 1976; L. Veres to Kabel-und Metallwerke Gutehoffnungshuette A.-G.), Stabilized Polyolefin Mixtures.
35. U.S. Patent 4,088,589 (May 9, 1978; A Rossi and W. J. Fernandez to Exxon Research and Engineering Co.), Dual Pour Depressant Combination for Viscosity Index Improved Waxy Multigrade Lubricant.
36. U.S. Patent 4,275,253 (June 23, 1981; M. Takahashi and A. Ito to Kureha Chemical Industry Co., Ltd.), Radiation-Resistant Oil and Method of Lubricating for Atomic Power Facilities.
37. U.S. 4,368,343 (Jan. 11, 1983; I. L. Kotlyarevsky, N. I. Myakina, M. A. Kamkha, I. M. Ikryanov, and S. A. Glyadinskaya), Process for Producing High-Vacuum Oils.
38. Belov, P. S., E. N. Grigoreva, and M. M. Fernandes-Gomes (1983). Selectivity of the alkylation of naphthalene by higher olefins. *Neftepererab. Neftekhim. (Moscow),* 4, 37–38.
39. Jpn. Kokai Tokkyo Koho JP 59,147,096 [84,147,096] (to Nippon Oil Co., Ltd.), Synthetic Oxidation-Resistant Lubricant.
40. U.S.S.R. Patent 1,002,279 (March 7, 1983; O. N. Tsvetkov), Preparation of Higher Alkylnaphthalenes.
41. Belov, P. S., E. N. Grigoreva, S. G. Gukasyan, N. B. Nagaev, and O. A. Stulova (1985). Synthesis of an alkylnaphthene heat transfer agent, *Neftepererab. Neftekhim. (Moscow),* 12, 10–12.
42. Japan Patent 61,056,138 (March 20, 1986; to Idemitsu Kosan KK), High-Purity Tetralin Derivatives Production.
43. U.S. Patent 4,604,491 (Aug. 5, 1986; H. Dressler and A. A. Meilus to Koppers Co., Inc.), Synthetic Oils.
44. Jpn. Kokai Tokkyo Koho JP 62,138,440 [87,138,440], (June 22, 1987; T. Yamashita, T. Inagaki, T. Sato, and M. Kojima to Lion Corp.), Alkylation of Aromatic Hydrocarbons.
45. Japan Patent 62,263,133 (Nov. 16, 1987, to Idemitsu Kosan KK), Di-aralkyl Aromatic Hydrocarbon Preparation.
46. Japan Patent 62,294,630 (Dec. 22, 1987; T. Minoe, Y. Saito, and T. Tsubochi to Idemitsu Kosan KK), Preparation of Di-aralkylaromatic Hydrocarbons.
47. Japan Patent 62,289,533 (Dec. 16, 1987; to Idemitsu Kosan KK), 1-(1-Tetralyl)-2-phenylpropane.
48. Japan Patent 87,009,093 (Feb. 26, 1987; to Exxon Research and Engineering Co.), Alkylation of Aromatic Hydrocarbons With Alkene Mixture.
49. U.S. Patent 4,714,794 (Dec. 22, 1987; T. Yoshida and H. Watanabe to Nippon Oil Co. Ltd.), Synthetic Oils.
50. Dressler, H. Unpublished results obtained in the Koppers Co., Inc., Research Department.
51. Japan Patents 63,014,738 and 63,014,739 (both Jan. 21, 1988 to Shin-Daikyowa Sekiyu), both titled Preparation of Mono- or Di-alkylnaphthalenes.
52. German (East) Patent 121,631 (Aug. 12, 1976; W. Faeder, W. Hoertzsch, and H. Uhlig), Composition Useful as Pour Point Depressant and Filtration Aid.

53. U.S.S.R. Patent 635,122 (Nov. 30, 1978; A. M. Kuliev, I. I. Namazov, N. G. Abdulaev, F. N. Mamedov, R. J. Rzaev, and F. N. Nuriev), Depressants for Lubricatiang Oils.
54. Romanian Patent 66,669 (Oct. 30, 1978; I. Scheianu, M. Mihailescu, and G. Moisescu to Intreprinderea Rafinaria Teldajen), Fluid Pour-Point Depressant Additive for Mineral Oils and Diesel Fuels.
55. U.S. Patent 3,426,076 (Feb. 4, 1969, H. Mertwoy and H. Gissen to the U.S.A. as represented by the Secretary of the Army), Oxidatively Stable Alkylbiphenyls and Terphenyls as Non-Spreading Lubricants.
56. U.S. Patents 3,640,870 (Feb. 8, 1972) and 3,931,334 (Jan. 6, 1976), both by R. M. Gemmill, Jr. and J. W. Schick to Mobil Oil Corp., both titled Lubricant Compositions Comprising Substituted Indans.
57. U.S. Patent 4,480,142 (Oct. 30, 1984; R. L. Cobb to Phillips Petroleum Co.), Catalytic Alkylation.
58. U.S. Patent 4,506,107 (March 19, 1985; A Sato, K. Endo, S. Kawakami, H. Yanagishita, and S. Hayashi to Nippon Petrochemical Co.), Electrical Insulating Oil and Oil-filled Electrical Appliances.
59. Kosswig, K. (1982). Tenside, in *Ullmann's Encyclopedia of Technical Chemistry,* 4th ed., vol. 22, p. 455, Verlag Chemie, Weinheim.
60. Takoaka, S. (1974). SRI-PEP (Process Economics Program) Report No. 59A Supplement, Aliphatic Surfactants, Menlo Park [the given process is based on U.S. 3,355,508 (Nov. 28, 1967; H. N. Moulden to Chevron Research), Continuous Process for Alkylating Aromatic Hydrocarbons.]
61. *Oil Gas J.* (1990). July 16, p. 48, news item.
62. Scott, W. P., and A. P. McCloud (1977). *J. Nat. Lub. Grease Inst., NLGI Spokesman,* Nov., 260–264.
63. *Chem. Week* (1991). Jan. 16, p. 16, news item.
64. U.S. Patent 1,815,022 (July 14, 1931; G. H. B. Davis to Standard Oil Development Co.), Lubricating Oil.
65. U.S. Patents 1,963,917 and 1,963,918 (June 19, 1934; F. H. MacLaren to Standard Oil Co. of Indiana), Pour-Point Depressors.
66. U.S. Patent 2,030,307 (Feb. 11, 1936; F. H. MacLaren to Standard Oil Co. of Indiana), Pour-Point Depressors for Lubricating Oils.
67. U.S. Patent 2,174,264 (Sept. 26, 1939; E. Lieber and M. M. Sadlon to Standard Oil Development Co.), Method for Producing Wax Modifying Agents.

6
Perfluoroalkylpolyethers

Thomas W. Del Pesco
E.I du Pont de Nemours and Co., Inc.
Deepwater, New Jersey

I. INTRODUCTION

Synthetic lubricants are a broad class of lubricants that are derived by chemical synthesis rather than being refined from petroleum oils. There are many types of synthetic lubricants developed for a range of uses requiring properties such as high-temperature stability, low-temperature performance, fire resistance, long service life, chemical inertness, radiation resistance, and high viscosity index, to name just a few.

Perhaps the most common use for synthetic lubricants is in high-temperature applications. Most petroleum oils and greases will oxidize and rapidly lose their efficiency above 250°C, while below 10°C they become too viscous to be serviceable. When a combination of high- or low-temperature properties, chemical or oxidative stability, low volatility, materials compatibility, inertness, and nonflammability or concombustibility is all needed simultaneously, a good lubricant is very difficult to find.

A class of lubricants that can meet all these requirements is the perfluoropolyethers, also known as PFPE, perfluoropolyalkylethers, PFPAE, or perfluoroalkylethers, PFAE. These names are synonyms for the same chemical. The PFPE fluids are composed entirely of carbon, fluorine, and oxygen. They are colorless, odorless, and completely inert to most chemical agents including oxygen.

This chapter describes the process for the manufacture of four types of perfluoropolyether lubricants, their unique physical and chemical properties that make them useful as lubricants, and applications that demonstrate some of the lubrication problems that they are helping to solve in various industries.

II. PFPE CLASSIFICATION

Gumprecht (1) first disclosed the use of the perfluoroalkylpolyethers (PFPE) as lubricants in an ASLE/ASME lubrication conference in the fall of 1965. He described a new class of polymer that exhibits broad liquid range, good thermal stability, outstanding chemical stability, good material compatibility, and low wear in lubricant tests.

There are four distinct types of PFPE oils. Each type is chemically similar but structurally different. Each PFPE type exhibits similar physical and chemical properties. In fact, it is often quite difficult to tell them apart without sophisticated analytical equipment.

There are four types of PFPE polymers. Each type is differentiated by its chemical structure. The chemical stability and physical properties of each structural type is slightly different.

PFPE-K $CF_3CF_2CF_2O\text{-}[CF(CF_3)CF_2\text{-}O\text{-}]_nCF_2CF_3$
PFPE-Y $CF_3O\text{-}[CF(CF_3)CF_2\text{-}O\text{-}]_y\text{-}[CF_2\text{-}O\text{-}]_mCF_3$
PFPE-Z $CF_3O\text{-}[CF_2CF_2\text{-}O\text{-}]_z\text{-}[CF_2\text{-}O\text{-}]_nCF_3$
PFPE-D $CF_3CF_2CF_2\text{-}O\text{-}[CF_2CF_2CF_2\text{-}O\text{-}]_aCF_2CF_3$

A. Linear versus Nonlinear PFPE

Both PFPE-K and PFPE-Y polymer chains contain pendent trifluoromethyl groups ($-CF_3$) and are nonlinear molecules, while PFPE-D and PFPE-Z contain no pendant groups and are linear. The linear PFPE structures show less change of viscosity with temperature and pressure when compared to the nonlinear PFPE.

B. Shielded versus Nonshielded PFPE

Pendant trifluoromethyl groups adjacent to the ether (-O-) linkage shield that linkage from acid-catalyzed cleavage. PFPE-K has a fully shielded polymer chain, PFPE-Y has a partially shielded polymer chain, and PFPE-D and PFPE-Z have nonshielded polymer chains. PFPE-K contains a trifluoromethyl group adjacent to every ether linkage and is the most shielded of the four types of PFPE polymers.

C. Homopolymer versus Copolymer PFPE

PFPE-K and PFPE-D contain one linkage group while PFPE-Y and PFPE-Z contain two types of linkage groups. PFPE-Y and PFPE-Z both contain the difluoroformyl ($-CF_2O-$) linkage as well as the perfluoropropyl or perfluoroethyl group, respectively. Even though PFPE-Y and PFPE-Z appear to be copolymers, they are prepared using only one monomer.

III. PREPARATION OF PFPE TYPES

PFPE-K perfluoroalkylpolyethers are prepared by the anionic polymerization of hexafluoropropylene epoxide (HFPO) at low temperatures. The preparation of HFPO has been described (2,3). Polymerization of HFPO can be carried out with solvents such as aliphatic hydrocarbon polyethers or nitriles using cesium fluoride as the source of fluoride ions. The reaction temperature is the boiling point of HFPO (–24°C).

The polymer has an acid fluoride end group that is much too reactive for use in lubricants. Stabilization can be achieved (4) by reaction with elemental fluorine. The degree of polymerization, n, varies from approximately 1 to 80. That corresponds to a molecular weight range of 435 to 13,500.

PFPE-Y is obtained by the photochemical catalyzed polymerization of perfluoropropylene in the presence of oxygen at low temperatures (5). The crude reaction mass must be treated to eliminate unstable peroxide linkages. This can be accomplished by treatment with elemental fluorine or ultraviolet light. Either method causes partial decomposition of the peroxide linkages, giving two types of perfluoroalkyl linkages in the polymer chain, [$-OCF_2-$] and [$-OCF(CF_3)CF_2-$]. The molecular weight range of the polymer produced is 1,000 to 10,000.

When tetrafluoroethylene is used instead of hexafluoropropylene (6) the polymer obtained is PFPE-Z. The polymer is treated with elemental fluorine to stabilize the structure by destroying the peroxide sites. PFPE-Z crude polymer has a higher molecular weight distribution than PFPE-Y, ranging from 8,000 to 70,000. During the stabilization process the molecular weight decreases because chains are cleaved into smaller pieces. The reaction mechanism and kinetic equation for the photooxidation of fluoroolefins to PFPE-Y and PFPE-Z is described by Sianesi et al. (7).

PFPE-D is obtained by the Lewis acid catalyzed ring opening polymerization reaction of 2,2,3,3-tetrafluorooxetane (8). The hydrogen-carbon bonds are subsequently converted to fluorine-carbon bonds by direct fluorination with elemental fluorine. Ultraviolet light can catalyze the direct fluorination.

The crude polymers are purified by contact with absorbing agents to remove polar materials and distilled under reduced pressure. The reaction mass is fractionated by distillation into specific molecular weight or viscosity ranges. These ranges usually correspond to different grades or product types available commercially.

IV. PROPERTIES

PFPE oils are colorless and odorless fluids. Physical properties that change with molecular weight are pour point, viscosity, viscosity index, and volatility. Chemical properties are dependent on chemical structure and not on molecular weight. Perhaps the most important feature of PFPE lubricants is their thermal and oxidative stability.

A. Thermal and Oxidative Stability

The temperature at which thermal decomposition of the PFPE-K oils begins has been measured in our laboratory by differential thermal analysis. This method gives a value in excess of 470°C. PFPE impurities not removed during purification by the manufacturer can appreciably lower the temperature at which initial decomposition is observed. These impurities may be products of side or incomplete reactions during the manufacturing process. For this reason, a more useful value is obtained by determining the rate of degradation over a long period of time at a high temperature. Such an experiment was carried out by heating a sample of PFPE-K oil in an Inconel tube under an atmosphere of argon at 452°C for 31 days. The loss of oil averaged 1.1 wt% per day and the loss in viscosity at 38°C was 3%. The oil as recovered was colorless and suffered little change in molecular weight. Table 6.1 summarizes available data on the temperature stability of PFPE by type.

PFPE oils are about as stable in pure oxygen as they are in inert systems. Sianesi et al

Table 6.1 Stability of PFPE in Inert Atmosphere and Air[a]

PFPE type	Stability limit (inert)	Stability limit (air)
PFPE-K	>450°C	400°C
PFPE-Y	<316°C	<316°C
PFPE-Z	NA	<288°C
PFPE-D	NA	370°C

[a]Source: Individual manufacturer's reference literature. NA indicates that the data was not available.

(9) showed that the decomposition temperature for PFPE-Y oils in air was very close to the thermal decomposition temperature in an inert atmosphere.

Tests with PFPE-K (Table 6.2) under a variety of conditions demonstrate the inertness of the PFPE oil to reaction with oxygen. Shock loading tests of PFPE-K with gaseous oxygen ranging from 93°C and 7,500 psi to 277°C and 6,000 psi showed no reaction.

The thermal-oxidative stability of the PFPE polymer depends on its structure. The relative stability of each type of PFPE appears to be related to the degree of shielding and to the presence of the difluoroformyl group on the polymer chain. Jones et al. (10) studied the effect of branching on the thermal oxidative stability of low-molecular-weight PFPE in the gas phase in glass ampoules. Oxidative stability measurements seemed to indicate that highly branched PFPE oils are much less stable than PFPE molecules containing a repeating, single trifluoromethyl pendant group.

Impurities in the product can play an important role in the initial, apparent thermal-

Table 6.2 Oxygen Compatibility Tests with PFPE-K

Test type	Temperature	Oxygen pressure, MPa(psi)	Test agency or method	Results
Ignition in O_2 pressure bomb	>400°C	13 (1886)	BS3100[a]	No ignition
Pressure drop in O_2 bomb	99°C	0.7 (100)	ASTM D-942[b]	No pressure drop, 600 h
Mechanical impact LOX	25°C	98 J	MSFC spec.[c] 106B	20 Trials, no reaction
Mechanical impact LOX	25°C	98 J	NASA HB 8060.LB 13.1[d]	20 Trials, no reaction
Mechanical impact LOX	25°C	98 J	ASTM D-2512	20 Trials, no reaction
Mechanical impact LOX	25°C	122 J	BS3100	10 Trials, no reaction
Mechanical impact LOX	25°C	736 J	BAM 8104-411[e]	Multiple trials, no reaction

[a]British Specification, BS.
[b]American Society for Testing and Materials, ASTM.
[c]Marshall Space Flight Center Specification, MSFC.
[d]National Aeronautics and Space Administration, NASA.
[e]West German Federal Institute for Materials Testing, BAM.

oxidative stability of the PFPE. Paciorek et al. (11) found that PFPE-K, PFPE-Y, or PFPE-Z (PFPE-D was not tested) oils that contained impurities initially appeared to be have poor thermal-oxidative stability. However, once the impurities were removed or reacted away, the oil appeared significantly more stable until the upper temperature limit was reached. If these impurities are not removed during manufacturing, the apparent upper temperature limit of thermal-oxidative stability can be much lower than that expected. The decomposition products do not appear to promote or catalyze further degradation of the PFPE when the reaction is performed in glass.

Gumprecht (12,13) reported that $CF_3CF{=}CF_2$, CF_3COF, and COF_2 were the main products in the thermal decomposition of PFPE-K. Paciorek et al. (14) found SiF_4, CO_2, and some BF_3 when oxygen was present. SiF_4 and BF_3 are formed by reaction with the glass ampoule walls in the presence of oxygen.

Other studies confirm the effect of structure on the thermal oxidative stability of PFPE oils. Paciorek et al. (14) found PFPE-K to be stable in the presence of oxygen to the highest temperature tested, >340°C. However, Jones et al. (15) found that PFPE-Z is inherently unstable at 316°C in oxidizing atmospheres. Neither hydrogen end-capped polymer nor polymer with remaining peroxide linkages is the cause of this apparent instability. The instability has been attributed to the presence of the difluoroformyl (-CF_2O-) and tetrafluoroethylene oxide (-CF_2CF_2O-) units in the polymer chain. These same difluoroformyl, (-CF_2O-) groups are also present in PFPE-Y. In fact, Pearson (16) reports that in the temperature range 325–405°C in Monel vessel, PFPE-Y, which contains the difluoroformyl (-CF2O-) group, decomposes 20–40 times faster than PFPE-K. These observations are consistent with the theory that the difluoroformyl groups contribute to instability of the PFPE at higher temperatures.

PFPE lubricants leave no deposits when they volatilize. Even when PFPE molecules thermally decompose, tars or residues are not observed. Generally the volatile decomposition products are removed as they are formed. PFPE decomposition products can react with the surface in contact with the PFPE if the decomposition products are not removed. In contrast, hydrocarbon lubricants thermally form gums or oxidatively decompose or polymerize to form tars, or deposits. These hydrocarbon-derived deposits then cause problems that can lead to apparent lubricant failure. The oxidation stability of PFPE oils is perhaps best demonstrated on a practical basis by the fact that samples of the oil contaminated with traces of hydrocarbon oils can be purified by blowing air through the mixture at high temperatures (300°C and up) for several hours. Under these conditions, the hydrocarbons are oxidized and the PFPE oils can be recovered in a pure state by filtration.

B. Lewis Acids

Metal fluorides can catalyze the decomposition of PFPE oils, especially those PFPE oils that contain the difluoroformyl (-CF_2O-) group. PFPE lubricants will react with Lewis acids, such as aluminum trichloride, at temperatures normally considered safe for PFPE oils (17). The apparent catalytic activity of the Lewis acid depends on the PFPE structure type. Polymers that contain the difluoroformyl group appear less stable than those without it (11). PFPE-Z is stable up to 260°C in contact with metals but not stable at temperatures even less than 100°C in the presence of metal halide Lewis acids (18). The effect of aluminum trichloride on the stability of different PFPE is shown in Fig. 6.1.

Sainesi et al. (9) show that metal oxides can also lower the decomposition tempera-

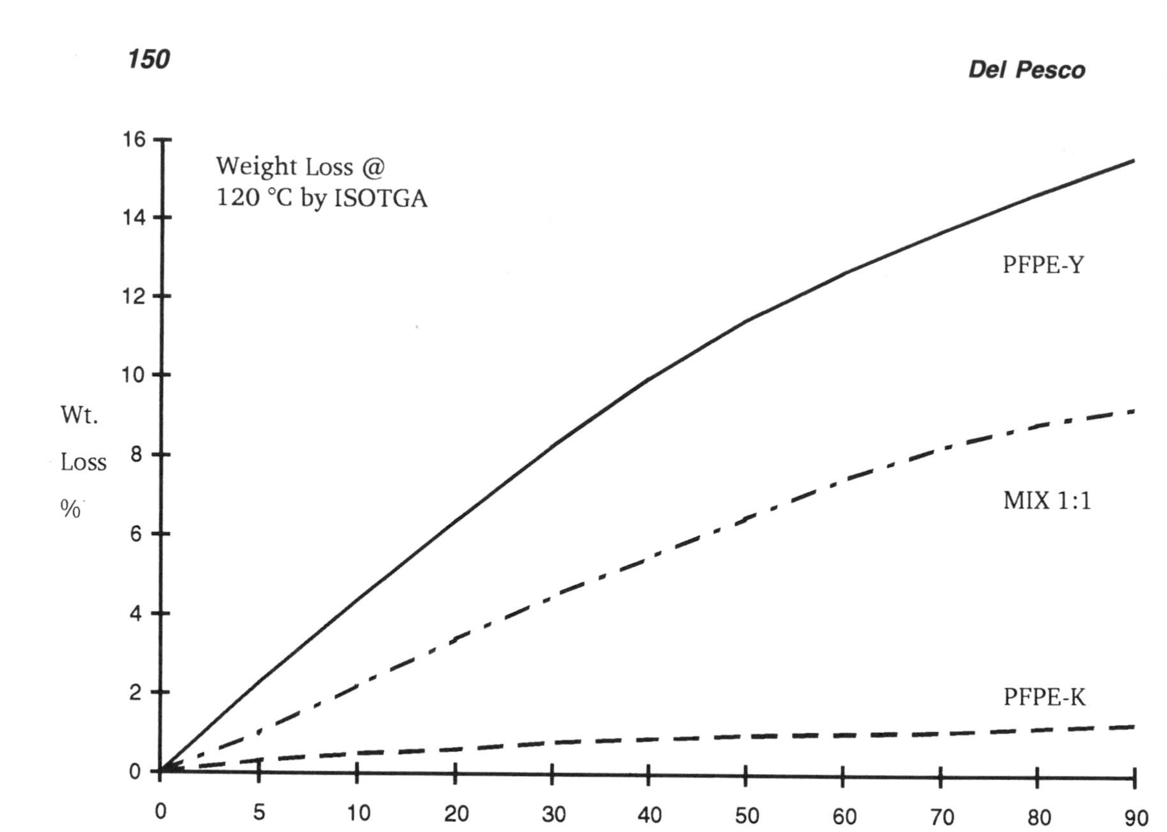

Figure 6.1 Effect of Lewis acids on PFPE degradation.

ture of PFPE-Y. Cuprous oxide had a small effect on the rate of decomposition at 360°C, whereas aluminum oxide had 100 times that rate. The data are summarized in Table 6.3

Carre and Markowitz (19) demonstrated that iron fluoride can form in the raceways of bearings lubricated with PFPE and that the presence of FeF_3 can lower the temperature at which the polymer begins to decompose. Unfortunately, the PFPE type is not disclosed and the magnitude of this effect is difficult to evaluate.

C. Compatibility with Metals

PFPE oils can also react with nascent metal surfaces. Mori and Morales (20) studied the reactions of the PFPE-Z, PFPE-D, and PFPE-K in sliding contact with stainless steel, type 440C, specimens under ultra-high-vacuum conditions. All three fluids appeared to react with the steel specimens during sliding because the surface composition had changed. PFPE-Z, which has difluoroformyl linkages, decomposed under the sliding conditions and generated COF_2 and fluorinated carbons. PFPE-D and PFPE-K, however, gave no gaseous products. The oxide layer on the stainless steel surface was removed during sliding, and metal fluorides were formed with all three oils. Mori and Morales concluded that the decomposition reaction of PFPE-Z began when the fluid contacted the fresh metal surface formed during sliding and that the metal fluorides catalyzed the decomposition of the PFPE-Z. The PFPE-K and PFPE-D showed no gross decomposition.

Table 6.3 Percent Weight Loss of PFPE-Y per Minute, Exposure at 360°C[a]

Metal oxide	Rate (%wt loss/min)
ZrO_2, CrO_3, SnO_2, WO_3, ThO_2, Al_2O_3	>5
CeO_2, TiO_2	1–5
Fe_2O_3, NiO	0.5–1.0
V_2O_5, MgO	0.3–0.5
Cr_2O_3, Sb_2O_3, Co_3O_4, $Ga_2O_3 \cdot 12\% H_2O$	.15–.3
CuO, MoO_3, HgO, ZnO, SnO, MnO_2, TeO_2, BaO, CaO, SiO_2, In_2O_3, GeO_2, Bi_2O_3, P_2O_5, PbO	0.09–0.15
Tl_2O_3	
Cu_2O	0.05
Inert gas, N_2	<0.02

[a]Source: Sianesi et al. (9).

PFPE oils can also react with metal surfaces at high temperatures. PFPE stability toward metal surfaces will depend on the PFPE type and structure. Jones et al. (21) examined the effect of the pure metals, titanium and aluminum, and their alloy, Ti(4Al,4Mn), on the degradation of linear PFPE-Z and found that the pure metals were much less effective at decomposing PFPE-Z than the alloy. Therefore, the activity of the alloy cannot be predicted by the activity of the individual metals. PFPE-Z degradation with M-50 steel correlated with the depletion of chromium and vanadium on the surface. On the other hand, PFPE-K was stable to much higher temperatures, >310°C (11).

At temperatures up to approximately 288°C, PFPE-K oils are inert to most metals. Where the oil temperature is likely to exceed 288°C, metals, and alloys used for construction become important. Extensive studies of the compatibility of PFPE-K oils with many metals and alloys have been made using the "Micro Oxidation-Corrosion Test" developed by the Air Force Materials Laboratory. Results obtained for PFPE-K at various temperatures with a number of different metals are shown in Fig. 6.2.

In general, nickel and cobalt alloys exhibit the greatest resistance to corrosion and are suitable for use with the PFPE oils up to at least 370°C. Ordinary steels are not suitable for temperatures above 288°C. Certain stainless steels are satisfactory at 316°C. A summary of metals compatible with PFPE oils at various temperatures is given in Table 6.4.

Certain alloys have been found to cause catalytic depolymerization of the PFPE oils at high temperatures. Titanium alloys that contain aluminum decompose PFPE oils above 136°C. Aluminum 2024 appears to do the same thing at 370°C. These problems are greatly minimized in the absence of oxygen, indicating that the reactions involved are between the oil and the oxide coating on the metals.

A number of additives are under development that can markedly decrease the reaction rate of PFPE oils with many metals at high temperatures (22). Some of these also minimize the catalytic depolymerization of the oils by titanium alloys (19,23–27).

D. Hydrolytic and Chemical Stability

Hydrolytic stability of PFPE oils is excellent. Long-term contact with steam or boiling water produces no adverse effects on the oils.

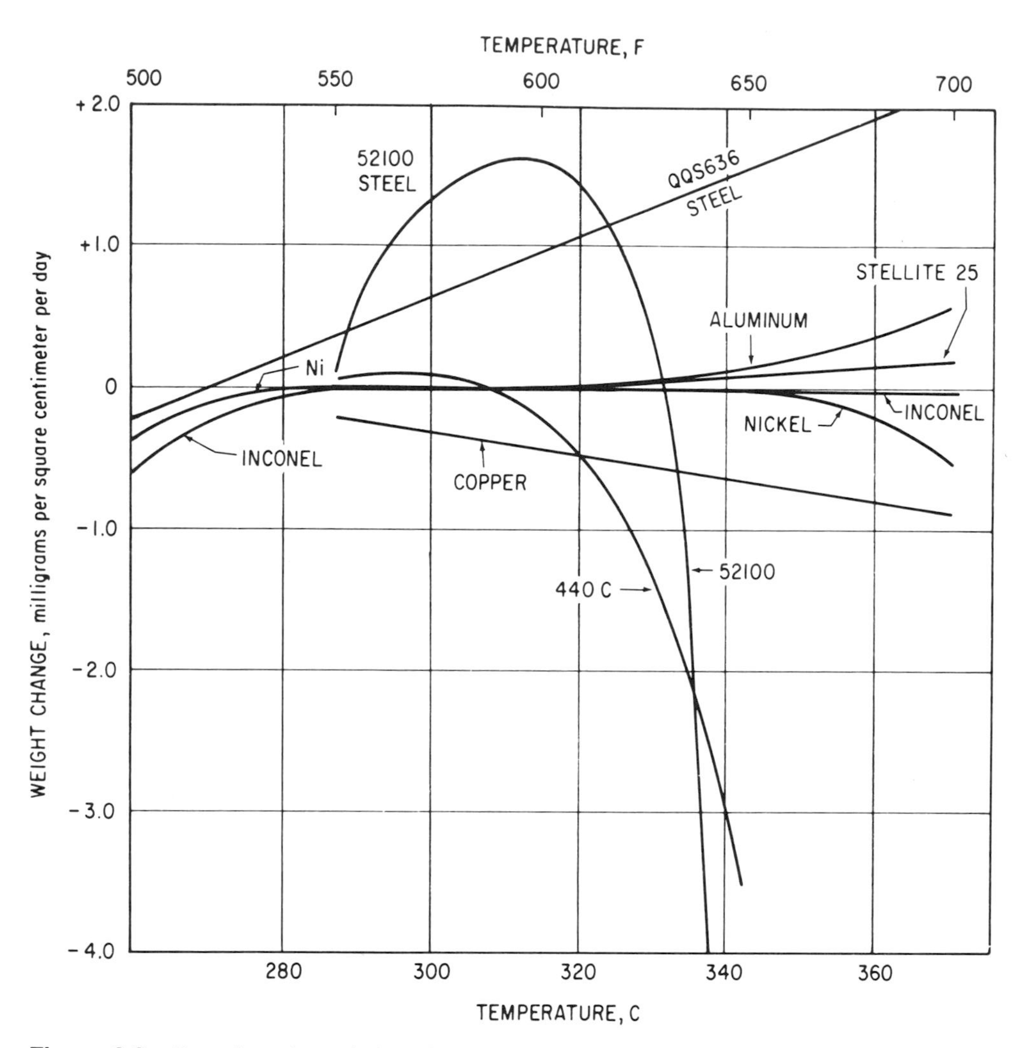

Figure 6.2 Corrosion of metals by PFPE-K at elevated temperatures. Oxidation-corrosion test conditions: 20 ml sample in Inconel test tube, 20 liters/h dry flow, 24 and 48 h test duration, no reflux condenser used.

All PFPE oils show remarkable chemical inertness. No evidence of reactivity was observed between the oil and boiling sulfuric acid, fluorine gas at 200°C, chlorine trifluoride at 10–50°C, uranium hexafluoride gas at 50°C, molten sodium hydroxide, or any of the following materials at room temperature: ethyl alcohol, JP-4 turbine fuel, hydrazine, unsymmetrical dimethyl hydrazine, diethylenetriamine, aniline, 90% hydrogen peroxide, inhibited red fuming nitric acid, and nitrogen tetroxide. The PFPE-K oils are insoluble in these chemicals with the exception of a slight solubility in hydrazine and a moderate (25–30%) solubility in nitrogen tetroxide. PFPE-K does not react with nitrogen tetroxide, or inhibited red fuming nitric acid in impact tests.

Again, there may be stability differences depending on the type and structure of the PFPE. PFPE-Y is reported to react with liquid (100%) and gaseous ammonia, alkaline

Table 6.4 Metals and Alloys Suitable for Use with PFPE-K Oils at Elevated Temperatures[a]

	Air	Inert gas
371°C (700°F)	Nickel alloys	Nickel alloys
	Cobalt alloys	Cobalt alloys
	AMS-5547 steel	AMS-5547 steel
343°C(650°F)	AMS-5525 steel	304, 446 stainless steel
	Ti (6Al, 6V, 2Sn)	
	Mg, Ag, Cr, V	
316°C(600°F)	301, 304, 310, 316, 321, 446	440C
	N-155	High-speed stainless steel
	Ti (13V, 11Cr, 3Al)	
	Ti (6Al, 4V)	QQ-S-636, M-1, M-50, WB-49, 52100 steels
	Al (QQ-A-355)	
	Bearing bronze	Ti(8Mn), Cu
288°C (550°F)	405, 410, 440 Stainless steels	Ti (7 Al, 4 Mo)
	High-speed stainless steel (14.5Cr, 4.OMo, 1. 2V, 1. 15C, balance Fe)	
	QQS-636, M-1, M-50, WB-49, 52100 steels	
	Ti (8 Mn), Cu	
Below 288°C	Most metals and alloys show little or no evidence of corrosion	

[a]Test conditions: 72 h at 5 liters dry gas/hr, qualifying corrosion rate <0.4 mg/cm·day. Any metal or alloy suitable at one temperature is also suitable at all lower temperatures. Any metal suitable in air is also suitable in inert atmosphere. Source: DuPont literature.

metals, and finely divided powder of light metals such as aluminum, magnesium, and their alloys (28). The stability of PFPE-Y toward reactive chemicals is shown in Table 6.5. PFPE-K has shown no reactivity toward ammonia (personal observation). Although there are no available data on the reactivity of PFPE-K with alkaline metals, all PFPE molecules will probably react with nascent metal surfaces and unpassivated, high-surface-area metals.

Table 6.5 Chemical Resistance of PFPE-Y Oils[a]

Chemical type	Chemical name	Temperature, °C
Organic solvents	All common	350
Organic acid	Carboxylic acids	350
Organic base	Tributyl amine	200
Inorganic acid	HF, H_2SO_4, HCl	300
Inorganic base	KOH, NaOH,	300
Inorganic salt	NaCl, KF	350
Oxidizing agent	Br_2, Cl_2, F_2	250
Oxidizing agent	$KMnO_4$, K_2CrO_4	220
Lewis acid	$AlCl_3$	160
Lewis acid	SbF_5	100

[a]Source: Montedison literature.

E. Solubility

PFPE oils are insoluble in most common solvents, acids, and bases. However some solvents will dissolve in PFPE oils. Solubility data for PFPE-Y is shown in Table 6.6. PFPE oils are completely miscible in highly fluorinated solvents such as trichlorotrifluoroethane, hexafluorobenzene, perfluorooctane, hexafluoropropylene dimer, and some, but not all, of the new HCFC fluorocarbons currently being considered for commercialization.

The solubilities of gaseous oxygen and nitrogen in PFPE-K at 100°C are 192 and 114 ppm, respectively. Table 6.7 summarizes available data for other gases.

F. Compatibility with Elastomers

Elastomer compatibility with PFPE-K oil has been evaluated by immersion for 168 h at various temperatures (29). Results, summarized in Table 6.8, show that many elastomers are affected only slightly by contact with the oil at 93°C. At higher temperatures the inherent deficiencies in the elastomers themselves and not the compatibility with PFPE determines the maximum use temperature. At 204°C only Viton A fluoroelastomer and Teflon® TFE fluorocarbon resin, of those tested, are relatively unchanged. Other data obtained at 260°C show only triazine elastomer to have promise. In general, PFPE oils do not appreciably change the volume of elastomers at 93 or 149°C.

In a separate series of tests (30), one investigator immersed a number of elastomers in PFPE oil at 70°C for 6 months. Elastomers used were neoprene, butyl, Buna N and natural rubber, polyfluorosiloxane, polychlorotrifluoroethylene, and vinylidene fluoride-hexafluoropropylene copolymers. No significant changes in dimension, hardness, or color were noted.

Table 6.6 Solubility of Common Solvents and PFPE Oils[a]

Solvent	Solubility of PFPE in solvent	Solubility of solvent in PFPE	Solvent, g in 100 g PFPE
Hexane	SS	PS	2.0
n-Heptane	PS	SS	1.6
n-Octane	I	I	—
Cyclohexane	I	I	—
Benzene	I	SS	1.3
Toluene	I	I	—
Xylene	I	I	—
Ethyl ether	SS	PS	2.6
Tetrahydrofuran	I	SS	1.4
Ethyl acetate	I	I	1.5
Methyl formate	I	I	—
Dimethyl ketone	I	I	—
Methyl ethyl ketone	I	I	—
Trichloroethylene	I	PS	3.0
Monochlorobenzene	I	I	—
Chloroform	SS	PS	4.1
Carbon tetrachloride	SS	PS	2.9
Perfluoroalkanes	Miscible	Miscible	—

[a]Source: Montedison literature.

Table 6.7 Gas Solubility in PFPE at 20°C[a]

Type of gas	ml Gas/ml PFPE-Y	ppm Gas
Helium	0.08	8
Hydrogen	0.09	4
Oxygen	0.290	218
Nitrogen	0.190	125
Carbon dioxide	1.30	1,344
Chlorine	3.189	5,320
Fluorine	0.197	176
Hydrogen fluoride	0.806	379
Water Vapor	0.04[b]	17

[a]Source: Montedison literature.
[b]Estimated from water solubility.

G. Compatibility with Plastics

PFPE oils do not have any significant effect on plastics. The following were treated with PFPE-Y for 1,000 h at 70°C: phenylene-oxide based resins, polyethylene terephthalate, polybutylene terephthalate, polystyrene, polystyrene impact-resistant, polyethylene low density, polyethylene high density, polypropylene, acrylonitrile-styrene copolymer, polymethylmethacrylate, acrylonitrile-butadiene-styrene copolymer, polyamide 66, polyvinyl chloride, acetal copolymer, and polycarbonate. There were no significant changes.

Table 6.8 Elastomer Compatibility with PFPE-K Oil[a]

Elastomer	93°C	149°C	204°C
Fluorosilicone	S	S	NG
Ethyl acrylate	S	S	NG
Methyl silicone	S	S	NG
Viton A fluoroelastomer	S	S	S
Urethane	S	–	–
Hypalon synthetic rubber	S	NG	–
Butyl 325	S	NG	–
VPA	S	NG	–
Natural rubber	NG	–	–
Neoprene, WRT	S	NG	–
Hycar 100 (Buna N)	S	–	–
Polysulfide (Thiokol)	S	–	NG
cis-1,4-Polybutadiene	NG	–	–
Synpol 1013, 5BR	NG	–	–
EPT, sulfur cure	S	–	NG
EPT, peroxide cure	S	S	NG
Polypropyleneoxide. XP-139	–	NG	–
Cyanacryl, acrylic rubber	–	S	NG
Teflon	S	–	S

[a]Based on 168-h tests at temperatures indicated. Code: S = only minor changes in elastomer properties, NG = major, generally harmful, effect on elastomer properties. Source: Personal data and DuPont literature.

H. Flammability

PFPE-K oils are nonflammable under practically all conditions likely to be encountered. They show no autogenous ignition, and flash or fire points up to 649°C (1,200°F) in standard ASTM tests. A sample of PFPE-K did not flash or burn when allowed to contact a manifold at temperatures in excess of 649°C (1,200°F). In a high-pressure spray ignition test (MIL-F-7100), PFPE-K oil did not flash or fire at up to 2 f from the spray orifice.

I. Radiation Resistance

PFPE-K oils are quite stable to radiation when compared with many materials used as lubricants or power fluids. In general, irradiation of PFPE-K oils causes minor depolymerization with a consequent reduction in viscosity and the formation of volatile products but no insoluble solids or sludge. In one test, exposure of a sample of PFPE-K to 108 rads of electron bombardment at ambient temperatures in air resulted in a decrease in viscosity of 21%. The irradiated sample contained no sludge and was unchanged in appearance (personal observation).

J. X-Ray Stability

Mori and Morales (31) studied the degradation of three types of commercially available perfluoroalkyl polyethers, PFPE-D, PFPE-Z, and PFPE-K, by x-ray irradiation. The products formed were COF_2 and low-molecular-weight fluorocarbons. PFPE-Z, which has difluoroformyl linkages (-OCF_2-), produced a large quantity of COF_2 gas. Liquids became more tacky and the molecular weight distribution became broader and higher-molecular-weight polymers were formed. They concluded from these results that degradation and cross-linking took place simultaneously. PFPE-D cross-linked more easily than the other fluids. There was no substrate effect on the degradation reaction. The rates of degradation were similar when stainless steel (440C) and gold-coated surfaces were used. Metal fluorides were formed on stainless steel during the reaction.

K. Shear Stability

Although PFPE-K oils are polymeric in nature, they do not break down when subjected to high rates of shear. Irradiation at 10 kHz in a sonic shear tester for 1 h resulted in viscosity changes of less than 0.5% (personal observation).

V. PHYSICAL PROPERTIES

Physical properties that show little change with molecular weight are summarized in Table 6.9.

A. Heat-Transfer Properties

The thermal conductivity of PFPE-K varies only slightly over a wide temperature range as shown by the data in Table 6.9. These values are somewhat lower than those of many other hydrocarbon lubricants with similar viscosity.

B. Specific Heat

The specific heat of PFPE-K with number average molecular weight of 6,000 is a linear function of temperature as shown in Fig. 6.3.

Table 6.9 Typical Physical Properties of PFPE Oils[a]

	PFPE-K	PFPE-Y	PFPE-Z	PFPE-D
Density, g/ml	1.86–1.91 at 24°C	1.86–1.91 at 20°C	1.82–1.85 at 20°C	1.86–1.89 at 20°C
Refractive index	1.29–1.301 nD24	1.29–1.304 nD20	1.29–1.294 nD20	1.29–1.298 nD20
Surface tension	16–20 mN/m	18–21 mN/m	23–24 mN/m	17.7–19.1 mN/m
Isothermal secant bulk modulus	1,034 MPa/150,000 PSI at 38°C, 34.5 MPa (5,000 psi)	9,650 kg/cm^2 at 25°C, 100 kg/cm^2		
Coefficient of thermal expansion, /°C	0.00095–0.001	0.00092–0.00109		
Specific heat, cal/g·C, 38°C	0.23–0.24	0.24	0.20–0.23	
Thermal conductivity	0.0831–0.0934 W/m·K at 38°C 0.0692–0.0883 W/m·K at 260°C			

[a]Source: Individual manufacturer's literature.

C. Thermal Coefficient of Expansion

Information on the thermal coefficient of expansion for two PFPE-K oils is given in Fig. 6.4 over the temperature range from 40 to 200°C. The thermal coefficient of expansion and its rate of change with temperature decrease with increasing molecular weight.

D. Density

The densities of PFPE-K oil as a function of temperature are shown in Fig. 6.5. The densities of PFPE fluids are nearly twice that of hydrocarbon lubricants. At a given temperature, density increases slightly with increasing molecular weight.

E. Electrical Properties

Electrical properties are significantly affected by the presence of trace amounts of moisture from exposure to a humid atmosphere or other sources of water contamination. The data in Table 6.10 were obtained on routinely produced oil with no extraordinary drying techniques employed.

F. Volatility

The volatility of PFPE varies inversely with number average molecular weight, and thus the higher-viscosity grades have lower volatility losses. Volatility losses, as measured by ASTM D-972 Modified, are shown in Table 6.11. This test employs a steady stream of hot air blown across the surface of the hot oil and is useful in determining the light-end

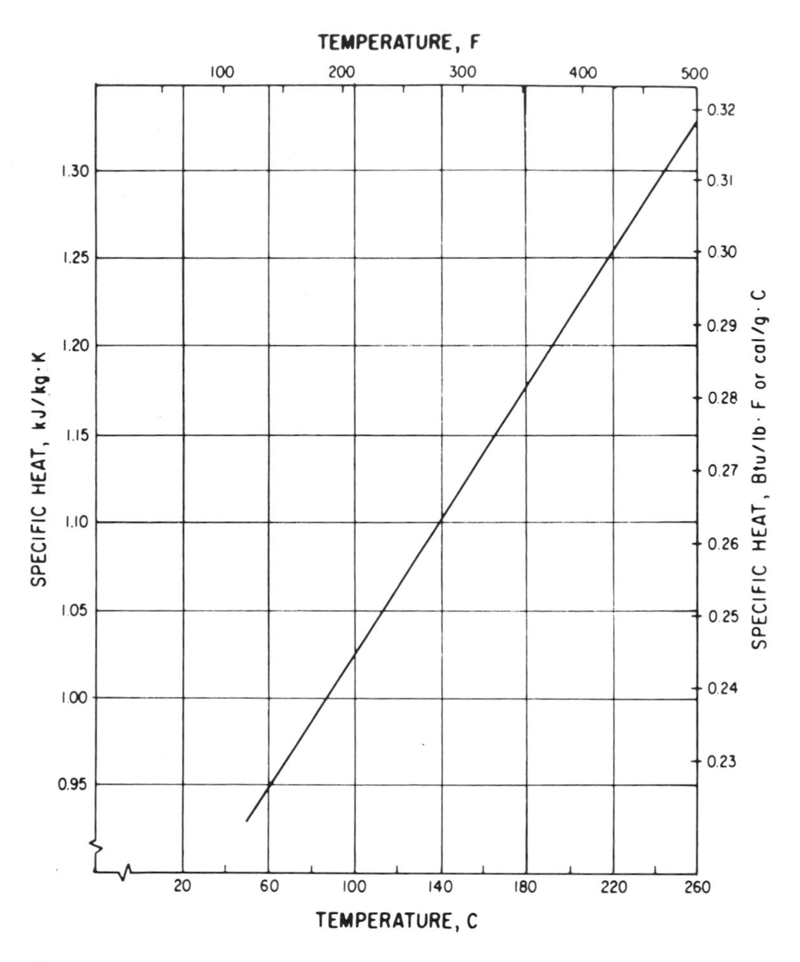

Figure 6.3 Specific heat of PFPE-K.

content of the particular oil. The effect of molecular weight on the vapor pressure of homologues (PFPE-K) is shown in Fig. 6.6.

G. Viscosity

Viscosity is one of the most useful properties of the PFPE oils. Manufacturers differentiate their product grades by viscosity. Lubricating oils are chosen for their viscosity and viscosity index. PFPE viscosities and their change with temperature for each type of PFPE are shown in Fig. 6.7. The absolute or centipoise (mPa·s) viscosities will be considerably higher than the centistoke values because the density of PFPE oils is about 1.9 g/ml. The temperature dependence of viscosity is similar to that of high-grade petroleum oils and is much superior to that of conventional fluorocarbons.

PFPE oil viscosity increases with an increase in the molecular weight. Commercial PFPE polymers are fractionated by distillation to produce a series of grades, based on viscosity. The viscosity is a function of average molecular weight, and the same viscosity can be obtained from a very narrow or a very broad molecular weight distribution.

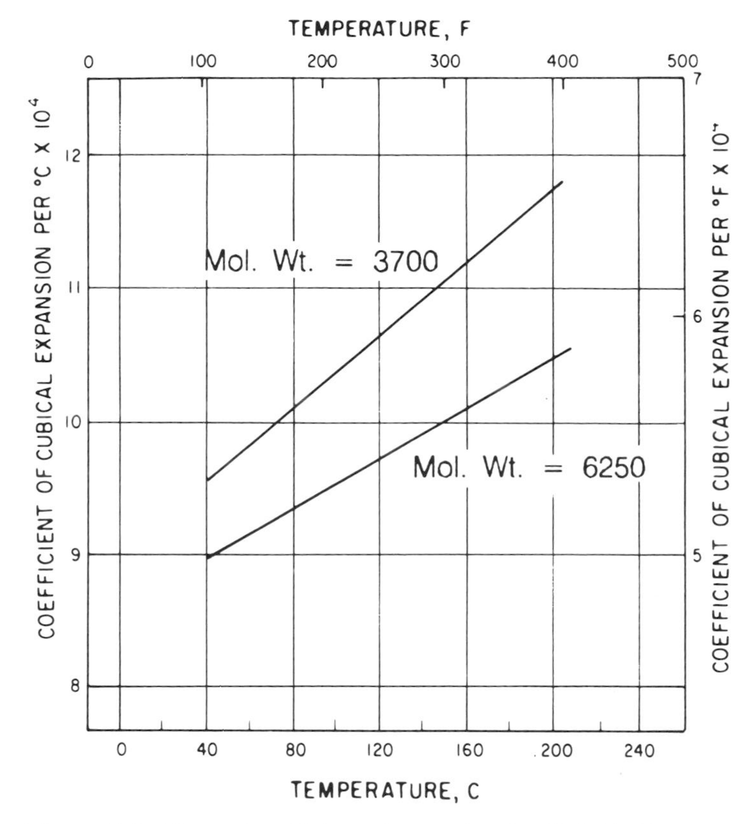

Figure 6.4 Thermal coefficient of expansion for PFPE-K oils.

Viscosity data and its change with temperature for seven different molecular weight fractions of the PFPE-K oils are shown in Fig. 6.8

H. Compressibility

The compressibility of a fluid is an important consideration in certain uses such as hydraulic systems. A useful measure of compressibility relating to the performance of fluid systems is the fluid bulk modulus, the reciprocal of compressibility. Fluid bulk modulus is determined by measuring the velocity of sound through a fluid under the desired conditions of temperature and pressure. Adiabatic tangent bulk modulus data (obtained by sonic methods) for PFPE-K are given in Fig. 6.9 for pressures up to 34.5 MPa (5,000 psig) and temperatures to 204°C (400°F). PFPE oils are considerably more compressible than conventional hydraulic fluids and thus show lower bulk modulus test results. Figure 6.10 compares the compressibility of PFPE-K fluorinated oil with that of a typical hydrocarbon-based hydraulic fluid. At 38°C (100°F) and an applied pressure of 34.5 MPa (5,000 psi), the volume of the hydrocarbon oil is reduced by about 2% while PFPE-K is compressed almost 3½%.

Cantow and Barrall (32) examined the dependence of the viscosities of PFPE fluids on temperature and pressure and found that differences in viscosities could be attributed to structural differences. The bulky CF_3 groups are responsible for the high pressure sensitivity of the nonlinear PFPE.

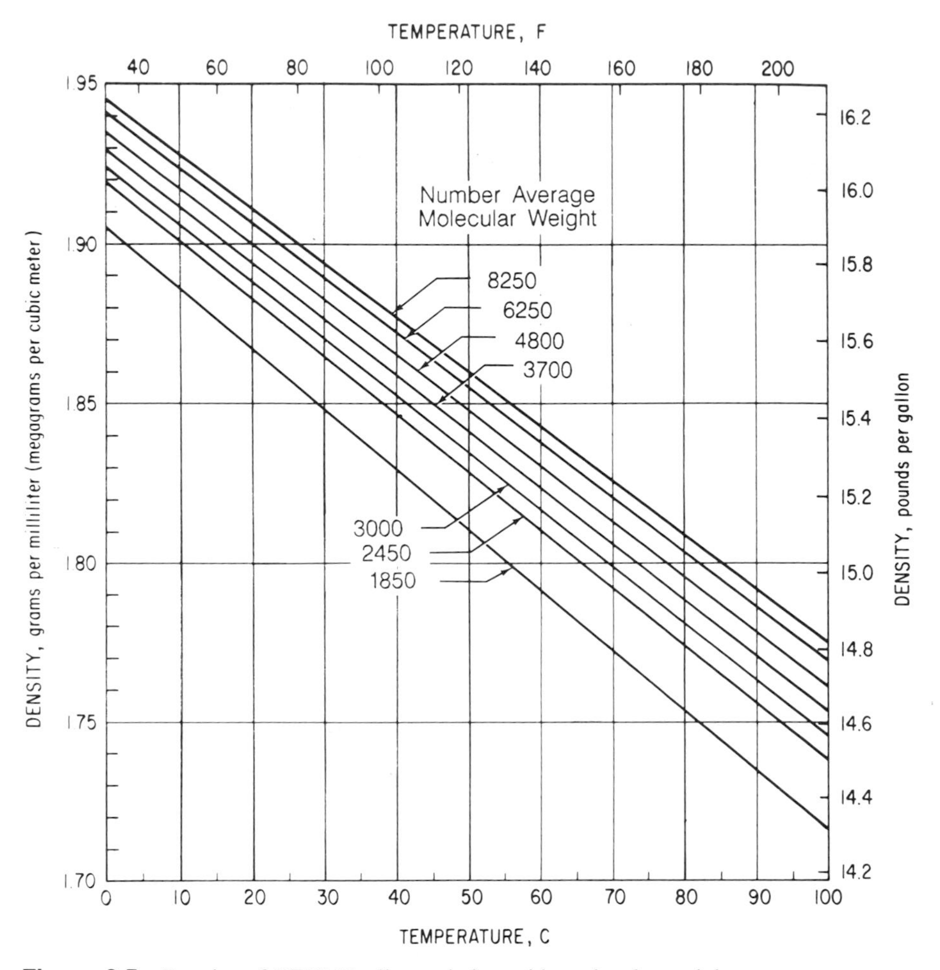

Figure 6.5 Density of PFPE-K oils, variation with molecular weight.

I. Vapor Pressure

Vapor pressure of PFPE fluids is inversely proportional to molecular weight. The vapor pressures of the individual homologues of PFPE-K are shown in Fig. 6.11.

The vapor pressure of a PFPE fluid can be very sensitive to the presence of small amounts of impurities, such as very-low-molecular-weight fractions or solvents. The

Table 6.10 Electrical Properties of PFPE Oils at 25°C[a]

Property	PFPE-K	PFPE-Y
Dielectric breakdown voltage, ASTM D-877, kV/0.1 in	38.8–41.1	40 kV/100 MIL
Specific resistivity, ASTM D-257,ohm–cm	6.4×10^{13}–4.1×10^{14}	1015
Dielectric constant, ASTM D-150 at 102–105 Hz	2.10–2.15	No data
Dissipation Factor, ASTM D-150, % at 102–105 Hz	<0.003–<0.007	No data

[a]Source: Manufacturer's literature.

Table 6.11 Typical Volatility of PFPE-K Oils (ASTM D-972)

	Wt% loss in 6½-h test at				Wt% loss in 22-h test at		
MW	99°C	149°C	204°C	260°C	149°C	204°C	260°C
1,850	<1	10	60	—	20	80	—
2,450	<1	5	30	—	5	40	—
3,000	<1	1	10	45	3	20	60
3,700	<O.I	1	5	30	1	5	40
4,800	<O.I	<1	1	15	—	2	20
6,250	<O.I	<1	1	4	—	1	6
8,250	<O.I	<1	<1	2	—	—	3

Source: DuPont data

Isoteniscope method is very sensitive to the presence of small amounts of high-vapor-pressure impurities whereas the Knudsen method is not. Comparing the vapor pressure obtained by each method indicates the level of these impurities. The Knudsen method is normally reported in the literature and manufacturer's information bulletins.

The vapor pressure and viscosity of specific homologues of PFPE will be dependent on its molecular weight. Commercial PFPE products may appear to contradict this correlation because different grades with the same viscosity have different vapor pressures. Commercial PFPE oils are composed of distributions of different molecular weight polymers, and the fraction of lower-molecular-weight polymer will determine the vapor pressure of that grade. Figure 6.12 shows three series of PFPE-K oils with different molecular weight distributions.

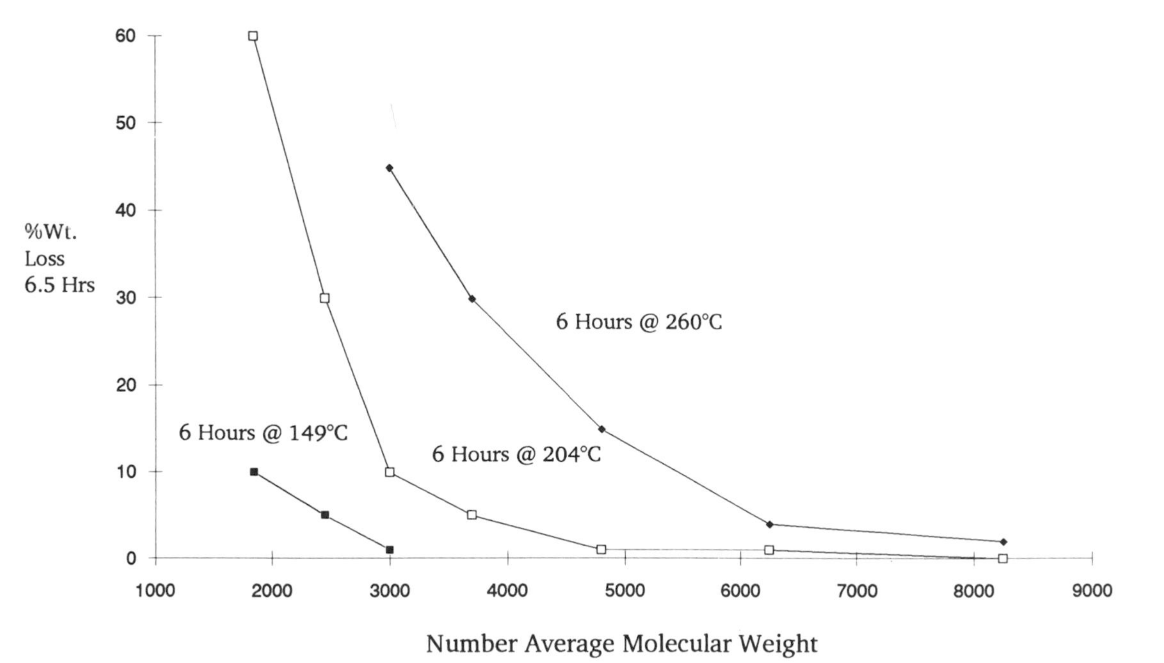

Figure 6.6 Volatility of PFPE-K vs. molecular weight, ASTM D-972.

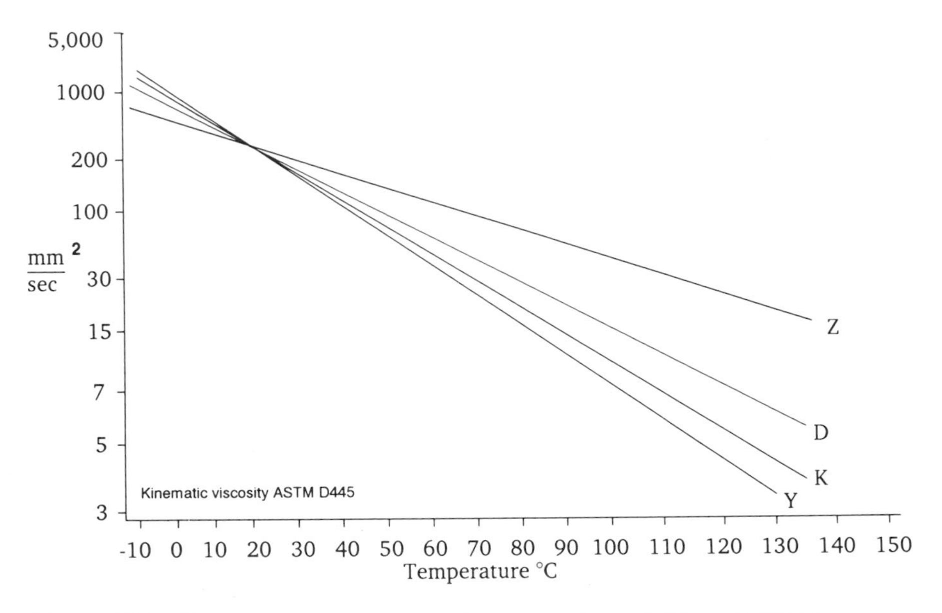

Figure 6.7 Viscosity-temperature properties: comparison of different PFPE types.

VI. LUBRICATION

PFPE oils are excellent lubricants under normal, severe, and starved operating conditions. PFPE oils have excellent lubricating power under heavy loads, at high speeds, and at elevated temperatures. In general, when failure does occur, it does happens over time, giving the operator warning of a failure.

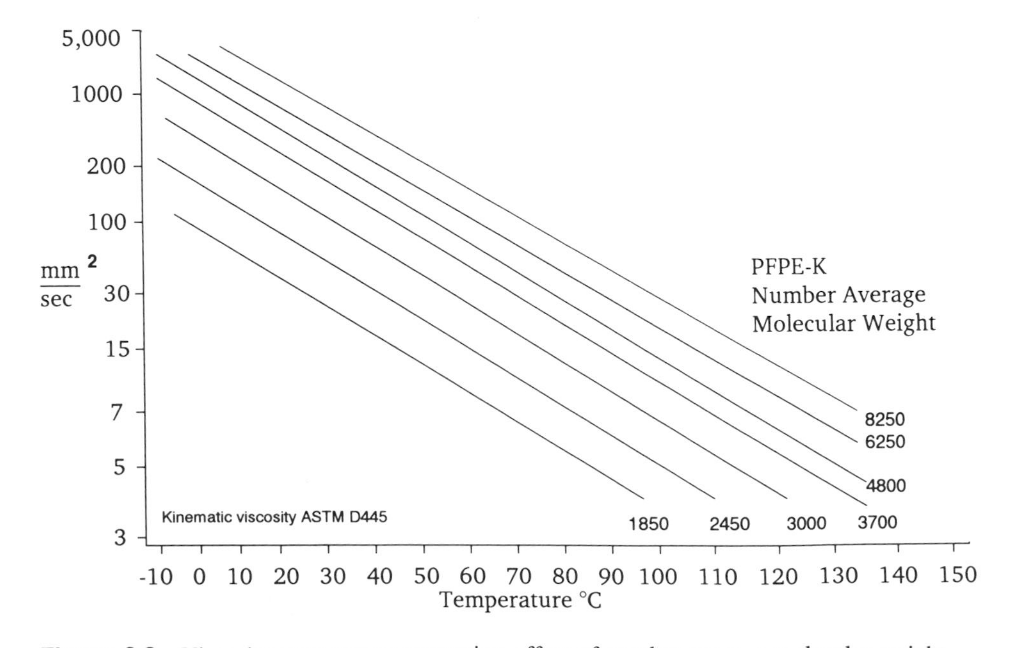

Figure 6.8 Viscosity-temperature properties: effect of number average molecular weight.

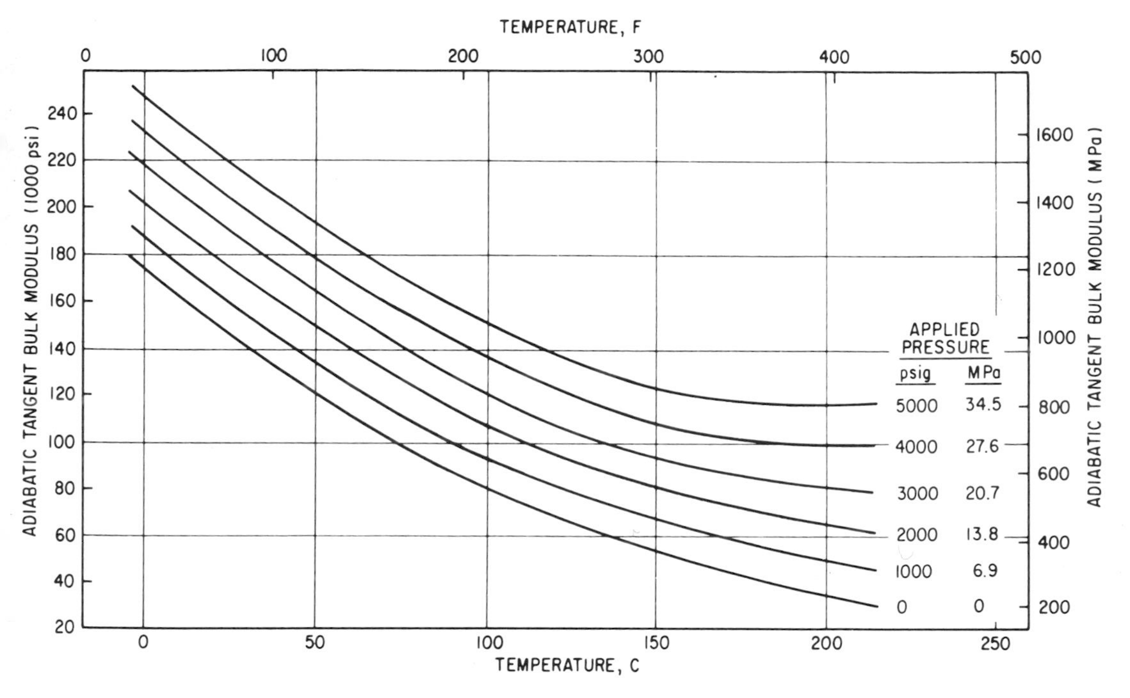

Figure 6.9 Adiabatic tangent bulk modulus of PFPE-K fluorinated oil.

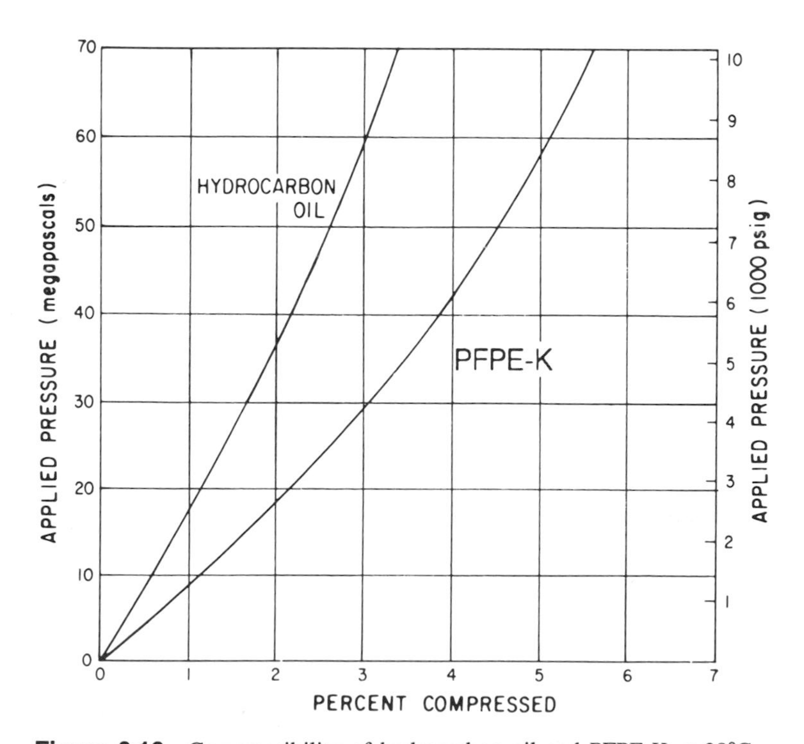

Figure 6.10 Compressibility of hydrocarbon oil and PFPE-K at 38°C.

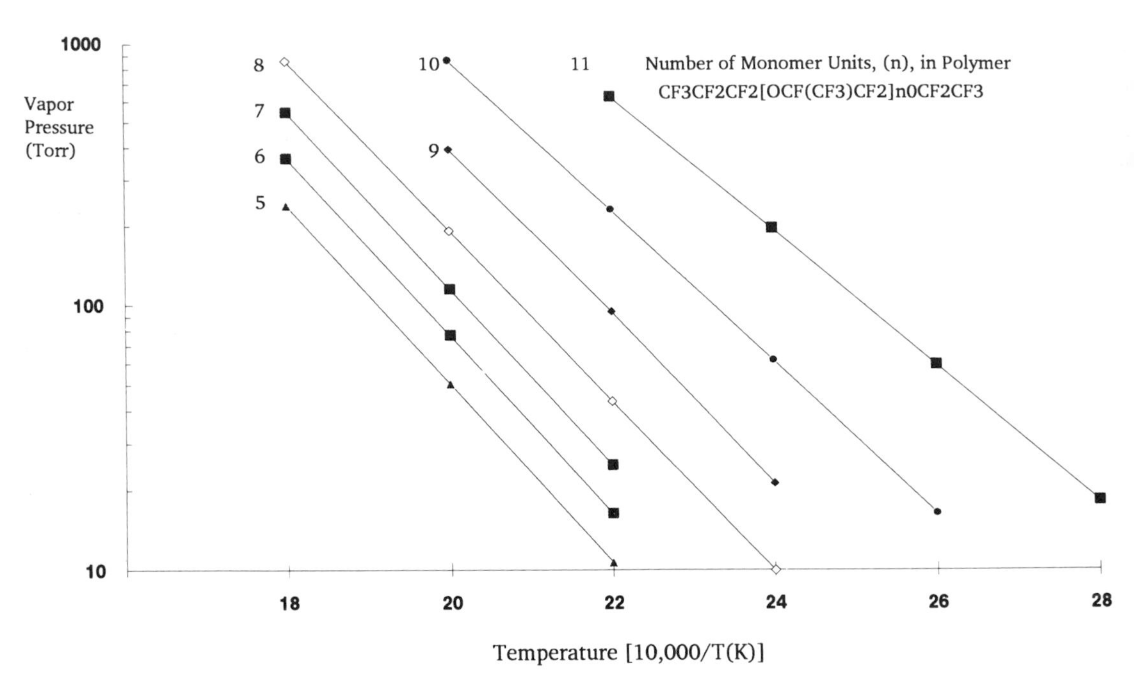

Figure 6.11 Vapor pressure, temperature properties: effect of molecular weight.

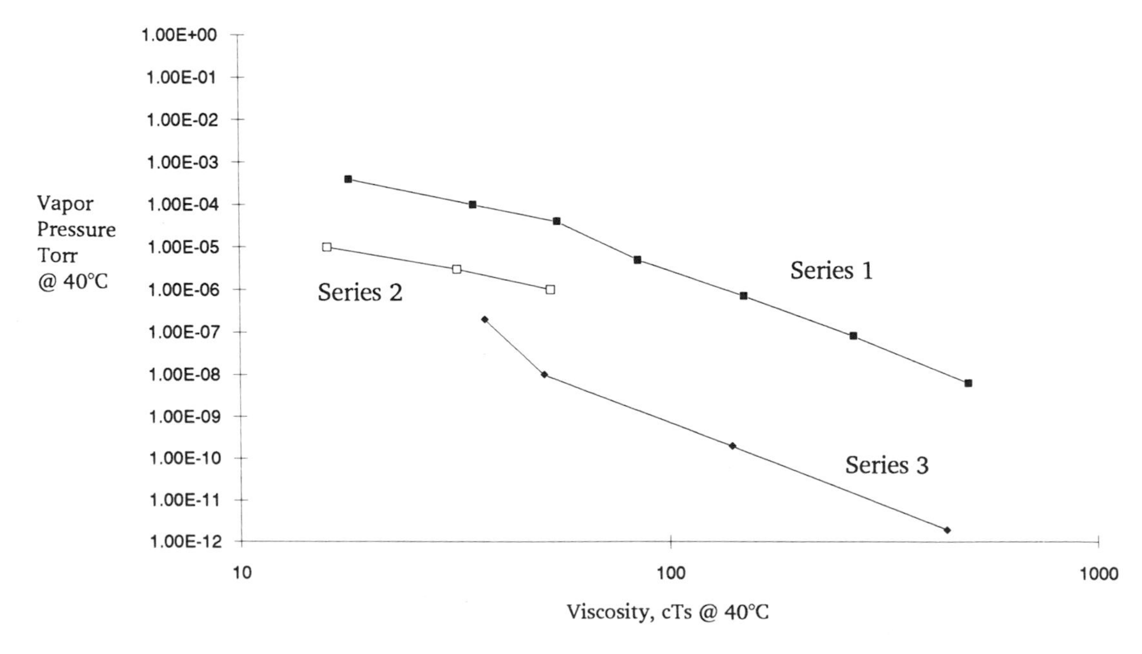

Figure 6.12 Effect of molecular weight distribution on vapor pressure.

The test data in the following lubricity tests were obtained on PFPE-K oil which contained no additives.

A. Four-Ball Wear Tests

Four-ball wear tests were conducted on PFPE-K at 75 and 316°C and at loads of 1–40 kg using balls made of different alloys. Typical results at 620 rpm and 1,280 rpm are given in Fig. 6.13. In general, the wear scars increase in diameter with increasing temperature, load, and speed, as expected. In most cases, lower wear occurs with type M-10 steel than with SAE 52100 steel. Wear scars produced with PFPE oils show less surface roughness than other oils.

B. Four-Ball Extreme Pressure Tests

The load-carrying ability of PFPE oils has been compared with that of various other oils in the Shell four-ball extreme pressure tester. As can be seen from the results in Table 6.12, the load-carrying ability of these oils increases with viscosity. The values of PFPE-K oils compare favorably with those for a diester base stock that contains no additives, with two finished synthetic hydraulic oils, and with a petroleum-based extreme pressure type gear oil.

C. Falex Extreme Pressure Tests

Extreme pressure properties of PFPE oil have also been obtained using the Falex pin & V-block tester. The results, shown in Table 6.13, indicate outstanding properties. Using

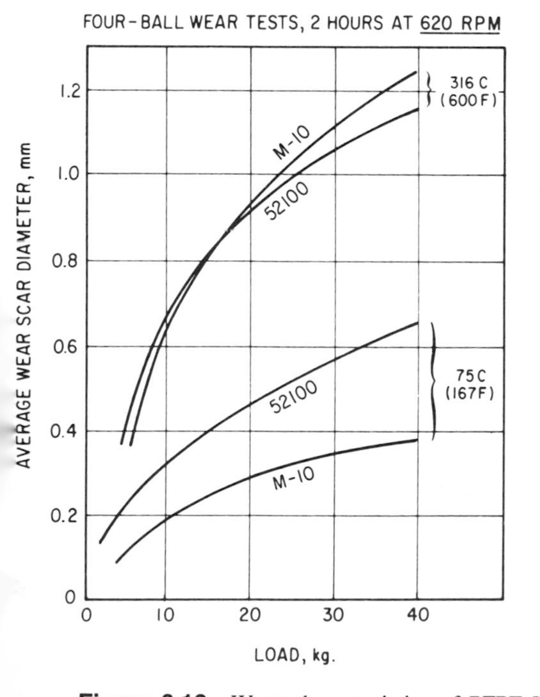

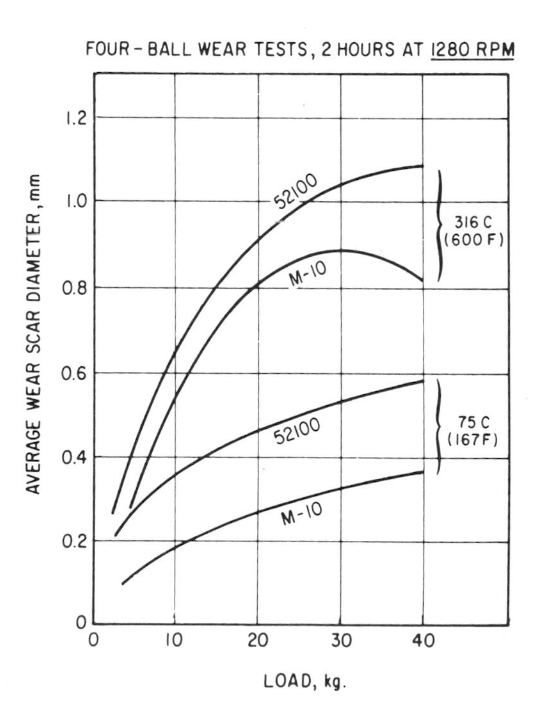

Figure 6.13 Wear characteristics of PFPE-K oil.

Table 6.12 Extreme Pressure Lubricating Characteristics of PFPE-K Oils, Four-Ball Test[a]

Oil	Viscosity, 38°C	Mean hertz load, kg	Incipient seizure, kg	Weld point, kg
PFPE-MW=1,850	18.1	45.9	141	200
PFPE-MW=3,700	87	62.3	200	224
PFPE-MW=6,250	280	98.0	200	398
Di-2-ethylhexyl sebacate	12.3	15.1	40	89
Silicate ester hydraulic fluid	33	20.9	63	100
Phosphate ester hydraulic fluid	11.5	25.3	79	112
EP petroleum gear oil	35.3	64.4	89	398

[a]Four-Ball Load Carrying Capacity Test (mean hertz load) by FTMS-791 method 6503.1. Source: DuPont data.

shafts and V-blocks made of 52100 or M-10 steels, the load-carrying properties of PFPE oil exceeded the limits of the test equipment. Using the standard shaft and V-block metals, the load at failure for PFPE oil far exceeded that of the other test oils.

D. Bearing Fatigue Tests

Fatigue tests in a rolling contact bearing rig showed PFPE oil to have excellent fatigue life at room temperature, 218°C, 260°C, and 316°C. The test pieces were made from M-50 steel; the stress was 4,826 MPa (700,000 psi) maximum Hz and the speed was 25,000 stress cycles per minute. Oil was fed at the rate of 20 drops per minute. Sufficient tests were run to permit calculation of B-10 and B-50 life stress cycles. Test results are summarized in Table 6.14.

Ball bearing performance tests conducted at high temperatures on the PTFE-K, PFPE greases using FTMS-791, method 333, show that PTFE greases give excellent perfor-

Table 6.13 Extreme Pressure Characteristics of PFPE-K Oil, Pin and V-Block Test[a]

Oil	Specimen material		Maximum normal conditions		Failure load (kg)
	Shaft	V-Blocks	Load, kg[b]	Torque, J	
PFPE-K MW = 6,250	SAE 3135	AISI 1137	1,814	5.88	1,928[c]
PFPE-K MW = 6,250	52100	52100	2,041	5.42	No failure
PFPE-K MW = 6,250	M-10	M-10	2,041	4.29	No failure
Deep dewaxed mineral oil	SAE 3135	AISI 1137	454	2.83	567[d]

[a]Using the Falex pin and V-block tester; load increased by 250-lb increments every minute. Source: DuPont data.
[b]2,041 kg Capacity gauge.
[c]Load dropped.
[d]Shaft fractured.

Table 6.14 Roller Contact Fatigue Tests[a]

Temperature °C (°F)	Bearing life	Cycles PFPE-K	MIL-L7808 oil
32 (90)	B10	3.3×10^6	2.65×10^6
32 (90)	B50	10.1×10^6	6.5×10^6
218 (425)	B10	7.2×10^6	2.8×10^6
218 (425)	B50	13.5×10^6	7.0×10^6
316 (600)	B10	8.4×10^6	—
316 (600)	B50	18.1×10^6	—

[a]Minimum 25,000 stress cycles, maximum 700,000 psi (4,826) MPa) Hertz stress, air. Test Specimens: M-50 steel bars and rollers, 65,000 psi (448 MPa) maximum Hertz load. Source: DuPont data.

mance. In this test the normal speed is 10,000 rpm and the bearing load is 3 lb radial and 5 lb thrust. The results of such tests are reported in hours to failure. The grease passes the test if there is no excessive bearing noise or screech, excessive running torque as evidenced by a large increase in spindle input power, or increase of more than 20°F in bearing temperature over the programmed bearing test temperature.

The results in Table 6.15 indicate that grease A performs satisfactorily at temperatures up to 260°C and that grease B is satisfactory to at least 288°C.

E. PFPE Grease

The PFPE oils can be used directly as lubricants, or they can be used as the base oils for making greases. The principal advantage of the greases over the oils is greases will "stay put" on surfaces where the oils might tend to run off or drain away rapidly.

A large number of solids are suitable thickening agents for the formation of greases from the PFPE oils. Many of the thickening agents used for PFPE greases are the same as those used for hydrocarbon greases. Hydrocarbon soaps are not used because their thermal and chemical stability is poor. Thickening agents commonly used are finely divided silica, "Attapulgus clay," montmorillonite, ammeline, boron nitride, copper phthalocyanine,

Table 6.15 Performance of PFPE-K Grease in Ball Bearings[a]

PFPE grease	Temperature, °C	Speed, rpm	Hours to failure
A	204	20,000	> 500
B	204	20,000	> 500
A	260	10,000	> 500
B	260	10,000	> 2,000
B	288	10,000	> 500

[a]Source: DuPont data.

metal-free phthalocyanine, polytetrafluoroethylene polymers and telomers, fluorinated ethylene-propylene copolymers, and zinc oxide.

Some thickeners or fillers have been described as imparting thermal conductivity to the grease. For instance, a grease made with PFPE-Y and boron nitride has a coefficient of thermal conductivity of 0.68 and also possess good heat and acid resistance (33). Some of the PFPE greases thickened with fluorinated ethylene-propylene copolymer or PTFE contain antioxidants (34) such as tris(fluoroalkoxyphenyl)phosphine or corrosion inhibitors such as benzimidazole derivatives (35), perfluoroalkyl- and perfluoroalkyl ether-substituted benzoxazoles, benzothiazoles, bis-benzoxazoles, and bis-benzothiazoles (36). Some PFPE greases are prepared by polymerizing tetrafluoroethylene in the perfluoroalkyl polyether (37).

The thickener used in the greases described herein, and the one chosen for use in greases made from the PFPE oils for commercial development, is a telomer of tetrafluoroethylene (PTFE) having an approximate molecular weight of 20,000 to 30,000. Typical properties of the telomer are softening point, 323°C; density, 2.1 g/ml; particle size, 100% less than 30 μm.

VII. APPLICATIONS

The chemical resistance of the PFPE lubricants (oils and greases) has led to their widespread use in many applications in the space programs where complete inertness to the fuels and oxidants is vital. Thus, both the oils and greases have been used to lubricate O-rings in many applications on the Apollo and other space projects. In general, a lighter oil is used if the O-rings are expected to operate after assembly. If the O-ring is static and lubricant is required solely to assist in assembly, then a grease made from one of the heavier oils is preferred because of its lower volatility. Both oils and greases have been used as lubricants and antiseize compounds in the assembly of threaded fittings on spacecraft.

PFPE oil is used on the fasteners and moving parts of astronauts' pressure suits, largely because of its nonflammability. This same oil has been used successfully on the bearings of antenna arrays on space craft. This oil has also proved satisfactory in lubricating bearings that permit extension of the paddle arms supporting solar cells on other space craft. PFPE is also used as the lubricant for the slide wire of potentiometers installed in space craft. The oil minimizes wear and does not migrate to other parts of the system.

In nuclear power plants, PFPE was used as a bearing lubricant for pumps operating in an area where it was exposed to radiation. In this case, the radiation rapidly decomposed ordinary petroleum oils but the PFPE oil operated for long periods of time without appreciable decomposition. In one instance, after 900 h of operation the viscosity had changed only 2% and no sludge or gum had formed.

The PFPE oils, because of their excellent stability to oxidation and the absence of gum formation, are used as lubricants for the bearings, jewels, and pivots of many different kinds of instruments. The shelf life for PFPE oils is nearly indefinite.

In an office machine a lightly loaded size R-6 bearing was placed near an electrical heating element and often reached temperatures greater than 260°C. This bearing, operating at 500 rpm, failed in less than 500 h with all MIL-G-25013 (dye-thickened, silicone-based) greases that were tested. This operating life was considered inadequate for the purpose. When PFPE grease was used as the lubricant, operating times in excess of 5,000 h were obtained between failures.

Premature failure of the bearings of a synchronous electrical motor in a plant that produces synthetic textile fibers occurred when a dye-thickened silicone-based grease was used. The size 210 shielded cartridge-width ball bearings operating at 3,200 rpm at a temperature of about 375°F failed in 2–6 months, for an average life of 2½ months. When PFPE grease was substituted for the silicone grease, bearing life increased to an average exceeding 1 year.

In a machine designed to orient polypropylene film at a temperature of 150°C, the film is held in position by stud-mounted, sealed roller bearings of ½ in bore. When these bearings were lubricated with a high-quality diester-type grease, the incidence of bearing failure reached an intolerable level, and production downtime became excessive. Introduction of a dye-thickened silicone grease improved but did not solve the problem. PFPE grease was then adopted with appreciable improvement in performance, and it has been in use at this plant for many years.

The sealed rolling element bearings on a track and chain conveyor in another plant, one that manufactures polyester film, operate in an oven at 260°C. The use of a carbon black-thickened silicone grease led to many premature bearing failures due to the tendency of the grease to run out of the bearings past the seal. This problem was solved by the use of PFPE grease, which has performed satisfactorily at this plant for many years.

A battery of 7½-hp electrical motors is used to drive centrifugal-type blowers circulating hot air in a chemical drying process unit. The ambient air temperature in the vicinity of the motor bearings is about 250°C. The bearing nearest the blower is a size 310 shielded ball bearing and the one on the opposite side is a 35 MM bore roller bearing. When these bearings were lubricated with a premium quality high temperature bearing grease, bearing failures reduced motor life expectancy to less than 12 weeks. Bearing failure appeared to be due to lubrication starvation caused by the grease melting and running out of the bearing and by grease deterioration products forming within the bearing. When PFPE grease was adopted, average motor life increased to over 36 weeks, and the failures were increasingly due to parts other than the grease.

Table 6.16 gives many applications by type, and the special properties of the PFPE oil that make the oil or grease especially suited for the application.

A. Economics

There are four companies that actually make all the PFPE products in the world: Du Pont in the United States, Montedison in Italy, Daiken in Japan, and NOK in Japan. The total production of PFPE products is very difficult to determine, but it should be more than 500 tons/year.

Information about the fraction that each company produces is not readily available. Within the last 10 years, Daikin reported a 60 metric tons per year fluorine oil plant and Asahi Glass reported a 50 tons per year fluorinated oil plant.

The notion that these lubricants are superexpensive and can only be used in very special situations is false. The lowest selling price is under US$100/lb. The highest is >US$3000/lb. PFPE oils and greases can be used for general purpose lubrication, especially in the presence of severely harsh and demanding environments.

Table 6.16 Applications for PFPE Lubricants

Application	Special properties
Seal lubricant in reactive chemical environments	Chemical inertness, high-temperature stability, insolubility
Lubricant for aircraft instrument bearings	Low vapor pressure, biostable fluid
Valve and O-ring lubricant in oxygen service	Stable to oxygen atmospheres
Base oil for specialty lubricants	Inertness, stability, nonflammability
Antiseize lubricants	Nonflammability
Electrical instrument oil	Insolubility, stability
Pump-working fluid in chlorine service	Chemical inertness
Chemical reaction medium	Chemical inertness, high-temperature stability
Electric motor bearing lubricant	High-temperature stability, excellent lubricity
Oven conveyor chain and bearing lubricant	High-temperature stability
Plug valve lubricant in corrosive chemical service	Chemical inertness, high-temperature stability
Plasma etching equipment lubricant	Chemical stability, low volatility
Computer gear drive grease	Biostability
Rack and pinion disk drive lubricant	Biostability, thermal stability, excellent lubricity
Seal lubricant for chemical reactor agitator shafts	Chemical inertness, insolubility, thermal stability
Lubricant for water-immersed valves and bearings	Insolubility, chemical inertness
Lubricant for pressure relief valves	Chemical inertness, insolubility in solvents

[a]Source: Personal data.

ACKNOWLEDGMENTS

The author thanks Robert Kelly, Jimmie Patton, John Graham, Gerry Madden, and Nandan Rao for many helpful conversations and contributions to the contents and for reading and proofreading this chapter.

REFERENCES

1. Gumprecht, W. H. (1966). PR-143—A new class of high-temperature fluids, *ASLE Trans.* 9, 24–30.
2. Carlson, D. P., and A. S. Milian (1967). Fourth International Symposium on Fluorine Chemistry, Estes Park, CO, July.
3. British Patent 904,877 (September 5, 1962).
4. Gumprecht, W. H. (1967). The preparation and thermal behavior of hexafluoropropylene epoxide polymers, Fourth International Symposium on Fluorine Chemistry, Estes Park, CO, July.
5. U.S. Patent 3,442,942 (May 5, 1969).
6. U.S. Patent 3,715,378 (Feb. 6, 1975).
7. Sianesi, D., A. Pasetti, R. Fontanelli, G. C. Bernardi, and G. Caporiccio (1973). La chimica e L'industria, *Wear*, 55(2), 208.

8. European Patent Application 0148482, Dec. 20, 1984.
9. Sianesi, D., V. Zamboni, R. Fontanelli, and M. Binaghi (1971). *Wear*, 18, 85.
10. Jones, W. R., T. R. Bierschenk, T. J. Juhlke, H. Kawa, and R. J. Lagow (1988). *Ind. Eng. Chem. Res.*, 27/8, 1497.
11. Paciorek, K. J. L., J. Kratzer, J. Kaufman, and J. H. Nakahara (1979). *J. Appl. Polym. Sci.*, 24, 1397.
12. Gumprecht, W. H. (1967). The preparation and thermal behavior of hexafluoropropylene epoxide polymers, paper presented at the Fourth International Symposium on Fluorine Chemistry, Estes Park, Col, July.
13. Gumprecht, W. H. (1968). The preparation, chemistry and some properties of hexafluoropropylene epoxide polymers, paper presented at the Gordon Fluorine Conference, New Hampshire.
14. Paciorek, K. J. L., J. Kratzer, J. Kaufman, J. H. Nakahara (1979). *J. Appl. Polymer. Sci.*, 24, 1397.
15. Jones, W. R., K. J. L. Paciorek, T. I. Ito, and R. H. Kratzer (1983). *Ind. Eng. Chem. Res. Dev.*, 22, 166.
16. Pearson, R. K., J. A. Happe, G. W. Barton (1982). Study of the degradation of two candidate diffusion pump oils, Krytox and Fomblin, Lawrence Livermore Laboratory report number UCID 19571.
17. Sianesi, D., and R. Fontanelli (1967). Perfluoro polyethers. Their structure and reaction with aluminum chloride, *R. Makromol. Chem.* 102, 115.
18. Montedison brochure.
19. Carre, D. J., and J. A. Markowitz (1985). The reaction of perfluoropolyalkyl oil with FeF_3, AlF_3, and $AlCl_3$ at elevated temperatures, *ASLE Trans.*, 28, 40.
20. Mori, S., and W. Morales (1989). Reaction of perfluoroalkyl polyether oils with stainless steel under ultrahigh vacuum conditions at room temperature, *Wear*, 132, 111.
21. Jones, W. R., K. J. L. Paciorek, D. H. Harris, M. E. Smythe, J. H. Nakahara, and R. H. Kratzer (1985). *Ind. Eng. Chem. Res. Dev.* 24/3, 417.
22. Dolle, R. E., and F. J. Harsacky (1966). *New High Temperature Additive Systems for PR*-143 Fluids*, U.S. Air Force Materials Laboratory Technical Report AFML-TR-65-349, January.
23. Corti, C., and P. Savelli (1989). Perfluoropolyether lubricants, *Proc. Conf. Synth. Lub.*, ed. A. Zakar, pp. 128–155, Hungarian Hydrocarbon Inst., Szazhalombatta, Hungary.
24. Strepparola, E., P. Gavezotti, and C. Corti (1989). Antirust additives for lubricants or greases based on perfluoropolyethers, Eur. Pat. Appl., 12 pp., EP 337425 A1 18 Oct, *Chem. Abstr.*, 111(26):236482v.
25. Jones, W. R., Jr., K. J. L. Paciorek, T. I. Ito, and R. H. Kratzer (1983). Thermal oxidative degradation reactions of linear perfluoroalkylethers, *Ind. Eng. Chem. Prod. Res. Dev.*, 22(2), 166–70.
26. Christian, J. B., and C. Tamborski (1980). Benzoxazole and benzothiazole antirust greases, *Lub. Eng.*, 36(11), 639–642.
27. Snyder, C. E., C. Tamborski, H. Gopal, and C. A. Svisco (1978). Synthesis and development of improved high-temperature additives for polyperfluoroalkylether lubricants and hydraulic fluid, *J. ASLE*, 35, 8, 451.
28. PFPE manufacturer bulletin MA-816E
29. Dolle, R. E., et al. (1965). *Chemical Physical and Engineering Performance Characteristics of a New Family of Perfluorinated Fluids*, U.S. Air Force Materials Laboratory Technical Report AFML-TR-65-358, September.
30. Messina, J. (1967). Perfluorinated Lubricants for Liquid Fueled Rocket Motor Systems, ASLE Preprint no. 67 AM8A-4, May 1–4.
31. Mori, S., and W. Morales (1989). NASA Technical Paper 2910. *Degradation and Crosslinking of Perfluoroalkyl Polyethers in Ultra Vacuum.*
32. Cantow, M. J. R., E. M. Barrall II, B. A. Wolf, and H. Geerissen (1987). Temperature and

pressure dependence of the viscosities of perfluoropolyether fluids, *J. Polym. Sci., Part B: Polym. Phys.*, 25(3), 603–609.

33. Mizushima, S., H. Nakahara, and H. Yamada (1988). Perfluoropolyether compound compositions with excellent thermal conductivity, Japan Patent 63251455, 18 October.
34. Christian, J. B. (1983). Oxidation stable polyfluoroalkyl ether grease compositions, U.S. Patent Application, 14 pp., avail. NTIS order no. PAT-APPL-6-418 106., US 418106 A 15 April 1983, *Chem. Abstr.*, 99(10), 73640n.
35. Christian, J. B. (1981). Grease compositions, U.S. Pat. Appl., 20 pp., avail. NTIS order no. PAT-APPL-225 546, US 225546 A0, 31 July, *Chem. Abstr.*, 96(4), 22263x.
36. Christian, J. B., and C. Tamborski (1980). Benzoxazole and benzothiazole antirust greases, *Lub. Eng.*, 36(11), 639–642, *Chem. Abstr.*, 94(8):49855n.
37. Tohzuka, T., Y., Kataoka, S. Ishikawa, and K. Fujiwara (1989). Fluorine-containing grease and its preparation, EP 341613, A1 15 November, *Chem. Abstr.*, 112(12), 101951x.

7
Chlorotrifluoroethylene

Dale A. Ruesch*
Halocarbon Products Corporation
River Edge, New Jersey

Chlorotrifluoroethylene (CTFE) oils, greases, and waxes are saturated, hydrogen-free lubricants that are chemically inert, nonflammable, and have high thermal stability, good lubricity, high dielectric strength, high density, low compressibility, and nonpolar characteristics.

I. HISTORICAL DEVELOPMENT

The CTFE lubricants were first synthesized in the 1940s. Scientists on the Manhattan Project were looking for an inert lubricant for mechanical equipment needed in uranium isotope separation using the extremely reactive uranium hexafluoride (UF_6). The CTFEs were found to make an excellent inert lubricant for this application, so they were utilized as oils, greases, and the plastic polychlorotrifluoroethylene (PCTFE).

Following World War II the demand grew rapidly for industrial chemicals and gases, among them many aggressive materials such as chlorine, fluorine, oxygen, hydrogen peroxide, nitric acid, and sulfuric acid. Thus, a growing market was generated for chemically inert lubricants, and CTFE lubricants were the materials of choice.

Though most applications for CTFEs are for aggressive service, a number of applications have developed using other unique characteristics such as high density (gyroscope flotation fluids), low compressibility (hydraulic fluids), low vapor pressure (inert vacuum

*With assistance on technical matters by Louis L. Ferstandig, Ph.D, Halocarbon Products Corporation.

pump oils), and extreme pressure lubrication characteristics. Newer applications including metal-working lubricants for exotic metals such as tantalum and molybdenum, and low-temperature inert bath fluids are beginning to expand significantly. In metal-working CTFEs extend cutting tool life appreciably.

II. CHEMISTRY

Chemically, CTFE oils and waxes are saturated low-molecular-weight polymers of chlorotrifluoroethylene and have the general formula -$(CF_2CFCl)_n$- with n varying from 2 to about 10 units. They are made by a controlled polymerization technique. The low polymer is then separated into various fractions from light oils to waxes. The oils are, of course, totally miscible with each other, and oils of intermediate viscosity can be prepared by blending those of higher- and lower-viscosity.

III. PROPERTIES AND PERFORMANCE CHARACTERISTICS

A. Chemical Properties

The most outstanding property of CTFE lubricants is their chemical inertness. This property, together with nonflammability, provides the rationale for the large majority of usage. Specifically, CTFE lubricants do not react with the following chemicals and many others not as widely used.

Ammonium perchlorate	Chlorine trifluoride (gaseous)	Oxygen (LOX and GOX)
Boron trichloride	Fluorine (gaseous)	Ozone
Boron trifluoride	Fuming nitric acid	Sodium chlorate
Bromine	Hydrogen fluoride	Sodium hypochlorite
Bromine trifluoride (gaseous)	Hydrogen peroxide	Sulfur hexafluoride
Carbon dioxide	Hydrogen sulfide	Sulfur trioxide
Calcium hypochlorite	Muriatic acid	Sulfuric acid
Chlorinated cyanurates	Nitrogen oxides (all)	Thionyl chloride
Chlorine	Nitrogen trifluoride	Uranium hexafluoride
Chlorine dioxide	Oleum	

CTFE lubricants are not recommended for contact with sodium or potassium metal, amines including amines additives, liquid fluorine, or liquid chlorine trifluoride. They should also not be used with aluminum and magnesium (and alloys of these metals) under conditions where galling or seizing may occur.

Silica-thickened and CTFE polymer-thickened greases are both produced commercially. The presence of silica as a thickener compromises the grease inertness with chemicals such as gaseous fluorine, halogen fluorides, hydrofluoric acid, or caustic solutions that react with silica. The CTFE polymer-thickened greases exhibit the same chemical resistance as the base stock CTFE oil.

Oxygen compatibility is a major characteristic of CTFE lubricants, and therefore many test procedures have been developed to determine the safety of lubricants in contact with oxygen. Several CTFE oils and greases, including those with rust inhibitors, were tested using American Society for Testing and Materials (ASTM) method G 72, Standard Test Method for Autogenous Ignition Temperature of Liquids and Solids in a High Pressure Oxygen-Enriched Environment. None ignited throughout the entire testing range, rising to 400°C (752°F) in the presence of 2,000 psig (138 bar) of oxygen. Another test with CTFE oils and greases, including those with rust inhibitors, exceeded the energy limits of ASTM D 2512, Test for Compatibility of Materials with Liquid Oxygen (Impact Sensitivity Threshold Technique). None of the lubricants tested showed any sensitivity at the highest impact loading of 114 ft·lb (36.9 g·cal).

A CTFE oil and grease recently passed NASA testing at 8,500 psi (586 bar) and 100°F (38.7°C).

United States government specifications that are met or exceeded by CTFE lubricants are:

DoD-L-24574 (SH), Lubricating Fluid for Low and High Pressure Oxidizing Gas Systems
MIL-G-47219A, Grease, Lubricating, Halogenated
NASA 79K22280, Lubricant for 1000-GPM LO_2 Pump Bearings, Specification for

These tests also serve as indirect measures of CTFE inertness to other strong oxidizing agents.

The CTFE oils decompose thermally to non-sludge-forming volatiles rapidly at 320°C (608°F), noticeably at 300°C (572°F), and in lesser extents at lower temperatures. The maximum safe recommended long-term operating temperature is 204°C (400°F) and the maximum short-term temperature recommended is 260°C (500°F) in scrupulously clean systems. Thermal stability is affected by the presence of metals, and such exposure to temperatures above 177°C (350°F) should be evaluated before field applications. Oils containing antirust inhibitors may have lower recommended operating temperature due to decomposition of the inhibitor, even though the oil would not be affected at the lower temperature. The thermal decomposition products contain toxic materials, which should be handled only in well-vented areas.

CTFE lubricants, especially at elevated temperature, affect some elastomer and plastic materials. Most gaskets and O-rings have proprietary compositions, so the prudent approach to elastomer selection should involve bench testing and, if possible, testing of the specific product under operating conditions. CTFE oils have been found compatible with specific formulations of the following materials:

Ethylene propylene rubber
Polyvinyl alcohol
Buna-N (butadiene/acrylonitrile), ambient to moderate temperatures
Neoprene
Teflon
Chlorinated polyethylene
Viton, Fluorel
Polyimides
Polycarbonates
Fluorosilicone
Cured epoxies
Urethanes
PNF (phosphonitrilic fluoroelastomer)
EPDM (ethylene propylene diene rubber)

Most solvent-resistant elastomers and plastics are unaffected by CTFE fluids, but at higher temperatures the fluids may dissolve in and seriously weaken the following materials:

Buna-N (butadiene/acrylonitrile).
Buna-S (butadiene/styrene) rubber.

Silicone rubbers.
Natural rubber.
Polymers or copolymers of chlorotrifluoroethylene.
PVC (polyvinyl chloride).

The CTFE materials wet metallic surfaces readily and form lubricating films similar to the more common lubricants. Steel parts that have been lubricated with CTFE oils and then disassembled for inspection appear to have benefited from the lubrication even in severe service. However, it has been reported that the disassembled, solvent-cleaned parts rust rapidly on exposure to air. Rusting can be inhibited by keeping a thin film of oil on the part and, where necessary, using oil with rust inhibitor.

Halocarbon oils and greases are noncorrosive toward metals at temperatures up to about 350°F (177°C) with the exception of copper and some of its alloys, which will discolor at temperatures over 120°F (49°C). Prior testing should be done on all metals for applications above those temperatures.

Fluoro- and chloro-organic compounds, such as CTFE oils, greases, and waxes, may react with aluminum and magnesium under conditions of high friction. For example, a reaction may occur from the tightening of a bolt of aluminum or magnesium in an aluminum or magnesium device with CTFE oil or grease in the threads. Such reactions are extremely localized and nonpropagating, even though they may be accompanied by a sharp noise. These reactions do not always occur, even under large frictional forces. No reaction was observed when ¾-in flared aluminum hydraulic tubing, lubricated with CTFE grease, was deliberately tightened with an aluminum coupling. The torque exerted (over 200 ft-lb) destroyed the threads, but the grease was not affected. CTFE oils and greases may be used routinely in aluminum and magnesium housings, tubing, containers, and other parts when galling possibilities do not exist.

The CTFE light oils are soluble in most organic liquids including aromatic and aliphatic hydrocarbons, chlorinated and fluorinated solvents, alcohols, ketones, and esters. The solubility decreases as the molecular weight increases. All CTFE fluids are insoluble in aqueous solutions, whether they are neutral, acidic, or alkaline.

CTFE fluids will dissolve halogens and volatile anhydrous inorganic salts (such as titanium tetrachloride). Typical organic materials with which various CTFE fractions are miscible are:

Acetone
Amyl acetate
Benzene
n-Butyl alcohol
Carbon tetrachloride
Carbon disulfide
Chloroform
Dibutyl phthalate
Dioctyl phthalate
Dioctyl sebacate
Ethanol
Ether
Glacial acetic acid
Hexachlorobutadiene
Hexane
Kerosene
Methanol
Methyl ethyl ketone
Methyl isobutyl ketone
Methylene chloride
Mineral oils
Silicone oils
Tetrachlorodifluoroethane (R-112)
Trichlorobenzene
Tetrachloroethylene
Trichlorofluoromethane (R-11)
Trichloroethane
Trichloroethylene
Trichlorotrifluoroethane (R-113)

CTFE fluids dissolve gases readily. Chlorine, for example, is soluble to the extent of several weight percent at ambient conditions. Air is about 15 volume percent soluble. Although both viscosity and density are changed by dissolved gas, most properties of the fluid itself are not affected.

Tests performed by the Institut Français de Petrole indicate that CTFE oils are excellent lubricants. On standard four-ball machines, seizure does not take place until 2.5 s under as much as 200 kg load. This value compares favorably with that of "extreme pressure oils" intended for gear lubrication. Other extreme pressure tests using ASTM method D 2783, Standard Method for Measurement of Extreme-Pressure Properties of Lubricating Fluids (Four-Ball Method), show that CTFE oils exhibit no seizure even at the final applied load of 800 kg. Low-viscosity base stock greases do show weld points but only at the highest load. Even those greases and all the oils show load wear indices that are appreciably higher than hydrocarbon oils. The same data show reasonable scar diameters increasing uniformly with load.

CTFE oils can be used interchangeably with hydrocarbon oils to lubricate a wide range of standard equipment such as bearings, compressors, gear boxes, and oil pumps. Occasionally some equipment modification may be required because of density, viscosity, and vapor pressure differences from hydrocarbon lubricants.

The CTFE oils have been used successfully in all types of equipment for over 40 years. However, if there is any question of direct interchangeability with conventional lubricants, a monitored test in the piece of equipment may be advisable.

B. Physical Properties

The CTFE oils, greases, and waxes have no flash and fire points. They increase in viscosity, density, and pour point, and they decrease in vapor pressure as the average molecular weights increase. Molecular weights for the oils range from 300 to 1,000. Waxes typically have molecular weights of 1,100 to 1,300.

The viscosity of the oils ranges from less than 1 cSt to 1400 cSt at 100°F; (38.7°C). Because the viscosity of CTFE oils drops significantly with increased temperature, thought must be given to insure proper viscosity at operating conditions. Viscosity versus temperature charts are readily available.

Oil density varies from 1.71 to 1.97 g/ml at 100°F (37.8°C). Greases and waxes have densities of approximately 1.9–2.0 g/ml at 160°F (71.1°C). Since the viscosity in centipoises is determined by multiplying the centistoke viscosity by density, CTFE oils have centipoise viscosities almost twice the centistoke value. In conventional fluids centistoke and centipoise values are usually similar (density about 1) or the centipoise value is lower (density less than 1). When changing from a conventional oil to a CTFE the viscosities of the oils in centipoises should be used to determine the grade.

The oils show no breakdown under shear: no drop in viscosity of CTFE oils was noticed after 20 min in a Durel & Sausse microgrinder running at high speed.

Various physical properties are listed in Table 7.1. Certain of the properties are especially noteworthy compared to conventional lubricants. The surface tension is very low, contributing to better wettability of most surfaces. The lighter oils have very low pour points so they are still useful at cryogenic temperatures. The oils are not very compressible, a property required for quick hydraulic response.

Table 7.1 Physical Property Data

Property	Value
Bulk modulus	Over 200,000 psi [13.8 × 10^8 Pa at 100°F (37.8°C) with applied pressure of 10,000 psi (6.9 × 10^7 Pa)]
Coefficient of cubic expansion	7.6–9.6 × 10^{-4} $cm^3/cm^3/°C$
Dielectric constant	2.25–4.0, varies with frequency (Hz) and oil temperature
Distillation range	Initial boiling points at 1 atm (760 mm Hg) range from 130°C (260°F) to above 300°C (572°F)
Heat of vaporization	20–23 cal/g
Pour point	–129°C (–200°F) to 21.2°C (70°F) as molecular weight increases
Refractive index	$n^{25}{}_D$ = 1.380–1.415 as molecular weight increases
Specific heat	0.2–0.25 cal/g
Surface tension	20–30 dyn/cm at 77°F (25°C)
Vapor pressure	Regular oils: 0.004–40 mm Hg at 122°F (50°C) Vacuum pump oils: < 0.001 to 0.0035 mm Hg at 122°F (50°C)
Volume resistivity	10^{13} to 10^{14} ohm-cm

C. Comparative Performance

Some of the synthetic lubricants with low vapor pressures may have very high flash points, but they are still flammable and would explode with liquid oxygen and similar chemicals, whereas CTFEs are completely nonreactive.

Applications where the CTFEs are used based on comparative performance characteristics and cost-effectiveness are:

1. Aerospace industry.
   a. Lubricant in oxygen delivery system to space shuttle oxidizer tanks.
   b. Nonflammable hydraulic fluid for future aircraft.
2. Biological science.
   a. Immersion oil for embryo studies.
3. Chemical industry.
   a. Compatible with chemicals previously mentioned and many others not as widely used.
   b. Bromine pump oil.
   c. Sulfur trioxide spill control mixture. (This mixture, when spread over a sulfur trioxide spill, stops fuming completely, and further cleanup can be done on a nonemergency basis.)
   d. Lubricant for equipment used in fluorination process for blow-molding polyethylene bottles and gasoline tanks.
   e. Sealant for flange faces.
   f. Mechanical seal barrier fluid for the following operations: bromination, chlorination, fluorination, nitration, oxidation, and sulfonation.
   g. Chlorine service, as pump oils (vacuum and pressure pump applications), oil-injected helical screw compressors, valve and plugcock grease, chlorine vaporizer lubrication, valve-stem lubricant, assembly and repair of chlorine cylinder valves, tank car maintenance (chlorine valve and pressure relief valve lubricant), thread lubricant, and instrument fill fluid.

4. Cryogenic gas industry.
   a. Oxygen service, as lubricant for remote control solenoid valves, thread lubricant, instrument fill fluid, rotary meter lubricant, diaphragm compressor oil, vacuum pump oils for evacuating oxygen cylinders and bulk storage tanks, vacuum pump oils for oxygen plasma cleaning, bearing grease for liquid oxygen (LOX) pumps, lubricant for compressors in portable oxygen plants, and carbon dioxide pump oil.
   b. Welding supply, as lubricants for bearings in LOX pumps, and vacuum pump oils for evacuating oxygen cylinders.
   c. Helium, as oil for helium compressors, and lubricants for helium regulators.
5. Electronics industry.
   a. Vacuum pump oil for semiconductor manufacturing equipment.
   b. Vacuum pump oil for equipment used to plasma desmear multilayer printed circuit boards.
   c. Vacuum pump oil in equipment used to plasma clean electronic and medical devices.
   d. Inert grease for semiconductor processing equipment.
6. Instrument fill fluids. Strong oxidizing agents such as oxygen, chlorine, fluorine, nitric acid, and hydrogen peroxide preclude the use of glycerine or silicone fill fluids. For those agents inert CTFE fluids are used in the following areas:
   a. Diaphragm seals.
   b. Pressure gauges.
   c. Manometers.
   d. Dead weight testers.
   e. Sensors.
7. Laboratory apparatus.
   a. Cold temperature bath fluid (nonflammable).
   b. Stopcock lubricant.
   c. Ground-glass joint lubricant.
   d. Wax coating to protect glass from attack by aggressive species.
   e. Vacuum pump oil for mass spectrometer equipment.
8. Life support systems. CTFE oils and greases are used to lubricate life support systems where oxygen-enriched atmospheres (>25%) or high-pressure air is required. They include the following:
   a. Diving gear/U.S. Navy.
   b. Hyperbaric oxygen chambers.
   c. Hospital nitrous oxide systems.
   d. Home oxygen units.
   e. Liquid oxygen respiratory equipment.
   f. Anesthesia machines.
   g. Portable oxygen-generating plants.
   h. Systems for evacuating and refilling oxygen bottles.
   i. Breathing systems in airplanes and submarines.
9. Lubricant industry.
   a. Base stock for specialty greases.
   b. Base stock for antiseize compounds and thread sealants.
   c. Extreme pressure additives for lubricating oils.
10. Metalworking industry.
    a. Cutting oil for machining tantalum, molybdenum, and niobium.

   b. Drawing of tantalum wire and stainless steel tubing.
   c. Forming of tantalum parts.
   d. Manufacture of woven wire and cable in aggressive services.
   e. Additive to other cutting oils for enhanced tool life.
   f. Machining of high-nickel alloys.
11. Nuclear industry.
    a. Lubricant in process to convert UF_6 to UO_2.
    b. Hydrogen-free oil for use in nuclear service.
    c. Greases to lubricate controls for nuclear applications.
12. Paper industry. Chlorine, sodium chlorate, chlorine dioxide, oxygen, and hydrogen peroxide are widely used as pulp bleaching chemicals by the paper industry. CTFE lubricants are compatible with all of these chemicals and are finding increased use in this industry.
13. Petroleum industry.
    a. Antiseize for drilling tools in hydrogen sulfide environments.
    b. Alkylation lubricant (compatible with HF and sulfuric acid).
    c. Instrument fill fluid for oil exploration equipment.
14. Steel industry.
    a. Grease for swivel joints in oxygen delivery systems and oxygen heating systems.
15. Water and wastewater treatment.
    a. Water treatment. CTFE lubricants are compatible with water-treating chemicals such as oxygen, ozone, hydrogen peroxide, chlorine, calcium hypochlorite, sodium hypochlorite, and chlorinated cyanurates. They are used in chlorinators, pumps, valves, etc.
    b. Swimming pool chemicals. Lubricants in compacting equipment for calcium hypochlorite and chlorinated cyanurates, and die release in tableting swimming pool chemicals.
16. Miscellaneous.
    a. Index matching fluids.
    b. Potting and sealing waxes.
    c. Damping fluids.
    d. Heat-transfer fluids.
    e. Release agent for molding of plastics and elastomers.
    f. Plasticizer for polychlorotrifluoroethylene and epoxy resins.

IV. MANUFACTURE, MARKETING, AND ECONOMICS

Active companies in the CTFE field are Atochem S.A., Paris, France; Daikin Industries, Ltd., Osaka, Japan; Halocarbon Products Corporation, River Edge, N.J.; Occidental Chemical Corporation, Niagara Falls, N.Y. All of the aforementioned companies market the oils and some formulated products such as greases, special waxes, vacuum pump oils, and oils and greases with performance enhancement additives such as rust inhibitors or polytetrafluoroethylene.

Many specialty lubricant producers, process equipment manufacturers, and government branches purchase the base stock oils and formulate their own finished products for sale or internal consumption.

Even though they are more expensive on a volume basis than ordinary hydrocarbon oils and the synthetic hydrocarbons, the CTFEs are generally less expensive than the competitive perfluoropolyether (PFPE) lubricants. In all applications CTFE oils are cost-effective as determined by both tangible benefits (reduced downtime, superior product quality, increased equipment life, reduced maintenance) and intangible benefits (employee and plant safety, reduced company liability).

V. OUTLOOK

Both the near- and long-term outlook for these lubricants appear to be of growth along with industry in general.

Although new applications are constantly developing and growing rapidly, their total volume is small when compared with the major applications. The use of inert lubricants in some major applications such as oxygen, chlorine, and the electronics industry is expected to show little or no growth. The newer applications will grow to provide long-term growth. Factors that could significantly affect the long-term outlook for CTFE lubricants are:

Full development of nonflammable hydraulic fluids for military aircraft and land vehicles.

Replacement of phosphate esters as vacuum pump fluids in oxygen service. The U.S. Navy is already converting to CTFE vacuum pump oils.

Reclamation of CTFE oils, which would make them even more cost-effective.

Replacement of environmetally unacceptable chlorinated solvent-type cutting oils with CTFE oils.

Replacement of hydrocarbons by inert, nonflammable lubricants as safety is emphasized more. Preventative maintenance techniques are becoming more sophisticated, enhancing equipment life, and complying with more stringent environmental regulations. Inert lubricants will be utilized more as the high initial cost is offset by higher waste disposal and maintenance costs associated wtih standard lubricants.

Although the use of oxygen is growing, there are many new customer-site noncryogenic oxygen plants. These plants use relatively low-pressure compressed air and do not require inert lubricants. It is projected by one industry expert that noncryogenic plants could account for 30% of the total oxygen and nitrogen production in less than 10 years.

The chlor-alkali industry is projected to grow no more than 1% per year or even to decline in demand in a few years. Chlorine use is declining because of the regulation of CFCs, chlorinated solvents, and reduced use in the pulp and paper industry (due to the formation of dioxin) in chlorine bleaching operations. This decline in usage could lead to plant shutdowns and reduce the use of inert lubricants.

The overall outlook is for slow but continued growth for CTFEs.

REFERENCES

1. Atochem S. A. Brochure, Voltalef^{R} Oils and Greases.
2. Daikin Industries Ltd. Technical information bulletin, Daifloil^{R} and Daiflon Grease.
3. Halocarbon Products Corporation. Brochure, Chlorotrifluoroethylene (CTFE) Oils, Greases & Waxes.
4. Occidental Chemical Corporation. Data Sheet 184H386, Fluorolube^{R}.

8
Silicones

Donna H. Demby
GE Silicones, Waterford, New York

Stanley J. Stoklosa
GE Silicones, Waterford, New York

Allen Gross
GE Silicones, Overland Park, Kansas

I. INTRODUCTION

Silicones are a broad family of synthetic polymers that are partly inorganic and partly organic. Their structure consists of alternating silicon and oxygen atoms rather than the carbon-to-carbon backbone that characterizes organic materials. Typically, one or more organic side groups are attached to the silicon atoms, imparting properties such as chemical resistance, lubricity, improved thermal and oxidative stability, and reactivity with organic chemicals and polymers. Additionally, these materials are characterized by chemical inertness, low surface tension, excellent water repellency, good electrical properties and weatherability, and a high degree of slip on most rubber or plastic surfaces.

Silicones are grouped into three major categories: fluids, resins, and elastomers (Fig. 8.1).

Silicone fluids in commercial use often have side groups such as methyl, trifluoropropyl, phenyl, vinyl, longer alkyl, and other organic groups. In addition, they may be modified with the use of fillers, solvents, antioxidants, or lubricity additives for use in a wide variety of applications.

II. HISTORICAL DEVELOPMENT

Organosilicone chemistry had its beginnings about 100 years ago. Kipping of England was the first to name silicones. As early as 1904, he synthesized a number of R-Si-X

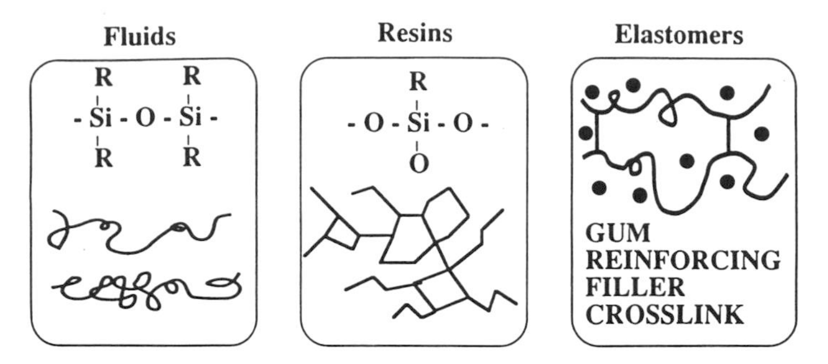

Figure 8.1 Forms of silicone.

compounds using Grignard reagents. As scientific knowledge about the formation of large polymer molecules increased in the early 1930s, industrial development of silicones began.

Early work focused on resinous polymers for use as insulating materials that would withstand temperatures above the 105°C limit of Class A insulation (2). In 1940, through research done at its corporate research and development laboratory in Schenectady, NY, General Electric became the first company to successfully develop an economical procedure for silicone production. Both Dow Corning Corporation and General Electric began commerical development of silicone polymers during the 1940s. The Dow Corning plant opened in 1943; General Electric began commercial production in 1944 and completed the initial phase of its Waterford facility in 1947.

Silicones were initially used for military applications during World War II, followed by use in the aerospace industry. Silicones' success in meeting the demanding environmental conditions characteristic of these applications, together with their thermal and chemical stability, led to increased usage in the United States.

Typical military applications for silicones include damping fluids for aircraft instruments, antifoams in petroleum oils, and greases used as ignition sealing compounds. After World War II, civilian uses expanded beyond these applications to include release agents for molding rubber, water repellents, and ingredients for paints, lubricants, and polishes.

In recent years, silicones have grown from a chemical rarity into a widely used family of products. Silicone fluids have proved extremely versatile in solving many industrial problems and have been key ingredients in a number of new products. Probably no other synthetic fluid has found such a variety of uses in so many different industries.

III. CHEMISTRY

A. Product Structure

Silicone fluids are unique polymers that combine an inorganic silicon-oxygen (siloxane) backbone with organic side chains: R groups such as methyl (CH_3) and phenyl (C_6H_5), as shown in Fig. 8.2. This hybrid nature explains why silicones behave somewhat like organic polymers, yet retain important inorganic properties like heat resistance.

Many of the unique properties of silicone fluids are due to the free rotation of molecules along the Si-O and Si-C bond axes and the flexible nature of the siloxane

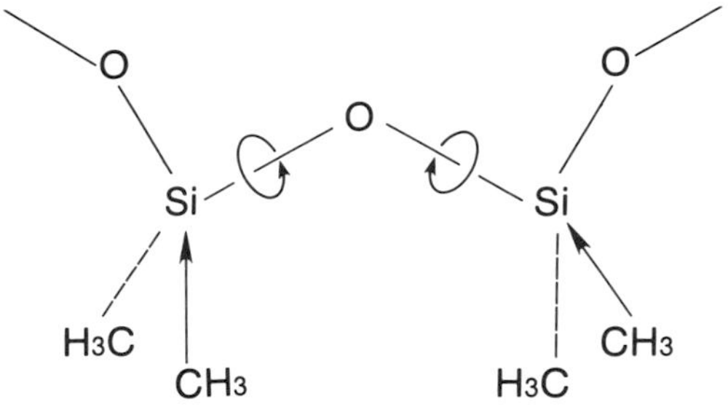

Figure 8.2 Silicone polymer structure.

backbone. This freedom of motion leads to greater intermolecular distances and, therefore, to lower inter-molecular forces. These factors explain the low modulus, low glass-transition temperatures (T_g), and high permeability of silicones (3).

Structure also explains the slight temperature dependence of other properties, such as viscosity. At lower temperatures, the siloxane has short chain extension and low molecular entanglement of the R groups. As the chain extends with higher temperatures, the backbone must shift to a higher-energy configuration where the R groups are closer together. The greater molecular mobility due to temperature is thus negated by increased molecular entanglement. Therefore, the viscosity of the flexible silicone has much less temperature dependence than does the stiff-chained hydrocarbon (5).

B. Nomenclature

Silicone terminology is a blend of "official" International Union of Pure and Applied chemistry (IUPAC) rules, American Chemical Society (ACS) rules, and commonly used shorthand.

The terminology borrows from organic chemistry nomenclature and defines SiH_4 as silane, analogous to CH_4, methane. Hence, H_3SiCl is chlorosilane; $(CH_3)_2SiCl_2$ is dimethyldichlorosilane; and H_3SiOH is silanol. The molecule $H_3SiOSiH_3$ is disiloxane; polymers are polyorganosiloxanes (a more descriptive term than silicones). Polydimethylsiloxanes (PDMS) are the predominant commercial polymers.

Strictly proper nomenclature can be awkward to use, especially with polymer chemistry and in the spoken language. A useful shorthand has been developed, and it reflects the predominance of PDMS polymers. The letters, M, D, T, and Q are used to represent mono-, di-, tri-, and quadrifunctional monomer units: that is, monomer units with the silicon bonded to one, two, three, and four oxygen atoms, respectively. (Keep in mind that each O is actually a bridge in the siloxane chain, and is "shared" by two Si atoms.) The remaining substituents are assumed to be methyl groups; monomer units with alternate substituents are labeled with a prime, such as M' (3). Tables 8.1 and 8.2 show common examples of this nomenclature and other common shorthand terms.

The D chain-propagating units are used to build up the polymer, since they can form one Si-O bond to the growing chain and still have a bond site available to add the next monomer unit. The M units are used as chain terminators, since simply attaching to the chain uses their only Si-O bond site. The T and Q units are used to add branching and cross-linking, especially in resins.

Table 8.1 Siloxane Structures

Chemical name	Structural formula	MDT formula	Common name
Hexamethyldisiloxane	$(CH_3)_3SiOSi(CH_3)_3$	MM	Mono
Octamethyltrisiloxane	$(CH_3)_3SiOSi(CH_3)_2OSi(CH_3)_3$	MDM	Linear trimer
Decamethyltetrasiloxane	$(CH_3)_3SiO]Si(CH_3)_2O]_2Si(CH_3)_3$	MD_2M	Linear tetramer
Octamethylcyclotetrasiloxane	$(CH_3)_2Si—O—Si(CH_3)_2$ / O O / $(CH_3)_2Si—O—Si(CH_3)_2$		
Octaphenylcyclotetrasiloxane	$C_6H_5)_2Si—O—Si(C_6H_5)_2$ / O O / $(C_6H_5)_2Si—O—Si(C_6H_5)_2$	D'_4	Cyclic phenyltetramer
2,4,6,8-Tetramethyl-2,4,6,8-tetraphenylcyclotetrasiloxane	$(CH_3)(C_6H_5)Si—O—Si(CH_3)(C_6H_5)$ / O O / $(CH_3)(C_6H_5)Si—O—Si(CH_3)(C_6H_5)$	D'_4	Cyclic methylphenyltetramer
Methyltris(trimethylsiloxy)silane	$[(CH_3)_3SiO]_3SiCH_3$	M_3T	
1,1,1,3,5,5,5-Heptamethyltrisiloxane	$(CH_3)_3SiOSi(H)(CH_3)OSi(CH_3)_3$	MD'M	
1,1,3,5,5-Pentamethyl-1,3,5-triphenyltrisiloxane	$(CH_3)_2(C_6H_5)SiOSi(CH_3)(C_6H_5)$ / $OSi(C_6H_5)(CH_3)_2$	M'D'M'	

C. Synthetic Methods

Oxygen (50%) and silicon (25%) are the most abundant elements in the earth's crust. Silicon is not found pure in nature, but rather in combination with oxygen in a wide variety of three-dimensional crystalline networks called silicates (Fig. 8.3).

The substitution of atoms such as iron, aluminum, calcium, sodium, and potassium in

Table 8.2 Silicone Functionality Shorthand Notation

Formula	Functionality	Symbol
$(CH_3)_3SiO_{0.5}$	mono	M
$(CH_3)_2SiO$	di	D
$(CH_3)SiO_{1.5}$	tri	T
$(CH_3)(C_6H_5)SiO$	di	D'
$(C_6H_5)_2SiO$	di	D'
$(CH_3)(H)SiO$	di	D'
SiO_2	quadri	Q

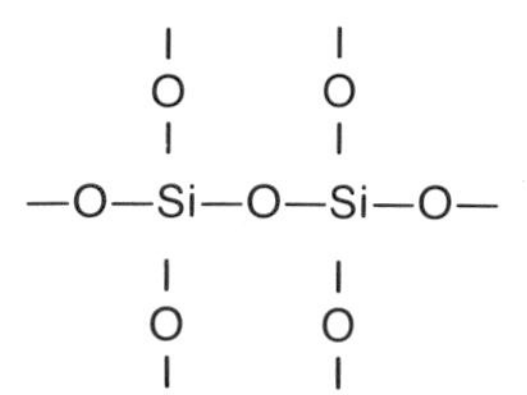

Figure 8.3 Silicate structure.

the matrix gives rise to a wide variety of rocks and minerals. Through the craft of ceramics, humans have been using practical silicate chemistry for thousands of years (6,7). However, due to the stable nature of silicates, silicon was not isolated and recognized as an element until 1824.

Since silicon and carbon are both tetravalent (have four bonding electrons available), much of the research in the 1800s focused on synthesizing silicon analogs of organic compounds (Fig. 8.4 and 8.5).

However, there was little success. Due to silicon's more complex electron orbital structure, silicon chemistry differs from carbon (organic) chemistry in several important ways (6,7):

1. Si-Si and Si-H bonds react rapidly with oxygen and water, making them unstable under normal conditions (they will react with air).
2. Silicon has a tremendous affinity for (and is difficult to separate from) oxygen. Put another way, silicon builds polymers with oxygen, while oxygen tends to degrade hydrocarbons to single molecules (CO_2 and CO).
3. Silicon does not form double bonds, while C=C and C=O bonds are important in organic chemistry.
4. Silicon compounds generally require higher energy (higher temperature) to react, making them more stable.

The potential usefulness of the Si-O bond was not realized until the 1930s, when organic resins used for electrical insulation were found to be inadequate for new higher temperature applications. It was thought that a resin siloxane backbonc with small organic side chains might have the proper temperature resistance (5).

Early synthetic schemes used Grignard reagents and silicon tetrachloride. The Grignard reagent is prepared as follows:

$$Mg + RBr \xrightarrow{\text{ether}} RMgBr$$

The silicon tetrachloride is derived from quartzite rock, oil coke (carbon), and chlorine gas:

```
  H  H  H
  |  |  |
—C—C—C—
  |  |  |
  H  H  H
```

Figure 8.4 Hydrocarbon.

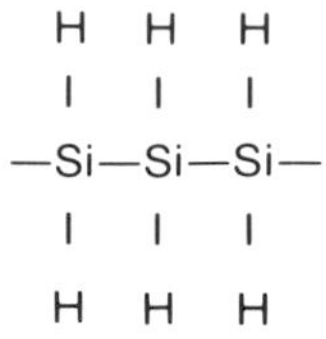

```
H   H   H
|   |   |
—Si—Si—Si—
|   |   |
H   H   H
```

Figure 8.5 Polysilane.

$SiO_2 + 2C \rightarrow Si + 2CO$ (electric furnace)
$Si + 2Cl_2 + \text{heat} \rightarrow SiCl_4$

The reactants are combined:

$$2RMgBr + SiCl4 \xrightarrow{\text{ether}} R_2SiCl_2 + 2MgBrCl$$
$$nR_2SiCl_2 + (2n)H_2O \rightarrow (2n)HCl + nR_2Si(OH)_2$$
$$nR_2Si(OH)_2 \rightarrow nH_2O + (R_2SiO)_n \text{ (linear and cyclic)}$$

This method proved costly and difficult on a production scale. Safety is also a concern with large quantities of reactive materials such as ether and magnesium.

A direct process was developed by Rochow of General Electric Company that allowed the chlorosilane intermediates to be produced in one step:

$$Si + 2RCl \xrightarrow{\text{catalyst}} R_2SiCl_2$$
$$nR_2SiCl_2 + (2n)H_2O \rightarrow (2n)HCl + nR_2Si(OH)_2$$
$$nR_2Si(OH)_2 \rightarrow nH_2O + (R_2SiO)_n \text{ (linear and cyclic)}$$

This process made the commercialization of silicones possible, and it remains the standard of the industry.

D. Current Commercial Routes

Commercial silicone manufacture starts with the production of chlorosilane intermediates, especially methylchlorosilanes.

Typically, methyl chloride vapor is passed at high velocity through a fluid-bed reactor containing finely ground silicon metal and copper catalyst. The silicon is converted (at up to 90% efficiency) to a crude mix of dimethyldichloro-, methyltrichloro-, methyldichloro-, and trimethylchlorosilanes, as well as other monosilanes and higher-boiling residues (3) (Fig. 8.6). Conditions are generally set to yield at least 50% dimethyldichlorosilanes.

Fractional distillation columns are used to separate the desired components from the crude mix. The task is somewhat difficult since the boiling points of the chlorosilanes are similar (Table 8.3).

Methylchlorosilanes are converted to silicone fluids in a two-step process. In the hydrolysis step, dimethyldichlorosilane is converted to cyclic and linear dimethylsiloxanes (D units):

$$(CH_3)_2SiCl_2 + H_2O \rightarrow \underset{\text{cyclic}}{[(CH_3)_2SiO]_n} + \underset{\text{linear}}{HO[(CH_3)_2SiO]_mH} + HCl$$

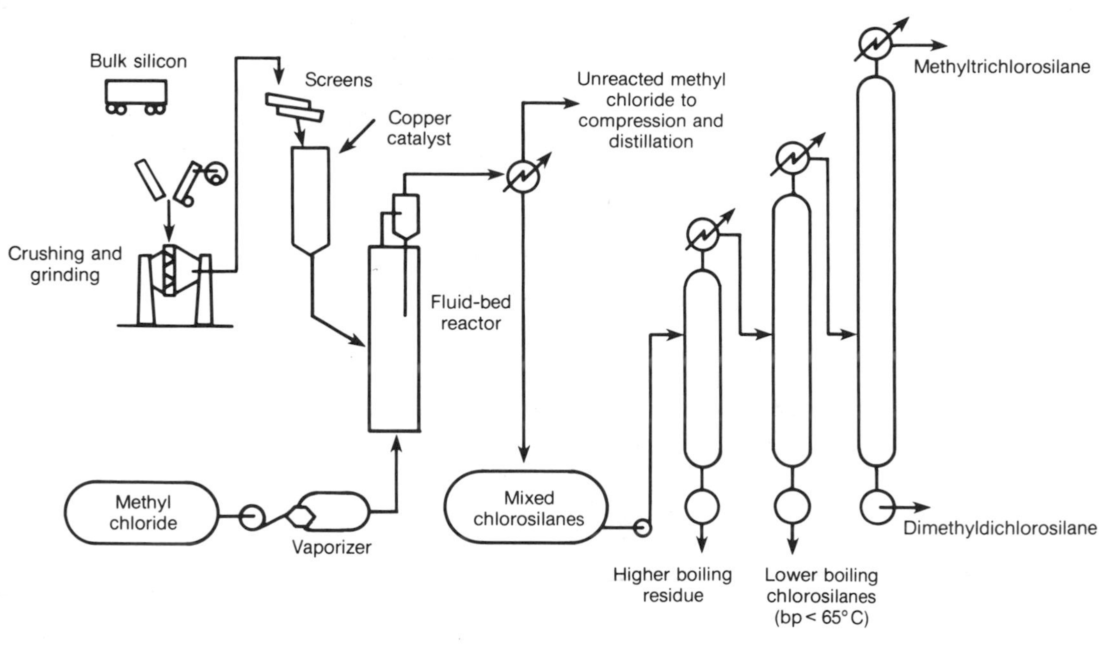

Figure 8.6 Conversion of silicon ore to methylchlorosilane.

Likewise, trimethylchlorosilane is converted to hexamethyldisiloxane, the source of chain-terminating M units:

$$(CH_3)SiCl + H_2O \rightarrow (CH_3)_3SiOSi(CH_3)_3 + HCl$$

Siloxane polymer (MD_XM) is formed in the equilibration reaction:

$$[CH_3)_2SiO]_n + HO[(CH_3)_2SiO]_mH + (CH_3)_3SiOSi(CH_3)_3 \xrightarrow{H^+ \text{ or } CH^-} [(CH_3)_2SiO]_n + (CH_3)_3SiO\text{-}[\text{-}(CH_3)_2SiO\text{ -}]_p{=}Si(CH_3)_3 + H_2O$$

The average chain length can be controlled reliably by choosing the ratio of D to M. (However, note that some cyclic siloxanes remain at equilibrium with the mix of polymer chains.) The equilibrated fluid is washed with water to kill residual active groups, neutralized, dried, and devolatilized to yield the final silicone fluid product.

Table 8.3 Chlorosilane Properties

Compound	Boiling point, °C	Density d^{20}, g/cm^3	Refractive index, n_D^{20}	Assay, %
$(CH_3)SiCl_3$	66.4	1.273	1.4088	95–98
$(CH_3)_2SiCl_2$	70.0	1.067	1.4023	99–99.4
$(CH_3)_3SiCl$	57.9	0.854	1.3893	90–98
$(CH_3)SiH(Cl)_2$	41.0	1.110	1.3982	95–97
$(CH_3)_2SiH(Cl)$	35.0	0.854	1.3820	

Commercial products can be pure or modified dimethylsilicone fluids. Fluids can be modified chemically by copolymerization with methylalkyl, methylphenyl, diphenyl, methyltrifluoropropyl, or other organofunctional oligomers. The addition of silica fillers to fluids gives greaselike silicone compounds, used for electrical insulation and waterproofing or as antifoam compounds. The use of soap as a filler results in a lubricating grease.

IV. PROPERTIES AND PERFORMANCE

A. Structural Considerations

Silicones are used for lubrication in both critical (metal-to-metal) and noncritical applications, such as plastic-to-plastic. Critical lubrication conditions can be divided into three categories: hydrodynamic, boundary, and extreme pressure. These characteristics define the mechanisms whereby damage is prevented as one metal surface slides across another.

Early application testing soon made it clear that the most basic silicone, polydimethylsiloxane fluid (Fig. 8.7), in its unmodified and uncompounded state, lacked the ability to lubricate sliding metal-to-metal in any of the three categories outlined.

Dimethyl fluids as well as molecular modifications with phenyl were ideal in thermal and oxidative stability. However, in viscosity and shear stability, they lacked the essential load-bearing characteristics necessary for critical lubricant use.

Molecular modifications utilizing phenylalkyl, fluoro, nitrile, methyl alkyl, and chlorophenyl functionalities, and/or compounding with fillers such as soaps, graphite, treated or untreated silicas, molybdenum disulfide, polyethylenes, carbon blacks, clays, or combinations of the preceding are used to achieve the variety of lubrication properties necessary for effective critical lubrication.

Compounding lubricating fluids produces either greases or compounds. A grease is a "semi-structured lubricant that is utilized to lubricate or protect metal-to-metal contact surfaces" (13). A compound is a fluid that has been thickened and stabilized to a greaselike consistency. Compounds are used for applications other than metal-to-metal.

B. Chemical and Related Properties

Generally speaking, silicones are chemically inert, self-extinguishing, resistant to oxidation and thermal attack, hydrolytically stable, nonstaining or noncorrosive, and, as a class of materials, very low in toxicity. The physiological properties of dimethyl silicone oils are presented in Table 8.4.

It is the Si-O linkage that contributes to the outstanding high-temperature characteristics and general inertness of silicone fluids. The Si-O linkages are comparable to linkages found in similar high temperature materials (e.g., quartz, glass, and sand). The carbon-to-hydrogen bonds of hydrocarbon fluids (Fig. 8.8) lack this relationship.

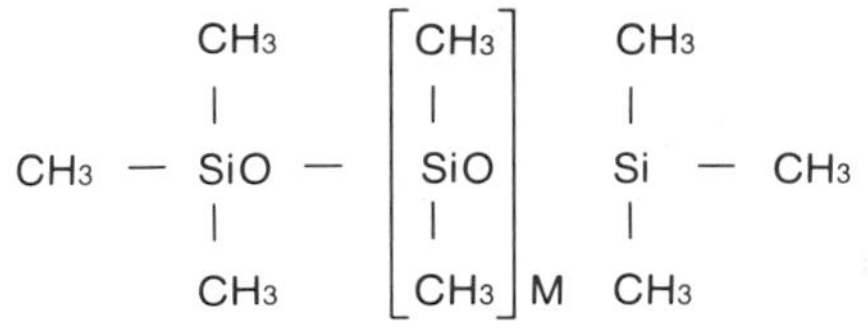

$$CH_3 - \begin{matrix} CH_3 \\ | \\ SiO \\ | \\ CH_3 \end{matrix} - \left[\begin{matrix} CH_3 \\ | \\ SiO \\ | \\ CH_3 \end{matrix} \right]_M \begin{matrix} CH_3 \\ | \\ Si \\ | \\ CH_3 \end{matrix} - CH_3$$

Figure 8.7 Polymethylsiloxane fluid structure.

Table 8.4 Physiological Properties of Dimethyl Silicone Oils

Skin	No reaction
Ingestion	LD50 15 g/kg
Inhalation	No injuries known (low vapor pressure)
Eyes	Transient irritation
Patch testing	No skin irritation

1. Chemical Inertness

Silicone fluids are inert to most common materials of construction. Generally, silicone polymers are unaffected by rubber, plastic, and most metals. However, the metals outlined in Table 8.5 either cause or inhibit gellation of silicones.

Typically, water and ordinary aqueous solutions of inorganic acids or bases will not react with siloxanes intended for mechanical applications. However, strong acids or alkalis will cause molecular rearrangement and speed up gellation of silicone fluids under oxidative conditions.

2. Self-Extinguishing

Silicone polymers produce a white silica ash if burned and will eventually self-extinguish. See Section IV, C, 7 (Flammability) for additional information.

3. Oxidation Stability

In oxidative breakdown, oxygen reacts with the organic groups of the molecules, causing the fluids to lose volatiles and increase in viscosity until gellation occurs. The reaction is dependent on the temperature and supply of air present (Fig. 8.9).

At the point called the oxidation threshold temperature, a significant amount of oxidation by-products starts to appear. Table 8.6 compares the oxidation thresholds of several silicone and organic fluids.

Although silicone lubricants are characterized as having a high oxidation threshold, they have poor tolerance for oxygen. Antioxidants are sometimes added to silicone lubricants to extend their lives at high temperature.

4. Thermal Stability

The best thermal stability for silicones is achieved in the absence of air or in an inert atmosphere such as nitrogen or carbon dioxide.

Under strictly thermal conditions where oxidation is not a factor, the bonds linking

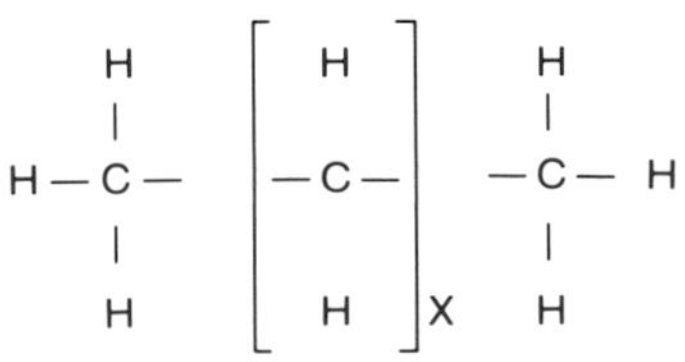

```
     H        ┌   H   ┐      H
     |        │       │      |
H — C —       │ — C — │    — C — H
     |        │       │      |
     H        └   H   ┘X     H
```

Figure 8.8 Hydrocarbon fluid structure.

Table 8.5 Effect of Metals on Silicone at High Temperatures[a]

Metal	Dimethyl (400°F)	Methyl phenyl (450°F)
Copper	Inhibits gellation[b]	None
Phosphor bronze	Inhibits gellation[b]	None
Lead	Inhibits gellation	Volatilization
Yellow brass	Causes gellation	None
Selenium	Causes gellation	Causes gellation
Tellurium	Causes gellation	causes gellation

[a]*Silicones S-9B Technical Data Book, Silicone Fluids*, General Electric Silicones, Waterford, NY.
[b]Copper and phosphor bronze inhibit gellation up to 400°F. At higher temperatures little effect is noted.

silicon and oxygen can be broken at very high temperatures. The result is the forming of lower-molecular-weight, volatile silicones. An activation temperature of approximately 316°C (600°F) is typically observed.

Thermal stability is typically improved by functionalizing basic dimethyl siloxanes with phenyl groups.

The decomposition products of basic siloxanes are noncorrosive, nonabrasive, and essentially nontoxic. However, halogenated silicones (e.g., halophenyls and haloalkyls) decompose into acidic products at temperatures above 316°C (600°F).

5. Hydrolytic Stability

When exposed to water at temperatures that will hydrolyze many other high-temperature hydraulic fluids, silicone lubricating or hydraulic fluids will remain unaffected. Problems like gel precipitation and viscosity reduction will not be encountered.

6. Corrosion and Staining

Pure silicones contain no acid-producing chemicals to cause staining or corrosion. However, chlorophenyl polysiloxanes have been known to produce negligible to slight staining, with the exception of copper. For this reason, the use of pure copper is not recommended with chlorophenyl polysiloxane. Structural surfaces have been known to discolor at 204°C (400°F).

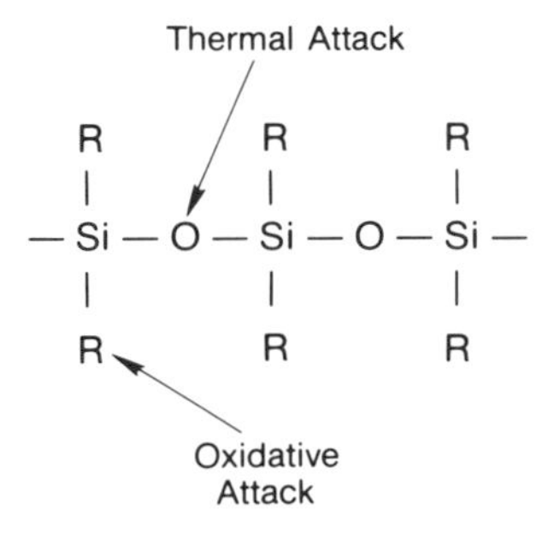

Figure 8.9 Thermal and oxidative attack mechanisms.

Table 8.6 Oxidation Threshold

Fluid	Oxidation Threshold Temperature, °C (°F)
Methyl phenyl siloxane (45 wt% MePH)	271 (520)
Chlorophenyl polysiloxane	221 (430)
Dimethyl polysiloxane	204 (400)
Dibasic acid ester	66 (150)
Petroleum oil	66 (150)

7. Effect on Plastics and Rubber

Although silicones are generally not affected by rubber or plastics, these materials can be affected by long immersion in silicone fluids. Very low-molecular-weight fluids (10 cSt or less) have been shown to have the most adverse effects, acting as solvents with plastics, and leaching out plasticizer in rubbers. Table 8.7 shows the effect on plastic after 30-day immersion in dimethyl and methylphenyl silicones.

Plastics, coatings, and resins are generally not affected by moderate-viscosity silicone lubricants at ambient temperatures. However, silicone fluids, like many organic materials, may cause stress cracking in polyethylene. Therefore, this plastic should be stress-relieved if it is to operate reliably in contact with silicone fluids for long periods of time. Cellulose acetate butyrate is stiffened and crazed by dimethyl fluids. Polyacetal (sold under the trade name Delrin) is stiffened and crazed by both dimethyl and methylphenyl fluids.

Table 8.7 Effect on Plastics after 30-Day Immersion

Plastic[a,b]	Dimethyl SF96 (350 cSt)	Phenyl containing SF
Nylon	No effect	No effect
Polystyrene	No effect	No effect
Methacrylics	No effect	No effect
Modified methacrylics	No effect	No effect
Polycarbonates (Lexan)	No effect	No Effect
Phenolics	No effect	No effect
Cellulose acetate butyrate	Stiffened	No effect
Polyacetal (Delrin)	Stiffened and crazed	Stiffened and crazed
Polyethylene	Some stress cracking	Some stress cracking
Linear polyethylene	Some stress cracking	Some stress cracking
Linear polypropylene	Some stress cracking	Some stress cracking
Polyvinyl chloride	Shrinks and hardens	Shrinks and hardens
PTFE (Teflon)	No effect	No effect

[a]Linear polyethylene and linear polypropylene are not as susceptible to stress cracking or crazing as ordinary polyethylene.
[b]Teflon is a registered trademark of E.I. duPont de Nemours & Co., Inc.; Lexan is a registered trademark of GE Company; and Delrin is a registered trademark of Dow Chemical Co.

When silicone fluids are used to provide surface coatings on rubber in order to impart slip to rubber parts, no effect is noted. Rubber containing little or no plasticizer compatible with silicone fluids is unaffected. Table 8.8 lists the temperature ranges recommended for various rubber uses with silicone.

When silicones are used with these rubbers outside the recommended temperature range, the effect is usually a reduction of weight and volume and an increase in hardness caused by leaching of the plasticizer.

C. Physical and Related Properties

Silicones' excellent viscosity-versus-temperature characteristics, low-temperature properties, and shear stability along with their high compressibility, good radiation resistance, and low surface tension are the physical properties that ensure their usefulness as lubricants.

1. Viscosity-Temperature Properties

Silicones experience a relatively small change in viscosity with regard to temperature. Other fluids like petroleum oils and dibasic acid esters exhibit larger changes in viscosity with temperature than most silicones (Fig. 8.10).

The viscosity–temperature coefficient (VTC) is an indication of how much the viscosity changes with temperature. The lower the value, the less the viscosity changes with temperature. The VTC is defined as

$$1 - \frac{\text{Viscosity at } 99°\text{C } (210°\text{F})}{\text{Viscosity at } 38°\text{C } (100°\text{F})}$$

Silicones generally have a VTC of only 0.6. This is different from organic fluids, which typically have a value of 0.8 or higher. Figure 8.10 shows the viscosity–temperature relationship of various fluids.

2. Low-Temperature Properties

Low-temperature fluidity is an important advantage for silicones over conventional oils. It is measured by pour point: the temperature at which a fluid is so cold that it loses its ability to flow. Although analogous to the freezing point of a pure compound, the pour point of a polymeric fluid is not sharply defined and can vary with the test method. The most common method is ASTM D-97.

Table 8.8 Rubber in Silicone Fluid Systems

Type	Product[a]	Recommended service temperature range, °F
Chloroprene	Neoprene	–40 to 200
Isobutylene isoprene	Butyl	–40 to 200
Nitrile-butadiene	Nitrile Buna N	–40 to 300
Fluororubber	Viton Fluorel	–20 to 450

[a]Neoprene and Viton are registered trademarks of E.I. duPont de Nemours & Co., Inc., and Fluorel is a registered trademark of Minnesota Mining & Manufacturing Co.

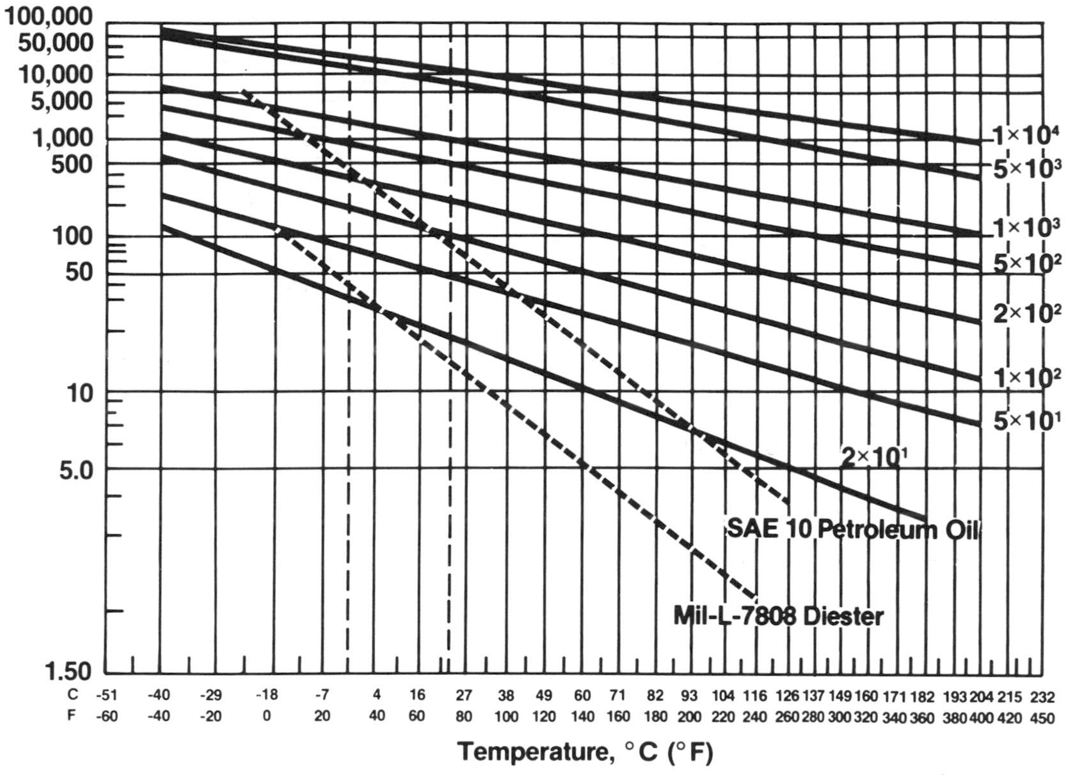

Figure 8.10 Viscosity–temperature relationship of dimethyl fluids (20 to 10,000 cS at 25°C) versus selected hydrocarbon oils.

Basic dimethyl polysiloxanes above 50 cSt have characteristic pour points of –50 to –54°C (–58 to –65°F). By specially formulating fluids, pour points down to extremely low values can be achieved. Typically, branched siloxanes remain pourable at the extremely low temperature of –84°C (–120°F).

3. Shear Stability

Low-molecular-weight or low-viscosity silicone fluids are essentially Newtonian in behavior. This means the measured viscosity does not change with different rates of shear. Dimethyl polysiloxanes of 40–1,000 cSt at 25°C (77°F) are typical examples of this category.

Fluids with viscosities above 1,000 cSt at 25°C (77°F) exhibit pseudo-plastic flow. This means that there is deviation from linearity when the apparent viscosity is plotted against the rate of shear. Further, as the viscosity increases, the deviation becomes more pronounced (Fig. 8.11).

Additionally, the loss of viscosity is much smaller than that of organic fluids, and silicone fluids will return to their original absolute viscosity as long as the temperature remains below 316°C (600°F).

Table 8.9 shows the stability of a methyi chlorophenyl siloxane in a pumping test. Compared with MIL-H-5606 in the same type of pump, the methyl chlorophenyl fluid was pumped hotter and longer, with a resulting negligible viscosity change and pump wear.

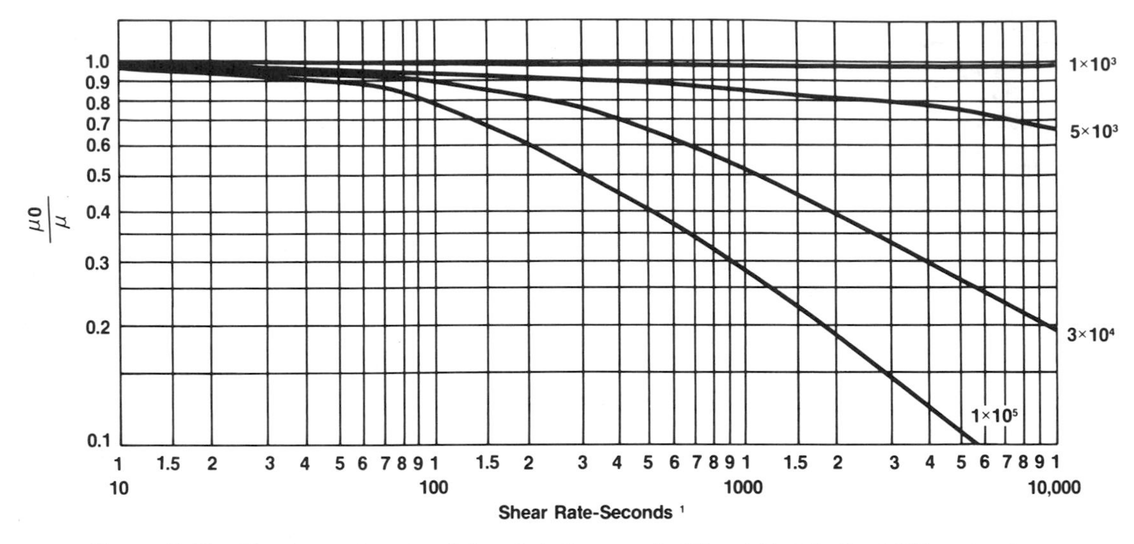

Figure 8.11 Transitory viscosity of dimethyl siloxanes (1,000 to 100,000 cS at 25°C) under shear conditions.

4. Compressibility

All fluids show some tendency to decrease in volume when subjected to high pressures. Silicone fluids are considerably more compressible than natural oils like petroleum products.

It has been stated that a siloxane bond is like a ball and socket—free to move in any direction. When a load is applied to the "chain," the linkage deflects, compressing the chain. The compression here is greater for siloxane because the spacing between the silicone and oxygen is greater than that of a hydrocarbon bond. Temperature has only a small effect over a range of –40 to 149°C (–40 to 300°F).

The low-viscosity silicones have been found to be most compressible. For example, a dimethyl siloxane of 0.65 cSt was found to offer a 17% volume reduction at 50,000 psi. However, the low volatility and effect on materials of construction make fluid of this

Table 8.9 Shear Stability in Pumping Tests[a]

Fluid	Methyl chlorophenyl siloxane	MIL-H-5606
Pressure, psig	4,000	4,000
Temperature, °F	215	175
Test duration, h	1,200	500
Recirculation, cycles	44,000	20,000
Flow rate during test	Constant	Dropped, >18%
Fluid replaced	None	>50% (Leakage)
Pump wear	Nil	Serious
Viscosity change, cSt, at 38°C (100°F)	–3.5%	–25%

[a]Pump used: Lockheed MK 7 radial piston.

viscosity inappropriate for many lubrication applications. A dimethyl siloxane of 1,000 cSt was found to offer only a 15% decrease in volume reduction under a pressure of 50,000 psi.

Figure 8.12 and Table 8.10 provide pressure versus percent compressibility data for selected silicone fluids.

5. Radiation Resistance

It has been shown that radiation resistance is a function of aromatic content for both silicones and organic fluids. The dimethyl siloxanes contain no aromatic groups in their structure. Methyl and phenyl siloxanes and other aromatic siloxanes, therefore, demonstrate a much greater radiation resistance.

Dimethyl silicone fluids show relatively poor resistance to gamma radiation. Tests have shown that a dose of 1×10^7 roentgens will produce a large increase in the viscosity of dimethyl fluids. A dosage of 1×10^8 roentgens is usually sufficient to cause gellation. A methyl phenyl siloxane with 40–50 mol% phenyl would show minimum change at 1×10^8 roentgens. A methyl chlorophenyl siloxane of 8–13.5 mol% tetrachlorophenyl would demonstrate better resistance than a dimethyl siloxane, but not as good resistance as that of a methyl phenyl siloxane, methyl alkyl siloxane, petroleum oil, dibasic acid ester, silicate, or disiloxane ester.

Theoretical treatment of the effects of radiation on silicone can be found in references 20 and 21.

6. Surface Tension

Surface tension of silicone fluids is usually low, making them appropriate for applications where high surface activity and great spreading power are necessary.

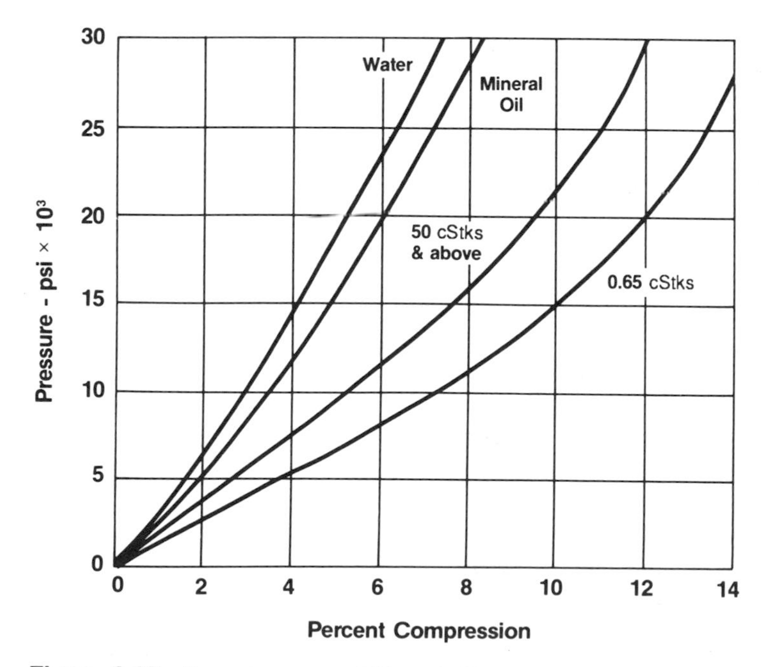

Figure 8.12 Percent compressibility of silicone fluids versus mineral oil and water.

Table 8.10 Percent Compressibility of Selected Silicone Fluids

Fluid	7,100 psi	35,500 psi	284,000 psi	568,000 psi
100-cSt Polydimethylsiloxane	4.5	12.7	28.6	34.0
1,000-cSt Polydimethylsiloxane	4.6	12.7	28.2	33.5
12,500-cSt Polydimethylsiloxane	4.5	12.5	28.1	33.5

Dimethyl silicones have the lowest surface tension values, and these are largely independent of viscosity [about 21 dyn/cm at 25°C (77°F) over a viscosity range of 20–100,000 centistokes]. Methyl phenyl fluids have slightly higher surface tension values [about 24–25 dyn/cm at 25°C (77°F)], but these values are still much lower than those of organic materials.

The surface tension of organic fluids is typically in the range of 30–40 dyn/cm. The value of water at room temperature is about 72 dyn/cm.

7. Other: Flammability

Characteristic of silicones are high flash, fire, and autoignition temperatures. Conventional silicone fluids of 50 cSt or more in viscosity have flash points of 238–302°C (460–575°F) when measured by the conventional "closed-cup method."

The fire points of silicones are significantly higher than their flash points. It has been estimated that their closed-cup fire points can be as much as 93°C (200°F) higher than their flash points. The difference between the flash and fire points accounts for the self-extinguishing characteristics of nonvolatile, high-molecular-weight silicone fluids. Conventional nonsilicone fluids frequently have flash and fire points within a few degrees of each other, reducing the likelihood that they will self-extinguish.

The autoignition temperatures for conventional silicones are estimated to be in the 438–460°C (820–860°F) range.

D. Performance Data

Typically, silicones operate at temperatures between –73 and 232°C (–100 to 450°F). Specially formulated silicones like the chlorophenylmethyl silicones have been known to operate at temperatures up to 315°C (600°F). Other modified silicones like the methyl alkyls cannot operate continuously above 350°F because of the limited oxidative stability inherent in the fluid.

In one of the most popular lubricating applications for high-performance silicones (ball bearing lubrication), the concept of "speed factor" or DN values (bearing bore diameter in mm, times speed in rpm) is used to define the operating capability of silicone and other bearing lubricants. Silicone fluids used as bearing lubricants are recommended for use between 200,000 and 350,000 DN at light to medium loads. This recommendation is also true for most silicone greases that were designed primarily for antifriction bearings.

Silicone greases are known to provide extended bearing life over a wide operating temperature range. They are limited by the physical properties of the base silicone oil, in addition to the thickener chosen.

While the DN value defines the operating capability of lubricants, tests like the Shell four-ball wear test, the Falex test, and the Ryder gear test are widely accepted screening and performance tests for lubricants. Others, like the Navy gear test, have poor reproducibility but have been used in the past to characterize the performance of lubricants.

It has been demonstrated that chlorophenylmethyl siloxanes and methyl alkyl siloxanes, when properly formulated, compare favorably to traditional lubricants. Fluids or compounded methyl alkyl products have been known to perform even better than traditional lubricants in applications requiring aluminum-to-aluminum contact or lubrication of dissimilar metals. Conventional silicones compare poorly to traditional metal-to-metal lubricants in these tests.

Other specialty formulations like trifluoropropyl methyl fluids or even silicone glycol fluids have improved lubricating properties under certain conditions. However, physical properties or the inherent stability of the siloxane chain is often compromised.

1. Four-Ball Wear Test

Performance under sliding contact is measured by the four-ball wear test. Table 8.11 compares typical wear scar data for a number of fluids in the temperature range of 25–75°C (77–167°F) in steel-to-steel applications.

Methyl alkyl silicones are not included in the data in Table 8.11. However, in other studies, methyl alkyl fluids were compared to SAE 30, a traditional petroleum lubricant.

Table 8.11 Shell Four-Ball Wear Test—Comparison of Typical Data[a]

Test fluid	Test conditions	Wear scars, mm at	
		10 kg	50 kg
Petroleum base			
Mineral oil	167°F, 600 rpm	0.44	0.59
Mil-H-5606	Ambient, 600 rpm	0.30	0.55
SAE 10 engine oil	Ambient, 600 rpm	0.37	0.50
Dibasic Acid Esters			
Uncompounded			
Dibasic ester	Ambient, 600 rpm	0.52	0.79
Mil-L-6085A	Ambient, 600 rpm	0.60	0.85
Mil-L-7808	167°F, 600 rpm	0.22	0.40
Orthosilicate ester	167°F, 600 rpm	0.71	1.10
Phosphate ester	Ambient, 600 rpm	0.46	0.57
Polyphenyl ether	167°F, 600 rpm	0.56	1.25
Silicones			
Chlorophenylmethyl siloxane (8–13.5 mol% Cl_4PH)	Ambient, 600 rpm	0.39	0.53
Dimethyl silicone	Ambient, 600 rpm	0.50	1.83
Dimethyl silicone	400°F, 1,200 rpm	0.91	—
Methyl phenyl silicone (45 mol% MePH)	Ambient, 600 rpm	1.39	4.18
Chlorinated methyl phenyl silicone	Ambient, 600 rpm	0.80	2.32
Trifluoropropyl methyl silicone	400°F, 1,200 rpm	0.54	—
Trifluoropropyl methyl-dimethyl copolymer	400°F, 1,200 rpm	0.78	—

[a]Steel balls: AISI 52-100. Time: 1 h, except 400°F/1200 rpm tested fluids, which are 2 h.
[b]Tests performed at ambient temperature are initiated at approximately 25°C (77°F) but no effort at temperature control is attempted. Temperatures will rise 10–50°F (depending on the fluid); such data are roughly comparable to that run at 75°C (167°F).

The fluids were used to lubricate sliding aluminum S2 versus stationary tool steel with a load of 40 kg. The test temperature is 75°C (167°F) at speeds of 600 rpm. The methyl alkyl siloxane wear scar measured only 0.55 mm, as compared to a 0.70 mm scar for SAE 30 petroleum lubricant.

2. Falex Test

This test is designed to measure the antiweld properties of lubricants and their resistance to wear under extreme pressure conditions. Table 8.12 shows data on various lubricants.

3. Ryder Gear Test

The Ryder Gear Test (data generated using F.S. 791, method 6508) measures gross surface damage between case-hardened spur gears. Ratings are in terms of "scuff load," the load at which 22.5% of the gear tooth area is scored. Methyl phenyl silicones show some advantages in these tests. Performance is primarily dependent on the lubricant's "pressure-viscosity" properties: the greater the viscosity increase under pressure, the better the gear lubricant. Table 8.13 presents data on selected fluids.

E. Applications

Silicone fluid applications vary widely. They are used as a base fluid for a variety of products such as emulsions, solutions, greases, and compounds. Although there are certain uses for silicone lubricating oils, more often the lubricant is in the form of a grease. Silicone fluids have found limited use as lubricants in applications calling for a material that can withstand high temperatures and moderately loaded conditions, but are generally ineffective as lubricants for heavily loaded metal surfaces.

However, advances in technology have improved the load-carrying capabilities and lubricity of silicones; for example, fluorosilicone lubricants are effective in applications such as pumps and valves for fuel or solvent tanks where the metal surface and lubricant are exposed to hydrocarbon solvents.

Another drawback of silicone fluids, their adverse effect on the adhesion of postdecorative coatings such as paint, has also been addressed by technology. Methyl alkyl-substituted polysiloxanes possess unique properties that allow them to be used in areas where soldering and painting are being done.

A number of silicone fluids have been blended successfully with synthetic, organic fluids to combine the best performance qualtities of each. An example that has been marketed successfully is a lithium soap-based diester-polysiloxane blend. This product provides enhanced lubricating properties and lower cost when compared with the polysiloxane equivalent.

Silicone fluids are also used extensively as release agents in a variety of molding

Table 8.12 Falex Test: Loads to Seizure

Dibasic acid ester	1,150–2,000 psi
Chlorophenylmethyl siloxane (8–13.5 mol% Cl_4PH)	1,100–1,200 psi
Petroleum oil	600–900 psi
Methyl alkyl silicone	To 750 psi
Methyl phenyl silicone	50–100 psi
Dimethyl silicone	60–80 psi

Table 8.13 Ryder Gear Test: Scuff Loads

Methyl phenyl silicone	4,000 psi
Dibasic acid ester	2,800 psi
Versilube F-50®	2,400 psi
Petroleum oil	1,200 psi
Dimethyl silicone	1,000 psi

operations. They are desirable due to their resistance to high temperatures, smoking, and fuming, and to the small amount of product needed, which offsets their higher price.

Other applications include personal care products, paper coatings, hydraulic fluids, damping fluids, and polishing fluids. Table 8.14 lists the various types of silicone fluids, their properties, and the applications in which they are used.

Table 8.14 Applications for Common Silicone Fluids

Silicone type	Properties	Applications
Dimethyl	Excellent viscosity temperature characteristics, hydrolytic stability, reduces surface tension, imparts excellent slip to rubber and plastic, acts as a water repellent	Plastic bearings, sheeting, cutting tools, molded and extruded parts, sewing thread, base fluid for compounds, hydraulic fluids, damping fluids
Fluoro	Good lubrication and resistance to chemicals and solvents, extended bearing life, excellent high-temperature properties, high load-carrying characteristics	Base fluid for greases, hydraulic fluids, bearings, chemical process compressors, vacuum pumps, other chemical and corrosive environments
Methyl phenyl	Increased thermal stability, good high- and low-temperature stability, excellent radiation resistance, improved oxidation resistance	Base fluid for greases (maintenance and lube for life applications), impart slip to rubber and plastic, hydraulic fluid, thread and fiber
Methyl alkyl	Develops thick films under dynamic conditions, most compatible with organic materials, does not contaminate surfaces to be painted	Base fluid for greases, difficult metal combinations, die casting, metal working, cutting oils, penetrating oil
Chlorophenylmethyl	Greatly improved high-temperature lubricity; chlorine provides chemical reactivity with metal necessary for effective boundary lubrication, improved pour point	Miniature bearings, base fluid for greases, bearings in high ambient temperature industries, clocks and timing devices, hydraulic systems, tape recorders, vacuum pumps

V. MANUFACTURE, MARKETING, AND ECONOMICS

There are seven fully integrated competitors in the global silicones market (Table 8.15). Although the sizes of these producers vary widely, each has a fairly complete product line. These are also a large number of specialty finishing companies who purchase siloxane intermediates and process them into different elastomers, coatings, greases, compounds, and adhesives.

Overall demand for silicone fluids in the United States, Europe, and Japan grew at a rate of 10–15% during the 1970s as market potential and value in a wide range of applications were being realized. As these markets became more fully exploited, growth slowed due to maturing applications and displacement by lower cost materials.

However, during this period, new, more complex organomodified silicone polymers were introduced, increasing the overall use of silicone fluids as lubricants. In addition, a host of specialty silicone elastomers, coatings, and adhesives is expected to drive growth through the 1990s and beyond, with an anticipated growth rate of 8–12% for the next few years.

There is generally strong price competition among manufacturers of basic silicone fluids due to the similarity of their products. However, in the case of more complex, specialty fluids, the issue is not competition, but rather price sensitivity on the part of consumers. Because of their higher cost compared to competing organic chemicals,

Table 8.15 Major Global Producers of Siloxane Fluids

Company	Manufacturing sites
North America	
Dow Corning Corporation (a joint venture of the Dow Chemical Co. and Corning Glass Works)	Midland, MI Carrolton, KY
GE Silicones	Waterford, NY
Union Carbide Corporation[a] (Specialty Chemical Division)	Sistersville, WV South Charleston, WV
Europe	
Dow Corning Ltd.	Barry, Wales
Wacker-Chemie GmbH	Burghausen, Germany
Rhone-Poulenc Specialites Chimique SA	Saint-Fons, France
Bayer AG	Leverkusen, Germany
Japan	
Dow Corning Toray Silicones Co. Inc.	Ichihara, Chiba Prefecture
Toshiba Silicone Co. (a joint venture between Toshiba Corp. and GE Company)	Ohta, Gumma Prefecture
Shin-Etsu Chemical Co.	Annaka, Gumma Prefecture Takefu, Fukui Prefecture Naoetsu, Niigata Prefecture

[a]No longer producing methyl chlorosilanes.

silicones are considered high-end products and are selected for applications demanding their unique properties. Successful marketing and sales of these silicone fluids focuses on their special qualities or value-added economics.

Within the past few years, an increasing share of silicone fluid production has been dedicated to specialty products. This trend drives new technology and product development.

The highly engineered nature of specialty silicone products demands an investment in technology. The high cost of this technology has led to different product marketing strategies among the producers. Some have bundled specialty silicone products with organic specialties and targeted them at specific lubricant markets. Others have consolidated, formed joint ventures, or sold their interests to competitors. However, all these approaches are based on bringing value to the customer in specialty products like specialty lubricants.

The silicones industry has experienced a major shift from growth within traditional geographic boundaries to a global marketing approach. Silicone producers seek to improve existing product lines, develop new products, and enter new markets.

REFERENCES

1. Miller, J. W. (1984). Synthetic lubricants and their industrial applications, *Appl. Rev. J. Syn. Lub. 1*, No. 2 (July), 136–152.
2. *GE Silicones Technical Data Book S-9.*
3. Hardman, B. and A. Torkelson (1989). Silicones, Reprinted from *Encyclopedia of Polymer Science and Engineering*, vol. 15, pp. 204–303, John Wiley & Sons, New York.
4. *Synthetic Lubricants* (1962). Reinhold Publishing Co.
5. Liebhafsky, H. A. (1978). *Silicones Under the Monogram*, John Wiley & Sons, New York.
6. Noll, W. (1968). *Chemistry and Technology of Silicones*, Academic Press, New York.
7. Rochow, E. G. (1987). *Silicon and Silicones*, Springer-Verlag, Berlin.
8. Burkhard, C. A., E. G. Rochow, H. S. Booth, and J. Hartt (1947). *Chem. Rev.*, 97.
9. U.S. Patent 2,380,995 (August 7, 1945), E. G. Rochow (to General Electric Co.).
10. Rochow, E. G., and W. F. Gilliam (1945). *J. Am. Chem. Soc.*, 67, 963.
11. Smith, R. E. (1975). Silicone lubricants for the chemical processing industry, *J. Am. Soc. Lub Eng.*, ASLE 30th Annual Meeting, Atlanta, GA (May 5–8).
12. *Versilube Silicone Lubricants*, Historical Technical Data Book S-10B, GE Silicones Products Dept.
13. Barnes, J. E., and J. H. Wright (1988). Silicone greases and compounds: Their components, properties and applications, Presented at the 55th NCGI Meeting.
14. Lonsky, P. (1985). Some characteristics of silicones developed as lubricants, *J. Syn. Lub.*, *1*, No. 4. (January), 302–313.
15. C. Gunderson and A. Wihart (eds.) (1962). *Synthetic Lubricants*, Dow Chemical Co., pp. 299–305.
16. Wilcock, D. F. (1946). *General Electric Rev.*, 49, 11, 1228.
17. Brigeman, P. W. (1949). *Proc. Am. Acad. Sci.*, 77, 115.
18. Dugan, W. J. (1951). General Electric Viscasil Silicone Fluids for Mechanical Operations, Report No. CHSD-49, Waterford, NY.
19. Smart, M. (1988). Silicone fluids, *Chemical Economics Handbook: Chemical Products Group, Plastics.*
20. Miller, A. A. (1960). Radiation chemistry of polydimethylsiloxane: I. Crosslinking and gas yields *J. Am. Chem. Soc.*, *82*, (July 20), 3519–3523.
21. Miller, A. A. (1964). Radiation stabilities of arylmethylsiloxanes. *Ind. Eng. Chem. Prod. Res. Dev.*, *3*, No. 3 (September), 252–256.

9
Silahydrocarbons

F. Alexander Pettigrew and Gunner E. Nelson
Ethyl Corporation
Baton Rouge, Louisiana

I. INTRODUCTION

The term silahydrocarbons is used to describe a class of compounds that is more correctly termed tetraalkylsilanes (SiR_4). The alkyl groups on silicon can be the same or different. Additionally, the alkyl groups can be straight-chained (i.e., *n*-alkyl) or branched. For reasons discussed later, the silahydrocarbons of most interest have been those that contain *n*-alkyl groups. For this discussion the term silahydrocarbon will be used to generically refer to the class of compounds. The term tetraalkylsilane will be used as appropriate to clarify discussions of the chemistry.

II. HISTORICAL DEVELOPMENT

While the history of silahydrocarbons goes back well over 100 years, it was in the 1950s that the modern history of their development as functional fluids began. The role of the U.S. Air Force, for the most part, defines that history up until the very recent past.

The U.S. Air Force Materials Laboratories began a general investigation into the synthesis and properties of tetraalkylsilanes in the 1950s (1–3). These fluids were called silahydrocarbons (4) to emphasize their hydrocarbonlike behavior. The driving force for this research was the recognition that new high-performance aircraft would require lubricants and fluids capable of operating in a high-temperature/high-load environment. Over the years, the interest in silahydrocarbons rose and fell, finally reaching a peak in the

1980s. During this same period, target specifications evolved for candidate fluids for use as hydraulic fluids in high-performance aircraft. These target specifications eventually evolved to incorporate the low-temperature requirements of MIL-H-5606 (mineral oil) (5) with the high-temperature requirements of MIL-H-83282 (synthetic polyalphaolefins, PAO) (6). A partial listing of these hybrid target properties is shown in Table 9.1. Kinematic viscosity and most of the other properties were targeted to those of MIL-H-5606D fluid at low temperature and MIL-H-83282 at high temperature.

The early front-runner was a PAO-based fluid. However, it was recognized that the flash point of this new PAO would be lower than the MIL-H-83282 specification (205°C minimum). The lower flash point was a consequence of the lower molecular weight required to obtain the desired kinematic viscosities at lower temperatures. This compromise in flash point, in large part, spurred further development of silahydrocarbons as candidate hydraulic fluids. The inherent differences between silahydrocarbans and PAOs made it possible to produce a silahydrocarbon based fluid with a flash point matching MIL-H-83282 PAO and having the low-temperature viscosity of the more volatile mineral oil. The subject of silahydrocarbon properties versus those of hydrocarbons is discussed later. By the late 1970s, the Air Force had successfully developed a silahydrocarbon composition nearly meeting these specifications (7). It possessed excellent thermal stability, viscosity index, and low-temperature flow characteristics. The properties of the Air Force silahydrocarbon fluid are compared to the target properties in Table 9.2.

III. CHEMISTRY

The literature contains numerous reports on methods of preparation of silahydrocarbons, some dating back over a century. The types of reactions that appear to be useful include alkylation of silicon halides and hydrosilation of olefins. Alkylation can be affected by lithium, magnesium, zinc, and aluminum alkyl compounds. Each of these will be discussed in turn. A note regarding nomenclature is appropriate at this point. The valence of silicon in silane-type compounds is always four. When naming silanes, it is customary to leave out any hydrogens directly bonded to the silicon. Thus, methyldichlorosilane is CH_3Cl_2SiH.

A. Magnesium

The first synthesis of tetraalkylsilanes (silahydrocarbon) was reported by Friedel and Crafts in the mid 1800s (8,9). They reported the reaction of alkyl magnesium halides

Table 9.1 Original Target Properties of Low-Temperature MIL-H-83282

Fluid property	Target	MIL-H-5606	MIL-H-83282
Kinematic viscosity, cSt			
At –54°C, maximum	2,500	2,500	N.A.
At –40°C, maximum	500	600	2,600
At 100°C, minimum	3.5	4.9	3.5
Point point, °C, maximum	–59.4	–60	–55
Shear stability, MIL-H-5606D (% viscosity change)	0	0	0
Flash point, °C (open cup), minimum	163	82	205
Fire point, °C, minimum	191	N.A.	245

Table 9.2 Silahydrocarbon Performance versus Target Properties for Low-Temperature MIL-H-83282

Fluid property	Determined	Target
Kinematic viscosity, cSt		
At –54°C	2,410	2,500 maximum
At –40°C	564	500 maximum
At 100°C	2.58	3.5 minimum
Pour point,°C	<–65	–59.4 maximum
Shear stability, % viscosity loss	0	0
Flash point, °C	227	163 minimum
Fire point, °C	238	191 minimum

(Grignard reagent) with tetrachlorosilane to give low-molecular-weight symmetrical tetraalkylsilanes such as tetraethylsilane.

$$4RMgX + SiCl_4 \rightarrow R_4Si + 4MgXCl$$

Since then many workers have studied the preparation of tetraalkylsilanes via magnesium reagents. Gilman and Clark (10) studied the reaction of alkyl magnesium halides with alkyl trichlorosilane.

$$3R'MgX + RSiCl_3 \rightarrow RSiR'_3 + 3MgXCl$$

The Air Force researchers (4,11) developed a route based on the reaction of mixed alkyl magnesium halides or mixed alkyl lithium reagents with tetrachlorosilane or alkyltrichlorosilane. The case for the mixed magnesium reagent reaction with methyltrichlorosilane is shown.

$$RBr + R'Br + 2Mg \rightarrow RMgBr + R'MgBr$$
$$CH_3SiCl_3 + 3[RMgBr + R'MgBr] \rightarrow$$
$$[CH_3SiR_3 + CH_3SiR_2R' + CH_3SiRR'_2 + CH_3SiR'_3] + 3MgBrCl$$
$$R = n\text{-octyl} \qquad R' = n\text{-decyl}$$

A key to obtaining desirable low-temperature properties in the final fluid was the fact that the product was a mixture of compounds of close molecular weight. This suppresses solidification due to crystallization at low temperature. For the fluid whose properties are reported in Table 9.2, R and R' are a 50:50 mixture of *n*-octyl and *n*-decyl.

Lennon at Monsanto developed two routes to methyltrialkylsilanes based on the catalyzed reaction of dialkyl magnesium with methyltrichlorosilane (12–14). The preparation of magnesium hydride and its subsequent reaction with α-olefins to give dialkyl magnesium was previously known (15).

$$Mg + H_2 \rightarrow MgH_2$$
$$MgH_2 + \alpha\text{-olefin} \rightarrow MgR_2$$
$$3MgR_2 + 2CH_3SiCl_3 \text{ (cat)} \rightarrow 2CH_3SiR_3 + 3MgCl_2$$

Both cyanide compounds, such as cuprous cyanide (12), and thiocyanate salts (13) are used as catalysts. With either class of catalyst, Grignard reagents may take the place of the dialkyl magnesium. If a mixture of α-olefins is used, then a mixture of products, such as that described by the Air Force researchers, results.

B. Lithium

Prior to the work by the Air Force discussed above, where an alkyl lithium was shown to be able to take the place of an alkyl magnesium halide (10), Gruttner and Wiernik reported the reaction of an alkyl lithium with a trialkylsilane to give the first unsymmetrically substituted tetraalkylsilane (16).

$$R'Li + R_3SiH \rightarrow R_3SiR' + LiH$$

Gilman and Massie reported the reaction of alkyl lithiums with tetrachlorosilane to give symmetrical tetraalkylsilanes (17).

$$4\ RLi + SiCl_4 \rightarrow R_4Si + 4\ LiCl$$

C. Zinc

There is a report by Bygden of the reaction of dialkyl zinc with tetrachlorosilane to give symmetrical tetraalkylsilanes (18).

$$2ZnR_2 + SiCl_4 \rightarrow R_4Si + 2ZnCl_2$$

D. Aluminum

Jenkner's route involves the reaction of a trialkyl aluminum with methyltrichlorosilane catalyzed by a metal chloride to give low-molecular-weight tetraalkylsilanes (19,20).

$$CH_3SiCl_3 + AlR_3 + MCl \rightarrow CH_3SiR_3 + MAlCl_4$$
$$R = CH_3 \text{ or } CH_3CH_2$$

Bakshi et al. later developed a similar route that gave higher-molecular-weight products (21). For example,

$$CH_3SiCl_3 + Al(C_8H_{17})_3 + NaCl \rightarrow CH_3Si(C_8H_{17})_3 + NaAlCl_4$$

Depending on the exact conditions, various levels of tetra-*n*-octylsilane and dimethyl-di-*n*-octyl silane could be produced as coproducts. The use of one-half equivalent of sodium chloride based on the tri-*n*-octyl aluminum was found to maximize the yield of the desired product, methyl-tri-*n*-octyl silane.

Nelson and co-workers at Ethyl developed a route to tetraalkylsilanes based on the alkylation of chlorosilanes with sodium aluminum alkylates (22–26). The route involves first the preparation of the sodium aluminum tetraalkylate from the reaction of sodium aluminum hydride with α-olefins. Then the intermediate tetraalkylate is next used to alkylate a chlorosilane such as methyltrichlorosilane.

$$NaAlH_4 + 4(C_8/C_{10}\ \alpha\text{-olefin}) \rightarrow NaAl(C_8H_{17})_x(C_{10}H_{21})_{4-x}$$
$$3NaAlR_4 + 4CH_3SiCl_3 \rightarrow 4CH_3SiR_3 + 3NaAlCl_4$$

The product is a mixture of the four components described by the Air Force workers. Product distribution can be readily altered by adjusting the C_8/C_{10} olefin ratio, or, if another composition is required, olefins of other carbon numbers can be used. In an interesting modification of this process, olefin/aluminate interchange is used.

$$NaAlH_4 + 4(C_8\ \alpha\text{-olefin}) \rightarrow NaAl(C_8H_{17})_4$$
$$3NaAl(C_8H_{17})_4 + C_{10}\ \alpha\text{-olefin} + 4CH_3SiCl_3 \rightarrow 4CH_3Si(C_8H_{17})_x(C_{10}H_{21})_{3-x} + 3NaAlCl_4$$

Presumably, a rapid exchange of alkyl groups on the aluminate occurs during the alkylation process, yielding again the four-component mixture, with the relative mole fractions of each calculable on the basis of the C_8/C_{10} ratio as before.

E. Hydrosilation

Hydrosilation has also been utilized in the preparation of tetraalkylsilanes. While it is likely not of preparatory significance, Austin et al. reported an interesting route to tetraalkylsilanes (27). They studied the photocatalyzed reaction of alkenes with alkylsilanes using trinuclear metal carbonyl catalyst precursors. Thus, irradiation of $M_3(CO)_2$ (M=Fe, Ru, or Os) in the presence of 1-pentene and $HSiEt_3$ gave n-$C_5H_{11}SiEt_3$.

Another interesting approach was studied by El-Durini and Jackson (28). In their studies of free radical reactions, they reported the reaction of tripropylsilane with 1-octene in the presence of a free-radical initiator. A high yield of the product, octyltripropylsilane, was obtained if the starting silane were present in excess.

Onopchenko and Sabourin developed a route based on catalytic hydrosilation of α-olefins by mono-, di-, or trialkylsilane (29,30). Various platinum-based catalysts are claimed. For example,

$$(C_8\ \alpha\text{-olefin}) + C_6H_{13}SiH_3 + H_2PtCl_6{\cdot}6H_2O\ (\text{cat}) \rightarrow (C_8H_{17})_3(C_6H_{13})Si$$

A key to high conversion to the tetraalkylsilane is exposure of the catalyst to oxygen. Under similar conditions, using a rhodium-based catalyst, the inventors were able to prepare mixtures of tetraalkylsilanes that contained various levels of unsaturation in the side chains (31). The mixture can be hydrogenated to provide a saturated tetraalkylsilane. The unsaturation can also be sulfurized to provide a compound that can act as a lubrication additive.

Onopchenko and Sabourin developed another route to tetraalkylsilanes based on hydrosilation of α-olefins (32). The scheme involves hydrosilation of an α-olefin by methyldichlorosilane followed by reduction of the resultant methylalkyldichlorosilane to the methylalkylsilane by lithium aluminum hydride. This product is, in turn, reacted with another α-olefin via catalytic hydrosilation. This reaction scheme allows the synthesis of a series of methyldialkylalkylsilanes.

$$CH_3SiCl_2H + \alpha\text{-olefin} + \text{catalyst} \rightarrow CH_3SiCl_2R$$
$$2CH_3SiCl_2R + LiAlH_4 \rightarrow 2CH_3SiH_2R + LiAlCl_4$$
$$CH_3SiH_2R + \alpha\text{-olefin} + \text{catalyst} \rightarrow CH_3SiR'_2R$$

Malcolm et al. developed a route to ethyltrialkylsilanes starting from silane (SiH_4) itself (26,33). This route involves the partial alkylation of silane by a sodium aluminum tetraalkylate. The olefin employed in the last step can be ethylene. The resulting product would thus be an ethyltrialkylsilane, in contrast to the methyltrialkylsilane produced by the methods starting from methyltrichlorosilane.

$$SiH_4 + NaAlR_4 \rightarrow R_2SiH_2 + R_3SiH$$
$$R_2SiH_2 + R_3SiH + Co_2(CO)_8 \rightarrow R_3SiH$$
$$R_3SiH + \alpha\text{-olefin}' + \text{catalyst} \rightarrow R_3SiR'$$

IV. PERFORMANCE

As mentioned earlier, the term silahydrocarbon was chosen to describe this class of compounds because of their hydrocarbonlike character. Indeed, in many ways they are similar to hydrocarbons. It has been shown that silahydrocarbons are compatible with mineral oils and PAOs (32). This compatibility is one aspect of the fluids that made them attractive to the Air Force as drain and fill replacements for existing hydraulic fluids. There are, however, some inherent differences, and this is the basis for the interest in these materials. Most of the work cited in the literature on the properties of silahydrocarbons centers around their potential as hydraulic fluids. Therefore, a property-by-property comparison to mineral oil and PAOs is warranted. Chapter 14 contains a more general discussion of the properties of a wider variety of fluids.

A. Viscosity

As seen in the discussion of the chemistry, it is possible to synthesize a large number of tetraalkylsilanes of the type SiR_4. Since R can be any hydrocarbon group from one carbon up, an enormous number of possibilities exist. A good starting point in understanding the differences between silahydrocarbons and hydrocarbons is a review of physical properties of compounds with similar structures. The melting points of several tetraalkylmethanes and tetraalkylsilanes are shown in Table 9.3 (4). From the data, it is clear that, for a given structure, the substitution of silicon for the quaternary carbon lowers the melting point dramatically. In practice, this translates to a beneficial decrease in pour point for fluids of similar molecular weight.

There are reports in the literature of the synthesis of silahydrocarbons over a wide range of molecular weights (3,26,32–34). A review of the data shows that the viscosity is a linear function of the molecular weight and is not influenced greatly by the molecular structure. The viscosity at 100°C for 59 compounds is plotted versus the number of carbons in the silahydrocarbon in Fig. 9.1. The linearity of the relationship is particularly impressive considering that some of the fluids are pure compounds and some are mixtures. Also, the fluids are composed of four classes of tetraalkylsilanes: dodecyltrialkylsilanes, didodecyldialkylsilanes, methyltrialkylsilanes, and ethyltrialkylsilanes. It is important to note that the viscosity correlation does not necessarily translate to low temperature. As the temperature is lowered, the fluids comprised of single compounds suffer from the problem of crystallization (i.e., freezing). A fluid that is a mixture of compounds of similar but different molecular weights will have a pour point below that at which the single-component fluid would solidify.

A discussion of viscosity is not complete without taking viscosity index and pour point into account. The viscosity indices of a mineral oil, a PAO, and a silahydrocarbon of the same 100°C viscosity are 94, 127, and 151, respectively (35). A high viscosity index

Table 9.3 Comparison of Melting Points of Silicon and Carbon Analogs

Carbon analog	m.p., °C	Silicon Analog	m.p., °C
$(CH_3)_4C$	–16	$(CH_3)_4Si$	–95
$(CH_3CH_2)_4C$	–31	$(CH_3CH_2)_4Si$	–69
$(n\text{-}C_3H_7)_4C$	–26	$(n\text{-}C_3H_7)_4Si$	–46
$(n\text{-}C_4H_9)_4C$	–6	$(n\text{-}C_4H_9)_4Si$	–46
$(n\text{-}C_4H_9)_3C(CH_2CH_3)$	–66	$(n\text{-}C_4H_9)_3Si(CH_2CH_3)$	–91
$(n\text{-}C_4H_9)_3C(n\text{-}C_6H_{13})$	–51	$(n\text{-}C_4H_9)_3Si(n\text{-}C_6H_{13})$	–72

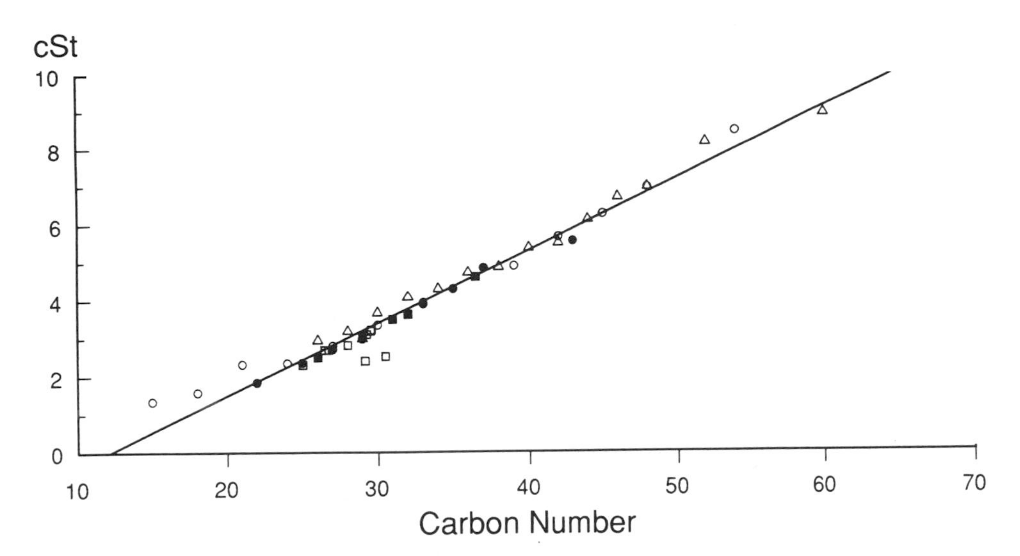

Figure 9.1 Viscosity of silahydrocarbons at 100°C: dodecyltrialkylsilanes ○ (3), didodecyldialkylsilanes △ (3), methyltrialkylsilanes ● (26), methyltrialkylsilanes □ (34), methyltrialkylsilanes ▲ (32), ethyltrialkylsilanes ■ (3,34).

is favorable, indicating that the viscosity does not show large variations with temperature. The pour point of the mineral oil is only –15°C while the PAO and silahydrocarbon pour points are both below –65°C.

B. Lubricity

The lubricity of silahydrocarbons was recently compared to a mineral oil and a PAO (35). In the absence of additives, the lubricity of the mineral oil was slightly better than the PAO, as measured by the size of the wear scar in the four-ball wear test. Both were better than the silahydrocarbon. In the presence of 1% antioxidant and 3% of an antiwear additive, the mineral oil and PAO werc equivalent and were both better than the silahydrocarbon. It is significant that the performance of the silahydrocarbon was improved by the additives, because this is not the case for some classes of fluids. It may be that an additive package developed specifically for silahydrocarbons would give better performance. Earlier work by the Air Force showed that the lubricities of the three fluids were equivalent in the presence of an antioxidant and an antiwear additive (36).

C. Thermal Stability

The Air Force researchers reported the results of a systematic investigation into the relative performance of silahydrocarbons and several other candidate fluids as high-temperature lubricants (36). In regard to thermal stability, they found that silahydrocarbons were inherently more stable than mineral oil or PAO. Later work by researchers at Ethyl confirmed this finding (35). This difference in thermal stability is likely due to the different amounts of branching in the structures. The silahydrocarbon has the advantage of having a well-controlled structure with no branching other than the quaternary branch centered at silicon. It is well established that both mineral oils and PAOs used as hydraulic

Table 9.4 Source of Silahydrocarbons

Fluid	Carbon Number	Source[a]
Dimethyldiethylsilane	6	Fairfield
Tetraethylsilane	8	Aldrich, Fairfield, Fluka, K & K, Lancaster, Pfaltz & Bauer
Tetrabutylsilane	16	Huls, Lancaster
Tri(hexyl/octyl)methylsilane	22	Ethyl Corp.
Tri(octyl/decyl)methylsilane	25–31	Ethyl Corp., Hüls, Monsanto
Tridecylmethylsilane	31	Ethyl, Hüls
Tri(dodecyl/tetradecyl)methylsilane	40	Ethyl

[a]Sources: Ethyl Corporation, Commercial Development Department, 451 Florida, Baton Rouge, LA 70801; Hüls America, Bartram Road, Bristol, PA 19007; Aldrich Chemical Co., Inc., 940 West Saint Paul Ave., Milwaukee, WI 53233; Fairfield Chemical Co., P.O. Box 20, Bluthewood, SC 29016; Fluka Chemical Corp., 980 South Second St., Ronkonkoma, NY 11779; ICN K & K Laboratories, Inc., P.O. Box 28050, 4911 Commerce Parkway Cleveland, OH 44128; Lancaster Synthesis Ltd., P.O. Box 1000, Windham, NH 03087; Pfaltz & Bauer, Inc., Division of Aceto Corporation, 172 East Aurora St., Waterbury, CT 06078; and Functional Products Research, Mail Zone R4B, Monsanto Co., 800 N. Lindbergh Ave., St. Louis, MO 63167.

fluids contain considerable amounts of secondary and tertiary branch points. It has been argued that hydrocarbons of the same structure should be no less thermally stable than a given silahydrocarbon, provided that both materials are pure (37). Other work has shown that the situation may be more complex. Researchers at Chevron found that incorporation of a tertiary hydrogen near the end of an alkyl chain in a silahydrocarbon decreased the thermal stability, as expected (32). Incorporation of a tertiary hydrogen on the carbon alpha to the silicon actually made the molecule more stable.

D. Oxidative Stability

Results cited by both the Air Force (36) and Ethyl (35) researchers show that the oxidative stability of formulated silahydrocarbons and PAOs are equivalent and that both are clearly superior to mineral oils.

E. Volatility and Flammability

The volatility of nonpolar compounds is a function of molecular weight, as is the viscosity. The volatility of a silahydrocarbon, a mineral oil, and a PAO of the same viscosity at 100°C has been measured by thermogravimetric analysis (35). The temperature required to induce 5, 50, and 95% weight loss is cited for each fluid. These data clearly show the silahydrocarbon is the least volatile. This difference in volatility translates to higher flash point as measured by ASTM D-92. The flash points are silahydrocarbon, 234°C; PAO, 218°C; and mineral oil, 197°C.

F. Summary of Properties

The key features of silahydrocarbons are low-temperature fluidity, high thermal stability, low volatility (and hence high flash point), and high viscosity index. In comparison to

PAOs, the superior thermal stability, high viscosity index, and low volatility (high flash point) seem to be the advantages that would lead one to consider a silahydrocarbon as a candidate fluid.

V. COMMERCIAL STATUS

At present, there is no readily available information that indicates significant commercial applications for silahydrocarbons. There is one patent in the literature on a use that falls outside of the hydraulic fluid data. This is the use of silahydrocarbons as a textile lubricant (38). Silahydrocarbons with compositions of potential interest as fluids are available from two companies. Several pure, low-molecular-weight tetraalkylsilanes are available as well. The compounds available and the suppliers are listed in Table 9.4.

REFERENCES

1. Rosenberg, H., J. D. Groves, and C. Tamborski (1960). *J. Org. Chem.,* 25, 243.
2. Tamborski, C., and H. Rosenberg, *J. Org. Chem.,* 25, 246.
3. Baum, G., and C. Tamborski (1961). *J. Chem. Eng. Data,* 6, 142.
4. Snyder, C. E., L. J. Gschwender, C. Tamborski, G. J. Chen, and D. R. Anderson (1982). *ASLE Trans.* 25(3), 299–308.
5. Military Specification MIL-H-5606E, 29 August 1980, Hydraulic Fluid, Petroleum Base; Aircraft, Missile, and Ordnance.
6. Military Specification MIL-H-83282C, 25 March 1986, Hydraulic Fluid, Fire Resistant, Synthetic Hydrocarbon Base, Aircraft, Metric, NATO Code Number H-537.
7. Gschwender, L. J., C. E. Snyder Jr., and G. M. Fultz (1986). *Lub. Eng.,* 485–490.
8. Friedel, C., and J. M. Crafts (1863). *Ann. Chem.,* 127, 28.
9. Friedel, C., and J. M. Crafts (1870). *Ann. Chem.,* 259, 334.
10. Gilman, H., and R. N. Clark (1946). *J. Am. Chem. Soc.,* 68, 1675.
11. Tamborski, C., and C. E. Snyder (1983). U.S. Patent 4,367,343 to the United States of America represented by the Secretary of the Air Force.
12. Lennon, P. J. (1987). U.S. Patent 4,650,891 to Monsanto.
13. Lennon, P. J. (1987). U.S. Patent 4,672,135 to Monsanto.
14. Lennon, P. J., D. P. Mack, and Q. E. Thompson (1989). *Organometallics,* 8(4), 1121–1122.
15. B. Bogdanovic (1982). U. S. Patent 4,329,301.
16. Gruttner, G., and M. Wiernik (1915). Ber. 48,1474.
17. Gilman, H., and S. P. Massie (1946). *J. Am. Chem. Soc.,* 68, 1128.
18. Bygden, A. (1911). Ber. 44B,2640.
19. Jenkner, H. (1961). U.S. Patent 3,103,526.
20. Jenkner, H. (1957). U.K. Patent 825,987 to Kali-Chemie.
21. Bakshi, K. R., A. Onopchenko, and E. T. Sabourin (1986). U.S. Patent 4,595,777 to Gulf Research and Development Co.
22. Nelson, G. E. (1987). U.S. Patent 4,711,965 to Ethyl Corp.
23. Nelson, G. E. (1987). U.S. Patent 4,711,966 to Ethyl Corp.
24. Nelson, G. E. (1989). U.S. Patent 4,845,260 to Ethyl Corp.
25. Nelson, G. E. (1990). U.S. Patent 4,916,245 to Ethyl Corp.
26. Pettigrew, F. A., L. Plonsker, G. E. Nelson, A. J. Malcolm, and C. R. Everly (1989). Paper presented at the Society of Tribologists and Lubrication Engineers, Atlanta, GA, May 4.
27. Austin, R. G., R. S. Paonessa, P. J. Giordano, and M. S. Wrighton (1977). U.S. NTIS, AD

Report Ad-A044506, 45 pp., Available from Gov. Rep. Announce. Index (U.S.) 1977, 77(25),92 74–1. *Chem. Abstr.* 88(20),144251x.
28. El-Durini, N. M. K., and R. A. Jackson (1982). *J. Organomet. Chem.,* 232(2), 117–121.
29. Onopchenko, A., and E. T. Sabourin (1986). U.S. Patent 4,578,497 to Gulf Research and Development Co.
30. Sabourin, E. T., and A. Onopchenko (1989). *Bull. Chem. Soc. Jpn.,* 62,3691–369b.
31. Onopchenko, A., and E. T. Sabourin (1986). U.S. Patent 4,572,791 to Gulf Research and Development Co.
32. Onopchenko, A., and E. T. Sabourin (1988). *J. Chem. Eng. Data,* 33, 64–66.
33. Malcolm, A. J., C. R. Everly, and G. E. Nelson (1987). U.S. Patent 4,670,574 to Ethyl Corp.
34. Tamborski, C., G. J. Chen, D. R. Anderson, and C. E. Snyder, Jr. (1985). *Ind. Eng. Chem. Prod. Res. Dev.,* 22, 172–178.
35. Thomas, S. G., G. E. Nelson, K. C. Lilje, M. S. Casserino, and R. C. Reid, Jr. (1991). Paper presented at the Society of Tribologists and Lubrication Engineers, Montreal, Quebec, Canada, April 29.
36. Snyder, C. E., and L. J. Gschwender, Liquid Lubricants for Use at High Temperatures.
37. See the discussion by B. Cupples following reference 4.
38. Plonsker, L. (1990). U.S. Patent 4,932,976 to Ethyl Corp.

10
Phosphazenes

Robert E. Singler
Army Materials Technology Laboratory
Watertown, Massachusetts

Mary Jo Bieberich
Naval Surface Warfare Center
Annapolis, Maryland

I. INTRODUCTION

Phosphazenes are ring or chain compounds consisting of alternating phosphorus-nitrogen atoms with two substituents attached to phosphorus. Representative structures are shown below. In these structures, R can be a halogen, organo, or organometallic substituent; X is generally a halide or metal halide counterion. The physical properties of phosphazenes vary considerably with molecular weight and choice of substituents (R). Many of the cyclic phosphazenes are either liquids or low-melting crystalline solids. As the molecular weight is increased, increasing either by the size of the substituent or the number of P-N repeat units, one can obtain oils and greases; ultimately, elastomers and thermoplastics are formed (1).

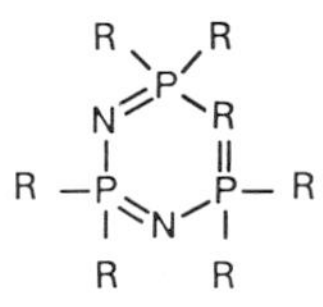

$[R_3P{=}N\text{-}PR_2{=}NPR_3]^{(+)}X^{(-)}$

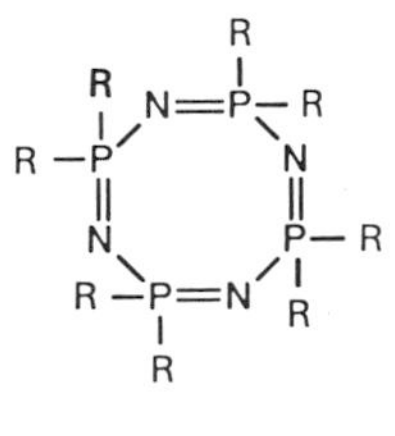

$\left[P(R)_2{=}N \right]_n$

1 2 3 4

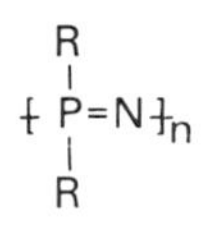

Phosphazenes are of interest in applications where fire resistance and thermal stability are important considerations. Due to the presence of phosphorus and nitrogen, phosphazenes are inherently fire resistant. Halogen-containing substituents further enhance fire resistance. With the proper selection of substituents, thermally and hydrolytically stable ring and chain compounds, including fluids with low pour points and good thermal stability, have been prepared.

Phosphazene fluids are based on structures **1**, **2**, and **3**. Occasionally in the literature these structures are referred to as "polymers"; however, we will reserve the term "polymer" for structure **4**, $n > 100$. Although they are of interest in their own right, high polymers such as **4** will not be described in this chapter. For a description of polyphosphazenes, including the commercial elastomers, the reader is referred to a recent symposium on inorganic and organometallic polymers (2).

II. CHLOROPHOSPHAZENES

A. Cyclic Chlorophosphazene Synthesis

The chlorophosphazenes form the basis for the preparation of phosphazene fluids and lubricants. Ammonium chloride and phosphorus pentachloride react to form a mixture of hexachlorocyclotriphosphazene **5** and octachlorocyclotetraphosphazene **6**, along with smaller amounts of higher cyclics and open-chain oligomers [equation (1)]. This process has been known for over 100 years and has been extensively studied (3). The chlorotrimer **5** and tetramer **6** are available from commercial sources.

$$NH_4Cl + PCl_5 \longrightarrow \underset{\mathbf{5}}{(NPCl_2)_3} + \underset{\mathbf{6}}{(NPCl_2)_4} \qquad (10.1)$$

B. Linear Chlorophosphazene Oligomer Fluids

One important modification of equation (1) is the use of excess PCl_5 with NH_4Cl or with chlorotrimer to obtain linear chlorophosphazene oligomers [equation (2)]. These open-chain oligomers are generally either oils or greases.

$$\left.\begin{array}{l} NH_4Cl + PCl_5\ (xs) \\ \text{Trimer} + PCl_5 \end{array}\right\} \longrightarrow [PNCl_2]_nPCl_5 \quad (n = 3 - 10) \qquad (10.2)$$

Over a period of approximately 15 years, the Air Force Materials Laboratory sponsored the development of linear chlorophosphazenes for high-temperature fluid applications (4–6). Oligomers such as $[Cl(PCl_2{=}N)_3PCl_3]PCl_6$ and $[Cl(PCl_2{=}N)_n PCl_3]Cl$ ($n \approx 10$) undergo further reactions above 300°C resulting in chain extension or cyclization. However, if these oligomers are end-capped with certain metal halides, the thermal stability is markedly enhanced. Oligomers, $[Cl(PCl_2{=}N)_3PCl_3]BCl_4$ and $[Cl(PCl_2{=}N)_3PCl_3]AlCl_4$, are reported to be stable up to 400 and 700°C, respectively. The thermal stability of the products seems to depend on the stability of the metal halide anion, which effectively end-caps the reactive $PNCl_2$ chain. Although these oligomers show excellent thermal stability, they are very sensitive to hydrolysis and thus are limited in their applications. Replacement of hydrolytically unstable chlorine substituents with organic substituents might increase hydrolytic stability of these oligomers. To date, little additional progress has been made to develop stable fluids from the linear chlorophosphazene oligomers.

III. CYCLIC PHOSPHAZENE FLUIDS

A. Synthesis

Most of the research and development of phosphazene fluids has centered on the alkoxy- and aryloxycyclophosphazenes, which are derived from the chlorotrimer and chlorotetramer. The cyclic chlorophosphazenes are versatile substrates for nucleophilic displacement processes (7). Using the chlorotrimer as an example, both single-substituent and mixed-substituent cyclophosphazenes can be obtained (e.g., **7**, **8**), depending on the selection of alcohols and/or phenols in the reaction [equation (3)]. It should be noted that structure **8** represents only one of the possible isomeric products from a two-substituent process. The differences between **7** and **8** will be discussed below.

$$N_3P_3Cl_6\ (\underline{5}) \xrightarrow[\text{Base}]{\text{ROH}} N_3P_3(OR)_6\ (\underline{7}) \qquad N_3P_3Cl_6\ (\underline{5}) \xrightarrow[\text{Base}]{\text{ROH/R'OH}} N_3P_3(OR)_3(OR')_3\ (\underline{8}) \tag{10.3}$$

A suitable base is required either to form the anion of the alcohol or phenol or to remove hydrogen chloride and thus drive the reaction to completion. A number of modifications of the process of equation (3) have been described, including the use of sodium metal or sodium hydride in various inert organic solvents, using a tertiary organic amine or sodium carbonate as the hydrogen chloride acceptor, use of sodium hydroxide in xylene followed by azeotropic distillation of water, and phase-transfer catalysis (8–10). In these reactions, it is important to obtain complete substitution of chlorine and to remove by-products in order to maximize the thermal and hydrolytic stability of the products. Partially substituted products, $N_3P_3Cl_x(OR)_{6-x}$ are generally thermally and hydrolytically less stable than fully substituted products (11).

While nonfluorinated alkoxycyclophosphazenes are prone to rearrangement and decomposition below 200°C (12), fluoroalkoxy- and aryloxycyclophosphazenes are exceptionally stable to hydrolysis and heat (8, 12). They are also fire resistant, being self-extinguishing in a flame. Whereas trifluoroethoxy and phenoxy-substituted trimers, $N_3P_3(OCH_2CF_3)_6$ and $N_3P_3(OC_6H_5)_6$, are crystalline solids, using long-chain fluoroalcohols and substituted phenols can result in fluid products. Also, if mixtures of alcohols and/or phenols are used in the substitution reaction [equation (3)], mixed-substituent products such as **8** are formed, which often are fluids over a wider temperature range. The ability to attach an almost endless variety of side chains onto a phosphazene ring in a controlled fashion, to produce materials with a wide range of physical properties, has led to the development of several promising fluids, which are described below.

B. Cyclic Phosphazene Fluids—Early Work

A number of references exist dated prior to 1980, mainly in the patent literature and in government reports, that describe the development of cyclic phosphazenes as fluids, lubricants, plasticizers, or as additives for lubricating oils and greases (1, 13–20). No attempt will be made to describe all of the early work, but several of the reports will be discussed in some detail, since they form the basis for current activity in this area. Structures given below are all derived from the same general reaction process shown in equation (3).

Nichols (14, 15) synthesized a series of aryloxy-substituted cyclophosphazene trimers and tetramers, using *m*-trifluoromethylphenol, *m*-trifluoromethoxyphenol, and phenol combinations. Both single- and mixed-substituent trimers (**7**, **8**) and higher cyclic oligomers were prepared and characterized. Use of fluorosubstituted phenols tended to suppress crystallinity, lower pour points, and enhance thermal stability of the products. Fluid **9** had a pour point of –31°C, and was reported to be stable up to 385°C (15). Other properties are listed in Table 10.1.

$[NP(OC_6H_4\text{-m-}OCF_3)_2]_3$

9

During the 1960s, the U.S. Navy supported the development of cyclophosphazene as fire-resistant, compression-ignition-resistant hydraulic fluids to meet the requirements for MIL-H-19457A (21). Selected examples are included in Table 10.1. Initial work (19) involved the preparation of fluoroalkoxycyclophosphazenes. Fluoroalcohols with terminal fluorine (e.g., CF_3CH_2OH, $C_3F_7CH_2OH$) generally gave crystalline products or had undesirably high vapor pressures. Hydrogen-terminated fluoroalcohols reduced the vapor pressure, but many of the products (**10**) were also crystalline.

$[NP(OCH_2(CF_2)_xH)_2]_y$ x = 2, 4, 6, or 8; y = 3 or 4

10

Attempts to eliminate crystallinity and improve other properties led to the synthesis of cyclophosphazenes with mixed fluoroalkoxy substituents (**11**). Thirteen products were prepared, 12 of which were liquid at room temperature, and 7 of which had pour points ranging from –23 to –48°C. Viscosities for these liquids, however, were too high for the intended use (21). Most of the densities exceeded 1.8 g/ml. Tetramer products like

Table 10.1 Properties of Selected Cyclic Phosphazene Fluids[a]

Property	**9**	**10**	**11**	**11A**	**12**	**13**[b]	**13**[b]
Substituent[c]							
Ar	$-C_6H_4$-*m*-OCF_3				$-C_6H_5$	$-C_6H_4$-*m*-Cl	$-C_6H_4$-*m*-CF_3
R		$-CH_2(CF_2)_4H$	$-CH_2(CF_2)_2H$	$-CH_2CF_3$	$-CH_2(CF_2)_4H$	$-CH_2CF_3$	$-CH_2CF_3$
R'			$-CH_2(CF_2)_4H$	$-CH_2C_3F_7$			
Viscosity,[d] cSt							
38°C	44.3	100.5	96.7	10.6	159	61.3	53.9
99°C	5.84	10.6	9.20	2.04	11.4	6.54	5.58
Density, 25°C	1.5[e]	1.85	1.71	1.7	1.60	1.51	1.52
Pour point,[d] °C	−31	−46	−48	<−40	−23	−20	−20
S.I.T.,[f] °C	—	566	538	—	635	663	677

[a]Data sources were **9** (14, 15); **10**, **11**, **12**, (19); **11A** (R. E. Singler, G. L. Hagnauer, and C. E. Snyder, unpublished results, 1977); **13** (20).
[b]Synthesized from a commercial chlorotrimer-tetramer mixture (80/20).
[c]R, R' = perfluoroalkyl groups, H or F terminated, structures **10–13**; Ar = phenyl or substituted phenyls in structures **9**, **12**, **13**.
[d]MIL-H-19457 target goals (21): viscosity (cSt) 38°C, 43–50; 99°C, 4.8 minimum; pour point = −18°C maximum.
[e]Estimated based on example 2 (14).
[f]S.I.T. = spontaneous ignition temperature (19, 20). Modification of ASTM D-2155-66.

11, y = 4, gave higher viscosities and lower ASTM (viscosity-temperature) slopes (22) than the trimers.

$$[NP(OCH_2(CF_2)_xH)_a(OCH_2(CF_2)_zH)_b]_y$$

y = 3 or 4
a + b = 2
x = 2, 4, or 6
z = 4, 6, 8, or 10
$x \neq z$

11

A series of mixed-substituent cyclophosphazenes with hydrogen-terminated fluoroalkoxy and aryloxy side chains, **12**, was also prepared and characterized in this study (19).

$$[NP(OAr)_a(OCH_2(CF_2)_xH)_b]_y$$

y = 3 or 4
a + b = 2
x = 2, 4, 6, or 8

12

Using both phenol and substituted phenols as the aryloxy group, some products were obtained that were liquid at room temperature. *Ortho*-substituted phenols tended to give products that failed hydrolytic stability tests, which was probably due to the presence of unreacted chlorine (P-Cl units). Pour points and viscosities were higher, and the densities were lower, than for the mixed-fluoroalkoxyphosphazenes (**11**) described previously.

In an attempt to obtain products with lower viscosities and densities than **12**, efforts were then directed toward the synthesis of mixed aryloxyfluoroalkoxycyclophosphazenes, **13**, using fluorine-terminated fluoroalcohols (20).

$$[NP(OAr)_a(OCH_2(CF_2)_xF)_b]_y$$

y = 3 or 4
a + b = 2
x = 1, 2, 3, or 7

13

Replacement of the terminal hydrogen on the fluoroalcohol reduced viscosity, and the use of short-chain fluoroalcohols reduced the density. Properties of the mixed-substituent derivatives **13** were dependent on the length of the fluoroalkoxy chain, the aryloxy/fluoroalkoxy ratio (a/b), and the substituents on the benzene ring. Trends were, in general, independent of ring size. A direct comparison of the physical properties of the trimer and tetramer products with identical a/b ratios showed that the tetramers generally had higher boiling points, viscosities, autogenous ignition temperatures (AITs), and thermal stabilities, as well as lower ASTM slopes and frequently lower pour points (20). Properties of selected examples are given in Table 10.1.

B. Cyclic Phosphazene Fluids—Recent Developments

1. Fire-Resistant Hydraulic Fluids—Naval Ship Applications

While some of the fluids prepared in the earlier Navy program (19, 20) showed promise for MIL-H-19457, they did not have the requisite balance of physical properties and systems compatibility for sustained development. Some of the fluids crystallized in storage or failed more severe hydrolytic stability testing. Part of the problem may have been the result of the synthetic procedures employed in the early work or the difficulty in characterizing complex product mixtures (23). Mixed-substituent products, such as **10–13**, are complex mixtures with different chemical compositions and isomer distributions; this often made it difficult to determine if a given composition was inherently unacceptable, or perhaps there was a minor component in the mixture that adversely affected properties. These issues, as well as projected high costs of phosphazene fluids, also discouraged commercial development.

In 1980, the Navy took a renewed interest in phosphazene fluids, after a consideration of alternate synthetic procedures that might be used to optimize fluid composition, and availability of modern characterization techniques that could more accurately analyze product composition.

Singler (24, 25), participating in a new program with the Navy to develop nontoxic, fire-resistant phosphazene hydraulic fluids, used a different process to synthesize aryloxy/trifluoroalkoxyphosphazenes. Unlike the earlier work (20), which used an azeotrope to remove water formed during the reaction, sodium salts of trifluoroethanol and the phenols were first formed with sodium hydride in anhydrous dioxane and tetrahydrofuran. The sodium salts were then added sequentially, starting with the least reactive nucleophile, to the $PNCl_2$ trimer in an aromatic solvent. Both two-substituent and three-substituent cyclophosphazenes were prepared [equation (4)].

$$[NPCl_2]_3 \xrightarrow[CF_3CH_2OHa]{ArONa,\ Ar'ONa} \underset{\underline{14}}{N_3P_3(OAr)_a(OCH_2CF_3)_b(OAr')_c} \quad a + b + c = 6 \qquad (10.4)$$

The mode of addition—that is, adding the sodium aryloxide(s) followed by sodium trifluoroethoxide to the trimer—changed the substitution pattern and improved the yield and homogeneity of the product, **14**. Viscosity and pour point combinations were optimized and crystallization tendency was minimized by varying the aryloxy/fluoroalkoxy substituent ratio through the sequential addition process. Substituent ratios and product distributions were monitored by proton nuclear magnetic resonance (^1H-NMR), high-performance liquid chromatography (HPLC), and gas chromatography/mass spectrometry (GC/MS).

Of the various multisubstituent fluids prepared in this study (25), those that exhibited the best combination of properties were in the range of $a + c = 1–4$ [**14**]. The removal of $b = 6$, $N_3P_3\ (OCH_2\ CF_3)_6$, was considered essential in order to eliminate crystallization of fluids in cold storage. The most promising fluids are listed in Table 10.2, along with property data and target property requirements for MIL-H-19457C hydraulic fluid replacements. While all of the fluids prepared in this study were not acceptable in every respect, this work demonstrated that trisubstituted aryloxy/fluoroalkoxyphosphazenes exhibit the best range of properties for hydraulic fluid performance, compatibility, and fire resistance (25).

Under Navy contract (26, 27), Borg-Warner pursued the Navy's interest in multisubstituent aryloxy/fluoroalkoxycyclophosphazenes, using a synthetic process that involved phase-transfer catalysis (PTC). Patents were also filed by Borg-Warner (9, 28). Using PTC, Carr reacted both the trifluoroethanol and the phenolic constituents, in stoichiometric quantities, directly with the $PNCl_2$ trimer in a medium comprising water, a base, a water-immiscible solvent, and a phase-transfer catalyst. This synthetic procedure eliminated the preparation, in advance, of the corresponding sodium salts, and thus the

Table 10.2 Cyclic Phosphazene Esters: Fluid Evaluation[a]

$N_3P_3(OC_6H_4CF_3)_a(OCH_2CF_3)_b(OC_6H_4CH_3)_c$				MIL-H-19457C target properties
Value of a	2.2	3.3	1.4	
Value of b	3.8	2.7	3.2	—
Value of c	—	—	1.4	
Viscosity, cSt				
40°C	22.8	54.0	42.9	43–50
100°C	3.5	4.7	4.8	4.8 minimum
Pour point, °C	–30	–16	–21	–18 maximum
Density, g/ml	1.5	1.5	1.4	2.0 maximum
Total acid number (TAN), mg KOH/g	0.02	0.02	0.01	0.1 maximum
High-pressure AIT[b] °C	332	337	280	235
Flash point, °C	250	[c]	245	275
Demulsibility (40/40 in 30 min)	Pass	Pass	Pass	Pass
Hydrolytic stability				
ΔTAN, mg KOH/g	0.01	0.01	0.01	0.2 maximum
H_2O acidity, mg/cm^2	0.58	1.11	0.42	5.0 maximum
Copper weight loss, mg/cm^2	0.02	0.03	0.03	0.3 maximum
Cu appearance	Dark tarnish	Corrosion	Light tarnish	No corrosion
Insolubles, wt%	0	0	0	0.5 maximum
Elastomer compatability, % volume swell				
Buna-N[d]	0.8	7.9	[c]	None
Viton, Class 1[e]	18.9	11.4	7.2	±5
Viton, Class 2[e]	15.0	12.2	–3.3	±5
EPR[f]	0.1	[c]	–3.3	±5
Paint compatibility[g]	No change	No change	No change	None

[a]Reference 24.
[b]Autogenous ignition temperature, ASTM G-72-82.
[c]Not determined.
[d]MIL-P-25732.
[e]MIL-R-83248.
[f]MIL-G-22050.
[g]MIL-P-24441.

direct handling of the hazardous materials sodium metal, sodium hydride, or sodium trifluoroethoxide. Significant reductions were made in the preparation and reaction times, while achieving high yields of phosphazene esters. The PTC bulk addition process, which involved premixing the phenols and trifluoroethanol with caustic before addition of $(PNCl_2)_3$, in contrast to a PTC sequential addition process, gave a broader distribution of components in the product mixture. Carr experimented with both pure and crude trimer in his PTC studies and concluded that crude or partially purified trimer, which contained a small amount of tetramer, could be used in place of pure trimer to yield fluids with a broader distribution of products and lower pour points.

In his investigations (26, 27), Carr found through duplicate synthesis that the reproducibility of the PTC process was excellent. He also identified several potential problems, which were eventually resolved by modifying the PTC process. Initial experiments showed that the products contained a higher amount of residual chlorine than products made by the sodium salts method. To achieve complete substitution, the organic phase was postreacted with sodium or potassium aryloxide. The PTC process also yielded crude products with higher than expected viscosities. This was attributed to residual catalyst and higher-molecular-weight species. Passage of the crude product through the ion-exchange resin removed the residual catalyst and part of the high-molecular-weight components. Acid numbers and color were slightly improved through distillation with about 1% by weight anhydrous calcium carbonate. Carr also investigated several other fluid purification techniques, which were deemed necessary to more effectively remove acid impurities in the crude product and thus improve the demulsibility and hydrolytic stability. Repeated vacuum and centrifugal molecular distillation were ineffective in removing all the acid impurities; however, treatment of both distilled and crude product with activated alumina reduced the acid number and the tendency of the fluid to emulsify. The effectiveness of this treatment eliminated the distillation process.

Of the many phosphazenes prepared by Carr, the most promising fluid, **15**, was prepared using PTC, bulk addition, and a mixture of alkylphenols (*m*- and *p*-methylphenol), phenol, and trifluoroethanol.

$$N_3P_3(OCH_2CF_3)_{3.5}(OC_6H_5)_{1.25}(OC_6H_4\text{-m-}CH_3)_{0.87}(OC_6H_4\text{-p-}CH_3)_{0.38}$$

15

The aryloxy/fluoroalkoxy substituent ratio was established at 2.5 : 3.5 to achieve the required viscosity and pour point. A copper corrosion inhibitor, tolyltriazole, was formulated into the fluid at a concentration of 100 ppm. Fluid analysis indicated better fire resistance, thermal and hydrolytic stability, and paint compatibility when compared with MIL-H-19457 phosphate ester. The low-temperature properties, elastomer compatibility, and wear characteristics were comparable. The property data are listed in Table 10.3 for the phosphazene ester fluid representative of a 30-gal pilot plant production (27).

Extensive toxicity evaluations, including acute toxicity tests, inhalation, and skin absorption kinetic studies, and 21-day repeated inhalation and dermal exposure testing of the Navy's cyclic phosphazene hydraulic fluid, **15**, were conducted by the Toxicology Detachment of the Naval Medical Research Institute (29). The most significant routes of exposure to ship board hydraulic fluids are dermal, due to spills or leaks, and aerosol inhalation from pressurized system leaks. Tests that simulated these routes of exposure showed no evidence of toxicity as a result of exposure of test animals to the phosphazene fluid. Test data clearly indicated that the phosphazene fluid is poorly absorbed into the body by any route and appears to produce little or no effect when artificially introduced into the body. Such materials are therefore unlikely to have more than minimal health consequences if used in shipboard hydraulic system applications.

The advances made in the state of the art by Singler and Carr encouraged the Navy to continue their pursuit of phosphazene hydraulic fluids. In 1986, the Ethyl Corporation produced 280 gallons of the Navy's phosphazene ester hydraulic fluid, **15**, using a sodium salt procedure developed by Kolich (30). The product met the Navy's physical and chemical property specifications.

The cyclophosphazene fluids prepared by Carr and Kolich via two distinctively

Table 10.3 Fluid Evaluation[a]
$N_3P_3(OCH_2CF_3)_{3.5}(OC_6H_5)_{1.25}(OC_6H_4\text{-}m\text{-}CH_3)_{0.87}(OC_6H_4\text{-}p\text{-}CH_3)_{0.38}$

Properties	PN fluids	Fluid acceptance criteria
Viscosity, cSt		
40°C	35.2 (50.7 cP)	43–50 (50–58 cP)
100°C	4.2 (6.1 cP)	4.8 minimum (5.5 cP)
Density, g/ml	1.45	1.5 maximum
Pour point, °C	–21	–18 maximum
Total acid number (TAN), mg KOH/g	0.02	0.1 maximum
Demulsibility	40/40; 5 minimum	40/40; 5 minimum
Hydrolytic stability		
Δ TAN, mg KOH/g	0.83	0.2 maximum
H_2O layer, mg KOH	0.03	5.0 maximum
Cu weight loss, mg/cm^2	0.02	0.3 maximum
Cu appearance	Slight tarnish	No corrosion
Insoluble, wt%	Nil	0.5 maximum
Flash point, °C	288	275 maximum
High-pressure AIT,[b] °C	285	235 maximum
Foam tendency	Nil	Nil
Bulk modulus, isothermal secant at 10,000 psig	266,000	Report
Four-ball wear, mm (wear scar diameter)	0.7	0.5–1.0
Material compatability		
% Volume swelling		
Viton, class 1[c]	8.7	±5
Viton, class 2[c]	6.6	±5
EPR[d]	–1.7	±5
Paint compatibility[e]	No change	No change

[a]Reference 27.
[b]Autogenous ignition temperature, ASTM G72-82.
[c]MIL-R-83248.
[d]MIL-G-22050.
[e]MIL-P-24441.

different synthetic procedure were characterized and compared using GC/MS, HPLC, and ^1H-NMR (31). Molecular weights of individual components and isomer distributions were determined for both products. Although the aryloxy/fluoroalkoxy ratio for both fluids was approximately the same (2.5 : 3.5) and the isomeric components were identical, the amounts of these components differed. Both fluids met the Navy's target property requirements.

The Navy conducted a 1,000-h performance test of the Ethyl-produced cyclotriphosphazene fluid using a high-pressure, variable-displacement piston pump (Sundstrand series 27) at simulated shipboard operating temperature, pressure, and flow rate to evaluate the lubricating ability of the phosphazene fluid (32). Measured dimensional changes of critical pump parts showed no significant wear or unusual wear patterns. The volumetric efficiency remained fairly constant throughout the test and was in the expected range, indicating no degradation to the pumping components. The torque efficiency dropped from 89 to 78% within 100 h, but remained constant for the remainder of the test. The results of the analysis of fluid samples drawn at 100-h intervals during the test indicated that there was no physical or chemical degradation (32).

2. Other Applications—Military and Commercial

Other researchers have continued to explore phosphazene functional fluids in the United States and Japan (33–49). These developments are also related to the earlier work in this field.

Researchers at Dow Chemical Company have synthesized a series of single- and mixed-substituent aryloxycyclotriphosphazenes intended for use as high temperature lubricants for aircraft gas turbine engines (33). Through careful choice of substituents, the Dow group prepared an aryloxycyclotriphosphazene, **16**, with optimal low temperature properties, oxidative stability and lubricity. Representative properties are shown in Table 10.4. This compound, **16**, which was designated as X-1P, also improved the friction and wear properties of a commercial pentaerythritol tetraester (PET) fluid.

$N_3P_3(O\text{-}C_6H_4\text{-}F)_2(O\text{-}C_6H_4\text{-}CF_3)_4$

16

Otsuka Chemical Company has developed a series of fluoroalkoxyphosphazene fluids, Phospharol NF fluids, that exhibit excellent lubricating ability, oxidative stability, chemical compatibility with plastics (particularly thermoplastics), and fire resistance (34–41). Two examples, illustrated in structures **17** and **18**, have properties similar to the fluids described in Table 10.1. Phospharol fluids were developed for uses in electronic devices containing plastic components and in the manufacture of semiconductors. Fluid **17** was reported to be stable in the presence of $AlCl_3$ (34). Because of its low vapor pressure and fire resistance, **18** is claimed to be suitable for use as a vacuum pump oil (37). Phospharol fluids were also reported as comprising the liquid phase of some electrorheological fluid compositions (40, 41).

$[NP(OCH_2(CF_2)_4H)_2]_3$

17

$[NP(OCH_2CF_2CF_3)(OCH_2(CF_2)_4H)]_3$

18

Dekura (48) has prepared liquid and solid (greaselike) fluorine-containing phosphonitrile amides, **19**, that exhibit favorable lubricating ability, enhanced adsorption on metal surfaces, corrosion inhibition, and fire resistance. These fluorine-substituted aminophosphazenes can be used on their own as synthetic lubricants, or as lubricity additives in other oils.

$[NP(NHR_f)NHR'_f)]_3$

19

The viscosity and lubricating properties of compounds having formula **19** are adjusted to suit different applications by selection of R_f and $R_f^!$ radicals. For example, if R_f

Table 10.4 Selected Properties of X-1P[a]

$$N_3P_3(O\text{-}C_6H_4\text{-}F)_2(O\text{-}C_6H_4\text{-}CF_3)_4$$

Property	Value
Viscosity[b]	
25°C	1.401 Pa·s (934 cSt)
100°C	0.0156 Pa·s (11.2 cSt)
Pour point[c]	–15°C
Low-temperature fluidity[d]	–11°C
Specific gravity at 25°C	1.50
Vapor pressure	
60°C	1.17×10^{-7} Pa (10^{-9} mm Hg)
330°C	2.66×10^{3} Pa (27 mm Hg)
Autoignition temperature[e]	>650°C
Oxidative stability[f]	429°C
Volatility[g]	
10% weight loss	314°C
50% weight loss	355°C
Lubricity behavior[h]	
Wear-scar diameter	0.51 mm
Coefficient of friction	0.03
Lubricity enhancement of commercial petroleum fluid,[i] wear-scar diameter	
0% X-1P	1.28 mm
1% X-1P	0.45 mm
Acute toxicity[j]	Low

[a]Reference 33.
[b]ASTM D-445-88.
[c]ASTM D-97-87.
[d]Pumpability temperature for viscosity of 20,000 cP.
[e]ASTM E=659-78.
[f]Pressure differential scanning calorimetry.
[g]Thermogravimetric analysis: same in air or nitrogen flow of 50 cm^3/min at a rate of 15°C/min.
[h]Four-ball method, 300°C, 10 lb load, 1200 rpm, 1 h, steel balls.
[i]Pentaerythritol tetraester. Same conditions as in *h* except 200°C, 120 lb load.
[j]LD50 in rats 2000 mg/kg. Dermal LD50 in rabbits 2000 mg/kg.

and R'_f are fluoroalkylcarbonyl and/or fluoroalkylalkylcarbonyl radicals, the compounds tend to be solid with superior load-carrying capacity. In contrast, lower-molecular-weight products exhibit good low-temperature properties but poor load-carrying capacity. Nibert (49) has also disclosed aryloxyphosphazene-containing lubricating grease compositions.

IV. CONCLUSION

When one considers the rich chemistry of the phosphazenes, it is somewhat surprising that there are only a few phosphazene fluids that have reached the commercial stage (39). Phosphazenes are inherently fire resistant, and they have other properties necessary for high-performance applications. The main factor that has discouraged extensive development of phosphazene fluids (and polyphosphazenes) is their high cost relative to commercial fluids. The Navy currently has curtailed its development program of phosphazene hydraulic fluids due to the projected cost ($12/lb, $144/gallon). This high cost is due to the limited demand for phosphazene fluids and the high cost of the starting materials, trifluoroethanol and chlorotrimer. The Navy's interest in phosphazene fluids as well as polyphosphazene materials, however, has not waned. Recently the Navy has requested participation in a Department of Defense program designed to stimulate the commercialization of materials, such as cyclic phosphazene fluids and phosphazene polymers, required for military applications. The chemistry of phosphazene fluids and polyphosphazenes is closely related, and large-scale development in one of these areas could be the stimulus for commercial development of all phosphazene technology.

ACKNOWLEDGEMENTS

The authors would like to acknowledge Charles Kolich, Ethyl Corporation, Ted Morgan and Bassam Nader, The Dow Chemical Company, Harry Allcock, Pennsylvania State University, Selwyn Rose, Himont, Inc., and Satoru Maki, Otsuka Chemical Co., for their advice and assistance during the preparation of this manuscript.

REFERENCES

1. Allcock, H. R. (1972). *Phosphorus-Nitrogen Compounds*, Academic Press, New York.
2. Zeldin, M., K. J. Wynne, and H. R. Allcock (eds.) (1989). *Inorganic and Organometallic Polymers*, ACS Symposium Series no. 360, American Chemical Society, Washington, D.C., Chapters 19–21.
3. Allcock, H. R. (1972). *Phosphorus-Nitrogen Compounds*, Academic Press, New York, Chapter 4.
4. Nichols, G. M. (1962). Synthesis of Inorganic High Temperature Fluids, AF33 616 7158, February, reported by C. D. Schmulbach in *Progress in Inorganic Chemistry* (F. A. Cotton, ed.), vol. IV, p. 275, Wiley Interscience, New York.
5. Moran, E. F. (1968). *J. Inorg. Nucl. Chem.*, 30, 1405; contract AF 657 10693 (AD 478 112L), 1966.
6. Murch, R. M. (1969). Thermally Stable Hydraulic Fluids, AFML-TR-68-337 (AD 849 271).
7. Allcock, H. R. (1972). *Phosphorus-Nitrogen Compounds*, Academic Press, New York, Chapters 6 and 7.
8. Allcock, H. R. (1972). *Phosphorus-Nitrogen Compounds*, Academic Press, New York, Chapter 6.
9. Carr, L. J., and G. M. Nichols (1986). U.S. Patent 4,600,791.
10. Wang, M. L., and H.-S. Wu (1990). *Ind. Eng. Chem. Res.*, 29, 2137.
11. Allcock, H. R. (1972). *Phosphorus-Nitrogen Compounds*, Academic Press, New York, Chapter 5.
12. Allcock, H. R. (1972). *Phosphorus-Nitrogen Compounds*, Academic Press, New York, Chapter 13.
13. Lipkin, D. (1940). U.S. Patent 2,192,921.
14. Nichols, G. M. (1966). U.S. Patent 3,234,304.
15. Nichols, G. M. (1967). U.S. Patent 3,316,330.
16. Kober, E. H., H. F. Lederle, and G. F. Ottmann (1966). U.S. Patent 3,291,865.
17. Kober, E. H., H. F. Lederle, and G. F. Ottmann (1967). U.S. Patent 3,304,350.
18. Drysdale, J. J., R. E. Le Bleu, and J. H. Fassnacht (1965). U.S. Patent 3,201,445. (1965).

19. Kober, E., H. Lederle, and G. Ottmann (1963). *ASLE Trans.*, 7, 389; U.S. Navy Contract NObs-86482, AD 432,367 (1963).
20. Kober, E., H. Lederle, and G. Ottman (1966). *J. Chem. Eng. Data*, 11, 221; U.S. Navy Contract NObs-90092, AD 608,144 (1964).
21. Military specification MIL-H-19457A (Ships) (1963). Superceded by MIL-H-19457D (Ships) (1989).
22. ASTM Standards (1967). Designation D-341-43, Viscosity Temperature Charts for Liquid Petroleum Products, American Society for Testing Materials, Philadelphia. Superceded by ASTM D-341-87 (1987).
23. Singler, R. E. (1976). Potential of phosphazenes as hydraulic fluids, *Conference on Hydraulic Fluids*, NASA Ames Research Center, February, NASA TM X-73,142.
24. Singler, R. E., T. N. Koulouris, A. J. Deome, H. Lee, D. A. Dunn, P. J. Kane, and M. J. Bieberich (1982). Preparation and properties of phosphazene fire resistant fluids, *Army Science Conf. Proc.*, 3, 297 (AD A117298).
25. Singler, R. E., A. J. Deome, D. A. Dunn, and M. J. Bieberich (1986). *Ind. Eng. Chem. Prod. Res. Dev.*, 25(1), 46.
26. Carr, L. J., G. M. Nichols, and S. H. Rose (1983). Phosphazene Base Hydraulic Fluid Development, Navy contract N00167-82-C-0168, Phase I, DTNSRDC/SME-CR-03-84.
27. Carr, L. J., and S. H. Rose (1984). Phosphazene Base Hydraulic Fluid Development, Navy contract N00167-82-C-0168, Phase II, DTNSRDC/SME-CR-18-84.
28. Carr, L. J., G. M. Nichols, and S. H. Rose (1986). U.S. Patent 4,601,843.
29. Kinkhead, E., E. Kimmel, H. Wall, and J. Grabau (1990). *Am. Ind. Hyg. Assoc. J.*, 51(11), 583.
30. Kolich, C. H., and W. D. Klobucar (1987). U.S. Patent 4,698,439.
31. Deome, A. J., and D. A. Bulpett (1986). Thermal Decomposition Product Analysis of Phosphazene and Phosphate Ester Fluids, U.S. Army Materials Technology Laboratory TR 86-35.
32. Barnes, R. (1988). Navy Test of Cyclotriphosphazene Hydraulic Oil in a 27 Series Pump, Sundstrand Sauer Letter Report to David Taylor Research Center.
33. Nader, B. S., K. K. Kishore, T. A. Morgan, C. E. Pawlowski, and W. L. Dilling (1991). Presented at the 46th STLE Annual Meeting in Montreal, Quebec, April, STLE preprint no. 91-AM-7D-1.
34. Tada, J., and H. Muramatsu (1985). *J. Vac. Soc. Jpn*, 28, 706.
35. Muramatsu, H., H. Baba, T. Nakanaga, and Y. Tada (1987). *Chem. Abstr.*, 106, 159363.
36. Nakacho, Y., Y. Tada, and T. Yagi (1987). *Chem. Abstr.*, 106, 87465.
37. Nakacho, Y., and Y. Tada (1988). U.S. Patent 4,724,264.
38. Nakanaga, T., Y. Tada, S. Yamada, M. Hirohama, and T. Akata (1988). *Chem. Abstr.*, 108, 115566.
39. Maki, S. (1991). Otsuka Chemical Company Co. Ltd., Product Bulletin.
40. Nakanaga, T., and Y. Tada (1991). *Chem. Abstr.*, 114, 27073.
41. Nakanaga, T., and M. Yasuki (1991). *Chem. Abstr.*, 114, 27074.
42. Morimoto, T., T. Nakanaga, and Y. Tada (1987). *Chem. Abstr.*, 106, 122909; 107, 99590.
43. Tsubokawa, M., and S. Mori (1988). *Chem. Abstr.*, 109, 173357; 109, 193589.
44. Dekura, T., and J. Endo (1989). *Chem. Abstr.*, 110, 60946.
45. Kamijama, S., K. Fujikawa, Y. Yoshikawa, T. Okamoto, and T. Nishikawa (1989). *Chem. Abstr.*, 111, 233197.
46. Mori, S., and M. Tsubokawa (1989). *Chem. Abstr.*, 26144, 111.
47. Amasaka, T., M. Ikeda, and H. Tsutsumi (1990). *Chem. Abstr.*, 112, 182729.
48. Dekura, T. (1990). U.S. Patent 4,898,683.
49. Nibert, R. K. (1988). U.S. Patent Application 851,635; *Chem. Abstr.*, 108, 8693.

11
Dialkylcarbonates

Giuseppe Fisicaro and Giampaolo Gerbaz
AgipPetroli
Rome, Italy

I. INTRODUCTION

Dialkylcarbonates represent a new class of synthetic fluids obtained by transesterification of dimethylcarbonate (DMC). The characteristics of the products strongly depend on the alcohols, which can be chosen among a wide variety of natural and synthetic products.

The performance and application fields of dialkylcarbonates are similar to those of traditional esters, with improvements in terms of toxicology, seal compatibility, and absence of acid compounds from decomposition.

Dialkylcarbonates are expected to gain increasing importance and application, both as lubricants base stocks and as performance fluid components, thanks to their characteristics, performance, and economics.

II. CHEMISTRY

A. Chemical Structure

The structure of dialkylcarbonates basically consists of a polar group ($>CO_3$) with the four atoms on the same plane, and two alkylic chains bonded to two oxygen atoms, free to rotate around the axis of the carbon-oxygen bond:

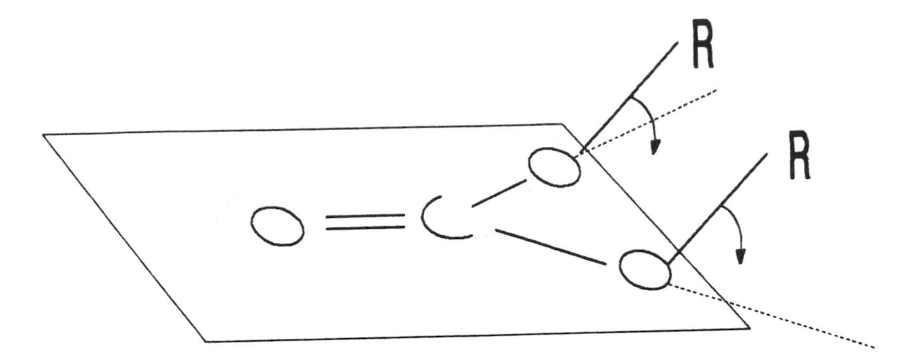

This chemical structure determines the physicochemical properties of dialkylcarbonates, as well as their chemical behavior; therefore, both reactions of the carbonic group and of the alkylic groups are expected to take place.

B. Synthetic Methods

Dialkylcarbonates are diesters of carbonic acid, deriving from the condensation of the acid with alcohols:

$$\mathrm{HO{-}C({=}O){-}OH} + \mathrm{ROH} \xrightarrow{-H_2O} \mathrm{HO{-}C({=}O){-}OR} + \mathrm{ROH} \xrightarrow{-H_2O} \mathrm{RO{-}C({=}O){-}OR}$$

In practice this route is not feasible because carbonic acid, as such, has not sufficient chemical stability to allow the condensation of alcohols; moreover, the monoester formed in the first condensation step is unstable and cannot undergo any further condensation reaction. Conversely, carboalcoxides can be directly obtained by reaction of carbon dioxide and inorganic alcoholate in alcoholic solution;

$$CO_2 + \mathrm{ROMe} \xrightarrow{\mathrm{ROH}} \mathrm{MeO{-}C({=}O){-}OR}$$

and subsequently converted to dialkylcarbonates. Carboalcoxides are quite unstable and can be readily hydrolyzed to inorganic carbonate and alcohol.

$$2\ \mathrm{MeO{-}C({=}O){-}OR} + H_2O \xrightarrow{-CO_2} Me_2CO_3 + 2ROH$$

Other laboratory and industrial routes have been investigated and developed to synthetize dialkylcarbonates with a wide range of performance and characteristics, starting from different raw materials. The following routes can be considered.

1. From alkylhalides.

$$Ag_2CO_3 + 2C_2H_5I \xrightarrow[-2AgI]{} C_2H_5O{-}\overset{\overset{\large O}{\|}}{C}{-}OC_2H_5$$

2. From carboalcoxides.

$$MeO{-}\overset{\overset{\large O}{\|}}{C}{-}OR + R_1X \xrightarrow[-MeX]{} RO{-}\overset{\overset{\large O}{\|}}{C}{-}OR$$

3. From phosgene.

$$Cl{-}\overset{\overset{\large O}{\|}}{C}{-}Cl + 2ROH \xrightarrow[-2HCl]{(NaOH)} RO{-}\overset{\overset{\large O}{\|}}{C}{-}OR$$

Actually, the reaction takes place in two separate steps:

$$Cl{-}\overset{\overset{\large O}{\|}}{C}{-}Cl + ROH \xrightarrow[-HCl]{(NaOH)} Cl{-}\overset{\overset{\large O}{\|}}{C}{-}OR$$

$$Cl{-}\overset{\overset{\large O}{\|}}{C}{-}OR + R_1OH \xrightarrow[-HCl]{(NaOH)} R_1O{-}\overset{\overset{\large O}{\|}}{C}{-}OR$$

The two-step process allows the formation of asymmetrical diesters without any further separation if the starting reactant is an alkylchloroformate instead of phosgene (step 2) (1–3). To recover high-purity free-chlorine carbonates, the reaction product can be distilled in the presence of inorganic carbonates.

4. From carbon dioxide, with an adequate hydrophilic system as a catalyst, such as ethylene oxide (6) or zeolites (7).

$$CO_2 + 2ROH \xrightarrow[-H_2O]{CAT.} RO{-}\overset{\overset{\large O}{\|}}{C}{-}OR$$

5. From urea, with high-boiling alcohols, at high temperature (>200°C) (8–10).

$$\mathrm{NH_2{-}C(=O){-}NH_2} + 2\,\mathrm{ROH} \xrightarrow[-2\mathrm{NH_3}]{\mathrm{CAT.}} \mathrm{RO{-}C(=O){-}OR}$$

6. From carbon monoxide, via oxidation in the presence of complex palladium (12) or copper catalyst (13, 14).

$$2\,\mathrm{ROH} + \mathrm{CO} + \tfrac{1}{2}\mathrm{O_2} \xrightarrow[-\mathrm{H_2O}]{\mathrm{CAT.}} \mathrm{RO{-}C(=O){-}OR}$$

7. Synthesis by transesterification.

$$\mathrm{RO{-}C(=O){-}OR} + 2\,\mathrm{R_1OH} \overset{\mathrm{OH^-}}{\rightleftharpoons} \mathrm{R_1O{-}C(=O){-}OR_1} + 2\,\mathrm{ROH}$$

This reaction, basically being an equilibrium reaction, is governed by the following main factor: the more nucleophilic the alcohol, the easier the substitution with a less nucleophilic alcohol. This means, as a consequence, that arylic alcohols are substituted by alkylic alcohols, that lower-molecular-weight alcohols are substituted by higher-molecular-weight alcohols, and that the transesterification reaction of asymmetric carbonates more easily leads to the substitution of the lower-molecular-weight alcohol.

C. Current Commercial Routes

The most effective commercial route for the production of higher dialkylcarbonates is the transesterification of dimethylcarbonate (DMC):

$$\mathrm{CH_3O{-}C(=O){-}OCH_3} + 2\,\mathrm{ROH} \xrightarrow[-\mathrm{CH_3OH}]{} \mathrm{RO{-}C(=O){-}OR}$$

Consequently, the most important step of the whole commercial route is now the synthesis of DMC.

1. Production of DMC

Presently there are three commercial routes for the production of DMC, all involving methanol as methylating agent:

From Phosgene. This synthesis route has been widely applied in the past to industrial-scale production as the sole way to produce DMC.

$$Cl-\overset{\overset{O}{\|}}{C}-Cl + 2CH_3OH \xrightarrow[-HCl]{} CH_3O-\overset{\overset{O}{\|}}{C}-OCH_3$$

To obtain quantitative yields (>99%), excess alcohol is generally required, and it is subsequently recovered by distillation. There are problems with this process, mostly related to handling of phosgene and chlorine compounds.

From Carbon Monoxide. An important industrial route to produce DMC starts from methanol and carbon monoxide (see Section II, B), in the presence of oxygen and a catalytic system based on Cu^+/Cu^{2+} (11). The reaction takes place in two separate steps.

Step 1 (oxidation):

$$2CH_3OH + 2CuCl + \tfrac{1}{2}O_2 \xrightarrow[-H_2O]{} 2Cu\begin{matrix}OCH_3\\Cl\end{matrix}$$

Step 2 (reduction):

$$2CH_3OH + CO + \tfrac{1}{2}O_2 \xrightarrow[-H_2O]{CAT.} CH_3O-\overset{\overset{O}{\|}}{C}-OCH_3$$

The overall reaction is

$$2CH_3OH + CO + \tfrac{1}{2}O_2 \xrightarrow[-H_2O]{CAT.} CH_3O-\overset{\overset{O}{\|}}{C}-OCH_3$$

This synthesis route is currently increasing in importance because of the availability of raw materials, simplicity of the process, and economics.

From Carbon Dioxide. By direct reaction of carbon dioxide and ethylene oxide, or propylene oxide, the corresponding cyclic five-member carbonates are easily obtained with high conversion and yield (4).

$$CO_2 + \text{(epoxide, R)} \xrightarrow{P} \text{(cyclic carbonate, R)}$$

R = H, CH_3

In a second step the cyclic carbonate is transesterified with methanol to DMC in a standard transesterifiction plant.

Actually, the synthesis route from ethylene oxide (5) is not economically competitive with that from carbon monoxide already described (15). It has been used in the past by Dow Chemicals in an industrial-scale plant.

The synthesis route from propylene oxide has not been equally investigated. It is likely to could exhibit valuable economics related to the cost of raw materials (propylene oxide) and the value of the coproducts (propylene glycol).

Moreover, dialkylcarbonates of high-molecular-weight alcohols can be directly obtained by transesterification of the cyclic carbonates, thus eliminating DMC as an intermediate, resulting in a neat benefit for the overall production economics.

III. PROPERTIES AND PERFORMANCE CHARACTERISTICS

Unsaturated dialkylcarbonates are widely used for the production of polycarbonates; these are of course beyond the scope of our work and will not be further considered.

The basic structure of the molecule, the carbonic group, is fixed, so properties and characteristics mostly depend on the alcohol(s) and on the relevant synthesis process. Other aspects, more related to product quality and some physicochemical characteristics, depend on the process and particularly on the finishing steps (alcohols and solvent recovery, catalyst elimination, etc.).

The viscosity of dialkylcarbonates produced by transesterification ranges from <1 to 40 cSt at 100°C.

Alcohols capable of covering this viscosity range can be chosen among the following main groups:

1. Natural, linear, short-chain C_{6-8} and long-chain C_{12-18}.
2. Synthetic, branched, long-chain C_{12-18}.
3. Polyols, synthetic: trimethylol propane (TMP), neopentyglycol (NPG), neopentaerythritol (NPE).

This choice is not at all restrictive; it is likely that a wider range of viscosities and characteristics will be met using alcohols and technologies different from those now used.

The first commercial production of carbonates suitable for application in the lubricant field was started in 1985 by Enichem Synthesis (Italy), with a capacity of 3000 tons/year. Starting raw materials were DMC (dimethylcarbonate) and a mixture of synthetic alcohols chosen in the C_{12-16} range.

The process involves three main steps: transesterification of DMC, purification of dialkylcarbonates, and methanol recovery.

A. Chemical Properties

Dialkylcarbonates represent an important class of reactants and intermediates in organic chemistry through the following main reactions:

To give urethanes and ureas, with ammonia and amines.
As alkylating agents, for amines, phenols, and acids.
As oxyalkylating agents, in the case of five-membered cyclic carbonates.
To give malonates, by Claissen condensation.

As carbonylating agents (DMC) for ketones, nitriles, hydroxy amides, and amino alcohols.
As methylating agents (DMC).

Apart from such applications, the particular structure of the carbonic group involves the fact that dialkylcarbonates undergo thermal and/or catalytic decomposition to alcohols, carbon dioxide, and olefins, without acid formation.

Similarly, in the presence of water and/or hydrolyzing agents, hydrolysis can lead to the back formation of alcohols, carbonic acid, and, consequently, carbon dioxide. For this case, as a general rule, the hydrolytic stability increases with the molecular weight of the alkylic groups.

B. Physical Properties

All the industrially important dialkylcarbonates are colorless liquids with a specific gravity lower than 1 and a high boiling point. Some of them exhibit a pleasant odor. The physical state depends on the type and molecular weight of the alcohols. High-molecular-weight ($>C_{10}$) linear alcohols easily lead to solid products. Low-molecular-weight carbonates with highly symmetrical structures, such as dimethylcarbonate and ethylenecarbonate, exhibit melting points above 0°C.

Dialkylcarbonates are generally soluble in organic solvents, especially in polar solvents (alcohols, ethers, ketones, esters, etc.). In water, only the cyclic five-member carbonates (ethylene and propylene) present good solubility. Other low-molecular-weight carbonates (dimethyl and diethyl) present very limited solubility. High-molecular-weight carbonates are essentially insoluble.

Lower dialkylcarbonates generally form azeotropic mixtures with water and/or organic solvents; this aspect increases the number of operation related to the production of dialkylcarbonates via transesterification.

Physical properties of the most important dialkylcarbonates are listed in Tables 11.1 (low-molecular-weight carbonates) and in 11.2 (high-molecular-weight carbonates).

C. Comparative Performance Data

1. *Application as Lubricant Component*

The first reported applications as a lubricant component are listed next.

1941 (16). U.S. patent claims the use of small amounts of carbonates (1–5%), mostly from low-molecular-weight alcohols, to enhance the spreading and penetrability of mineral lubricants.
1945 (17). U.S. patent claims the use of minor proportions of dialkylcarbonates in mineral oil formulations to improve the load carrying capacity of the lubricant. Alkylic groups employed range from ethyl to lauryl alcohol, including aromatic alcohols.
1953 (18). U.S. patent claims the use of dialkylcarbonates of C_{1-10} alcohols as synthetic lubricants with low pour points and high viscosity indices.
1954 (19). U.S. patent claims the use of polymers derived from monomers of carbonic ester as pour-point depressants.
1955 (20). As for preceding point.
1956 (21). U.S. patent claims the use of carbonates of alcohols C_{10-20} obtained from oxosynthesis of branched olefines.

Table 11.1 Physical Properties of Low-Molecular-Weight Dialkylcarbonates

Carbonate	d_4^{20}	b.p.,°C (atmospheric)	m.p.,°C	Flash point,[a] °C	N_D^{20}	Viscosity, cP, 25°C	CAS registry no.
Dimethyl	1.073	90.2	4	14[c]	1.3687	0.664^{20}	616.38.6
Diethyl	0.976	125.8	–43	33[c]	1.3843	0.868^{15}	105.58.8
Diallyl	0.994	—	—	—	1.4280	—	15022.08.9
Allyldiglycol	1.143	—	–4	177[o]	1.4503	9.0	142.22.3
Ethylene	1.322^{39}	248	39	—	1.4158^{31}	—	96.49.1
Propylene	1.207_{20}	242	–49	—	1.4189	—	108.32.7
Di-*n*-propyl	0.941	166	<–40	62[c]	1.4022	1.27	623.96.1
Diisopropyl	0.920	147	<–40	46[c]	—	1.14	6482.34.4
Di-*n*-butyl	0.924	208	<–40	92[c]	1.4099^{25}	1.72	542.52.9
Di-2-ethylhexyl	0.897_{20}	—	—	—	1.4352^{25}	—	14858.73.2
Methylisopropyl	0.969_{20}	120	<–40	33[c]	—	0.77	—
Methyl-*n*-propyl	0.983	131	<–40	37[c]	—	0.84	—
Methyl-*n*-butyl	0.964	162	<–40	54[c]	—	1.01	—

[ac]Closed cup, [o]open cup.

Table 11.2 Physical Properties of High-Molecular-Weight Dialkylcarbonates

Carbonate	Viscosity at 100°C, cSt	Viscosity at 40°C, cSt	Viscosity index	Pour point, °C	Volatility (Noack %)
Oxo C_{12-15}/Oxo *iso*-C_{13}, 50/50	3.9	17.2	115	−21	16
Oxo C_{14-15} branched	4.1	18.0	126	−36	12
Oxo C_{12-15}	3.8	15.7	144	+5	15
Oxo C_{12-15}/C_4 Diol, 90/10	5.2	23.9	153	0	10.8
iso-C_{10}	2.5	9.4	86	−57	—
iso-C_{13}	3.9	19.8	81	−42	—
iso-C_{18}	11.0	180	<0	−18	—
iso-C_8/C_6 diol	4.7	24.5	110	−34	—
iso-C_8/TMP	12.4	151	62	−30	—
n-C_{12}	3.4	12.2	169	+5	—

Other than these U.S. patents, there was no indication of commercial applications of dialkycarbonates during this time. Subsequently, other preparations and applications have been claimed in Italian patents:

1982 (22). Italian patent claims lubricant formulations containing dialkylcarbonates obtained by transesterification of DMC with high-molecular-weight alcohols. The resulting products exhibit improved characteristics and performance.

1989 (23). Italian patent claims lubricant formulations containing organic carbonates as improved lubricants for cold steel rolling.

1990 (24). Italian patent claims lubricant formulations containing dialkylcarbonates obtained by transesterification of DMC with high-molecular-weight branched alcohols. Particularly, these alcohols are obtained by oxosynthesis of linear olefins, followed by cryogenic separation of the linear fraction.

The first commercial applications of dialkylcarbonates as lubricant components dates to 1987, when AgipPetroli introduced such components as a new synthetic base stock in the formulation of semisynthetic gasoline engine oils. After this, AgipPetroli extended the application of dialkylcarbonates to the formulation of other types of automotive lubricants, diesel and super high performance diesel (SHPD) engine oils, and gear and two-stroke oils. In the field of industrial lubricants dialkylcarbonates have been used in the formulation of both molding and rolling oils.

The quantity of dialkylcarbonates as base stock in the finished product ranges from 5 to 30% (automotive) and from small amounts to 100% (industry).

2. Performance

The performance of dialkylcarbonates in engine oil applications has been presented elsewhere (25). The main conclusions can be summarized as follows:

Tribology. The presence of the carbonic group in the molecule of dialkylcarbonates imparts oiliness properties to these products, owing to the interaction between the carbonic group and the metallic surfaces. Tribological tests (SRV, α-LFW-1), carried out in comparison with other mineral and synthetic base stocks having the same viscosity at 100°C, confirm excellent oiliness properties of dialkylcarbonates, better than those of mineral oil and current synthetic base stocks.

Viscometrics. Viscometrics data indicate that the amount of dialkylcarbonates necessary to meet the requirement of low viscosity multigrade engine oils (5W/XX, 10W/XX) is comparable with the amount of other synthetic base stocks (PAO, esters, etc.) of the same viscosity.

Oxidation the Thermal Stability. Oxidation and thermal stability tests show outstanding performance for dialkylcarbonates. The absence of acid products formed at the end of the tests has already been outlined. This aspect may affect the overall behavior of fully formulated lubricants in that corrosive wear phenomena can be positively controlled.

Elastomer Compatibility. Elastomer compatibility, a critical performance of synthetic base stocks, is generally satisfactory for dialkylcarbonates, even with respect to silicon elastomers, which are the most sensitive to the basestock polarity. In particular, semisynthetic multigrade engine oils formulated with dialkylcarbonates exhibit a better performance as far as silicone rubber swelling is concerned than the similar formulations made with PAO or polyol esters.

Wear Protection. The engine performance of multigrade oils containing dialkylcarbonates shows that the same excellent antiwear properties exhibited by this component are maintained in the finished oil formulation, without any interference with antiwear additives, as exhibited by other ester type base stocks. Results have been obtained on the OM 616 Kombi, CRC L-38, Petter W1 (extended to 144 h) and VW cam and tappet tests.

Sludge Protection. The formulations containing dialkylcarbonates give good sludge control, better than that exhibited by a pure mineral-based formulation containing the same package at the same concentration. Results are obtained from the VE and M102E black sludge tests. This performance can be related to the particular chemical structure of dialkylcarbonates, with a particularly strong polar group and two alkylic chains in the same molecule, resulting in a synergistic effect with traditional dispersant additives in the control of sludges of different nature.

IV. TOXICOLOGY, HANDLING, AND ENVIRONMENTAL ASPECTS

A. Toxicology

Dialkylcarbonates can be classified as not dangerous substances for humans and for the environment, according to the relevant tests and experiments that have been carried on to assess this aspect. Among others, the effect of contact, ingestion, and inhalation, under acute, subacute, and chronic conditions, has been investigated and tested, allowing the classification of dialkylcarbonates as not dangerous for humans.

As far as the environment is concerned, dialkylcarbonates are not soluble in water, exhibit a very low volatility, can be recollected from water and the ground if any loss occurs, have a low degradation rate, and do not form harmful products by degradation. On the basis of these considerations, dialkylcarbonates can be considered not dangerous for the environment.

B. Biodegradability

The presence of the carbonic group, CO_3, does not by itself affect the biodegradability of the whole molecule, as can be concluded from the following data (MITI Test Mod.):

Carbonate of	Biodegradability[28D]
Oxo C_{12-15} / oxo i-C_{13}	75%
Oxo C_{14-15} branched	65%
TMP oxo C_{12-15} / i-C_8	15%
Oxo i-C_{13}	60%
i-C_{18}	Negligible
CH_3OH	90%

Actually, the biodegradability of dialkylcarbonates strongly depends on the biodegradability of the alkylic groups present in the molecule.

C. Handling

No special procedures and care are required in handling dialkylcarbonates; only standard hygiene and safety procedures are recommended.

Dialkylcarbonates are stable at ambient temperature and atmospheric pressure, do not require any special precautionary measures for storage, and can be safely stored in carbon-steel tanks, even for long periods.

ACKNOWLEDGMENT

The authors acknowledge Enichem Synthesis and Euron for their activity in the field of synthesis, evaluation, and application of dialkylcarbonates, and AgipPetroli for permission for publication.

REFERENCES

1. Pennwalt Corp. (1978). DE-OS 2926354 (J. R. Angle, U. D. Wagle, D. C. Reid).
2. Matzner, M., R. R. Kurkjy, and R. J. Cotter (1964). *Chem. Rev.*, 64, 645.
3. Babad, H., and A. G. Zeiler (1973). *Chem. Rev.*, 73(1), 81.
4. Union Carbide Corp. (1980). EP-A 47474 (C. H. McMullen).
5. Peppel, W. J. (1958). *Ind. Eng. Chem.*, 50, 767–770.
6. Bayer (1977). DE 2748718 (H.-J. Buysch, H. Krimm, H. Rudolph).
7. Bayer (1982). EP-A 85347 (J. Genz, W. Heitz).
8. Bayer (1979). EP 13957 (W. Heitz, P. Ball).
9. Bayer (1979). EP 13958 (W. Heitz, P. Ball).
10. BASF (1981). EP 41622 (W. Harder, F. Merger, F. Towae).
11. *Ing. Chim. Ital.* (1985). 21(1–3), Gen.-Mar. (M. Massi Mauri, U. Romano, F. Rivetti).
12. Rivetti, F., and U. Romano (1979). *J. Organomet. Chem.* 174, 221–226.
13. Shell Int. Res. (1981). EP-A 71286 (E. Drent).
14. Bayer (1981). DE-OS 3016187 (G. Stammann, R. Becker, J. Grolig, H. Waldmann).
15. SRI (1988). PEP Review 87-1-4 (Y. R. Chin, N. F. Shih).
16. Standard Oil (1941). U.S. Patent 2,263,265 (M. F. Fincke, J. H. Bartlett).
17. Lubrizol Corp. (1945). U.S. Patent 2,387,999 (A. T. Knutson, E. F. Graves).
18. Standard Oil (1953). U.S. Patent 2,651,657 (L. A. Mikeska, L. T. Eby).
19. Standard Oil (1954). U.S. Patent 2,673,185 (J. H. Bartlett).
20. Esso (1955). U.S. Patent 2,718,504 (J. H. Bartlett).
21. Esso (1956). U.S. Patent 2,758,975 (D. L. Cottle, F. Knoth, D. W. Young).
22. Assoreni (1982). Italian Patent 20264 A/82 (P. Koch, U. Romano).
23. Euron (1989). Italian Patent 20191 A/89 (E. Brandolese).
24. AgipPetroli (1990). Italian Patent 21812 A/90 (G. Fisicaro, G. P. Gerbaz).
25. Fisicaro, G., and S. Fattori (1989). Conference on Synthetic Lubricants, Sopron (Hungary).

12
Cycloaliphatics

Clifford G. Venier and Edward W. Casserly
Pennzoil Products Company
The Woodlands, Texas

I. INTRODUCTION

Naturally occurring cycloaliphatic materials have been generally recognized as important constituents of petroleum for a century (1–3). "Naphthenes" (4) contain at least one ring while "paraffins" contain none. The different properties, both good and bad, of comparable products derived from naphthenic crude oil and paraffinic crude oil have been ascribed to the relative proportion of cyclic and acyclic molecules. Aromatic compounds also occur naturally in crude oils, and, in general, mineral oil base stocks contain all three classes of compounds in varying proportions.

Synthetic hydrocarbon base materials can be classified into the same categories as conventional mineral oil: paraffinic, naphthenic, and aromatic. For example, polyalphaolefins (PAOs), the hydrogenated oligomers of l-alkenes (see Chapter 1), are analogous to the paraffinic materials in petroleum. Dimers of internal olefins of the type used in detergent manufacture are also analogs of paraffinics, and the commercial development of them under the name "polyinternalolefins" is under way in Europe. Polybutenes are a noncyclic aliphatic product with no analog in natural petroleum.

The aromatic constituents of petroleum find analogy in synthetics as well. Synthetic alkylated aromatic compounds have found limited use as synthetic lubricating oil base stocks. Dialkylated aromatics have enjoyed the most commercial success.

No example of a synthetic analog of the third major structural class of petroleum hydrocarbons, the naphthenes or cycloaliphatics, has yet been successfully commercialized.

The scope of this chapter is limited to a description of the preparation and properties of cycloaliphatic materials that may have potential use as lubricant base stocks and that have been reported in the literature and patent art. Because of their inherent stability, only five- and six-membered ring compounds have been considered.

The organization of the rest of this chapter is by the number of condensed rings in the system and by ring size: that is, cyclopentanes will be followed by cyclohexanes, hydrindanes and decalins, etc., in that order. When an example contains two different ring systems, it will be mentioned in both places.

Several general studies on structure-property relationships have included molecules containing saturated rings. The largest of these is the API Project 42 report (5), which compiles the physical properties of 321 pure hydrocarbons. The properties measured in the project include viscosities, densities, boiling and melting points, heats of vaporization, and light refraction properties. Other properties of selected sets of the compounds are also included: viscosities and densities at elevated pressures, vapor pressures, heats of combustion, heats of fusion, and thermal conductivities. No discussion of the relations of structure to properties is presented in the API Project 42 Report, however.

Using some of the data from Project 42, Denis (6) has described the effect of five- and six-membered rings on tribologically relevant properties, in particular the pour point and viscosity index (VI). He found that rings centrally located within the longest chain, for example in 1,4-dialkylcyclohexanes, have little effect on pour point or VI. On the other hand, rings have the same effect as other alkyl groups as side chains, namely, the lowering of both the pour point and the VI.

In general, the discussion of cycloaliphatics in this chapter is limited to compounds with 20 or more carbon atoms. Compounds of lower molecular weight, in general, have too low viscosities and too high volatilities for successful use as lubricants, but may be included to illustrate a special point or to discuss special applications.

II. CYCLOPENTANES

Two basic types of cycloaliphatics containing cyclopentane rings are discussed: (a) compounds in which a single cyclopentane ring bears one or, more often, more than one alkyl substituent, and (b) compounds containing two or more terminal cyclopentane rings, that is, $-C_5H_9$ substituents.

A. Multiply-Alkylated Cyclopentanes

1. Preparation

The preparations of a large number of examples of the first type of cycloaliphatics containing cyclopentane rings in which two or more alkyl groups are attached to a central ring were reported by Venier and Casserly (8, 9). Multiply-alkylated cyclopentanes (MACs) are available from the hydrogenation of the corresponding multiply-alkylated cyclopentadienes, as in equation (1). These, in turn, are readily available by the alkylation of cyclopentadiene, methylcyclopentadiene, etc., by two fundamentally different routes (8, 10, 11).

$$\text{C}_5\text{-diene}\text{-}(CHR_2)_m + 2\,H_2 \xrightarrow{\text{Catalyst}} \text{C}_5\text{-ring}\text{-}(CHR_2)_m \quad (1)$$

Both preparative routes take advantage of the extraordinarily high acidity of simple cyclopentadienes. The very low pK_a of cyclopentadiene ($pK_a = 18$) compared to that of other simple hydrocarbons ($pK_a = 30$ or greater) arises from the aromaticity of the cyclic six-electron periphery of the cyclopentadienide anion, as in equation (2).

$$C_5H_6 \xrightarrow{\text{Base}} C_5H_5^{-} \tag{2}$$

This relatively high acidity allows the cyclopentadienide anion to be usefully accessible using ordinary bases, such as alkoxides and hydroxides. The use of expensive air- and moisture-sensitive bases, such as alkyl lithium reagents, to generate the nucleophilic carbanion necessary for carbon-carbon bond formation is easily avoided in the synthesis of higher-molecular-weight multiply-alkylated cyclopentadienes.

Phase-Transfer Alkylation. In 1968, Makosza (12) reported that cyclopentadiene could be alkylated under phase-transfer conditions using alkyl halides, as in equation (3). Venier and Casserly (8, 9) extended the earlier phase-transfer alkylation work to synthesize multiply-alkylated cyclopentadienes and, subsequently, alkyl cyclopentanes in the useful lubricating oil range.

$$C_5H_6 + m\ R_2CH\text{-}X \xrightarrow[\text{KOH (aq)}]{\text{Catalyst}} C_5H_{6-m}(CHR_2)_m + m\ KX \tag{3}$$

Multiple alkylations of cyclopentadiene, in which two to six alkyl groups are added to the cyclopentadiene ring, are easily carried out in a single pot using greater than 40% sodium or potassium hydroxide as the aqueous phase, an alkyl halide as both the alkylating agent and organic phase solvent, and a quaternary ammonium salt as the phase transfer catalyst. Unfortunately, the high cost of most alkyl halides and the large excess of potassium hydroxide required may make this route prohibitively expensive, even though the reaction is virtually quantitative.

Alkylation with Alcohols. Bailey and Hirsch (13) and Fritz and Peck (14) reported the alkylation of cyclopentadiene using alcohols in the presence of base, as in equation (4). Venier and Casserly (8, 9) extended the alkylation of cyclopentadiene by alcohols to the synthesis of products in the useful lubricating oil range.

$$C_5H_6 + m\ R_2CH\text{-}OH \xrightarrow[190^{\circ}C - 250^{\circ}C]{\text{KOH}} C_5H_{6-m}(CHR_2)_m + m\ H_2O \tag{4}$$

Cyclopentadiene is conveniently generated in situ by the efficient cracking of commercially available dimer, dicyclopentadiene, at the elevated reaction temperature (190–250°C). Because it suppresses the concentration of ions more basic than hydroxide ion, water inhibits the reaction. The water produced in the reaction is therefore removed as it is formed during the course of the reaction.

Multiple alkylations, in which two to five alkyl groups are added to the cyclopentadiene ring, are easily carried out in one step. Venier and Casserly (8) have used this reaction procedure to react alcohols with eight or more carbon atoms with cyclopentadiene. In principle, alcohols of lower molecular weight could be used; however, the high reaction temperature (190–250°C) would necessitate the use of an inert high-boiling solvent or a closed system under the autogenous pressure of the alcohol. The reaction is virtually quantitative. This method has been used to produce one specialty lubricant marketed to the aerospace industry.

Friedel-Crafts Oligomerization of 1-Alkenes with Halocyclopentanes. The Friedel-Crafts oligomerization of 1-alkenes in the presence of halocyclopentanes gives products of lower viscosity and more uniform molecular weight than oligomerization in the absence of the halocyclopentane (15, 16). Yields were in the range of 40–60% based on starting 1-alkene. No structural data to suggest how much, if any, of the cyclopentyl derivative was incorporated into the product were presented in the 1961 patent.

2. Physical Properties

The discussion of cycloaliphatics containing cyclopentane rings will be limited to compounds with 20 or more carbon atoms. Compounds of lower molecular weight will, in general, have low viscosities and high volatilities.

The reader should explicitly note that many of the examples used to illustrate physical properties are *mixtures* of two, and occasionally more, materials of different molecular weight. A weight average carbon number of a sample is defined to allow materials to be compared. The weight average carbon number of a sample, C_{av}, is defined as

$$C_{av} = \Sigma\ [(\text{weight fraction of the } i\text{th component}) \times (\text{carbon number of the } i\text{th component})]$$

Viscosity. The viscosity of alkylcyclopentanes depends primarily on the molecular weight of the compound, increasing regularly as molecular weight increases. The kinematic viscosities (ASTM D-445) at 100°C for a variety of examples are given in Appendix 12.1 and Fig. 12.1.

Equations of the form log(vis) versus log(C_{av}) can be used to describe the dependence of viscosity on molecular weight. The equations relating log(vis) to log(C_{av}) for multiply-alkylated cyclopentanes, 1-decene PAOs, and normal alkanes are

MACs: $\log(\text{vis}) = 1.71[\log(C_{av})] - 1.92$
PAOs: $\log(\text{vis}) = 1.77[\log(C_{av})] - 2.06$
n-Alkanes: $\log(\text{vis}) = 2.03[\log(C_{av})] - 2.36$

Branching of the alkyl groups gives products that are more viscous than the products prepared from the corresponding linear alkyl groups of the same molecular weight. The viscosities of cyclopentanes prepared from the highly branched oxo alcohols, such as isodecanol and isotridecanol, lie above the regression line in Fig. 12.1. This is to be expected since these alcohols are prepared from propene oligomers by hydroformylation. Isodecanol and isotridecanol are best characterized as mixtures of trimethylheptanols and tetramethylnonanols, respectively.

Table 12.1 lists the properties of several cyclopentanes with a total of 29 carbon atoms. The alkyl groups vary from normal and branched octyls to normal decyls and normal dodecyls. Note that the branched alkyl cyclopentanes are more viscous than the normal alkyl cyclopentanes.

Viscosity Index. Viscosity index (VI; ASTM D-2270) is the measure of the change of viscosity with temperature. The higher the viscosity index, the less the viscosity of the fluid changes with temperature. For lubricants, a high viscosity index is usually desirable.

In the multiply-alkylated cyclopentanes in which two or more alkyl groups are

Appendix 12.1 Lubricant-Related Properties of Multiply-Alkylated Cyclopentane Synthesized Hydrocarbon Fluids

Alkyl, R[a]	m, Number of Groups	C_{av}[b]	Kinematic viscosity, 100°C, cSt	Kinematic viscosity, 40°C, cSt	Viscosity index	Pour point, °C	Reference
n-octyl	2	21	2.18	6.49	158	–24	8,9,22
n-octyl	3	29	3.68	15.58	124	<–57	8,9,22
n-octyl	3,4	34	4.96	24.70	128	<–57	8,9
n-octyl	4	37	5.99	33.43	125	<–48	8,9,22
n-octyl	4,5	41	7.40	45.33	127	<–54	8,9
n-decyl	2	25	3.03	10.37	161	+6	8,9,22
n-decyl	2,3	29	3.78	14.44	161	–9	8,9
n-decyl	3	35	5.15	23.99	151	–27	8,9,22
n-decyl	3,4	41	6.90	37.67	144	–45	8,9
n-decyl	4	45	7.99	46.70	143	<–60	8,9,22
n-decyl	4,5	48	8.92	54.07	144	<–45	8,9
n-dodecyl	2	29	4.13	15.64	178	+21	8,9,22
n-dodecyl	3	41	6.99	35.26	164	–9	8,9,22
n-dodecyl	3,4	46	8.18	45.17	157	–15	8,9
n-dodecyl	4,5	59	11.91	75.55	153	–9	8,9,22
2-octyldodecyl	2,3	62	14.56	109.30	137	–57	8,9
2-ethylhexyl	3	29	4.21	26.44	22	<–48	[d]
2-ethylhexyl	3,4	31	5.35	40.60	40	<–54	8,9
isodecyl	3,4,5	46	11.68	119.60	83	–33	8,9,22
isotridecyl	3,4,5	56	20.09	310.22	71	–27	8,9,22
2-octyl	3	29	4.71	28.23	74	<–54	8
2-decyl	2,3	32	4.07	18.34	123	—	8
n-octyl/*n*-decyl		35[c]	5.23	26.28	134	<–50	9
n-octyl/*n*-decyl		41[c]	6.91	38.67	139	<–50	9

[a]R is the alkyl group in ⬠–R_m.

[b]$C_{av} = \sum_i [(\text{weight fraction})_i \times (\text{carbon number})_i]$.

[c]Estimated from kinematic viscosity at 100°C.

[d]Previously unpublished data.

attached to a single cyclopentane ring, the viscosity index varies with both the number of substituents and the length of the alkyl groups. Therefore, a separate curve relating average carbon number to viscosity index is generated for each alkyl group length. Representative examples are given in Table 12.2 and Fig. 12.2.

The viscosity index depends most strongly on the average length of the alkyl groups and less on the number of alkyl groups attached to the cyclopentane ring. For a given degree of substitution, the longer the average length of the alkyl groups, the higher the viscosity index. For example, tri(*n*-octyl)cyclopentane has an average alkyl length of 8 and a VI of 124. Tri(*n*-decyl)cyclopentane has an average alkyl length of 10 and a VI of 151. Tri(*n*-dodecyl)cyclopentane has an average alkyl length of 12 and a VI of 164.

For a given average alkyl length, the greater the degree of substitution, the lower the viscosity index. For example, in progressing from di(*n*-decyl)cyclopentane to tri(*n*-decyl)cyclopentane to tetra(*n*-decyl)cyclopentane, the substitution increases from two to three to four alkyl groups, while the VI decreases from 161 to 151 to 143.

For a given carbon number, the VI increases as the length of the alkyl groups

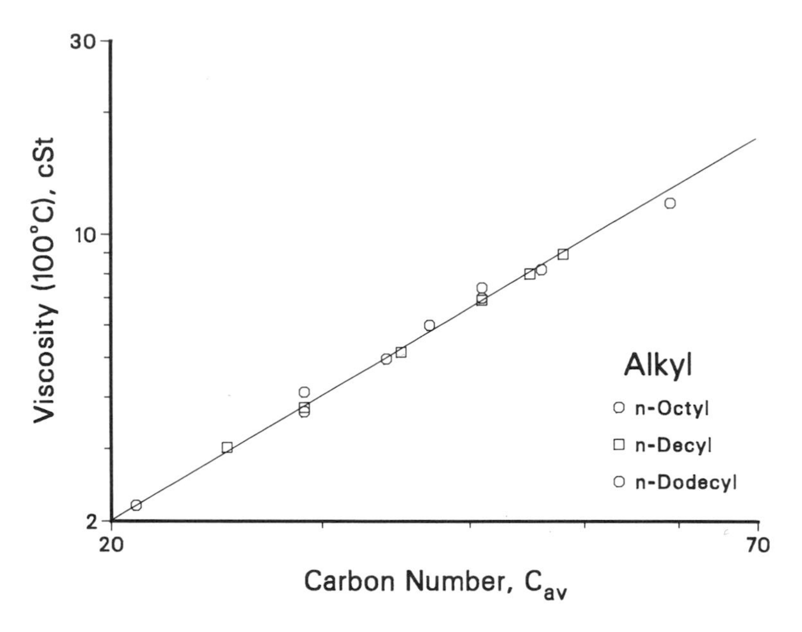

Figure 12.1 Kinematic viscosity as a function of carbon number for alkylated cyclopentanes.

increases or as the degree of substitution decreases. The following examples from Table 12.1 all have a total of 29 carbon atoms. Di(*n*-dodecyl)cyclopentane has a VI of 178, the mixture of di- and tri(*n*-decyl)cyclopentanes has a VI of 161, and tri(*n*-octyl)cyclopentane has a VI of 124.

Branching decreases the VI by effectively reducing the length of the longest alkyl group. In Table 12.1, examples of normal and branched octyl-substituted cyclopentanes are given. The tri(*n*-octyl)cyclopentane has a VI of 124, while the tri(2-octyl)cyclopentane has a VI of 74, and the tri(2-ethylhexyl)cyclopentane has a VI of 22. The length of the longest uninterrupted alkyl chain decreases from eight to six to four carbon atoms in this series.

The viscosity indices of blends of MACs with different alkyl groups depend most

Table 12.1 Lubricant-Related Properties of Multiply-Alkylated Cyclopentane Synthesized Hydrocarbon Fluids of 29 Carbon Atoms

Alkyl, R[a]	m, Number of groups	C_{av}[b]	Kinematic viscosity, 100°C, cSt	Kinematic viscosity, 40°C, cSt	Viscosity index	Pour point, °C	Reference
n-Octyl	3	29	3.68	15.58	124	<-57	8,9,22
2-Octyl	3	29	4.71	28.23	74	<-54	8
2-Ethylhexyl	3	29	4.21	26.44	22	<-48	23
n-Decyl	2,3	29	3.78	14.44	161	−9	8,9
n-Dodecyl	2	29	4.13	15.64	178	+21	8,9,22

[a]R is the alkyl group in (cyclopentane)-R_m.

[b]$C_{av} = \sum_i [(\text{weight fraction})_i \times (\text{carbon number})_i]$.

Table 12.2 Viscosity Index of Multiply-Alkylated Cyclopentane Synthesized Hydrocarbon Fluids

Alkyl, R[a]	m, Number of groups	C_{av}[b]	Kinematic viscosity, 100°C, cSt	Viscosity index	Pour point, °C	Reference
n-Octyl	3	29	3.68	124	<–57	9
n-Octyl	3,4	34	4.96	128	<–57	9
2-Octyl	3	29	4.71	74	<–54	9
2-Ethylhexyl	3	29	4.21	22	<–48	9
n-Decyl	2	25	3.03	161	+6	9
n-Decyl	2,3	29	3.78	161	–9	9
n-Decyl	3	35	5.15	151	–27	9
n-Decyl	4	45	7.99	143	<–60	9
n-Dodecyl	2	29	4.13	178	+21	9
n-Dodecyl	3	41	6.99	164	–9	9
n-Octyl/*n*-decyl		35[c]	5.23	134	<–50	9
n-Octyl/*n*-decyl		41[c]	6.91	139	<–50	9

[a]R is the alkyl group in ⌂–R_m.

[b]$C_{av} = \sum_i [(\text{weight fraction})_i \times (\text{carbon number})_i]$.

[c]Estimated from kinematic viscosity at 100°C.

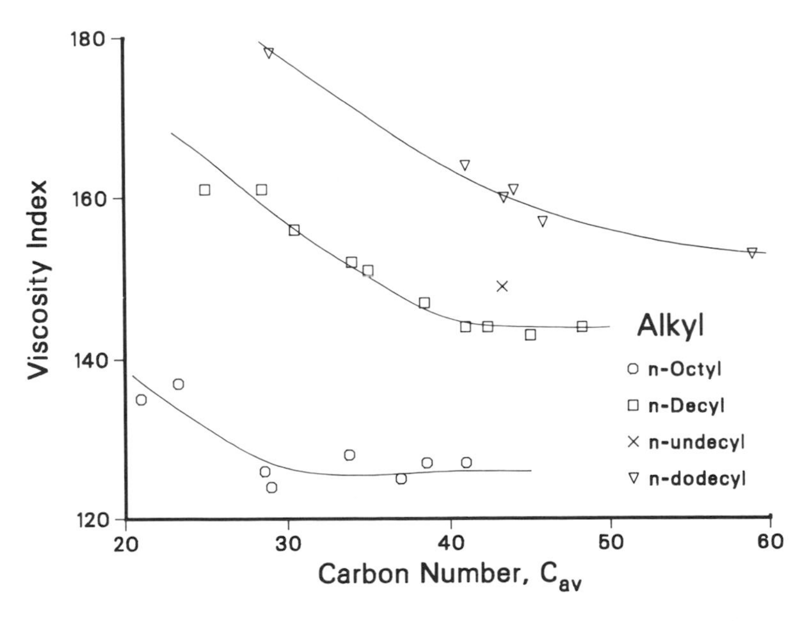

Figure 12.2 Viscosity index as a function of alkyl group and carbon number for alkylated cyclopentanes.

strongly on the length of the average alkyl group as well. Tri(*n*-octyl)cyclopentane has a VI of 124 and tri(*n*-dodecyl)cyclopentane has a VI of 164. A 50:50 mixture of the two has an average chain length of 10 and a VI of 140, which is between the VI of the components. Tri(*n*-decyl)cyclopentane has an average chain length of 10 by construction and exhibits a VI of 151.

Multiply-alkylated cyclopentanes that are prepared from mixtures of alcohols also exhibit viscosity indices characterized by an average chain length. Two examples serve to illustrate this. For MACs with an average total carbon number of 35 the VI values are *n*-octyl, 128; *n*-octyl/*n*-decyl, 134; and *n*-decyl, 151. For C_{av} = 41, the VI values are *n*-octyl, 127; *n*-octyl/*n*-decyl, 139; *n*-decyl, 144.

Low-Temperature Properties. The pour point (ASTM D-97) of cycloaliphatics with two or more alkyl groups attached to a single cyclopentane ring depends on a number of factors. First, for a given number of substituents, for example, di(*n*-alkyl)cyclopentanes, the pour point increases dramatically with increasing length of the alkyl group. Second, with a constant *n*-alkyl group, the pour point decreases as the number of substituents increases, that is, as the molecular weight increases. Finally, MACs with *n*-alkyl groups have higher pour points than cyclopentanes with branched alkyl groups. Representative data from Appendix 12.1 are given in Table 12.3 and illustrated in Fig. 12.3. As with viscosity index, a separate curve is generated for each alkyl group.

For the disubstituted cyclopentanes, the pour point increases from –24°C to +6°C to +21°C as the alkyl group is changed from *n*-octyl to *n*-decyl to *n*-dodecyl. The same trend is clear in the tri(*n*-alkyl) series: tri(*n*-octyl), <–57°C; tri(*n*-decyl), –27°C; and tri(*n*-dodecyl), –9°C. All known examples of *n*-octyl substituted cyclopentanes with greater than 29 carbon atoms and *n*-decyl substituted cyclopentanes with greater than 41 carbon atoms have pour points lower than –50°C.

The molecular weight of cyclopentanes with *n*-decyl substituents increases 140 units for each *n*-decyl group. The pour point, however, decreases with each additional substituent, that is, each increase in molecular weight. Thus, di(*n*-decyl)cyclopentane has a molecular weight of 350 and a pour point of +6°C; tri(*n*-decyl)cyclopentane has a

Table 12.3 Low-Temperature Properties of Multiply-Alkylated Cyclopentane Synthesized Hydrocarbon Fluids

Alkyl, R[a]	m, Number of groups	C_{av}[b]	Kinematic viscosity, 100°C, cSt	Viscosity index	Dynamic viscosity, −30°C, cP	Pour point, °C	Reference
n-Octyl	2	21	2.18	158	—	–24	8,9,22
n-Octyl	3	29	3.68	124	700	<–57	8,9,22
n-Octyl	4	37	5.99	125	3,700	<–57	8,9,22
n-Decyl	2	25	3.03	161	—	+6	8,9,22
n-Decyl	2,3	29	3.78	161	—	–9	8,9
n-Decyl	3	35	5.15	151	—	–27	8,9,22
n-Decyl	4	45	7.99	143	4,500	<–60	8,9,22
n-Dodecyl	2	29	4.13	178	—	+21	8,9,22
n-Dodecyl	3	41	6.99	164	—	–9	8,9,22
2-Octyldodecyl	2,3	62	14.56	137	20,500	–57	8,9,22

[a]R is the alkyl group in ⬠–R_m.

[b]$C_{av} = \sum_i [(\text{weight fraction})_i \times (\text{carbon number})_i]$.

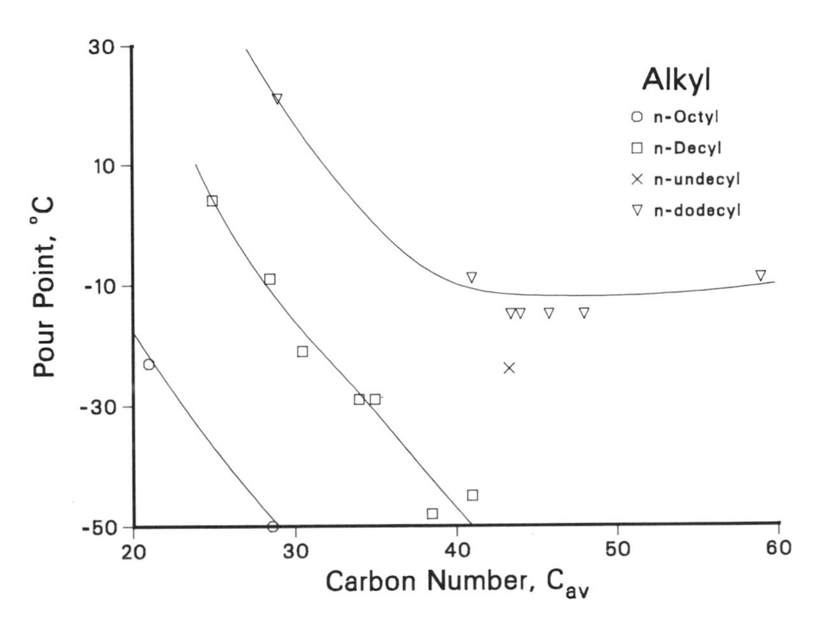

Figure 12.3 Pour point as a function of alkyl group and carbon number for alkylated cyclopentanes.

molecular weight of 490 and a pour point of –27°C; and tetra (*n*-decyl)cyclopentane has a molecular weight of 630 and a pour point of <–50°C. Since the molecular weight correlates with the 100°C viscosity, the more viscous examples of this series have lower pour points.

The effect of alkyl group chain length is greater than the effect of the number of substituents, as illustrated by the pour point of examples with the same molecular weight, that is, the same C_{av}. Tri(*n*-octyl)cyclopentane and di(*n*-dodecyl)cyclopentane each contain 29 carbon atoms. The tri(*n*-octyl)cyclopentane has a pour point of <–50°C while the di(*n*-dodecyl)cyclopentane has a pour point of +21°C. A mixture of di- and tri(*n*-decyl)cyclopentanes, with an average of 29 carbon atoms, has a pour point of –9°C. Thus, the pour point is very sensitive to the exact structure of the compound.

Isoparaffins in the lubricating oil range are liquids, whereas the paraffins are solids. Thus, the pour point of MACs with branched side chains would be expected to be lower than for unbranched examples. One example in particular illustrates this point well. A mixture of di- and tri(2-octyldodecyl)cyclopentane has an average carbon number of 62 and a pour point of –57°C. If the average carbon number versus pour point curves in Fig. 12.3 were extrapolated to alkyl chains of 20 carbon atoms, the series would all be solids. The introduction of a single branch in each alkyl group decreases the pour point dramatically.

High-Temperature Properties. As is the case with most alkanes, the flash and fire points are a function primarily of the molecular weight. Table 12.4 lists the flash and fire points of a number of MACs. The evaporation loss of representative MACs is also given in Table 12.4. Once again, as molecular weight increases, the evaporation loss goes down. The last entry in the table, tris(2-octyldodecyl)cyclopentane, is particularly noteworthy. Besides the low volatility and high flash and fire points, this material has a pour point of less than –55°C (see Table 12.3).

Table 12.4 High-Temperature Properties of Multiply-Alkylated Cyclopentane Synthesized Hydrocarbon Fluids

Alkyl, R[a]	m, Number of groups	C_{av}[b]	Kinematic viscosity, 100°C, cSt	Kinematic viscosity, 40°C, cSt	Viscosity index	Pour point, °C	Flash point, °C	Evaporation loss,[c] weight %	Reference
n-Decyl	3	35	5.15	23.99	151	−27	260	6.1	22
n-Decyl	4,5	48	8.92	54.07	144	<−45	288	1.3	22
n-Dodecyl	3,4	47	9.98	55.70	167	−15	294	0.4	22
2-Octyldodecyl	2,3	62	14.56	109.30	137	−57	307	0.1	8,9

[a]R is the alkyl group in ⬠–R_m.

[b]$C_{av} = \sum_i [(\text{weight fraction})_i \times (\text{carbon number})_i]$.

[c]Modified ASTM D-972, 250°C for 6.5 h with a nitrogen flow of 2 liters/min.

B. Alkanes with Cyclopentyl Substituents

API Project 42 reported physical properties for 321 hydrocarbons, including paraffins, isoparaffins, cycloaliphatics, and aromatics. In the normal paraffin series, the greater the number of carbons, the higher the viscosity, the higher viscosity index (VI), and the higher pour point. Branching decreases the viscosity slightly, but the VI and the pour point for hydrocarbons with a given number of carbon atoms are markedly decreased by branching. The API Project 42 report includes data for several multiring compounds and several compounds in which a single alkyl group is attached to a single cyclopentane ring. All of these data are compiled in Appendix 12.2. The number of carbon atoms in these examples range from 22 to 27. Additional data for some of these compounds have been reported by Denis (6) and Reith (7).

Appendix 12.2 Properties of Cyclopentanes from API Project 42 Report

Compound name	API Project 42 number	Carbon number	Viscosity, 100°C, cSt	Viscosity, 40°C cSt	Viscosity index	Pour point, °C
One cyclopentyl ring						
9(3-Cyclopentylpropyl)heptadecane	110	25	2.84	10.72	112	–21
11-Cyclopentylheneicosane	64	26	3.04	11.64	119	–13
1-Cyclopentylheneicosane	117	26	3.72	13.08	188	+45
11-Cyclopentylmethylheneicosane	74	27	3.26	12.79	125	–21
11(2,4-Dimethylcyclopentylmethyl)heneicosane	180	28	3.33	13.62	116	—
Two cyclopentyl rings						
1-Cyclopentyl-4(3-cyclopentylpropyl)dodecane	111	25	3.49	14.82	114	–40
1-Cyclopentyl-2-hexadecylcyclopentane	15	26	3.74	14.21	162	+19
1,1-Dicyclopentylhexadecane	202	26	3.81	16.12	130	+12
9[alpha(*cis*-0.3.3-Bicyclooctyl)methyl]heptadecane	178	26	3.64	17.30	87	—
Three cyclopentyl rings						
1,5-Dicyclopentyl-3(2-cyclopentylethyl)pentane	553	22	3.88	20.84	58	—
1,7-Dicyclopentyl-4(3-cyclopentylpropyl)heptane	113	25	4.52	23.43	105	–24
1,3-Dicyclopentyl-2-dodecylcyclopentane	199	27	4.75	26.78	93	0

Table 12.5 Selected Cyclopentanes from API Project 42 Report

Compound name	API Project 42 number	Carbon number	Viscosity, 100°C, cSt	Viscosity, 40°C, cSt	Viscosity index	Pour point °C
One cyclopentyl ring						
9(3-Cyclopentylpropyl)heptadecane	110	25	2.84	10.72	112	–21
11-Cyclopentylheneicosane	64	26	3.04	11.64	119	–13
11-Cyclopentylmethylheneicosane	74	27	3.26	12.79	125	–21
Cyclopentylheneicosane isomers						
1-Cyclopentylheneicosane	117	26	3.72	13.08	188	+45
11-Cyclopentylmethylheneicosane	74	27	3.26	12.79	125	–21
Two cyclopentyl rings						
1-Cyclopentyl-4(3-cyclopentylpropyl)dodecane	111	25	3.49	14.82	114	–40
1,1-Dicyclopentylhexadecane	202	26	3.81	16.12	130	+12

Selected data from Appendix 12.2 are given in Table 12.5. The structure/property relationships, described already for the multiply-alkylated cyclopentanes, hold true for the alkanes with cyclopentyl substituents. The viscosity follows molecular weight. For example, the first three entries are structurally related: there is one branched alkyl group attached to one cyclopentane ring. As the molecular weight increases, the kinematic viscosity at 100°C increases from 2.84 to 3.26 cSt.

The position of the cyclopentane ring on the alkyl group affects the viscosity index and pour point. For example, the isomers 1-cyclopentylheneicosane and 11-cyclopentylheneicosane differ only by the placement of the cyclopentane ring. The one branch point reduces the length of the longest unbranched chain from 21 to 10. The viscosity index is reduced from 188 for the 1-cyclopentylheneicosane to 119 for the 11-cyclopentylheneicosane. In the same way, the pour point for the 1-isomer is +45°C and is –12.7°C for the 11-isomer. 1-Cyclopentylheneicosane behaves more like the normal paraffin *n*-hexacosane (6) (kinematic viscosity at 100°C = 3.24 cSt, VI = 188, pour point = +56°C).

In examples containing two cyclopentyl groups, 1,1-dicyclopentylhexadecane has a higher molecular weight and a longer unbranched chain than 1-cyclopentyl-4(3-cyclopentylpropyl)dodecane. Hence, it has a higher kinematic viscosity at 100°C, a higher viscosity index, and a higher pour point.

III. CYCLOHEXANES

A. Introduction

A large number of individual examples of cyclohexanes of sufficiently high molecular weight to be in the lubricating oil boiling range are known. The API Project 42 Report lists many examples illustrating a wide variety of structural features. Perhaps the most promising molecules for commercial exploitation are alkylcyclohexanes that can be prepared by hydrogenation of alkylbenzenes [equation (5)], where Ar-H is any aromatic hydrocarbon, R is an alkyl group, and L is a leaving group such as halide. Acid catalysts are protonic (e.g., sulfuric) or Lewis (e.g., aluminum chloride) acids.

$$Ar\text{-}H_x + x\,R\text{-}L \xrightarrow{\text{Acid}} Ar\text{-}R_x + x\,H\text{-}L \qquad (5)$$

Hydrogenation of the corresponding alkylbenzenes is the most straightforward route to synthetic alkylated cyclohexanes [equation (6)]. As a consequence, no alkylated cyclohexane fluid will ever be less expensive than its parent alkylbenzene. Alkylated cyclohexanes would be viable commercial products *only* if unique properties increased the benefit side of the cost/benefit ratio, or if a route to them independent of the alkylation of aromatics were developed.

$$C_6H_{6-x}R_x + 3\,H_2 \xrightarrow{\text{Catalyst}} C_6H_{12-x}R_x \qquad (6)$$

B. Physical Properties

Appendix 12.3 is a compilation of the properties of all cyclohexanes in the boiling range of lubricant base stocks that are in the literature. The effect of structure on properties is very similar to that discussed in the cyclopentane section. Viscosity follows molecular weight, viscosity index reflects the average length of ring substituents, and pour point is a function of branching. Table 12.6 gives examples in which a long alkyl chain attached to a cyclohexane ring dominates the properties.

The effect of inserting rings into paraffinic chains can be illustrated by comparing the viscosity indices of eicosane, 1,3-didecylcyclohexane, and 1,4-didecylcyclohexane.

$CH_3(CH_2)_{18}CH_3$ VI = 177

$CH_3(CH_2)_9$–C_6H_{10}–$(CH_2)_9CH_3$ (1,4) VI = 168

$CH_3(CH_2)_9$–C_6H_{10}–$(CH_2)_9CH_3$ (1,3) VI = 147 (7)

1,4-Disubstituted cyclohexanes should have an essentially linear structure, altered little from that of the eicosane, whereas the 1,3-disubstituted compound of necessity will be bent.

The data on cyclohexanes are complementary to those on cyclopentanes in many ways, because different kinds of structural variations have been reported for the two different classes of fluids. For example, specific isomers of cyclohexyleicosane have been prepared and their properties determined. Table 12.7 lists the properties.

Figures 12.4 and 12.5 show how viscosity index and pour point vary with the position at which the cyclohexyl rings are attached to the 20-carbon chain. As was noted with cyclopentanes, both VI and pour point are largely governed by the length of the longest unbranched chain. As the average number of carbons in the longest unbranched chain decreases, both the viscosity index and the pour point fall.

For comparison, the properties of a mixture of eicosylcyclohexanes prepared by the Friedel-Crafts alkylation using a statistical mixture of monochloroeicosanes are also shown in both Figs. 12.4 and 12.5. Note that the viscosity index of the statistical mixture is higher than for substitution near the middle of the chain. The product is an equimolar

Appendix 12.3 Properties of Cyclohexanes in Lubricating Oil Range[a]

Compound	API Project 42 number	Number of R groups	Carbon number	Viscosity, 100°C, cSt	Viscosity, 40°C, cSt	Viscosity Index	Pour point, °C	Reference
1-Hexadecylcyclohexane		1	22	2.87	9.60	159	—	19
2-Methyl-5-isopropyl-1-dodecylcyclohexane		3	22	2.57	9.24	108	—	17
2-Methyl-5-isopropyl-1[2(3-methylbutyl)-6-methylhexyl]cyclohexane		3	22	2.72	14.76	–58	—	17
2-Methyl-5-isopropyl-1[2(1-decalyl)ethyl]cyclohexane		3	22	5.44	53.84	–39	—	17
1,2-bis(2-methyl-5-isopropylcyclohexyl)-ethane		3	22	3.92	29.13	–87	—	17
9-Heptadecylcyclohexane	509	1	23	2.69	10.77	79	–57	5
1-Octadecylcyclohexane		1	24	3.33	12.37	148	+40	18
3-Octylundecylcyclohexane	88	1	25	3.21	13.58	99	—	5
1,1-Di(cyclohexylmethyl)octane	19	1	25	4.85	30.23	70	—	5
3,3-Di(cyclopentylpropyl)propylcyclohexane	129	1	25	5.54	36.52	82	–46	5
4,4-Di(cyclohexylethyl)butylcyclopentane	127	1	25	7.11	65.47	49	–34	5
Tri(2-cyclohexylethyl)methane	90	1	25	9.75	44.88 (60°C)	–7	+41	5
1,4-Di(4-cyclohexylbutyl)cyclohexane	207	2	26	9.10	4.35 (135°C)	90	—	5
1,1-Dicyclohexyltetradecane	11	1	26	5.18	32.81	80	+38	5
1,1-Di(4-methylcyclohexyl)dodecane	139	1	26	5.34	42.69	50	–37	5
1,3-Didecylcyclohexane	209	2	26	3.59	13.83	147	—	5
1,4-Didecylcyclohexane	153	2	26	4.01	15.43	168	—	5

1-Eicosylcyclohexane	100	1	26	4.06	15.34	176	+48	5,7
2-Eicosylcyclohexane	102	1	26	4.06	16.05	162	+13	5
3-Eicosylcyclohexane	75	1	26	3.77	14.82	151	+23	5
4-Eicosylcyclohexane	104	1	26	3.72	15.80	125	+16	5,7
5-Eicosylcyclohexane	76	1	26	3.69	16.16	114	–2	5,7
7-Eicosylcyclohexane	77	1	26	3.58	15.71	108	—	5,7
9-Eicosylcyclohexane	78	1	26	3.44	14.95	105	—	5
4-Methyl-8-nonadecylcyclohexane	162	2	26	3.51	16.14	91	<–57	5
2,5-Dimethyl-1-octadecylcyclohexane	159	3	26	3.71	14.34	154	—	5
11-Heneicosylcyclohexane	60	1	27	3.68	16.16	114	–7	7
2,4,6-Trimethyl-1-octadecylcyclohexane	157	4	27	4.08	16.37	158	—	5
1-Docosylcyclohexane		1	28	4.61	18.43	179	+49	18
5-Docosylcyclohexane		1	28	4.09	18.12	128	<+23	18
Tri(3-cyclohexylpropyl)methane	172	1	28	9.59	38.74 (60°C)	46	—	5
1,1-Di(cyclohexylmethyl)dodecane	65	1	28	5.49	36.21	81	+2	5
2-Decyldodecylcyclohexane	91	1	28	3.78	17.00	111	–4	5
2,5-Dimethyl-11-heneicosylcyclohexane	169	3	29	4.32	22.92	90	—	5
1,1,6,6-Tetracyclohexylhexane	186	1	30	7.06 (150°C)	2.85 (200°C)	–76	+131	5
13-Pentacosylcyclohexane	69	1	31	4.66	22.08	131	+1	5
5-Hexacosylcyclohexane		1	32	5.48	25.97	155	+30	18
4-Cyclohexyl-(1-butyloctadecyl)-cyclohexane		2	34	8.86	69.00	101	<+23	18
15-Nonacosylcyclohexane	136	1	35	6.04	31.13	144	+13	5
17-Triacontylcyclohexane	138	1	39	7.45	40.79	151	+25	5

[a]Data in the range of room temperature to 150°C were extrapolated to 100°C and 40°C, temperatures more commonly used today.

Table 12.6 Cyclohexanes with Long Normal-Alkyl Chains

Compound	API Project 42 number	Number of R groups	Carbon number	Viscosity, 100°C, cSt	Viscosity, 40°C, cSt	Viscosity index	Pour point, °C	Reference
1-Hexadecylcyclohexane		1	22	2.87	9.60	159	—	19
1-Octadecylcyclohexane		1	24	3.33	12.37	148	+40	18
1,3-Didecylcyclohexane	209	2	26	3.59	13.93	147	—	5
1,4-Didecylcyclohexane	153	2	26	4.01	15.43	168	—	5
1-Eicosylcyclohexane	100	1	26	4.06	15.34	176	+48	5,7
2,5-Dimethyl-1-octadecyl cyclohexane	159	3	26	3.71	14.34	154	—	5
2,4,6-Trimethyl-1-octadecyl cyclohexane	157	4	27	4.08	16.37	158	—	5
1-Docosylcyclohexane		1	28	4.61	18.43	179	+49	18

Data in the range of room temperature to 150°C were extrapolated to 100°C and 40°C, temperatures more commonly used today.

Table 12.7 Properties of x-Eicosylcyclohexanes

x=	API number	Viscosity, 100°C, cSt	Viscosity, 40°C, cSt	Viscosity index	Pour point, °C	Reference
1	100	4.06	15.34	176	+48	5,7
2	102	4.06	16.05	162	+13	5
3	75	3.77	14.82	151	+23	5
4	104	3.72	15.80	125	+16	5,7
5	76	3.69	16.16	114	−2	5,7
7	77	3.58	15.71	108	—	5,7
9	78	3.44	14.95	105	—	5

Data in the range of room temperature to 150°C were extrapolated to 100°C and 40°C, temperatures more commonly used today.

mixture of the 10 isomers, 1- to 10-eicosylcyclohexanes, and the average point of attachment is between carbons 5 and 6. Thus, the viscosity index of the mixture (120) is close to that of the 5-eicosylcyclohexane (115), and the pour point (−15°C) is slightly lower than that of the pure material due to the natural melting point lowering observed in mixtures.

Table 12.8 shows the properties of a series of alkylcyclohexanes in which the alkyl group is a secondary linear alkyl chain.

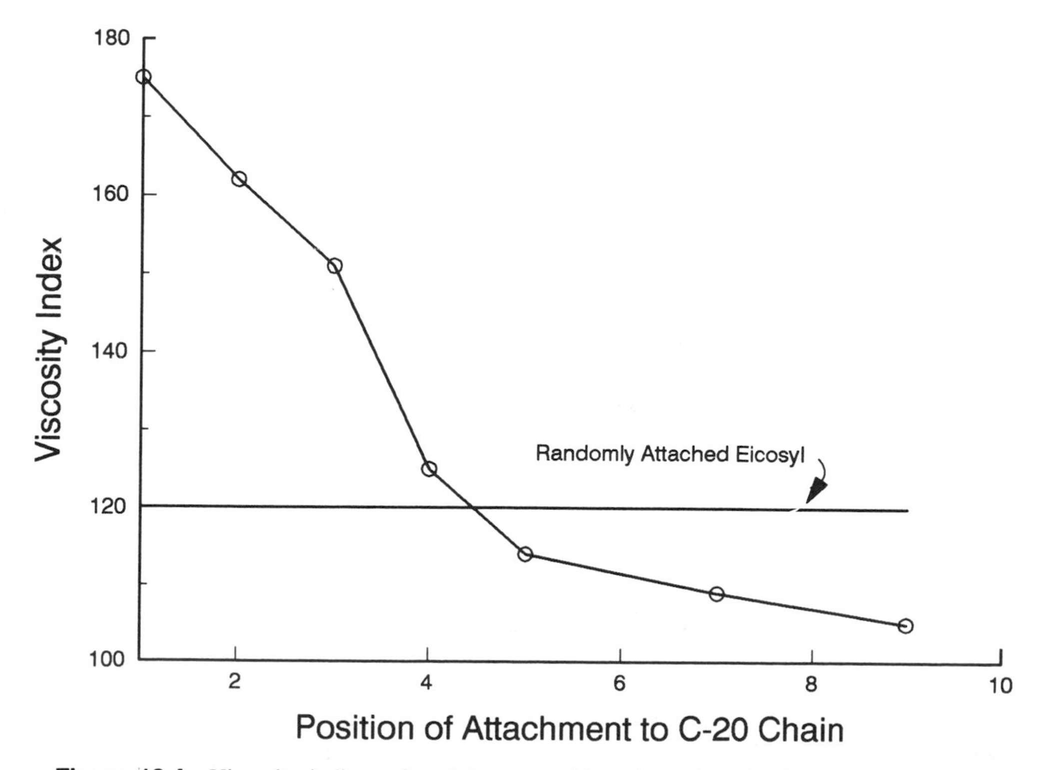

Figure 12.4 Viscosity indices of cyclohexanes with various eicosyl substituents.

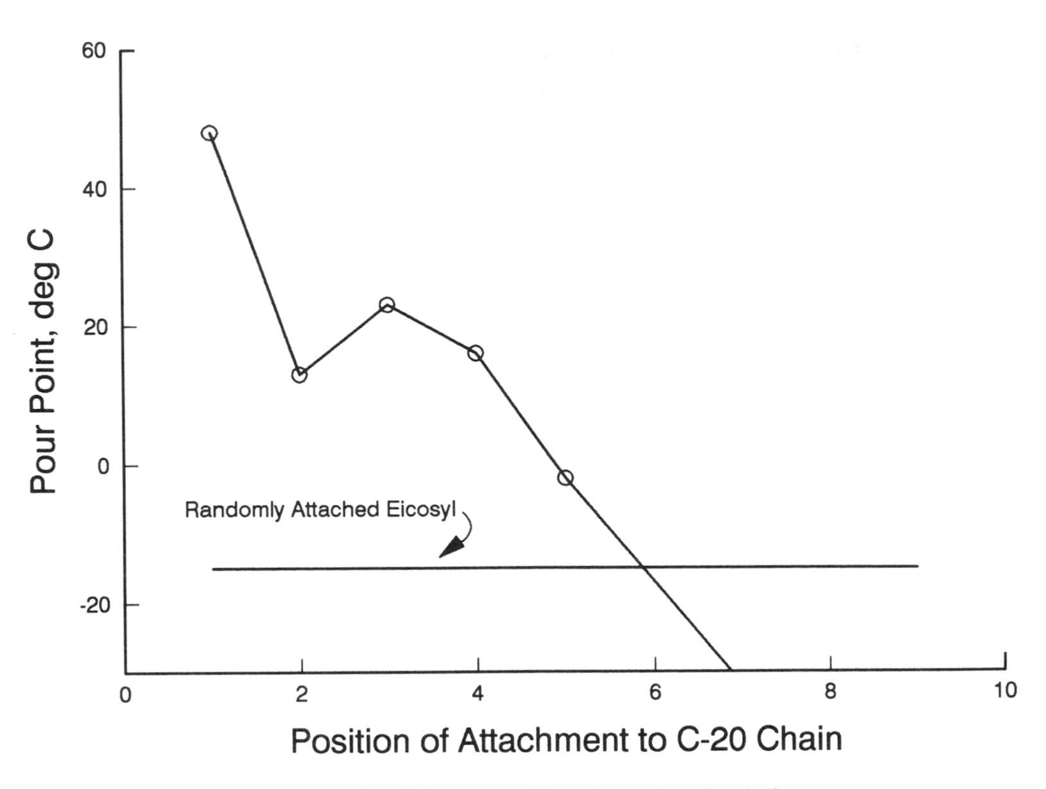

Figure 12.5 Pour points of cyclohexanes with various eicosyl substituents.

One set is a series of alkylcyclohexanes in which the alkyl group is a symmetric secondary substituent, RRCH-, in which the two R groups are both the same normal alkyl substituent. The dependence of viscosity index and pour point on the length of the *n*-alkyl group are shown in Figs. 12.6 and 12.7.

When the R group is large, the viscosity index approaches that of an alkylcyclohexane in which the alkyl group is a normal chain with the same number of carbons as the R group. For example, when the R of the RRCH- substituent is $C_{16}H_{33}$, the VI is 151. By way of comparison, hexadecylcyclohexane has a VI is 159. When the length of the R group is less, the two normal alkyl R-groups of the secondary substituent seem to interfere with the VI-enhancing properties of one another. For example, when R is $C_{10}H_{21}$, the VI is 114. For comparison, the VI values of 1,3- and 1,4-didecylcyclohexanes, in which the two decyl groups cannot interfere with one another, are 147 and 168, respectively. Unfortunately, pour points also increase dramatically with increasing *n*-alkyl group length.

A second set of alkylated cyclohexanes has one, two, or three linear alkyl substituents of the same carbon number as substituents. Each alkyl group is, however, independently attached to the cyclohexane at a randomly determined site on the linear alkyl chain. These materials were prepared by hydrogenation of so-called linear alkylbenzenes in which the substituents were introduced into the molecule by Friedel-Crafts alkylations of benzene by randomly chlorinated paraffinic substrates. The attachment to the ring is statistically distributed along the chain. The dependence of the viscosity index and pour point on the overall chain length of the substituents is also shown in Figs. 12.8 and 12.9. Both

Table 12.8 Cyclohexanes with Linear (Secondary) Alkyl Substituents

Compound	API Project 42 number	Number of R groups	Carbon number	Viscosity, 100°C, cSt[a]	Viscosity, 40°C, cSt[a]	Viscosity index	Pour point, °C	Reference
9-Heptadecylcyclohexane	509	1	23	2.69	10.77	79	–57	5
(Random)-octadecylcyclohexane[b]		1	24	3.07	12.10	112	–30	20,21
4-Methyl-8-nonadecylcyclohexane	162	1	26	3.51	16.14	91	<–57	5
2-Eicosylcyclohexane	102	1	26	4.06	16.05	162	+13	5
3-Eicosylcyclohexane	75	1	26	3.77	14.82	151	+23	5
4-Eicosylcyclohexane	104	1	26	3.72	15.80	125	+16	5,7
5-Eicosylcyclohexane	76	1	26	3.69	16.16	114	–2	5,7
7-Eicosylcyclohexane	77	1	26	3.58	15.71	108	—	5,7
9-Eicosylcyclohexane	78	1	26	3.44	14.95	105	—	5
(Random)-eicosylcyclohexane		1	26	3.65	15.57	120	–15	20,21
11-Heneicosylcyclohexane	60	1	27	3.68	16.16	114	–7	7
5-Docosylcyclohexane		1	28	4.09	18.12	128	<+23	18
2,5-Dimethyl-11-heneicosylcyclohexane	169	1	29	4.32	22.92	90	—	5
13-Pentacosylcyclohexane	69	1	31	4.66	22.08	131	+1	5
5-Hexacosylcyclohexane		1	32	5.48	25.97	155	+30	18
(Random)-ditridecylcyclohexane		2	32	6.46	42.33	102	–51	20,21
(Random)-ditetradecylcyclohexane		2	34	6.89	44.72	110	–47	20,21

Table 12.8 *(continued)*

Compound	API Project 42 number	Number of R groups	Carbon number	Viscosity, 100°C, cS[a]	Viscosity, 40°C, cS[a]	Viscosity index	Pour point, °C	Reference
15-Nonacosylcyclohexane	136	1	35	6.04	31.13	144	+13	5
(Random)-dipentadecylcyclo-hexane		2	36	7.72	51.70	114	–45	20,21
(Random)-dihexadecylcyclo-hexane		2	38	8.62	58.93	120	–39	20,21
17-Triacontylcyclohexane	138	1	39	7.45	40.79	151	+25	5
(Random)-dioctadecylcyclo-hexane		2	42	9.97	68.75	128	–15	20,21
(Random)-tritetradecylcyclo-hexane		3	48	13.52	116.10	113	–47	20,21
(Random)-trihexadecylcyclo-hexane		3	54	16.40	138.50	126	–45	20,21
(Random)-trieicosylcyclo-hexane		3	66	19.70	158.20	148	–30	21

[a]Data in the range of room temperature to 150°C were extrapolated to 100°C and 40°C, temperatures more commonly used today.
[b]Random means that the listed properties are of a mixture in which the linear alkyl group is attached to the ring through a random distribution of carbons in the chain.

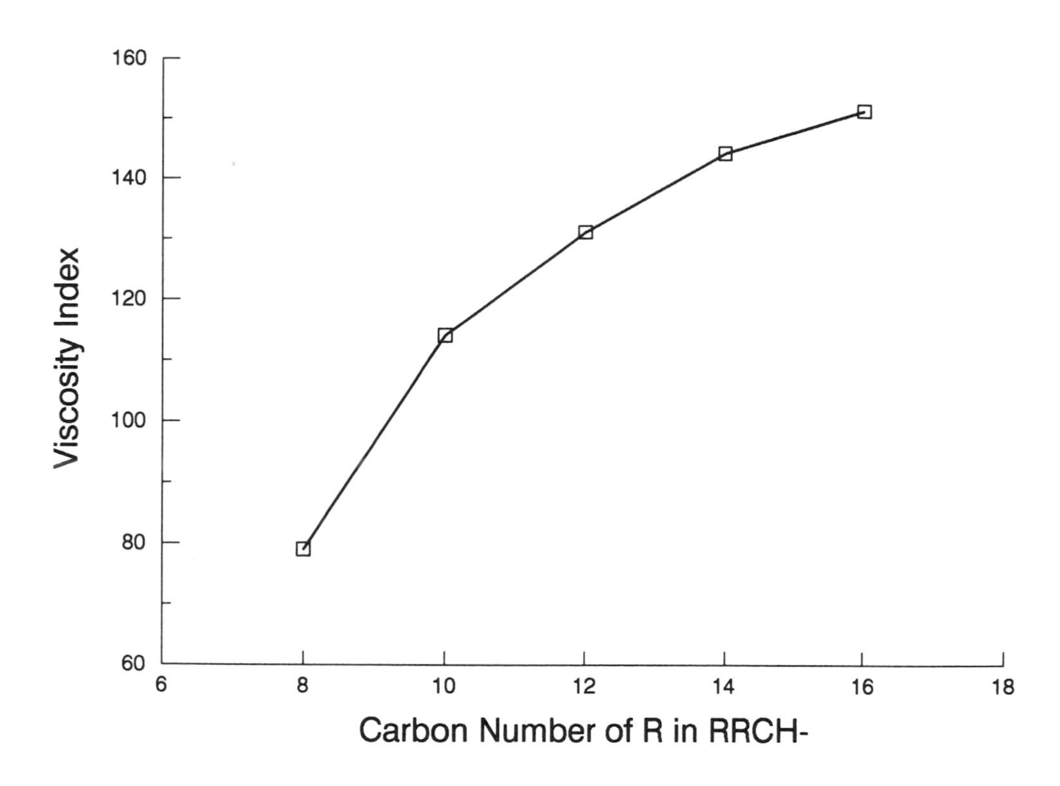

Figure 12.6 Viscosity indices of cyclohexanes with symmetric secondary alkyl substituents (RRCH-).

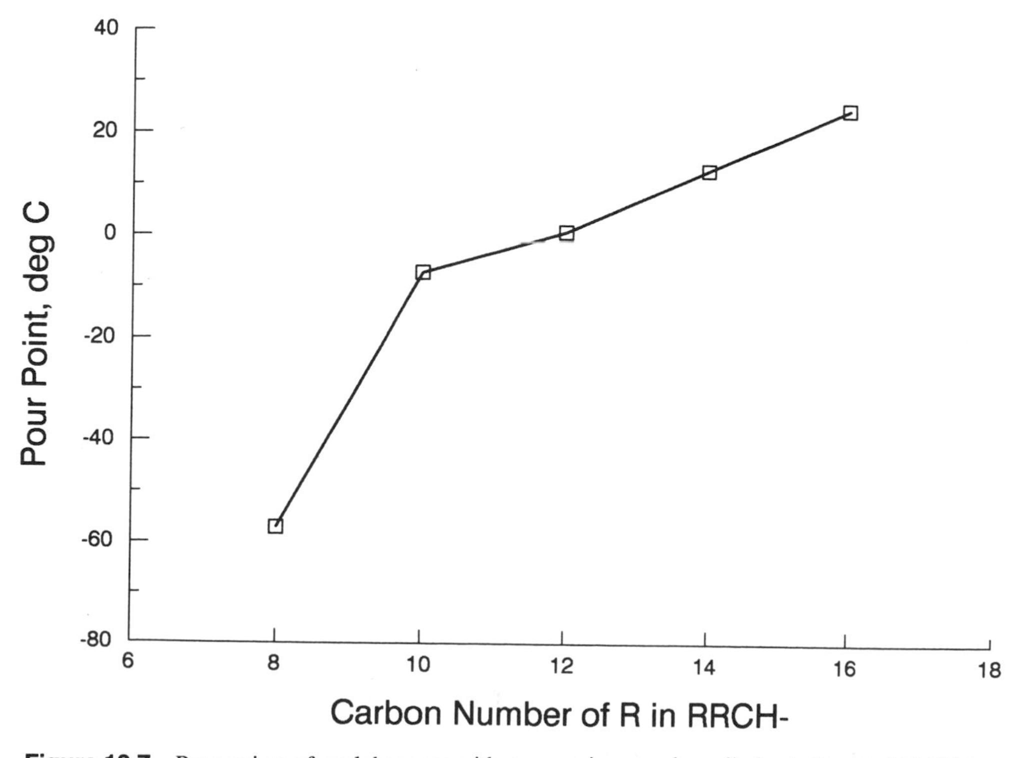

Figure 12.7 Pour points of cyclohexanes with symmetric secondary alkyl substituents (RRCH-).

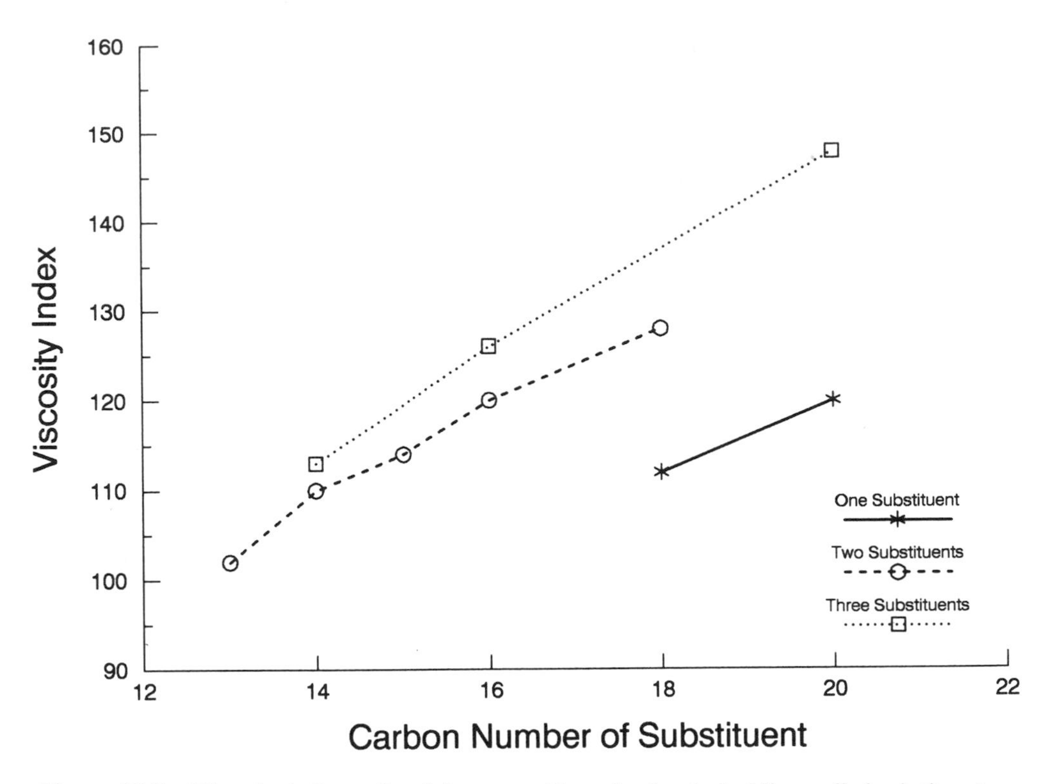

Figure 12.8 Viscosity indices of cyclohexanes with randomly attached linear alkyl substituents.

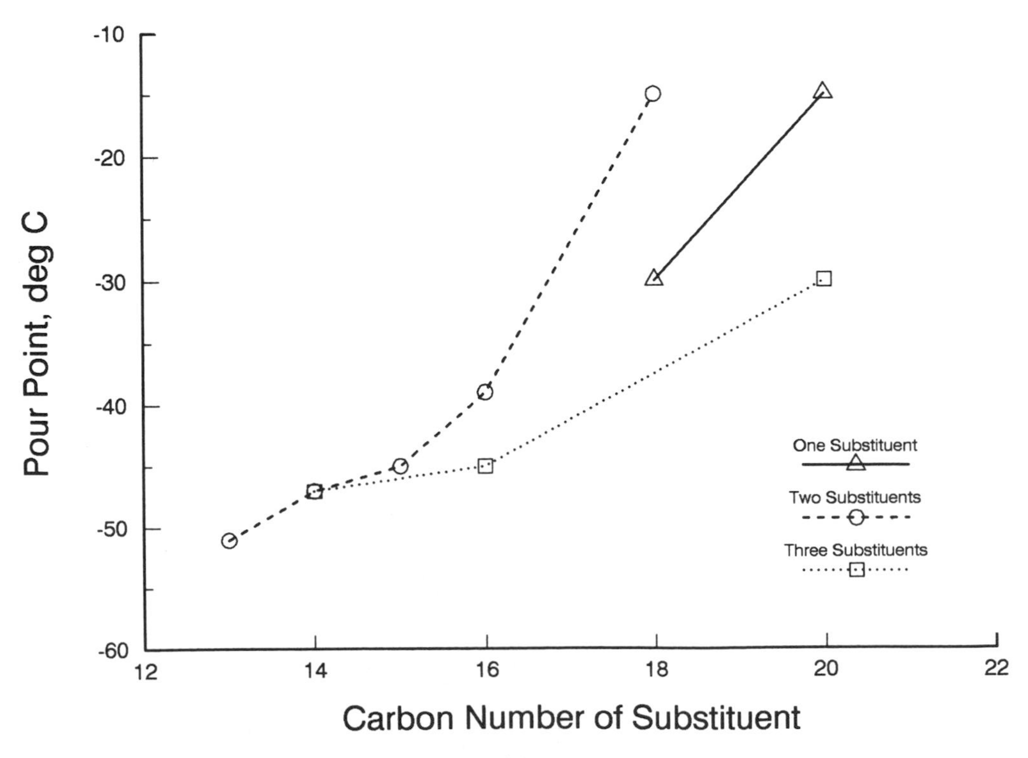

Figure 12.9 Pour points of cyclohexanes with randomly attached linear alkyl substituents.

viscosity index and pour point of these randomly alkylated cyclohexanes behave in the usual fashion. Both increase with increasing chain length.

Alkylated cyclohexanes that are mixtures of mono-, di-, tri-, and tetra-substituted products are shown in Table 12.9. These materials are more representative of materials that could be produced commercially. Their 100°C viscosities, for example, could be adjusted to commercially useful values, as is done with the trimer, tetramer, and pentamer of 1-decene.

IV. MULTIPLE FUSED RING COMPOUNDS

A. Decalins and Hydrindanes

Decalins and hydrindanes are both two-ring compounds in which the rings share two carbons in common. Decalins (decahydronaphthalenes) have two six-membered rings and hydrindanes (perhydroindenes) have a six- and a five-membered ring. Appendix 12.4 shows data for compounds with two fused rings.

Although there has not been a systematic study of decalins, enough data are available to make some generalizations about their properties (see Appendix 12.4 for all data available for decalins). Table 12.10 summarizes the pertinent data from Appendix 12.4. The now familiar increase in viscosity index with increasing *n*-alkyl side chain length is observed (Fig. 12.10). In this case, there are two isomeric series to consider, and, as can be seen in Fig. 12.8, the 2-decalin series has higher viscosity indices for a given alkyl group than the corresponding 1-decalin series. This could be viewed as arising from the more linear shape of the 2-decalin series.

The decalin series provides an example of an incredibly viscous material. Tri(1-decalyl)methane (API number 203) has a kinematic viscosity of 91 cSt at 150°C! If extrapolated to 100°C, a viscosity of about 10,000 cSt is expected. Such a material would be more viscous than polybutenes of 3,000 molecular weight.

The thickening effect of placing decalin moieties close together in the same molecule is also seen in the di(decalyl) examples. Di(1-decalyl)methane, the di(1-decalyl)ethanes, and 1,1-di(1-decalyl)undecane are all much more viscous than other decalins of comparable carbon number.

Table 12.9 Properties of Mixtures of Linear-Alkyl (Secondary) Cyclohexanes, R_x-cyclo-C_6H_{12-X}

R group	x =	Carbon number	Viscosity, 100°C, cSt	Viscosity, 40°C, cSt	Viscosity index	Pour point, °C	Reference
(Random)-undecyl[a]	2.4	32.0	5.67	34.88	101	–55	20,21
(Random)-tridecyl	2.4	36.7	8.00	56.42	110	–50	20,21
(Random)-tetradecyl	2.2	36.7	8.49	58.70	117	–45	20,21
(Random)-pentadecyl	2.2	38.2	7.56	48.82	119	–39	20,21
Methyl-(random)-pentadecyl	2.0	36.2	7.82	57.15	100	–38	21
(Random)-hexadecyl	2.3	43.3	11.24	83.53	123	–27	20,21
(Random)-octadecyl	2.0	42.2	10.33	72.74	127	–5	20,21

[a]Random means that the listed properties are of a mixture in which the linear alkyl group is attached to the ring through a random distribution of carbons in the chain.

Appendix 12.4 Properties of Cycloaliphatics with Two Fused Rings: Decalins and Hydrindanes[a]

Compound	API Project 42 number	Number of R groups	Carbon number	Viscosity, 100°C, cSt	Viscosity, 40°C, cSt	Viscosity Index	Pour point, °C	Reference
Alkyldecalins								
2,6-Dimethyl-3-octyldecalin	629	3	20	2.65	10.88	65	—	5
1,4-Dimethyl-5-octyldecalin	640	3	20	2.60	10.22	79	—	5
2-Decyldecalin	594	1	20	2.84	10.44	120	—	5
2-Butyl-3-hexyldecalin	615	2	20	2.79	13.95	–7	—	5
7-Butyl-1-hexyldecalin	612	2	20	2.74	12.07	44	—	5
1-Undecyldecalin	544	1	21	2.98	11.53	112	—	5
Di(1-decalyl)methane	586	1	21	10.51	379.8	<–200	—	5
1-Dodecyldecalin		1	22	3.39	14.27	111	—	17
1[2(2-methyl-5-isopropylcyclohexyl)ethyl]-decalin		3	22	5.44	53.84	–38	—	17
1,2-Di(1-decalyl)-ethane	562	1	22	11.77	321.9	–146	+97	5
1,1-Di(1-decalyl)-ethane	563	1	22	15.75	160.2 (60°C)	<–200	+21	5
1-Pentadecyldecalin	175	1	25	4.24	18.62	136	—	5
(Random)-octadecyldecalin[b]		1	28	6.05	36.59	110	–27	5
2(1-octadecyl)decalin		1	28	5.74	28.32	150	+43	18
1,10-Di(1-decalyl)-decane	132	1	30	12.63	140.6	77	—	5
1[4(1-cyclohexyl)-tetradecyl]decalin	192	1	30	9.63	110.2	47	—	5
Tri(1-decalyl)methane	203	1	31	97.99 (149°C)				5
1,1-Di(1-decalyl)-undecane	122	1	31	19.43	699.4	–96	–1	5
1(11-Heneicosyl)-decalin	62	1	31	6.19	41.07	95	—	5
2(1-Docosyl)decalin		1	32	7.63	40.56	159	—	18
2(5-Docosyl)decalin		1	32	7.43	47.90	118	<23	18

Appendix 12.4 *(continued)*

Compound	API Project 42 number	Number of R groups	Carbon number	Viscosity, 100°C, cS	Viscosity, 40°C, cS	Viscosity Index	Pour point, °C	Reference
Alkylhydrindanes								
5-Decylhydrindane	598	1	19	2.32	7.84	110	—	5
2-Decylhydrindane	596	1	19	2.46	8.23	129	—	5
2-Butyl-5-hexylhydrindane	603	2	19	2.24	7.91	87	—	5
2-Butyl-1-hexylhydrindane	601	2	19	2.32	10.18	–2	—	5
5-Butyl-6-hexylhydrindane	605	2	19	2.14	8.43	25	—	5
1-Hexadecylhydrindane	108	1	25	4.02	16.33	151	+2	5
2-Hexadecylhydrindane	118	1	25	4.23	17.27	157	—	5
1,10-Di(5-hydrindanyl)decane	145	1	28	9.44	4.68 (135°C)	102	—	5

[a]Data in the range of room temperature to 150°C were extrapolated to 100°C and 40°C, temperatures more commonly used today.
[b]Random means that the listed properties are of a mixture in which the linear alkyl group is attached to the ring through a random distribution of carbons in the chain.

Even fewer examples of hydrindanes are available. However, a series of C_{19} hydrindanes with various distributions of the 10 alkyl carbons have been reported (see Appendix 12.4). The data collected in Table 12.10 show the most interesting series. The relatively linear, 2-butyl-5-hexylhydrindane has a viscosity index of 87, closer to those of the decylhydrindanes (110 and 129) than to those of the other butylhexylhydrindanes (–2 and 25). As was observed in the cyclohexane series, interruption of a chain by a ring, in which the resulting structure is still more or less linear, has minimal effect on the viscosity index, while structures in which the ring structure biases the chain to a nonlinear conformation greatly reduce the viscosity index.

B. Miscellaneous Ring System with More Than Two Fused Rings

As can be seen from the compilation in Appendix 12.5, fusing three rings together tends to yield a compact molecule. As we have seen with substituted cyclopentanes and cyclohexanes, compactness leads to low viscosity indices. Thus, only those with a very long alkyl chain have viscosity indices above 100. Although melting points are not reported for the examples in Table 12.10, usually molecules with chains of 12 or more carbons tend to have higher than useful pour points.

Reinforcing the hypothesis that relative linearity leads to higher viscosity indices, the relatively linear 2-dodecylperhydrophenanthrene has a VI of 125, while the more compact 9-dodecylperhydrophenanthrene has a VI of only 84.

Table 12.10 Properties of Cycloaliphatics with Two Fused Rings: Decalins and Hydrindanes[a]

Compound	API Project 42 number	Number of R groups	Number of carbons	Viscosity, 100°C, cSt	Viscosity, 40°C, cSt	Viscosity index	Pour point, °C	Reference
Alkyldecalins								
1-Undecyldecalin	544	1	21	2.98	11.53	112	—	5
1-Dodecyldecalin		1	22	3.39	14.27	111	—	17
1-Pentadecyldecalin	175	1	25	4.24	18.62	136	—	5
2-Decyldecalin	594	1	20	2.84	10.44	120	—	5
2-Octadecyldecalin		1	28	5.74	28.32	150	+43	18
2-Docosyldecalin		1	32	7.63	40.56	159	—	18
Di(1-decalyl)methane	586	1	21	10.51	379.8	<–200	—	5
1,2-Di(1-decalyl)ethane	562	1	22	11.77	321.9	–146	+97	5
1,1-Di(1-decalyl)ethane	563	1	22	15.75	160.2 (60°C)	<–200	+21	5
1,10-Di(1-decalyl)decane	132	1	30	12.63	140.6	77	—	5
1,1-Di(1-decalyl)undecane	122		31	19.43	699.4	–96	–1	5
Tri(1-decalyl)methane	203	1	31	97.99 (149°C)				5
Alkylhydrindanes								
5-Decylhydrindane	598	1	19	2.32	7.84	110	—	5
2-Decylhydrindane	596	1	19	2.46	8.23	129	—	5
2-Butyl-5-hexylhydrindane	603	2	19	2.24	7.91	87	—	5
2-Butyl-1-hexylhydrindane	601	2	19	2.32	10.18	–2	—	5
5-Butyl-6-hexylhydrindane	605	2	19	2.14	8.43	25	—	5

[a]Data in the range of room temperature to 150°C was extrapolated to 100°C and 40°C, temperatures more commonly used today.

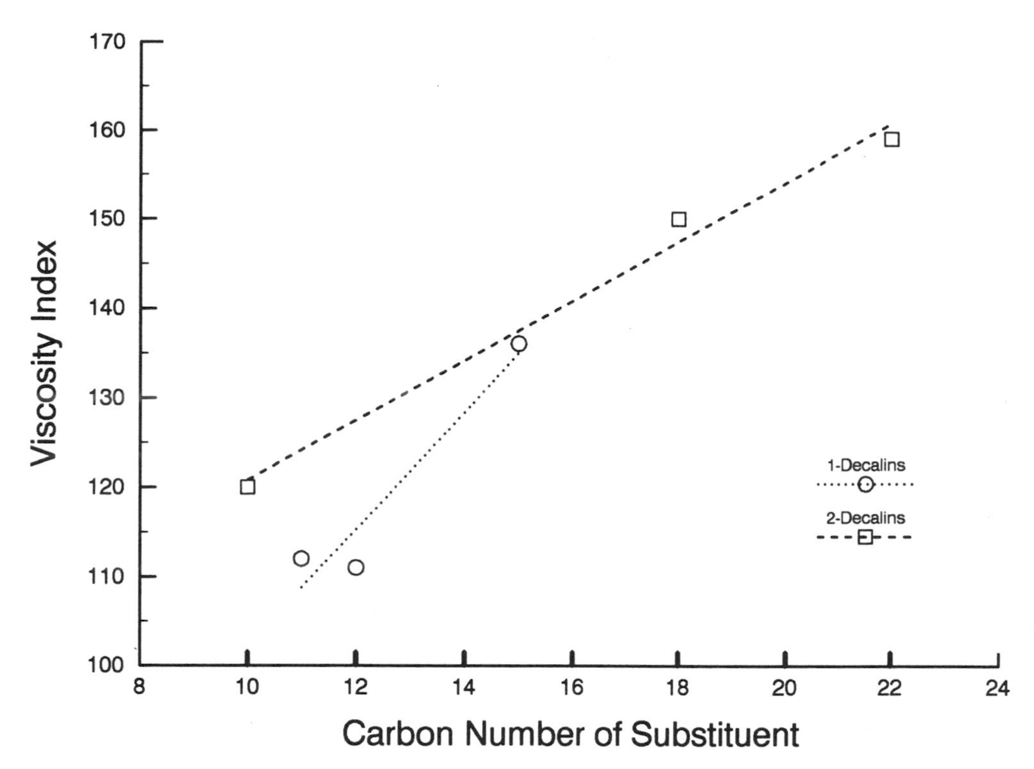

Figure 12.10 Viscosity indices of decalins with n-alkyl substituents.

V. MANUFACTURE AND ECONOMICS

At this time, no synthetic cycloaliphatic material has been successfully commercialized as a lubricating oil base stock.

The only viable route to alkylated cyclohexanes is the hydrogenation of alkylbenzenes. Therefore, cyclohexanes, and decalins, hydrindanes, and compounds with more than two rings, will always be more expensive than their precursor aromatic compounds. Since the physical properties are not changes significantly by the hydrogenation, there is little to recommend alkylated cyclohexanes as lubricants. Chemically, the hydrogenation should increase oxidative stability. However, in most cases oxidation is controlled by additives, and there is little to recommend alkylcyclohexanes compared to alkylbenzenes. The only scenario in which alkylated cyclohexanes might compete with alkylated aromatics is if the aromatics have toxicological properties or environmental effects that prevent their use.

Alkylated cyclopentanes, on the other hand, represent an independent class of potential lubricant base materials. They are derived from cheap starting materials in two high-yield steps (8, 9). Cyclopentadiene is a by-product of ethylene production, and primary alcohols are readily available from ethylene oligomerization, hydroformylation, and natural sources.

One multiply-alkylated cyclopentane, tris(2-octyldodecyl)cyclopentane, has been commercialized as a special lubricant for the aerospace industry under the name Pennzane® Synthesized Hydrocarbon Fluid X2000. It is a low-volume, high-priced product with especially low volatility and wide liquid range (8, 10).

Appendix 12.5 Properties of Miscellaneous Ring Systems[a,b]

Compound	API Project 42 number	Number of rings	Carbon number	Viscosity, 100°C, cSt	Viscosity, 40°C, cSt	Viscosity Index	Pour point, °C
Perhydroperylene	636	5	20	1.46 (200°C)	2.40 (150°C)	—	—
9-Cyclohexylperhydroanthracene	579	3	20	1.74 (200°C)	3.84 (150°C)	—	—
Perhydrodibenzo(a,i)fluorene	587	5	21	10.26	290.3	−199	—
6-Octylperhydrobenz(d,e)anthracene	196	4	25	7.68	82.52	28	—
9-Dodecylperhydrofluorene	237	3	25	4.34	23.84	81	—
1,1-Di(5-perhydroacenaphthyl)ethane	200	3	26	99.5	591.2 (80°C)	<−200	—
2-Octylperhydrotriphenylene	228	4	26	8.66	103.3	25	—
2-Decylperhydroindeno(2,1-a)indene	231	4	26	6.96	48.06	100	—
9-Octylperhydronaphthacene	166	4	26	11.85	261.1	−69	—
3-Decylperhydropyrene	216	4	26	6.10	18.3 (60°C)	93	—
2-Octylperhydrochrysene	225	4	26	9.90	156.7	−18	—
4-Decylperhydropyrene	219	4	26	6.47	51.03	66	—
9-Dodecylperhydroanthracene	125	3	26	5.87	41.39	76	—
2-Dodecylperhydrophenanthrene	143	3	26	6.00	33.70	125	—
9-Dodecylperhydrophenanthrene	141	3	26	5.83	39.38	84	−7
Cholestane	155	4	27	20.28	47.99 (80°C)	−90	+80
5-Pentadecylperhydroacenaphthene	193	3	27	5.59	28.55	138	—
4(9-Heptadecyl)as-perhydroindacene	177	3	29	6.44	48.42	75	—

[a]All data are from reference 5.
[b]Data in the range of room temperature to 150°C were extrapolated to 100°C and 40°C, temperatures more commonly used today.

VI. CONCLUSIONS

Both alkylcyclopentanes and alkylcyclohexanes with a wide variety of tribologically useful properties can be prepared by simple chemical reaction sequences. In the former case the facile alkylation of cyclopentadienide anion by alkyl halides and alcohols followed by hydrogenation of the intermediate alkylcyclopentadiene provides the products. In the latter, hydrogenation of alkylbenzenes prepared by Friedel-Crafts alkylation is the best route.

The properties of single-ring cycloaliphatics are determined by the number and nature of the substituents. Except for very compact or highly branched structures, the kinematic viscosity is determined almost exclusively by molecular weight.

The other main viscometric properties, viscosity index and pour point, are governed by the length of the average alkyl chain and by the extent of branching. As the average uninterrupted normal alkyl chain gets longer, the VI and the pour point will both get higher, all other things being equal. Branching drives both VI and pour point down. However, if a molecule is large enough to have both extensive branching and long uninterrupted alkyl chains, high viscosity index and low pour point are achievable. For example, tris(2-octyldodecyl)cyclopentane has VI = 136 and a pour point of <–57°C. Neither class has been commercialized in substantial quantities. However, alkylcyclopentanes are a newly reported class of synthetic hydrocarbons that show promise for economical production in the future.

Since alkylcyclohexanes are prepared from alkylbenzenes, the saturated material will be more expensive than the aromatic. Thus, if alkylcyclohexanes are to be commercially viable, they must exhibit significant performance advantages over their precursor alkylbenzenes, or alternative preparations must be developed.

REFERENCES

1. Engler, C. (1888). *Z. Angew. Chem.*, 1, 73.
2. Warren, C. M. (1891). *Proc. Am. Acad. Arts Sci.*, 27, 56.
3. Markownikoff, W., and J. Spady (1887). *Ber.*, 20, 1850.
4. The term "naphthenic" was first used in Markownikoff, W., and W. Oglobin, (1884). *Bull. Soc. Chim.*, 41, 258.
5. American Petroleum Institute (1967). *Properties of Hydrocarbons of High Molecular Weight Synthesized by Research Project 42 of the American Petroleum Institute*, American Petroleum Institute, New York.
6. Denis, J. (1985). *J. Synth. Lub.*, 1, 201.
7. Reith, H. (1973). *Schmierungstechnik*, 4, 48.
8. Venier, C. G., and E. W. Casserly (1991). *Lub. Eng.*, 47, 586.
9. Venier, C. G., and E. W. Casserly (1990). *Prep. Paper, Am. Chem. Soc. Div. Petrol. Chem.*, 35, 260.
10. Casserly, E. W., and C. G. Venier (1990). *Prepr. Paper, Am. Chem. Soc. Div. Petrol. Chem.*, 35, 265.
11. Venier, C. G., and E. W. Casserly (1990). *J. Am. Chem. Soc.*, 112, 2808.
12. Makosza, M. (1968). Polish Patent 55,571, 30 May.
13. Hirsch, S. S., and W. J. Bailey (1978). *J. Org. Chem.*, 43, 4090.
14. Fritz, H. E., and D. W. Peck (1966). U.S. Patent 3,255,267, 7 June.
15. Favis, D. V. (1961). U.S. Patent 3,000,981, 19 September.
16. Favis, D. V. (1990). *Chem. Eng. News*, 24 September, p. 2.
17. Neyman-Pilat, E., and S. Pilat (1941). *Ind. Eng. Chem.*, 33, 1382.
18. Mikeska, L. A. (1936). *Ind. Eng. Chem.*, 28, 970.
19. Evans, E. B. (1938). *J. Inst. Petrol. Technol.*, 24, 321.
20. Appelt, W., R. Baur, K. Bronstert, J. Nickl, and K. Oppenlaender (1983). Deutsches Patentamt DE 3133559 A1, 10 March.
21. Bronstert, K., J. Nickl, W. Ochs, and K. Starke (1981). Europaische Patentanmeldung EP 81 10 0018, 5 January.
22. Venier, C. G., and E. W. Casserly (1988). U.S. Patent 4,721,823, 26 January.

13
Polybutenes

John D. Fotheringham
BP Chemicals Ltd.
Grangemouth, Stirlingshire, Scotland

I. INTRODUCTION

A. History

Butlerov and Gorianov have been attributed with preparing the first polymers of isobutylene in the 1800s. This followed work by Berthelot and others who studied the polymerization of various olefins using Friedel-Craft catalysis systems (1). During the 1930s the first commercial processes for the manufacture of solid polyisobutylene rubbers were introduced first in Germany (2) and then in the United States (3). It was not until later in the 1940s and 1950s that commercial routes were developed for the preparation of liquid polybutenes by Chevron (4), the Standard Oil Company of Indiana (5), and the Cosden Petroleum Corporation (6) in the United States.

B. Manufacturing and Isobutylene Supply

The current commercial routes to polybutenes are believed to be still very much based on the original technology developed by Cosden and others. However, improvements have been introduced and continue to be introduced to improve the manufacturing process and the quality of the finished product. Specific details of the different manufacturing technologies that are in use are covered adequately and in some detail elsewhere (7, 8) and in patent literature (9, 10).

Polybutenes are produced by the polymerization of a hydrocarbon stream containing a high proportion of isobutylene. There are three main sources of isobutylene feedstock. These are refinery catalytic crackers producing petrol/gasoline, steam crackers producing ethylene, and from the dehydration of tertiary butyl alcohol produced as a by-product in the manufacture of propylene oxide.

Isobutylene is also used in the manufacture of several other important materials such as butyl rubber, methyltertiarybutyl ether (MTBE), alkylation products for gasoline antiknock, and the polyisobutylene rubbers as shown in Fig. 13.1. The rapid growth in the use of MTBE as an octane enhancer for gasoline (11) has placed heavy demands on isobutylene supplies. At present there appears to be sufficient isobutylene available to satisfy demand. However, future projections, particularly if MTBE production continues to grow, show tightness of supply and possible shortages.

The manufacturing process for polybutene is illustrated using an isobutylene stream derived from the steam cracking of naphtha in Fig. 13.2.

Crude oil is separated by distillation into its many fractions, of which the Naphtha cut is fed to a steam cracker. The steam cracker breaks down the medium-chain paraffins into short-chain olefins such as ethylene and propylene. A C_4 olefin cut is also produced, which is then processed in a unit to extract the diolefin butadiene. The butadiene raffinate, or raffinate 1 as it is often referred to after butadiene extraction, normally contains isobutylene, 40–50 wt%; *n*-butenes, 20–30 wt%; and butanes 20–30 wt%. The raffinate 1 is then desulphurized and dried prior to its introduction into the reactor section. The isobutylene in the raffinate 1 stream is polymerized selectively using a Lewis acid catalyst to produce the polybutene polymers. The molecular weight of the polybutene can be controlled by changing the reaction conditions, such as the temperature at which the polymerization is undertaken. The polymerization is an exothermic process, so cooling of the reactor is necessary to maintain the desired temperature of reaction. After completion of the polymerization, catalyst residues, unreacted butenes, and the butanes are removed from the product. The remainder is then distilled to remove light polymer, leaving the final polybutene product. This is then transferred to storage. After quality control it is sent out in bulk road cars, drums, or other suitable containers to the customer. The light polymer can be further processed to yield low-viscosity grades of polybutene. The unreacted butenes and the butanes (raffinate 2) can be used as a feed for processes requiring high levels of *n*-butenes or returned to the cracker for further processing.

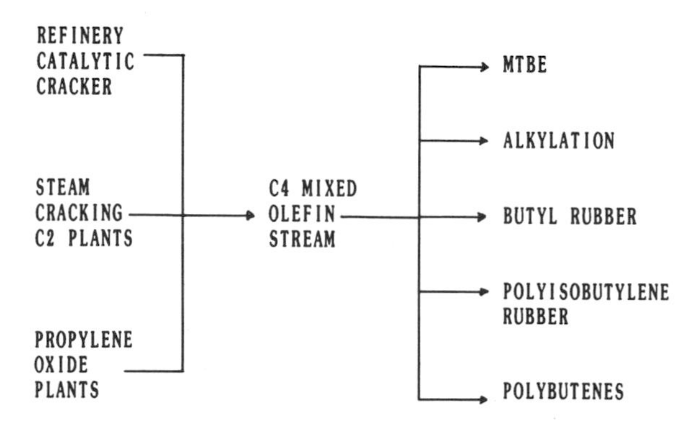

Figure 13.1 Main sources and major products derived from isobutylene feedstock.

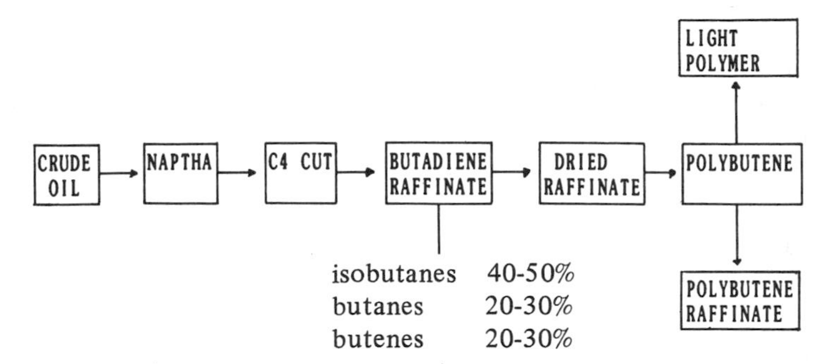

Figure 13.2 Schematic representation of the production of polybutenes from crude oil.

C. Nomenclature

As described, polybutenes are derived from a feed stream containing a mixture of butene monomers. The balance of monomers is rich in isobutylene (**1**) but contains proportions of butene-1 (**2**) and butene-2 (**3**). Polymerization of this stream yields a polymer of essentially isobutylene that also contains a certain level of the other monomer structures. The ratio of isobutylene/butene monomer incorporation increases as the molecular weight of the polybutene increases. That is, the higher-molecular-weight polybutenes contain less butene-1 or butene-2 structure in the polymer backbone. Nevertheless, to be strictly correct, these polymers should be described as copolymers of isobutylene/butene. Historically, for ease of reference the term isobutylene/butene copolymer has been replaced by the nomenclature polybutenes or PIB for short, and this nomenclature has been retained throughout the chapter. The polybutenes are polymers that are liquid in nature, although the higher-molecular-weight polybutenes possess some semisolid rubbery character.

$$CH_3 \cdot C(=CH_2) \cdot CH_3 \qquad [1]$$

$$CH_3 \cdot CH_2 \cdot CH{=}CH_2 \qquad [2]$$

$$CH_3 \cdot CH{=}CH \cdot CH_3 \qquad [3]$$

The nomenclature polyisobutylenes is normally reserved for polymers derived from a feed stream containing only the isobutylene monomer. The polyisobutylene polymers are normally high-molecular-weight ($M_n > 10{,}000$) rubbers. However it is known that a limited range of liquid polymers of polyisobutylene has been produced (12). The manufacturing process used for these liquid polyisobutylenes is unclear. One proposal is that they are the products from the sonic degradation of higher-molecular-weight polymer (7).

Another material that is sometimes confused with the polybutenes is the thermoplastic polybutene-1. Polybutene-1 is produced by a transition metal (Ziegler-Natta) catalyzed polymerization of butene-1. Isotactic polybutene-1 has good high-temperature properties and has found use in applications such as plastic hot water pipes and in heat-sealing plastic films (13).

Finally, another class of material containing high levels of isobutylene monomer is the butyl rubbers, which are copolymers of isobutylene and isoprene. The incorporation of isoprene at low levels introduces carbon-carbon double-bond functionality into the polymer structure, which allows butyl rubber to undergo conventional cross-linking reactions.

From this discussion, it should be clear that when the nomenclature polybutenes is used, it is referring to the liquid polymers made up of predominantly the isobutylene monomer but containing a level of other butene structures in the polymer chains. Polybutenes are available in a wide choice of different viscosities or grades, with appearance ranging from that of a free-flowing light oil to a semisolid rubber at 25°C. There is little concensus among manufacturers on a standard system of nomenclature for grades of polybutene. An approach that has been used historically and the one that will be adopted throughout this chapter is to name the polybutene grade on the basis of its viscosity measured in SSU (Saybolt viscosity) at 100°C divided by 100. For example, using this system a polybutene with viscosity of 2960 SSU at 100°C would be defined after rounding to the nearest whole number as a 30 grade.

D. Applications

The major applications following commercialization of liquid polybutenes made use of the many attractive physical properties that the polymers possessed. Polybutenes found use as components of adhesives and sealants, as electrical oils, and as modifiers of rubbers (14). Later the chemical reactivity of the polymer was recognized, with polybutenes being used as a starting material for the manufacture of ashless detergents (15).

Today about 70% of world production of polybutene is used in the manufacture of polybutene derivatives. The main class of derivatives is the polybutenyl succinimides, used as ashless detergent additives to combat wet sludge formation in crankcase engine oils and now also finding an important role as detergent additives for petrol/gasoline and diesel to prevent carburetor fouling and engine deposits (16). Derivatives of polybutene are also manufactured for use as corrosion inhibitors and as antiwear additives for lubricating oils and greases.

Of the remaining 30% of world production, between 5 and 10% is used in electrical applications where the excellent insulating properties, high quality, and consistency of polybutene are used to advantage to replace mineral oil in the formulation of filling compounds for electrical, telecommunication, and fiber optic cables and for the impregnation of metallized paper capacitors.

Industrial applications of polybutene that make use of the physical properties of the liquid polymers account for 20–25% of world production. The major industrial application for polybutenes remains in adhesives and sealants. However, in the period since their commercialization the useful range of physical properties that the polymers possess has led to their introduction into many other everyday applications. The main properties of the polybutene polymers are summarized in Table 13.1. Polybutenes are now widely used in oils and lubricants, agrochemicals, tackified polyethylene, bitumen modification, concrete mold oils, and in putties, anticorrosion coatings, masonry coatings, paints, inks, and dispersion aids. The main industrial applications, excluding oils and lubricants, which are covered later, are highlighted in Table 13.2.

It is estimated that between 25 and 30% of polybutene sold into industrial applications is for oils and lubricants.

Table 13.1 Summary of the Main Properties of Polybutene Polymers

Stable to light and air
Effective tackifiers/plasticizers
Colorless
High viscosity indices
Hydrophobic
Compatible with organic materials
Practically nontoxic
Noncorrosive
Evaporate or burn without residue
Inherently good lubricants

E. Synthetic Methods

The cationic polymerization of streams containing isobutylene is typically achieved using Lewis acid type catalyst systems. For example, the Cosden process describes the use of an anhydrous aluminium trichloride slurry with either gaseous HC1 or chloroform as cocatalyst (17). Many other combinations of catalyst and cocatalyst are capable of performing the polymerization (18). Cationic polymerization remains an active field of research, as evidenced by the considerable quantity of scientific publications that deal with all aspects of the polymerization process (19a–e).

II. CHEMICAL AND PHYSICAL PROPERTIES

A. Chemical Structure

The structure of polybutenes is essentially that of the isobutylene (**4**) repeat unit with incorporation of low levels of other butene (**5**, **6**) structures. The majority of polybutene polymer molecules contain a carbon-carbon double bond at the end of the polymer chain. A minority of polymer chains are terminated with a carbon-halogen bond. The position of the double-bond end group is important in determining the ease with which it may undergo chemical reaction. From carbon-13 NMR studies it has been shown that the positioning of the double bond in most polybutene grades gives rise to the *cis*- and *trans*-trisubstituted end-group structure (**7**). From studies of model compounds it has been shown that a number of other positions for the double bond at the end of the polymer chain are possible. An internal double bond may be present in some polymer chains, although these prove difficult to characterize. As polybutenes contain only one double bond per polymer molecule they cannot undergo conventional cross-linking reactions with, for example, sulfur or peroxides. The actual composition of a given polybutene in terms of isobutylene and butene content and the positioning of the double bond at the end of the polymer chain can be influenced by the manufacturing conditions but are perhaps mostly dependent on feedstock composition and the overall manufacturing process.

Table 13.2 Major Industrial End Applications With Properties and Grades of Polybutene of Interest

Application	Polybutene properties	Main grades of polybutene used											
Adhesives													
Hot melt	Plasticizer, tackifier												
Solvent based	Viscosity modifier	Very											
Water based	Colorless, nontoxic	Light	03	04	07	3	5	10	30	150	200	600	2000
Pressure-sensitive hot melt	Extender, stable				X			X	X		X		X
Sealants													
Tape, gun, knife grades	Nondrying binder												
Oleoresinous	Tackifier, plasticizer	Very											
Rubber based	Modifier, extender	Light	03	04	07	3	5	10	30	150	200	600	2000
Putties	Stable, inert					X	X	X	X	X	X	X	X
Tackified polyethylene													
Palletwrap	Cling agent or	Very											
Silagewrap	tackifier	Light	03	04	07	3	5	10	30	150	200	600	2000
Domestic cling films	Nontoxic							X	X	X	X		
Agrochemicals													
Ultra low volume spraying/	Viscosity modifier,												
controlled droplet application	adjuster, nonphototoxic	Very											
Rainfastness, sticker	Nontoxic	Light	03	04	07	3	5	10	30	150	200	600	2000
Physical trapping medium	Adhesive, hydrophobic	X	X	X			X		X		X		X
Bitumen modification													
Special road bitumens	Low temperature	Very											
Roofing felts	Improvements	Light	03	04	07	3	5	10	30	150	200	600	2000
Bitumen sealants, adhesives	Better aging, adhesive				X				X	X	X		
Concrete mold oils		Very											
High performance	Nonstaining, nontoxic,	Light	03	04	07	3	5	10	30	150	200	600	2000
release oils	better mold performance	X	X	X			X						
Process oils													
Rubber modification	Tackifier, plasticizer,												
Coatings, masonry,	binder, hydrophobic,	Very											
anticorrosion	nontoxic, colorless,	Light	03	04	07	3	5	10	30	150	200	600	2000
Inks, paints, paper	stable, inert, consistent quality			X	X		X	X	X		X		
Dispersion aids													

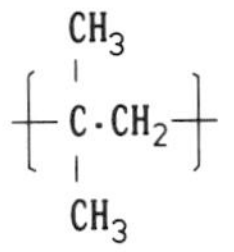

$$-\left[C(CH_3)_2 \cdot CH_2\right]- \qquad -\left[CH_2 \cdot CH_2 \cdot CH(CH_3)\right]-$$

[4] 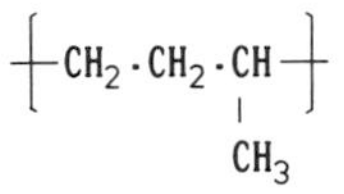 [5]

$$-\left[CH_2 \cdot CH(CH_2CH_3)\right]-$$

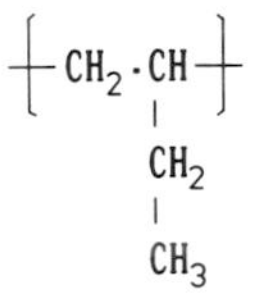

[6]

Polybutenes having the predominantly disubstituted vinylidene structure (**8**) at the end of the polymer chain are becoming commercially available (20a–c). The prepondrance of the vinylidene end-group structure is extremely useful in conferring improved chemical reactivity to the polybutene in, for example, the malenization reaction with maleic anhydride to produce the polybutenylsuccinic anhydride derivative.

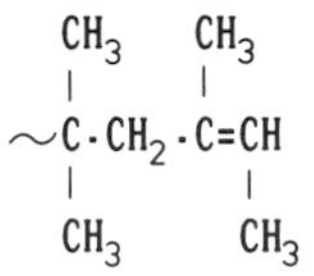

$$\sim C(CH_3)_2 \cdot CH_2 \cdot C(CH_3){=}CH(CH_3)$$

[7]

$$\sim CH_2 \cdot C(CH_3)_2 \cdot CH_2 \cdot C(CH_3){=}CH_2$$

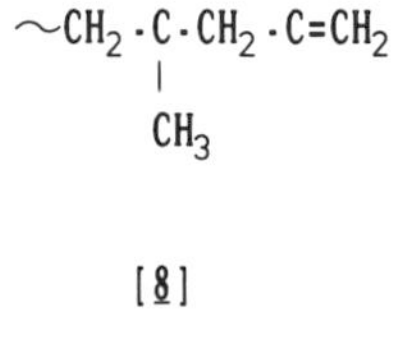

[8]

Polybutenes are made available on the market as grades that are usually defined by a viscosity range. The nature of the polymerization process results in each grade being made up of a distribution of polymer chains of different chain lengths or molecular weights. Polybutenes produced by cationic polymerization have relatively narrow molecular weight distributions. The shape of the distribution curve for a manufactured polybutene will approach to a good approximation that of a Gaussian distribution. For reasons of plant operation, economics, or to satisfy customer end-use requirements, polybutenes of different viscosities can be blended to provide a polybutene grade with an intermediate viscosity range.

B. Chemical Properties

Polybutenes are straight-chain, aliphatic polymers made up of predominatly the isobutylene repeat unit. The main hydrogen types in the polymer are secondary and primary. Incorporation of other butene structural types in the polymer chain introduces a small proportion of the more reactive tertiary hydrogen type. However, the main focus for any reactivity in the polybutene structure is the olefinic bond at the end of the polymer chain. Reaction of this olefinic group is only normally accessible under certain contrived conditions. Under normal storage conditions and for industrial applications not involving prolonged exposure to temperatures in excess of 100°C, polybutenes can be considered chemically stable to atmospheric oxidation. The chemical stability of the polybutene polymers is well demonstrated by their retention of viscosity, tackiness, and failure to harden, to become waxy, or to show any deterioration in color on storage for many years at ambient temperatures.

As produced, polybutenes are very pure materials containing extremely low levels of unintentional additional chemical species such as water, chlorine, and metals such as iron. The polymers are also free from nitrogen and sulfur species at detectable levels. Sensitive tests recognized for the detection of polycyclic aromatic hydrocarbons have failed to detect the presence of aromatic compounds in polybutenes (26).

Being constructed of carbon and hydrogen, polybutenes are nonpolar in character. As such they are in general soluble in nonpolar solvents and insoluble in polar solvents. Thus polybutenes are soluble in hydrocarbon solvents such as benzene, toluene, heptane, and kerosenes; in common halogenated hydrocarbons such as chloroform, methylene chloride, and carbon tetrachloride; and in certain oxygenated solvents like tetrahydrofuran and diethyl ether. Polybutenes are insoluble in simple alcohols, esters, and ketones, but solubility does increase with higher homologues and as the molecular weight of the polybutene decreases.

Polybutenes are fully compatible at all concentrations with low-, medium-, and high-viscosity mineral oils of varying aromatic, paraffinic, and naphthelinic contents. Full compatibility has also been demonstrated with polyalphaolefins and alkyl benzenes. In most but not all cases polybutenes have been found to be compatible with synthetic ester oils. Incompatibility has not unexpectedly been found with silicone oils and polyalkylene glycols. The compatibility of polybutene with mineral oils and synthetic oils is summarized in Table 13.3.

Polychloroprene and nitrile rubbers show good resistance to polybutenes, as do fluoroelastomers such as Teflon and Viton. These materials are most suitable for pump packings and seals of equipment for handling polybutenes.

C. Chemical Reactions

The major end use for polybutenes makes use of the reactivity of the polymer afforded through the double bond at the end of the polymer chain. This involves reaction of a polybutene with maleic anhydride at elevated temperatures to produce a polybutenylsuccinic anhydride, which is of importance in its own right as an oil soluble corrosion inhibitor. Further reaction of the polybutenylsuccinic anhydride with a polyethylene polyamine forms the important imide derivatives used as ashless detergent additives in automotive crankcase motor oils to disperse sludge, which forms in the engine at low engine temperatures. Polybutenyl succinimides are also used as petrol/gasoline additives; their use at a very low concentration in the gasoline improves cleanliness in the

Table 13.3 Compatibility of Polybutene with Mineral and Synthetic Oils

Oil	50:50 Oil/polybutene[a,b]
Mineral oil	
Paraffinic	C
Aromatic	C
Naphthelinic	C
Synthetic oils	
Polyalphaolefins	C
Alkyl benzenes	C
Esters[c]	C
Polyalkylene glycols	I
Silicone oils	I

[a]Polybutene 10 grade
[b]C, Compatible; I, incompatible.
[c]Esters need to be examined individually to ensure compatibility

engine carburetor, in valve ports, and on valve stems and helps to prevent formation of deposits on pistons.

Under suitable conditions polybutenes can be made to take part in many of the typical reactions of simple olefins as indicated in Fig. 13.3 (21–25).

In these reactions, the position of the olefin bond at the end of the polymer chain plays an important role in determining the extent of the reaction or yield that is possible. With the availability of polybutenes containing high levels of the vinylidene end-group structure, which show improved reactivity compared with conventional polybutenes, it is expected that these will provide routes to more efficient and cleaner chemical reactions. These more reactive polybutenes give the prospect of much higher yields of polybutene derivatives from chemical reactions.

D. Physical Properties

The main physical properties used on a routine basis to characterize a polybutene are its viscosity and its flash point. A typical product specification might also include properties such as color, water content, acid number, and appearance. Other polymer properties such as density, molecular weight, and refractive index are used less frequently for product specifications, as the possible range of values for these properties are to a large degree defined by the viscosity and flash point of the polybutene.

Polybutene grades that are representative of the range of polybutenes currently available on the market are given along with their typical physical properties in Table 13.4. The full range of grades from the low-viscosity 03 grade to the high-viscosity, semisolid 2000 grade find use in synthetic and/or semisynthetic lubricant applications.

1. Viscosity

The kinematic viscosity of polybutenes is normally measured using a suspended level viscometer under conditions set out in the ASTM D-445 method. The viscosity of polybutenes increases with the molecular weight of the polymer. Polybutenes are avail-

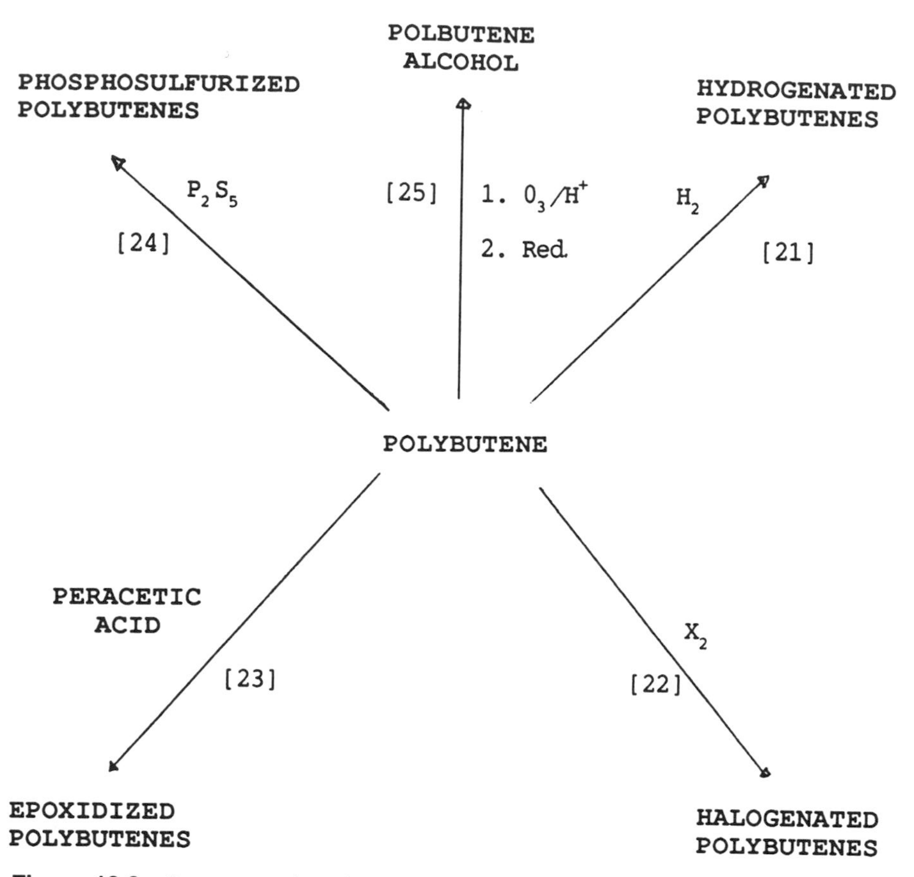

Figure 13.3 Some examples of the chemical reactions of polybutene.

able commercially with viscosity from about 1 to 45,000 cS at 100°C corresponding to a molecular weight range from 180 to 5,800. The change in viscosity with temperature is illustrated in Fig. 13.4. The rate of change in polybutene viscosity with temperature becomes less as the molecular weight of the polymer increases. To facilitate the handling and pumping of polybutene of 30 grade and above, heating to between 60 and 100°C is normally used to lower the viscosity of the material. Polybutene 10 grade and below offers a viscosity index comparable to that offered by mineral oil products. A viscosity index in excess of 170 and up to 380 is available from 30 grade and above, but associated with these grades are relatively high viscosities (>225 cS at 100°C) and pour points (>4°C). The viscosity of polybutenes up to 30 grade shows little dependence on shear rates, and these could be classed as near-Newtonian fluids. Above 200 grade, viscosity begins to show some dependence on shear rate at high shear rates, suggesting some pseudoplastic behavior.

The effect on the viscosity of polybutenes of increasing pressure is to increase the viscosity. This is discussed in more detail in Section III,C,3.

2. Flash Point

The flash point of polybutenes is normally measured using Pensky Martin closed cup (PMCC) or Cleveland open cup (COC) apparatus under conditions set out in the ASTM

Table 13.4 Polybutene Grades with Typical Physical Properties

Property	Polybutene grade										
	03	04	07	3	5	10	30	150	200	600	2000
Viscosity at 100°C (cS)	2	4	13	57	103	225	635	3,065	4,250	12,200	40,500
Viscosity at 100°C (SSU)	32	39	70	270	480	1,050	2,960	14,300	20,000	57,000	190,000
Flash point, PMCC[a] (°C)	95	120	130	140	155	165	170	175	175	180	190
Flash point, COC[b] (°C)	105	135	145	155	190	210	240	250	270	275	280
Pour point (°C)	–60	–60	–30	–21	–12	–7	4	18	24	35	50
Relative density at 15.5°C (g/cm^3)	0.824	0.830	0.851	0.869	0.884	0.894	0.902	0.911	0.914	0.918	0.921
Color (Hazen)	50	50	50	50	50	50	50	50	50	50	50
Viscosity index	—	—	95	98	100	128	181	246	264	306	378
Refractive index	1.461	1.468	1.474	1.487	1.490	1.494	1.498	1.503	1.504	1.505	1.508
Bromine number (g Br_2/100 g)	—	—	40	27	20	16	12	8	6	4	3
Acid number (mg KOH/g)	—	—	0.03	0.03	0.03	0.03	0.03	0.03	0.03	0.03	0.03
Water content (ppm)	40	40	40	40	40	40	40	40	40	40	40
Molecular weight (M_n)	270	300	440	620	780	955	1,250	2,100	2,400	3,800	5,800

[a]PMCC—Pensky Martin closed cup.
[b]Cleveland open cup.

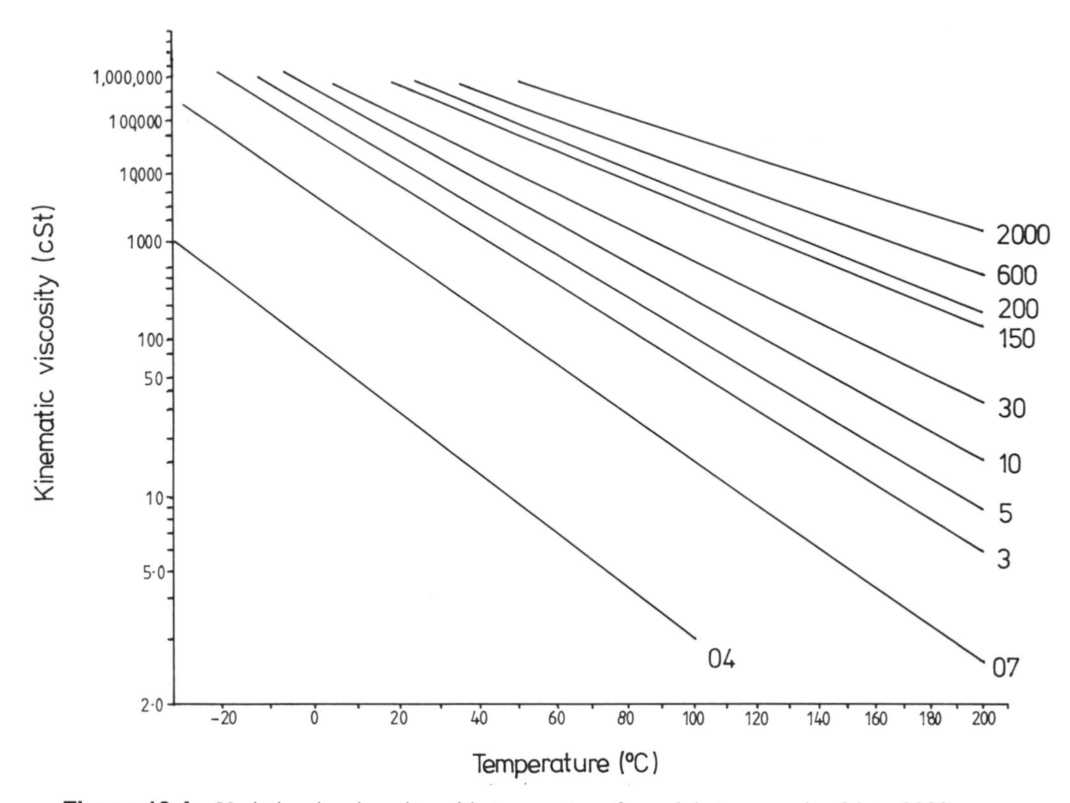

Figure 13.4 Variation in viscosity with temperature for polybutene grades 04 to 2000.

D-93 and ASTM D-92 methods, respectively. The flash point of polybutene gives an indication to the proportion of low-molecular-weight polymer in the bulk material. The level of low-molecular-weight polymer present in the polybutene is influenced by the extent of light polymer stripping by distillation that the manufacturer employs during production. Polybutene grades sold commercially are manufactured to a guaranteed flash-point specification. In general, although it need not be the case, flash points increase with the polybutene grade or molecular weight.

3. Color and Appearance

The color of polybutene is normally assessed by a visual examination comparing a sample of polybutene against a series of recognized color standards. This forms the basis for the ASTM D-1209 method for determination of color in hazen units. Polybutenes are extremely low-colored materials and can be described as essentially water white in nature. Polybutenes should contain no particulate matter or other visible impurities. The bright, clear, and water-white appearance of polybutene grades is a useful indicator of the quality of the material.

4. Water Content

The water content in polybutenes can be determined by a titrimetric technique using Karl-Fisher reagent in accordance with the ASTM D-1744 method. The water content of polybutenes is low and generally found in the region of 20–40 ppm. Water levels in the region of 100–150 ppm cause a cloudiness to appear in the material. Due to the hydrophobic nature of polybutene, higher levels of water are not tolerated, with the water separating to the bottom of the material as a discrete phase.

5. Acid Number

The acid number for polybutenes can be determined using the ASTM D-974 method and is typically of the order of 0.03 mg KOH/g.

6. Bromine Number

Bromine number determination by IP 129 method gives a measure of the amount of carbon-carbon double-bond unsatuation per unit weight of material. Polybutenes contain one double bond at the end of each polymer chain. The drop in bromine number with increasing molecular weight of polybutene reflects the lower level of double bonds per unit weight of material. As a simple argument 1 mole, based on number average molecular weight (M_n), of each polybutene grade should contain an equal amount of unsaturation. In the case of polybutene 2000 grade it is simply that unsaturation is more "dilute."

7. Density

The density of a polybutene can be measured using a hydrometer following the ASTM D-1298 procedure. The density of polybutene at 15.5°C increases with the polybutene grade or molecular weight and varies from 0.82 for 03 grade up to 0.92 for 2000 grade. As manufactured there is minimal batch-to-batch variation in density within each polybutene grade at a given temperature. However, the density of polybutene is a function of temperature and decreases with increasing temperature. This variation in density with temperature is shown for several grades of polybutene in Fig. 13.5.

8. Molecular Weight and Molecular Weight Distribution

The number average molecular weight (M_n) of polybutene can be determined by vapor pressure osmometry (VPO) using the method ASTM D-2503 and by gel permeation chromatography (GPC) following the method ASTM D-3593. Gel permeation chromatography also gives information on the weight average molecular weight (M_w) and

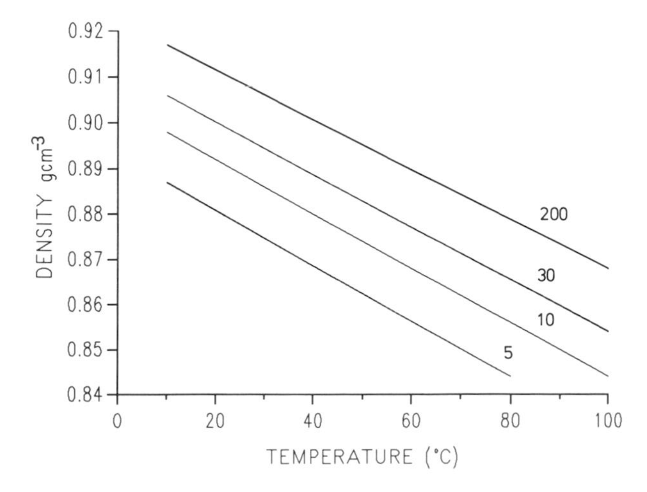

Figure 13.5 Variation in density with temperature for polybutene grades 5, 10, 30, and 200.

on the molecular weight distribution of the polymer. The dispersion index (DI), which is a measure of the broadness of the molecular weight distribution, can be calculated from the quotient M_w/M_n. Each polybutene grade contains a range of polymer molecules of different chain lengths spread around the number average.

The DI of polybutenes tends to increase with the molecular weight (M_n) of the polymer, with, for example, 04 grade having a typical DI of 1.2 while 2,000 grade has a typical DI of 2.4. The shape of the molecular weight distribution for polybutene from a manufacturing plant is normally unimodal and approximates that of a Gaussian distribution. Typical GPC traces for 04, 200, and 2,000 grades are shown in Fig. 13.6. The traces illustrate the increase in the broadness of the molecular weight distribution with grade M_n.

Intentional blending to form intermediate grades or poor manufacturing control can produce polybutenes with relatively high dispersion indices and molecular weight distributions that contain shoulders or distinguishable peaks.

9. *Volatility*

Each polybutene grade is made up of a distribution of polymer chains of different molecular weight (Section II,D,9). A consequence of this distribution is that at temperatures in excess of about 60°C polybutenes can lose a proportion of their mass through evaporation of the lowest-molecular-weight polymer molecules. The rate of loss of material decreases with increasing polybutene grade or number average molecular weight of the grade. The evaporation loss expressed as a percentage weight loss is shown for 07 to 2,000 grade polybutene in Fig. 13.7. Evaporation loss was determined on a small

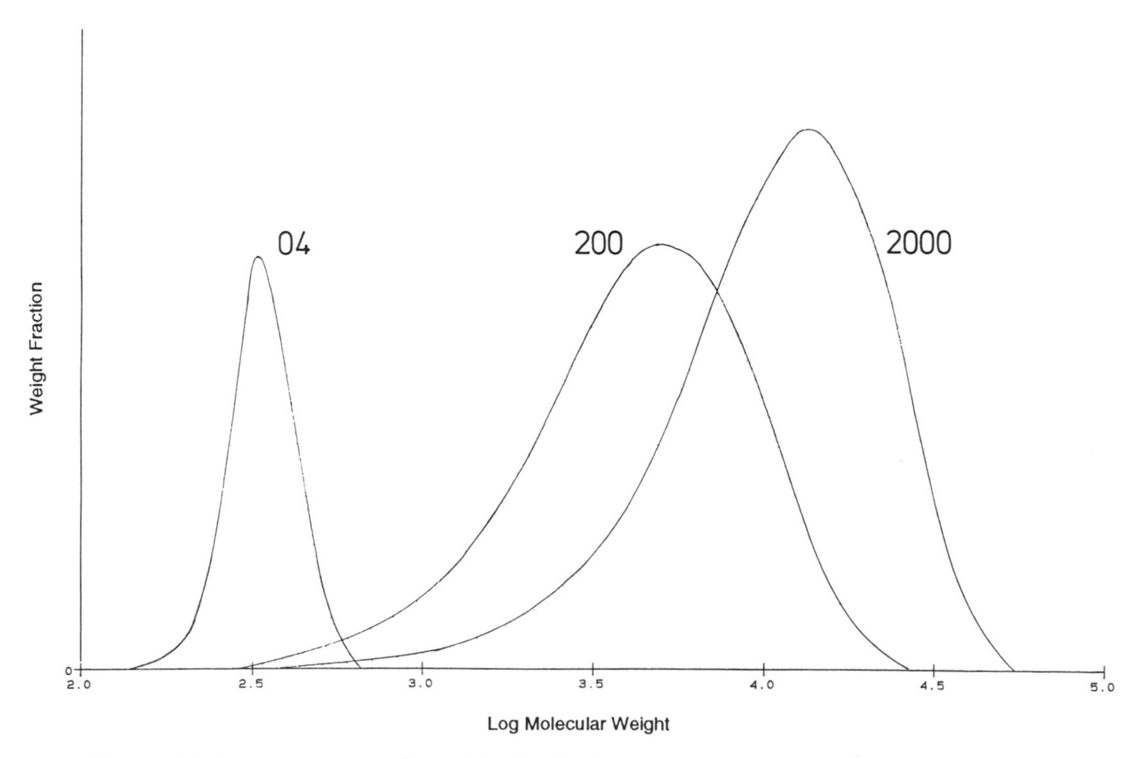

Figure 13.6 Typical molecular weight distribution curves for polybutene grades 04, 200, and 2000.

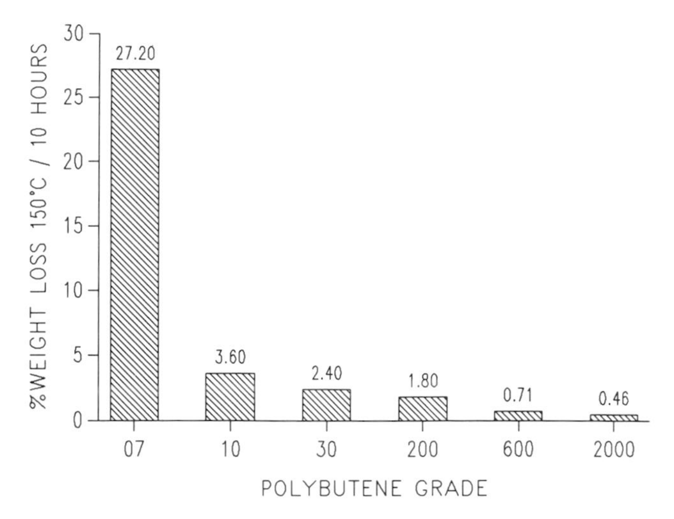

Figure 13.7 Volatile loss for polybutene grades 07 to 2000 when held at 150°C for 10 h following ASTM D972.

sample of polybutene held at a constant temperature of 150°C for a 10-h period following the method ASTM D-972.

10. Decomposition

Polybutenes differ from most other oils by decomposing at elevated temperatures through a depolymerization or unzipping mechanism. The decomposition products are of considerably lower molecular weight than the number average molecular weight of the grade from which they are derived. Under conditions of rapid depolymerization it has been shown that the main product of decomposition is the isobutylene unit. Depolymerization of the polymer chain does not proceed at a fast rate below temperatures of 200–250°C. At lower temperatures it is often difficult to distinguish between evaporation loss of low-molecular-weight polymer molecules and loss of polymer fragments from depolymerization.

Thermogravimetric analysis of several grades of polybutene is shown in Fig. 13.8. The similarity in the curves for 10 and 200 grade at temperatures above 250°C would appear to indicate the onset of depolymerization of the polybutene polymer in both grades. The curve for 04 grade most probably represents weight loss through evaporation of the polymer molecules. Polybutenes stored at high temperatures for any length of time are normally maintained under an inert gas atmosphere. This prevents the formation of potentially explosive mixtures of low-flash-point material over the liquids and also avoids slow oxidation and discoloration of the polybutene.

The depolymerization mechanism of polybutenes is very valuable in allowing the higher-molecular-weight grades to volatilize before rapid combustion takes place. This type of decomposition takes place cleanly and completely without the formation of carbon or tarry materials and is in contrast to mineral oils, which decompose to leave carbon residues. However, if a polybutene is heated to above its flash point and then ignited, carbon and smoke will be generated as combustion products. In the case of the higher-molecular-weight grades, combustion is not self-sustaining unless the body of the liquid is raised to a temperature such that rapid depolymerization also occurs.

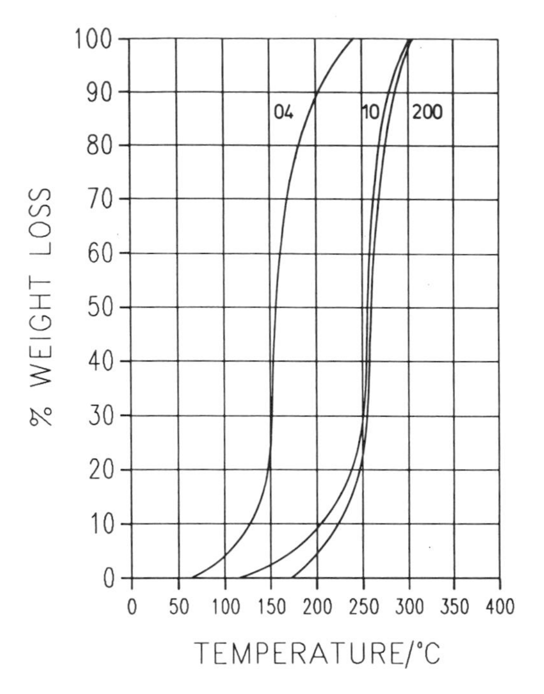

Figure 13.8 Thermogravimetric analysis of polybutene grades 04, 10, and 200, heating at 180°C/h or 3°C/min in dry air with flow rate of 25 ml/min.

11. Toxicity

Polybutene polymers are materials of very low biological activity, as evidenced by their accepted use for many years as components of cosmetics, surgical adhesives, and pharmaceutical preparations. The polymers are products of extremely low oral toxicity. The acute oral toxicity of polybutenes is extremely low, with LD_{50} values quoted in the range of greater than 15.4–34.1 g/kg in rats these being the maximum dose levels used. Long-term chronic oral studies employing 2 wt% polybutene in the diet of rats, and up to 1000 mg/kg/day in dogs for 2 years produced no treatment-related effects (27, 28). These findings indicate that polybutenes can be considered as practically nontoxic by ingestion, although intake of significant doses may give rise to gastrointestinal disturbance. Also, sensitive tests recognized for the detection of polycyclic aromatic hydrocarbons have failed to detect the presence of aromatic compounds in polybutene (26).

Polybutenes are virtually insoluble in water, and their biotoxicity in an aqueous medium is therefore difficult to assess. The International Marine Organisation (IMO) rates polybutenes as unlikely to bioaccumulate and as nonhazardous to marine species.

12. Biodegradation

In studies based on the EEC respirometric test for theoretical oxygen demand (ThOD), comparative tests revealed polybutene levels of biodegradation to be less than that for a polyalphaolefin or mineral oil of similar viscosity. Under the test conditions, none of the materials attained the necessary level of ThOD to be regarded as readily biodegradable as accepted by the Organisation for Economic Co-operation and Development (OECD) (29). However, it is clear that under more favorable conditions, such as higher levels of active sludge acting for a longer period, the levels of biodegradation for the oils including polybutene would be increased.

The highly branched structure of the polybutene polymer is thought to be responsible for its resistance to biodegradation (30).

E. Very-Low-Viscosity Polybutenes

In addition to the grades of polybutene described so far, there are also polybutenes commercially available with viscosity below that of the 03 grade. These grades are commonly referred to as very light polybutenes and are one of the products isolated from the light polymer stream by distillation (Section I,B). The very light polybutenes contain mainly trimers and tetramers of isobutene and n-butenes. Grades are available with viscosity in the range 1.3–14 cS at 20°C and flash points of between 35 and 60°C by Luchaire closed cup.

The very light polybutenes are solventlike in character and find use in applications as diluents and replacements for hydrocarbon solvents.

F. Polyisobutylene Rubbers

Polyisobutylene rubbers (Section I,C) are produced with molecular weights from 10,000 up to several million. They are used as components in adhesives and sealants, in coatings, and in oil and lubricant applications for thickening and for producing tacky or antithrow oils.

III. PERFORMANCE CHARACTERISTICS OF POLYBUTENES AS SYNTHETIC LUBRICANTS

A. Polybutenes: Advantages and Limitations

Synthetic oils are used in applications where a mineral oil cannot give the level of performance required of the lubricant or when the use of a synthetic provides a performance/cost advantage over a conventional mineral oil lubricant. Synthetic lubricants offer a number of advantages over mineral oil-based products. These are well known and are summarized in Table 13.5. The major synthetic oils, polyalphaolefins, and esters will offer most or all of the benefits described in Table 13.5. In comparison with polyalphaolefins or esters, polybutenes are more volatile, less resistant to oxidation, and give improvements in viscosity index only through use of the viscous, high-pour-point grades. A

Table 13.5 Performance Advantages Available for Synthetic Oils over Mineral Oils

Reduced	Increased
Pour point	Oxidation stability
Volatility	Viscosity index
Toxicity	Dispersancy
Deposits	Lubricity
	Flame resistance

comparison of the basic physical properties of polybutene, polyalphaolefin, ester, and mineral oils of similar viscosity is shown in Table 13.6. This illustrates the limitations of the polybutene polymer in terms of volatility, flash point, and viscosity index/pour point relationship.

It should be clear therefore that polybutenes cannot be considered as true synthetic base oils in the same sense as, for example, polyalphaolefins and esters are for automotive engine oils and aviation applications. Polybutenes find use where the special properties of the polymer, such as very low deposit formation, low toxicity, thickening power, and polymer shear stability, are of prime concern and other properties such as volatility and high-temperature oxidation resistance are of secondary importance. Such applications are in one-pass total-loss systems such as two-stroke oils and polyethylene compressor lubricants and in metal working, special greases, oil thickening, and as viscosity index improvers for gear and hydraulic oils.

B. Tests for Deposit Formation

An illustration of the tendency of polybutenes to form very low levels of deposits is given by the Conradson carbon residue test. In this test oils are destructively heated in accordance with the ASTM D-189 method and the level of carbon deposits remaining is then determined. The results of this test comparing polybutenes, esters, polyalphaolefins, and mineral oils are shown in Fig. 13.9. It can be seen that the full range of polybutenes from low to high viscosity gives rise to extremely low levels of carbon deposits. In the test polybutenes gave superior results to esters and mineral oils, with brightstock high-viscosity mineral oil giving the most deposits of all the oils tested. Polyalpholefins also give rise to low levels of carbon deposits, although slightly in excess of that given by polybutenes.

Under oxidation conditions as measured by the total oxidation products (TOPs) following IP 280 and IP 306 test methods it can be seen from Fig. 13.10 to 13.13 that the oxidation products of polybutenes are predominantly those of volatile acids. Oxidation products in the form of soluble acids or sludge that would be expected to remain with the lubricant are lower for polybutene than for esters, mineral oils, and polyalphaolefins. This again confirms the low deposit-forming tendencies of polybutene.

In the section that follows, the individual lubricant applications where polybutenes find use are reviewed in detail.

Table 13.6 Comparison of Physical Properties of Polybutene with Other Oils of Similar Viscosity

Oil	Viscosity, 100°C,cS	Viscosity, 40°C,cSt	VI	Pour point, °C	Flash point, PMCC,°C	Volatility
Polybutene 07	12.1	113.5	95	–30	130	Poor
Polybutene 5	103	3,250	100	–12	155	Fair
Polybutene 10	225	8,360	128	–7	165	Fair
Mineral oil SN150	5.2	30.5	100	–12	208	Fair
Brightstock 470/95	32.0	470.0	95	–9	257	Very good
Naphthelinic oil	2.0	7.7	62	–54	137	—
Polyalphaolefin	4.0	17.5	128	–68	213	Very good
Ester	12.8	88.0	144	–40	252	Good

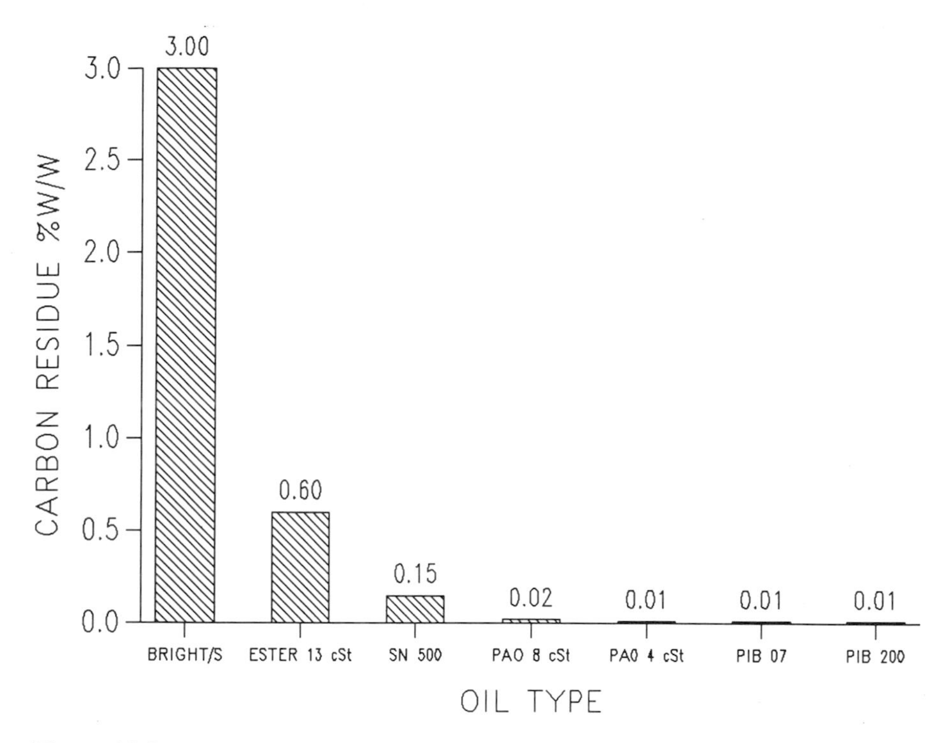

Figure 13.9 Results of Conradson carbon residue analysis of synthetic and mineral oils following ASTM D-189.

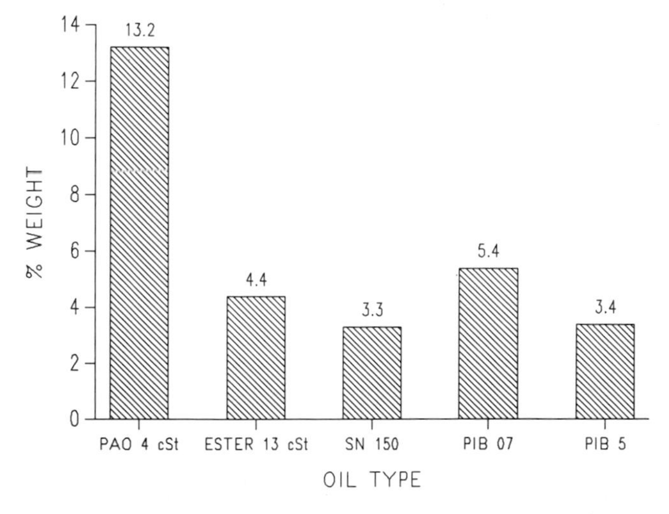

Figure 13.10 Total oxidation products (TOP) expressed as weight percent of oil for synthetic and mineral oils following IP280 and IP300, heating at 120°C for 48 h, pure oxygen flow with copper catalyst.

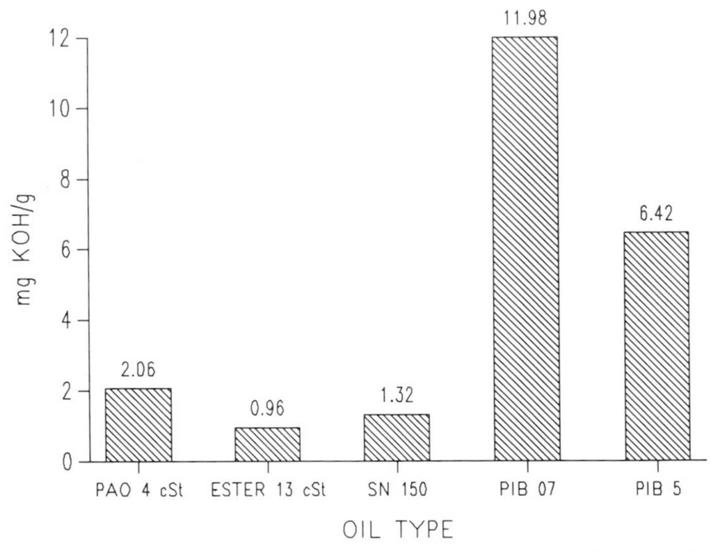

Figure 13.11 Proportion of total oxidation products (TOP) as volatile acids for synthetic and mineral oils.

C. Applications

1. Polybutenes in Oil Thickening

Thickening agents are often required to achieve the correct viscosity characteristics for automotive and industrial oils after refining or recovery processes. The high-viscosity mineral oil grades of brightstock are the traditional products used for thickening low-viscosity oils. Normally, large amounts are required to achieve the required viscosity

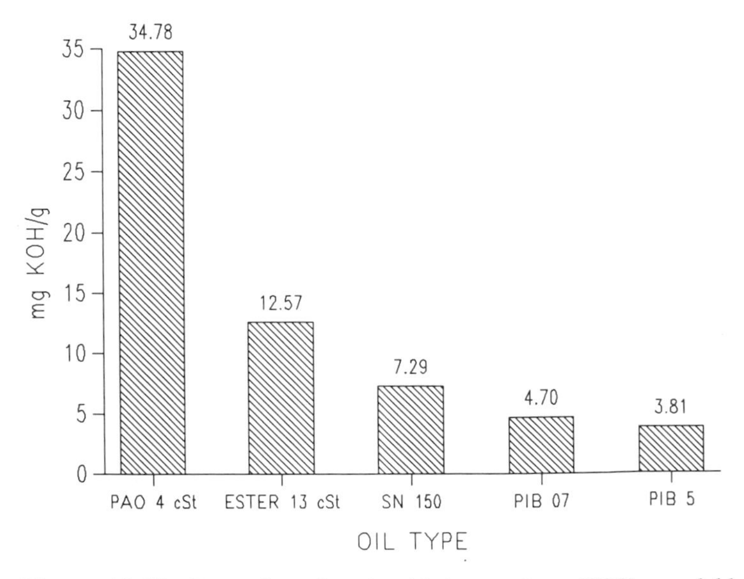

Figure 13.12 Proportion of total oxidation products (TOP) as soluble acids for synthetic and mineral oils.

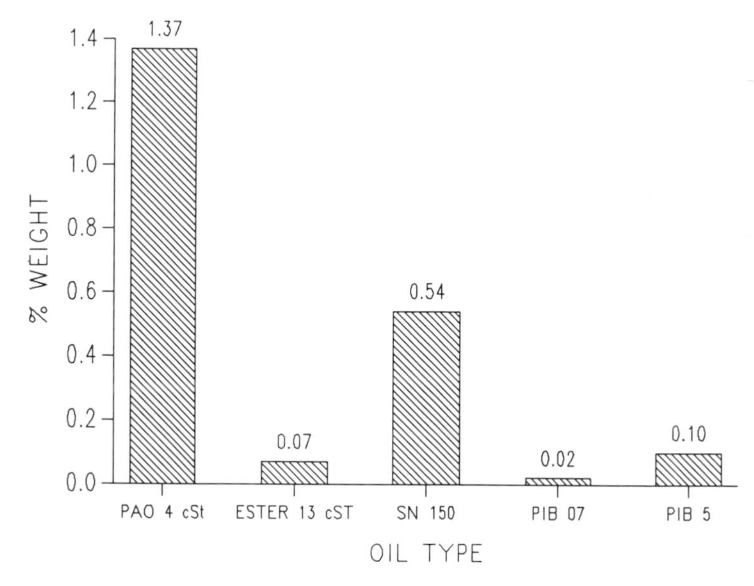

Figure 13.13 Proportion of total oxidation products (TOP) as sludge formation for synthetic and mineral oils.

adjustment. Polybutenes are very efficient synthetic alternatives to brightstock and have the advantage of raising the viscosity index of the thickened oil. In addition, the use of polybutenes improves the low-temperature properties of the thickened oil by lowering the pour point and lowering the –18°C viscosity. The exellent shear stability of the polybutene polymer (Section III,C,4) ensures that the viscosity of the thickened oils is retained under normal service conditions.

Figures 13.14 and 13.15 compare the thickening efficiency of several grades of polybutene and brightstock (33.1 cS at 100°C; VI = 96) when used to thicken solvent-neutral 150 and 500 base oils. As can be seen, significantly less polybutene is required to give the same thickening effect as brightstock. The improvements in formulated oil viscosity index, pour point, and –18°C viscosity obtained by using polybutenes are clearly demonstrated for the oils blended to SAE 30, 40, and 50 viscosity specifications shown in Table 13.7.

The main polybutene grades of interest for oil thickening and viscosity adjustment are shown in Table 13.8.

2. Polybutenes in Two-Stroke Oils

Two-stroke engines are used to power motorcycles, scooters, outboard motors for boats, snowmobiles, chainsaws, and small agricultural equipment. The advantages that two-stroke engines have over four-stroke crankcase engines lie in their high power to weight ratio, simplicity of design, and lower production costs. The market for two-stroke oil is very dependent on the fortunes of the motorcycle industry, which accounts for the majority of two-stroke oil consumption. While the market for two-stroke oil in Western Europe and Japan is seen as mature and in slow decline, growth in demand from the emerging countries in the Far East is maintaining the size of the total market. Within the total market, it has been predicted that the use of synthetic oils will grow steadily in the

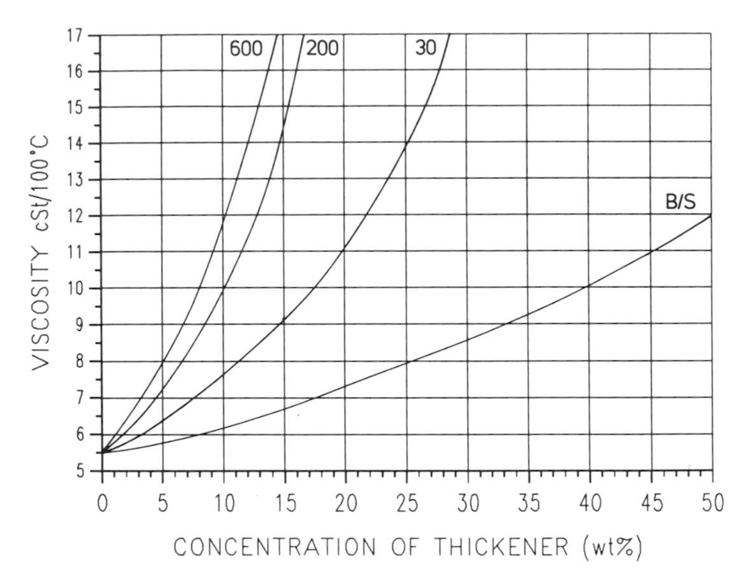

Figure 13.14 Thickening efficiency of polybutene grades 30, 200, and 600 in comparison with brightstock 470/95 for SN150 mineral oil at 100°C.

next few years as greater emphasis is placed on meeting and improving the performance of two-stroke engines (31).

Unlike four-stroke crankcase lubricants where the oil is held in the sump and is pumped continuously to lubricate the engine, two-stroke oils are required to operate on a one pass total loss basis. The oil is introduced into the engine either premixed with the fuel or by direct injection using a demand pump. The oil functions as a lubricant for the crank, bearings, and combustion cylinder and is then consumed along with the fuel. Products of decomposition are lost through the exhaust system. Oil is therefore used on a continuous basis while the engine is in operation.

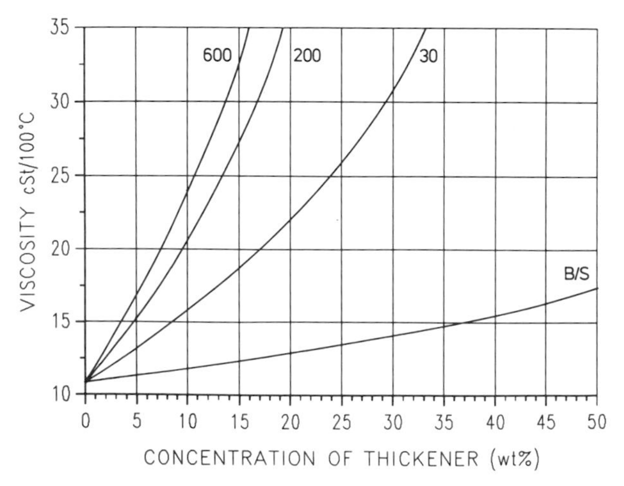

Figure 13.15 Thickening efficiency of polybutene grades 30, 200, and 600 in comparison with brightstock 470/95 for SN500 mineral oil at 100°C.

Table 13.7 Formulations for SAE 30, 40, and 50 Oils with Polybutenes and Brightstock

Formulation/performance	SAE 30				SAE 40				SAE 50		
Polybutene 600	9.0				12.5				14		
Polybutene 200		11.5				14.5				18.5	
Polybutene 30			18				24				31
Brightstock 470/95				47.5				60			
Solvent-neutral 150	91.0	88.5	82	52.5	87.5	85.5	76	40	86	81 5	69
Viscosity, 100°C, cS	11.7	10.8	10.1	11.3	15.7	13.1	12.8	14.2	17.7	16.7	17.0
Viscosity, 40°C, cS	83.7	78.5	78.0	98.3	121	101	111	140	141	141	170
Viscosity, –18°C, P	70	60	68	120	110	76	115	210	145	124	200
Viscosity index, VI	104	124	110	101	137	126	109	98	139	128	107
Flash point, PMCC, °C	207	205	207	213	211	200	200	221	207	196	196
Pour point (upper), °C	–15	–12	–12	–9	–15	–15	–15	–9	–15	–12	–15

Table 13.8 Main Polybutene Grades of Interest for Oil Thickening Applications

Grade	Very light	03	04	07	3	5	10	30	150	200	600	2000
Use							X	X	X	X	X	X

The important performance requirements for a two-stroke oil are summarized in Table 13.9. These relate to miscibility with the fuel, lubrication and corrosion protection of the engine, and control of deposits in the engine and exhaust system. More recently, greater concerns for the environment and the public image of two-stroke engines have placed an additional demand on the oil as a method of reducing visible smoke emissions from the two-stroke exhaust system.

Conventional two-stroke lubricants are based on mineral oil with a detergent additive and in some cases a hydrocarbon diluent. The mineral oil is present usually as a mixture of a brightstock high-viscosity oil and a low-viscosity oil. The final viscosity of commercial two-stroke oils is normally in the range 7–12 cS at 100°C. Brightstock is used to reduce scuffing and wear in the engine. However, it is now accepted that brightstock oils are the major contributors to deposit formation in the combustion chamber and exhaust system and that their use is linked to high levels of visible smoke, often seen as plumes of pungent blue smoke from the exhaust.

The property of the polybutene polymer to decompose cleanly at engine temperatures makes it well suited to the one-pass total-loss requirement of the lubricant. In addition to giving a reduction in carbon deposits in the engine and exhaust system, the use of polybutene has been shown to improve oil film strength, reduce engine wear, piston seizure, and corrosion (32), and maintain the antiscuff performance of the oil (33). Recently, public concern over the relatively high levels of visible smoke emitted from two-stroke engines has put pressure on the engine manufacturers in particular, to reduce smoke emissions from their engines. The manufacturers have responded by modifying the design of engines, such as fitting demand pumps for the oil and incorporating features such as special grove sections in the exhaust port (34). It is recognized, however, that the best opportunity to bring about reductions in visible smoke emission lies with the formulation of the two-stroke oil (34, 35). There is now a wide concensus among authors that the inclusion of polybutene in place of mineral oil in the two-stroke oil can give a significant reduction in the level of visible smoke (32–34, 36). Figure 13.16 is typical of the data obtained comparing the levels of visible smoke when using a conventional

Table 13.9 Important Performance Requirements for a Two-Stroke Oil

Prevent engine wear and piston scuffing
Form minimal deposits in the combustion chamber and exhaust port and cause minimal spark-plug fouling
Maintain piston cleanliness and prevent ring sticking
Maintain crankcase cleanliness and prevent corrosion
Have good low-temperature fluidity and fuel miscibility
Clean-burning oils giving little visible smoke emission

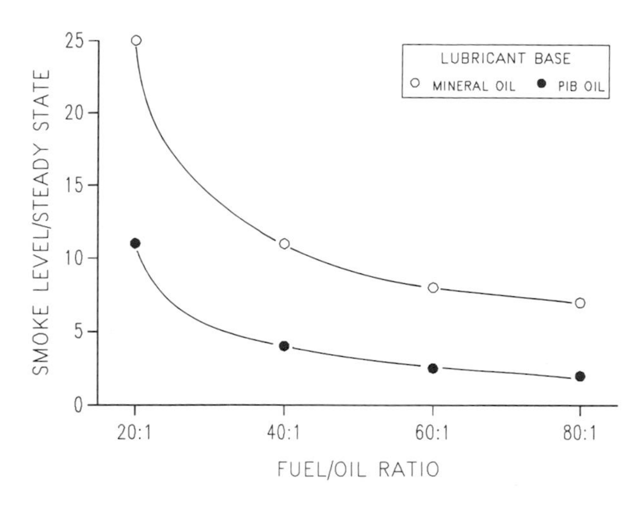

Figure 13.16 Comparison of smoke levels for polybutene and mineral oil formulated two-stroke lubricants under steady-state conditions and different fuel/oil ratios; 60% load, 4,500 rpm (36).

mineral oil lubricant and the improvements gained when a polybutene-based lubricant is used (36).

Recently a relationship between the "reaction energy" of an oil and its tendency to produce visible smoke has been proposed; "reaction energy" is a measure of the energy involved in oil decomposition (34). Polybutene-containing oils were shown to have a low reaction energy and correspondingly low smoke-producing tendencies. The low value of reaction energy is clearly linked to the facile depolymerization of the polymer at temperatures found in the two-stroke engine.

Polybutene is normally employed to replace the polluting brightstock component of a two-stroke oil formulation. However, its use can be extended to replacement of the lower-viscosity mineral oil and even the hydrocarbon diluent component of the oil formulation. Two-stroke lubricants containing up to 97 wt% of a polybutene or blends of polybutene are claimed in the patent literature (37–40). In practice, however, polybutene concentrations are lower and typically in the region of 15–70 wt%. At low concentrations the polybutene is used primarily as an antiscuff agent while at higher concentrations benefits in the reduction of visible smoke are achieved with the lubricant. It is also normal practice to retain a proportion of mineral oil in the lubricant formulation, as this prevents the problem of lacquer formation, which has been associated with oils formulated with very high levels of polybutene.

Polybutenes or a combination of polybutenes with viscosity in the range 4–635 cS at 100°C corresponding to grades 04–30 are used to replace brightstock and lower-viscosity mineral oils. In addition the solventlike polybutenes with viscosity 1–14 cS at 20°C can give improvements in lubrication and fuel miscibility when used to replace any hydrocarbon diluent in the two-stroke oil formulation.

Polybutenes are not readily biodegradable (Section II,D,1) by virtue of their branched structure, but have been shown to be essentially nontoxic in animals (27) and in tests gave no evidence of hazard to aquatic life (29). Fully synthetic oils based on esters find application where legislation specifies that oils must be biodegradable, for example, in two-stroke oils for outboard motors used on Swiss lakes. Esters normally make up 100%

of a two-stroke formulation. Oils based on esters provide effective lubrication, and very lean-burning oil/fuel ratios have been reported for their use (41, 42). However, there is some evidence to suggest that esters may be less effective in controlling deposits (e.g., Section III,B) and visible smoke emissions. Results of smoke levels for a range of conventional and synthetic oils in the West Bend smoke test are shown in Table 13.10 (33). As expected, polybutene-based lubricants gave a lower level of smoke than conventional mineral oil lubricants. Lubricants formulated with ester, polyalphaolefin, and polypropylene synthetic oils gave higher levels of smoke than the mineral oil lubricants.

The use of polybutene-based two-stroke lubricants should continue to increase where improvements in two-stroke engine performance are sought. The low visible smoke and particulate oil emission that characterize the exhaust effluent leave polybutene-based lubricants well positioned to meet the increasing demands on environmental grounds for clean-burning oils. In addition, polybutene will continue to be used for the control of deposits in the engine and exhaust system and to improve the lubrication and engine life. Only where legislation calls for the oil to be biodegradable will polybutene lubricants not find application.

The main polybutene grades of interest for two-stroke lubricants are shown in Table 13.11.

3. Polybutenes as Compressor Lubricants

Polybutenes are employed alone or in combination with other oils as lubricants for the barrels of compressors that are used to generate the high pressures of ethylene gas required for reaction in the manufacture of low-density polyethylene (LDPE). A simple scheme that illustrates the steps involved in the production of LDPE is shown in Fig. 13.17.

Ethylene gas is supplied to a primary compressor, where it undergoes a first stage compression to between 200 and 300 bar. It is then fed into the secondary compressor, where the ethylene gas is compressed to the final reaction pressure, which is normally in the range 2,000–3,000 bar. A peroxide initiator is pumped into the reactor, where, at reactor temperatures of between 180 and 200°C, it undergoes a thermal decomposition to free radical species, which catalyze the polymerization of the ethylene gas to LDPE. After leaving the reactor the unreacted ethylene gas is returned to the compressors for recycling

Table 13.10 Smoke Levels for Various Oils in West Bend Smoke Test (33)

Base oil formulation[a]	Average smoke level (%)
SN 650 oil and brightstock	55
SN 500	54
Polyalphaolefin (low viscosity)	64
Polyalphaolefin (high viscosity)	81
Synthetic ester	82
Polypropylene	63
SN 150 + 1,000 MW PIB	23
Polybutene 10 + Solvent	21
Polybutene 10 + No Solvent	35

[a]25 : 1 Fuel/oil ratio used in West Bend smoke test with constant performance additive package.

Table 13.11 Main Polybutene Grades of Interest for Two-Stroke Lubricants

Grade	Very light	03	04	07	3	5	10	30	150	200	600	2000
Use	X		X	X	X	X	X	X				

to the reactor, while the polyethylene passes to the polymer finishing section where it is prepared for sale (43). There are two designs of reactor, an autoclave reactor, which is a large stirred vessel, and a tubular reactor, which is basically a long high-pressure pipe (44). The reactor pressure employed in the autoclave reactor is generally 1,500–2,500 bar, and in the tubular reactor the pressure is normally higher, at 2,500–3,000 bar.

Conventional mineral oils are mostly used and are adequate for the lubrication of the compressor crankcase and bearings and for the hydraulic and gear oils of the initiator pumps. However, for the lubrication of the compressor barrels and in particular the secondary compressor barrel additional demands are made on the lubricant that cannot be satisfied by conventional mineral oil products. During the lubrication of the compressor barrel, the lubricant comes into contract with the compressed ethylene gas. Some of the lubricant is dissolved in the gas and is carried over continuously into the reactor area. A constant supply of lubricant to the cylinder is required to maintain adequate lubrication. After polymerization the lubricant is present at low levels as an intimate mix with the LDPE product. It is important therefore that the compressor barrel lubricant must not hinder the polymerization reaction, nor must its low concentration in the finished product have a detrimental effect on the performance of the LDPE in its intended application. A list of properties required of the cylinder lubricant are shown in Table 13.12.

Polybutenes provide the necessary high-pressure sealing, continuous film forming, oxidation resistance, and limited volatility and meet in full the requirements regarding purity, low moisture content, low deposit formation, inertness, and food contact approval for the lubricant.

Of importance in this application is the effect that pressure has on the viscosity of the lubricant. The viscosity of the lubricant will increase with increasing pressure. Its viscosity must not increase to the extent that the operation of the lubricant delivery system for the compressor barrel would be compromised, resulting in failure. Polybutenes thicken rapidly at high pressures, and for this reason their use tends to be found in the

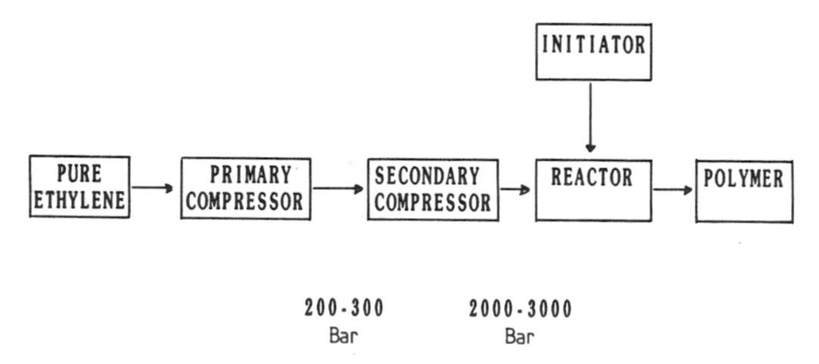

Figure 13.17 Schematic representation of the manufacture of low-density polyethylene from ethylene gas.

Table 13.12 Important Properties for the Lubricant Used in Secondary Compressor Barrel During LDPE Manufacture

Viscosity in the range 220–330 cS at 38°C
High-pressure sealing capacity
Continuous film forming up to 120°C.
Low deposit forming
Resistance to oxidation
High purity
Low volatility
Low moisture content
Inert to process gas and equipment
No deleterious effect on end properties of low-density polyethylene
Food contact approval

lower-pressure autoclave LDPE manufacturing process. Thickening under pressure is less of a problem if a blend of polybutene and a mineral white oil is used. In the higher-pressure tubular process polyalkyleneglycols are most often used to lubricate the secondary compressor barrel. Polyalkyleneglycols thicken less than either polybutenes or polybutene/white oil blends under pressure. This can be seen in Fig. 13.18.

In general, polyalkyleneglycols are considered to offer better lubrication for the secondary compressor barrel. Evidence for this is the lower usage rate of lubricant for quantity of LDPE produced and the increase in the packing life of the cylinder that have been reported for polyalkyleneglycols compared to that offered by polybutene-based lubricants. If required, however, the use of additives such as fatty acids can boost the antiwear performance of polybutene based lubricants. Polybutene lubricants are used in preference to polyalkyleneglycols in the manufacture of ethylene/vinyl acetate copolymers where the solubility of vinyl acetate in polyalkyleneglycols precludes its use. Polybutene lubricants are also employed in the manufacture of LDPE that is destined for

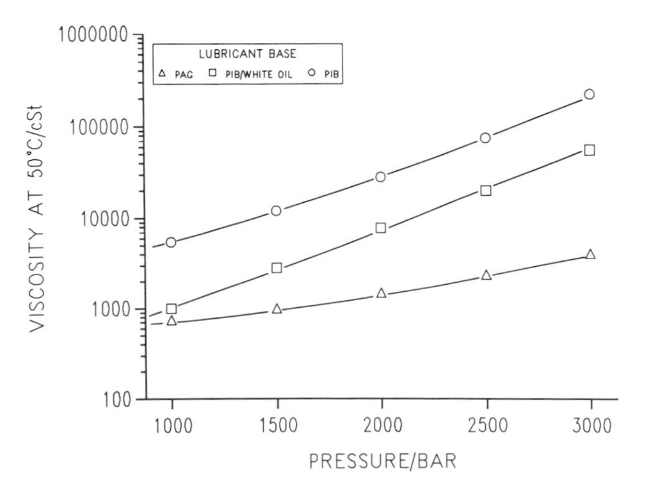

Figure 13.18 Variation in viscosity with pressure for polybutene, polybutene/white oil, and polyalkylene glycol lubricants.

use either in electrical applications such as wire and cable or coating applications such as card lamination. The residual low level of polyalkyleneglycol in the LDPE appears to have a detrimental effect on the electrical properties of the polyethylene and also adversely affects the adhesion quality of coating grades.

The use of polybutene has also been reported for use as the lubricant in compressors for other nonoxidizing gases such as hydrogen, nitrogen, and carbon dioxide (45).

The main polybutene grades of interest for use as compressor barrel lubricants are shown in Table 13.13.

4. Polybutenes in Gear and Hydraulic Oils

The selection of an oil for gear lubrication involves careful consideration of the gear design, the materials involved in the construction of the gears, and the likely conditions of operation. The viscosity of the oil has a key role in providing the correct thickness of oil to provide hydrodynamic lubrication of the gears, determine the load-carrying capabilities of the gears, control leakage at seals, and limit noise generation of the unit. The viscosity of the oil at low temperatures is of importance because the flow properties will determine the ease of start-up in cold climates and the ease of gear shifting. Gear systems are usually classified as automotive gears for cars and commercial vehicles and industrial gears in manufacturing and processing industries.

Automotive Gear Oils. For automotive gear oils an SAE classification system has been adopted to define the performance of the oil in terms of viscosity at high and low temperature (Appendix 13.1). Historically in Europe monograde oils have been used in cars (SAE 80 or SAE 90) and commercial vehicles (SAE 140). More recently, demands for improvements in the low-temperature properties of gear oils and the realization that gear oil performance could make a sizable contribution to fuel economy (46) have led to the development of multigrade gear oils. Multigrade oils are applicable for use over a wider temperature range than monograde oils, and the low-temperature properties of the former are better defined.

Initially 80W/90 and 85W/140 multigrade oils were introduced, which could be formulated using conventional mineral oils. However, there is now a trend to even wider multigrade oils such as 75W/90 for cars and 80W/140 for commercial vehicles, which require the use of synthetic base oils and/or VI improvers. In Europe, demand for the wider multigrade oils is growing in areas such as Scandinavia, which experiences severe winter climates. In the rest of Europe, where monograde oils at present dominate the market, the use of multigrade oils should continue to advance as wider acceptance of advantages of synthetic oils is gained and the drive to improve fuel efficiency intensifies. Multigrade oils have found greater use in countries such as the United States, Canada, and Japan where there is at present more emphasis on improving fuel economy and where there can be a wide variation in climatic conditions.

As indicated, the viscometrics for 80W/90 and 85W/140 oils can be met by blending brightstock oils with pour-point depressants and extreme-pressure additives, although

Table 13.13 Main Polybutene Grades of Interest for Compressor Barrel Lubricants

Grade	Very light	03	04	07	3	5	10	30	150	200	600	2000
Use			X	X			X	X				

Appendix 13.1 Lubricant Viscosity Classification for Axle and Manual Transmission Oils—SAE J306c

SAE viscosity grade	Maximum temperature for 150 Pa·s, °C[a]	mm^2/s (= cS) at 100°C	
		Minimum	Maximum
75W	–40	4.1	
80W	–26	7.0	
85W	–12	11.0	
90		13.5	<24.0
140		24.0	<41.0
250		41.0	

[a]150 Pa·s = 1500 P.

some companies will include a viscosity index (VI) improver to enhance the low- or high-temperature properties of the oil. The wider multigrade oils 75W/90 and 80W/140 cannot be blended using mineral oils alone, and synthetic base oils, blends of synthetic and mineral oils, or a VI improver in conjunction with synthetic or mineral oils must be used.

The selection of the correct VI improver is most important, because under the severe mechanical shearing that exists in gear configurations the polymeric VI improver must be very shear stable in order to maintain effective VI contribution to the oil during service. Problems of VI improver breakdown were experienced during the initial stages of multigrade oil commercialization. The early VI improvers, such as olefin copolymers or polymethacrylates with molecular weights in excess of 40,000, developed for crankcase engine oil were found to undergo catastrophic mechanical breakdown when used in gear oils. It is now well recognized that much lower-molecular-weight polymers are required to withstand the greater demands made on polymer shear stability while operating in a gear-box arrangement. Bench tests with diesel injector, FZG gear, and sonic equipment are now used on a routine basis to assess the shear stability of prospective VI improvers and lubricant formulations.

Polybutenes are shear-stable polymers that are effective as VI improvers in the formulation of high-quality multigrade gear oils. Polybutenes are efficient oil thickeners at high temperatures and at the same time offer improved low-temperature properties over oils thickened with brightstock (Section III,C,1). The best combination of VI improvement and polymer shear stability has been found for polybutene grades 150, 200, and 600. The high viscosity of polybutenes at the high-molecular-weight end of the grade range—for example, 600 grade has a viscosity of 12000 cS at 100°C—has led some manufacturers to offer special products that have the high-viscosity polybutene cut back with mineral oil to facilitate handling and blending (47). Table 13.14 illustrates the excellent shear stability of the polybutene polymer when tested in diesel injector and FZG gear rigs. For comparison, the results of a commercial polymethacrylate VI improver are shown. The results show clearly the relationship between the molecular weight and the shear stability of the polymers. The trade-off that can be made between treatrate/cost, which favors the use of a relatively high-molecular-weight polymer, and the required polymer shear stability, which favors the use of a relatively low-molecular-weight polymer, will be dependent on the particular demands placed on the oil in the end application.

Table 13.14 Shear Stability of Polybutene and Polymethacrylate Viscosity Index Improvers in Different Tests for Solvent Neutral Base Oil Thickened with PIB and PMA to Give a Similar Viscosity at 100°C

Thickener, PIB grade/PMA	150	200	600	PMA
Molecular weight (M_n)	2,100	2,400	3,800	9,000
Initial viscosity, cS, 100°C	19.12	18.75	17.72	19.60
Viscosity after diesel injector,[a] cS, 100°C	19.00	18.44	17.35	19.01
Viscosity loss, %	0.6	1.4	2.1	3.0
Initial viscosity, cS, 100°C	19.12	18.24	17.72	
Viscosity after FZG shear,[b] cS, 100°C	17.57	16.52	15.65	
Viscosity loss, %	8.1	9.4	11.7	
Initial viscosity, cS, 100°C	18.44	18.24	17.72	19.13
Viscosity after FZG shear,[c] cS, 100°C	16.40	15.29	14.39	14.68
Viscosity loss, %	11.0	16.2	18.9	23.3

[a]Diesel injector rig (IP 294; 250 cycles).
[b]FZG gear rig; proposed CEC TLPG7 conditions; load stage 5; EP additive present.
[c]FZG gear rig; IP 351 (B); load stage 6; EP additive present.

Typical automotive multigrade gear oil formulations for 80W/90, 85W/140, and 80W/140 oils making use of polybutene as a VI improver are shown in Table 13.15. The polybutene grades PB800 and PB600/45 referred to in Table 13.15 are examples of specially produced polybutene/mineral oil blends that facilitate the use of higher-viscosity polybutenes in lubricant formulations. In the 80W/90 and 85W/140 oils the use of polybutene as a brightstock replacement produces a lubricant with superior low- and high-

Table 13.15 Formulations for 80W/90, 85W/140, and 80W/140 Oils Using Polybutene as a Viscosity Index Improver

Composition[a]/performance	80W/90		85W/140		80W/140		
SN 100, wt%					40.0	64.5	52.0
SN 150, wt%		76.0					
SN 300, wt%					26.0		
SN 500, wt%	62.5		6.5	76.0			
Brightstock 470/95, wt%	30.0		86.0				8.0
PB 800, wt%							32.5
200 Grade PIB, wt%						28.0	
PB 600/45, wt%		16.5		16.5	26.5		
Viscosity, 100°C, cS	15.1	16.2	28.0	29.1	25.6	26.4	25.2
Viscosity, 40°C, cS	150	130	339	310	219	243	245
Viscosity index	101	133	96	127	148	140	131
Pour point, °C	–18	–27	–18	–21	–27	–27	–27
Viscosity at –10°C, P			387	177			
Viscosity at –18°C, P			2,320	599			
Viscosity at –26°C, P	948	482			1,400	1,460	1,450
Viscosity at –30°C, P	7,560	1,060					

[a]All formulations contain 6.5 wt% EP additive and 1.0 wt% pour-point depressant.

temperature properties. For the wider multigrade 80W/140, the use of polybutene as a VI improver in combination with mineral oils is sufficient to achieve the viscosity specifications of the oil without recourse to the use of special mineral oils or synthetic base oils. Polybutenes can be used as a VI improver in the formulation of 75W/90 oils as shown in Table 13.16. However, without access to hydrocracker base oils, the use of a synthetic oil such as polyalphaolefin or ester is required to achieve the viscosity specification. The quantity of synthetic oil in the formulation can be reduced by including a low-pour-point base oil.

An extended road-car trial carried out with a 75W/90 gear oil illustrates the excellent shear stability of the polybutene VI improver, with the oil showing a viscosity loss of only 16.8% after 40,000 km of service (Fig. 13.19). The trial used a VW Passat with a transaxle gear arrangement that is known to provide a severe test of the shear stability of the VI improver.

Industrial Gear Oils. Industrial gear units represent perhaps 60–75% of the total market for gear oils. Gears are classified as open and enclosed, with the enclosed gears gradually replacing open gears as the preferred design. The ISO viscosity system given in Appendix 13.2 is used to specify industrial fluids. The requirements of an industrial gear oil are similar to those for an automotive gear oil. In addition, as water is often a contaminant in industrial processes, the oil must show good water separation characteristics.

Polybutenes are polymers of high shear stability (Table 13.14) and can be used to adjust the viscosity of mineral base oils to the required viscosity specification for industrial gear lubricants. Polybutenes produce lubricants with superior viscosity index and low-temperature properties over brightstock-thickened oils and can therefore be used to formulate more energy-efficient gear oils. In addition, the tacky nature of the high-molecular-weight polybutenes can improve the adhesiveness and antithrow properties of the oil, thereby increasing the retention time of the lubricant on the gear surface. Polybutenes show good water separation superior to solvent-neutral 100 and 500 mineral oils but not as good as polyalphaolefins. Figures 13.20 to 13.22 compare the thickening efficiency of polybutenes and brightstock for adjusting the visocity of solvent-neutral 150 and 500 base oils and brightstock 470/95 oil to ISO specifications.

Table 13.16 Formulations for a 75W/90 Oil Using Polybutene as a Viscosity Index Improver

Composition[a]/performance				
6 cS Polyalphaolefin, %wt	50.0	57.0	30.0	40.0
SN100, %wt	23.5	14.0		
Naphthelinic oil, %wt			39.5	25.5
PB 600/45, %wt	19.0		23.0	
PB 800, %wt		21.5		27.0
Viscosity at 100°C, cS	16.4	15.3	15.8	16.4
Viscosity at 40°C, cS	112	110	101	119
Viscosity index	158	145	167	149
Pour point, °C	–39	–39	–39	–39
Viscosity at –40°C, P	1,250	1,450	1,220	1,400
Viscosity at –18°C, P	51.9	56.7	52.6	57.6

[a]All formulations contain 6.5 wt% EP additive and 1.0 wt% pour-point depressant.

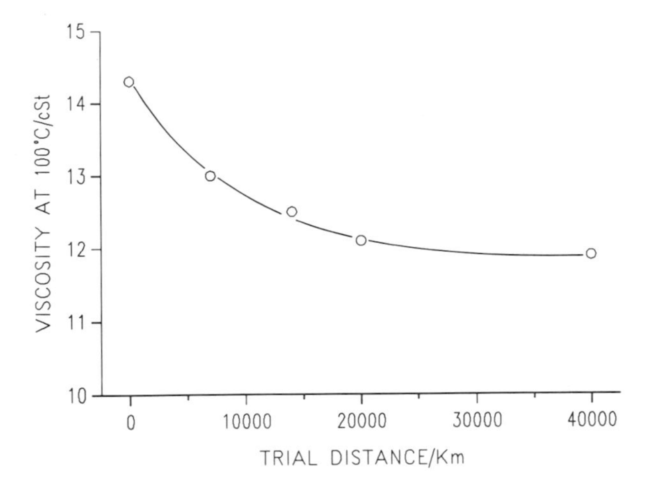

Figure 13.19 Extended road-car trial showing loss in viscosity/shear stability of 75W/90 oil formulated with polybutene as the viscosity index improver.

Appendix 13.2 ISO VG Viscosity System for Industrial Fluid Lubricants

Viscosity system grade identification	Midpoint viscosity, cS (mm^2/s) at 40.0°C	Kinematic viscosity limits, cS (mm^s/s) at 40.0°C	
		Minimum	Maximum
ISO VG 2	2.2	1.98	2.42
ISO VG 3	3.2	2.88	3.52
ISO VG 5	4.6	4.14	5.06
ISO VG 7	6.8	6.12	7.48
ISO VG 10	10	9.00	11.0
ISO VG 15	15	13.5	16.5
ISO VG 22	22	19.8	24.2
ISO VG 32	32	28.8	35.2
ISO VG 46	46	41.4	50.6
ISO VG 68	68	61.2	74.8
ISO VG 100	100	90.0	110
ISO VG 150	150	135	165
ISO VG 220	220	198	242
ISO VG 320	320	288	352
ISO VG 460	460	414	506
ISO VG 680	680	612	748
ISO VG 1000	1000	900	1100
ISO VG 1500	1500	1350	1650

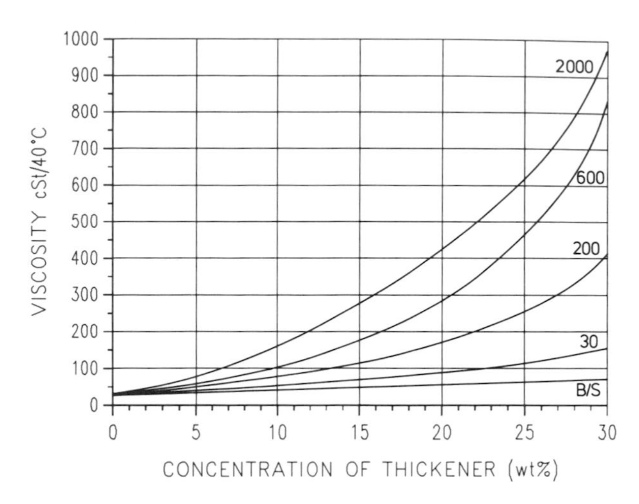

Figure 13.20 Thickening efficiency of polybutene grades 30, 200, 600, and 2000 in comparison with brightstock 470/95 for SN150 mineral oil at 40°C.

The trend in the industrial sector is to the use of multipurpose and more energy-efficient oils. This suggests a move to the use of more synthetic oils such as polyalphaolefins and an increased use of the cost-effective approach of employing shear-stable viscosity index improvers with conventional or special base oils. In addition, the demand for safer, nontoxic lubricants for use in the industrial sector will perhaps focus more attention on materials such as polybutenes, which are essentially nontoxic and which are recognized as versatile materials for the formulation of lubricants.

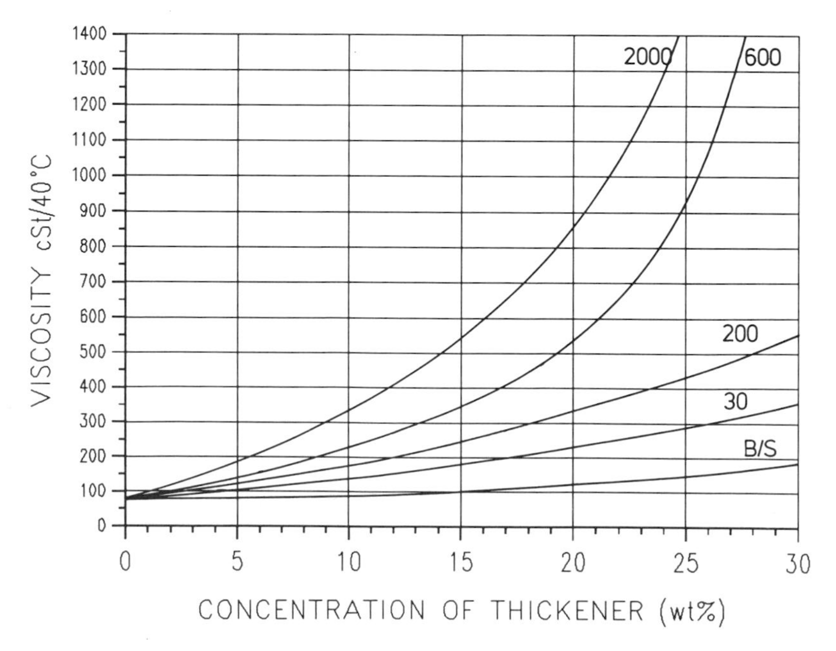

Figure 13.21 Thickening efficiency of polybutene grades 30, 200, 600, and 2000 in comparison with brightstock 470/95 for SN500 mineral oil at 40°C.

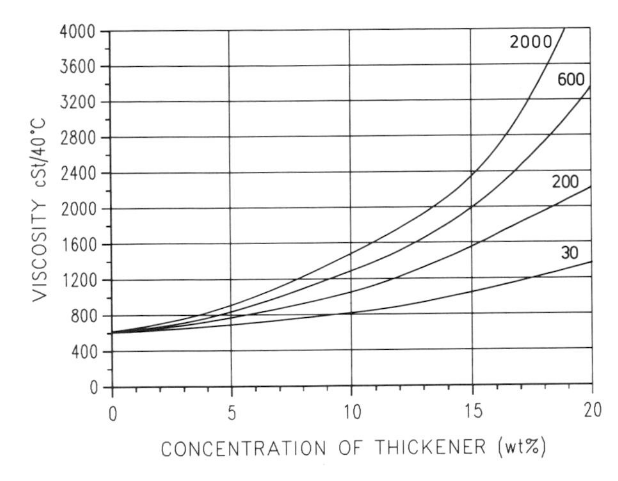

Figure 13.22 Thickening efficiency of polybutene grades 30, 200, 600, and 2000 for Brightstock 470/95 at 40°C.

Hydraulic Oils. The vast majority of hydraulic oils are based on mineral oils (48). Polybutenes find use as a viscosity adjuster and as a shear-stable viscosity index improver in the formulation of mineral oil-based hydraulic fluids. Polybutenes are especially useful where there is concern over contamination of the working area or the materials under construction by oil leakage from the hydraulic systems. This is the case, for example, with the lubricant used for the hydraulic system in an aluminum rolling plant (Section III,C,5).

The main polybutene grades of interest for use as viscosity index improvers for high-quality multigrade and energy-efficient oils are summarized in Table 13.17.

5. *Metal Working*

In metal-working applications such as cutting, rolling, stamping, drawing, and pressing, a lubricant is used to maintain a protective film between the die or tool and the metal billet. This acts to reduce the frictional heat generated during the operation. The lubricant is expected to prolong the die or tool life, reduce energy requirements, and produce a smooth, stain-free surface on the metal article.

Neat oils or soluble oils are suitable for applications where the loads between the die or tool and the metal being worked are light. High-viscosity oils are preferred as these loads increase. In very demanding or severe operations it is normal to include an extreme pressure additive in the oil. For example, fatty acid, fatty alcohol, or glycerine additives might be chosen with sulfurized oils, chlorinated compounds, phosphates, or borates being reserved for the most severe operating conditions.

If heat generation during the metal-working process is substantial, there are advantages in using the lubricant in the form of an emulsion. The larger volume of fluid that can

Table 13.17 Main Polybutene Grades of Interest for Use as Viscosity Index Improvers in Gear and Hydraulic Oils

Grade	Very light	03	04	07	3	5	10	30	150	200	600	2000
Use									X	X	X	X

be used and the higher heat capacity of water compared to that for an oil both contribute to more efficient removal of heat.

Polybutenes have a number of advantages over mineral oils when used as lubricants in certain nonferrous metal-working applications. These properties are summarized as follows.

1. Polybutenes do not contain aromatic compounds and are considered to be practically nontoxic by all routes of exposure. Many metal-working operations involve the worker in close proximity with the lubricant. With the growing concern over the safety of the workplace, the nature, composition, and hence safety implications for lubricants is being reassessed.
2. Polybutenes are aliphatic hydrocarbon polymers. They contain no sulfur or nitrogen species at detectable levels, which if present could lead to staining of the metal.
3. Polybutenes on the surface of a metal will undergo a rapid and total decomposition if the metal is annealed or treated in excess of 300°C directly after forming. The decomposition of polybutene by depolymerization of the polymer chain leaves no carbon deposits that could be a source of corrosion problems. The clean decomposition also avoids the need to clean the metal prior to brazing, welding, or additional processing.
4. Polybutenes are available with viscosities ranging from solventlike to semisolid at 25°C. There is often a polybutene of suitable viscosity to provide a lubricant of high film strength between the die or tool and metal even under adverse operating conditions.
5. Polybutenes are typically more resistant to biodegradation than mineral oils. In emulsified form polybutenes show extremely good resistance, superior to that shown by mineral oil, to microbal attack during storage.

The following sections illustrate the metal-working applications where polybutenes are used to advantage over conventional mineral oil lubricants. Throughout, particular emphasis is placed on the ability of polybutene to decompose cleanly and totally at high temperature, to avoid staining and leaving carbon deposits on the metal article.

Metal Rolling. In the rolling of aluminum sheet or foil a lubricant is used to reduce the friction between the rollers and the metal sheet. This ensures a good surface finish. In the production of high-quality sheet or foil an important consideration is that any lubricant remaining on the aluminum from the rolling process must not give rise to surface staining or blemishes during the subsequent annealing of the aluminum.

In the cold rolling of aluminum, desulfurized gas oil or odorless kerosene is often used as the lubricant. However, some manufacturers consider that the staining that results from annealing the aluminum in the presence of residual mineral oil-based lubricant is unacceptable and instead choose to use polybutene as a part or total replacement for the mineral oil. A typical lubricant used for the cold rolling of aluminum might contain 0.5–2 wt% of oleic acid, lauryl alcohol, butyl stearate, or palm oil as a lubricity additive and a polybutene with viscosity in the range 2–4 cS at 100°C (49–51). In the hot rolling of aluminum the lubricant is often used in emulsified form at 5–10 wt% solids (oil) content. As in the cold rolling process, a fatty additive can be added to improve the lubrication. Polybutenes with viscosity in the range 2–4 cS at 100°C can be used (52). However, emulsified lubricants based on polybutenes of higher viscosity such as 200 grade have been claimed for the hot rolling of aluminum (53).

In the rolling of aluminum the need to produce stain-free sheet or foil also places certain demands on the lubricants that serve the bearing and hydraulic systems in the mill (54). In normal operation there are frequent oil leaks or contamination of these systems, with the leaked oils entering the rolling oil or coming into contact with the metal sheet. If the bearing or hydraulic oils are formulated on mineral oils, this can lead to staining of the aluminum sheet on annealing. The problem of staining is overcome if solutions of polybutene in the rolling oil are used, together with the normal wear-reducing and viscosity-stabilizing additives. Oil leaks that contaminate the rolling oil or metal sheet do not then give rise to staining, due to the clean decomposition of polybutene at annealing temperatures (55–57). A more viscous lubricant with higher film strength than the rolling oil is required for the bearings that support the rollers and for the hydraulics that are used to provide the movements on the millstand. Figure 13.23 shows that by varying the quantity of polybutene added to a typical rolling oil, lubricants with viscosity suitable for the bearing and hydraulic systems can easily be manufactured. In addition to the nonstaining properties of polybutene, their performance in the bearing and hydraulic systems makes use of the excellent shear stability and good viscosity index properties that are available from the polymers. In the production of aluminum foil or deep-drawn containers that are used to wrap or contain foodstuffs the nontoxic nature of polybutene is also of importance if there is any concern over the possibility that residual lubricant could remain on the metal article (58).

In the rolling of steel, polybutenes are also used to eliminate discoloration or blemishes produced by the residual lubricant after tempering at high temperatures of 540–980°C. Polybutenes with viscosity in the range 13–225 cS at 100°C have been employed either alone or in emulsified form together with additives to improve lubrication (59–61).

Tube Drawing. Polybutenes with viscosity in the range 13–225 cS at 100°C have been shown to be suitable lubricants for copper tube drawing when used either alone or in combination with an extreme pressure additive (62). The use of polybutenes in place of mineral oil lubricants is found to give improved die life and, importantly, eliminates the

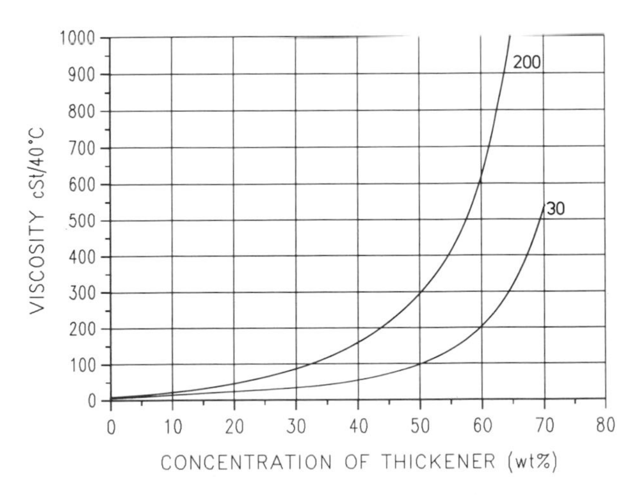

Figure 13.23 Thickening of typical rolling oil with polybutene grades 30 and 200 to provide lubricants for the bearing and hydraulic systems in aluminum rolling.

need for a cleaning stage that is required when mineral oils are used. At the annealing temperatures of 550–600°C used for copper tubing, the polybutene lubricant is removed quickly and completely from the surface of the metal (63).

Cutting Oils. As neat oils, polybutenes are only suitable for light-duty cutting operations. Problems of fuming and misting prevent their use in heavy-duty cutting applications. The polybutenes of interest for use as neat oils are generally those with viscosity in the range 4–13 cS at 100°C.

All grades of polybutenes can be emulsified to produce oil-in-water emulsions (64) and used as soluble oils for combined lubrication and cooling. Cutting fluids based on polybutene emulsions perform as well as conventional mineral oil products in terms of cutting requirements and in addition have a number of advantages. As polybutenes are not readily biodegradable, an increased service life of the emulsion coupled with less problems of unpleasant odor from lubricant breakdown can be expected (65). In addition, polybutenes are free from aromatic species and are recognized as being of very low toxicity, particularly following skin contact, which is a likely route of exposure for workers operating in close proximity to cutting fluids.

In other metal-working applications such as forging, welding, casting, and ironing, polybutene can be used in the lubricant to provide improvements in nonstaining properties, die or tool life, and lubricant service life, and can improve the safety of the lubricant in terms of toxicity (66–73).

Polybutenes can also be used in the temporary protection of metal articles while in storage or during shipment. The stable, adhesive, nonstaining, and water-resistant film of polybutene provides anticorrosion protection and can easily be removed if necessary by conventional solvent degreasing or by heating the metal piece to high temperature.

All grades of polybutene can be used either alone or in combination with other oils in metal working applications as shown in Table 13.18.

6. Polybutenes in Grease Manufacture

A grease is a stable mixture of a base oil and a thickening agent. The manufacturing process and the quantity, dimensions, and distribution of the thickener in the oil largely dictate the nature and properties of the finished grease. As with lubricating oils, antioxidants, pour-point depressants, rust inhibitors, antiwear additives, and other advantageous substances can be incorporated in the grease to enhance its properties or performance.

The use of a grease is favored where the lubricant is required to remain in position after application to a mechanism; if a lubricating oil were applied it would be lost through dripping or being flung during operation. The use of a grease is also preferred if physical opportunities to reintroduce a lubricant are limited by access or if the economics of relubrication are prohibitive. A grease is expected to perform all the functions of a lubricating oil with the exception perhaps of efficient cooling and cleaning of the contacting surfaces. A grease may often be expected to function as a sealant against dirt, water, or other contaminants and be the primary means of preventing corrosion (74).

Table 13.18 Main Polybutene Grades of Interest for Metal-Working Applications

Grade	Very light	03	04	07	3	5	10	30	150	200	600	2000
Use	X	X	X	X	X	X	X	X	X	X	X	X

Polybutenes have particular properties as base oils and as modifiers of conventional mineral oils that can help achieve the levels of performance demanded by automotive, industrial, and specialist greases. Table 13.19 gives an overview of these properties and lists the areas of current and potential interest for polybutene greases.

Polybutenes of low viscosity in the range 4–103 cS at 100°C can be used as the base oil for the manufacture of a grease. Traditional metal soaps such as aluminum, calcium and lithium stearates, bentonite clay, and silica thickeners have been used to produce polybutene greases (75). Some examples of these greases are given in Table 13.20. The results shown are used to provide a general guide to the physical properties and performance levels that can be expected. The composition and manufacture of the polybutene greases illustrated have not been optimized and performance-boosting additives have not been included. Polybutenes do not oxidize readily, but their resistance to oxidation can be improved by the addition of a proprietry antioxidant. Addition of antioxidants to polybutene greases is recommended if the grease is to be used at temperatures in excess of 100°C.

Polybutenes of medium to high viscosity in the range 4,200–12,200 cS at 100°C can be used in conjunction with a conventional mineral oil to produce a blended oil from which a grease can be manufactured (76). The polybutene is used to adjust the viscosity

Table 13.19 Important Polybutene Properties for Grease Manufacture

Property	Polybutene character	Applications
Pour point	Low for low-viscosity polybutenes. Improved performance over mineral oil bases under low-temperature conditions.	Low-temperature greases for automotive and industrial bearings and pinions.
Toxicity	No detectable aromatic content and practically nontoxic.	Incidental food contact greases.
Color	White or transparent when thickened with light-colored thickening agent.	Food or medical greases.
Water resistance	Hydrophobic with good water resistance and sealant properties for corrosion protection.	Greases for bottling and canning lines. Screw thread ubricants.
Inertness	Noncorrosive and no effect on nitrile, neoprene, or Viton rubber seals. Good chemical and physical stability. Permanently nondrying and contain no detectable sulfur or nitrogen.	Corrosion protection. Anti-seize pastes. Wire rope lubricants. Electrical greases.
Viscosity	Wide range to modify adhesiveness and temperature response of conventional greases. Excellent product uniformity.	Heavy-duty gears operated over a wide temperature range.
Volatility	Depolymerize to volatile compounds above 275°C without residue formation or staining.	Graphite or molybdenum disulfide suspensions for conveyor chains and roller bearings on ovens and kilns.

Table 13.20 Physical and Performance Characteristics of Some Polybutene-Based Greases

Polybutene grade	Thickening agent compound	wt%	Color	Drop point °C (ASTM D-566)	Worked penetration, mm/10 (ASTM D-127)	NGLI number	Oil separation, 164 h/40°C (IP121)	Water washout, % wt (IP215/ASTM D-1264)	Four-Ball wear test, mm (ASTM D-2266)
04	Al distearate	10	Clear	105	281	2	7.20	7.0[a]	0.53
07	Al distearate	10	Clear	108	313	1	4.25	4.0[a]	0.48
07	Li 12-OH stearate	11	White	197	267	2	0.95	6.0[b]	0.32
3	Bentone 34·MeOH	5	Light brown	>230	293	2	0.5	<1.0[b]	0.44

[a]IP 215 method, 38°C.
[b]ASTM D-1264 method, 79°C.

and viscosity index of the mineral oil to modify and improve the properties of the finished grease. An example of a polybutene 200 grade being used to adjust the viscosity of a white oil is shown in Fig. 13.24.

A number of advantages over grease manufacture from straight mineral oil have been claimed for grease manufactured from blends of polybutene with mineral oil. These advantages relate to improvements in the viscosity/temperature properties, in the adhesive and cohesive strength, in reduced oil bleeding (77–82), in the physical and chemical integrity (83, 84), and in the low-temperature properties of the grease. In addition, polybutenes can be used to manufacture greases for industries such as those involved in food processing (85) or in pharmaceutical production that require the lubricant to have incidental food contact approval. Being water-white, polymers containing no detectable aromatic species, polybutenes produce in conjunction with light colored thickening agents white or even transparent greases that are well placed to meet the exacting standards demanded in such industries.

Polybutenes can also be used in high-temperature specialty greases as carriers for suspensions of solid lubricants such as graphite or molybdenum disulfide. These types of lubricants are normally applied to conveyor chains and roller bearings or used as release agents in the steel and glass industries. At the normal high temperature of operation in these applications, the polybutene decomposes by depolymerization of the polymer chain to leave the solid lubricant in position. Polybutenes, in contrast to mineral oils, do not form deposits on thermal decomposition that could give rise to problems with wear, seizure, and corrosion.

The main polybutene grades of interest for use in the manufacture of greases are shown in Table 13.21.

7. Polybutenes as Wire Rope Lubricants

Wire ropes are used in a variety of industrial applications including cranes, bridge construction, elevators, mine haulage, guy ropes, and drilling. In its simplest form a wire rope is made up of a collection of wires twisted together. In practice, wire ropes are produced to sophisticated designs to meet the performance requirements demanded of the

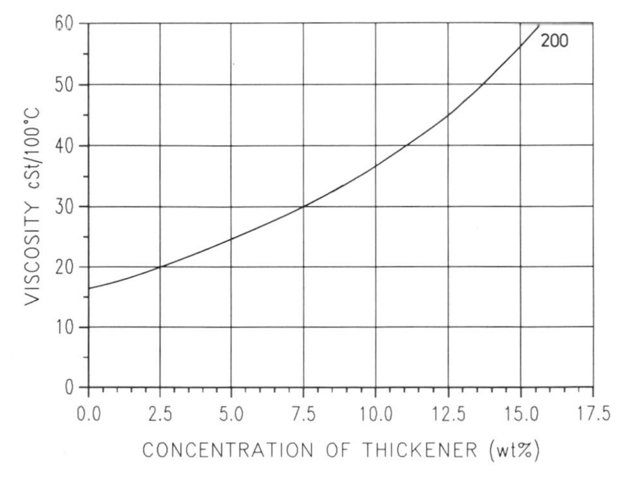

Figure 13.24 Viscosity adjustment of a white oil using polybutene grade 200 to provide a blended base oil for grease manufacture.

Table 13.21 Main Polybutene Grades of Interest for Grease Manufacture

Grade	Very light	03	04	07	3	5	10	30	150	200	600	2000
Use			X	X	X	X				X	X	

rope in different end applications. The two major designs for a wire rope are the strand rope and the locked coil rope. In a strand rope, individual wires are wound into strands, which are then twisted around a central core. In locked coil ropes the individual wires are wound together and special shaped wires in the outer layers interlock to enclose the rope structure. The locked coil types are characterized as having exceptional strength for a given rope thickness.

When in use a wire rope flexes and the wires and strands in its construction rub together. Also, wire ropes are often used in corrosive environments such as dockside cranes or subsea cables. Therefore a wire rope lubricant must not only minimize internal wear between the wires and strands and external wear between the rope and pulley or drum arrangement but must also provide the rope with a high level of corrosion protection.

The application of the wire rope lubricant is normally undertaken during the spinning or construction of the rope. Lubrication at the manufacturing stage is especially important for locked coil ropes, as their design makes relubrication difficult. The central core of the rope can also be impregnated with lubricant to provide a reservoir of lubricant to maintain the rope while in use. It is also common to relubricate a rope at regular intervals during its service life.

Polybutenes exhibit the physical characteristics required of a wire rope lubricant. They are inert, nondrying, and adhesive materials that provide effective lubrication and corrosion protection for the wire rope. The medium- to low-viscosity polybutenes appear to have the best combination of properties for lubrication. However, the use of a high-viscosity polybutene has also been claimed in a lubricant formulation for a locked coil rope (86). Paraffinic, microcrystalline, and polymer wax can be used in a blend with polybutene to give control over the drop point of the lubricant. Fatigue and friction testing of wire rope lubricated with polybutenes, polybutene/wax blends, mineral oil, and bitumen-based lubricants has been carried out. The results shown in Table 13.22 indicate that polybutene and polybutene/wax lubricants offer an improved performance in rope life and give considerably higher (30% plus) coefficients of friction than mineral oil-derived products. The high coefficient of friction is thought to be related to the branched structure of the polybutene polymer (87). A high coefficient of friction or traction is desirable for applications that rely on friction to drive the ropes. Polybutene lubricants have therefore been claimed for use in applications such as friction-driven elevator ropes (88).

For general-purpose wire rope lubrication, bitumen, waxes, and high-viscosity mineral oils are most often used. However, in higher performance applications such as in the lubrication of ropes for bridges, subsea cables, and elevator ropes, polybutenes are finding use. In addition, polybutenes are well positioned to replace bitumen or mineral oils in wire rope applications now demanding safer, nontoxic lubricants.

The polybutene grades of interest for use as wire rope lubricants are summarized in Table 13.23.

Table 13.22 Wire Rope Fatigue, Friction, and Corrosion Protection Data for Various Wire Rope Lubricants

Lubricant base	Relative[c] fatigue life	Friction tests[d]		Salt spray[e] (ASTM B-117)
		Wear area, mm^2	Coefficient of friction	
Polybutene[a]	263–303	0.07–0.16	0.10–0.12	—
Polybutene/wax[b]	188–230	0.22–0.27	0.09–0.11	Pass
Mineral oil	163	0.07	0.07	—
Bitumen	—	4.12	0.54	Pass

[a]Polybutenes grades 5 and 10.
[b]Polybutene content in excess of 90%.
[c]For 3-mm Rope, 1 cycle/second, reverse bend under load. Dry rope life = 100.
[d]Cameron-Plint machine. Load 30 N, reciprocation rate 50 Hz, duration 2.5×10^5 cycles, stroke length 2.3 m.
[e]Mild steel panel, dip coated 20°C above drop point, 720 h at 35°C, no corrosion for Pass.

IV. MANUFACTURE AND ECONOMICS

A. Manufacturers and Capacities

The main centers for the production of polybutenes are in the United States and Europe. Production units also exist in Japan—Idemitsu Petrochemical Co., Nippon Petrochemical Co. (Nisseki), and Nippon Oil and Fats Co. (Nichiya)—and in South America—Polybutenos Argentenos (Argentina) and Polybutinos Industrios Quimicas (Brazil). Elsewhere there are small polybutene production facilities in the USSR, China, and India. In 1990 there was an estimated worldwide nameplate capacity for the production of polybutene of around 640 kTes. However, the nameplate capacity is based on the manufacturer producing a single grade of polybutene of normally around 1,000 molecular weight. The actual production figure would therefore be considerably less for reasons of plant operation, plant reliability, and the manufacture of a range of grades of polybutene. During the period 1984–1990 new production capacity for polybutene came on stream but supply remained constant as inefficient and aged plants were shut down by Cosden, Lubrizol, Chevron, and Petrofina in the United States and Canada together with the closure of the Amoco-Fina plant in Belgium in early 1989. For the period 1990–1993 production capacity is expected to expand modestly. This is shown in Table 13.24.

In Europe BP Chemicals produces polybutenes on two production sites, at Grange-

Table 13.23 Main Polybutene Grades of Interest for Wire Rope Lubricants

Grade	Very light	03	04	07	3	5	10	30	150	200	600	2000
Use						X	X	X				X

Table 13.24 Estimated World wide Polybutenes Nameplate Capacity (kTes), 1984–1993

Location	1984	1987	1990	1993
North America	380	345	370	370
Europe	195	195	180	180
Japan	30	30	30	30
South America	10	10	25	25
Others	25	30	35	40
Total	640	610	640	645

mouth in Scotland and Lavera in the south of France. Exxon produces at Koln in Germany and Lubrizol produces at Le Havre in France. The polybutene production of Lubrizol and the majority of the production of Exxon are used captively for the production of lubricant and fuel additives. Exxon markets a limited number of polybutene grades under the trade name Parapol. BP Chemicals is the major Western European producer of polybutenes and is committed to supplying the open market. It offers a wide range of standard polybutene grades of molecular weight 200–6,000 under the trade names Hyvis and Napvis and has recently commercialized a new, highly reactive polybutene tradenamed Ultravis. Ultravis differs from conventional polybutenes in that it has a much higher proportion of terminal vinylidene double bonds.

In the United States, Lubrizol at Deer Park, Texas and Exxon at Bayway, New Jersey and Baytown, Texas manufacture polybutenes. As in Europe, the production by Lubrizol and the majority of production by Exxon are for use in-house. Amoco Chemicals is the largest worldwide producer of polybutenes and manufactures on two production sites at Texas City, Texas and Whiting, Indiana (see Table 13.25).

At Whiting, Amoco also have the capability to produce MTBE from isobutylene feedstock. Amoco offer a wide range of polybutene grades on the open market under the tradename Indopol.

Table 13.25 Estimated Nameplate Capacities of Manufacturers in Europe and the United States

	1990 Capacity (kTes)
Europe	
BP Chemicals, UK and France	100
Exxon, Germany	45
Lubrizol, France	35
Total	180
United States	
Amoco	200
Exxon	95
Lubrizol	75
Total	370

Polyisobutylene rubbers (ultra-high-molecular-weight polyisobutylenes) are available from BASF under the tradename Oppanol and from Exxon under the tradename Vistanex.

B. Price

It is difficult to be definitive about pricing. At the present time it is fair to say that polybutenes are less expensive than other synthetic oils such as polyalphaolefins and esters. The information detailed in Table 13.26 has been proposed as a rough guide to the pricing of synthetic oils relative to mineral oil (89, 90). On this basis polybutene would be priced in the range 3–4 depending on volume and grade of material requested.

V. OUTLOOK

The use of polybutene in oils and lubricant and related applications is expected to show modest growth over the next few years. Areas where growth is anticipated are in transport fuel additives, clean burning two-stroke oils, specialist metal-working and grease applications, and in energy efficient oils for industrial and automotive uses. In these applications the special performance characteristics of polybutene will continue to produce a cost/performance advantage over the use of other types of oils. In addition, with lubricants now under greater scrutiny for evidence of toxicity or hazard, formulators will increasingly seek safe, cost-effective starting materials for use as components of lubricants. This should result in synthetics such as polybutene being investigated in greater detail in future years.

In the longer term, as synthetic oils become more widely accepted there will be the opportunity for polybutene to be used in synergy with other synthetic oils. For example, this is already being seen with the early stages of development of polybutene/ester blends for specialist two-stroke lubricants. These blended oils aim to combine the clean-burning properties of polybutene with the superior lubricity characteristics of esters to produce a high-performance lubricant. For other applications polybutenes may find use as a thickener, viscosity adjuster, or cost-saving diluent for more expensive synthetic oils in high-performance lubricant applications.

ACKNOWLEDGMENTS

I would like to thank BP Chemicals, Polybutenes Business Management for permission to publish this chapter. My thanks extends to many colleagues, present and past, who

Table 13.26 Guide to the Relative Cost of Synthetic Oils Compared with Mineral Oil

Oil	Relative cost
Mineral oil	1
Polybutene	3–4
Polyalphaolefin	5
Diester	5
Polyalklene glycol	4–5
Phosphate ester	8
Silicon fluids	10–50

contributed in different ways to the work presented. I would also like to thank Miss M. Martin for her considerable assistance in the preparation of the chapter manuscript.

REFERENCES

1. Mark, D., and A. R. Orr (1956). *Petroleum Refiner*, 35, 7.
2. Otto, M., and M. Muller-Cunradi (1933). GP 641,284 to I. G. Fabrin.
3. Standard Oil (New Jersey) (1943). U.S. Patent 2,311,567.
4. California Research Corporation (1946). U.S. Patent 2,484,384.
5. Standard Oil (Indiana) (1960). U.S. Patent 2,677,002.
6. Cosden Petroleum Corporation (1960). U.S. Patent 2,957,930.
7. Kennedy, J. P., and E. Marechal (1982). *Carbocationic Polymerisation*, Chapter 10, Wiley and Sons, New York.
8. Frederickson, M. J., and A. J. Simpson (1980). In *The C4 Hydrocarbons and Their Industrial Derivatives*, Chapter 19, ed. E. Hancock, Ernest Benn.
9. Cosden Petroleum Corporation (1964). U.S. Patent 3,119,884.
10. Standard Oil (Indiana) (1961). U.S. Patent 3,121,125.
11. Mills, G. A., and E. E. Ecklund (1989). *Chemtech*, 19, 626.
12. Low Molecular Weight "Oppanol," BASF.
13. I. D. Rubin (1968). *Poly (1-Butene)—Its Preparation and Properties*, Macdonald Technical and Scientific.
14. Ornite Chemical Company Industrial Technical Bulletin Booklet, circa 1947.
15. Cosden Petroleum Booklet, circa 1960.
16. Herbstman, S., and K. Virk (1990). *Chemtech*, 20, 243.
17. *Petroleum Refiner* (1959). 38, 6.
18. Guterbock, H. (1959). *Polyisobutylen*, Springer-Verlag, Berlin.
19a. Kennedy, J. P. (1975). *Cationic Polymerisation of Olefins*, Wiley Interscience, New York.
19b. Russell, K., and G. Wilson (1977). *Polymerisation Processes*, Chapter 10.
19c. *Encyclopedia of Polymer Science and Engineering* (1988). vol. 2, p. 729, John Wiley and Sons, New York.
19d. Nuyken, O., and S. D. Pask (1989). In *Comprehensive Polymer Science*, p. 619, Pergamon Press, New York.
19e. Sauvet, G., and P. Sigwalt (1989). In *Comprehensive Polymer Science*, p. 579, Pergamon Press, New York.
20a. *European Chemical News* (1990). 17/24, 12.
20b. BP Chemicals Press Release Ref. 42/BP/125.
20c. *The Chemical Engineer* (1990). 13/12.
21a. Labrofina (1976). Patent 1,513,853.
21b. Cosden Petroleum Corporation (1961). U.S. Patent 3,100,808.
22a. Lubrizol. U.S. Patent 3,252,908.
22b. Esso. French Patent 1,446,344.
23. Chisso Corporation (1968). Japan Patent 13,218.
24a. Esso (1958). Great Britain Patent 847,339.
24b. Esso (1957). Great Britain Patent 838,928.
25. Monsanto. U.S. Patent 3,429,936.
26. Internal BP Chemicals Correspondence regarding Hyvis/Napvis Polybutenes.
27. BP Chemicals Technigram PB102 (1990). *Polybutenes—Health, Safety and Environmental Information*, Ref. 197/500.
28. BP Chemicals Technigram PB103 (1990). *Polybutenes—Food Contact Status*, Ref. 198/500.
29. Internal BP Chemicals Correspondence.
30. Higgins, I. J., and P. D. Gilbert (1978). In *The Biodegradation of Hydrocarbons*, ed. H. J. Somerville, Heyden, London.

31. College of Petroleum Studies (1990). Synthetic Lubricants Course Code SP5, Oxford.
32. Souillard, G. J., F. Van Quaethoven, and R. B. Dryer (1971). SAE Paper 710,730.
33. Fog, D. A., R. M. Brown, and D. H. Garland (1989). *I. Mech. Eng.*, C372/040.
34. Yashiro, Y. (1987). SAE Paper 871,216.
35. Kovacs, A., and D. Olagos (1990). *J. Synth. Lub.*, 7, 47.
36. Sugiura, K., and M. Kagaya (1977). SAE Paper 770,623.
37. Labrofina (1965). Great Britain Patent 1,162,157.
38. British Petroleum Company (1969). Great Britain Patent 1,287,579.
39. SFBP (1968). French Patent 1,597,015.
40. Labrofina (1972). Great Britain Patent, 1,340,804.
41. Kenbeek, D., and G. Van der Waal (1988). *J. Synth. Lub.*, 5, 215.
42. Unichema International (1987). Technical Bulletin 2k/5.87.
43. Wikelski, K. W. (1981). *J.A.S. Lub. Eng.*, 37, 203.
44. Sittig, M., ed. *Chemical Technology Review* number 70, p. 48, Noyes Data Corporation.
45. Chevron Chemical Company. Technical Data Sheet, *Chevron Polybutenes as Lubricants.*
46. O'Connor, B. M., and Ross, A. R. (1989). *J. Synth. Lub.*, 6, 31.
47. BP Chemicals and Lubrizol Products.
48. College of Petroleum Studies (1990). Synthetic Lubricants Course SP5 Section 1.2, Oxford.
49. Chisso Co. Ltd. Japan Patent 15,387.
50. Standard Oil. U.S. Patent 2,899,390.
51. Bucsi, W. G. (1990). *J.A.S. Lub. Eng.*, 46, 186.
52. Swiss Aluminium. European Patent Application 0048216.
53. Agnum, A. I. Great Britain Patient, 1,052,652.
54. Schimon, W. (1984). *J.A.S. Lub. Eng.*, 40, 471.
55. Aluminium Laboratories. Great Britain 964,268.
56. Guminski, R. D. U.S. Patent 3,298.951.
57. Swiss Aluminium Ltd. U.S. Patent 4,488,979.
58. Swiss Aluminium Ltd. U.S. Patent 4,228,217.
59. Esso. Netherlands Application 6503934.
60. Standard Oil Co. Great Britain Patent 1,109,304.
61. W. R. Grace and Co. European Patent Application 0375412.
62. Akzo Chemicals. Technical Bulletin G-4/1, *Chemicals for Metalworking.*
63. Armour Industrial Chemical Company. U.S. Patent 2,990,943.
64. BP Chemicals (1989). Technigram PB107, *Polybutene Emulsions*, Ref. 215/2000.
65. Toho Chemical Industries Ltd. Japan Patent 184,375.
66. Shell. U.S. Patent 2,621,159.
67. Comp. Gen. Du Duralumin. Belgian Patent 678,683.
68. Soc. de Prod. Chem. et de Synthese. Great Britain Patent 1,060,114.
69. Smallman Lubricants Ltd. Great Britain Patent 2,185,996.
70. National Distillers and Chemicals Co. European Patent Application 0206280.
71. Standard Oil Company. U.S. Patent 3,397,734.
72. Nalco Chemical Co,. U.S. Patent 4,260,502.
73. Van Straaten Corp. U.S. Patent 4,758,358.
74. SAE Information Report J310a on Automotive Lubricating Grease.
75. BP Chemicals (1989). Technigram PB20, *Polybutene Greases*, Ref. 218/500.
76. Labofina S. A. Belgian Patent 779,526.
77. Cities Service Oil Co. U.S. Patent 3,663,726.
78. Cato Oil and Grease Co., Inc. French Patent 2,045,577.
79. Chem. Fabrik Rhenus. German Patent 1,955,951.
80. Chevron Res. Co. U.S. Patent 3,472.770.
81. Cato Oil and Grease Co. CA 936,353.
82. Southwest Petroleum Chemicals Inc. U.S. Patent 379,972.

83. Suwa Sickosha. Japan Patent 7,232,873.
84. Esso Research and Engineering Co. Great Britain Patent 853,751.
85. Birko Corporation. U.S. Patent 4,828,727.
86. Shell International Research. Great Britain Patent 2,095,696.
87. Muraki, M. (1987). *Tribol. Int.*, 20, 347.
88. Mitsubishi Denki Kabushiki Kasisha. Great Britain Patent 2,118,195.
89. *Industrial Lubrication and Tribology* (1989). Nov./Dec., 17.
90. College of Petroleum Studies (1989). Synthetic Lubricants Course SP5, Section 3, Oxford.

14
Comparison of Synthetic Fluids

Wilfried J. Bartz
Technische Akademie Esslingen
Ostfildern, Germany

I. INTRODUCTION

The conditions of the special frictional contacts as well as those of the environment have to be taken into account for the application of lubricants. Both influencing factors result in the total complex of requirements to which the lubricants are exposed and that have to be met in considering an adequate service life. Ranges and limitations of application of lubricants are controlled by the extent and importance of the single requirements. If a single requirement exceeds the limits of application of a certain fluid, its use will be generally questionable, regardless of whether all other requirements might be met easily.

II. REASONS FOR CHOOSING SYNTHETIC FLUIDS

Regarding the limits of application of fluids, physical and chemical effects have to be distinguished. The physical effects are characterized by temperature and pressure. Figure 14.1 shows temperature and pressure ranges for certain fields of application for lubricants and operational fluids. Both factors particularly control the so-called "liquid range." The solidification of a fluid at low temperatures and high pressures limits its application, as does its evaporation at high temperatures and low pressures.

The chemical effects are characterized by oxidation and radiation influences, both affected by temperature. Figure 14.2 shows the oxidation, radiation, and temperature ranges for some fields of application for lubricants and operational fluids.

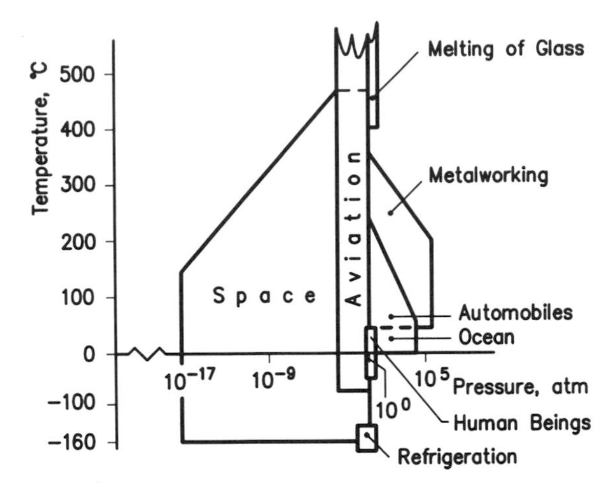

Figure 14.1 Temperature and pressure ranges for several applications of lubricants and operational fluids.

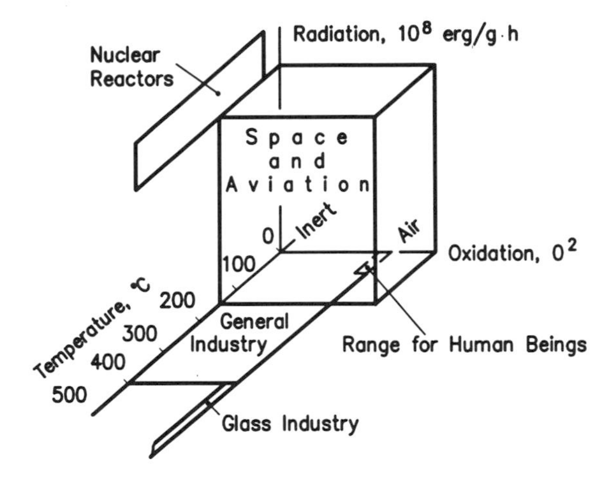

Figure 14.2 Oxidation, radiation, and temperature ranges for several applications of lubricants and operational fluids.

We cannot expect one special or single fluid, such as a mineral oil, to meet the ranges of requirements shown in Figs. 14.1 and 14.2. Therefore, it becomes a technical necessity to use synthetic fluids as lubricants and operational fluids.

There are two reasons for a decision in favor of a synthetic fluid instead of a mineral oil:

1. The desired or required property cannot be obtained by mineral oils, even those containing additives.
2. The extent of the desired or required property cannot be obtained by mineral oils, even those containing additives.

Mineral oils can be inferior to certain synthetic fluids regarding the following properties or property complexes:

Thermal stability
Oxidation stability
Viscosity temperature behavior
Flow behavior at low temperatures
Volatility at high temperatures
Temperature range of application
Radiation stability
Ignition resistance

No synthetic fluid combines all properties in a superior manner to those of mineral oils. In general, some inferior properties have to be taken into account. Possibly existing inferior properties of synthetic fluids compared with mineral oils may include:

Hydrolytical behavior
Corrosion behavior
Toxicological behavior
Compatibility with other design materials
Miscibility with mineral oil
Compatibility with seal materials
Additive solubility
Availability, in general or in certain viscosity grades
Price

Again, it must be stated that no synthetic fluid combines all properties in an inferior manner to those of mineral oils.

This results in the fact that some synthetic fluids might exhibit inferior performance in some specific applications while they have superior performance in other specific areas.

Therefore, the following point of view must also be considered. Certain advantageous properties may be linked inevitably with less sufficient properties. For instance, the behavior under mixed film lubrication conditions is strongly controlled by mutual interactions between the lubricant molecules and the surfaces of the frictional contact, characterized by the chemical and physical reactivity capability of the fluid. But many synthetic fluids stand out for good chemical and physical stability resulting in good oxidation and thermal stability. Rather moderate or even poor behavior at mixed film lubrication conditions can be the result.

III. CLASSIFICATION OF SYNTHETIC FLUIDS

One possibility for the classification of synthetic fluids is their distinction regarding their chemical composition. As Table 14.1 reveals, the following basic types can be distinguished (1):

C, H
C, H, O
C, H, O, Si
C, H, O, P
C, F, O
C, F, Cl

Another criterion of classification, especially for the distinction between mineral oils and synthetic fluids, might be the production process (1). Mineral oils are produced by

Table 14.1 Classification of Synthetic Fluids According to Chemical Composition

Synthetic fluids	Composition
Synthetic hydrocarbons	C, H
Polyalphaolefins	
Alkylated aromatics	
Monoalkylbenzenes	
Dialkylbenzenes	
Polyalkylene glycols	C, H, O
Carboxylic acid esters	C, H, O
Dicarboxylic acid esters	
Neopentyl polyesters	
Phosphoric acid esters (phosphate esters)	C, H, O, P
Silicone oils	C, H, O, Si
Silicone oils	
Polysilicone oils (siloxanes)	
Silicate esters	
Polyphenylether	C, H, O
Polyfluoroalkylether (alkoxyfluoro oils)	C, F, O
Chlorofluorocarbons	C, F, Cl
Chlorotrifluoroethylenes	
Polymethacrylate/polyalphaolefine-cooligomers	C, H, O

distillation and raffination, whereas synthetic fluids are obtained by chemical reactions (Table 14.2).

A third possibility for the systematic classification of synthetic fluids is based on their chemical structure (2) (Table 14.3).

IV. ADVANTAGES AND DISADVANTAGES OF SYNTHETIC FLUIDS

The advantages and disadvantages of some important synthetic fluids are summarized in Tables 14.4–14.12.

Table 14.2 Classification of Mineral Oils and Synthetic Fluids According to Production Process

Mineral oils, produced by distillation and refining		Synthetic fluids, produced by chemical reactions	
Conventional technologies	Modern technologies	Synthetic hydrocarbons	Other synthetic fluids
Acid refining	Hydrotreating	Polyalphaolefins	Dicarboxylic esters
Solvent extraction	Hydrocracking	Polyisobutenes	Neopentyl polyesters
Dewaxing		Dialkylbenzenes	Polyalkylene glycols
			Phosphate esters
			Silicone oils
			Polyphenylether
			Perfluoroalkylether
			Chlorofluoroalkylether
			PAMA/PAO-Cooligomers

Table 14.3 Classification of Synthetic Fluids According to Chemical Structure

1. Synthetic hydrocarbons
   - 1.1. Polymers (of olefins)
     - 1.1.1. Ethylene polymers
     - 1.1.2. Propylene polymers
     - 1.1.3. Polybutenes (e.g., polyisobutenes)
     - 1.1.4. Polymers of higher olefins (e.g., polyalphaolefins)
   - 1.2. Chlorinated hydrocarbons
   - 1.3. Condensation products (of aromates with olefines) (e.g., polyalkylaromates, dialkylaromates)
   - 1.4. Hydrocracked oils

2. Polyether oils
   - 2.1. Aliphatic polyether (e.g., polyalkylene glycols)
   - 2.2. Perfluoroalkylether
   - 2.3. Polythioether
   - 2.4. Polyphenylether

3. Esters, carboxylic acid esters
   - 3.1. Ordinary and complex esters (e.g., dicarboxylic acid esters, complex esters)
   - 3.2. Esters of neopentylpolyols (e.g., neopentyl polyesters)
   - 3.3. Fluorine-containing esters (e.g., fluoroesters, perfluorodialkylether)

4. Phosphoric acid esters, phosphate esters

5. Oils containing silicon
   - 5.1. Silicone oils, siloxanes
   - 5.2. Silicic acid esters (e.g., silicate esters)
   - 5.3. Silahydrocarbons (e.g., tetralkylsilanes)

6. Halogen hydrocarbons, halogen carbons
   - 6.1. Chlorohydrocarbons (e.g., chlorodiphenyls, Polychloroparaffins)
   - 6.2. Aliphatic fluoro- and chlorofluorocarbons (e.g., chlorotrifluoroethylene)
   - 6.3. Hexafluorobenzene

7. Others
   - 7.1. Ferrocene derivatives
   - 7.2. Aromatic amines (e.g., triarylamines)
   - 7.3. Heterocyclic N, B, and P compositions
   - 7.4. Urea derivatives
   - 7.5. PAMA/PAO-Cooligomers

1. Synthetic hydrocarbons.
   Polyisobutenes (Table 14.4)
   Polyalphaolefins (Table 14.5)
2. Polyether oils.
   Polyalkylene glycols (Table 14.6)
   Perfluoroalkylether (Table 14.7)
   Polyphenylether (Table 14.8)
3. Ester oils, carboxylic acid esters.
   Diesters and polyesters (Table 14.9)

Table 14.4 Advantages and Disadvantages of Polyisobutenes

Advantages
Available in many viscosity grades
Good corrosion behavior
Nontoxic
Clean burning without residues
Good lubricating properties
Miscible in mineral oils and in synthetic hydrocarbons
Disadvantages
Moderate oxidation stability
Higher volatility
Moderate cold flow behavior
Poor viscosity temperature behavior

4. Phosphate esters (Table 14.10)
5. Oils containing silicone
   Silicone oils (Table 14.11)
6. PAMA/PAO-Cooligomers (Table 14.12)

Table 14.5 Advantages and Disadvantages of Polyalphaolefins

Advantages
Available in many viscosity grades
Very good cold flow behavior, pour points <–60°C
Very low volatility up to +160°C, even at low viscosities
High oxidative and thermal stability with oxidation inhibitors
Very good viscosity temperature behavior
Compatible with mineral-oil-resistant paints
Higher viscous types compatible with mineral-oil-resistant seal materials
Good friction behavior
Unlimited miscibility with mineral oils and esters
Good hydrolytic stability
Good corrosion behavior
No toxic potential owing to absence of aromatic contained in mineral oils
Moderate costs
Disadvantages
Low viscous types have moderate compatibility with a lot of seal materials; compatible with fluorine rubber (FPM) materials
Scuffing and wear protection properties less good than mineral oils, polyglycols, and esters
Moderate solubility of extreme pressure and antiwear additives
Biological degradation moderate for low-viscosity grades—poor for higher-viscosity fluids

Table 14.6 Advantages and Disadvantages of Polyalkylene glycols

Advantages
Producible in all viscosity grades desired
Very good viscosity temperature behavior, high viscosity index
High load-carrying capacity (i.e., good scuffing and wear-protecting properties)
Excellent friction behavior, especially with steel/phosphor bronze contacts
With inhibitors, very good oxidation stability
High service temperatures up to 250°C
Good cold flow properties, low pour point
Good corrosion behavior
Nontoxic
Unbranched polymers with molecular weights up to 1,500
Biologically degradable
Disadvantages
In general not miscible with mineral oils, esters, and synthetic hydrocarbons (but there are some more expensive types miscible with mineral oils)
Moderate solubility and response for additives
Less pronounced viscosity pressure behavior compared to mineral oils
Fire resistant only in water solutions
Compatible only with paints based on epoxy resin and polyurea
Compatible only with seal materials based on fluorine rubber (FPM) and polytetraethylene (PTFE) materials; limited compatibility with acryle butadiene rubber (NBR) and methyl vinyl silicone rubber (MVQ) materials

Table 14.7 Advantages and Disadvantages of Perfluoroalkylethers

Advantages
Extraordinary high thermal and oxidative stability
Highest chemical stability of all lubricating oils
Very wide service temperature range
Very low volatility
Very good cold flow behavior
Compatible with seal materials, plastics, and paints
Fire resistant
High radiation stability
Good wear and scuffing protecting behavior
Low surface tension; good wetting properties
Disadvantages
Moderate viscosity temperature behavior
Low corrosion protection
No solubility for additives
Not miscible with any other oil
Nontoxic up to decomposition temperature (280–350°C); at higher temperatures toxic decomposition vapors are formed
Extraordinarily high costs

Table 14.8 Advantages and Disadvantages of Polyphenylethers

Advantages
Highest thermal and oxidative stability of all lubricating oils
Highest stability against high-energy radiation
High chemical stability; highest stability against acids
Excellent lubricating properties, even at mixed film conditions
Low volatility
Excellent hydrolytical stability
Good miscibility for mineral oils and additives
Disadvantages
Available only in limited viscosity grades
Poorest cold flow behavior of all lubricating oils
Poorest viscosity temperature behavior (negative VI values)
Moderate compatibility with paints
Moderate compatibility with seal materials
Not miscible with perfluoroalkylethers, silicone oils, and polyalkylene glycols
Moderate corrosion protection properties
High costs

V. COMPARISON OF CHEMICAL, PHYSICAL, AND TECHNOLOGICAL PROPERTIES

An overall review of several chemical and physical properties and some selected technological properties (e.g., indications regarding toxicology, behavior against seal materials and metals, and stability against chemicals) is summarized in Table 14.12 (3).

Table 14.9 Advantages and Disadvantages of Diesters and Polyolesters

Advantages
Containing inhibitors: better oxidative and thermal stability compared with mineral oils
High service temperatures
Low pour point, good cold flow behavior
Good viscosity temperature behavior
Unlimited miscibility with mineral oils and most other synthetic oils
Good wear and scuffing protection and friction behavior
Low volatility
Nontoxic
Many types biologically degradable
Mean costs
Disadvantages
Available only in low viscosity grades
Problematical compatibility with seal materials; compatible only with FPM, PTFE, and MFQ materials
No compatibility with paints
Poor hydrolytical stability
Moderate corrosion protection

Table 14.10 Advantages and Disadvantages of Phosphoric Acid Esters (Phosphate Esters)

Advantages
Fire resistant
Containing inhibitors: good oxidation stability
Good cold flow behavior
Excellent scuffing and wear protection, good friction behavior
Good radiation stability
Triarylester types not toxic
Biologically degradable
Disadvantages
Poor viscosity temperature behavior
Poor hydrolytical stability
Moderate corrosion protection
Compatible only with FPM seal materials
Not miscible with mineral oil

Table 14.11a Advantages and Disadvantages of Silicone Oils

Advantages
Available in many viscosity grades
Best viscosity temperature behavior of all lubricating oils
Very good oxidative and thermal stability
Excellent cold flow behavior
Low volatility, even at low viscosities
High flash points
Compatible with seal materials, plastics, and paints
Corrosion protection properties and hydrolytical stability similar to mineral oil
Good cold flow properties
High chemical stability
Good electrical properties, such as high specific resistance and high dielectric strength (electronic insulation value)
Water insoluble
Disadvantages
Low surface tension, good wetting capacity
Very poor lubricating properties at mixed film conditions
Lowest load-carrying capacity (wear and scuffing protection), which cannot be improved by additives
Not miscible with mineral oils, synthetic hydrocarbons, esters, polyphenylethers, and perfluoroalkylethers
High costs

Table 14.11b Advantages and Disadvantages of PAMA/PAO-Cooligomers

Advantages	Disadvantages
High viscosity index	Low fire resistance
Low evaporation losses	Not biologically degradable
Excellent low temperature behavior	Limited viscosity grades
Good viscosity temperature behavior	Higher costs
Good oxidation stability	
Excellent compatability with seal material	
Good miscibility with mineral oils and additives	
Nontoxic	

Table 14.12 Comparison of Properties of Synthetic Lubricating Oils

Property	Diester	Neopentyl polyol (complex) esters	Typical phosphate ester	Typical methyl silicone	Typical phenyl methyl silicone	Chlorinated phenyl methyl silicone	Polyglycol (inhibited)	Chlorinated diphenyl	Silicate ester or disiloxane	Polyphenyl ether	Fluorocarbon	Mineral oil (for comparison)	Remarks
Maximum temperature in absence of oxygen (°C)	250	300	120	220	320	305	260	315	300	450	300	200	For esters this temperature will be higher in the absence of metal
Maximum temperature in presence of oxygen (°C)	210	240	120	180	250	230	200	145	200	320	300	150	This limit is arbitrary. It will be higher if oxygen concentration is low and life is short.
Maximum temperature due to decrease in viscosity (°C)	150	180	100	200	250	280	200	100	240	150	140	200	With external pressurisation or low loads this limit will be higher.
Minimum temperature due to increase in viscosity (°C)	–35	–65	–55	–50	–30	–65	–20	–10	–60	0	–50	0 to –50	This limit depends on the power available to overcome the effect of increased viscosity.
Density (g/ml)	0.91	1.01	1.12	0.97	1.06	1.04	1.02	1.42	1.02	1.19	1.95	0.88	
Viscosity index	145	140	0	200	175	195	160	–200 to +25	150	–60	–25	0 to 140	A high viscosity index is desirable

Flash point (°C)	230	255	200	310	290	270	180	180	170	275	None	150 to 200	Above this temperature the vapor of the fluid may be ignited by an open flame.
Spontaneous ignition temperature	Low	Medium	Very high	High	High	Very high	Medium	Very high	Medium	High	Very high	Low	Above this temperature the fluid may ignite without any flame being present.
Thermal conductivity (W/M °C)	0.15	0.14	0.13	0.16	0.15	0.15	0.15	0.12	0.15	0.14	0.13	0.13	A high thermal conductivity and high thermal capacity are desirable for effective cooling.
Thermal capacity (J/kg °C)	2,000	1,700	1,600	1,550	1,550	1,550	2,000	1,200	1,700	1,750	1,350	2,000	
Bulk modulus	Medium	Medium	Medium	Very low	Low	Low	Medium	Medium	Low	Medium	Low	Fairly high	There are four different values of bulk modulus for each fluid but the relative qualities are consistent.
Boundary lubrication	Good	Good	Very good	Fair but poor for steel on steel	Fair but poor for steel on steel	Good	Very good	Very good	Fair	Fair	Very good	Good	This refers primarily to antiwear properties when some metal contact is occuring.
Toxicity	Slight	Slight	Some toxicity	Nontoxic	Nontoxic	Nontoxic	Believed to be low	Irritant vapor when hot	Slight	Believed to be low	Nontoxic unless overheated	Slight	Specialist advice should always be taken on toxic hazards.
Suitable rubbers	Nitrile, silicone	Silicone	Butyl, EPR	Neoprene, Viton	Neoprene, Viton	Viton, fluorosilicone	Nitrile	Viton	Viton nitrile, fluorosilicone	(None for very high temperatures)	Silicone	Nitrile	
Effect on plastics	May act as plasticizers		Powerful solvent	Slight, but may leach out plasticizers	Slight, but may leach out plasticizers	Slight, but may leach out plasticizers	Generally mild	Powerful solvent	Generally mild	Polyimides satisfactory	Some, softening when hot	Generally slight	

Table 14.12 *(continued)*

Property	Diester	Neopentyl polyol (complex) esters	Typical phosphate ester	Typical methyl silicone	Typical phenyl methyl silicone	Chlorinated phenyl methyl silicone	Polyglycol (inhibited)	Chlorinated diphenyl	Silicate ester or disiloxane	Polyphenyl ether	Fluorocarbon	Mineral oil (for comparison)	Remarks
Resistance to attack by water	Good	Good	Fair	Very good	Very good	Good	Good	Excellent	Poor	Very good	Excellent	Excellent	This refers to breakdown of the fluid itself and not the effect of water on the system.
Resistance to chemicals	Attacked by alkali	Attacked by alkali	Attacked by many chemicals	Attacked by strong alkali	Attacked by strong alkali	Attacked by alkali	Attacked by oxidants	Very resistant	Generally poor	Resistant	Resistant but attacked by alkali and amines	Very resistant	
Effect on metals	Slightly corrosive to nonferrous metals	Corrosive to some nonferrous metals when hot	Enhance corrosion in presence of water	Noncorrosive	Noncorrosive	Corrosive in presence of water to ferrous metals	Noncorrosive	Some corrosion of copper alloys	Noncorrosive	Noncorrosive	Noncorrosive, but unsafe with aluminium and magnesium	Noncorrosive when pure	
Costs (relative to mineral oil)	5	10	10	25	50	60	5	10	10	250	300	1	These are rough approximations, and vary with quality and supply position.

VI. COMPARISON OF CERTAIN PROPERTIES

Figure 14.3 reveals the temperature limits of mineral oils, which are between 100 and 150°C for oxidation and between 350 and 400°C considering the thermal stability.

The comparatively small influence of oxidation inhibitors in order to increase the service temperatures can clearly be recognized. If oxidation were avoided, much higher service temperatures could be applied to mineral oils. The limiting temperatures are characterized by the limited thermal stability of the oxidation inhibitors.

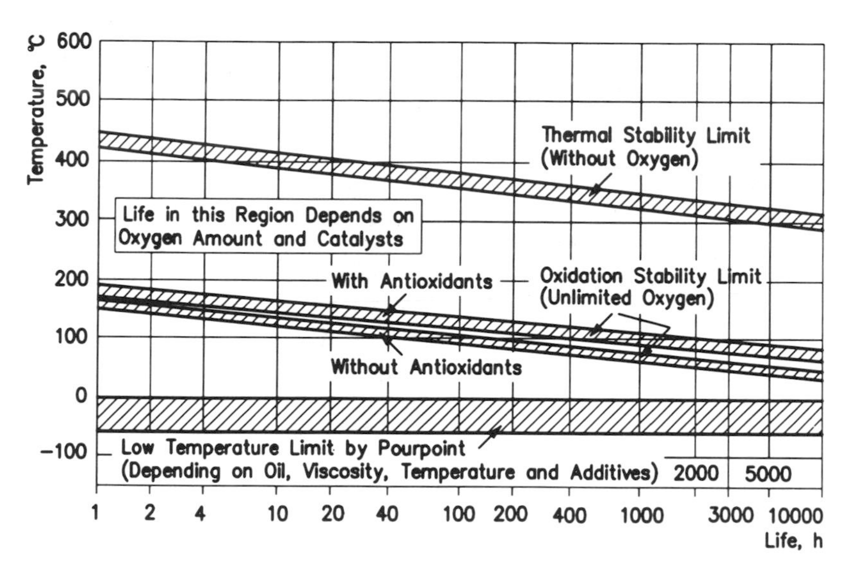

Figure 14.3 Temperature limits for mineral oils (3).

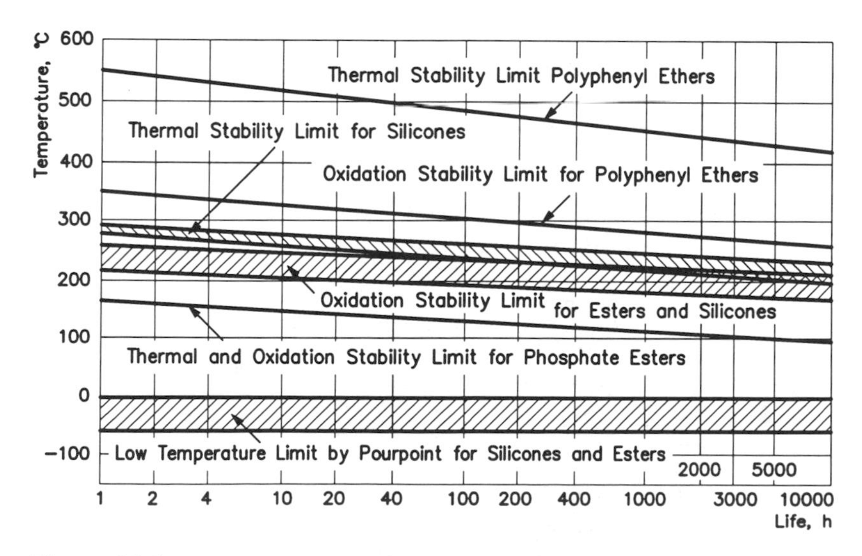

Figure 14.4 Temperature limits for some synthetic oils (3).

The temperature limits of some synthetic fluids are listed in Fig. 14.4. The temperature limiting the oxidation and the thermal stability was plotted against the service life of the fluid. Obviously, the higher oxidation limit and thermal stability limit of some synthetic fluids result in a better high-temperature behavior compared to mineral oils. Comparing the behavior of polyphenyl ether with silicone and ester oils illustrates the problematic nature of a situation that can combine excellent high-temperature properties with insufficient low-temperature behavior. The extraordinary high-temperature performance of polyphenyl ether becomes obvious, whereas the very low pour points of silicone oils and esters dedicate these fluids for low-temperature application. These relations result in a maximum service temperature for endurance and temporary service periods according to Table 14.13, and it becomes obvious that some synthetic fluids are superior to mineral oils. In principle, the same statement is valid with regard to low service temperatures (Table 14.14). Of course, any direct comparison of fluids of different classes can only be made for fluids having the same viscosity.

These results can be summarized by comparing the service temperature ranges of some synthetic fluids (Fig. 14.5). Their superiority to mineral oils can clearly be recognized (4). The temperatures that in general can be applied to synthetic fluids containing adequate oxidation inhibitors are remarkably higher than those for mineral oils (Fig. 14.6).

Figure 14.7 reveals the comparison of the viscosity temperature behavior of some synthetic fluids and formulations based on mineral oil. Comparing curves E and F reveals the extent of improvement by using additives in mineral oils. For further improvements other fluids have to be selected, such as in curve B. It is obvious that the excellent viscosity temperature behavior of a silicone oil cannot be realized with mineral oils, whether they contain VI improvers or not.

Table 14.13 Maximum Service Temperatures for Synthetic Lubricating Oils

Fluid	Permanent (°C)	Temporary (°C)
Mineral oils	90–120	130–150
Synthetic hydrocarbons	170–230	310–340
Carboxylic acid esters	170–180	220–230
Polyalkylene glycols	160–170	200–220
Polyphenylether	310–370	420–480
Phosphoric acid esters (alkyl)	90–120	120–150
Phosphoric acid esters (aryl)	150–170	200–230
Silicic acid ester (silicate esters)	180–220	260–280
Silicone oils, siloxanes	220–270	310–340
Silahydrocarbons	170–230	310–340
Halogenated polyphenyls	200–260	280–310
Perfluorohydrocarbons	280–340	400–450
Perfluoropolyglycols	230–260	280–340

Table 14.14 Flash Points and Pour Points of Some Synthetic Lubricating Oils

Fluid	Flash point, °C	Pour point, °C
Mineral oils	200–300	0 to –60
Polyalphaolefins	200–350	–20 to –60
Polyalkylene glycols	200–260	–30 to –50
Perfluoralkylether	—	–20 to –75
Polyphenylether	230–350	+20 to –20
Diesters	200–270	–50 to –80
Phosphate esters	100–260	–10 to –60
Silicone oils	230–330	–10 to –100
Silahydrocarbons	270–290	–5 to –45
Silicate esters	180–280	–50 to –70

VII. OVERALL COMPARISON OF SYNTHETIC FLUIDS

Table 14.15 shows an overall comparison of some important physical, chemical, and technological properties of the most important synthetic fluids (2, 5, 6). While desirable, the realization of such an overall comparison is difficult, considering the aim of a justified evaluation. Therefore some introductory remarks seem to be necessary. As several fluids are commercial while some others are development fluids, any comparison might be misleading at the present time. In addition, there are various subclasses within some general classes, a fact that could not be considered.

The same restriction applies to the influence of additives on the changes of some properties. Last but not least, it should be mentioned that the ratings between "excellent" and "poor" need some explanation. For instance, a fluid was rated as "excellent" if it was nontoxic. To give another example, the cyclophosphazene fluids could not be rated regarding biodegradability. In most cases the authors of the specific chapters of this book gave valuable assistance for the making of this table, which should be used to obtain a first impression of the properties of synthetic fluids relative to each other. As no information about the alkylated cyclopentanes could be obtained, these fluids have been omitted from this table. On the other hand the properties of rapeseed oils have been included. It can be seen that in regarding all compared properties, no fluid is superior to mineral oil or would

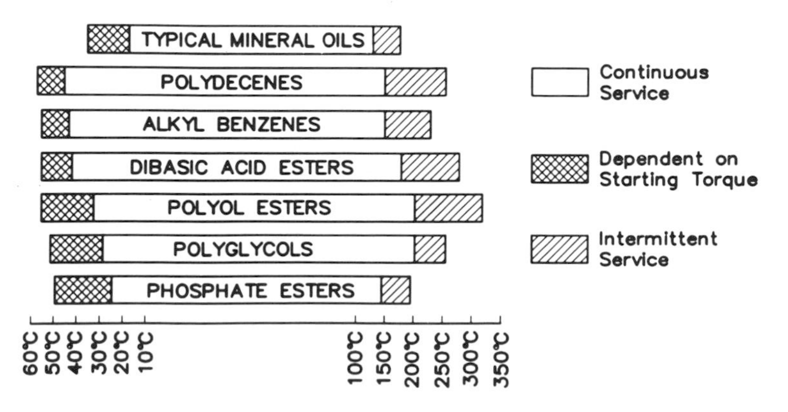

Figure 14.5 Comparison of service temperature ranges of synthetic fluids (3).

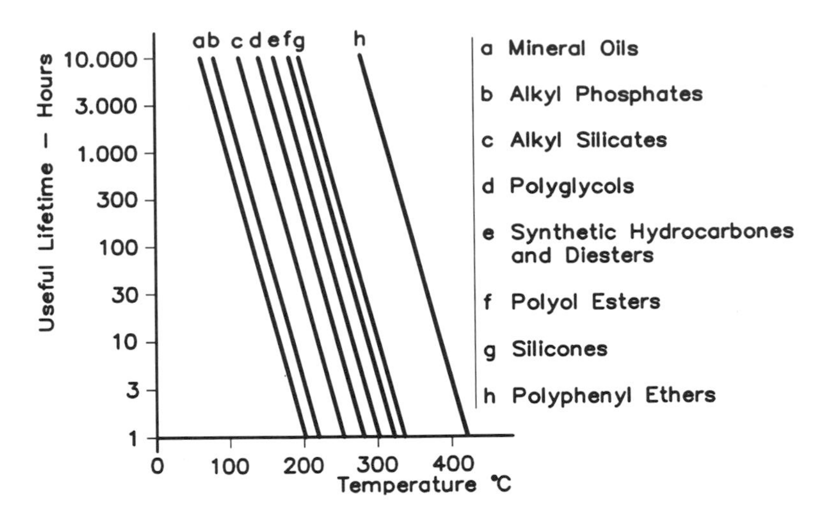

Figure 14.6 Service life depending on temperature for different synthetic oils containing inhibitors (3).

obtain the best evaluation for every property. Particularly when compared with mineral oils containing additives, synthetic fluids are not superior in all cases.

The choice of a synthetic fluid is usually the result of the synthetic fluid having particular properties of performance characteristics not obtainable with mineral oils.

Lubricating and functional fluids are chosen to cover the requirements of specific applications. Therefore, the fluid of choice for one application may be totally inadequate for another.

If certain cases of application require special properties that cannot be covered by mineral oils, for example, a certain synthetic fluid will have to be selected and applied in spite of some less good or even disadvantageous properties which have to be taken into account. Every single property, therefore, has to be supplemented by a weight factor depending on the special application.

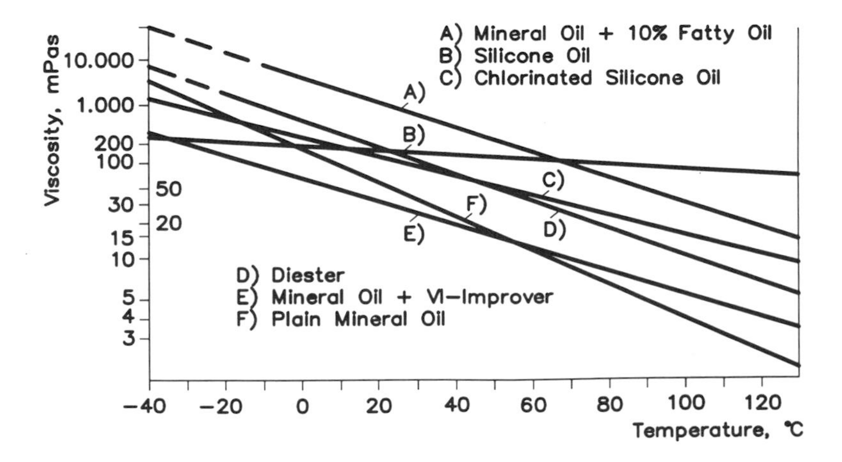

Figure 14.7 Viscosity temperature behavior of some lubricating oils.

Table 14.15 Comparison of Properties of Important Synthetic Fluids

Evaluation 1–Excellent 2–Very good 3–Good 4–Moderate 5–Poor	Mineral oils	Polyisobutenes	Polyalphaolefins	Alkylated aromates	Polyalkylene glycols	Perfluoroalkylethers	Polyphenylethers	Dicarboxilic acid esters	Neopentyl polyesters	Triaryl phosphate esters	Trialkyl phosphate esters	Silicone oils	Silicate esters	Silahydrocarbons	Chlorofluorocarbons	Cyclophosphazene fluids	Dialkylcarbonates	Alkylated cyclopentanes	PAMA/PAO-Cooligomers	Rapeseed oils
Viscosity temperature behavior (VI)	4	5	2	4	2	4	5	2	2	5	1	1	1	2	4	5	3	3	2	2
Low-temperature behavior (pour point)	5	4	1	3	3	3	5	1	2	4	1	1	2	3	3	3	3	3	2	3
Liquid range	4	5	2	3	3	1	5	2	2	2	3	1	1	2	5	4/5	2	1	2	3
Oxidation stability	4	4	2	4	3	1	2	2/3	2	2	4	2	2	3	1	3	2	2	2	5
Thermal stability	4	4	4	4	3	1	1	3	2	2	3	2	3	2	2	3/4	3	4	3	4
Evaporation losses, volatility	4	4	2	3	3	1	3	1	1	2	2	2	3	2	3	3	4	1	1	3
Fire resistence, inflammability temperature	5	5	5	5	4	1	4	4	4	1/2	1/2	3	4	4	1	1/2	3	5	4/5	5
Hydrolytical stability	1	1	1	1	3	1	1	4	4	4	3	3	4	1	2	3	3	1	2	5
Corrosion protection properties	1	1	1	1	3	5	4	4	4	4	4	3	5	1	5	3	1	1	2	1
Seal material compatibility	3	3	2	3	3	1	3	4	4	5	5	3	3	2	4	3/4	3	2	1	4
Paint and lacquer compatibility	1	1	1	1	4	2	4	4	4	5	5	3	4	1	3	3/4	2	1	1	4
Miscibility with mineral oil	—	1	1	1	5	5	3	2	2	4	4	5	4	1	5	5	2	1	1	1
Solubility of additives	1	1	2	1	4	5	2	2	2	1	1	5	3	3	5	4	2	3	1	3
Lubricating properties, load-carrying capacity	3	3	3	3	2	1	1	2	2	1	3	5	4	3	1	2/3	2	3	2	1
Toxicity	3	1	1	5	3	1	3	3	3	4/5	4/5	1	4	2	2	2	1	1	1	1
Biodegradability	4	5	5	5	1/2	5	5	1/2	1/2	2	2	5	4	5	5	—	1	5	4/5	1
Price compared with mineral oil	—	3–5	3–5	3–5	6–10	500	200–500	4–10	4–10	5–10	5–10	30–100	20–30	30–70	300–400	30–50	4–10	3–8	5–10	2–3

VIII. CONCLUSION

Mineral oils cannot meet all requirements on lubricants for frictional contacts or on operational fluids. Therefore, certain synthetic fluids have gained extraordinary importance. These fluids can be classified by the production process, the composition, or the chemical structure. After comparing their chemical, physical, and technological properties, the advantages and disadvantages of some important synthetic fluids are discussed.

An overall comparison with reference to mineral oils shows that synthetic fluids are superior to mineral oils only with regard to certain characteristic properties. Therefore, so-called weight factors have to be applied to the special properties in order to obtain a practically orientated overall comparison.

REFERENCES

1. Bartz, W. J. (1987). Notwendigkeiten für den Einsatz synthetischer Schmierstoffe und Betriebsflüssigkeiten. *Tribol. Schmierungstechnik* 34, 262–269.
2. Klamann, D. (1982). *Schmierstoffe und verwandte Produkte*, Verlag Chemie, Weinheim.
3. Neale, M. J. (ed.) (1973). *Tribology Handbook*, Butterworth, London.
4. Williamson, G. I. (1988). Synthetic Lubricants—The State of Market Development and Inter-Product Competition, Course SP5, The College of Petroleum Studies, Oxford, 5 and 6 December.
5. Schmid, W. (1980). *Synthetische Industrieschmierstoffe*. Mobil Oil, Hamburg.
6. Möller, U. J., and U. Boor (1987). *Schmierstoffe im Betrieb*. VDI-Verlag, Düsseldorf.

15
Automotive Crankcase

Alan J. Mills
Castrol International, Ltd.
Pangbourne, Berkshire, England

I. INTRODUCTION

Engine oils can play a major role in protecting the engine. This is indisputable and is clearly demonstrated by the proliferation of new performance requirements that have been demanded by the engine and vehicle manufacturers (1). Synthetic engine oils represent the freedom to formulate oils that are better for the engine, better for the environment, and better for the driver.

Synthetic engine oils have been marketed for many years and have been a small but growing part of the market. More recently these oils have come of age and have moved away from being "generic" to being designed with specific performance benefits for different applications. Synthetic engine oils must now be considered as a part of each and every engine oil market. It is necessary to compare different synthetic engine oils with their counterpart mineral engine oils in each market, rather than to consider the performance of a single synthetic engine oil. In doing this it is most relevant to consider the potential that synthetic engine oils have in each aspect of performance, rather than exact results obtained from a single oil.

As synthetic engine oils move from being generic to being specific, their formulations become more commercially sensitive. The cost of developing, proving, and marketing a synthetic engine oil is as high as or higher than for a mineral engine oil. Formulation details are therefore no longer being publicized. The most that can be revealed are some of the commerical forces that are shaping the developments and some of the concepts and directions that synthetic engine oil technology can take. After this it is down to the vision of the synthetic oil marketer and the skill of the synthetic oil formulators. The chemistry is the tool that they employ.

II. WHAT IS A SYNTHETIC ENGINE OIL?

This is simple question, but one that has generated much debate over recent years. As with many simple questions, there is a complex answer.

First, it is necessary to define what synthetic means. In simple terms synthetic means put together from something different, typically from simpler, more basic components. Classically this includes esters, which are manufactured from the reaction of acid and alcohol, and synthetic hydrocarbons such as polyalphaolefins (PAOs), which are manufactured by polymerizing olefins and further processing. Other synthetic base stocks would include polyalkylene glycols, silicones, etc. (1, 2). Conventional mineral oil base stocks are clearly not synthetic, as they are essentially fractions obtained from crude oil. Similarly, it s difficult to justify that "modified" mineral oils are synthetic. The refinery stream used to produce these base stocks has many of the properties of a lubricating base stock even though it would for several reasons be unacceptable in terms of performance. These "modified" mineral oil base stocks are treated as "unconventional" mineral oils.

Having defined what synthetic means it is necessary to decide how much synthetic. Fully synthetic should be the easiest to define. However, even in a fully synthetic oil, there is almost certainly some mineral oil present. The chemical components used to manufacture the additive package contain mineral oil as a carrier or processing aid to make the chemical reactions operate efficiently. Similarly, the additive package needs to be of a handleable viscosity, and a further addition of mineral oil may be included. Then of course there is the viscosity index improver (VII), which also requires a base stock to make it handleable. When all these aspects are considered, it is possible for a "fully synthetic" engine oil to contain over 10% mineral oil. If the mineral oil is of good quality, at such a level it is unlikely to cause any serious derating of performance. However, when the level increases to above 20%, the claim to be fully synthetic becomes more difficult to maintain.

There are of course benefits that can be built into an engine oil with lower levels of synthetic base stock. However, the law of diminishing returns may start to apply (3). In general, when the level of synthetic base stock falls much below 20%, it is difficult to claim that the oil is "synthetic based" unless the synthetic component makes a major modification to the performance of the oil. This may be the case with synthetic base stocks that are polar and can affect aspects such as surface protection and cleanliness. Under these conditions the whole formulation may be directed by the inclusion of the synthetic component, and the development programme will be extensive.

There is a totally different way of answering the question of "What is a synthetic engine oil?" and this is to consider it from a commercial point of view. Clearly the oil is going to be expensive and it will occupy a premium position in the market. On this basis the synthetic engine oil must have excellent performance that is clearly superior to standard mineral-oil-based engine oils. This need to be better, and to be seen to be better, is a commercial need. It is of course both hypothetically and practically possible to formulate a synthetic engine oil that has performance weaknesses. Commercially this is unacceptable. Such an oil could cause engine damage and in this way destroy the position of synthetic engine oils as providing the best possible protection.

A synthetic engine oil therefore represents the ultimate in performance for two reasons: first, the "freedom to formulate" that is available when synthetic components can be used, and second, the commercial need for excellent performance. This combination of freedom and need makes the formulation of synthetic engine oils a complex and stimulating area to work in.

III. THE BENEFITS FROM USING SYNTHETIC ENGINE OILS

Synthetic engine oils cost considerably more than conventional mineral-oil-based engine oils. How can this be justified? And who does it need to be justified to?

The ultimate consumer must be convinced that he has made a wise choice despite the higher expenditure. The workshop manager will find it necessary to justify the increased cost to the car owner. The engineer who specifies a synthetic engine oil for initial fill by the vehicle manufacturer will have to defend his position on a strict cost-performance basis to the management. The synthetic engine oil must provide clear performance benefits that will be understood by some extremely sceptical audiences.

There is one further justification that is necessary and that predates any of the purchasing decisions. This is the need to justify the development of the synthetic engine oils themselves. To develop and prove (and redevelop and reprove) a new synthetic engine oil costs millions of dollars. One reason for this is that all current approval systems established over the years have been based on conventional mineral oil technology. "Atypical" formulations require fleet tests to demonstrate that there are no unexpected performance side effects (4). In addition to the technical development cost there is also the commercial development cost. Packaging, launch, and promotion of a new engine oil are again extremely expensive. The decision to develop a new synthetic engine oil is not taken lightly, and the decision must be based on the benefits that the oil can provide. The question of performance benefits should be addressed a thousand days and millions of dollars before the new synthetic engine oil reaches the marketplace. So what can these benefits be?

A benefit is irretrievably linked to a person (or group of people) and their particular situation. What is seen as a benefit by one person may not be seen as a benefit by another person. One motorist may want to run the vehicle for 20 years but another may plan to replace it after 1 year. One motorist may desire improved fuel economy but another may want increased engine power and acceleration. One motorist may be interested in extending the oil drain interval but another may prefer to change the oil every season.

The desires of the workshops may be different again. Their main concerns are likely to be the retention of existing customers, the recruiting of new customers, and compliance with their supplier's warranty requirements. Different workshops will have different priorities and they will position themselves differently. Some may aim to provide a low-price service. Some may aim to provide a high-security, minimum-breakdown-risk service. Some may aim to provide a "sporty" service. There are many other options available. Each option may be marketed differently.

It is obvious that there are many different desires from the many different purchasers and users of engine oils and that it is not possible to produce one synthetic engine oil that meets everyone's requirements. In any single synthetic engine oil there will inevitably be some compromises. For example, a compromise must be reached between oil consumption control and low-temperature oil flow rates. Both may be superior to conventional mineral oil technology, but there will be some compromise nonetheless. Every synthetic engine oil has its own balance of benefits.

The benefits that can be engineered into a synthetic engine oil fall into several main aspects. These are engine protection, fuel economy, and environmental protection. The following sections describe how these aspects can be improved by the use of synthetic basestocks.

IV. IMPROVED ENGINE PROTECTION

A. The Need for Improved Engine Protection

Engines built today are more durable than ever before. If cars are serviced as specified by the vehicle manufacturer, then the majority of cars should be able to put at least 50,000 miles "on the clock" without any major engine repairs being required. So where is the need for improved performance? The answer is that the expectations of the vehicle owner and the legislator are increasing just as fast as, and sometimes faster than, engine durability. Many cars do need major repairs and do go out of permissible emission control limits. The engine oil can play a critical role in protecting the engine and keeping it operating satisfactorily.

The last few years have seen many new specification and approval tests being developed for engine oils by vehicle manufacturers around the world (4, 5). Such new tests are established so as to overcome field problems that have been admitted by the vehicle manufacturers. One, several, or all vehicle manufacturers may have suffered frequent (or potential) engine failure in the field based on a particular problem. Respecifying the engine oil has been identified as the most attractive option by the vehicle manufacturers. The vehicle manufacturers have then required the lubricant industry to run (and often also to help develop) tests to identify engine oils that will overcome the problem. The end result is that many millions of dollars are spent in proving, and to a lesser extent improving, engine oil quality.

There are three important features of this system. The first is that the vehicle manufacturers are clearly convinced that engine oils make a major difference to engine performance. The second is that the current system requires a "basketful of broken parts" to be produced before there is any recognised improvement in standard (mineral) oil performance as defined by new approvals or classifications. The third feature is that this is an ongoing situation with new field problems occurring when using mineral engine oils meeting the most modern approvals and classifications. Each successive new lubricant specification and classification fails to provide adequate protection. The solution has always been to provide a new "basketful of broken parts" and to develop a new test for a new specification or classification to allow engine oils to be proved or improved. The system as it stands only responds after field problems. There is no element of performance enhancement before a problem occurs.

There is an alternative to this approach, and that is to formulate stronger oils that exceed the performance required by the standard industry tests. The potential areas of weakness of mineral engine oils can be anticipated in advance and extra protection can be built into a properly designed and formulated synthetic engine oil before a field problem occurs.

The typical causes of engine damage are components that have become severely worn or excessive deposits that have built up. If wear and deposits can be reduced, then the useful life of the engine will be increased.

B. Improved Wear Protection

Wear is essentially caused by surface-to-surface contact. Wear can be increased by weakening the surface (e.g., by corrosion), or it can be reduced by strengthening the surface [e.g., by using antiwear or extreme pressure (EP) chemistry]. Base stocks and additive selection will also affect wear protection (6). The best way to minimize wear is to reduce surface-to-surface contact. This is particularly important when considering the protection of nonferrous materials, because these are not easily protected by antiwear or EP chemistry. If a full lubricant film is provided by the engine oil then wear is reduced to a minimum. As the engine oil becomes hotter it becomes thinner. As the lubricant film becomes thinner the surface-to-surface contact is increased. This is the origin of high-temperature wear. Different wear phenomena also occur at the engine start-up and during engine warm-up.

1. High-Temperature Wear Protection

High-temperature wear is initially determined by the physical attributes of the engine oil. If the high temperature viscosity under the experienced speed/load conditions is inadequate, there will be a breakdown of the lubricant viscosity film. Some parts of the engine are designed to operate continually with full film lubrication. Metal-to-metal contact between these parts will cause rapid wear. Typical examples of this are engine bearings. Bearings are designed to operate on oils of a minimum in-bearing viscosity. This is currently defined by the SAE (Society of Automotive Engineers) J300 classification (7). This relates to the kinematic viscosity at 100°C (kv100). Other measurements, such as high-temperature, high-shear (HTHS) viscosity and measurement of bearing oil film thickness (BOFT), can be used in an attempt to more closely reproduce bearing conditions. It is essential that the viscosity of the oil does not fall below that required by the bearing. This can occur due to the oil having too low an initial viscosity for the conditions, due to the oil shearing down in the engine, or due to fuel dilution reducing the oil viscosity. Synthetic engine oils can provide protection against all these concerns.

The superior performance of synthetic base stocks such as PAOs means that wider viscosity range oils can be produced without compromising other performance properties. The kinematic viscosity of a synthetic 5W-50 oil is approximately 75% higher than a mineral 5W–30 oil at 100°C. The same figure applies for a synthetic 10W–60 oil compared to a mineral 10W–40 oil (7). The benefits of this higher viscosity are seen under extreme driving conditions. Such extreme conditions include high temperature, high load, low-speed driving (slow speeds only generate thin films), and changes in engine conditions such as acceleration, deceleration, or changing gear.

In addition to producing a synthetic engine oil with a wider viscosity range than a VI-improved mineral oil, it is also possible to produce the same SAE viscosity oil by using higher-viscosity synthetic base stocks with a lower treat of VII. For any given VI this means that the synthetic oil is more shear stable than the mineral oil, and therefore the synthetic oil viscosity will be higher than the mineral oil after shearing by the engine. The freedom to select the polymer type according to performance rather than cost also allows the use of more shear-stable, possibly higher treat rate, polymers that further improve the shear stability of the synthetic oil. For example, fully formulated synthetic 5W–50 or

10W–60 oils shear down by less than 5% in the CEC L-14-A-78 shear stability test, whereas mineral 5W–30 or 10W–40 oils may shear down by 10% or more (1).

Fuel dilution can be a concern, and dilutions in excess of 10% can be observed in vehicles used for cold-climate, short-trip travel. By the time that the oil has reached normal operating conditions the fuel dilution is unlikely to be much above 2% due to the volatilization of the gasoline. High-speed driving can increase this back up to 5%, and under racing conditions fuel dilutions of up to 15% can occur in the hot oil. The viscosity of the oil can be reduced by 50%. The use of high-viscosity synthetic engine oils provides an important safety margin for high-temperature wear protection.

2. Start-Up Wear Protection

If you ask people who are involved in automotive lubrication "When does the majority of engine wear occur?" they will almost certainly tell you that it occurs when the engine is started and while it warms up. You will not have to ask many people before you will actually hear hard numbers quoted. "Every time you start a car it in effect puts 100 miles of driving wear on the engine." "One cold start at –20°C is equivalent to 1000 miles of driving." Trying to track down the origin of these numbers can be frustrating, as the source often disappears into the realms of hearsay. A few examples are traceable, however, and some detailed studies have confirmed the effect in the particular vehicles tested. Several SAE papers have been published on this topic (8, 9). The practical problem is that what happens with one engine design may be totally different in another engine design. Each engine needs to be considered in its own right. In terms of the engine oil, the performance can be addressed by two basic concerns: how fast does the new oil flow around the engine, and what happens to the engine oil after it has been in the engine for some time?

The rate at which different engine oils flow around an engine can be measured by a variety of techniques. In most engines the act of cranking and then idling the engine at start-up should get the oil to the camshaft (which is almost always the furthest point in the lubrication route) as quickly as possible (10, 11). In practice, when an engine oil is used at its minimum recommended temperature, as defined by its SAE viscosity grade, it can take up to 30 s to lubricate even a well-designed lubrication system. If the lubrication system is not optimal then it can can take a lot longer for the oil to arrive (Castrol Ltd., unpublished data). The faster the oil arrives, the sooner full-flow lubrication can begin. Using the appropriate viscosity grade is essential to obtaining rapid full-flow lubrication.

There are two problems in using the appropriate viscosity-grade engine oil. First, the choice has to be predictive in anticipating the temperature conditions that will be encountered over the next drain interval. Sometimes the weather, or the altitude on a trip, may take the driver by surprise. Second, what will happen to the oil as it ages? The driver may have filled his engine with a certain viscosity engine oil, but what will it actually be after 5,000 miles of driving?

Mineral oil base stocks contain a significant proportion of linear paraffinic molecules or molecular segments. At low temperatures these can interact to form microcrystalline regions that hold the oil in a matrix and resist movement. This will slow down the flow of oil or may even prevent oil flow altogether. The result is lubricant starvation until the oil arrives. This leads to extensive metal-to-metal contact and, potentially, wear. Synthetic base stocks can be selected to have excellent low-temperature properties. Pour points of synthetic oils are typically below –45°C (2), and it is possible to design 0W–X oils that have good cold cranking performance down to below –30°C without having high volatility or low viscosity at high temperatures. Even 0W–60 oils can be formulated to meet the most severe physical requirements. The 5W–X mineral engine oils are limited to a viscosity range of 5W–30 if volatility is to be controlled. Similarly, 10W–X and 15W–X mineral engine oils can go no wider than 10W–40 and 15W–50 viscosity ranges. Under the most severe winter conditions a synthetic 0W–X engine oil may allow oil to flow and the engine to start whereas a 5W–X or 10W–X mineral engine oil may not. At more common low temperatures, synthetic engine oils typically pump three or four times as quickly as mineral engine oils. At temperatures above the freezing point synthetic engine oils provide almost instant lubrication (Castrol Ltd., unpublished data). An oil that gets to the camshaft four times as quickly will get full-flow camshaft lubrication going four times as quickly.

The second aspect of start-up wear protection is what happens to the oil when it has been in the engine (2). Engine oils are not totally shear stable, and therefore their viscosity tends to reduce as soon as the engine is run. Initially therefore there will be a reduction in low-temperature viscosity and an increase in low-temperature flow rates (combined, of course, with a reduction in high-temperature viscosity protection). This has already been discussed. There are, however, two effects that operate in the opposite direction. Engine oils contain low-viscosity components that are volatile under high-temperature driving conditions. The loss of these components makes the oil thicker at low temperatures and reduces flow rates. Engine oils also oxidize under exposure to heat, air, and assorted catalytic materials. Oxidation causes an increase in oil viscosity and again reduces the flow rate of the oil.

Commercially available 5W–30 mineral oils have a high volatility at high temperatures. Exact values depend on formulation strategies and test methods, but it is not unusual to see 25% volatility in the European NOACK test (CEC L-40-T-87) or 20% in the ASTM simulated distillation (ASTM D-2887) test (Castrol Ltd., unpublished data). These tests are designed to mimic the effect of high-temperature driving conditions on the oil under European and North American driving conditions, respectively. Oil consumption under severe driving conditions is a continual concern. If 20% of an oil is lost by volatilization then its viscosity increases dramatically, especially at low temperatures. A 5W–30 mineral oil can easily become a 15W–40 oil. The oil is no longer suitable for use under the intended low-temperature conditions. Fully formulated synthetic engine oils can possess superior volatility to equiviscous mineral engine oils for two reasons. First, synthetic base stocks are normally better controlled in terms of volatility range and therefore have inherently lower volatility than mineral oils. This is because synthetic base stocks are more tightly defined chemically than mineral base stocks. Second, synthetic base stocks can also offer higher viscosity at high temperature compared to mineral base stocks, and this means that less VII is required to achieve the high-temperature viscosity. Less VII means less thickening at low temperatures as well, and therefore low-temperature performance is derated less. The consequence is that higher-viscosity, lower-volatility base stocks can be used, and the fully formulated synthetic engine oil can be designed to have extremely low volatility. It is therefore possible to produce synthetic engine oils with the same viscosity as a mineral engine oil but with minimal volatility loss at high temperatures.

Exposure to high temperatures causes not only volatilization but also oxidation. The difference between mineral and synthetic engine oils can be seen in the ASTM sequence IIIE test (4). The maximum allowed viscosity increase in this test to meet the requirements of API SG is 375% as determined at 40°C. The viscosity of the oil will also increase at

100°C and subzero. A simple parallel wih a 375% viscosity increase at 40°C would suggest that a 5W–30 oil could end up at a considerably higher and perhaps even undefined SAE viscosity like a "30W–70." Start-up lubrication times could be excessive as soon as the temperature falls to freezing. The used engine oil will be a very long way from meeting the engine's start-up lubrication requirements. In contrast, synthetic engine oils can be designed to have better resistance to oxidative thickening due to their selected and controlled molecular structure. Results from ASTM IIIE tests on synthetic engine oils give viscosity increases that are typically less than 50% and may be less than 10% (1). Synthetic engine oils can provide stay-in-grade performance in the test. The used synthetic engine oil provides the same low-temperature flow rates as it did when new.

A used 5W–50 synthetic engine oil should provide rapid lubrication down to its design temperature of –25°C (–13°F). A used mineral oil may be so viscous as to be unsuitable for use below 0°C (32°F). The synthetic engine oil would under these conditions provide an extra 25°C (45°F) of engine protection compared to an oxidized 5W–30 mineral oil.

3. *Warm-Up Wear Protection*

When the lubricant has been pumped around the engine and the engine has started, the lubricant has to provide protection against wear. This protection is a combination of physical and chemical performance. The high viscosity of an oil at very low temperatures provides a thick film of lubricant that minimizes surface-to-surface contact (when the oil has arrived). This protects against wear. As the temperature increases and the oil thins down, the lubricant film reduces. Significant surface contact can occur and "antiwear" chemistry may be required to protect the surfaces. The classical antiwear chemistry that is utilized is zinc dialkyldithiophosphate (referred to as zddp or zdtp). However, this chemistry is not normally activated at ambient temperatures. Even the lowest-temperature industry engine tests do not simulate the level of wear protection that is offered by an oil at oil temperatures less than 60°C. It can take 15 min or more for engine oil to warm up from extreme cold to 60°C when it is left to idle (Castrol Ltd., unpublished data). The engine may not warm up at all on short-trip driving. Many vehicles spend a high proportion of their time operating under such warm-up conditions.

Modern engine oils are designed to be lower viscosity for rapid lubrication, and the benefits of this are clear. There is, however, the aspect that low-temperature oil films are thinner and that the potential for metal-to-metal contact is increased. Traditional antiwear chemistry, as explained above, is not designed to be active under these conditions. Increased protection can be built into the oil by the use of synthetic base stocks with extra film-strength performance. These base stocks have a high affinity for metal surfaces and can provide extra wear protection where traditional mineral oil lubricants do not. Such chemistry and measurement technology are regarded as proprietary to individual oil companies and are not published in the literature.

Increased film-strength performance may become increasingly important as lower-viscosity oils are used in greater volumes.

C. Improved Deposit Control

When fuel is burned in the engine it forms water and carbon dioxide. However, there are significant levels of other compounds formed that are unstable and can form deposits in the engine. The high temperatures and partially combusted fuel gasses can also degrade the engine oil, and this degraded engine oil can also lead to engine deposits. The engine is designed to operate satisfactorily in the presence of a moderate amount of deposits. However, if the deposits become excessive, then engine performance and efficiency will suffer. The driver will notice a loss of power and acceleration, and an increase in fuel consumption. The environment will experience an increase in emissions.

Deposits can form in different parts of the engine, and these have different origins, different effects, and different methods of control. Ultimately the engine will require stripping and rebuilding or, if the deposits lead to excessive engine wear, the engine will need to be scrapped. The use of a synthetic engine oil can reduce engine deposits and extend efficient engine performance.

1. Piston Deposits

Partially degraded fuel and engine oil components can polymerize and degrade on hot surfaces to form tars, varnishes, and carbonaceous deposits. The area where deposits cause the most noticeable and irreversible effects on engine performance is on the pistons in the piston ring grooves. Excessive deposit buildup can restrict gas flow, and this will slow down the response time of the piston rings. This leads to loss of compression, increased combusted gas blow-by, and increased oil consumption. The effect of this is to increase the tendency to form deposits, and a vicious circle is developed. Deposits in piston grooves cause nipping and sticking of the piston rings.

The use of synthetic engine oils with increased resistance to breakdown and increased cleanliness can significantly reduce deposits in the piston ring grooves. Engines that have run on suitably formulated synthetic engine oils give considerably reduced deposits (1). In some extremely severe engine tests using the latest fuel-efficient technology, typical mineral engine oils suffer piston ring stick after the equivalent of approximately 1,000 miles of high-speed operation. In contrast, synthetic engine oils can survive for the equivalent of 4,000–5,000 miles of continuous high-speed driving without problems (Castrol Ltd., unpublished data). In more conventional engine tests, the use of synthetic engine oils can allow the duration of the test to be multiplied by a factor of two, three, or more while maintaining the same piston ring cleanliness obtained when using a conventional engine oil. This effect has also been observed in field trials where vehicles have been run for three times the recommended engine oil drain intervals while maintaining piston ring cleanliness superior to that obtained from traditional mineral engine oils under standard drain intervals (Castrol Ltd., unpublished data).

2. Inlet Valve Deposits

Another area of the engine that causes concern with deposits is the inlet valve. Deposits on the back of the inlet valve absorb fuel from the injectors and cause problems with engine starting and idling. The deposits also strangle the airflow into the engine at high engine speeds, reducing power output and increasing emissions. Major concerns have occurred in the field and have caused much debate over fuel quality and the benefits from the use of fuel detergents. The engine oil can also play a role in either contributing to or reducing inlet valve deposits. Engine oil consumed by passing through the piston rings into the combustion chamber or from mist generated in the crankcase and vented back to the engine air intake can form deposits on the engine intake valves.

Synthetic engine oils can be formulated to have low oil consumption. This can be achieved by controlling volatility and other oil properties. The aspect of volatility is simple to understand and communicate. Clearly, an oil with a lower volatility will give less oil consumption, all other things being equal. Other important oil properties include the "viscoelastic" properties of the oil. For example, multigrade engine oils are preferred by some commercial vehicle engine manufacturers due to their reduced oil consumption. The excellent low-temperature performance of synthetic base stocks allows the formulation of wide-viscosity-range, low-volatility engine oils with a high VII content that provides high viscoelasticity. Synthetic engine oils can also be formulated to be cleaner burning than mineral oils, and this too can contribute to reduced inlet valve deposit formation. Tests have shown that inlet valve deposits on the Mercedes Benz M102E inlet valve deposit test [the test has provisional Coordinating European Council (CEC) status] can be significantly reduced by the use of a fully formulated synthetic engine oil compared to a mineral engine oil (Castrol Ltd., unpublished data).

3. Turbocharger Deposits

Another part of the engine that can clearly benefit from the use of synthetic engine oils is the turbocharger. When a turbocharged engine is switched off immediately after a high-speed run, "heat soak" occurs from the engine, and it considerably increases the oil temperature in the turbocharger. This can lead to deposits that can block the oil flow to the turbocharger bearings, leading to turbocharger seizure. Synthetic engine oils that are inherently more thermally stable than mineral engine oils have been shown to keep turbochargers considerably cleaner than mineral engine oils. A poor mineral engine oil will give turbocharger seizure, caused by blocked oilways, in under 100 h of a severe "heat soak" test cycle. A good mineral engine oil will survive for 100 h but with considerable deposits and incipient blocking. A good synthetic engine oil will maintain the turbocharger in almost "as fitted" condition (1).

4. Combustion Chamber Deposits

The contribution of engine oil to combustion chamber deposits that cause preignition is well known. The use of high "sulfated ash" oils can lead to high levels of combustion chamber deposits. However, different formulations have been shown to perform differently, and the correlation between sulfated ash level and preignition is not exact. One reason for this is that the rate of oil consumption into the combustion chamber is also a relevant factor.

Preignition problems have been greater in Europe than elsewhere, due to the higher ash levels typical of European passenger car engine oils. In the past the problem caused by preignition was actual engine damage, as well as severe performance loss and fuel consumption increase. With the increase in the proportion of cars that are fitted with a knock sensor and correction mechanism, engine damage is now often avoided. However, the activation of the correction mechanism (typically retarding the ignition or reducing turbocharger boost) causes its own loss of power, performance, and efficiency, and an increase in fuel consumption and vehicle emissions is observed. The use of synthetic engine oils that reduce oil consumption into the combustion chamber can considerably reduce combustion chamber deposits and minimize engine damage and or performance loss. A fully formulated synthetic 5W–50 engine oil has been shown to have only 50% of the oil consumption loss of a commercially available 5W–30 mineral engine oil under a severe driving cycle on a test bed (Castrol Ltd., unpublished data).

5. Oil System Deposits

If deposits form in the oil system, lubrication times are increased and lubricant starvation occurs. Serious engine damage is possible. For example, "black sludge" has caused many problems around the world. Black sludge is primarily caused by a combination of changing fuel chemistry and new engine designs. The problem has been attacked by modifying engine design, by improving fuel quality, and by modifying the engine oil to control the deposition of the sludge. To an extent, the problem has been thought to be solved three times, and now it is only with specially produced artificial fuels and severe engines that significant levels of sludge can be created.

Concern over black sludge has, over the past few years, prevented vehicle manufacturers from increasing their recommended engine oil drain intervals. The improvements that have been made in terms of fuel quality and engine design should make black sludge less of a concern, and consideration can again be given to increasing drain intervals with appropriate-quality engine oils. The cleanliness and durability provided by synthetic engine oils make them the most appropriate engine oils to use once black sludge as a major problem can be consigned to the history books.

Overall, the benefits of synthetic engine oils are focused on keeping the engine running under "design" conditions. Extensive testing and field trialing of synthetic engine oils has shown that they can maintain the whole engine in a far cleaner condition than realized using mineral engine oils. This keeps engine efficiency to a maximum, keeps emissions to a minimum, and minimizes engine damage.

V. IMPROVED FUEL ECONOMY

Fuel economy and energy conservation are major issues in terms of resource conservation, economics, and pollution. The engine oil, as with all other constituent parts of the engine, can contribute to fuel economy. In order to minimize fuel consumption due to the engine oil it is necessary to minimize viscous drag and to minimize surface-to-surface contact (i.e., oil film breakdown). These two requirements are in conflict, in that a "thick" oil will have a high resistance to flow but will form a strong hydrodynamic "wedge" helping to keep surfaces apart, while the reverse is true for a "thin" oil. Somewhere between these two, and with the assistance of surface-active chemicals and base stocks, there is an optimum formulation. Obviously, this balance will be different for different vehicles and different driving conditions.

Some parts of the engine operate under conditions where some degree of surface contract is accepted and can be tolerated by the oil's antiwear chemistry, preventing serious damage. Examples are the valve train and the piston rings. Since the realms of chemistry are being entered, it is necessary to have an engine test. No type or number of viscometers will anticipate the behavior of surface-active chemistry. Due to the considerable differences between engine designs, different results will be obtained with different engines. "Fuel economy" is therefore a difficult claim to justify for an engine oil. What is likely to be true, however, is that if a large number of different vehicles are compared, there will be a significant reduction in fuel consumed. Therefore, it is justifiable to describe an engine oil as "energy conserving" based on the results from engine tests.

The API sequence VI test is the only accepted test method for measuring engine-oil-derived fuel economy. The test has the benefits of being relatively easy and inexpensive to operate under statistical control. This is no small achievement, and these were key objectives of the task force developing the engine test. Many engine oils are now formulated to achieve good ratings from this test. The system allows a definition of "energy-conserving" engine oils that can be made widely available in the U.S. marketplace and therefore allows the car manufacturers to carry out their U.S. Environmental Protection Agency (EPA) vehicle fuel economy determinations using such an oil and to specify such an oil for service fill. The test is important to the car manufacturers since it allows them to use such oils to help them to achieve improved CAFE (corporate average fuel economy) figures for their vehicles. There is, however, one major aspect of fuel consumption that is not taken account of by the sequence VI test. This is the fuel consumption over the life of the engine.

Some vehicle road tests have demonstrated that a commercially available 5W–50 synthetic engine oil can give considerably lower fuel consumption than a commercially available 10W–30 mineral engine oil over a distance of 30,000 miles. Figures in excess of 5% were obtained (Castrol Ltd., unpublished data). Not surprisingly, the 10W–30 oil gave a better result in the sequence VI test. This brings the role of the API sequence VI test into question. Is it preferable to use an engine oil that gives better results over the road or one that gives better results in the API sequence VI test? It all depends on whether the object of using the oil is to obtain a better EPA vehicle fuel economy result or reduced over-the-road fuel consumption. Unfortunately, EPA vehicle fuel economy testing is itself only a short-duration test. No account is taken of long-term fuel economy due to long-term lubricant effects on engine performance.

Much more information on the real fuel economy of cars over the road is required. If the benefits of synthetic engine oils over mineral engine oils can be substantiated, the current "energy-conserving" system will need to be reconsidered. It may be appropriate to take a fresh look at defining energy-conserving oils so that long-term effects are given due credence. Perhaps the CAFE figures should be based on more realistic, long-distance fuel economy testing. Such fuel economy testing could be made a part of the emissions compliance evaluation for the vehicle.

It is of course still possible to demonstrate good results for synthetic engine oils in the API sequence VI test. Results at least comparable with mineral engine oils can be achieved, and particularly low-viscosity oils can be formulated to take advantage of the test conditions. Engine Fuel Efficiency Index (EFEI) values of 3% can be achieved (Castrol Ltd., unpublished data).

VI. REDUCED ENVIRONMENTAL DAMAGE

The impact of any product on the environment should be known and steps taken to minimize any negative impact. Products need to be considered in terms of their environmental impact "from the cradle to the grave." This involves their manufacture, their performance, and their disposal.

The manufacturing processes involved in the production of the lubricant and its components can be easily identified. The common synthetic base stocks are produced by chemical processes that are typically no more damaging to the environment than the production of mineral oil base stocks. Some synthetic base stocks, such as esters, can be derived from renewable resources. This can be regarded as environmentally beneficial as the source completes the carbon cycle and goes some way toward counteracting the effect of base-stock combustion on atmospheric carbon dioxide levels. The use of any synthetic base stock that is manufactured by processes that are environmentally damaging should be reviewed in detail. This of course applies to any raw material used in any process and is not unique to synthetic lubricants.

The performance obtained from synthetic engine oils can be clearly beneficial to the environment. Synthetic engine oils can:

Extend the useful life of an engine before a high-resource rebuild or replacement.

Maintain the engine at peak performance so as to minimize emissions and maximize fuel conservation. (Reducing fuel consumption is a critical part of global environmental strategy).

Reduce oil consumption so as to minimize damage to exhaust after-treatment systems and reduce lubricant-derived pollutants.

Potentially extend the drain interval of the engine oil and so reduce the quantity of used engine oil that needs to be disposed of.

Environmentally considerate disposal of used engine oil is viewed with increasing concern. Used synthetic engine oil disposal needs to be considered in some depth. Collection and recycling (or controlled burning at high temperatures) are the environmentally preferred methods of used engine oil disposal. The first thing that is achieved is the discouragement of environmentally inconsiderate disposal such as dumping. It is important that used synthetic engine oils do not make an increased environmental demand on the recycling process. When used synthetic engine oil is drained and enters the collection system there is no way of preventing it from entering the recycling system. The addition of suitable used synthetic engine oils to the recycling pool can help to produce a better-quality recycled base stock in terms of viscometrics and stability. It is clear, however, that if the used synthetic engine oil interferes with the recycling process, then it may cause major environmental concerns.

Ultimately, the ability of used synthetic engine oil to be recycled mixed with used mineral engine oil may put a constraint on the use of some synthetic base stocks in the consumer market. There does not appear to be a problem at present. However, this needs to be considered in some detail if novel synthetic (or natural) base stocks are to be used.

VII. CONCLUSIONS

Synthetic engine oils have ceased to be a novelty and are now a small but well-accepted part of the engine oil market in most parts of the world.

Synthetic engine oils represent an attempt to formulate the best possible engine oils that will avoid potential field problems before they occur.

Synthetic engine oils can provide many different benefits, and combinations of benefits, and these can be chosen according to the target application and its needs. Synthetic engine oils should therefore be viewed as a series of different products rather than as a group of similar products. This optimization of performance for applications can make synthetic engine oils even more valuable.

Engine tests used by the lubricant industry are defined around mineral engine oils. As a consequence, they may not allow the full benefits of synthetic engine oils to be demonstrated. New test methods may need to be considered.

One area where synthetic engine oils can provide significant benefits is in terms of reducing the contribution of engine oils to environmental damage. This can be achieved by better protecting and maintaining the engine and by reducing the direct impact of the consumed and used engine oil on the environment.

The knowledge base for synthetic engine oils is continuing to grow, and there is much more to learn and benefit from. As this knowledge base develops into new products, it is reasonable to expect the market penetration of synthetic engine oils to increase far above its current level.

REFERENCES

1. Coffin et al. (1989). *The Application of Synthetic Fluids to Automotive Lubricant Development, Trends Today and Tomorrow*, IIIe Symposium CEC, Paris.
2. Hammond, B. (1987). Synthetic lubrication, *Motor Ind. Rev.*, 2, 132–147.

3. Bartz, W. J. (1984). Synthetic lubricants and operational fluids, *Proc. 4th Int. Colloquium*, Esslingen.
4. SAE J183 (1990). *Engine Oil Performance and Engine Service Classification* (Other than Energy Conserving), June.
5. Cahill, G. F. (1989). Evolution of the CCMC Engine Lubricant Sequences, *CEC IIIrd Int. Symp. Automotive Fuels and Lubricants*. Paris, France.
6. Van der Waal, G. (1987). Improving the performance of synthetic base fluids with additives, *J. Synth. Lub.*, 4, 267.
7. SAE J300 (1989). *Engine Oil Viscosity Classification*, June.
8. May, C. J., and J. J. Habeeb. *Lubricant Low Temperature Pumpability Studies—Oil Formulation and Engine Hardware Effects*, SAE 890037.
9. Stewart et al. *Summary of ASTM Activities on Low Temperature Engine Oil Pumpability*, SAE 821206.
10. Richman, W. H., and J. A. Keller *An Engine Oil Formulated for Optimised Engine Performance*, SAE 750376.
11. Zurner and Gotre (1987). *Mineraloel-Technik*, *32*, No. 12.
12. Didot F. E., and T. F. Lonstrup *Low Temperature Viscosity Characteristics of Used Engine Oils*, SAE 800366.

16
Automatic Transmission Fluids

Douglas R. Chrisope
Ethyl Petroleum Additives, Inc.
St. Louis, Missouri

James F. Landry
Castrol Specialty Products
Irvine, California

I. INTRODUCTION

The main objective of this chapter is to detail performance differences between synthetic automatic transmission fluids (ATFs) and those based on mineral oils. In addition, the performance areas little affected by base fluid are also discussed. Following a brief introduction and historical review, various types of fluid are discussed and their performance levels compared. Then commercial practices are touched on before a brief summary.

A. Applications

Automatic transmissions are present in about 80% of all passenger cars in the United States. They are also prevalent in light trucks and buses. Automatic transmission use in Japan is rapidly approaching the U.S. level, while in Europe only about 15% of passenger cars are so equipped. ATFs are often used in manual transmissions in cars and light trucks because they flow better at lower temperatures than do typical gear oils.

Specialized vehicles, such as tractors, earth-moving equipment, garbage trucks, and cement trucks, often have hydraulic systems and various wet clutch and wet brake components requiring a specialized fluid. Some ATFs can qualify for this type of use by meeting the manufacturer's specifications. ATFs are sometimes recommended for use in the transmissions of agricultural implements, snowmobiles, helicopters, military tanks and personnel carriers, and various marine applications. While automatic transmissions have a greater presence than is commonly recognized, roughly one-half of domestic ATF production finds its way into nonautomotive applications.

B. General Performance Requirements for ATF

ATFs are the most complex functional fluids known to the automotive lubricant industry. Modern automatic transmissions demand a fluid that functions as a torque transfer medium, a hydraulic fluid, a gear lubricant, and a heat-transfer agent, while resisting oxidation from exposure to air. Other properties needed are compatibility with seal materials and resistance to rust, corrosion, and foaming. Lastly, but of primary significance, the fluid must provide the proper frictional profile in a wet clutch system. A wet clutch system usually involves single or multiple steel plates interacting with clutch plates made of asbestos, paper or other fibrous compositions, bronze, or graphite. We will show that synthetic ATFs can provide significant performance advantages over ATFs based on mineral oils.

C. Evolution of Modern Fluids

1. Historical Development

The evolution of automatic transmission fluids to the sophisticated, modern types follows closely the development of the transmissions that depend on them (1, 2). Until the 1930s, driving the typical automobile required considerable skill and physical strength, a considerable disincentive to universal use of the passenger car. As better roads and increasing vehicle speeds emphasized the problems in conventional transmission designs, the car makers became active in developing a self-shifting transmission. This effort culminated in the Hydra-matic model 180 in the 1940 Oldsmobile.

Almost all of the elements of modern automatic transmissions were incorporated in the Hydra-matic design: fluid coupling (although without torque multiplication), three planetary gear sets, logic and control by hydraulic computer, and torque transmission through multiplate clutch packs. The modern driver merely had to start the engine, place the transmission in Drive, and drive away.

This admirable feat of engineering was, however, initially dependent on the same lubricant as the engines of the day, straight mineral oil. GM realized that the success of this most complex mechanism relied on the lubricant quality. Car owners were told reliability could only be ensured by the use of a special fluid (mineral oil with antioxidants and friction modifiers) provided exclusively by GM dealers. This rather narrow distribution network lasted a few years before GM instituted a qualification system that made appropriate fluids widely available.

In 1949, the Type A transmission fluid was introduced, along with an approval system and endorsements by GM and Ford for its use in their transmissions. The Type A specification set minimum requirements in various bench, rig, and vehicle tests; viscosities, oxidative stability, and durability in repetitive clutch-plate engagement tests ensured satisfactory performance in service. Additional requirements caused GM to revise the specification in 1956, to Type A, Suffix A. Ford and Chrysler introduced their own oxidation tests in this period, which caused some divergence in the specifications.

Along with a specification revision in 1967, GM introduced the DEXRON trademark (3). Fluid approval now involved licensing to display the trademark on the ATF container. GM's next revision in 1973, called DEXRON-II, involved a contemporary transmission for oxidation and friction cycling tests (4). A wear test and a more strenuous friction retention test were also introduced.

Ford introduced its own trademark, MERCON, for ATF qualification purposes in 1987 (5). GM's latest revision, in October 1990, called DEXRON-IIE, required improved oxidation resistance and low-temperature fluidity, primarily for electronically controlled transmissions (6). Modern ATF formulations must satisfy these two specifications to compete in the North American marketplace.

Japanese and European original equipment manufacturers (OEMs) emphasize different properties than do the U.S. car makers. Chief among these are more stringent frictional requirements and better shear stability. The viscosity loss due to mechanical shearing of the polymeric thickeners is less acceptable to non-U.S. manufacturers. They do not, in general, follow the U.S. model regarding fluid qualifications.

The continuous evolution of the automatic transmission and the ever greater demands on the fluids have resulted in sophisticated, multifunctional fluids that bear little resemblance to the mineral oils first used in the 1940 Hydra-matic.

2. OEM Requirements

In general, ATFs must have the DEXRON-IIE and MERCON approvals to be commercially viable. These approvals require submission of a candidate fluid to one of two independent testing facilities recognized by Ford and GM. The fluids are tested according to published procedures and the results are submitted to ATF committees at the OEMs. Successful candidates are assigned a unique identification number and a license to use the DEXRON or MERCON trademarks in connection with the fluid. This process is required for every combination of additive package and base oil desired.

The tests required by the DEXRON-IIE and MERCON procedures are rather similar and include chemical and bench tests as well as rig and vehicle tests, which are summarized in Table 16.1.

The Allison Transmission Division of GM builds heavy-duty transmissions that are used mainly in buses, tanks, and personnel carriers. Its most recent specification is called C-4. The C-4 fluids are similar to DEXRON-II fluids but have additional elastomer compatibility and frictional requirements. This qualification is very important commercially.

Caterpillar TO-4 is another heavy-duty specification for off-highway applications. Frictional properties with various clutch-plate materials are significant with these fluids.

Mercedes-Benz accepts candidate fluids for their own in-house testing program and field tests, as does Zahnradfabrik Friedrichshafen AG (ZF). Satisfactory completion allows listing in the service manuals of fluids containing the passing additive system.

Hydraulic fluid approvals involve pump tests specified by the manufacturers. Vickers, Hägglunds-Denison, and Sundstrand each have their own procedures that test for wear, flow degradation, corrosion, and general parts condition.

3. Increasing Performance Requirements

There is a trend in automotive drive trains toward higher loading and temperatures. Automotive transmissions keep growing smaller and lighter while engine power and efficiency are increasing. Four-speed automatics may be replaced by five- and six-speed models, maintaining the pressure to downsize all components. Continuously slipping torque converters represent a new design, which "locks up" at lower speeds. This provides better fuel economy and less noticeable lock-ups but demands more thermally stable fluids. Advances in aerodynamic body design mean less air flow around transmissions for cooling. Increasing warranty periods are demanding greater reliability in the drive train, and automatic transmissions are historically a most troublesome item.

Competitive pressures keep the auto makers looking far into the future for new designs. Often, in the development of a new transmission, the fluid is found to be the

Table 16.1 Tests Used in ATF Qualifications

Test[a]	DEXRON-IIE	MERCON
Miscibility	Miscible with reference fluid	
Kinematic viscosity, D-445	Report at 40 and 100°C	6.8 cSt minimum at 100°C
Brookfield viscosity, D-2983	3,000 cP maximum at −20°C	1,700 cP maximum at −18°C
	20,000 cP maximum at −40°C	50,000 cP maximum at −40°C
Flash point, D-92	160°C min	177°C min
Copper corrosion, D-130	No blackening with flaking	1b
Rust, D-665-A	No rust	No rust
Rust, D-1748 Mod	No rust	—
Wear test, D-2882 Modif	15 mg weight loss	
Color, D-1500	6.0–8.0 (red)	
Foaming	No foam	D-892 no foam
Elastomer compatibility	7 Materials	3 Materials
Oxidation	Turbo Hydra-matic oxidation test (THOT)	Aluminum beaker oxidation test (ABOT)
	Fluid analysis	
	Parts rating	
Friction durability	Clutch plate test, 24,000 cycles	Clutch plate test, 4,000 cycles
	Band clutch test, 24,000 cycles	
Transmission cycling	THCT 20,000 cycles	
	Stable shift times, fluid and parts analysis	
Shift feel	In-vehicle test vs. a reference fluid	

[a]ASTM methods are given where applicable.

limiting factor in the system. Then a new level of fluid performance is required to advance the technology.

Fleet operators are under constant pressure to optimally maintain an expensive set of vehicles. A balance is sought between more frequent maintenance with standard lubricants and less frequent service with higher-cost, extended-drain lubricants. A comparison of total cost for each program is often surprising by favoring a more costly synthetic ATF.

D. Synthetic ATFs

Synthetic ATFs are best understood by considering the components of a conventional ATF. The typical ATF on a store shelf bearing the DEXRON and MERCON trademarks is 10–15% additives in a high-quality base oil. One-third to one-half of the additive is a viscosity index (VI) improver to increase the fluid viscosity at higher temperatures and to improve fluidity at low temperatures. The remainder of the additive package consists of dispersants to suspend carbonaceous oxidation products, antioxidants, antiwear components, agents to resist rust and corrosion of metals, antifoaming components and additives to adjust friction levels on the wet clutches, and perhaps diluent oil to aid in the blending of the additive concentrate.

A synthetic ATF may or may not contain VI improver, depending on the viscosities of the base fluid at high and low temperatures. In general, synthetic ATFs will require less VI improver than mineral oil-based fluids because of the intrinsically higher VI of the synthetic base fluids. The remaining components listed above will be necessary, although some adjustments may be required for the change in base fluid. Depending on the properties needed, the synthetic base fluid may be more polar or less polar than mineral oil. Antioxidant response often changes with base fluid and must be considered.

1. Part-Synthetic ATFs

Low-temperature fluidity of conventional ATFs can be improved considerably by adding 10–20% polyalphaolefins (PAOs) to the fluid. This inhibits the formation of a wax gel at temperatures down to –40°C. The cost of this method is generally not competitive with other means of improving the low-temperature performance of mineral oils. Other advantages accompanying the small amount of PAO in these mixed fluids, such as improved oxidative resistance or flash points, are small or negligible (7, 8).

2. All-Synthetic ATFs

Effects on elastomeric seal materials make most synthetic base fluids unsuitable if used alone. PAOs often cause seal shrinkage and can be poor solvents for some additives. Seal swelling additives can be used at 3–5% doses to achieve the required seal compatibilities. Esters of various classes (adipates, azelates, sebacates, dimerates, and phthalates) range from being too highly swelling to swell-neutral, but few have viscometric properties appropriate for ATF. Ester mixtures of appropriate viscosities usually fail seal tests or tests dependent on surface-active components, such as rust, copper corrosion, or wear tests. Esters are much more polar and surface-active than are mineral oils or PAOs and compete with current antiwear and antirust/anticorrosion agents. These additives, developed for mineral oils, perform nicely in the even less polar PAOs. Esters, however, can disrupt the delicate balancing of surface activities required to simultaneously deal with wear, corrosion and friction issues (7). A PAO–ester mixture often provides the best combination in terms of seal compatibility, antagonism toward surface-active agents, and cost.

II. COMPARATIVE PERFORMANCE

A. Viscometric Properties

1. High-Temperature Viscosity

Viscometric behavior is one area where synthetic ATFs are clearly superior to their mineral oil counterparts. If a single fluid is to meet both the DEXRON-IIE and MERCON specifications it must have a kinematic viscosity at 100°C of at least 6.8 cSt and a –40°C Brookfield viscosity of no more than 20,000 cP. The way this is achieved can be quite different for synthetic and mineral oil-based fluids.

The high-temperature viscosity of a functional fluid is dictated by application requirements, and the specification limits were chosen with these in mind. To the extent that wear is controlled by hydrodynamic mechanisms, higher viscosities are preferable. The pumps used to generate pressure for hydraulic control and circulation in automatic transmissions are designed for a certain viscosity range, and leakage around vanes or other surfaces leads to insufficient pressures. Simple gear rattling and vibration can be damped somewhat by a sufficiently viscous fluid. Shear-heating is worse with higher-viscosity fluids, and viscous drag on cold start-ups places an upper limit on viscosity.

In mineral oil formulations, the base oil selection is controlled by the low-temperature properties. Neither the high- nor the low-temperature viscosities are suitable without extensive modification by additives. Once the low-temperature issues are known, the viscosity at 100°C is adjusted by the use of polymeric VI improvers. Achieving viscometric targets will be discussed later in conjunction with low-temperature behavior.

Viscosity index improvers in ATF are generally mixed polymethacrylates or styrene–maleic ester copolymers. Conventional ATFs contain 1–5% of such polymers, which act as both a thickener at operating temperatures and a pour-point depressant at low temperatures. For synthetic base fluids a pour-point depressant is not necessary, but olefin copolymers (OCPs) or styrene–diene copolymers can be useful. The mechanism of viscosity increase by polymers has been examined and involves thermally induced extension of the backbone and side chains to maximize the volume occupied by the molecule. This increased molecular cross-section resists the flow of the base oil (9–12). Under high-shear conditions the apparent viscosity increase is somewhat smaller.

Polymethacrylates and styrene–maleic ester copolymers have varying degrees of resistance to mechanical shearing. A polymer of relatively high molecular weight, 100,000–200,000, will lose viscosity faster during use than one of a lower molecular weight, say 30,000–60,000. A typical North American ATF might lose roughly 20% of its viscosity in service or testing (13). European and Japanese ATFs, because of OEM requirements, have markedly better shear stabilities, with only 5% losses being common. If the fluid is used long enough, some thickening due to oxidation will occur and counteract this shear loss.

Another result of high-shear conditions is temporary viscosity loss. Base fluids, in general, do not suffer any viscosity loss under high-shear conditions. Fluids containing VI improvers, however, exhibit lower high-shear viscosities than one might predict from low-shear measurements. That is, the viscosity increase due to VI improver is partly invisible at high shear rates, which are common in automatic transmissions. This effect is related to molecular weight, so polymers that suffer less permanent viscosity loss from shearing also exhibit less temporary viscosity loss under high-shear conditions (14).

Data demonstrating permanent and temporary viscosity loss for some ATFs are shown in Table 16.2 (15). The fuel injector shear stability test (FISST) forces fluid through a nozzle multiple times to shear down the polymer permanently. The tapered

Table 16.2 Kinematic and HTHS Viscosities of Various ATFs before and after Mechanical Shearing[a]

	Before FISST[a]		After FISST	
ATF	KV100[b]	HTHS[c]	KV100	HTHS
Synthetic A, no VI improver	7.19	2.39	7.16	2.39
Synthetic B, no VI improver	5.68	1.90	5.69	1.93
Synthetic C, VI improver	7.55	2.12	5.83	1.97
Mineral oil D	7.19	2.11	5.67	1.97
Mineral oil E	7.32	2.10	5.50	1.90

[a]Fuel injector shear stability test, ASTM D-3945-B.
[b]Kinematic viscosity at 100°C, ASTM D-445.
[c]Tapered bearing simulator, 150°C and 10^6 s^{-1} shear rate, ASTM D-4683.

bearing simulator measures high-temperature, high-shear (HTHS) viscosities at 150°C and a shear rate of 10^6 s^{-1}. This is also called dynamic viscosity. The results, not surprisingly, show a greater stability in viscosity for the fluids containing no VI improver. Fluids A and B are both synthetic ATFs containing no VI improver and do not change their kinematic or dynamic viscosities after shearing in the 20-pass FISST procedure. Fluid C, a synthetic fluid with VI improver, loses over 22% of its 100°C kinematic viscosity in the FISST. In fact, fluid C has a much higher kinematic viscosity than does B before shearing, but afterward they have about the same kinematic and HTHS viscosities. As expected, the conventional mineral oil-based fluids D and E lose 21% and 25% of their kinematic viscosities due to the breakdown of the VI improver. The performance advantage clearly lies with synthetics containing no VI improver.

2. Low-Temperature Performance

As mentioned before, the low-temperature properties of a conventional ATF are controlled by the wax content of the mineral oil base. Wax is simply the fraction of a paraffinic oil with little or no branching, allowing a stable solid phase. The more highly branched molecules of oil and PAOs solidify at somewhat lower temperatures. The detrimental effect of the wax is to form a three-dimensional network throughout the oil, practically solidifying it. The additives to counteract this are polymers called pour-point depressants, which limit the size of the wax crystals. This results in suspension of a microcrystals in a separate liquid phase.

The apparent viscosity of this mixture must be no more than 20,000 cP at –40°C to meet the DEXRON-IIE specification (16). The recent change in the Brookfield viscosity requirement has resulted in greater use of light oils and partial naphthenic oils. Specially dewaxed base oils require less of the pour-point depressing additives but still solidify without them.

In contrast, PAOs and ester base fluids are wax-free and easily meet the DEXRON-IIE requirements. Table 16.3 shows the viscosities of commercial PAOs and blends useful for ATFs. By using VI improvers, one can obtain lower Brookfield viscosities and maintain high-temperature viscosities, but at the cost of some shear stability.

The impetus for better low-temperature performance, reflected in the recent DEX-

Table 16.3 Viscosities of Commercial PAOs used in ATF

PAO or 50:50 blend	Kinematic viscosity[a] 100°C, cSt	Brookfield viscosity,[b] –40°C, cP
PAO 4	4.06	2,042
PAO 4/6	4.82	3,680
PAO 6	5.91	6,538
PAO 6/8	6.79	10,550
PAO 8	7.78	15,105
PAO 8/10	8.72	21,440
PAO 10	9.87	—

[a]ASTM method D-445.
[b]ASTM method D-2983.

RON-IIE specification, is the requirement for reduced viscosity in the new electronically controlled transmissions. Low-temperature viscosity affects automatic transmission performance in two major ways. Viscous drag in the torque converter during starting is greatly affected by fluid viscosity, and a more viscous fluid requires larger starter motors, batteries, and cables (16). This is counter to the trend toward weight reduction for improved fuel economy. The second area is simply poor operation of the transmission due to sluggish flow of the ATF (17). Some investigators discuss this in terms of a maximum viscosity consistent with good operation, about 5000 cP (16). With various ATFs, this translates into different minimum temperatures for satisfactory operation. The authors concluded that a synthetic ATF provides an operability gain of 10–11°C over a fluid of 50,000 cP at –40°C.

B. Oxidative Stability

The improved resistance to oxidation of PAOs and PAO–ester blends over mineral oils is well known and often discussed (18–21). The myriad of test types, conditions, and procedures in the literature makes comparisons or extrapolations from one fluid type to another tenuous at best. Here we will compare a competent synthetic ATF to conventional ATFs in oxidation tests used for qualifications.

The severe oxidation test defined in the DEXRON-IIE procedure is known as a THOT (tee-hot) or Turbo Hydra-matic oxidation test. This is simply a model 4L60 transmission driven in third gear at 1755 rpm by an electric motor but with no other load applied. Excess heat is generated in the torque converter, and cooling water through a standard in-radiator cooler maintains a constant temperature of 163 ± 1°C. Air is bubbled through the fluid at 90 ml/min for 300 h. Various pressures and temperatures are maintained within defined limits and reported. Fluid samples are taken at specified intervals throughout, and the condition of certain parts is rated at the end of the test. Fluids may fail due to fluid condition or parts ratings.

Typical indicators of oxidative stability are increases in viscosity, total acid number (TAN, ΔTAN), and infrared absorbance (IR, ΔIR) in the carbonyl region at 1725 cm^{-1}. Free radicals formed by oxidation generally react with base fluid molecules to crate larger molecules than were originally present. These larger molecules increase the viscosity. Because oxidation forms carboxylic acids, ΔTAN and ΔIR are good guides to the degree of oxidation suffered by the fluid.

The fluid comparison in the THOT is provided in Table 16.4 (22). Mineral oil A was the passing reference fluid for the test. Mineral oil B contains the same additive package (and VI improver) as the synthetic fluid to provide a fair comparison. The synthetic ATF clearly resists oxidation better than the mineral oil-based fluids, as shown by its lower ΔTAN and ΔIR values. Parts that are rated for sludge display less sludge with the synthetic ATF than the others. The change in viscosity of the synthetic fluid after 300 h is almost negligible. The results for the synthetic ATF at 600 h are very similar to mineral oil B at 300 h, superiority that must be attributed to the base fluid.

Also notice that the effects of oxidation on the –40°C Brookfield viscosities vary greatly. The synthetic fluid was initially about 11,000 cP and increased only 3,000 cP during the test, even to 600 hours. Mineral oil B started at 30,000 cP and suffered a 50% increase at 300 h. Mineral oil A was 35,000 cP when fresh, but >270,000 cP at –40°C after oxidation, beyond the point of usefulness. Some samples of A were too viscous to measure, and hence only a lower limit can be given.

Table 16.4 Comparison of Synthetic and Mineral-Oil-Based ATF in the DEXRON-IIE THOT[a]

Parameter	Requirement	Mineral A[b] (reference), 300 h	Mineral B,[c] 300 h	Synthetic 300 h	Synthetic 600 h
Pentane insolubles	No limit	0.57	0.00	0.05	0.00
ΔTAN	4.5 Maximum	4.20	1.78	0.73	2.55
ΔIR	0.55 Maximum	0.50	0.41	0.14	0.38
Viscosity, 100°C (cSt)					
Fresh	—	7.05	7.22	7.18	
Used	—	6.80	7.78	7.34	7.93
–40° Brookfield	Fresh	35,000	30,000	11,000	
viscosity (cP)	Used	>270,000	45,500	13,800	14,240

[a]Run in model 4L60 transmission as per reference 6.
[b]Average of 21 tests.
[c]Same additive as synthetic ATF.

The variation in high-temperature viscosity due to shear-down of the VI improver and subsequent oxidation is more dramatic when examined graphically. In Fig. 16.1 we compare kinematic viscosities during a THOT of the synthetic fluid tested above against a single run (not an average) of a typical mineral oil-based ATF. The graph shows that the mineral oil reaches its minimum viscosity at about 60 h, giving a viscosity loss of 11%. This is followed by an increase from the low point of 37% at 300 h. In contrast, the synthetic fluid hardly varies for 200 h and then slowly increases by 2.2% at 300 h and by 10% at 600 h. Similarly, a synthetic fluid containing VI improver will suffer a viscosity loss much like mineral oil-based fluids, but better oxidative stability should limit the subsequent viscosity increase. A synthetic ATF without VI improver would therefore exhibit the most stable viscosity during use.

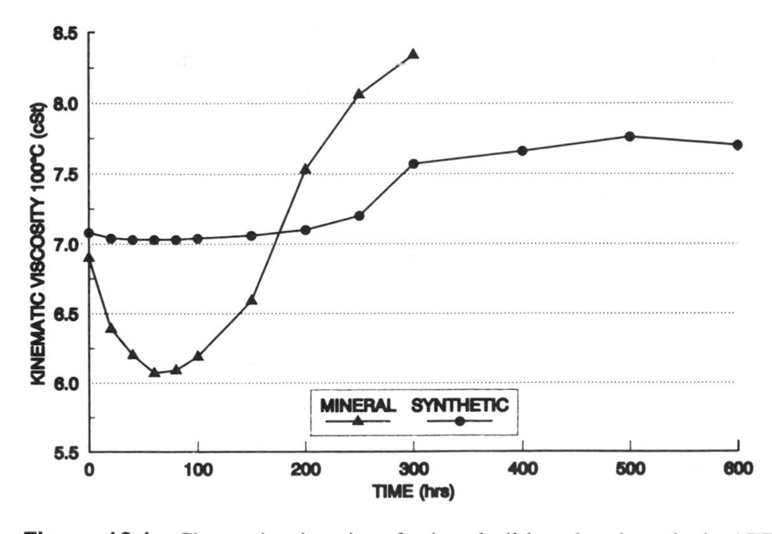

Figure 16.1 Change in viscosity of mineral-oil-based and synthetic ATFs during THOT.

The MERCON-required oxidation test is the aluminum beaker oxidation test or ABOT (ay-bot). This test is perhaps less severe than the THOT but probably more reproducible, since the testing apparatus is less complex. A fluid sample of 300 ml is heated at 155 ± 1°C for 300 h. Aeration at 5 ml/min is provided with agitation from a gear pump. Copper and aluminum strips are immersed in the fluid to check for sludge and deposit formation. Roughly the same tests for fluid oxidation as in the THOT are performed.

Comparison of a synthetic to mineral oil-based ATFs in the ABOT is made in Table 16.5. Again, mineral oil A is the reference fluid for this test and mineral oil B is based on the same additive package as the synthetic fluid. The single-length test results show that the synthetic fluid has no peer among mineral oils in resisting oxidation. The synthetic ATF must go 600 h to appear comparable to the mineral oils in a single-length test in viscosity increase and pentane insolubles. The ΔTAN and ΔIR are better for the synthetic fluid at 600 h than for the mineral oil-based ATF at 250 h. The difference in oxidative stability is again attributable to the base fluid.

C. Antiwear Performance

In a device as complex as an automatic transmission there is a variety of lubrication regimes. Depending on the application, wear is controlled by additive chemistry or fluid viscosity. The ATFs contain additives to control wear of metal parts, and the base fluid (provided it is of low polarity) exerts little influence on the outcome. One report demonstrates that a single ATF additive package in four different base fluids, including a synthetic with no VI improver, gave essentially identical results in both the FZG low-speed wear test and the MERCON vane pump wear test (13).

Conversely, other authors emphasize the importance of high-temperature high-shear (HTHS) viscosities as the factor controlling the hydrodynamic film strength (16). In an automatic transmission cycling test that is part of the DEXRON-II and MERCON specifications, the engine is brought from idle to nearly full throttle 20,000 times, causing many shifts in the transmission. The investigators found that pinion pins of the planetary gears showed differing amounts of wear depending on the HTHS viscosities of the ATFs, which all contained the same additive package. These pins cycle between boundary and hydrodynamic lubrication. There is an expectation that a greater film strength allows the pins to spend a greater fraction of their time in the hydrodynamic mode, leading to less

Table 16.5 Comparison of Synthetic and Mineral-Oil-Based ATF in the MERCON ABOT[a]

Parameter	Requirement	Mineral A, reference, single-length	Mineral B,[b] single-length	Synthetic	
				Single	600 h
Pentane insolubles	1% maximum at 200 h	0.24	0.16	0.17	0.26
ΔTAN	5.0 Maximum at 250 h	2.80	2.33	0.59	1.50
ΔIR	50 Maximum at 250 h	33.99	25.78	7.56	11.75
Viscosity increase 40°C, %	50% Maximum at 250 h	8.52	8.08	0.67	13.88

[a]Test procedure described in reference 5.
[b]Same additive as synthetic ATF.

wear. The ATF with the highest HTHS viscosity displayed the least wear on the pinion pins.

Hammond has also addressed the issue of film strength with synthetic base fluids (23). He claims unspecified synthetic base stocks have inherent film strengths of 22 MPa and above, while mineral oils average about 3.4–4.8 MPa. This larger effort required to remove a layer of synthetic lubricant is obviously very important when operating under hydrodynamic conditions.

D. Operating-Temperature Reduction

The literature contains several references to decreased operating temperatures and increased gear efficiency from the use of synthetic gear lubricants. Laukotka has shown a 20°C lower temperature in the FZG rig for a polyglycol-based lubricant relative to a mineral oil of equivalent viscosity at 40°C (24). Coffin and co-workers report a temperature decrease of 30°C in a manual transmission comparing a synthetic lubricant to a mineral oil (25). Temperature reductions and gear efficiency improvements are given by Jordan (26). O'Connor and Ross investigated axle efficiency increases from the use of PAO-based synthetic lubricants (27). These observations are usually explained in terms of improved traction, which is a lower coefficient of friction in the load zone between gear surfaces (26).

One might expect that similar temperature reductions or efficiency improvements could be observed in automatic transmissions. Very little work with automatics, however, has been undertaken. Such temperature reductions would be very welcome for their beneficial effects on the durability of parts and fluid. As a significant fraction of the heat generated in an automatic transmission is from the torque converter, the potential temperature reduction might be somewhat smaller than predicted from the manual gearboxes and differentials that have been studied. We must await further work in this area to progress beyond speculation.

E. Friction Retention

A typical clutch engagement involves rotating clutch plates pressing against stationary steel plates. By increasing the applied pressure, the torque is transmitted through the coupling, and the stationary plates begin to rotate. Toward the end of the engagement, or lock-up, the sliding speed difference between the mating plates approaches zero, and the friction changes from dynamic to static. The tendency for static friction to be higher than dynamic friction leads to a harsh or severe shift-feel. The ATFs contain additives to reduce the level of static friction, or more properly, the relative levels of static and dynamic friction. If these friction levels are very close together, the driver can experience a smoother or unnoticeable shift.

The base fluids typically play no direct role in the relative friction levels of wet clutches. The friction-modifying additives developed for mineral oils work just as expected in PAOs. It is possible that some esters, because of their greater polarity, could interfere with the friction modifiers. Likewise, severe oxidation of the base fluid creates polar compounds that might compete for the surfaces modified by these additives. More importantly, oxidation of the fluid directly oxidizes the friction modifiers to other compounds, which are no longer effective. The longer the fluid resists oxidation, the longer the original frictional properties remain. The superior oxidative stability demonstrated for synthetic ATFs thereby leads to extended retention of frictional properties.

F. Miscellaneous Issues

1. Seal Compatibility

As mentioned before, PAOs tend to shrink some seal materials unless blended with appropriate esters or additives to counteract this behavior. Formulators are accustomed to seal compatibility dictating the omission of certain components or chemical classes of additives, but the inclusion of seal-swelling additives can increase the complexity of the formulating task. The seal swellers must cause the proper amount of swelling in some materials without detriment to others. The types of compounds that swell elastomers are often antagonists for antiwear, antirust, and anticorrosion agents. A trend exists toward fluids that are compatible with an increasing number of elastomers.

2. Miscibility

Miscibility of mineral oil and synthetic fluids is a common concern. The OEM qualification tests require that candidate fluids be miscible with a mineral-oil reference fluid. Thus, a synthetic ATF can be topped off with a mineral-oil ATF and vice versa. Addition of mineral oil to a synthetic ATF, however, will increase its susceptibility to oxidation.

3. Foaming

An automatic transmission is particularly susceptible to generating excessive foam under certain conditions. There is greater possibility of foaming when the transmission is hot and filled above a certain level. Rotating gears, drums, and clutch packs can whip considerable amounts of air into the fluid. Unless the foam breaks down immediately, there will be sponginess in the hydraulic logic and control circuits and possible spewing of fluid out the filler tube. The tendency to foam is greater with increasing amounts of naphthenic oils in the fluid (16). Naphthenic oils are commonly added to mineral oil mixtures to achieve the new DEXRON-IIE maximum viscosity target of 20,000 cP at –40°C. A synthetic ATF, on the other hand, should achieve superior low-temperature performance without an inordinate tendency to foam. Few systematic studies in this area have been published (28).

4. Volatility

The PAOs and esters have a narrower distribution of molecular sizes than do typical mineral oils. This means that at a given temperature, a larger fraction of mineral oil will distill away than will a synthetic base fluid of equivalent viscosity. Considering that the base fluids in synthetic ATFs will probably have *higher* viscosities than a mineral oil base, the difference in volatility will be dramatic. Table 16.6 shows some simulated distillation data illustrating this difference.

5. Hydrolytic Stability

Hydrolysis is not a possibility with PAOs but is of some concern with ester-containing fluids. The typical base-stock-type esters will hydrolyze given water, acidic or basic catalysts, and suitable temperatures. Fortunately, transmissions usually achieve temperatures sufficiently high to drive off any water that gets into the system. Qualification for use in transmissions does not involve tests for hydrolytic stability, but approval for use as hydraulic fluids generally will require demonstration of hydrolytic stability.

III. COMMERCIAL PRACTICE

Although ATF manufacturers can supply data on volumes of ATF produced, they can not supply data on the great diversity of end uses. The 1990 worldwide demand for ATF was

Table 16.6 Simulated Distillation Data for Some PAOs and Typical Mineral-Oil-Based ATFs[a]

Percent distilled	Temperature (°C) for percent distilled				
	PAO 4[b]	PAO 6[b]	PAO 8[b]	DEXRON-II[c]	DEXRON-IIE[c]
5	391	391	402	334	329
10	402	402	432	346	345
20	412	412	468	363	361
50	422	476	491	396	388
90	468	525	537	449	434

[a]ASTM method D-2887.
[b]Data from Ethyl Corp., unpublished results.
[c]Data from reference 29.

estimated at about 210 million gallons of blended ATF. Less than 1% of this is synthetic ATF. The entire market is understood to be growing at about 1–2% per year, while the fraction of the market that is synthetic is growing considerably more quickly, perhaps 15% a year. There is a modest but growing recognition of the superior performance offered by synthetic ATF. The OEM interest in this area could spark further development and faster growth in this segment of the market.

IV. SUMMARY

We have demonstrated that synthetic ATFs have several performance advantages relative to mineral-oil-based fluids. The major ones are:

Viscometrics. The low-temperature characteristics of appropriate PAOs and esters are outstanding, easily meeting the 20,000 cP maximum Brookfield viscosity at –40°C of the DEXRON-IIE specification. While achieving this, or even 10,000 cP or less, no compromise is necessary in high-temperature viscosities, even without the use of VI improvers. The shear stability of fluids containing no VI improver is unmatched, as are the HTHS viscosities, leading to better lubrication and less wear of critical parts.

Oxidative Stability. In oxidation tests used for qualification, synthetic ATFs were shown to be superior to mineral-oil-based fluids, even with the synthetic fluids subjected to double length tests. Among other changes, the mineral-oil-based fluids suffer considerable increases in Brookfield viscosity at –40°C after oxidation, while the synthetic ATF examined had inconsequential thickening at this temperature. The better oxidative stability of synthetic fluids protects the additive package as well, maintaining the performance of friction modifiers and of antiwear and anticorrosion components much longer.

Operating Temperature Reduction. Lower temperatures in gearboxes using synthetic oils offer longer life for fluids and parts. The amount of the temperature reduction in automatic transmissions has not yet been established but will offer the same advantages.

REFERENCES

1. Gott, P. G. (1991). *Changing Gears: The Development of the Automotive Transmission*, Society of Automotive Engineers, Warrendale, PA, Chapters 1–6.
2. Deen, H. E., and J. P. Szykowski (1984). The evolution of today's versatile, multifunctional automatic transmission fluids, *Natl. Petrol. Refin. Assoc.*, Tech. Paper FL-84-82.
3. General Motors Passenger Car Automatic Transmission Fluid Bulletin (1967). General Motors Corporation, Ypsilanti, MI.

4. DEXRON®-II Automatic Transmission Fluid Specification, GM 6137-M (1973). General Motors Corporation, Ypsilanti, MI.
5. MERCON® Automatic Transmission Fluid Specification, Specification WSP-M2C185-A (1987). Ford Motor Company, Livonia, MI.
6. DEXRON® Automatic Transmission Fluid Specification, GM 6137-M (1990). General Motors Corporation, Ypsilanti, MI.
7. Boylan, J. B., and J. E. Davis (1984). Synthetic basestocks for partially synthetic motor oils, *Lub. Eng.*, 40, 427–432.
8. Willermet, P. A., C. C. Haakana, and A. W. Sever (1985). A laboratory evaluation of partial synthetic automatic transmission fluids, *J. Synth. Lub.*, 2, 22–38.
9. Flory, P. J. (1953). *Principles of Polymer Chemistry*, Cornell University Press, Ithaca, NY.
10. Port, W. S., J. W. O'Brien, J. E. Hansen, and D. Swern (1951). Viscosity index improvers for lubricating oils, *Ind. Eng. Chem.*, 43, 2105.
11. Alfrey, T., A. Bartovics, and H. Mark (1942). The effect of temperature and solvent type on the intrinsic viscosity of high polymer solutions, *J. Am. Chem. Soc.*, 64, 1557.
12. Evans, H. C., and D. W. Young (1947). Polymers and viscosity index, *Ind. Eng. Chem.*, 39, 1676.
13. Hartley, R. J., J. P. Sunne, and D. R. Chrisope (1990). *The Design of Automatic Transmission Fluid to Meet the Requirements of Electronically Controlled Transmissions*, Society of Automotive Engineers, Warrendale, PA, Tech. Paper 902151.
14. Alexander, D. L., and S. W. Rein (1980). *Temporary Viscosity Loss in Shear Stability Testing*, Society of Automotive Engineers, Warrendale, PA, Tech. Paper 801390.
15. Chrisope, D. R. (1991). Ethyl Petroleum Additives, Inc., unpublished results.
16. Watts, R. F., and J. P. Szykowski (1990). *Formulating Automatic Transmission Fluids with Improved Low Temperature Fluidity*, Society of Automotive Engineers, Warrendale, PA, Tech. Paper 902144.
17. Linden, J. L., and S. P. Kemp (1987). *Improving Transaxle Performance at Low Temperature with Reduced Viscosity Automatic Transmission Fluids*, Society of Automotive Engineers, Warrendale, PA, Tech. Paper 870356.
18. Blackwell, J. W., J. V. Bullen, and R. L. Shubkin (1990). Current and future polyalphaolefins, *J. Synth. Lub.*, 7, 25–45.
19. van der Waal, G. B. (1989). Properties and application of ester base fluids and PAOs, *Natl. Lub. Grease Inst. Spokesman*, 53, 359–368.
20. Gunsel, S., E. E. Klaus, and J. L. Bailey (1987). Evaluation of some poly-alpha-olefins in a pressurized Penn State microoxidation test, *Lub. Eng.*, 43, 629–635.
21. Gunsel, S., E. E. Klaus, and J. L. Duda (1988). High temperature deposition characteristics of mineral oil and synthetic lubricant basestocks, *Lub. Eng.*, 44, 703–708.
22. Chrisope, D. R. (1991). Ethyl Petroleum Additives, Inc., unpublished results.
23. Hammond, B. (1987). Synthetic lubrication, *Motor Industry Rev.* 2(Nov.), 132–147.
24. Laukotka, E. M. (1985). Lubrication of gears with synthetic lubricants, *J. Synth. Lub.*, 2, 39–62.
25. Coffin, P. S., C. M. Lindsay, A. J. Mills, H. Lindenkamp, and J. Fuhrmann (1990). The application of synthetic fluids to automotive lubricant development: Trends today and tomorrow, *J. Synth. Lub.*, 7, 123–143.
26. Jordan, G. R. (1983). Electric bills and the traction fraction (synthetic vs mineral lubricants), *Lub. Eng.*, 39, 491–495.
27. O'Connor, B. M., and A. R. Ross (1989). Synthetic fluids for automotive gear oil applications: A survey of potential performance, *J. Synth. Lub.*, 6, 31–48.
28. Hubman, A., and A. Lanik (1985). Air entrainment and foaming properties of synthetic lubricants at extreme temperatures, *J. Synth. Lub.*, 2, 121–142.
29. Kemp, S. P., and Linden, J. L. (1990). *Physical and Chemical Properties of a Typical Automatic Transmission Fluid*, Society of Automotive Engineers, Warrendale, PA, Tech. Paper 902148.

17
Gear Oils

Philip S. Korosec
Ethyl Corporation
Baton Rouge, Louisiana

I. INTRODUCTION

Gear oils are used to lubricate various types of "gears" designed to transmit power from one point to another in order to do work. The primary function of a gear lubricant is to provide a high degree of reliability and durability in the service life of gear equipment. Use of gear lubricants dates back to the early days of the industrial revolution when ways were being sought to generate and transmit more power, more efficiently and at higher loads than could normally be handled. Today the technology associated with gears, their function, and their lubrication has become very sophisticated. This chapter discusses the development and the use of synthetic gear oils and their role in both automotive and industrial gear applications.

To better understand the development and application of synthetics in gear lubrication, it is necessary to review some of the general principles of gears, gear lubrication, and how/why the technology developed. Some of the basic types of gears include:

Spur and helical.
Bevel.
Hypoid.
Worm.

These basic types of gears are described in Table 1. All gear "systems" are developed from these basic types of gears. A system, whether automotive or industrial, may contain

Table 17.1 Basic Types of Gears

Spur and helical gears
Spur and helical gears are in the form of a cylinder or disc on which the gears are cut. The spur gear has the teeth in a plane parallel to the axis, while the helical gear has the teeth cut in a spiral plane to the axis. The herringbone gear is a form of helical gear in that the teeth form a chevron pattern. The rack-and-pinion gear system has a spur gear running on a rack, giving a reciprocating motion.

Bevel gears
Bevel gears are gear systems where the axis centers intersect, transmitting power at almost any angle. The most common is 90°. These gears have teeth cut on a truncated cone and can be of the spur or helical principle, with the helical gear being a spiral bevel gear.

Hypoid gears
Hypoid gears are a form of spiral bevel gears. However, the axis of the pinion is below the axis of the ring gear. This type of gear system can withstand greater loads than the bevel gear system.

Worm gears
Worm gears consist of a worm and a worm gear. The worm and worm gear are special types of helical gear, with the worm resembling a screw.

only one type of gear or it may involve a combination of gears. However, most automotive applications use hypoid and spur gearing, while most industrial applications use bevel and worm gearing. The speeds may range from very low to very high under both light and heavy loads, depending on the type of work required from the gear system. These gear systems may be enclosed or open and exposed to the atmosphere. Speed, load, and climatic conditions all have an effect on the operating temperature of the particular gear system (3, 14).

Proper lubrication is essential to ensure optimum life of the gears. The gear lubricant should be of the type, grade, and quality to provide proper protection for the gears to function effectively for the intended application. Such lubricants should provide the following general performance characteristics (3, 12, 23):

Extreme pressure and antiwear protection.
Thermal and oxidative stability.
Foam suppression.
Ability to demulsify water.
Proper working viscosity.
Good low-temperature flow properties.
Seal compatibility.

In addition, gear oils need to be cost-effective, environmentally friendly, and in sufficient supply.

Technical developments and applications have occurred in both the automotive and industrial application sectors. I have chosen in writing this chapter to concentrate on automotive applications for synthetic fluids, since the bulk of technical development has occurred in this area (4, 32). The use of synthetic fluids in industrial gear applications is discussed later in the chapter as a separate section.

II. HISTORICAL DEVELOPMENTS

A. Automotive Gear Oils

Early automotive gear oils only met the minimum of performance requirements necessary to lubricate the various types of gear equipment. This requirement was generally based on providing a lubricant with the proper viscosity in the appropriate working temperature range. It was not until after World War II that the need for more formalized requirements was recognized. In the early 1950s the U.S. military developed the first automotive gear oil specification that defined the requirements of a lubricant suitable for use in specified military equipment (20, 21, 34). This was designated MIL-L-2105 and was the forerunner of the current technical requirements for automotive gear oils. Table 2 summarizes the development of this important specification from 1950 to present. As can be seen, the development and evolution of this specification defined certain performance criteria for gear oils, including test procedures for antiwear protection (L-37), extreme pressure (L-42), rust inhibition (L-33), and thermal and oxidative stability (L-60), as well as standardization of the chemical and physical properties for gear lubricants. It is important to realize that the military specifications and tests did not address the use of synthetic gear lubricants, although gear lubricants formulated with synthetic fluids must meet these requirements because the specifications are based on rigid performance criteria.

In 1960 the American Petroleum Institute (API) published a gear oil classification

Table 17.2 Performance Requirements of MIL-L-2105B, MIL-L-2105C, and MIL-L-2105D Military Specifications

Property	MIL-L-2105B	MIL-L-2105C	MIL-L-2105D
Moisture corrosion			
FTM 5326	(1-day test; 7-day test if required)	CRC L-33 (7-day test)	CRC L-33 (7-day test)
Copper strip corrosion (ASTM D-130)			
3 h at 121°C	2c maximum	3b maximum	3b maximum
Thermal oxidation stability (FTM 2504)			
Viscosity increase at 50 h, %	100% maximum	100% maximum	100% maximum
Pentane insolubles, %	3 maximum	3 maximum	3 maximum
Toluene insolubles, %	2 maximum	2 maximum	2 maximum
Foam suppression (FTM 3211)			
Sequence I, ml foam	300 maximum	20 maximum	20 maximum
Sequence II, ml foam	50 maximum	50 maximum	50 maximum
Sequence III, ml foam	300 maximum	20 maximum	20 maximum
High-speed shock loading axle test (two plain gears)	CRC L-42	CRC L-42	CRC L-42
High-speed/low-torque, low-speed/ high-torque			
Axle test (one plain gear, one lubricated gear)	CRC L-37	CRC L-37	CRC L-37
Heavy-duty field test, miles	100,000	100,000	100,000
Light-duty field test, miles	50,000	50,000	50,000
Rerefined base oils	Not allowed	Not allowed	Allowed

system that highlighted performance requirements based on criteria which all gear oils must meet (19). Table 3 summarizes the API classification system. Of the six classifications, only GL-4 for transmission oils and GL-5 for heavily loaded offset gears are of importance today. GL-1, GL-2, and GL-3 are of minor importance and can usually be satisfied by mineral oil alone or mineral oil containing small amounts of additive. GL-6 is obsolete. The API is currently working on definitions of two new classifications. These are PG-1 for modern manual transmissions and PG-2 for very heavy-duty offset gearing (22).

In 1987 the American Society for Testing and Materials (ASTM) published test procedures formalizing the testing required to meet both the military and API requirements (15). These test procedures are currently in use and are required for evaluation and qualification of all automotive gear oils, including those formulated with synthetic fluids.

B. Industrial Gear Oils

Lubrication specifications for industrial gear lubricants have historically been less formal than for the automotive sector. Gear systems are of many different types, used under a vast number of different conditions, and have been found not to be as critical regarding general lubricant performance under normal operating conditions (6). Currently, the standards of the industry are the American Gear Manufacturers Association (AGMA)

Table 17.3 API Gear Oil Categories

API category	Type of service	Type of gears and transmissions	Type of additive
GL-1	Mild conditions of low pressure and sliding velocity	Spiral bevel, worm Manual transmissions	Oxidation and rust inhibitors, defoamer, pour-point depressant; no extreme pressure (EP) or friction modifiers
GL-2	Worm gear axles and manual transmissions operating under such conditions of load, temperature; and sliding velocities that lubricants satisfactory for GL-1 service will not suffice		
GL-3	Moderately severe conditions of speed and load	Spiral bevel Manual transmissions	Mild EP
GL-4	Severe conditions of sliding speed and load (high-speed/low-torque and low-speed/high-torque)	Hypoid gears Truck and car manual transmissions	Medium EP; friction modifiers for limited-slip differentials
GL-5	High-speed shock load, high-speed/low-torque, low-speed/high-torque	Hypoid gears Truck and car manual transmissions	High EP; friction modifiers for limited-slip differentials
GL-6	High-speed, high-performance	High-offset hypoid gears	High EP; friction modifiers for limited-slip differentials

specifications for performance. The requirements for open and enclosed gear systems are listed in Tables 4 and 5 respectively (17, 18). The requirements define tests for measurement of the extreme pressure characteristics, oxidation protection, corrosion characteristics, water separation, and a variety of chemical and physical properties of the proposed lubricant. Most of the test procedures are defined by the ASTM. Again, I must emphasize that no specific specification exists for industrial gear lubricants formulated with synthetic fluids, but all such lubricants must meet these requirements. We discuss the use of synthetics in industrial applications later in the chapter.

C. Need for Synthetic Gear Lubricants

The needs for synthetic gear lubricants have only been identified recently. These needs have developed because of requirements from several original equipment manufacturers (OEM). These requirements are:

Table 17.4 AGMA Standard Specification 250.04, Lubrication of Industrial Enclosed Gear Drives

	Limits	
Property	R and O gear oils	EP gear oils
Viscosity index	90 minimum	90 minimum
Rust Protection		
ASTM D-665A	NR[a]	No rust
ASTM D-665B	No Rust	NR
Corrosion protection, ASTM D-130		
3 h at 100°C (212°F)	NR	1b maximum
3 h at 121°C (250°F)	1b maximum	NR
Oxidation stability, ASTM D-943, hours to TAN 2.0	500–1500 minimum[b]	NR
Oxidation stability, ASTM D-2893, viscosity increase at 99°C (210°F), %	NR	10 maximum
Foam suppression, ASTM D-892		
Sequence I volume of foam, ml[c]	75/10 maximum	75/10 maximum
Sequence II volume of foam, ml[c]	75/10 maximum	75/10 maximum
Sequence III volume of foam, ml[c]	75/10 maximum	75/10 maximum
Demulsibility, ASTM D-2711		
Total free water, ml	30 minimum	50–60 minimum[b]
Water in oil, %	0.5 maximum	1.0 maximum
Emulsion, ml	2 maximum	2–4 maximum[b]
Timken test, ASTM D-2782, OK load, 1b	NR	60 minimum
FZG, DIN 51-354, stages pass	NR	11 minimum
Cleanliness	Must be free from grit and abrasives	Must be free from grit and abrasives
Additive solubility	NR	Must be Filterable to 25 μm wet or dry without loss of EP additive

[a]NR, no requirement.
[b]Depends on AGMA grade of lubricant.
[c]After 5-min blow and 10-min rest.

Table 17.5 AGMA Standard Specification 251.02, Lubrication of Industrial Open Gearing

Property	Limits	
	R and O gear oils	EP gear oils
Viscosity index	90 minimum	90 minimum
Rust protection		
ASTM D-665A	No rust	No rust
ASTM D-665B	No rust	NR[a]
Corrosion protection, ASTM D-130		
3 h at 100°C (212°F)	NR	1b maximum
3 h at 121°C (250°F)	1b maximum	NR
Oxidation stability, ASTM D-943, hours to TAN 2.0	500–1500 minimum[b]	NR
Oxidation stability, ASTM D-2893, viscosity increase at 99°C (210°F), %	NR	10 maximum
Foam suppression, ASTM D-892		
Sequence I volume of foam, ml[c]	75/10 maximum	75/10 maximum
Sequence II volume of foam, ml[c]	75/10 maximum	75/10 maximum
Sequence III volume of foam, ml[c]	75/10 maximum	75/10 maximum
Demulsibility, ASTM D-2711		
Total free water, ml	30 minimum	50–60 minimum[b]
Water in oil, %	0.5 maximum	1.0 maximum
Emulsion, ml	2.0 maximum	2–5 maximum[b]
Timken test, ASTM D-2782[d], OK load, 1b	NR	45 minimum
FZG, DIN 51-354[d], stages pass	NR	9 minimum
Cleanliness	Must be free from grit and abrasives	Must be free from grit and abrasives
Additive solubility	NR	Must be filterable to 100 μm wet or dry without loss of EP additive

[a]no requirement.
[b]Depends on AGMA grade of lubricant.
[c]After 5-min blow and 10-min rest.
[d]Either 45 1b OK load or nine stages on FZG.

Improved thermal stability.
Lubrication over a wide temperature range.
Long drain or fill for life.
Fuel economy considerations.
Axle efficiency.
Quality.

Manufacturers, such as Mack Truck, Eaton Fuller, Navistar, Rockwell, Ford, General Motors, and several European companies, have developed or are in the process of developing specific criteria for synthetic gear lubricants (22). Synthetic gear lubricants are more suited for some of these requirements because of their superior properties and performance characteristics. We will discuss some of these in a later section.

As performance requirements increase for gear lubricants, their cost-effectiveness improves and they become more comparable with economics associated with their mineral-oil counterparts. Generally, synthetic gear lubricants are more expensive than conventional gear lubricants and their use becomes dependent on factors such as superior performance, quality, service life, and warranty (3).

Government regulations continue to be factor relative to the future of lubricating oils, including gear lubricants. Legislation mandating the registration of new chemicals makes new chemistry questionable economically and may require longer lead times for commercialization. Laws regarding the limits on certain elements that can be used in lubricating oils are pending. Certain base oils, such as those containing aromatics and high volatility characteristics, are being scrutinized by law makers. Biodegradability of lubricating oils is yet another environmental factor to consider, especially in Europe (12).

Quality is another important issue, both of the product as a lubricant and also its performance in the application intended.

All of these factors lend themselves toward the use of synthetic fluids for gear applications.

III. TYPES OF SYNTHETIC GEAR LUBRICANTS

Types of synthetic gear lubricants may include materials based on a wide variety of fluids and combinations including mineral-oil combinations. These fluids include the following (3, 33; also private communication):

Alkyl phosphates.
Aryl phosphates.
Alkylated benzenes.
Polybutenes.
Polyglycol ethers.
Polyol esters.
Esters of dibasic acids.
Synthetic hydrocarbons (PAO).
Silicates.
Polysiloxanes.
Silicones.

Most of these fluids are not discussed in this chapter, since for a variety of reasons they are not generally applicable to either automotive or industrial gear lubrication. The reasons may include general performance characteristics, operating temperature, cost, availability, environmental considerations, and the like. Of those fluids just listed, only the synthetic hydrocarbons, polyol esters, dibasic acid esters, and polybutenes have practical application to gear lubrication. Reference should be made to other chapters in this book for specifics on the use of other fluids as gear lubricants.

IV. PERFORMANCE PROPERTIES

A. Formulations

Synthetic automotive gear lubricants may be completely synthetic, combined with other synthetics, or mixed with conventional mineral oils to define the desired gear lubricant. Table 6 summarizes several combinations of synthetics used to formulate various grades of automotive gear lubricants. Note the use of synthetic hydrocarbons, dibasic acid esters,

Table 17.6 Part-Synthetic and Synthetic Gear Lubricant Formulations

COMPONENT	75W–90 mineral	75W–90 part-synthetic	75W–90 synthetic	75W–90 synthetic VII[a]	75W–140 synthetic	80W–140 part-synthetic	80W–140 Synthetic	80W–140 Mineral	80W–90 Mineral
EP additive	7.0	9.5	9.5	8.0	9.5	6.5	6.5	5.5	5.5
100 Neutral	55.0								
200 Neutral	18.0							64.5	43.0
600 Neutral		20.0				13.5			
150 Bright stock									51.0
PAO 4 cSt		24.0			23.0		26.0		
PAO 6 cSt				53.0					
PAO 8 cSt			24.0						
PAO 40 cSt			52.0			41.0			
PAO 100 cSt		24.5			51.0	17.0	50.5		
Diisodecyl adipate		20.0	15.0	9.0	15.0	20.0			
Ditridecyl phthalate							15.0		
Viscosity index improver	19.0			30.0				29.0	
Pour-point depressant	1.0	2.0			1.5	2.0	2.0	1.0	0.5
Viscosity at 100°C, cSt	14.20	14.45	15.48	16.81	26.9	25.62	26.29	25.9	14.2
Viscosity at 40°C, cSt	102.2	95.43	116.00	106.4	234.9	211.9	197.5	200.1	158.0
Viscosity index	138	157	140	172	148	153	168	141	101
Brookfield at –40°C, cP	138,000	128,800	71,000	143,000	—	—	127,000	145,000	—
Brookfield at –26°C, cP	—	—	—	—	38,300	47,900	—		—

[a]VII, viscosity index improver.

and polybutene in combination with mineral oils to meet the viscometrics required for a particular grade. Important basic features to consider are operating temperature, viscosity, general lubrication characteristics, oxidative stability, solubility and compatability, and the effect of these parameters on field service in terms of durability and efficiency of operation. We will refer to some of these lubricant grades in our discussion of some of these features.

B. Viscometrics

Synthetic gear lubricants operate over a different temperature range in terms of viscosity than do conventional mineral-oil-based gear lubricants. This range is important to the user, enabling the use of one lubricant for a variety of applications. The viscometrics for a PAO, for example, enable gear lubricants to be formulated for both low- and high-temperature performance without the use of polymeric viscosity index improvers, since the PAOs have a high "natural" viscosity index. Note the minimal use of viscosity index improvers in the PAO-based formulations in Table 6.

Among the most important factors of synthetics, especially PAOs, are their low-temperature viscosity properties. When oils are cold their viscosity may be many more times thick than at ambient termperature. This changes the lubricity of the lubricant and in extreme cases can cause damage to the gears. Synthetic gear lubricants have better low-temperature pumpability. Figure 1 illustrates this point. Note that a 75W–90 synthetic gear lubricant flows at temperatures as low as –40°C, while a conventional mineral-oil-based 80W–90 gear lubricant's flow is retarted at these very low temperatures (Eaton Fuller, 1989, private communication).

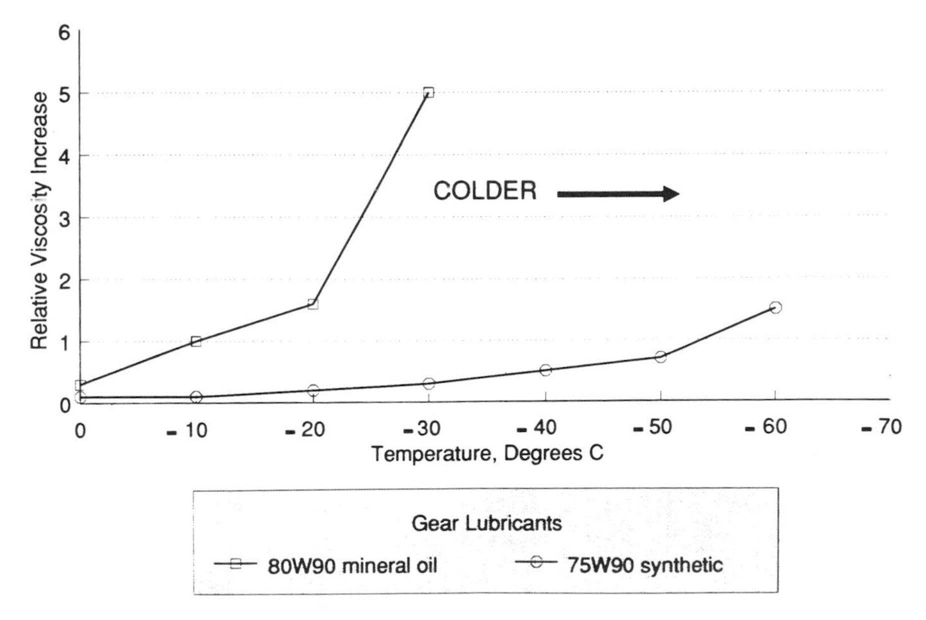

Figure 17.1 Low-temperature flow characteristics of synthetic vs. mineral-oil gear lubricants. (From Eaton Fuller, 1989, private communication.)

C. Shear Stability

Table 7 examines the shear stability of several gear lubricants as measured by an industry recognized test procedure (Uniroyal Chemical, 1980, private communication). This table shows that the viscosity of certain synthetic gear lubricants does not change as the gears churn and shear the lubricant. For example, comparison of a 75W–90 mineral oil and a synthetic gear lubricant shows different shear characteristics. The mineral-oil lubricant contains a viscosity index improver in order to meet the 75W–90 viscosity requirements. Note that the mineral-oil gear lubricant actually shears out of grade, while there is little shearing effect attributed to the synthetic gear lubricant. This is an important factor in the service life and durability of modern gear lubricants and one of the principal reasons for the increased use of synthetics in automotive applications. This property has been demonstrated in actual field service (2, 13, 25). Figure 2 shows what can happen to a viscosity-index-improved 75W–90 mineral oil when tested in two different trucks. The 75W–90 oil sheared out of grade, lowering the viscosity considerably. The gears from this test showed excessive wear indicative of the low viscosity. By comparison, the 75W–90 synthetic based on PAO showed very little shear loss and showed acceptable wear performance at the end of the test (13). This shear stability property also allows design of specific viscosities of gear lubricants, which allow the best axle efficiencies possible. We discuss the importance of this in a later section.

D. Oxidative and Thermal Stability

I have chosen to combine the discussion of these two properties since in practice their effect goes hand in hand. Because most synthetic stocks are synthesized (pure) compounds, their thermal and oxidative stability is usually better or can be more closely controlled than for mineral oils. This effect is translated into gear oils that can operate at higher temperatures and have potential for longer service life. The oils remain cleaner, free from oxidative sludge and grit, and more tolerant of seal deterioration. Many tests have been used to evaluate gear lubricants for their oxidative and thermal stability (1, 15, 16). Table 8 lists these tests, and Table 9 illustrates the various oxidative/thermal stabilities of several synthetic/mineral-oil gear lubricants. As the data show, the viscosity increase for the synthetic gear lubricants is generally less than for mineral oils indicating better temperature stability. Also, measurement of acidity, insolubles in certain hydrocarbons, and visual appearance of the oil and gear surfaces indicate the superiority of synthetic gear lubricants.

Table 17.7 Shear Stability of Gear Lubricants[a]

Fluid	Viscosity, KV 100		% Viscosity loss
75W–90 VII	14.45	11.25	22.1
75W–90 synthetic	14.25	14.05	1.4
75W–140 synthetic	24.93	24.84	0.4
80W–140 VII	29.63	26.48	10.6
80W–140 synthetic	25.71	25.54	0.7
80W–90 reference	13.80	13.59	1.5

[a]Cannon shear method, 3450 rpm, 8 h. VII, viscosity index improver.

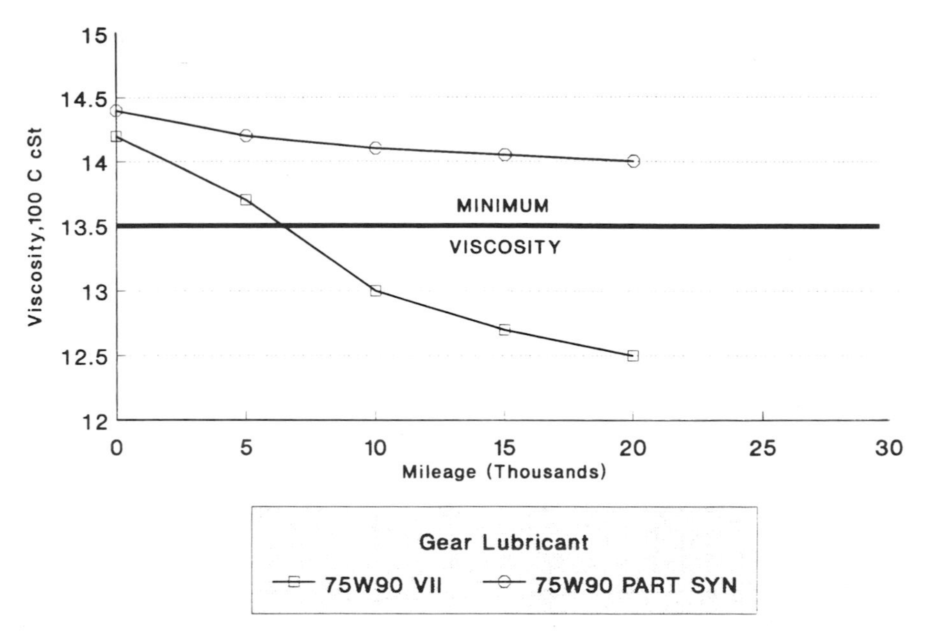

Figure 17.2 Effect of shearing on the viscosity of a part-synthetic vs. mineral-oil gear lubricant in actual field service (13).

E. Extreme Pressure Performance

In general, automotive gear lubricants based on synthetic fluids do not exhibit extreme pressure or antiwear properties any different than mineral oils. Most formulated gear lubricants contain extreme pressure/antiwear additives designed to meet the necessary performance requirements. There have only been a few reports of enhanced performance with the use of synthetic gear lubricants, whether they are based on PAO, diester, or polybutene chemistry (4, 5). It has been observed, however, that certain diesters in very high concentrations (50%) perform poorly in heavily loaded hypoid axles, and use of either special additives or higher dosages of standard extreme pressure additives is

Table 17.8 Comparison of Thermal Oxidation Stability Tests

Test	L-60	Indiana stirring	ZF/Renault/Peugeot (GFC)
Time (h)	50	96	200
Temperature (°C)	163	135	150
Air flow (liters/h)	1.1	0	10.0
Catalysts	Copper, steel	Copper, steel	None
Method of agitation	Large and small spur gears (1725 rpm)	Glass stirrer (1300 rpm)	Air turbulence
Equipment	Steel case with glass front	Glass beaker	Glass Erlenmeyer flask

Table 17.9 Thermal Oxidation Stability Test Results of Automotive Gear Lubricants

Lubricant	75W–90 mineral	75W–90 part-synthetic	75W–90 synthetic	75W–90 synthetic/VII	80W–90 mineral
L-60					
Viscosity inc, %	25.49	19.71	28.49	36.9	29.69
Pentane insolubles wt%	0.58	1.40	0.70	0.81	1.14
Toluene insolubles wt%	0.25	0.32	0.10	0.40	0.65
Deposit rating[a]	0.90	3.26	8.64	8.60	0.94
Indiana stirring					
Viscosity inc, %	0.2	2.8	4.5	9.1	-1.0
Acid number change, %	–22.0	15.1	56.9	8.1	2.4
Hexane insolubles wt%	0.30	0.23	0.07	0.81	0.06
Laquer rating[b]	6	5	0	4	5
ZF/Renault/Peugeot					
Viscosity inc, %	17.1	16.4	16.4	17.5	17.1
Acid number change, %	96	170	124	83	125
Hexane insolubles, wt%	0.30	0.10	0.01	0.16	0.07

[a]10 = Clean.
[b]0 = Clean.

required (13). The automotive gear lubricants discussed earlier (Table 5) all passed the required extreme pressure and antiwear gear tests.

One important aspect, however, that can effect extreme pressure and/or antiwear performance of synthetic gear lubricants is the solubility of certain additives in these fluids at the working temperature. Insolubility can, in extreme cases, lead to inadequate performance and system failure because the additive simply falls out of solution. This is a particular problem with PAOs.

F. Solubility and Compatibility

The solubility of certain synthetic gear lubricants in state-of-the-art antiwear/extreme pressure additives may be problematic. As discussed already, most of these additives are only marginally soluble in PAO fluids, and either esters or solubilizing mineral oils must be used in combination. Likewise, certain esters are not soluble in mineral oil, and solubilizing agents are required. This phenomenon is important, since among the important characteristics of synthetics used for gear lubricants are their low-temperature viscosity and lubricity advantages, as well as claims of improved axle efficiency. At low temperatures additives must still function to prevent wear of the gear surfaces under high loads. Much work is going on in the industry to improve the low-temperature additive solubility characteristics of synthetics. Also, new additives designed specifically for synthetic gear lubricant application are under development.

As long as all components of the lubricant are soluble, no other compatibility and/or miscibility concerns have been observed relative to combinations of various synthetic gear oils.

G. Operational Efficiency and Fuel Economy Considerations

Corporate Average Fuel Economy (CAFE) regulations have motivated manufacturers to investigate every possible means of improving vehicle fuel economy, including gear lubricant improvements. Many U.S. passenger car manufacturers have replaced hypoid rear-drive axles with transaxles employing helical gears, lubricated mostly with automatic transmission fluid. Hypoid or spiral bevel gears dominate in the remaining segment of passenger cars and light and heavy-duty trucks and are lubricated by GL-5 quality gear lubricants. These vehicles consume over 90% of the total U.S. gear lubricant production. Improvement in gear lubricant efficiency as a means to save fuel would make a significant technical contribution to this segment of the vehicle population. New lubricants based on a variety of synthetic combinations have been developed to provide this higher degree of efficiency (13, 25).

Synthetic gear lubricants can offer significant improvement in energy conservation and high-temperature/high-torque performance. We discussed earlier in this chapter the importance of the viscosity–temperature relationship influencing the proper lubrication of axles. This relationship also effects the efficiency of these axles. Previous research and development of automotive gear lubricants have been aimed at minimizing churning losses by reducing lubricant viscosity. While this approach is applicable to improvement in efficiency, it has certain limits relative to film thickness at the mating surface of the gears, which in turn is controlled by the viscosity–temperature relationship. If the viscosity is too low, the lubricant film is lost, leading to friction and heat buildup on the asperities of the mating gear surfaces. This can lead to poor durability and shortened service life.

Synthetic gear lubricants offer a solution to this problem because formulations can be structured with different and unique viscosity–temperature relationships compared to conventional mineral oils. The magnitude of the improvement may depend on the application and service cycle, but there is a growing bank of information suggesting the effect is always positive. Several test methods have been used to measure fuel economy effects of lubricants and are usually variations of the following procedures:

Axle efficiency test rigs (8, 26).
EPA 55/45 F.E.T. (21, 33).
Two truck tests, SAE/DOT/ATA type (10).
Extended duration fleet tests (2, 13).

One researcher (22), for example, describes a test apparatus and procedure to evaluate combinations of PAOs and PAO/ester gear lubricants compared to conventional mineral-oil gear lubricants. Details of their test rig and procedure, as well as those of other researchers, are found in Ref. 11, 13, 21, 22, and 33.

Gear lubricants based on various combinations of synthetic fluids similar to those described in Table 6 were evaluated. The reference oil was an SAE 80W–90 grade gear lubricant. All oils were dosed with good-quality GL-5 extreme pressure additive.

These lubricants were evaluated by two procedures: a heavy-duty cycle representing truck applications and a light-duty cycle representing passenger car operation. Combined heavy-duty and light-duty results are summarized in Fig. 3. As can be seen from these data, all the synthetic gear lubricants showed improvement in axle efficiency over the 80W–90 reference oil. The 75W–90 synthetic lubricants provided the largest gains, but the 75W–140 lubricant also showed positive gains in efficiency.

Speed is also a factor that can effect axle efficiency. This is typical of what is

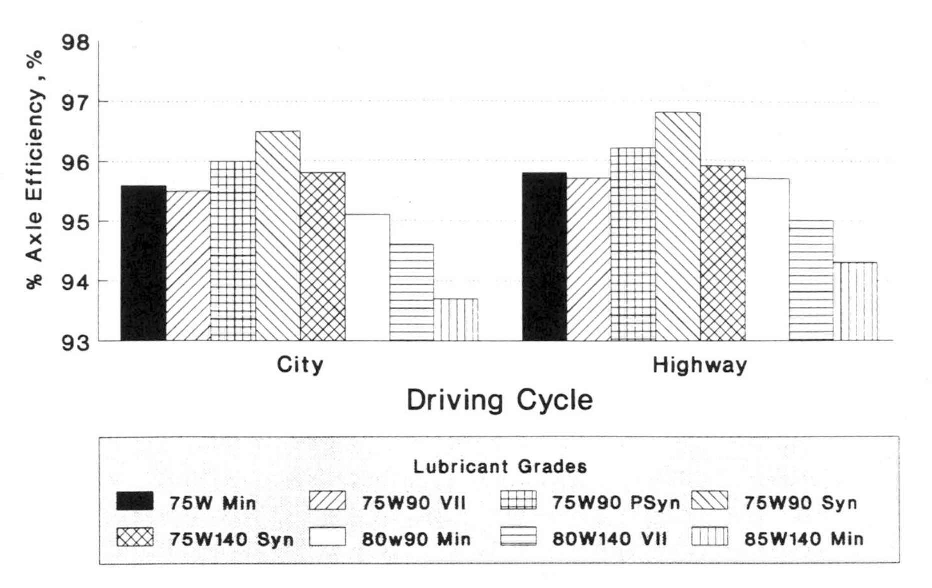

Figure 17.3 Axle efficiency of several synthetic and part-synthetic gear lubricants compared to conventional gear lubricants under city and highway driving conditions.

observed in actual field service. To show this effect, a typical duty cycle for both truck and passenger car was used in conjunction with an axle efficiency test. Each lubricant was evaluated at six road speeds. Lubricant response is shown in Figs. 4 and 5. The plot in Fig. 4 of the truck speeds shows that a 75W–90 synthetic lubricant exhibits better efficiency over the entire duty cycle. However, the data from the passenger car show a decrease in efficiency at high speeds, although the efficiency was positive over the entire cycle.

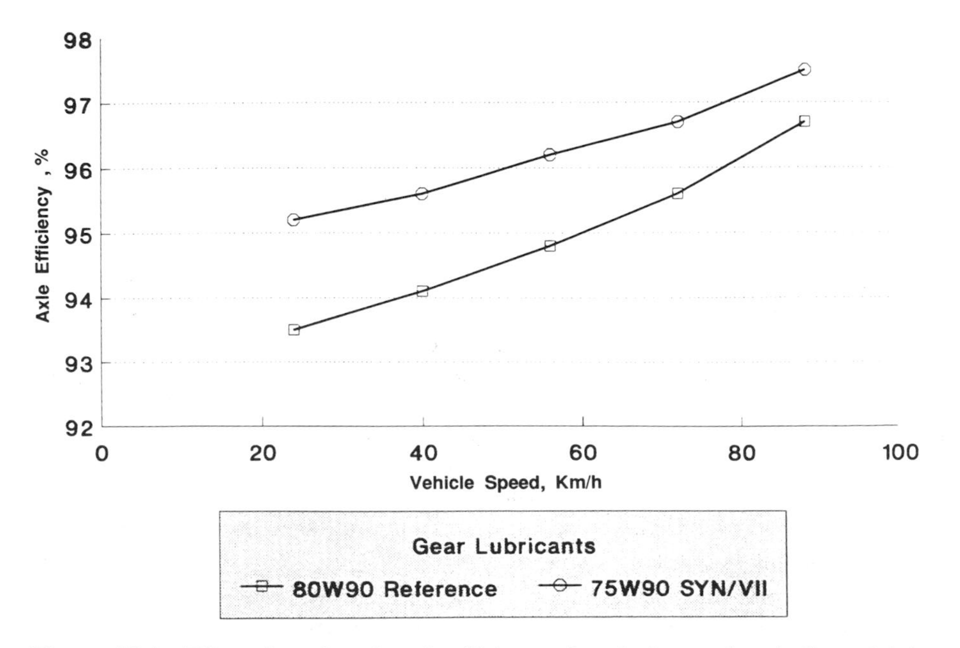

Figure 17.4 Effect of speed on the axle efficiency of synthetic vs. mineral-oil gear lubricants in heavy-duty trucks (22).

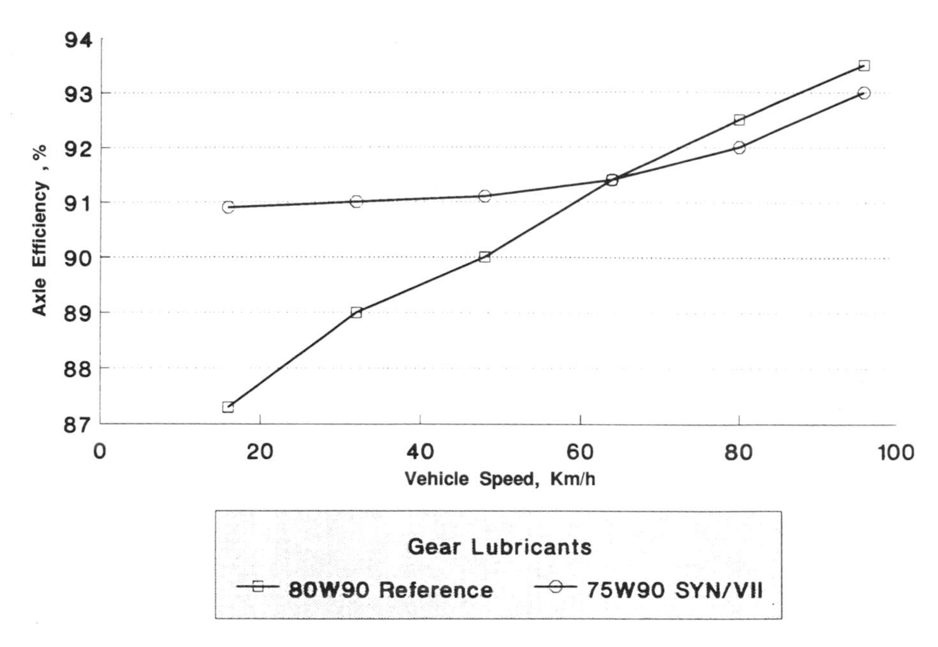

Figure 17.5 Effect of speed on the axle efficiency of synthetic vs. mineral-oil gear lubricants in passenger cars (22).

H. Field Performance

The earlier discussions of various gear oil properties and improvements seen with the use of synthetic lubricants must be demonstrated in actual field performance testing in order to demonstrate the improved durability, range of performance, and extended service life suggested.

Field tests have been conducted by several researchers on synthetic gear lubricants in automotive applications (2, 13). Results generally show that these synthetic lubricants offer the durability necessary for longer service life of the vehicles axles. This is shown in Fig. 6 by inspection of oils for wear by iron analysis after 50,000 and 100,000 miles of operations. Iron levels in a 75W–140 formulation were significantly less than those found in an 85W–140 mineral oil lubricant. The 75W–90 synthetic lubricant exhibited good wear protection as well (2). These results were confirmed by inspection of the gears at the end of the test and by performance against MIL-L-2105D.

Fuel economy has also been demonstrated in actual fleet service. Table 10 summarizes the fuel savings from three fleets showing values ranging from 1.3 to 8.3% in various types of operation for an extensive number of miles (2).

V. INDUSTRIAL GEAR LUBRICANTS

We have discussed and compared properties of automative gear lubricants formulated with synthetic fluids, but have not discussed their use in industrial gear lubricants. At present, this is a small application area for synthetic fluids but is expected to grow.

Most industrial gear oil specifications are based on performance criteria easily met by conventional gear lubricants. Refer to the AGMA specifications in Tables 4 and 5, as discussed previously. These lubricants are very cost-effective; therefore, incentive to

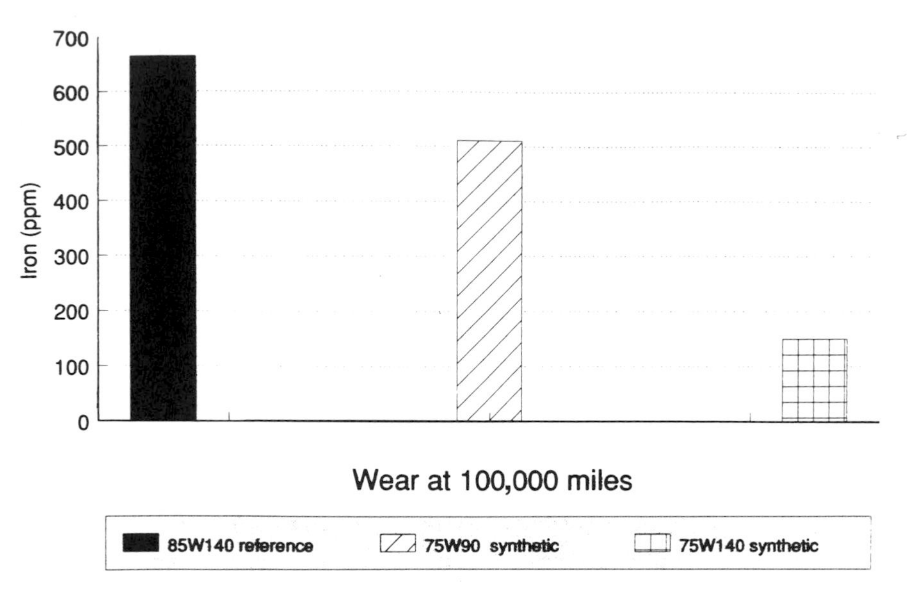

Figure 17.6 Wear characteristics of synthetic vs. mineral-oil gear lubricants in heavy-duty truck field testing (2).

change has been negligible. However, there is a growing amount of information regarding the energy savings afforded by using synthetic gear lubricants that could increase their use in a variety of industrial applications (6, 9). Most industrial plants have at least one gear system that operates beyond its original design specification. Sometimes these systems run continuously when they were designed to operate intermittently. At other times these systems run at higher temperatures or simply at higher loads. In each of these situations, however, the lubricant is usually stressed beyond its normal capabilities, and the result is frequently measured in terms of excessive gear wear and equipment downtime.

One solution to this problem is to run the equipment within original design. Alternatively, use of a synthetic gear lubricant in such systems is receiving more attention. Many plant operations also find that in addition to solving the tough lubrication problems, the fluids provide substantial cost saving in many less severe gear systems (31).

Operation over a wider temperature range can eliminate the seasonal changing of conventional lubricants from a thinner viscosity grade in winter to a thicker viscosity grade in the summer months. This feature alone can result in considerable savings in terms of labor and downtime costs, as well as eliminating the need to replace and inventory several grades of oil.

Table 17.10 Fuel Economy from Field Experience: Synthetic vs. Mineral-Oil Gear Lubricants[a]

Fleet	Miles	Fuel economy (%)
Line haul	1,300,000	1.3
Private delivery	113,000	8.3
Private delivery	221,000	6.1

[a]Reference lubricant 85W–140.

The higher viscosity index of the synthetic gear lubricants also means that it could still provide adequate lubrication should the system run hotter than normal for any reason. A comparable-viscosity mineral oil will thin out more quickly at higher temperatures and could fail to provide an adequate oil film to protect the gears from excessive wear.

Perhaps the most significant benefits of synthetic industrial gear lubricants are related to their improved thermal and oxidative stability. Not only will synthetics provide adequate lubrication at higher temperatures, but they also provide extended oil drain intervals at more moderate temperatures. For example, one power plant reported that it had increased the gear lubricant change interval on a fly-ash blower from weekly to once every 6 months, a 26-fold increase (9). Similarly, a nuclear power plant reported that it had increased the drain interval on their reactor coolant gear pump system from every 6 months to only once a year. That simple extension of gear oil life by 100% allowed the plant to eliminate downtime that cost $500,000 in the form of purchased electrical power (35).

Synthetic gear lubricants can also provide lower maintenance costs by reducing gear wear and the downtime required for gear replacement. The improved oxidative stability of synthetics means that substantially less sludge, varnish, and carbon deposits are formed in the gear box. Both laboratory and field performance has shown that reducing the levels of these deposits can decrease gear failures, sometimes by as much as 65% (35).

Other studies have examined the effects of viscosity and lubricant type on the formation of the oil film that keeps gear surfaces from rubbing against each other during operation. These studies have concluded that the wider temperature range of the synthetics helps provide better lubrication at very low temperatures and at very high temperatures. The studies also indicate that the chemical differences between conventional mineral oils and some synthetics may also affect the formation and thickness of this vital film, again to the advantage of the synthetic (32).

The improved viscosity–temperature characteristics of synthetic industrial gear lubricants can have a measurable effect on energy consumption, with savings of 1–3% reported (31). These savings are a result of the easier start-up at low temperatures and more adequate reduction of friction at high temperatures. The saving can also be attributed to the lower shear strength of synthetics, which makes it easier to move the lubricant through the gear system. These same characteristics also allow gear systems to operate cooler with a synthetic than with conventional mineral oil. This not only extends lubricant life, but can be a factor in extending gear and seal life as well.

VI. CURRENT COMMERCIAL PRACTICE

A. Synthetics in Use

As mentioned earlier in this chapter, although many synthetic fluids are available to formulate gear lubricants, only a few meet the general performance and economic targets associated with most commercial applications. For automotive gear lubricants these are:

Synthetic hydrocarbons (PAO).
Dibasic acid esters.
Polybutenes.

For industrial applications, polyalkylene glycols have been successfully used in addition to these three types of fluids listed (32).

Improved fuel economy requirements plus increasing gear-train operating temperatures in heavy-duty trucks and other off-road vehicles have resulted in the development and commercialization of several fully and part-synthetic gear lubricants. These synthetics offer significant improvements in energy conservation and high-temperature/high-torque performance. At the same time, these synthetics allow wide-range multigrade blending capabilities.

In Western Europe, 75W–90 gear lubricants containing up to 50% PAO are common. Full synthetics using PAO/diester combinations and PAO/polybutene combinations are used in small quantities as well.

B. Market for Synthetic Gear Lubricants

The market for gear lubricants world wide is 650,000 metric tons; of this, less than 5% is synthetic fluids of various types. Currently, PAO demand for synthetic gear lubricants is estimated to be 22,500 metric tons in the United States and 10,000 metric tons in Western Europe. Japan's use is much smaller (30).

C. A Trend Toward Higher Performance

Current trends toward increasing power and efficiency are requiring design engineers to more accurately determine the proper lubricant viscosity for maximum performance. Increased power output also implies higher gear surface temperatures, smaller lubricant film thicknesses, and an even greater demand on the lubricant.

Specifying a lower-viscosity conventional mineral lubricant may improve gear efficiency, but it may also result in increased gear wear at higher temperatures. Synthetic lubricants, with their improved viscosity–temperature characteristics, can provide greater efficiency without increased wear at higher temperatures. Synthetic gear lubricants should also continue to play an important role in critical or problematic gear systems that operate beyond their original design conditions. In summary, these observations and trends will expand the market for synthetic gear lubricants considerably (27–30). Manufacturers of gear equipment are beginning to demand better gear lubricants that meet tougher performance targets. These are:

Even better thermal stability.
Longer or no drain intervals (fill for life).
Heavier loads.
Quality.
Efficiency.
Extended warranty.

Of the synthetics available for use, the PAOs will dominate the market because of their combined superior performance, availability, and potential cost-effectiveness over other possible synthetic combinations.

REFERENCES

1. P. Barrand, J. C. Hypeaux, P. Marchand and A. Grocosa. An Appropriate Test Method for Evaluating the Thermal Stability of Transmission Lubricants, Third CEC Symposium, 1989.
2. B. J. Beimesch, M. A. Margeson and J. E. Davis. Field Performance of Synthetic Automotive Gear Lubricants, SAE Paper No. 831730, 1983.
3. C. J. Boner. *Gear and Transmission Lubricants*, Reinhold, New York, 1964.

4. R. H. Boehringer. Synthetic Lubricants for Industry, ASLE/ASME Section Meeting, St. Louis, MO, October 18, 1982.
5. J. G. Damrath and A. G. Papay. Gear Oils and the Function of E. P. Additives, SAE Paper No. 860757, 1986.
6. G. Daniel and W. R. Murphy. Practical Applications of Polyalphaolefin Based Industrial Gear Oils, Second World Congress on Gearing, 1984.
7. H. Gross. Formulating broadly cross-graded lubricants with high VII polyalphaolefins, *Plant Maintenence*, October 1983.
8. D. E. Hobson. Axle Efficiency—Test Procedures and Results, SAE Paper No. 790744, 1979.
9. How to determine if synthetic lubricants can solve critical equipment problems, *Fluid and Lubricant Ideas*, 1983.
10. R. A. Hunter. DOT/SAE fuel economy verification testing, SAE SP-488, *Truck and Bus Fuel Economy*, April 1981.
11. R. L. Johnson and D. L. Facchiano. An Examination of Synthetic and Mineral Based Gear Lubricants and Their Effect on Energy Efficiency, NLGI, Kansas City, MO, October 23–26, 1983.
12. P. S. Korosec, D. K. Walter, I. McPherson and M. J. Openshaw. Recent Developments in European Automotive Driveline Lubricants, MISR Conference, Cairo, March 1990.
13. P. S. Korosec, S. Norman and R. E. Kuhlman. Gear Oils Beyond GL-5, SAE Paper No. 821184, 1982.
14. P. S. Korosec and S. Norman. Gear Lubrication: Fifty Years of Progress, NLGI, October 1983.
15. Laboratory Performance Tests for Automotive Gear Lubricants Intended for API GL-5 Service, STP 512A, American Society for Testing and Materials, March 1987.
16. G. G. Lamb, C. M. Loane and J. W. Gaynor. Indiana stirring oxidation test for lubricating oils, *Ind. Eng. Chem. Anal. Ed.*, *13*(5): 317–321.
17. Lubrication of Industrial Enclosed Gear Drives, American Gear Manufacturers Association Standard Specification, September 1981.
18. Lubrication of Industrial Open Gearing, American Gear Manufacturers Association Standard Specification, November 1974.
19. Lubricant Service Designations for Automotive Manual Transmissions and Axles, API 1560, American Petroleum Institute.
20. Military Specification Mil-L-2105B, Lubricating Oil, Gear, Multipurpose, February 19, 1962.
21. Military Specification Mil-L-2105C, Lubricating Oil, Gear, Multipurpose, October 1976.
22. S. Norman, G. H. Kovitch and R. M. Klein. Industry Requirements for a New Generation of Automotive Gear Oils, NLGI, Kansas City, MO, October 1990.
23. S. Norman and R. M. Klein. The New Look in Automotive Gear Oils for Heavy Duty Axles and Transmissions, NLGI, St. Louis, MO, October 1989.
24. B. M. O'Conner, L. F. Schiemann and R. L. Johnson. The Relationship Between Laboratory Axle Efficiency and Vehicle Fuel Consumption, SAE Paper No. 811206, 1981.
25. B. M. O'Conner. Evaluation of Fuel Efficient Gear Oils, presented at the Institute of Petroleum Symposium of Performance and Testing of Gear Oils and Transmission Fluids, London, October 1980.
26. D. C. Porrett, S. D. Miles, E. F. Werderits and D. L. Powell. Development of a Laboratory Axle Efficiency Test, SAE Paper No. 800804, 1980.
27. M. Reisch. Lube additives face higher performance demands, *Chem. Eng. News*, January 9 (1989).
28. B. F. Stricker. North American Market for PAO's, private communication, 1988.
29. Synthetic Lubricants in the Automotive Market, Emery Industries, private communication, 1984.
30. Private communication, 1991.

31. Synthetic lubricants can reduce downtime and extend bearing life, *Pulp and Paper*, p. 127 (1989).
32. Synthetic gear lubricants—Application guide, *Fuel and Lubricant Ideas*, September/October (1981).
33. D. J. Taylor and W. F. Olszewski. Synthetic automotive gear lubricants, *Automotive Eng.*, November:49–53 (1988).
34. R. B. Whyte. History of CRC Activity in Automotive Gear Lubricants, paper presented to CRC Vehicle Fuel, Lubricant and Equipment Research Committee, November 25, 1969.
35. V. Villena-Denton. Synthetic oils carve out niche in gear oil market, *Automotive Lubricants, The Oil Daily*, October 10 (1988).

18
Grease

Paul A. Bessette
William F. Nye Company
New Bedford, Massachusetts

I. HISTORICAL PERSPECTIVES

Lubricants greatly reduce the frictional burden that nature levies against people and machines when two objects in contact are moved relative to each other. The recognition by ancient engineers that certain natural products were effective in facilitating the movement of massive objects and reduced the demand for muscle power was indeed the genesis of today's lubricant industry.

The archaeological record indicates that stone-building civilizations routinely used olive oil beneath wooden planks to reduce the friction and work needed to transport massive stones. Moreover, in the ancient transportation sector, animal fats were used in conjunction with inorganic fillers to reduce the frictional forces between the bearing surface and shaft on chariot wheels. These discoveries clearly indicate that lubricants were a viable engineering tool to many early societies; however, the original use of any lubricating substance probably has its roots in some forgotten cave. It may have been some prehistoric cave dweller who first realized that the slippery residue left on his hands after dinner would increase the effectiveness of his spear when applied to the point. Perhaps the first use of a lubricant was to bring down a woolly mammoth?

Thoughtful people throughout the ages have always sought various methods to reduce the effort required to accomplish a task. It is not uncommon even today to see a retired carpenter instructing an apprentice on the merits of applying wax to the tips of finish nails to successfully drive the fastener into an obdurate stair tread made of oak.

There is one other aspect pertaining to the discovery of lubricant in the hubs of ancient wheeled vehicles that is both technically interesting and fascinating. Along with animal fats, fats are mixed esters of glycerol, calcium oxide, and lime were also identified by chemical methods in chariot wheel hubs. Although the combination of ingredients may appear quite inconsequential a priori, closer scrutiny reveals some real magic. The ramifications of the serendipitous mixture can be explained as follows: Animal fat is added to the axle and wheel hub along with calcium oxide. During operation, frictional heating generates the necessary thermal energy to activate a chemical reaction between the lime and fat, producing a quasi in situ saponification. The resultant product has superior lubricating characteristics compared to the original ingredients and also has a higher dropping point than that of the fat—and all this was accomplished millennia before the great French chemist Lavoisier elucidated the chemistry involved in displacement reactions. For a chemist specializing in lubricants, the realization that a substance with modern greaselike attributes was discovered in chariot wheel hubs is tantamount to a computer scientist finding a 1950s vintage computer chip among the Dead Sea Scrolls.

II. THE NATURE OF SOLIDS AND THE UTILITY OF GREASE

Much of the material world that humans find suitable for fabricating useful products is solid. Spacecraft, automobiles, and bicycles are all made from solids. However, when two contacting solids are moved tangentially in relation to each other, frictional forces are encountered that act to oppose the motion—friction is nature's way of taxing us for moving something! Although friction is a fundamental property of all matter, it is particularly detrimental in the case of contacting solids since it usually leads to the unwanted removal of material from one or both of the contacting members. Solids, unlike liquids and gases, are held together by extremely strong forces, and as such these substances have a propensity to bond or adhere strongly when their cleaned surfaces come into contact. On a microscopic scale all surfaces appear rough.

Profilographic analysis reveals successive maxima and minima, with the peaks often referred to as asperities. Although it may be tempting to assume that the frictional force arises due to the interlocking of opposing asperities, this is clearly not the case. According to the advocates of the adhesion theory, the nature of the friction force is the product of the real area of contact between the contacting surfaces and the shear strength of the softer contacting member. The coefficient of friction can be expressed as the ratio of a material's resistance to plastic deformation due to shear and that due to compression. Mathematically, the coefficient of friction can be represented as follows:

$$F = S/P \qquad (1)$$

where S is the strength or the ability of the softer of two contacting materials to resist plastic flow due to shear and P is the compressive strength of the softer material or its ability to resist plastic flow due to compressive loads.

Greases, on the other hand, are semisolids. Since greases are held together by forces orders of magnitude weaker than the forces that bind solids, they are useful materials to place between solids. In the absence of lubricating greases, rigid materials tend to self-destruct. When solids are rubbed together, heat is generated, causing surface and subsurface electrons to vibrate with greater amplitude, and at some critical temperature metallic bonds are ruptured and a wear particle is created. Greases are useful as lubricants because they reduce friction and wear, dampen vibration and noise, seal out contaminants including water, remain where applied under rigorous temperature excursion, and serve as oil reservoirs, releasing oil as conditions warrant. Moreover, under conditions of low shear they exhibit a high apparent viscosity, but approach the viscosity of the base fluid component at sufficiently high rates of shear.

III. THE COMPOSITION OF SYNTHETIC GREASE

All lubricating greases, including synthetics, consist of two fundamental components: a base fluid representing the principal ingredient in the formulation, and a thickening agent that is used to immobilize the fluid. The concentration of thickener determines the consistency of the finished product; however, it is the nature of the oil that determines whether the grease will be classified as a synthetic. Although all greases contain an oil portion and a thickener, the possible oil and thickener combinations and the specific antioxidants and antiwear, anticorrosion, or extreme pressure agents employed provide manufacturers with a great deal of flexibility in formulating products with many different physical and chemical attributes. The lubricating oils commonly used to formulate synthetic lubricating greases are listed in Table 18.1 along with some of the notable characteristics of each fluid.

Table 18.1 Lubricating Oils Commonly Used in Synthetic Greases

Fluid	Characteristics
Diester	Suitable for making grease serviceable to −73°C.
Polyol esters	Higher viscosity than the diesters with better oxidative stability and less volatility.
Pentaerythritol esters	Primarily used to formulate synthetic greases requiring excellent lubricting properties from −40°C to over 177°C.
Polyglycols	These oils have a tendency to produce less carbonaceous residue upon degradation and find use in formulating greases used in arcing electrical equipment.
Polyalphaolefins	A relatively new class of synthetic fluid available in a range of viscosities. The oils exhibit improved compatibility with ester-vulnerable thermoplastics and elastomers.
Silicones	Outstanding thermooxidative stability, chemically inert, low- and high-temperature usefulness, and unmatched viscometric properties as a function of temperature. The viscosity index of one phenylsilicone is over 600.
Chlorofluorocarbons	Chemically inert toward oxygen and possessing excellent innate boundary and extreme pressure lubricating properties.
Aromatic ethers	These fluids make superior lubricants for gold electrical contacts. Greases made with these oils are extremely inert toward the ravages of ionizing radiation.
Phosphate esters	Can be used to manufacture relatively inexpensive greases that are able to resist ignition.
Perfluoropolyethers	Synthetic greases formulated from these fluids are inert toward oxygen and all but the most aggressive chemicals. Greases made from these oils have unsurpassed thermooxidative stability.
Alkylated cyclopentanes	The newest commercially available synthetic fluid, demonstrating exceptionally low volatility under hard vacuum and elevated temperature.

Although it is the oil that characterizes a grease as synthetic, the thickener type determines the identity of the grease. For example, a lubricating grease prepared from a polyalphaolefin oil (polyalphaolefins are commonly referred to as synthetic hydrocarbons, or abbreviated as SHCs or PAOs) thickened with lithium 12-hydroxystearate would be referred to as a lithium-soap thickened synthetic hydrocarbon. An ester thickened with an organomodified clay would be described as a clay-based synthetic ester.

Synthetic greases are prepared from both organic and inorganic thickeners. Organic thickeners are prepared from the reaction of a suitable alkali metal with either high-molecular-weight carboxylic acids or fats. When the chemical reaction takes place in the oil used in formulating the grease, it is referred to as in situ neutralization or in situ saponification, depending upon whether an acid or fat is the coreactant.

Inorganic thickeners, such as chemically modified clay, amorphous silica, and polytetrafluoroethylene, can also be used to form grease, but without the need for a chemical reaction for grease formation to occur. The efficacy of a particular thickener to convert a synthetic oil into grease is dependent on the ultimate surface area of the thickener, its ability to hydrogen bond, and its tendency to associate with the fluid on a molecular level. The thickener must have an affinity for the base fluid that is intermediate between the forces that lead to greater solubility and those forces tending to induce phase separation. This meso-solubility is a prerequisite for all successful grease formation. Table 18.2 identifies some of the commonly employed alkali metals used to make greases. The alkali metals are usually reacted with stearic acid, myristic acid, 12-hydroxystearic acid, or hydrogenated castor oil, a triglyceride that liberates 12-hydroxystearic acid during saponification.

IV. GREASE MANUFACTURE

Organic greases are usually manufactured in kettles. The size of the vessels ranges from laboratory units capable of manufacturing only 5–10 lb per batch to very large units capable of manufacturing 40,000 lb of grease in a single operation. Since heat is required to initiate the reaction of the ingredients used to manufacture soap-based grease, grease kettles are heated. Most are jacketed to accommodate either steam or hot oil. Steam is an advantageous thermal medium because cold water can be circulated through the same jacket to cool the batch on completion of the chemical reaction. The primary disadvantage of heating the grease charge with steam is that high pressure is required to attain temperatures above 450°F. Aside from steam, oil, electricity, and direct heating are also used to heat grease kettles. Generally, a soap-thickened grease is manufactured by adding a small portion of base oil to the kettle along with all of the fatty acid. At this stage, only

Table 18.2 Alkali Metals Used to Prepare Synthetic Lubricating Grease

Alkali	Formula
Lithium hydroxide monohydrate	$LiOH \cdot H_2O$
Calcium hydroxide	$Ca(OH)_2$
Sodium hydroxide	$NaOH$
Aluminum hydroxide	$Al(OH)_3$

enough heat is applied to melt the charge. Once the fatty acid has melted into the base oil, an aqueous solution of the alkali metal is added to the kettle incrementally. The addition is carried out in four steps to prevent the reaction mass, which is approximately at the boiling point of water, from unexpectedly discharging from the kettle due to the formation of copious quantities of steam. The kettle contents are continuously stirred, usually with counterrotating stirrers, to facilitate the dehydration of the soap mass as the reaction proceeds. After dehydration, additional base oil is gradually added to the kettle. The addition of oil must proceed slowly in order to maximize proper mixing of the intractable soap mass and the oil being added. After the addition of the required quantity of oil, the kettle contents are heated to some predetermined elevated temperature and maintained at that temperature for several hours. After the heating cycle, the grease is rapidly cooled to optimize the dispersion of the thickener. The rate at which the kettle contents is cooled has a pronounced effect on the finished consistency of the grease. Additives are usually added to the grease after the temperature of the batch has fallen below 100°C. When the kettle contents have reached room temperature, the grease may be either milled or homogenized. As a final step the grease may be filtered to remove contaminates. Quality control testing usually occurs prior to discharging the grease from the kettle for packaging.

V. GREASE CHEMISTRY

A. Preparation of a Synthetic Ester Grease

The following batch workup illustrates the preparation of 100 kg of a synthetic ester grease thickened with sodium myristate.

The finished product is to contain 12% thickener. The kilograms of thickener required equals the batch size multiplied by the desired thickener concentration, so that the amount of thickener required is given by

$$100 \text{ kg} \times 0.12 = 12 \text{ kg}$$

The balanced chemical reaction is

$$NaOH + H_3C(CH_2)_{12}COOH = NaOOC(CH_2)_{12}CH_3 + H_2O$$

Because the molecular weight of the thickener, sodium myristate, is 226 g per mole and 12 kg of thickener is needed, the number of moles of each reactant is equal to the number of moles of thickener, as there is a one-to-one stoichiometry indicated by the balanced chemical reaction. Therefore, the numbers of moles of sodium hydroxide and myristic acid required are determined as follows:

$$\frac{12{,}000 \text{ g}}{226 \text{ g/mol}} = 53.09 \text{ mol}$$

Ingredient	Number of moles	g/mol	Amount needed, kg
NaOH	53.09	23	1.221
Myristic acid	53.09	204	10.83

The sodium hydroxide would be dissolved in sufficient deionized water to completely dissolve the base, and the myristic acid would be added to the reaction vessel along with a small portion of nonsaponifiable petroleum oil (the reaction cannot be conducted in the presence of the ester base oil due to the susceptibility of esters to de-esterification in the presence of strong bases). After the neutralization reaction is completed and the soap mass has been adequately dehydrated, the thickener is then capable of absorbing the synthetic ester base fluid. The oil is added incrementally to effect proper mixing while maintaining the temperature of the kettle contents at approximately 150°C. When the total charge of oil has been added to the kettle, the temperature is reduced and the desired additives are added to the grease. The batch is further cooled and discharged from the kettle.

A primary advantage of synthetic lubricants over petroleum-based products is their improved serviceability at elevated temperatures. However, the ability of a lubricating grease to resist excessive softening or melting at extreme temperatures is dictated by the nature of the thickening agent. Most organic soap thickeners melt at temperatures below 450°F. To overcome the apparent thermal deficiency, complex thickeners are used that do not melt. Dropping points for many complex greases are usually above 500°F.

B. Preparation of an Aluminum Complex Grease

The preparation of 100 kg of an aluminum complex grease containing a synthetic hydrocarbon base oil can be illustrated as follows:

Batch size	100 kilograms
Thickener content	7.5%
Ratio of benzoic acid to stearic acid	0.8
Ratio of both acids to aluminum	1.9
Kilograms of thickener required = 100 kg × 0.075 = 7.5 kg	
Balanced chemical equation:	
$Al(OC_3H_7)_3 + C_{17}H_{35}CO_2H + C_6H_5CO_2H = (C_{17}H_{35}CO_2)AlOH(C_6H_5CO_2) + 3C_3H_7OH$	

The amount of each ingredient is determined based on the specified composition of the grease in molar ratios.

Ingredient	Moles	Molecular weight	Unit	Batch
Stearic acid	1.0	278.6	278.6	5.20
Benzoic acid	0.8	122.0	97.6	1.82
Aluminum	0.94	27.0	25.4	0.474
Total			401.6	7.49

Since the aluminum (as the isopropoxide) is supplied as a 12.7% mixture in oil, the 0.474 lb of aluminum would be available in 3.73 kg of the commercial product.

To manufacture the grease, the reaction of the ingredients would be carried out in a portion of the synthetic hydrocarbon fluid, after which the remaining oil and additives would be added. Kettles suitable for manufacturing complex aluminum greases must be designed to recover the alcohol liberated by the reaction.

C. Polyurea Grease

Although many polyurea greases contain petroleum base oils, these grease deserve mention in a text dedicated to synthetics because of their unique thickener chemistry. The alkylarylpolyurea thickening agent commonly used is a synthetic high-molecular-weight polymer that provides the following advantages: a dropping point of approximately 250°C, excellent water resistance, low oil separation, and an uncanny ability toward shear thinning that renders polyurea grease ideally suited for many rolling-element bearing applications.

D. Organomodified Clay-Thickened Synthetics

The ability of clay to function as an effective thickener for lubricating fluids is dependent on its surface pretreatment. The addition of a suitable organic compound to the surface of individual clay platelets transforms the material from a hydrophile to an oil absorber. Clay is useful as a thickening agent for grease due to its dispersible lamellar-type structure and ability to hydrogen bond. Individual platelets are only angstroms thick and possess tremendous surface area when efficiently dispersed. Moreover, since a chemical reaction is not required to form a synthetic grease from a clay thickener, these greases can be easier to produce and require less energy to manufacture than soap based greases. However, to achieve a viable product the grease formulator must optimize the amount of dispersant and water used to open the individual clay bundles and promote hydrogen bonding. Clays are suitable for thickening synthetic esters, synthetic hydrocarbons, polyglycols, and some silicones. However, the suitability of the clay thickener in a particular lubricating medium is predicated on the polarity of the oil and the clay's surface treatment. A clay (alkylaryl ammonium hectorite) thickened polyglycol grease can be prepared by simply adding approximately 7% clay to the polyglycol, stirring the mixture, then adding the optimized level of distilled water an propylene carbonate,* and finally passing the pregrease through a mill or high-pressure homogenizer to effect gellation.

Although clay-based greases are nonmelting and are often recommended for high-temperature applications, sustained usefulness at elevated temperature is limited in all greases by the thermooxidative stability of the base oil.

E. Polytetrafluoroethylene-Thickened Synthetic Grease

Polytetrafluoroethylene (PTFE) is one of two thickening agents used for synthetic oils that can be classified as a universal gellant. Polytetrafluoroethylene is capable of forming grease from all of the synthetic fluids mentioned in Table 18.1. The ability of PTFE to thicken oil is due to its low surface tension, surface area, and dispersability in organic fluids. Greases prepared from PTFE and synthetic fluids are serviceable at extremely low temperatures and are capable of continuous service above 200°C. Grease prepared from telomers of PTFE are probably the most suitable PTFE-thickened lubricants for long-term lubrication of rolling element bearings. Because an NLGI Grade 2 grease can be prepared from 20% PTFE derived from the telomer dispersion and 40% PTFE is required from the less expensive micropowders, the grease with the greater oil reservoir would be expected to function long after the lubricant with the higher concentration of thickener became

*Propylene carbonate is one of several dispersants used to manufacture clay based synthetic greases. Acetone and ethanol are also used.

intractable due to oil loss. The efficacy of grease is dependent on a favorable balance between the amount of oil and thickener. As the oil to thickener ratio changes because of oil separation, oil evaporation, and oil oxidation, the grease is gradually transformed into an intractable mass that fails to lubricate.

VI. ADVANTAGES OF SYNTHETIC GREASE

The primary advantages of synthetic grease over petroleum-based products are improved thermooxidative stability, wide temperature serviceability, and less change in apparent viscosity as a function of temperature. Each advantage can be elaborated. Oxygen is a pernicious molecule capable of rapidly degrading organic lubricants at elevated temperatures either by abstracting hydrogen from vulnerable sites along the lubricant molecule or by directly reacting with sites of unsaturation. Since oils used to formulate synthetic greases are virtually free of unsaturation and are synthesized to be resistant to oxidative attack, synthetic greases, as a class, greatly outperform their nonsynthetic rivals under severe oxidizing conditions. Moreover, silicones and perfluoropolyethers demonstrate superior resistance to oxidative degradation at temperatures above 150°C after years of exposure. Similar exposure would transform petroleum-based products into intractable polymers commonly referred to as sludge or varnish.

With selected oils, synthetic grease can be formulated with unsurpassed ability to remain pliable at –54°C while not deteriorating or excessively evaporating at temperatures above 177°C.

Adequate viscosity under operating conditions is, in this author's judgment, the most important property any lubricant can possess. Without it, moving surfaces are destined for self destruction regardless of how well the grease is fortified with special additives. Greases made from synthetic oils maintain their apparent viscosity as a function of temperature better than nonsynthetics.

Additionally, synthetic greases, as a class, are more compatible with various engineering thermoplastic and elastomers. A notable exception is the synthetic esters, which adversely affect certain vulnerable plastics such as acrylonitrile butadiene styrene (ABS), polyvinyl chloride (PVC), polystyrene, polysulfone, and polyphenylene oxide. It is prudent to always determine the compatibility on any lubricating fluid with elastomers due to the ability of low-viscosity oils to permeate into the rubber matrix and thus cause swelling. Special synthetic greases prepared from polyphenyl ethers demonstrate superior resistance to ionizing radiation.

VII. COST OF SYNTHETIC GREASES

The principal disadvantage of synthetic grease is their cost. Compared to many nonsynthetic products their prices are truly stratospheric; however, demands placed on modern devices have made the higher costs palatable to many firms. As an extreme example, a popular sodium complex high-speed spindle bearing grease formulated from a petroleum oil is commercially available at less than $1.00 per pound. A bearing grease formulated from a linear perfluoropolyether oil and thickened with polytetrafluoroethylene oligomers would command approximately $500.00 per pound. Such greases can only be considered where the operating requirements dictate the need for very specialized lubricants. However, most synthetic greases are typically priced between $5.00 and $50.00 per pound in 35-lb containers.

VIII. GREASE TESTING

The American Society of Testing and Materials (ASTM) along with its European counterparts has standardized a vast number of tests used to measure a specific chemical or physical property of grease. Some of these tests may be classified as those that measure characteristics particular to the composition, while other tests are more suitable for assessing batch-to-batch variability. Volatility per D-972, water washout per D-1264, and the four-ball wear test per D-2266 are examples of tests that measure properties of the grease inherent to the formulation. Some variation in these properties is possible, due to either the product or the test method, but the magnitude of the variation should be minor from one batch to the next. However, it is necessary and prudent to conduct tests that are sensitive to the composition of a synthetic grease at established intervals. The specific interval ought to be determined after consultation with customers.

Because grease manufacturers cannot exercise absolute control over the chemical reactions or processes that produce grease, certain physical tests should be conducted on each batch of grease to monitor important grease characteristics. These tests may be categorize as batch specific and can be used to monitor the homogeneity of the product as a function of time. Such tests are a benefit for statistical process control. Table 18.3 lists those tests that should be conducted on a batch basis.

The consistency of a grease measures its resistance to deformation under an applied force. Consistency attempts to quantify plastic behavior as viscosity tries to delineate fluidity. The National Lubricating Grease Institute (NLGI) has developed a numerical scale to classify the consistency of greases based upon their worked 60-stroke penetration values. Semifluid greases have a triple zero rating while the hardest greases would receive a rating of six. The nine NLGI categories or grades are:

Table 18.3 Synthetic Grease Properties That Tend to Vary Batchwise

Property	Method	Purpose
Unworked penetration	ASTM D-217	Measures the consistency of the grease prior to the input of mechanical energy. This is what the customer's pump is required to transfer.
Worked penetration, 60 strokes	ASTM D-217	Measures the mechanical stability of the greases. Excessive change may signal problems in applications imparting high shear to the grease.
Oil separation	FTM 791B, method 321.2	This test assesses the amount of oil released from the grease structure after 30 h at 100°C.
Evaporation	ASTM D-972	Measures evaporation of any volatile ingredients in the grease or residual by-products generated during manufacture.
Dropping point	ASTM D-2265	Determines the highest temperature attainable before a drop of oil separates from the grease.

NLGI grade	ASTM worked penetration, 60×
000	445–475
00	400–430
0	355–385
1	310–340
2	265–295
3	220–250
4	175–205
5	130–160
6	85–115

IX. APPLICATIONS FOR SYNTHETIC GREASES

A. Ball Bearings

Rolling-element bearings represent a major market for synthetic greases. Bearings today are expected to operate over temperatures that may range from –54°C to over 200°C. The application may dictate that the lubricant function during the entire life of the device without the opportunity for relubrication, and precision bearings demand synthetic greases with minimum particulate contamination. For example, ball bearings used for computer spindle applications require greases that provide adequate lubrication under high-frequency oscillatory motion, minimize viscous drag, have sufficiently low vapor pressure, are oxidatively stable, are free from particulate matter that may damage the bearing or jeopardize runout, and do not separate appreciable oil in either a dynamic or static mode.

Although all greases exhibit pseudoplastic flow behavior as a function of increasing shear, synthetic sodium-based ester greases become extremely soft during mechanical agitation, and as such this type of grease is ideally suited for bearing applications where motion of limited amplitude has a tendency to cause lubrication starvation.

Mechanically stable greases do not soften sufficiently to enter the contact zone and thus tend to allow metal-to-metal contact. Viscosity index is a dimensionless parameter used to assess a fluid's change in viscosity with temperature. Since most synthetic oils possess a higher viscosity index than nonsynthetics, greases made from high-viscosity-index oils are able to provide improved film strength over a broader temperature range than petroleum analogs.

Bearings used for computer disc drive applications must also be lubricated with greases that have a low vapor pressures. If the lubricant is not sufficiently nonvolatile, lubricant contamination of the storage media could occur with disastrous results. The vapor pressure of lubricating grease is determined by the molecular weight and molecular weight distribution of its base oil, the specific additives it contains, and its ability to resist degradation by oxygen. Under vacuum conditions, thermooxidative stability becomes irrelevant due to the absence of oxygen. However, in the presence of oxygen at elevated temperature, lubricants are susceptible to degradation, which during the initial stages liberates low-molecular-weight fragments that are highly volatile. If thermooxidative stability can be ignored, vapor pressure as defined by the Langmuir equation is proportional to surface area, temperature in degrees Kelvin, and molecular weight. Vapor pressure in torr can be expressed as follows:

$$P = 17.41G(T/M)^{1/2} \tag{2}$$

where G is the rate of evaporation in grams per centimeter squared per second, T is the temperature in degrees Kelvin, and M is the molecular weight of the material. Clearly, under vacuum conditions, the molecular weight of the ingredients in the lubricating grease determine its tendency to volatilize. At 100°C, synthetic hydrocarbon greases made from base oils with a viscosity of 5.8 cS at 100°C possess vapor pressures that approximate 10^{-6} torr. Synthetic greases made from perfluoropolyethers and thickened with polytetrafluoroethylene have vapor pressure below 10^{-10} torr at room temperature.

Dirt or contamination in a synthetic lubricating grease is undesirable. Greases employed as lubricants for precision bearings in guidance systems should contain minimum particulate contaminate in order to prevent race or ball damage caused by contaminate-induced lubricating film rupture. The sources of lubricating grease contamination are numerous. However, ultrafiltration is quite effective in removing contaminates greater than 35 μm from most synthetic greases. Some contamination in the 10–34 μm range is usually present in the grease after ultrafiltration and is not usually removed because of concern for the technical merit and the cost of doing so. Aside from removing contamination, the ultrafiltration of a grease improves its homogeneity by increasing the dispersion of the grease thickener. Ultrafiltration makes some grease slightly firmer and tends to reduce oil separation. Not long ago it was a popular misconception that grease filtration through micrometer-sized filters removes the thickener. However, hundreds of grease types have been ultrafiltered in the author's laboratory without a single incident of gross removal of the thickener. Qualitatively, the actual merits of grease ultrafiltration are somewhat limited; however, evidence is emerging substantiating the benefits of ultraclean greases in prolonging bearing life and improving performance. A substantial amount of grease intended for high-speed spindle applications and aircraft instrumentation is ultrafiltered.

Table 18.4 lists the physical properties of a synthetic ester ball-bearing grease used for high-speed applications.

For continuous duty at temperatures above 260°C, applications requiring a nonflammable grease, or chemical resistance toward aggressive chemicals like fuming sulfuric acid, only synthetic greases based on perfluoropolyethers and thickened with polytetrafluoroethylene will suffice. Because this author believes that more oil is better than less for most ball-bearing application requiring years of trouble-free operation, fluoroether greases thickened with polytetrafluoroethylene (PTFE) oligomers produce superior products. Using the PTFE oligomer, an NLGI Grade 2 grease can be prepared with less than 20% thickener, whereas with PTFE micropowder approximately 40% is needed to achieve the same consistency. Any grease containing 40% thickening solids can afford to lose little oil prior to becoming an intractable mass. Table 18.5 lists typical physical properties of a perfluoropolyether grease.

B. Electrical

One of the most rigorous applications for any synthetic grease is to effectively lubricate an electrical switch that produces an electrical arc during on–off cycling. The temperature of an electrical discharge is estimated to be above 1,000°C, and no present lubricant can cope with such extreme temperatures. However, polyglycol oils thickened with either organic or inorganic thickeners such as amorphous silica are usually the grease of choice for

Table 18.4 Physical Properties of Synthetic Ball-Bearing Grease

Physical property	Typical value
Viscosity of base oil at	
100°C	8.9 cSt
40°C	51 cSt
−40°C	2,550 cSt
Flash point	230°C
Pour point	−55°C
Unworked penetration	260
Worked penetration, 60×	300
Dropping point	202°C
Evaporation, 22 h at 135°C	1.5%
Oil separation, 22 h at 135°C	11.4%
Oxidation stability	
Conditions of test	100 h at 100°C
Pressure drop	Less than 68 kPa
Low-temperature torque at −54°C	
Starting	8,732 g-cm
After 10 min	2,330 g-cm
After 60 min	1,150 g-cm
Shell four-ball wear test	
Conditions	1,200 rpm, 40 kgf load for 1 h
Wear scar at 100°C	0.54 mm
Wear scar at 150°C	0.43 mm
Speedability or DN value	500,000

arcing electrical devices because of the tendency of the polyglycol to produce less carbonaceous residue upon thermooxidative degradation.

The operation of an electrical switch also places severe demands on grease characteristics. When closed, the voltage drop within a circuit, if the negligible contact resistance is ignored, is zero. However, as the contact begins to open during switch actuation, the

Table 18.5 Typical Properties of Perfluoropolyether Grease

Physical property	Typical value
Color	White
Appearance	Smooth
Unworked penetration	241
Worked penetration	269
Evaporation, 24 h at 204°C	0.79%
Oil separation, 24 h at 204°C	13%
Four-ball wear test	
Conditions	1,200 rpm, 40 kgf for 1 h
Wear scar at 75°C	0.85 mm
Wear scar at 204°C	1.26 mm
Low-temperature torque at −73.3°C	
Starting	2714 g-cm
After 1 h	590 g-cm
Solubility in fuel	None

grease must also function as a dielectric if arc damage to the contact is to be minimized. The great dilemma for manufacturers of synthetic grease for electrical switchgear is to produce products that have a tendency to remain between the moving contacts during operation while not impeding current flow while the switch is energized. To no small degree this simple requirement has remained quite elusive! Lubricants intended for electrical contacts service must protect the surface from oxidation, not form insulating films, and not be so viscous that electrical continuity cannot be established when the moving elements are at rest.

Electrical connectors represent an entirely different aspect of the successful lubrication of electrical components. Unlike electrical switches, which operate under sufficient voltage and current, electrical connectors may carry only a few milliamperes of current driven by millivolts of potential difference. These dry circuits require lubricants that do not impede current flow, that protect the connector interface from environmental corrodants, that ease the force required to make or break the connection, and that provide years of wear prevention under fretting conditions. Tables 18.6 and 18.7 list the properties of two synthetic greases intended for arcing switch and electrical connector applications, respectively.

C. Nuclear

Synthetic greases intended for nuclear service encounter three adversaries: alpha particles, beta particles, and x-rays. All three types of ionizing radiation if sufficiently abundant can destroy organic material in a matter of hours or days. Alpha particles, the doubly charged nuclei of helium atoms, are the most aggressive organocides because they are most capable of abstracting hydrogen atoms from the molecules that comprise lubricants. Beta particles possessing sufficient kinetic energy can also disrupt the covalent linkages that form the carbon to carbon chains in lubricating grease base fluids. On the other hand, x-rays are high-energy electromagnetic radiation and are also capable of disrupting the molecular structure of synthetic lubricants. If the flux of the radiation is sufficiently intense, no lubricant would be expected to survive indefinitely; however, silica-based greases prepared from polyphenylether are extremely durable toward exposure to ionizing radiation.

Table 18.6 Synthetic Grease for Arcing Electrical Switches

Physical property	Typical value
Viscosity of base oil at 99°C	30 cSt
Viscosity of base oil at 38°C	526 cSt
Pour point	−34°C
Grease dropping point	+225°C
Unworked penetration	334
Worked penetration, 60×	334
Worked penetration, 10,000×	351
Oil separation, 24 h at 100°C	0.8%
Evaporation, 24 h at 100°C	5.0%
Specific gravity at 25°C	0.9
NLGI grade	1

Table 18.7 Synthetic Grease for Electrical Connectors

Physical property	Typical value
Viscosity of base oil at 100°C	40 cSt
Viscosity of base oil at 40°C	400 cSt
Pour point	−34°C
Flash point	316°C
Grease dropping point	+260°C
Unworked penetration	310–340
Worked penetration	317

Polyphenyl ethers are completely aromatic fluids joined by an ether bridge. The resonance-stabilized aromatic structure of polyphenyl ether oils and a minimum number of vulnerable hydrogen sites are two possible explanations for the unsurpassed resistance of greases prepared from these fluids to withstand the ravages of ionizing radiation.

Grease intended for nuclear service must be virtually free of sulfur, chlorine, and other elements that tend to embrittle steel. Table 18.8 summarizes the physical properties of a nuclear-capable synthetic grease.

X. THE NLGI GREASE EDUCATION PROGRAM

The world of lubricating grease is diverse, and opinions regarding what constitutes a suitable lubricant for a particular application abound. The National Lubricating Grease Institute, headquartered in Kansas City, MO, serves as a vehicle bringing manufactures, users, and distributors together on a yearly basis for information exchanges pertaining to novel products, new performance requirements, and marketing trends. In conjunction with the general program, the institute conducts an education course where some of the country's recognized leaders in grease chemistry and technology serve as instructors. In addition, the institute publishes on a monthly basis the *NLGI Spokesman*, which is the only publication in the country solely dedicated to the understanding of the properties, behavior, and use of lubricating greases.

Table 18.8 Example of Synthetic Grease Intended for Nuclear Service

Physical property	Typical value
Viscosity of base oil at 99°C	13 cSt
Viscosity of base oil at 38°C	363 cSt
Flash point	288°C
Fire point	349°C
Dropping point of grease	+250°C
Unworked penetration	281
Worked penetration, 60×	287
Evaporation, 24 h at 150°C	0.08%
Oil separation, 24 h at 150°C	0.8%

Letters patents can also serve as a viable educational tool for those wishing to broaden their knowledge about greases.

REFERENCES

1. Fagan, G. (1989). *Polyurea and Complex Soap Greases*, National Lubricating Grease Institute Education Course, Kansas City, MO.
2. Frye, R. (1990). *An Introduction to Lubricating Grease*, National Lubricating Grease Institute Education Course, Kansas City, MO.
3. Kruschwitz, H. (1990). *Thickener Systems for Aluminum Complex Greases*, National Lubricating Grease Institute Education Course, Kansas City, MO.
4. Labude, K. (1989). *The Chemistry of Soap Base Grease*, National Lubricating Grease Institute Education Course, Kansas City, MO.
5. O'Connor and Boyd (1969). *Standard Handbook of Lubrication Engineering*, McGraw-Hill, New York.
6. Rabinowicz, E. (1965). *Friction and Wear of Materials*, John Wiley & Sons, New York.
7. Rymuza, Z. (1989). *Tribology of Miniature Systems*, Elsevier, Amsterdam.

19
Compressors and Pumps

Glenn D. Short and J. William Miller
CPI Engineering Services, Inc.
Midland, Michigan

I. INTRODUCTION

Compressors and pumps are used to move materials for a large variety of purposes. Applications for these machines are in industries such as steel, petroleum, chemical, mining, food, gas production and storage, energy conversion (refrigeration), etc. Their shutdown means lost production. Compressors and pumps transfer gases, liquids, or sometimes slurries of liquid solid mixtures. Lubricants in these applications lubricate moving parts such as bearings and gears, provide a liquid seal, and remove heat. This can be a difficult task as the lubricant often operates in a hostile environment.

There are many reasons for using a synthetic lubricant for almost any type of machine. An individual synthetic may have specific advantages, but none is superior in all respects. When considering compressors and pumps, some characteristic or group of desirable properties contributes to their performance advantage over mineral oils. Synthetic fluids can operate at higher temperatures, exhibit better low temperature fluidity and pour points, and are less volatile than their mineral oil counterparts.

Similarities exist between the design of compressors and the design of pumps. Classifications for compressors are dynamic or positive-displacement. Dynamic compressors include centrifugal compressors and axial compressors. These types of compressors develop pressure by the action of a rotating blade, which imparts velocity and pressure to the flowing medium. Examples of positive-displacement compressors include reciprocating, rotary screw, rotary vane, lobe, and scroll. Positive-displacement compressors confine successive volumes of gas within a closed space and elevate them to a higher pressure.

Vacuum pumps are compressors that operate with a suction pressure as a vacuum. Classifications for liquid pumps are similar to compressors. There are centrifugal pumps and positive-displacement types. The latter includes vane, screw, reciprocating, gear, and lobe.

II. COMPRESSORS

A brief description of these machines and their lubrication helps to provide an understanding of how synthetic lubricants improve their performance. Examples of these compressors appear in Fig. 19.1.

A. Dynamic Compressors

Dynamic compressors handle significant flows at relatively high speeds. Multistage compressors are required for higher pressures. These machines are usually driven through a speed-increasing gear. The lubricant reduces friction and prevents wear in bearings and gears. Oil is not meant to pass into the gas stream. Various methods of shaft sealing are used due to the variety of gases and applications. These include labyrinths, carbon rings, contact seals, and bushings. The lubricant may act as a sealing fluid or otherwise aid in the

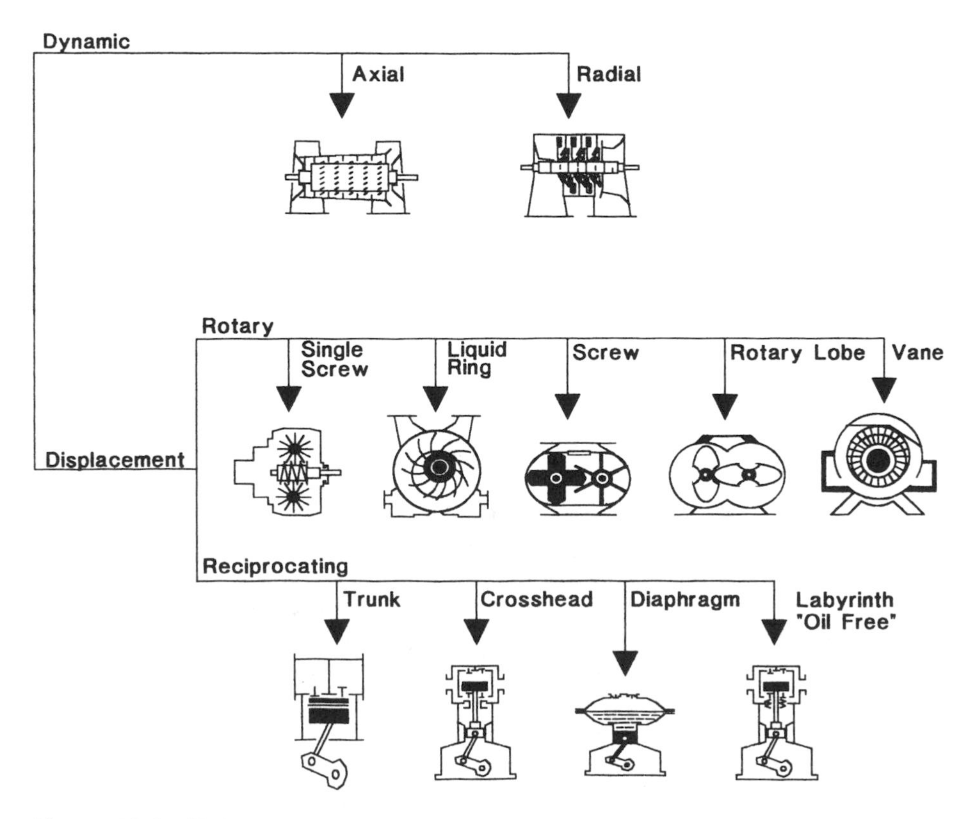

Figure 19.1 Various compresor types.

operation of these seals (i.e., hydraulic action). Small amounts may enter the gas stream through seal leakage. Conversely, gas may enter into the lubricant system. In some cases the gas provides a positive pressure to the lubricant reservoir to prevent leakage of air into the system (as with refrigeration and some chemical applications).

B. Positive-Displacement Compressors

1. Reciprocating Compressors

There are several types of reciprocating compressors. Lubrication points include cylinders, valves, pistons, piston rings, crankshafts, connecting rods, main and crank pin bearings, and other associated parts. Double-acting machines use crossheads and crosshead guides with connecting pins to join the crosshead to connecting rods. Most single-acting compressors use connecting rods attached directly to the pistons with wristpins or piston pins. "Oil-free" machines do not require lubrication in the compression area.

Cylinder lubrication includes cylinder parts, pistons, rings, valves, and rod packings. Crankcase parts include main and crankpin, crosshead (or wristpin) bearings, and crossheads and crosshead guides. Additionally, the lubricant may help the efficiency of seals and minimize their wear. The crankcase on reciprocating compressors may either be open to the cylinders (as in many vertical, V-type, and radial compressors) or sealed from the cylinder by a bulkhead and exposed to air (horizontal compressors). Bearing and other crankcase components require relatively large amounts of lubricant, which is usually supplied from the crankcase. Various methods are used, including "splash" utilizing dippers in the oil, "flooded" with devices to lift the oil such as disks, screws, grooves, or oil ring gears, and forced-feed systems.

Cylinder lubrication is either "splash" from the crankcase or force-fed from either the crankcase or a separate reservoir. A minimum amount of lubricant is used to provide a strong lubricant film to minimize wear and friction, to seal piston rings, valves, and rod packings, to remove heat, and to prevent corrosion.

When the crankcase is exposed to the cylinder, gas may leak past the oil control rings and contact the crankcase oil. When separated from the crankcase, oil is fed to the cylinder walls and piston rod packings with a force-feed lubrication system. In the latter case, essentially all the oil fed to the cylinder eventually leaves the compressor with the gas.

2. Rotary-Screw Compressors

Screw compressors are constant-volume and have a built-in compression ratio. Compression results in the single-stage double-helical type through the meshing of two rotors in a one-piece, dual-bore cylinder. The cylinder has air inlet passages, oil injection, a compression area, and discharge ports. Rotors are designated as male, with helical lobes, and female, with corresponding helical grooves. In oil-flooded machines, the lubricant is injected into the compression area to provide sealing, provide an oil film between the intermeshing screws, and remove the heat of compression. Oil separators are used to remove the oil from the discharge gas. "Dry screw" machines utilize timing gears to position the screws so that no internal lubrication is required.

Liquid-injected single-screw compressors are constant-volume, variable-pressure machines. Compression results from the intermeshing of a single screw with one or two gate rotors. The screw and casing combine to act as a cylinder. The gate rotor(s) acts as a piston. The screw also provides the action of a rotary valve, the screw and gate act as a suction valve, and the screw and casing (port) a discharge valve. There is a relatively low amount of friction between the screw and gate, as nearly all the compression is supplied by the screw. Bearings may be lubricated by grease or fluid, depending upon design.

3. Other Rotary Compressors

There are two common types of small vane compressors: fixed-vane and rotating-vane. Both types of compressors provide positive-displacement, nonreversing compression. The fixed-vane type uses a ring or roller, which rotates around an eccentric shaft. A single vane is mounted in a nonrotating cylinder housing. The rotating-vane compressor has a rotor concentric with the shaft and off-center with respect to the cylinder housing. The rotor is equipped with radially sliding vanes, which are forced against the cylinder walls by centrifugal force. Gas is trapped between the vanes and wall, where it is compressed due to a reduction in volume.

The lubricant in rotary-vane compressors helps to provide a seal between the sliding vanes and the cylinder (or ring) wall. Larger systems may use oil pumps. Adequate lubrication should be provided to the vanes, vane slots, bearings, and seal faces. The oil to the cylinders may be supplied later from the bearing lubricant discharge. The lubricant also prevents gas leakage in rotating shaft seals.

The basic compression unit in a scroll compressor is a set of two scrolls, one fixed and the other moving in a controlled orbit around a fixed point. Areas of lubrication include a short throw crank mechanism, bearings, and the scroll tip. Sealing is achieved through very accurate machining, various mechanisms to insure proper balancing of pressures between scrolls, linkage mechanisms, and sometimes a sealing element at the tip of the involute.

Other types of rotary compressor include the lobe and liquid piston or liquid ring, which require no internal lubrication.

C. Compressor Applications for Synthetic Lubricants

1. Air Compressors

The major reason for selection of synthetic oils for air compressors is their stability in the presence of air and moisture. Air contains about 21% by volume oxygen, which reacts with hydrocarbon oils (oxidation) to form organic acids, carbon oxides, varnishes, sludge, and hard, carbonlike deposits. Water can condense as air is compressed and later cooled. This may cause corrosion, solubilize, or form emulsions. This not only interferes with compressor lubrication but promotes more rapid deterioration of the oil. These reactions are catalyzed by certain metals such as iron and copper. The acids and sludge formed promote rapid deterioration of the oil (1).

Other considerations for selecting a synthetic lubricant are safety, maintenance, and the potential to reduce energy requirements. Each type of compressor can take advantage of the these properties.

Reciprocating Air Compressors. Conventional mineral oils produce carbonaceous deposits on valves, heads, discharge ports, and piping. The safety aspects of an air compressor lubricant are a major area of concern (2). Explosions and fires in the compressor and piping are not uncommon with oil-lubricated air compressors.

The following explanations have been developed over the past 50 years but remain essentially the same today (3). The majority of these occurrences are the result of oxidized oil residuals (as carbon–oxygen complexes) on compressor discharge components and in the piping. The deposit continues to oxidize, in the presence of air and iron oxides, to eventually produce enough heat to ignite (4). The temperature and oxygen partial pressure must be taken into account. Fires and explosions are more common when oxidation of carbon and oily deposits are accompanied by temperatures above 300°F (149°C) and pressures above 100 psig (68 kPa) (5). Carbon deposits on discharge valves can result in their malfunction and recompression of the gas, leading to excessive discharge temperatures. Deposits in intercoolers and aftercoolers reduce their efficiency.

Low volatility reduces the chance of a vapor fire and resulting explosion if excessive oil is present (6). The ignition source is usually caused by carbon deposits or valve failure. The low carbon-forming tendencies of many synthetics and higher flash and fire and autoignition temperatures reduce this hazard.

Figure 19.2 illustrates higher autoignition temperatures for synthetic fluids at elevated air pressures. This information represents common values for new formulated compressor oils. Deterioration (through oxidation) will lower the autoignition temperature as shown for the lubricants containing iron oxides (7). The discharge temperatures are for reciprocating compressors with inlet temperatures of 21°C (70°F) to all stages. Actual discharge temperatures will vary with cylinder size, degree of cylinder cooling, compression ratio, rpm, and other design variables (8).

Discharge temperatures for these reciprocating air compressors can easily reach 230°C (450°F), near the reduced autoignition of mineral oils at 135 psi (31 kPa) (9). Excessive oil accumulation in the air system can propagate a fire. The special case of detonation, caused by the development and propagation of shock waves, is attributed to this type of fire. A simplified explanation follows. As the fuel burns, hot vapors evolve and expand, and pressure waves push into the unburned gas. These waves cause the unburned gas to develop extreme pressures and heat. The flame that follows results in an explosion in the piping, coolers, or receiver.

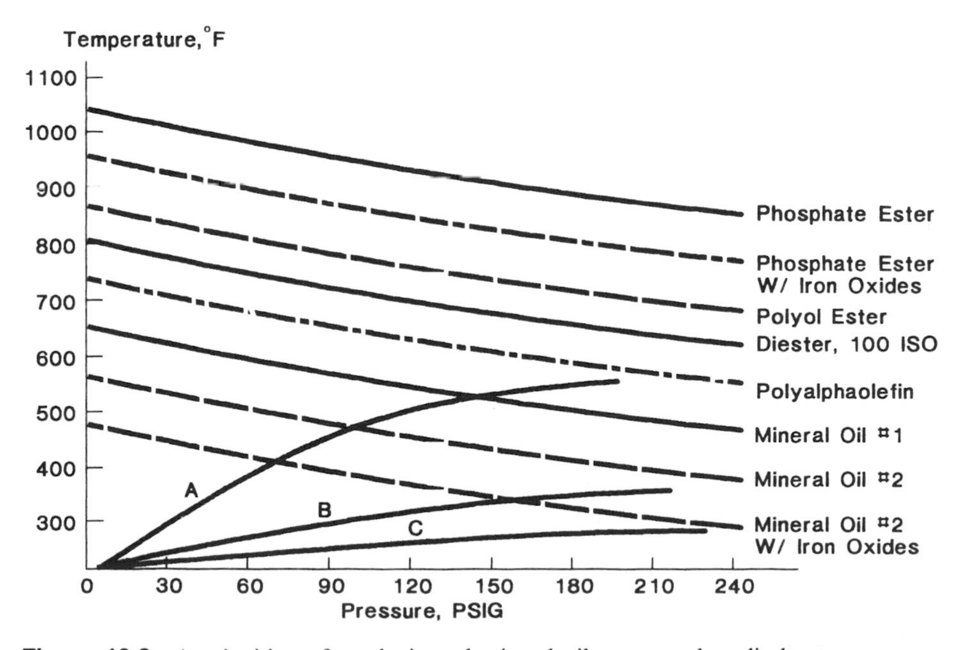

Figure 19.2 Autoignition of synthetic and mineral oils compared to discharge pressures and temperatures: (A) single stage, (B) two stage, (C) three stage.

Phosphate esters. Phosphate esters were originally selected for their fire-resistant nature (10). Problems occurred when the lubricant degraded to form by-products that were aggressive toward paints and elastomers in older machines. Higher feed rates may be required than with mineral oils, and there have been problems with close tolerance lubrication between aluminum and cast iron cylinders. Their high phosphorus content caused disposal problems. Even with these problems, phosphate esters were found to be helpful in high-pressure multistage machines (11).

Fires and explosions can occur even with the use of fire-resistant synthetic oils such as phosphate esters if they are allowed to accumulate in the air system (12). Lower feed rates to compressor cylinders and oil/air separation equipment may be required to reduce this hazard. Devices have been developed to deliver smaller and more accurate amounts of lubricants (4).

Di-, tri-, and tetraesters and polyol esters. These have low vapor pressures at elevated temperatures relative to mineral oils. This property, good oxidative stability, and an affinity to metal surfaces make it possible to use lower amounts of ester-type synthetics for cylinder lubrication.

The feed rate for a diester can be lowered to 50–65% that of a conventional mineral oil (5, 13). Polyol esters with low volatility, excellent stability, and high viscosity indices allow a feed rate of about 25% that of a mineral oil. The actual feed rate used should be determined through gradual reduction in feed rates combined with periodic inspections of the compressor cylinder to insure that there are no dry spots. The lower carbon-forming tendency of esters helps to keep cylinder walls, intercoolers, aftercoolers, and piping clean. This reduces the fuel source for fires. Another obvious advantage of reduced feed rates to compressor cylinders is the reduction lubricant usage.

The cleanliness and good solvent action of the esters help to prolong the maintenance intervals of piston rings and valves. Valve sticking can cause recompression, which results in higher discharge temperatures. The latter accelerates deposit formation. There have been several documented cases of prolonged maintenance intervals of two to four times that experienced with mineral oils (5, 14, 15). At least one investigator reported superior results with diesters as opposed to triesters (trimellitate) (16).

There have been many commercial claims that synthetic lubricants are more efficient and consume less energy than petroleum oils. This has been described as the result of the low coefficient of friction of esters, viscosity stability, and their cleanliness. One study showed an average savings of 6.6% in kilowatt-hours consumed for a group of 23 reciprocating compressors ranging in size from 25 to 100 hp (17).

Polyalkylene glycols. Polyalkylene glycols are not generally accepted as lubricants for air compressor applications. This is due to their relatively poor volatility, tendency to form volatile decomposition products, and poor compatibility with mineral oils. Blends of polyglycols and esters overcome these difficulties (18).

Polyalphaolefin-type synthetic hydrocarbon oils. These have been formulated with low volatility and low carbon-forming tendency. This has led to cleaner operation and reduction of fires in critical installations. Performance was good, with discharge temperatures at 392°F (200°C) and pressures at 100 psi (68 kPa) in a 120-hp two-stage reciprocating compressor after 16,000 h of operation (19). Other fully formulated PAO-based lubricants have surpassed the requirements of the ISO DP 6521 specifications (18). The PAOs have been blended with esters to improve their solvency and with oil-soluble silicones to reduce cylinder feed rates (4). When deposits are formed they are either very sticky (polymers) or hard varnishes (as with paraffinic oils). The major benefit of PAOs is their compatibility with elastomers and paints found in older machines, which are designed for use with mineral oils.

Other types. Other types of synthetic oils have been used for reciprocating air compressor applications. Fluorosilicone lubricants have been shown to reduce or eliminate air compressor explosions. Their high cost is offset by a reduction in feed rates to 5% of that with mineral oils (20). These oils have good lubricity with steel but may cause seizure with aluminum.

Rotary-Screw Air Compressors. The lubricant in a rotary-screw air compressor is continually circulated through a system where it is sheared by rotors while aggressively mixing with hot, wet air under pressure and in the presence of metals. Both thin-film and bulk oxidation conditions exist. The presence of moisture promotes hydrolysis and may promote the catalytic action of metals. Thermal stability is important as the lubricant is exposed to the heat of compression and then cooled. Conventional mineral oils deteriorate in a relatively short time and often require excessive makeup, as they volatilize and are inefficiently removed by oil separation equipment (commonly the coalescing type).

Table 19.1 lists the desirable and critical characteristics for these lubricants (21, 22). Most synthetic lubricants listed will provide drain intervals of 8,000 h when operated in a normal environment and with discharge temperatures of 180°F (or 90°C) and pressure of 68 kPa (100 psig). Differences are mainly with material compatibility, volatility, and operating life at higher temperatures or in high humidity.

Figure 19.3 shows the effect of high discharge temperatures for various synthetic lubricants. Most of these tests were conducted by a major compressor manufacturer in a 25-hp compressor operating with discharge temperatures from 220 to 230°F (104–110°C) and were reported in earlier literature (21–25). The expected life of these lubricants at standard temperatures of 180–190°F (or about 85°C) is about four times that shown for the elevated temperature.

Table 19.1 Properties of Rotary-Screw Air Compressor Lubricants

Property	Diester	Polyol ester	Polyglycol and ESTER	Polyalphaolcfin	Silicone
Oxidation resistance	Good	Excellent	Very good	Very good	Excellent
Flash point	Very good	Excellent	Very good	Very good	Excellent
Pour point	Very good	Excellent	Good	Very good	Excellent
Volatility	Very good	Excellent	Good	Very good	Very good
Lubricity	Good	Very good	Very good	Good	Fair
Demulsibility	Good	Good	Poor	Excellent	Excellent
Nonfoaming	Very good	Very good	Very good	Very good	Good
Rust and corrosion inhibited	Good	Good	Good	Very good	Fair
Hydrolytic stability	Good[a]	Good[a]	Good	Excellent	Excellent
Nontoxic	Good	Good	Good	Excellent	Good[b]
Material compatible	Fair[c]	Fair[c]	Fair[d]	Excellent	Excellent[e]

[a]Requires special additives.
[b]Chlorinated additives.
[c]Swells certain elastomers and may dissolve plastics.
[d]Not compatible with certain plastics.
[e]Silicone lost through separators may cause problems in painting operations.

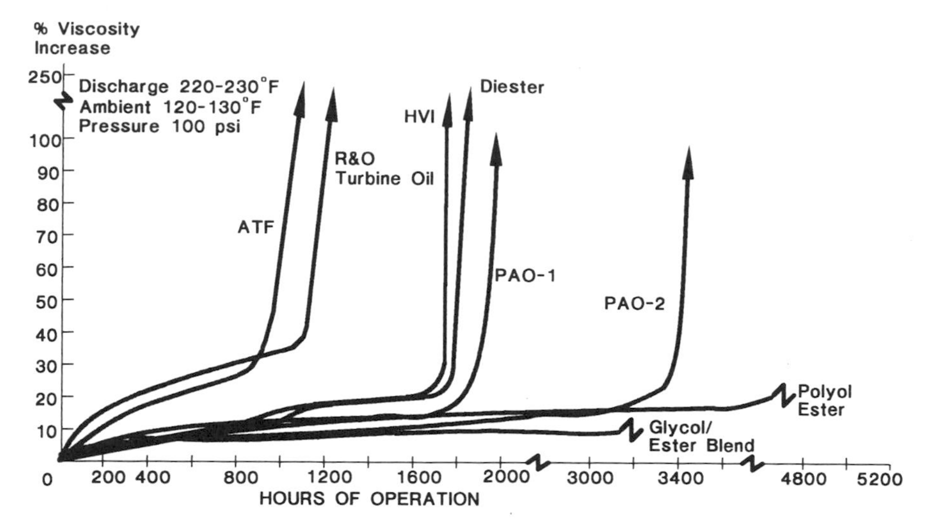

Figure 19.3 High-temperature compressor test results.

Energy savings are possible with synthetic lubricants. Conversations with original equipment manufacturers suggest the penalties attributed to using an unstable oil (Table 19.2). Careful selection of a synthetic lubricant will prevent these penalties in a properly maintained compressor. Most of these fluids can reduce or eliminate separator blockage and varnish formation. Heavy varnish increases energy consumption and can restrict rotor movement. The result can be air-end failures through excessive wear of rotors and bearings. A stable viscosity and good lubricity can reduce friction. Viscosity stability and cleanliness can improve oil-side heat transfer. Some of the synthetics have better thermal conductivity.

Diester and triester lubricants. Diester- and triester-based rotary-screw air compressor lubricants offer the advantages of being a natural detergent and not producing varnishes or heavy polymers. This detergency does not always end problems with separator blockage. One study showed that coalescer blockage with a diester was only slightly better than a mineral oil and significantly worse than a polyalphaolefin compressor oil (26). Esters dissolve varnishes leftover from mineral oils. This improves efficiency over the dirty machine as previously described.

One study of 11 machines, 40–300 hp, showed that rotary-screw compressors may

Table 19.2 Efficiency Loss in Rotary-Screw Compressors

Problem	Penalty
Viscosity increase or air-end varnish	3–5%
Blocked air/oil separator, P = 10 psi	3–5%
Excessive discharge temperature 18°F (10°C)	1–3%

benefit from reduced energy requirements averaging 9.3% (kilowatt hours consumed) (17). The study attributed energy saving to "friction reducing."

Depending on the type (adipate, phthalate, triester, etc.), these fluids usually increase in organic acid levels and gain or lose viscosity during extended use at higher air compressor discharge temperatures (above 210°F). This could lead to mechanical failures or corrosion problems (16, 21, 22). Both base fluid and additives must be carefully selected to avoid these problems.

The diesters have a wide range of viscosity indexes, depending on type and initial viscosity grade. They do not always meet rotary-screw compressor manufacturer minimum viscosity index requirements (usually at least 90). In other cases a viscosity requirement at 210°F (or 100°C) is specified. Diesters that have poor viscosity indexes may have too high a viscosity for cold starting conditions. An adipate-type diester (or a blend) meets both requirements.

Polyol esters. Polyol esters have not been used extensively for rotary-screw air compressors. This is most likely due to their higher cost. These fluids have good detergency and can provide excellent service at higher temperatures due to their excellent oxidative and thermal stability (22). Earlier formulations were limited by the availability of additives that would enhance the performance of these fluids. Long-term operation in rotary-screw compressors at higher temperatures requires antioxidants and rust and corrosion inhibitors that are not volatile and do not themselves deplete too rapidly. Newly developed additives and increased availability of these fluids (some improved) should result in their increased use. The polyol esters are the least volatile of the synthetics described (depending on type). This helps to limit their consumption through use of high-efficiency oil separators. They have excellent lubricity (elastohydrodynamic), viscosity indices, and low-temperature properties. Many of these fluids are biodegradable, an important factor when considering that water (and sometimes lubricant) is drained from aftercoolers and receivers.

Polyalkylene glycols. Polyalkylene glycols blended with polyol esters provide long life in compressors with discharge temperatures up to 230°F (110°C) (23). Polypropylene glycols by themselves were said to have deficiencies. They do not wet metal, and tend to run off ferrous metal surfaces and leave exposed metal, resulting in rust under humid conditions. The ester is added to compensate for this deficiency. Polypropylene glycol has excellent hydrolytic stability and less water adsorption than other types of polyglycols. Good vapor-pressure characteristics are achieved with the higher-viscosity fluid chosen for the blend (1,200 molecular weight). The pentaerythritol ester used in the blend was low viscosity to blend down the final lubricant viscosity. Like the esters, the polyglycol/ester blend was reported to remain clean upon failure, producing "volatile components rather than crosslinking to viscous gums."

Phosphate esters. Phosphate esters, except as additives, are not normally used in rotary-screw compressors due to their rapid breakdown and formation of acids. The exception is critical applications that require the use of a "fire-safe" lubricant. In these instances the phosphate ester must be changed more frequently than other types of synthetics, usually every 500–1,000 h. The alternative is an "oil-free" compressor where the phosphate ester may be used in anticipation of seal leakage.

Polyalphaolefin-type oils. Polyalphaolefin (PAO) type synthetic hydrocarbon oils have been used extensively for these compressors. They are often preferred over the use of esters for use in standard 100 psi (68 kPa), 180°F (or 90°C) applications. The PAOs have excellent hydrolytic stability and compatibility, not only with rubber and plastic materials in the compressor, but also in the compressed air system (24). They are also compatible with mineral oils and common additives. This makes compressor conversions from mineral oils simple and helps to prevent problems with materials and equipment used in the compressed air system. The PAOs have very little effect on swelling of rubber or elastomers. It is very common to add from 8 to 15% of a diester (or polyol ester) to increase seal swell and to help solubilize contaminants.

The PAO-based compressor oils have longer drain intervals than do diesters (usually about 30% longer). Their low water adsorption and rapid water separation result in improved rust and corrosion protection and help water to be easily drained from the oil reservoir during long periods of shut down. A high viscosity index and low volatility allow the use of lower-viscosity grades. This combined with excellent low-temperature fluidity reduces power consumption during start-up in cold environments.

Food-grade formulations of PAO air compressor oils are available that meet U.S. Department of Agriculture and Food and Drug Administration requirements for incidental food contact (27). It is not clear how these lubricants will be eventually treated in Great Britain, where white mineral oils have been eliminated for this use (28). The polyalphaolefins are not considered mineral oils and thus have not been banned for use with food by that government. Most other areas of the world accept the U.S. regulations, especially where food is imported into their country. These lubricants are recommended for drain intervals of 4,000–6,000 h for food applications and 8000 h for standard applications (24). The reduced hours for food applications are recommended as a preventative measure against unusual conditions (temperature, environment, etc.) that may result in early deterioration of the lubricant.

Dimethyl silicone has an excellent viscosity index, thermal and oxidative stability, and hydrolytic stability. Additives improve its otherwise poor lubricity with metals. A specially formulated dimethyl silicone rotary-screw air compressor lubricant has been marketed since the 1970s (21). Several compressors have been in operation for 40,000 h. In most cases the results have been favorable (24). Maintenance of the air oil separator is critical. Excessive oil carryover is expensive. In some cases, as with some painting operations, the dimethyl silicone can deposit on production parts and cause manufacturing problems.

Rotary-Vane Air Compressors. Flooded rotary-vane compressors benefit from the same properties of the synthetic lubricants described for rotary-screw compressors. Many of these machines use a higher viscosity grade than the screw. An ISO 100 blend of polyalphaolefin and phthalate esters was examined in two small (5.5 and 7.5 hp) air-cooled machines (29). Severe test conditions were used:

Insufficient air cooling (oil temperatures to 104°C).
High moisture.
Oxidants in air.
Stop-and-go operation.
Use of lime near by.

The results were determined to be favorable after 3,000–3,500 h of operation, with viscosity and acid number remaining relatively unchanged. The later was considered important to maintaining a constant seal for stable air pressure. Exceptional low oil consumption resulted in the lubricant not having a detrimental effect on ozone synthesis in a corona discharge process.

Diester compressor oils are commonly used in vane compressors. Their natural detergent action helps to prevent sticking caused by deposits in the rotor slots of sliding vane machines. Higher-viscosity polyol esters should be particularly useful in machines operating with high discharge temperatures. Some types of water-cooled machines have discharge temperatures above 325°F (163°C). Some vane machines are equipped with plastic or composite vanes. The manufacture should be consulted as to their compatibility with esters (most are compatible).

Centrifugal and "Oil-Free" Compressors. Synthetic lubricants provide a wide operating temperature range, better heat transfer, lower energy requirements, or reduced hazard if seal leakage occurs.

Polyol esters have been used for service in lobe compressors where high temperatures occur. One case involved the use of the lobe compressor as a blower for 600°F (315°C) air (24). The polyol ester reduced bearing deposits (and failures), provided adequate viscosity for lubrication, and provided more efficient heat removal.

The polyalphaolefins have very low exotherms when mixed with mineral acids. This is due to their completely saturated hydrocarbon structure. This property has made specially formulated polyalphaolefin air compressor lubricants permissible in chemical plant operations where leakage through seals could result in exothermic reactions. One such case is their use in ammunition plants where concentrated nitric and sulfuric acids are used.

2. Gas Compressors

The lubrication of gas compressors encompasses many different types of gas, which can be categorized as inert, highly soluble, and reactive. The International Standards Organization has suggested further classifications under ISO DP 6743/3 B as shown in Table 19.3. The recommendation is that these should be used for reference only and that the lubricant supplier and the equipment builder be consulted. In some cases these gases are divided into categories that relate to their industrial uses, oil refinery, chemical, halogen, polymer, landfill, etc. (30). In the case of discussing the use of synthetic lubricants, a variation of the ISO method of categorizing appears to be more useful, that is, to discuss the type of gas (hydrocarbon, carbon dioxide, etc.) and how synthetic lubricants help the compressor performance.

There appears to be three distinct areas of concern: solubility, reaction, and effect of the lubricant as a contaminant in the gas being compressed. The first two affect the compressor and the latter the gas application.

The solubility of a gas in the lubricant is a major concern. Dilution may cause a reduction in viscosity and a loss in the film thickness of the lubricant. The lubricant can be washed off by liquid components of the gas. Additional problems can occur when there is a reduction of pressure and degassing takes place (foaming, loss of lubricant film).

The solubility of the lubricant in the gas can result in loss of lubricant by absorption of the lubricant into the gas phase (31). The result can be the requirement of excessive amount of lubricant feed rates for compression and a source of contamination to the gas.

Reactions of the gas with the lubricant can result in premature failure of the compressor or, in more severe cases, fires and explosions. The lubricant or additives may react or inhibit catalysts, cause mechanical problems in the application (valves etc.), or plug areas of gas flow.

Hydrocarbon Gas Applications. Hydrocarbon gases are encountered in the collection and transmission of natural gas, in vapor recovery, and in the chemical industry. Synthetic oils are selected for their unique viscosity–temperature relationships or for their resistance to dilution by hydrocarbons.

Table 19.3 Draft ISO DP 6743/3 B Gas Compressor Lubricants

Symbol	Application	Application example	Lubricant type
DGA	Inert gas	< 1500 psi	Mineral oil (or synthetic)
		Nitrogen	
		Hydrogen	
		Ammonia	
		Carbon dioxide	
		Argon	
		All pressures	
		Helium	
		Sulfur dioxide	
		Hydrogen sulfide	
		< 150 psi	
		Carbon monoxide	
DGB	Inert gas moisture	As above	Above plus additives
DGC	High solubility	Hydrocarbons	Synthetic
		> 1500 psi	
		Carbon dioxide	
		Ammonia	
DGD	Chemical reaction	Hydrogen chloride	Nonhydrocarbon fluids
		Chlorine	
		Oxygen	
		> 150 psi	
		Carbon monoxide	
DGE	Dry inert plus reducing gas	Nitrogen	Synthetic
		Hydrogen	
		1500 psi Argon	

Compressed natural gas and other hydrocarbon gases are used to fuel gas turbines. The compressor supplies the gas at the flow rate and pressure needed for continuous operation of the turbine. Petroleum-based lubricants carried in the gas may produce carbonaceous deposits in the gas inlet nozzles of the turbine, restricting flow and causing flame-out.

Hydrocarbon gases are often the feedstock for a chemical process. Examples include the manufacture of polyethylene and polypropylene. Synthetic oils are often used for these applications due to their purity or lack of reaction with catalysts.

Reciprocating compressors applications. For pressures below 1000 psig, ISO 100–150 mineral oils may be used. Problems occur when the gas is wet or at increased pressures. The addition of fatty oils is common. These additives are difficult to pump at low temperatures, can cause damage to discharge valves, accumulate in aftercoolers and piping, and emulsify with water. Detergent-type heavy-duty engine oils have been used in sour gas (H_2S) applications. The additives in these oils can cause problems as described below.

High-pressure reciprocating compressors (5,000 psig) are used to re-inject natural gas into crude. Four basic problems have been identified (31):

1. Loss of lubricant viscosity.
2. Increased cylinder wear rate as a result of lubricant (680 ISO mineral oil) being washed off cylinder surfaces by liquid components in the gas.
3. Loss of lubricant to the high-pressure gas stream. This resulted in feed rates to rod packings of 10 times normal rates or up to a barrel of lubricant per day per compressor.
4. Reaction of additives with well-bore fluids, leading to permanent impairment to the well and resulting in reduced gas injection rates.

The study showed that the use of a polyglycol, 200 cSt at 40°C (104°F), would solve all of these problems. The higher viscosity index of the polyglycol resulted in a higher initial viscosity (undiluted) at compressor discharge temperatures. Resistance to dilution by hydrocarbon gases resulted in a viscosity in the cylinder of about twice that of the ISO 680 cylinder lubricant. It was shown that the polyglycol did not adsorb in the gas phase, while the mineral and acidless tallow is nearly completely adsorbed. Field trials showed a 20-fold increase in the life of pressure packings and a reduction in overhaul maintenance from eight times to one time per year.

High feed rates with mineral oils can cause problems in natural-gas pipeline booster compressors. Excessive oil downstream reduces the volume within the pipeline and leads to a significant reduction in pipeline efficiency. Silicone "semisynthetic" blends with white oils (sometimes polyalphaolefins) are used in combination with microlubrication hardware to lower cylinder feed rates as much as 95% from that with mineral oils (32–34). These systems have been in operation in more than 230 natural-gas pipeline booster compressors in the United States since 1981. The reduced feed rates and the use of the synthetic oil have resulted in longer cylinder ring life. One report showed no measurable wear after 2 years. Discharge valve life is improved through lower carbon buildup. The lower feed rates, using the microlubrication system, result in lower oil costs. The payback for the initial cost of the hardware and lubricant is less than 1 year, with greater savings after that time (35).

Rotary-screw compressors. Viscosity is the most critical lubricant requirement that must be met in the rotary-screw compressor when handling hydrocarbon gases. Viscosity may be lowered as the lubricant is diluted by the hydrocarbon gas as they mix in the compressor and pass into the separator. The final level of dilution is determined by the temperature and pressure in the separator, located on the discharge side of the compressor. Synthetic lubricants offer the advantage of being available in very high viscosity grades and in some cases are resistant to dilution through lower solubility with hydrocarbon gases.

Polyalkylene glycols have reduced solubility with hydrocarbon gases. Different types of polyglycols are available that offer unique solubility characteristics, dependent on their physical make-up and structure. Polypropylene glycols are used with lighter hydrocarbons (as with propane production) because of their resistance to dilution with hydrocarbons. Advantages such as higher volumetric efficiency and energy savings have been proved and are often the main criteria in their selection.

Inclusion of ethylene in the structure of the polyglycol reduces hydrocarbon solubility. A copolymer of ethylene and propylene oxide will allow 10–20% hydrocarbon solubility by weight before the fluid is saturated. Water solubility is increased but controllable, as these fluids exhibit an inverse solubility at temperatures above 140–160°F (60–70°C). Solubility with water at lower temperatures helps reduce corrosion, particularly where H_2S is present, through a combination of reducing water on the metal surface and the use of specially designed additives. Complete resistance to normal alkanes is possible with polymers based on ethylene oxide. Falling-ball viscosities carried out at 13,790 kPa (2,000 psi) with a typical wellhead gas proved no loss in viscosity. Although these fluids are water soluble, lubricity is not reduced (four-ball tests) with up to 7% water in the lubricant (36). Propane compressors may gain up to 18% improvement in volumetric efficiency with certain polyglycols (22).

Polyalphaolefins have also shown efficiency improvements. In one application compressing methane gas, a cogeneration plant measured a 2% increase in volumetric efficiency, without increased power consumption (36). The compressor was equipped with a slide valve allowing it to run in an unloaded mode when demand was less than output capacity. With the synthetic, less power was used to drive the compressor, with more frequent operation in the unloaded mode. The polyalphaolefin was originally chosen for improved water separation and low-temperature fluidity, as the oil reservoir was located outdoors.

Reactive Gases. The use of "nonlubricated" compressors is common in the chemical industry. There may still be some chance of the gas contacting the lubricant (as with a seal leak). Each type of synthetic lubricant has a unique chemical structure. This offers the opportunity to select a lubricant that will not chemically interact with the gas. This same property has led to the use of selected synthetics in lubricated compressors. In some cases the lubricant must provide a barrier to corrosive attack on the compressor materials. Another consideration should be the effect of the lubricant on the chemical process (as with catalysts).

1. Oxygen compressors are usually of the nonlubricated type. Highly fluorinated lubricants such as perfluoroethers are used.
2. Chlorine and hydrogen chloride compressors are usually nonlubricated types with halogenated lubricants used to lubricate mechanical parts. The lubricants are either the fluorinated types, perfluoroethers or fluorosilicones, or may be the chlorofluorcarbon lubricants.

   Other types of silicones have excellent chemical stability and protect metal surfaces from chemical corrosion. These have been combined with polyalphaolefins and severely hydrotreated "semisynthetic" hydrocarbons to provide the same metal protection as with the more expensive all-silicone fluids (22, 37). They are effective in many process gas applications involving chlorosilanes, methyl chloride, hydrogen chloride, and other chlorinated compounds. Applications such as vent gases, or "off gases," can contain water and can be extremely corrosive. Successful operation is achieved at temperatures above the dew point of water and purging with a dry, inert gas upon shutdown (36).
3. Hydrogen sulfide, nitrous oxide, and sulfur dioxide all require lubricants that are dry. The later two are soluble in mineral oil and therefore reduce its viscosity. Polyalphaolefins have a low tendency to absorb moisture and are available in high viscosity grades and are therefore suitable for these applications. The blended silicone described in the previous section has also been used for its additional corrosion protection.

Inert Gases. Inert gases are classified as such as they do not react with mineral oils or lead to general lubrication problems. Only highly purified mineral oils are used, as polar contaminants may react with contaminants in the gas or catalysts. Synthetics are preferred for improved stability and for prevention of catalyst poisoning. Additives must also be carefully evaluated for similar reasons, and, as with ammonia, may react with the gas (38). Esters are not used with ammonia or with gases containing ammonia as they react to form viscous polymers or solids.

The use of rotary-screw compressors with oil separators for these applications has resulted in the increased use of synthetics with low volatility. This helps reduce contamination of the gas. Food-grade polyalphaolefins are used in carbon dioxide (CO_2) compressors where the gas is later used in food products. The PAO has the additional advantage for some processes of only containing carbon and hydrogen. High-temperature catalytic processes remove organic contaminants by using oxygen to convert the carbon to CO_2 and the hydrogen to water.

Polyglycols have been commonly used in both hydrogen and helium applications. Polyglycols of the oxyethylene/oxypropylene copolymer types have shown good lubricity with reciprocating-compressor polytetrafluoroethylene (PTFE) packings and compression rings in hydrogen applications and are compatible with plant processes. Polyalphaolefins have replaced polyglycols in many hydrogen applications where rotary-screw compressors benefit from their lower volatility (oil/gas separation) in relationship to viscosity.

With the exception of ammonia, the esters have been used successfully in most inert gases.

Refrigeration. The choice of refrigerant will influence the compressor system design and the characteristics required of the lubricant. Centrifugal compressors present a simple problem. There is little contact between the refrigerant and the lubricant. The lubricant in these machines may be selected to prevent problems associated with seal leakage (lubricant into the system) or where the refrigerant has contact with the oil reservoir. The discussion that follows is primarily for positive-displacement compressors where there is interaction between the refrigerant and the lubricant.

3. Refrigeration

Refrigeration lubricants may be required to provide many years of service without makeup and with a minimum of maintenance. Final compression temperatures may reach 320°F (160°C) for some applications where unsuitable oils form carbonaceous deposits. In the special case of hermetic compressors the motor materials must not be adversely affected by the lubricant/refrigerant mixture or by-products from its deterioration. This requires a lubricant that has excellent thermal and chemical stability and produces a minimum of deposits.

Figure 19.4 shows a simple diagram of the refrigeration process. A liquid refrigerant evaporates to provide the required cooling, the vapors are compressed (to condensation pressure) and condensed to a liquid using a cooling medium such as water or air, and the liquid is sent through an expansion valve (to reduce pressure) and returned to the evaporator where the cycle starts again.

The lubricant in compression refrigeration systems has an influence on the operation and efficiency of the entire system. Some lubricant is carried out of the compressor and into the system. The lubricant must act as a compression sealing aid and reduce wear and friction in the compressor without adversely affecting the operation of the filter driers, condenser, expansion valve, or evaporator. The behavior of the oil/refrigerant pair is of major importance.

The solubility of the refrigerant gas in the lubricant and the miscibility of the liquid

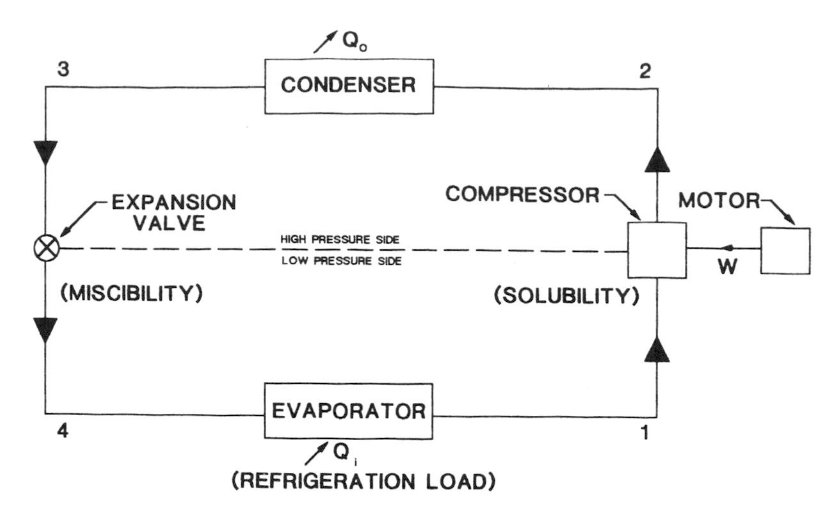

Figure 19.4 Refrigeration cycle.

refrigerant with the lubricant respectively affect compressor performance and system performance. Dissolved gas has the effect of reducing lubricant viscosity. Miscibility is considered for design of components and piping to promote uniform oil movement through the system and back to the compressor. Heat-transfer problems are more significant in systems where the oil is immiscible or partly miscible with the refrigerant.

The main differences between the screw compressor and the reciprocating compressor with respect to the oil system are:

The screw compressor has an oil separator and an oil sump situated on the high-pressure side.

The compression chamber is flooded with oil to seal the threads that are under compression.

The lubricant has more effect on performance than it does with the reciprocating compressor. To reach high performance, the screw compressor needs a lubricant with limited solubility of the refrigerant gas at discharge conditions (at the oil separator). Limited solubility will reduce or eliminate bypassing of refrigerant from discharge to suction or to a lower situated thread. External bypass caused by the refrigerant circulating with the oil is also reduced. This leads to both high volumetric efficiency and low torque. Most lubricants with low solubility also have low miscibility (liquid refrigerant in oil). In some cases a synthetic lubricant can meet the requirement of low solubility while maintaining good miscibility. This combined with low volatility reduces oil in the system to improve heat transfer (39).

Synthetic lubricants offer a wide range of properties and the opportunity to customize a lubricant for a particular refrigeration system. They have been considered for refrigeration applications since 1929 (40) and have continued to show improvements in performance properties when compared to mineral oils (41). Properties of synthetic base stocks as they relate to refrigeration compression appear in Table 19.4. Of particular interest are the miscibilities of these fluids with various refrigerants as shown in Fig. 19.5 and 19.6. Typical applications appear in Table 19.5.

Polyalphaolefins. Polyalphaolefins (PAO) have been extensively used in refrigera-

Table 19.4 Properties of Synthetic Refrigeration Lubricants[a]

	Synthetic hydrocarbons		Polyalkylene		Esters	
Properties	PAO	Alkyl benezene	Glycol	Dibasic	Polyol	Silicate
Chemical stability	E	VG	G	G[c]	G[c]	G[e]
Thermal stability	VG	VG	G[b]	G	VG[d]	G
Miscibility (polar refrigerant)	P	VG	E	VG	E	E
Volatility	E	G	G	VG	E	VG
Low temperature	VG	G	G	VG	VG	VG
Viscosity–temperature	VG	F	E	G	VG	VG
Water adsorption	E	G	P	F	F	F
Mineral oil compatibility	E	E	P	VG	G	P

[a]Key: P = poor, F = fair, G = good, VG = very good, E = excellent.
[b]Decomposes at 500°F, additives may be required.
[c]Additives may be required, reacts with ammonia (R-717).
[d]Additives required above 200°C.
[e]Hydrolyzes to form gels and solids.

tion compressor applications for their high viscosity index, low-temperature fluidity, and excellent thermal and chemical stability.

The PAOs have good miscibility with R-12. Superior chemical and thermal stability have reduced the risk of carbonizing at high temperatures in heat pumps with R-12 and R-114 (22, 42). Superior adiabatic efficiency of 3–10% is achieved in rotary-screw compressors using these refrigerants when compared to naphthenic refrigeration oils. This efficiency improvement is largely attributed to the higher viscosity of the PAOs at higher temperatures in the presence of the refrigerant. Superior performance and reliability have been achieved in reciprocating, twin-screw, and single-screw compressors (43).

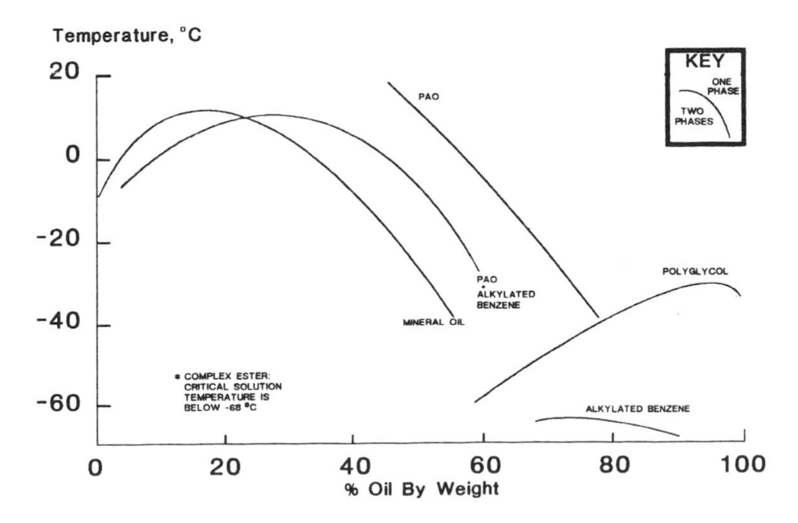

Figure 19.5 Miscibility characteristics for various fluids with HCFC-22.

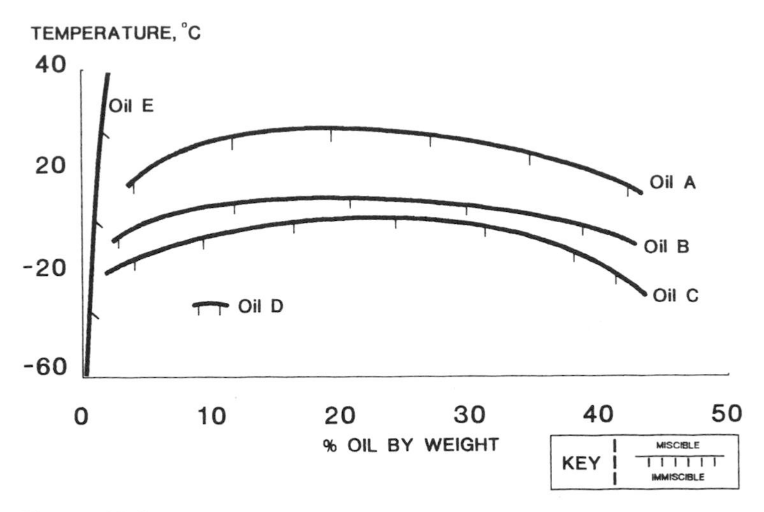

Figure 19.6 Miscibility of esters with HFC-134a. Oil A, polyol ester ISO 32 (35); oil B polyol ester ISO 22 (35); oil C polyol ester ISO 32 (27); oil D proprietary ester ISO 116 (37); oil E diester, ISO 32 (27).

Lower evaporator temperatures are permissible with PAOs than with mineral oils because PAOs are wax-free and have good low-temperature fluidity for oil return. Low-temperature fluidity is the major reason PAOs have been used in the United States for relatively insoluble refrigerants such as R-13 and R-503. The lubricant viscosity, ISO 15 or 32, is selected by considering operating viscosity in the compressor as well as low-temperature fluidity below –73°C (–100°F) in the direct expansion dry-type evaporator system (44).

Applications with R-22 are limited, by low miscibility, to systems that limit the amount of the oil. Dry expansion type evaporators are used (similar to R-13 applications) or flooded evaporators are equipped with oil-skimming devises.

Polyalphaolefins with low volatility have been used in very-low-temperature, –118°C

Table 19.5 Typical Applications for Refrigeration Lubricants

Refrigerant	Synthetic product
Chlorofluorocarbons	Polyalphaolefins
Hydrofluorochlorocarbons	Alkyl benzenes
	Complex esters
Highly fluorinated HCFCS	Polyol esters
	Polyalkylene glycols
Hydrofluorocarbons	Polyalkylene glycols
	Polyol esters
	Other esters
Ammonia	Polyalphaolefins
	"Semisynthetic" hydrotreated
Propane/hydrocarbons	Polyalkylene glycols
	Polyalphaolefins
	Esters

(–180°F), ethylene systems. Separators can effectively control the amount of lubricant in the low-temperature side of the system. Careful selection of oil viscosity provides enough lubricant film to lubricate compressor components in the presence of the gas. The small amount of miscible lubricant is diluted in the evaporator and can be easily returned. There is no problem of wax or viscous liquid formation as found with some mineral oils.

Polyalphaolefins and "semisynthetic" high viscosity index (HVI) hydrotreated isoparaffinic mineral oils have been found to provide several performance advantages in ammonia systems (45). Better thermal and chemical stability with ammonia results in reduced sludge and varnish and has resulted in extended drain intervals (46). Lower solubility helps improve lubrication and reduces foaming. Lower volatility reduces oil consumption and improves heat transfer by limiting the amount of oil in the system and on heat-exchanger tubing. The excellent low-temperature fluidity and high viscosity index of the PAO fluids allow evaporator temperatures below –46°C (–50°F) and maintain viscosity for higher compressor operating temperatures. Good low-temperature fluidity facilitates oil removal in most systems.

Alkyl benzenes. Alkyl benzene synthetic hydrocarbon oils have good chemical stability and good miscibility with R-22 and R-502. They have somewhat improved miscibility with more highly fluorinated chlorofluorocarbon (CFC) and hydrochlorofluorocarbons (HCFC) refrigerants such as R-13 and R-503. Their relatively low cost, compared to other types of synthetics, has made their use common for these moderately polar refrigerants.

The improved miscibility of the alkyl benzenes has led to their use with more environmentally safe chlorofluorocarbons (HCFC) such as HCFC 124 (47) and blends of HCFC with hydrofluorocarbons (HFC) (48). These refrigerants have been investigated for reduced ozone depletion and greenhouse warming potentials.

Alkyl benzenes are commercially available in ISO viscosity grades as high as 68 or 100. This has restricted their use where the solubility and viscosity relationships indicate a chance of overthinning in compressors where oil reservoirs are on the high-pressure side, such as rotary-screw types. The viscosity–temperature relationship of the oil may be improved by increasing the alkyl chains at the expense of reduced miscibility. Formulations have been made by blending an alkyl benzene with mineral oils or with a PAO to overcome these deficiencies. Private reports have indicated that blends with PAO may separate in the liquid refrigerant in the evaporator (44).

Polyalkylene glycols. Polyalkylene glycols (PAG) have excellent miscibility with polar refrigerants, good lubricity, and good low-temperature fluidity. Major drawbacks include their tendency to absorb water, low compatibility with mineral oils, and the requirement of additives for good chemical and thermal stability. New developments with structural changes show promise in overcoming these deficiencies.

Polypropylene glycols have been used in R-12 heat pump applications in rotary-screw compressors (40, 49). The PAGs have excellent viscosity temperature characteristics and reduced solubility with R-12. Efficiencies achieved were similar to those with PAO. The PAGs have also been used with R-22. When carefully selected, these oils have complete miscibility even at concentrations over 50% by weight with R-22 to temperatures below –73°C (–100°F). Higher-viscosity-grade PAG fluids have low miscibility with some highly fluorinated CFCs such as R-13 (chlorotrifluoromethane).

Polyalkylene glycols are miscible with hydrofluorocarbons (HFC) such as HFC-134a and also with blends of HFC and hydrochorofluorocarbons (HCFC). This has resulted in their initial use with HFC-134a in automotive air conditioners (50, 51) and rotary-screw compressors (39, 51). Increasing the viscosity of polyglycols reduces miscibility, especially at higher temperatures. There have been some lubricity problems with these highly fluorinated refrigerants, particularly with higher loads in reciprocating-type compressors and with aluminum surfaces. Problems associated with water and materials compatibility have limited their use in small hermetic machines. "Modified" polyglycols have provided somewhat better performance (53).

Rotary-screw compressors have shown improvements in volumetric efficiency (+2%), refrigeration capacity (+21%), and coefficient of performance (+12%) in constant-pressure-ratio comparisons with CFC-12 (39, 52). The CFC-12 data was obtained using an equivalent viscosity (at compressor discharge temperature) polyalphaolefin as the lubricant. The polyglycol has excellent miscibility with R-134a at condenser and evaporator conditions but becomes less soluble within the compressor rotor bores. A high viscosity during the compression cycle results in more efficient sealing between the rotors and between rotor and compression cylinder walls.

The polyglycols are used with hydrocarbon refrigerants for the properties described in the previous hydrocarbon gas section. These lubricants are easily drained from the bottom of evaporators due to their higher density.

Diesters. Diesters have been recognized for their potential application in R-22 applications, as they are widely available and have excellent miscibility. Major problems may exist with these materials due to poor stability with R-22 and a significant effect on the swelling of many elastomers (40). An additional restriction for the use of these fluids in rotary-screw compressors with R-22 is a limited range of viscosity grades available, up to ISO 100. Triesters tend to be more reactive with R-22 and less thermally stable than the higher viscosity polyol esters.

Neopentyl esters. Neopentyl esters (polyol esters) were reported for use in R-22 systems with evaporator systems of –80°C (–112°F) as long ago as 1968 (40, 54). Due to their polar nature (especially low-viscosity grades) these products tend to adsorb moisture. Additives may be required to inhibit chemical reaction and copper transfer (plating) at higher temperatures.

Lower-viscosity grades of neopentyl esters, to 32 ISO, are miscible with the more polar refrigerants such as HFC-134a (55). Higher-viscosity polyol esters with selected chemical structures have been described for HFC-134a (56, 57). These fluids have lower water absorption than the polyglycols.

High-viscosity, ISO 100–320, complex esters have improved adiabatic and volumetric efficiencies in twin-screw compressors using HCFC-22 (58). These oils were developed to have moderate to low dilution effects due to solubility while maintaining good miscibility. Their good miscibility (to below –90°F) has also led to their use in low-temperature applications.

Silicate esters. Silicate esters have been used to very low temperatures (with good miscibility) with highly fluorinated CFC refrigerants such as R-13. Problems have been reported that are associated with hydrolytic stability and the formation of varnishes or solids (59). Different structures may solve these problems (40).

Phosphate esters. Phosphate esters, such as tricresyl phosphate, are used as lubricity additives. Their use as primary lubricants has been eliminated due to poor viscosity–temperature relationships and the formation of acids at higher temperatures (40).

Fluorinated lubricants. Fluorinated lubricants such as perfluoroethers and fluorosilicone fluids are extremely miscible with nearly all halocarbon refrigerants. Their high cost has been a limiting factor for use in refrigeration compressors.

Fluorosilicone fluids are extremely miscible with HFC-134a. Compressor tests with 100 ISO fluorosilicone lubricants in automotive air-conditioning applications resulted in bearing failures with aluminum components (60).

There has been recent interest in the partial fluorination of at least one alkyl end-cap of polyglycols to improve their miscibility with HFC-134a (61). Another approach has been to blend a chlorotrifluoroethylene polymer with polyglycols (62). Either method is said to result in miscibility with HFC-134a throughout the operating range for air-conditioning applications.

III. VACUUM PUMPS

The same oils used for air compressors are suitable for most industrial vacuum pump applications. Synthetic oils with lower vapor pressures are used to achieve higher levels of vacuum without requiring substantial increase in viscosity. These oils include the esters and polyalphaolefins previously described. Other types of synthetic fluids are used for their chemical compatibility with the gases being pumped. Diffusion pumps require lubricants that have been molecularly distilled for ultrahigh vacuum.

Halogenated lubricants are used for their high chemical resistance to acids, alkalis, and corrosive gases, and their nonmiscibilities with many common solvents. Perfluoroethers and chlorofluorocarbons are used extensively for plasma etching in industries such as semiconductor manufacturing and for producing aluminum chloride. The perfluoroether can be used with pure oxygen and resist polymerization or the formation of nonvolatile residues when subjected to electron bombardment in sputter ion pump applications (63). The perfluoroethers can breakdown in the presence of certain Lewis acids at temperatures above 210°F (100°C) to form toxic fragments (64).

Polyphenyl ethers and alkyl silicones have been used for high-vacuum applications as they are thermally and chemically stable (65). Fluorosilicones have better chemical stability.

IV. LIQUID PUMPS

The use of synthetic lubricants for pump applications is limited to those applications either requiring compatibility with the material being pumped or having temperature requirements. The latter includes high-temperature applications where mineral oils may break down and applications where improved heat-transfer properties of selected synthetics help remove heat. A few examples of the variety of applications for these lubricants follows.

The superior lubricity of some synthetic fluids provides better lubrication to seals and packings. One example is the reciprocating pump. The biggest maintenance problem with these pumps is packing. There are many causes of short packing life; one of them is lack of lubrication. Mineral oils with tallow additives have been used to provide tenacity for the plunger surface to improve the lubricant film between plunger and packing. It was shown in previous sections that polyglycols, silicones, and halogenated lubricants provide this same type of action in the presence of hydrocarbons that would wash away the tallow. Other synthetics are selected to resist washing away by the specific material being pumped.

Centrifugal pumps are widely used in the chemical industry. Fluorosilicone lubricants have been applied successfully on several seal arrangements to eliminate product leakage and extend seal life. These lubricants are resistant to temperatures as high as 500°F, and to many chemicals and solvents that attack mineral oils (however, they are not suitable with ketones) (66).

Mineral oils and greases may break down at temperatures as low as 250°F (120°C), causing further temperature increases. Friction with packings rubbing on the shaft can easily generate these temperatures. The chemical being pumped is often at temperatures between 300 and 700°F (150 and 370°C). One method of heat removal is to use a double inside mechanical seal with a thermal convection lube system. The advantage of this system is that it normally requires very little attention. A molecular film of lube is maintained on the seal faces to provide lubrication. A low-viscosity polyalphaolefin has been used as the sealing and lubricating fluid. The fluid heats up in the pump stuffing box, rises, flows back to the supply tank, and is replaced by cool fluid from the supply tank. This is a thermal convection circulation system. A thin fluid is used for good circulation and to prevent openings in the seal faces.

Diaphragm pumps have been used for paint applications in the food industry. The fluid in one of these applications lubricates a reciprocating pump, which provides the pulsating action and acts against the diaphragm. There is the possibility that the diaphragm could develop a leak that could remain unnoticed for some time. Synthetic fluids provide a broad enough selection to insure that the fluid is compatible with the paint and also is "food grade."

A final example is the use of phosphate esters as lubricants for water-reactor coolant pump motors. Fires had occurred when the oils used to lubricate primary system pump motors leaked onto hot primary piping. The oil is pressurized in the oil lift, so it is sprayed on surrounding pipes if a leak occurs. Several lubricants were tested for the required fire-resistant properties and adequate stability including radiation resistance. These included mineral oils, diesters, methyl alkyl silicones, and phosphate esters. It was found that mineral oils soaked in pipe insulation and maintained at temperatures above 375°F (190°C) eventually undergo exotherms to temperatures as high as 580°F (304°C). Storage temperatures for this type of exotherm to occur for other fluids were: diester, 425°F (218°C); silicone, 400°F (204°C); and phosphate ester, 575–625°F (301–329°C). The temperature at which exotherms occurred was higher than the 550°F (288°C) operating temperature for the reactor's coolant pumps (67).

REFERENCES

1. Sugiura, K. M., and H. Nakano (1982). Laboratory evaluation and field performance of oil flooded rotary compressor oils, *Lub. Eng.*, 38(8), 510–518.
2. Thoenes, H. W. (1975). Safety aspects of selection and testing of air compressor lubricants, *Lub. Eng.*, 26(11), 409–411.
3. Majors, G. M. (1987). Air line fires, explosions, and detonations, Tutorial Course B, Gulf South Compressor Conference, Baton Rouge, LA, August.
4. Busch, H. W., L. B. Berger, and H. H. Schrenk (1949). *The 'Carbon–Oxygen Complex' as a Possible Initiator of Explosions and Formation of Carbon Monoxide in Compressed Air Systems*, U.S. Bureau of Mines, RI4465, June.
5. Miller, J. W. (1982). *Synthetic Microlube Systems Lower Explosion and Fire Hazards in Air Compressors*, STLE Annual Meeting, preprint 82-AM5E-1.
6. Loison, R. (1952). *The Mechanisms of Explosions in Compressed Air Pipe Ranges*, paper no. 26, Seventh Int. Conf. of Directors of Safety in Mine Research, Buxon, England, July.
7. McCoy, C. S., and F. J. Manley (1975). Fire resistant lubricants for refinery air compressors, National Meeting, Fuels and Lubricants, Houston, TX.

8. *Compressed Air and Gas Data*, 2nd ed., Ingersoll-Rand Company, Woodcliff Lake. NJ.
9. Zabetak,s, M. B., G. S. Scott, and R. E. Kennedy (1962). *Autoignition of Lubricants at Elevated Pressures*, report 6122, U.S. Bureau of Mines, p. 6.
10. Rekstad, G. (1965). Fire-resistant lubricants for air compressors, *Hydraulics Pneumatics*, 18, 100–105.
11. Arbocus, G., and H. Weber (1978). Synthetic compressor lubricants—State of the art, *Lub. Eng.*, 34(7), 372–374.
12. Ball, W. L. (1963). Explosion in aftercooler of a compressor using synthetic lubricant, *Safety Air Ammonia Plants*, 6, AIChE.
13. Product Literature, Anderol Synthetic Compressor Lubricants, Nuodex, Inc., Piscataway, NJ.
14. Wits, J. J. (1989). Diester compressor lubricants in petroleum and chemical plant service, *J. Synth. Lub.*, 5, 4. Original author: Fairbanks, D. R. (1975). Diester compressor lubricants in petroleum and chemical plant service, presented at ASLE Annual Meeting.
15. Application profile, synthetic air compressor oils (1981). *Fluid Lub. Ideas*, July/August.
16. Jayne, G. J., and A. P. Jones (19). Current progress in the development of synthetic air compressor oils, Synthetic lubricants and operational fluids, *Technische Akademie Essligen*, 4th Int. Colloquium, pp. 24.1–24.4.
17. Whiting, R. (1988). Monitoring energy savings of diester compressor oils, *Technische Akademie Essligen*, 6th Int. Colloquium, pp 14.4-1–14.4-3.
18. Ward, E. L., P. W. McGraw, and T. J. Appleman (1988). Lubricants for reciprocating air compressors, U.S. Patent no. 4,751,012, June.
19. Hampson, D. F. G. (1975). Reducing the risk of fires in large reciprocating air compressor systems, *Wear*, 34, 399–407.
20. Kivela, W. C. (1970). Fluorosilicone oil eliminates air compressor explosion hazard, *Environ. Control Safety Manage.*
21. Doperalski, E. J., and K. R. List (1979). Search for a practical extended coolant/lubricant for rotary screw air compressor, AIChE 86th National Meeting, paper no. 68E, April.
22. Short, G. D. (1984). Development of synthetic lubricants for extended life in rotary-screw compressors, *Lub. Eng.*, 40(8), 463–470.
23. Carswell, R., P. W. McGraw, and E. L. Ward (1983). A blend of a polyglycol and a tetra ester as a lubricant for rotary screw air compressors, *Lub. Eng.*, 39(11), 684–689.
24. Miller, J. W. (1989). Synthetic and HVI compressor lubricants, *J. Synth. Lub.*, 6(2), 107–122.
25. Cohen, S. C. (1988). Development of a 'synthetic' compressor oil based on two staged hydrotreated petroleum basestocks, *Lub. Eng.*, 44(3), 230–238.
26. Mills, A. J., M. A. Tempest, and A. S. Thomas (1986). Performance testing of rotary screw compressor fluids in Europe, *Lub. Eng.*, 42(5).
27. Miller, J. W. (1984). New synthetic food grade rotary screw compressor lubricant, *Lub. Eng.*, 40(7), 433–436.
28. Ministry of Agriculture, Fisheries & Food (1989). Government to ban use of mineral hydrocarbons in food, *MAFF News Release*, 53/89, February.
29. Legeron, J. P., and L. Beslin (1989). Rotary-vane compressors: Some technical aspects of long-life lubricants, *J. Synth. Lub.*, 6(4), 229–309.
30. Compressed Air and Gas Institute (1973). *Compressed Air and Gas Handbook,* 5th ed., Prentice-Hall, Englewood Cliffs, NJ.
31. Matthews, P. H. D. (1989). The lubrication of reciprocating compressor, *J. Synth. Lub.*, 5(4), 271–290.
32. Lyons, R. L., and D. L. Van de Bogert (1983). A positive microlube technique for industry, *Lub. Eng.*, 30(10), 648–650.
33. Miller, J. W. (1980). Microlubrication of natural gas compressors with synthetic lubricants, *Fluids Lub. Ideas*, September/October.

34. Van de Bogert, D. L. (1984). Microlubrication control, U.S. Patent 4,467,892, August 28.
35. Miller, J. W., and B. D. Krukowski (1989). Microlubrication cuts gas-transmission-line contaminants and improves flow, *Oil Gas J.*, July.
36. Tolfa, J. C. (1991). Synthetic lubricants suitable for use in process and hydrocarbon gas compressors, *Lub. Eng.*, 47(4), 289–295.
37. Miller, J. W. (1984). Synthetic lubricants and their industrial applications, *J. Synth. Lub.*, July.
38. Sumersmith, D. (1978). *Selection of Oils and Lubricant System Design for Ammonia and Methanol Synthesis Plant Compressors*, AMPO 78, Billingham Press Limited, Stockton, England.
39. Sjöholm, L. I., and G. D. Short (1990). Twin screw compressor performance and suitable lubricants with HFC-134a, *Proc. Int. Compressor Engineering Conf.*, Purdue, July.
40. Sanvordenker, K. S., and M. W. Larime (1972). A review of synthetic oils for refrigeration use, *ASHRAE Trans.*, 78, part 2.
41. Spauschus, H. O. (1984). "Evaluation of lubricants for refrigeration and air conditioning compressors, *ASHRAE J.*, 26(5), 59.
42. Kruse, H. H., and M. Schroeder (1984). Fundamentals of lubrication in refrigeration systems and heat pumps, *ASHRAE J.*, 26(5), 5–9.
43. Daniel, G., M. J. Anderson, W. Schmid, and M. Tokumitsu (1982). Performance of selected synthetic lubricants in industrial heat pumps, *J. Heat Recovery Systems*, 2(4), 359–368.
44. Short, G. D. (1990). Synthetic lubricants and their refrigeration applications, *Lub. Eng.*, 46(4).
45. Short, G. D. (1985). Hydrotreated oils for ammonia refrigeration, *Technical Papers, 7th Annu. Meeting, Int. Institute of Ammonia Refrigeration*, pp. 149–176, March 10–13.
46. Short, G. D. (1990). Refrigeration lubricants update: Synthetic and semi-synthetic oils are solving problems with ammonia and alternative refrigerants, *Technical Papers, 12th Annu. Meeting, Int. Institute of Ammonia Refrigeration*, pp. 19–53, March 4–7.
47. Reed, P. R., and H. O. Spauchus (1991). HCFC-124: Applications, properties, and comparison with CFC-114, *ASHRAE J.*, 33(2), 40–41.
48. Bateman, D. J., et al. (1990). Refrigeration blends for the automotive air conditioning aftermarket, SAE technical paper 900216, International Congress and Exposition, Detroit, MI, February/March.
49. Shoemaker, B. H. (1950). Symposium, synthetic lubricating oils, *Ind. Eng. Chem.*, 42(12), 2414.
50. El-Bourini, R., K. Hayahi, and T. Adachi (1990). Automotive air conditioning system performance with HFC-134a refrigerant, SAE technical paper no. 900214, International Congress and Exposition, Detroit, MI, February/March.
51. Struss, R. A., J. P. Henkes, and L. W. Gabbey (1990). Performance comparison of HFC-134a and CFC-12 with various heat exchangers in automotive air conditioning systems, SAE technical paper no. 900598, International Congress and Exposition, Detroit, MI, February/March.
52. Short, G. D., and R. C. Cavestri (1990). Selection and performance of synthetic and semi-synthetic lubricants for use with alternative refrigerants in refrigeration applications, *Proc. ASHRAE/Purdue CFC Conf.*, July.
53. Sundaresan, S. G. (1990). Status report on polyalkylene glycol lubricants for use with HFC-134a in refrigeration compressors, *Proc. ASHRAE/Purdue CFC Conf.*, July.
54. Altman, S. S., et al. (1969). Foreign Technology Division, Dept. of Commerce, Translation: FTD-HT-23-1489-68, April 22.
55. Sanvordenker, K. S. (1989). Materials compatibility of R-134a in refrigerant systems, Seminar 89–01 Ozone/CFC–CFC Alternative Studies, ASHRAE Annual Meeting, June 24–28, Vancouver, Canada, in *CFCs Time of Transition*, pp. 211–219, ASHRAE.
56. Jolly, S. T. (1990). New unique lubricants for use on compressors utilizing R-134a refrigerant, *Proc. ASHRAE/Purdue CFC Conf.*, July.

57. Kaimai, T. (1990). Refrigeration oils for alternative refrigerants, *Proc. ASHRAE/Purdue CFC Conf.*, July.
58. Sjöholm, L. I., and G. D. Short (1990). Twin screw compressor performance and complex ester lubricants with HCFC-22, *Proc. ASHRAE/Purdue CFC Conf.*, July.
59. Downing, R., and W. D. Cooper (1972). Operation of the Three Stage Cascade System, presented at ASHRAE Meeting, Nassau.
60. CPI Engineering Services, Inc. (1986). Internal correspondence on compressor testing with fluorosilicone lubricant and HFC-134a with air conditioning compressor builder, Midland, MI.
61. Short, G. D. (1990). Rotary displacement compression heat transfer systems incorporating highly fluorinated refrigerant—Synthetic oil compositions, U.S. Patent no. 4,916,914, April.
62. McGraw, P. W., and E. L. Ward (1989). Lubricants for refrigeration compressors, U.S. Patent no. 4,851,144, July.
63. Baker, M. A., L. Holland, and L. Laurenson (1971). The use of perfluoroalky ether fluids in vacuum pumps, *Vacuum*, 21(10).
64. Leybold-Heraeus Vacuum Products (1986). Engineering Notes, Vacuum Pump Fluids, revised, Export, PA.
65. Holland, L., et al. (1972). Pollution free high vacuum, *Physics*, 23.
66. Miller, J. W. (1973). Super-lube systems eliminate shaft-seal leakage, *Chem. Eng.*, 88–90.
67. Pietch, H. (1980). Evaluation and test of improved fire resistant fluid lubricants for water reactor coolant pump motors, Fluid lubricants for water reactor coolant pump motors, *Fluid Evaluation, Bearing Model Tests, Motor Tests, and Fire Tests*, prepared for Electric Power Research Institute, vol. 1, EPRI NP-1447.

20
Hydraulics

Andrew G. Papay
Ethyl Petroleum Additives, Inc.
St. Louis, Missouri

I. INTRODUCTION

A. Hydraulic Principles and Fluid Functions

Hydraulic systems offer a simple solution to the problems of transmitting and applying large forces while retaining high flexibility and control. That explains the fact that such systems are used in most forms of mechanized human endeavor to transmit, transform, and control mechanical work. They are part of a major technical discipline called fluid power technology. A typical hydraulic system, besides the fluid, includes the following main parts:

1. A force-generating unit that converts mechanical energy into hydraulic energy, such as a pump.
2. Piping for transmitting fluid under pressure.
3. A unit that converts the hydraulic energy of the fluid into mechanical work, such as an actuator or fluid motor. There are two types of motors, cylindrical and rotary (1).
4. A control circuit with valves that regulate flow, pressure, direction of movement, and applied forces.
5. A fluid reservoir that allows for separation of any water or debris before returning the clean fluid to the system through a filter.

In this setup, the pump can be considered the heart of the system and the hydraulic fluid the lifeblood of the equipment. The efficiency of the pump depends on good wear control, which the fluid must provide.

Wear causes internal slippage, that is, internal leakage from the high-pressure back to the low-pressure side of the pump. Leakage reduces the pump output, results in power loss, and increases the operating temperatures. Any deposits or varnish from fluid oxidation or any contaminants, such as rust or dust, can clog the generally fine clearances and cause problems.

Since hydraulic fluids are transmitters of power through pressure and flow hydrostatics, they must be practically incompressible and flow readily at all operating temperatures. They must also provide adequate seal, protect the working metal surfaces from wear and corrosion, and separate themselves easily from water and debris, while in the sump, before being recirculated into the pump. Of all the components in a hydraulic stem, the fluid is the most critical and multifunctional part (2).

The functions and requirements of a hydraulic fluid as indicated by the desired properties are summarized on Table 20.1.

Many of the properties shown on Table 20.1 affect not only the function indicated but most other functions to some degree or in a round-about way. A typical example is viscosity. If the fluid does not flow properly, heat transfer suffers, power transfer and control deteriorate, and all functions are put under stress.

Synthetic hydraulic fluids are similar to mineral oil fluids in most of these properties, with certain added advantages. As we are going to see later on, synthetic hydraulic fluids can be more durable under thermal/oxidative stress, cleaner in operation, and able to span wider areas of use. In the case of fire-resistant fluids, some synthetics are overwhelmingly superior to mineral hydraulic fluids, offering viable solution to hard problems of safety.

Table 20.1 Functions and Properties of Hydraulic Fluids

Function	Property
Power transfer and control medium	Low compressibility (high bulk modulus)
	Good foam release, low foaming tendency
	Low volatility
Heat-transfer medium	Good thermal capacity and conductivity
Sealing medium	Adequate viscosity and shear stability
Lubricant	Viscosity for film maintenance
	Low-temperature fluidity
	Thermal stability
	Oxidation stability
	Hydrolytic stability/water tolerance
	Cleanliness and system cleansing
	Filterability
	Demulsibility
	Antiwear characteristics
	Material compatibility
	Corrosion control
Special function	Fire resistance
	Friction modification
	Radiation resistance
Environmental impact	Low toxicity, when new or decomposed
Functioning life	Properties durability for long use

B. Historical Development

The earliest use of a non-petroleum-based hydraulic fluid is shrouded in the mists of history. What is well known is that water was used as a hydraulic fluid—within the broader sense of the term—for centuries. With Pascal's discovery of the fundamental law of physics on which hydraulic systems are based, hydraulic power development was placed on a scientific basis. One of the first documented uses of hydraulic power in industrial applications was that of London Hydraulic Power Co., which supplied power for the Tower Bridge, among other things. In 1906 oil-based hydraulic systems were used in warships to move guns. In the 1920s the use of water in hydraulic systems decreased and that of oil increased.

Synthetic hydraulic fluids were mentioned in the context of ester-based lubricants in the 1937 Zurich Aviation Congress. The largest push for the development of synthetic hydraulic fluids came during World War II by the military establishments, both Allied and German. The main driving force was the need for better low-temperature fluidity and fire resistance. Aviation has been perhaps the most demanding application for synthetic hydraulic fluids. Overall, the U.S. Military has played the leading role in opening the road to new and better lubricants and hydraulic fluids, and is generally given credit for the coordination and focusing of diverse industry efforts toward the highest lubricant and hydraulic fluid technology ever seen anywhere. An outgrowth of that is the development of a dynamic lubricant and hydraulic fluid additive industry in the United States that dominates this technology worldwide.

II. SYNTHETIC HYDRAULIC FLUIDS AND PROPERTIES

Generally, need generates demand, and application defines the type of product to be used. In the case of synthetic hydraulic fluids, there is a larger number of types and chemistries than in most other liquid lubricant categories combined. A partial list includes polyalphaolefins (PAOs), dicarboxylic acid esters (diesters), polyol esters, phosphate esters, polyalkylene glycols (polyglycols), alkylated aromatics, silicones, silicate esters, silahydrocarbons, perfluorocarbons, polyphenyl ethers, etc. Treating each type of synthetic hydraulic fluid separately would be a tedious and inefficient way to our goal. Therefore, the relevant types of fluids will be grouped around each main application area, expounding on individual chemistries versus performance when necessary to illuminate the application.

A. Why Synthetics

It is important to keep in mind that the main reason that synthetic hydraulic fluids have been displacing, in selected areas, the low cost and generally excellent quality mineral type is some extreme condition in each particular equipment. Some units of equipment are deliberately made to run at high temperatures or are compacted for space and weight or otherwise overstressed with respect to energy flow, and the result is a quick breakdown of the lubrication with a mineral fluid. Others are subject to hazardous conditions, such as fire. To improve reliability and extend maintenance-free life of the equipment, a well-chosen synthetic fluid is recommended. The choice is based on specific properties that address the particular extreme condition in each system.

B. Properties and Chemistry

When choosing a certain type of chemistry for hydraulic fluid, one has to know how chemistry relates to each fundamental property of the fluid. Following is a brief discussion of each property with respect to performance in a hydraulic system. The comparative performance in each property is given in the next section.

1. Viscosity

Viscosity is the most important property of any lubricant or hydraulic fluid, mineral or synthetic, and for all types of systems and applications. It is directly related to wear control (elasto-hydrodynamic lubrication film), liquid frictional losses, leakage, start-up ease, and efficiency. Of several possible viscosity grades for hydraulic fluids, the most popular are three: ISO 32, ISO 46, and ISO 68 (1).

2. Viscosity Index

High viscosity index (VI) is a very desirable property. It controls the temperature range within which a fluid can be used, and the higher, the better. The addition of special polymers called viscosity index improvers (VII) can raise the VI of a fluid and improve some parts of the performance, but not necessarily all of it (3).

3. Low-Temperature Fluidity

If the flow is poor at low temperatures, cavitation can occur in the pump intake, or the cylinder operation can be impeded. Pump manufacturers have set limits for the maximum viscosity at cold start. The limit recommended for vane pumps is 4,000 SUS, or about 860 cSt. For piston pumps, the limit is 7,500 SUS, or about 1600 cSt.

4. Viscosity–Pressure Characteristics

The measure of this property is the viscosity–pressure coefficient α. The higher the coefficient, the higher the viscosity increase with pressure. That defines the range of pressure within which a fluid can be used without exceeding the operating limits of the hydraulic system. Lower coefficients allow wider ranges of operating pressures and permit lower power losses.

5. Compressibility

As power transmitters, hydraulic fluids require very low compressibility. The latter is defined as the change in volume by a change in pressure. A low compressibility ratio, which is the reciprocal of the volume elastic modulus (or bulk modulus), translates into a fast response time, high pressure-transmission velocity, and low power loss. Materials with a high ratio or low bulk modulus would act as damping fluids. The importance of the high bulk modulus becomes greater when equipment size and weight are critical, as in aircraft. A poor bulk modulus will require increased line-sizes and actuator cross-sectional areas to compensate for the lower stiffness or higher compressibility of the fluid. These increases will also mean larger fluid volume and weight in the system (4).

6. Foam Release/Antifoam

Fast foam release and antifoaming properties are important for two reasons: to prevent "spongy" hydraulic controls and poor power transmission due to air, and also to avoid oil flow discontinuity that can affect lubrication and wear protection. Some antifoam additives, such as silicone oils, act on the surface tension to inhibit foaming, and they could be detrimental in some hydraulic fluid applications because they might weaken the air release properties.

7. Filterability

This is the ability of the fluid to pass through the filter without causing clogging. Filterability is influenced heavily by the overall hydrolytic stability of the fluid (5).

8. Hydrolytic Stability

A synthetic hydraulic fluid has to face the number one problem of most hydraulic oils, which is water. In this encounter, it must accomplish two things: retain its chemical integrity (not hydrolyze), and separate easily from the water soon after the encounter.

9. Thermal Stability

Synthetic hydraulic fluids require good thermal stability at high temperatures because they have to operate under severe conditions. Bond energies, position of an atom in a chain, and general molecular structure underlie thermal stability. This fact serves as a guide in designing molecules for use as synthetic hydraulic fluids. The thermal stability is frequently measured with the thermal degradation temperature (TDT). Since TDT is frequently adversely affected by the presence of metals, one must examine both numbers before making a choice of fluid.

10. Oxidation Stability

Good oxidation stability is one of the most important requirements for a synthetic hydraulic fluid. Actually, the property that is measured is a combination thermal/oxidation stability, since most of the tests run are high-temperature oxidation tests. Viscosity increase, total acid number (TAN), and sludge or deposits are the main criteria in such tests.

11. Antiwear Properties and Fatigue Life

The antiwear properties (AW) of a synthetic hydraulic fluid are not necessarily better than that of a mineral type. However, formulation technology can boost the AW performance of most synthetic hydraulic fluid types. Therefore, meaningful AW comparison and hence choice can be based only on finished fluids. The fatigue life, such as the rolling fatigue life of bearings, is an important consideration in most systems. Generally, it appears that the more polar or reactive a fluid, the worst the expected fatigue life. Thus AW and fatigue life do not always see eye-to-eye.

12. Friction Modification

This term usually implies a low static coefficient of friction or at least one that is not significantly bigger than the dynamic coefficient of friction. Most hydraulic fluids, both synthetic and mineral-based, do not require friction modification. Nevertheless, friction modification reduces significantly the "solid friction" (the one between boundary layers or solid surfaces) and increases efficiency in most mechanical systems. Automatic transmission fluids and tractor fluids that also function as hydraulic oils in their respective systems are friction modified, each within strict specifications. Those, however, are outside the scope of this chapter and are treated separately in this volume. If a hydraulic fluid needs friction modification, to avoid, for example, stick–slip phenomena, special additives are used. These are called friction modifiers, and they are mainly long-chain polar molecules (6).

13. Compatibility

Compatibility of the fluid with all materials of the system is always required. That includes seal compatibility. To avoid leaks, a hydraulic fluid must slightly swell a seal, but never overswell, shrink, harden, depolymerize, or otherwise harm the integrity of the seal. Synthetics with highly polar groups act as plasticizers and can pose a difficult challenge in that they overswell and attack conventional seals. Special seal and hose materials can solve this problem. Other materials that require fluid compatibility are packing, plastic parts, paints, and all types of metals in contact with the fluid.

14. Volatility

In choosing a high-temperature hydraulic fluid, one has to make sure that there will be no volatility problems. This is not only needed to avoid significant evaporation rates or low-flash-point-related situations, but also to avoid bubbles formation at low pressure points, and thus pump cavitation. In homogeneous fluids, such as synthetic bases, no volatile components are expected. The volatility of the fluid will be determined mostly by the molecular weight and the molecular cohesion forces. At low base viscosity, such as a 2-cSt (at 100°C) PAO fluid thickened with a viscosity index improver (VII) (7), one has to examine the volatility factor before deciding on an application.

15. Radiation Resistance

This is a special requirement of systems exposed to radiation.

16. Heat-Transfer Properties

The thermal capacity and conductivity of the fluid are design parameters in every hydraulic system. Preferred are fluids with a low coefficient of thermal expansion and low specific gravity.

17. Contamination

Contaminants from the environment can clog fine passages in the system and derate the flow and lubricating ability of the fluid, as well as negating several other desirable characteristics. The result is costly downtime and repairs. Seals, when they retain their integrity and function, can protect the system from much of the outside contamination.

18. Fire Resistance

Fire resistance is a major requirement that has been pushing large segments of industry and commerce toward the adoption of special, nonmineral hydraulic fluids made of synthetic bases. Fires have been caused by accidental leakage of hydraulic fluid onto hot areas, resulting in combustion. Rupture or puncture of a high-pressure hydraulic hose has been known to squirt fluid up to 12 m away in the form of fine mist, which is highly combustible, frequently explosively so, if made of mineral oil. Factories, ships, airplanes, military, and others all have experienced similar problems, and many have adopted special fire-resistant hydraulic fluids as the best practical solution. Fire resistance is a complex property measured by several different tests, and some of its aspects are not agreed on by some specification writers. Generally, values for spontaneous ignition, ignition point, combustion persistence, and spreadability are some of the criteria for fire resistance.

19. Additive-Based Properties

Additives are called to reinforce or supply a large number of properties in synthetic hydraulic fluids, as they do in most lubricants and fluids. These properties include cleanliness, demulsibility, and corrosion protection. Cleanliness, which in tandem with stability retards the formation of insoluble by-products (sludge, deposits), also helps in their removal from working surfaces, especially critical parts such as valves, and allows smooth movement. In addition to additives, filtration helps fluid circulation and performance by removing, mechanically, both deposits and solid contaminants. Demulsibility, a very important property in most hydraulic fluids, allows the easy and continuous separation of water from the fluid and return of the latter to the system in a satisfactory state for reuse. Corrosion protection for all the metals used in the equipment helps maintain the useful integrity of all working surfaces and mechanisms.

20. Toxicity and the Environment

Environmental considerations, including safety for people dealing with the hydraulic fluid through its course of life, are critical when choosing a product. Although this factor poses great restrictions in selecting a fluid, it cannot be overridden by any performance or cost considerations. In cases where a toxic fluid is mandated as the only practical answer, very severe system design safeguards are mandatory for allowing its use. Generally, a synthetic hydraulic fluid must be safe and benign to the environment, both when new and after decomposition.

21. Durability of Properties

Quality performance starts with a fresh, high-quality hydraulic fluid charge in a clean piece of equipment. It proceeds with the retention of all the good properties, and hence performance, over a long period of operation. A short-lived performance would necessitate frequent down-times for fluid change, at the very least, and sometimes costly equipment repairs. To assure durability of properties, a carefully crafted additive package is fitted to the proper base fluid. Such additives as antioxidants, detergent/dispersants, stabilizers, corrosion inhibitors, lubricity or antiwear agents, and demulsifiers are among those chosen to build up and sustain desirable fluid properties. The fluid durability and useful life will depend heavily on the proper choices.

III. COMPARATIVE PERFORMANCE

The relative performance of the various synthetic hydraulic fluids is the main guide in the selection of the best fluid for a specific application. In this section, the relative performance is presented in parts corresponding to each group of properties that pertain to hydraulics. These values, or ratings, are averages, because each type of fluid category contains several grades or variation of molecular structures. Therefore, properties can vary widely within each category, and actual selection for an application needs more specific information than what can be given here.

A. Viscometrics

Viscometrics includes high-temperature viscosity, low-temperature fluidity, viscosity index (VI), and pressure viscosity. Table 20.2 presents the relative performance in viscometrics for a variety of synthetic fluids. As seen in the table, some of the most expensive synthetics, such as fluorocarbons and polyphenyl ethers, show poorer viscometrics than a good-grade paraffinic mineral oil. Conversely, polyalphaolefin oligomers (PAOs), diesters, polyol esters, alkyl benzenes, polyglycols, silicones, and silicate esters are superior to mineral oil. This holds mostly true also for pressure–viscosity (8). Therefore, these synthetics would make good hydraulic fluids with regard to fluidics in a fairly wide range of temperature and pressures.

Table 20.2 Relative Performance in Viscometrics

Product	Low-temperature fluidity	VI	Pressure–viscosity
Mineral oil (paraffinic)	Fair–good	Good	Good
PAOs	Excellent	Very good	Good
Diesters	Very good	Very good	Very good
Polyol esters	Very good	Very good	Very good
Polyalkylene glycols	Very good	Very good	Very good
Phosphate esters	Fair[a]–good	Fair[a]–good	Very good
Silicones	Very good	Excellent	Excellent
Alkyl benzenes	Good	Very good	Good
Fluorocarbons	Fair	Fair	Fair
Polyphenyl ethers	Poor	Fair	Poor
Silicate esters	Excellent	Excellent	Very good
Silahydrocarbons	Excellent	Excellent	—

[a]Triaryl phosphates have poorer VIs and low-temperature fluidity than trialkylphosphates.

1. *Viscosity*

Most chemistries can provide products at low- and medium-range viscosities. Higher viscosities can be obtained with PAOs, polyglycols, silicones, and fluorocarbons in further polymerized forms.

2. *Viscosity Index*

Polyol esters, PAOs, diesters, polyglycols, silicate esters, and silicones usually have high VI. Polyaromatics, hologenated hydrocarbons, phosphate esters, and polyphenyl ethers have a lower VI, some of them even lower than mineral fluids.

3. *Low-Temperature Fluidity*

Good low-temperature fluidity is observed in synthetics having a large flexible molecular group or side chains. Fluorocarbons or polyphenyl ethers have therefore poor low-temperature fluidities. The PAOs and most esters, with the exception of aryl phosphate esters, have very good low-temperature fluidity.

Pumping efficiency is one area of superiority of some synthetics over mineral fluids. Thus, PAO has shown significant pumping efficiency improvement over mineral hydraulic oils, especially at lower viscosities (9).

4. *Compressibility*

The compressibility or bulk modulus of polyphenyl ethers is excellent, which means it has a very small compression ratio. It is followed by phosphate ester, which is superior to mineral oil. Generally, fluids with aromatic rings in their molecules have a smaller compression ratio, although the change in viscosity with temperature is quite large (10). Silicate esters and silicones are worse than mineral oils. Perfluorinated fluids, like fluoroalkylesters, show poor bulk modulus (4).

B. Stability

This group includes thermal stability, oxidation stability, hydrolytic stability, and volatility.

Table 20.3 Relative Stability of Synthetic Hydraulic Fluids

Product	Thermal	Oxidation	Hydrolytic	Volatility
Mineral oil (paraffinic)	Good	Fair	Excellent	Poor–fair
PAOs	Very good	Very good	Excellent	Very good
Diesters	Good	Very good	Good	Good
Polyol esters	Good	Very good	Good	Good
Polyalkylene glycols	Good	Good	Good	Good
Phosphate esters	Fair	Good	Fair	Fair–good
Silicones	Very good	Good	Excellent	Excellent
Alkyl benzenes	Good	Good	Excellent	Good
Fluorocarbons	Excellent	Excellent	Good	Poor
Polyphenyl ethers	Excellent	Good	Excellent	Good
Silicate esters	Very good	Good	Fair–poor	Good
Silahydrocarbons	Excellent	Very good	Excellent	—

Table 20.3 shows the relative performance in stability for several types of synthetic hydraulic fluids. These fluids are compared in their finished form, which means fully formulated with the proper additives.

1. Thermal Stability

The thermal stability of polyphenyl ether, fluorocarbons, and silicones is superior to that of mineral oil and even exceeds that of the various carboxylic acid esters. Silahydrocarbons also show superior thermal and storage stability (11–14), especially the saturated ones (15).

2. Oxidation Stability

While fluorocarbons and fluoroesters are top-rated in oxidation stability, PAOs and esters are performing very well also. Silicones, polyphenyl ethers, and others are achieving moderately good results. It must be emphasized here that antioxidants are practically always used in formulating synthetic hydraulic fluids with the type of antioxidant frequently adapted to the chemistry and solubility/compatibility of the base fluid. The useful oxidation stability of a fluid depends on the temperature of application.

Table 20.4 shows the tentatively recommended temperatures of operation for several

Table 20.4 Recommended Temperature of Operation

Product	Long service, °C	Short service, °C
Mineral oil	93–121	135–149
PAOs	149–232	288
Polyglycols	163–177	191–204
Diesters	177	204
Polyol estyers	191	218
Aryl phosphates	204	274
Silane	149–232	288
Silicate esters	191–204	246–274
Silicone	218–288	288–329
Polyphenyl ethers	316	427

synthetic hydraulic fluids (10). The numbers are only approximations, because variations in grades and above all in formulation can make a very big difference in the actual operation.

3. Hydrolytic Stability

Nothing can top hydrocarbons in hydrolytic stability. That includes PAOs, alkyl benzenes, and mineral oils. However, silicones and polyphenyl ethers are also excellent. Silahydrocarbons are hydrolytically stable, whereas those with silicon alkoxy bonds and silicate esters are unstable (16, 17). However, hydrolytic stability of silicate fluids can be considerably improved if the silicon atoms are shielded by carbon moieties, as in certain silicone clusters (18). All esters tend to succumb to hydrolysis, especially phosphate esters, whose decomposition by-products can be very corrosive. Traces of chlorine from chlorinated solvents accelerate the phosphate ester hydrolysis to form corrosive products (19). Such hydrolysis problems can explain why esters in general are infrequently used as hydraulic fluids in areas where water can be found in close proximity. The exception is when fire resistance is of paramount importance. In such cases, carefully designed hydraulic systems, to exclude water contamination, can use fire-resistant fluids, such as phosphate esters. In the case of carboxylic acid esters, hydrolysis can produce acids and alcohols or polyols. The acids will cause trouble with corrosion and other properties. Some polyhydric alcohols from the polyol esters, such as pentaerythritol, when falling out of solution can form lumps of jelly matter that can clog up the fine tolerances of a modern hydraulic system. This is one of the main reasons why carboxylic acid esters in general make a poor choice for hydraulic fluids. Consequently, such esters are used only rarely in hydraulic systems by themselves. Silicic acid esters also can hydrolyze and in the presence of acids form SiO_2, which, like sand, is abrasive. Polyglycols are more stable, but not as stable as hydrocarbon fluids.

4. Volatility

Volatility depends not only on the chemistry but also on the grade of the synthetic fluid, because most of them come in a range of molecular weights or viscosities. Overall, silicones are excellent having very low volatility. The PAOs are also very good, followed by most of the esters. Fluorocarbons and phosphate esters are no better than mineral oil. It is good to keep in mind that in very-low-viscosity grades volatility can usually be a concern for most types of fluids.

C. Lubricity and Wear Protection

This group of properties includes lubricity, antiwear, and fatigue life of the system as influenced by the fluid. Although all of these properties can be strongly affected by the additive package in the formulation, the base fluid itself has a significant impact. The base supplies something akin to a foundation on which the additives are building the final structure of properties. Thus, it is useful to examine the level of strength each fluid base brings with it.

Table 20.5 presents an approximation of those basic properties for several synthetic hydraulic fluid types, keeping in mind that there can be wide variations within each type.

1. AW and Lubricity

The natural lubricity or load-carrying ability of the hydrocarbons can only be exceeded by that of the phosphate esters, whose decomposition products (acid phosphate esters) can be strong AW and extreme pressure (EP) agents. On the other hand, the carboxylic acids esters are polar molecules that absorb onto metal surfaces, tending to displace many

Table 20.5 Relative Lubricity and Wear Protection

Product	Natural lubricity and AW	AW with additives	Fatigue life
Mineral oil (paraffinic)	Good	Excellent	Fair–good
PAOs	Good	Excellent	Good
Diesters	Fair	Good	Fair
Polyol esters	Fair	Good	Fair–good
Polyalkylene glycols	Good	Good	Fair
Phosphate esters[a]	Excellent	Excellent	Fair
Silicones	Poor	Fair	Fair–good
Alkyl benzenes	Good	Excellent	Good
Fluorocarbons	Fair–good	Good	Fair–good
Polyphenyl ethers	Good	Good	Good
Silicate esters	Good	Very good	Poor

[a]Acid phosphate esters have excellent AW properties but poor fatigue life.

beneficial polar molecules, such as those of lubricity and AW agents that are specifically included in the additive package. Therefore, the AW properties of finished ester-based hydraulic fluids are usually lower than those of comparable mineral- or PAO-based fluids. Naturally, all AW comparisons must be made on a level basis of viscosity and additive treat. Silicones are very nonpolar and have poor adsorbability on metal surfaces as well as poor load-carrying characteristics. This is attributed to the repelling interaction of the Si atoms, and it is difficult to overcome with conventional additives (20).

2. *Fatigue Life*

The fatigue life, such as the rolling fatigue life of bearing metal surfaces, varies among the different types of molecules composing the fluid. As Table 20.5 suggests, the polarity or reactivity of a fluid can be detrimental to the expected metal surface fatigue life. Among the various synthetic fluids, polyphenyl ether, PAOs, and alkyl benzenes have good fatigue lives. Most polar molecules like polyglycols and many esters, especially if they contain acidic species, or fluids containing reactive groups exhibit a shorter fatigue life (10). When estimating the fatigue life of a fluid, it is therefore important to consider the degree of purity or amount of decomposition products present in a fluid. A prime example is the phosphate esters. Such esters tend to hydrolyze in the presence of moisture, and the acid phosphate species, being very polar and reactive, will tend to shorten the fatigue life. This is more pronounced in the alkyl rather than the aryl phosphates. Therefore, in actual applications, the wise formulator of the finished phosphate fluid incorporates acid neutralizers, such as epoxides, in the formulation. Also, the hydraulic system engineer incorporates—besides strict exclusion of water, which by itself contributes to fatigue life deterioration (21)—a module for filtering the fluid through activated clay for the removal of the acidic species and the purification of the fluid. Finally, it is very important to keep sight of the fact that the reactivity and type of AW additives used in the fully formulated hydraulic fluid are also significant factors in actual fatigue life.

D. Compatibility

This group includes compatibility with seals and other organic parts, and compatibility with additives.

1. Seal Compatibility

Seal compatibility of PAOs is inadequate in that they display negative swelling power, causing shrinkage of many conventional types of seals. Formulation can correct this shortcoming through the incorporation of moderate amounts of seal swell additives, such as aromatics or esters. Conversely, esters, especially the more polar ones such as lower- and medium-molecular-weight diesters, and certainly phosphate esters, tend to overswell the conventional seals. Many esters need special elastomers. Some esters actually act as plasticizers, destroying the integrity of the seal—a dramatic reminder of why one must be very careful when mating a fluid to a seal material. Contrary to most other properties, there is no known commercial additive that can correct overswelling at reasonable concentrations. Fluorocarbons generally require special seals, such as phosphonitriles.

Table 20.6 shows the seal compatibility of selected hydraulic fluids.

2. Compatibility with Additives

Additives compatibility is or can easily be made good with all hydrocarbon bases. It is also fairly good with phosphate esters, polyphenyl ethers, carboxylic esters, and silicate esters, although special additives might be needed. It is not as good with polyglycols but it can be improved, and it is truly poor with fluorocarbons and silicones. This fact imposes a limit on the AW capabilities and in other desirable properties of the fluid, and that in turn limits the equipment design parameters.

E. Radiation Resistance

Radiation in all of its forms is an energy input on the fluid and, like heat or oxidation, constitutes a shock to the structures composing it. Resistance to radiation is one form of stability, although a special one. Most hydraulic fluids can skip this requirement because it pertains to a limited number of applications. Two of these are of special interest: nuclear radiation and sonic radiation environments.

1. Nuclear Radiation

Nuclear radiation can be destructive for many molecular structures. Of the synthetic hydraulic fluids, carboxylic acid esters, phosphate esters, and polyglycols are the most vulnerable. Hydrocarbon types, especially aromatic fluids, fare better. Polyphenyl ethers are among the best, perhaps 10 times more resistant than mineral oil (22). The explanation given is that when a neutron knocks out a hydrogen atom, the H^+ carries the entire kinetic energy of the neutron and breaks C–C bonds until its energy is dissipated and the proton cools off. Polymerization occurs when paraffins are used. However, when aro-

Table 20.6 Seal Compatibility of Selected Hydraulic Fluids

Compatible seals	Mineral oil	PAOs	Water/ glycol	Phosphate ester	Polyol ester
Buna N	Yes	Yes	Yes		Yes
Polychloroprene	Yes	Yes	Yes		Yes
Fluoroelastomer	Yes	Yes	Yes	Yes	Yes
PTFE	Yes	Yes	Yes	Yes	Yes
Butyl rubber				Yes	
Ethylene/propylene				Yes	

matics are used, fewer fragments are formed and less polymerization takes place. Therefore, less viscosity increase is caused by radiation. With gamma radiation, bonding electrons are kicked out and the molecules decompose into radicals, which leads to polymerization. Again aromatic systems are able to convert the absorbed energy, to a large extent, into resonance energy of the aromatic ring systems without causing bond cleavage.

2. Sonic Radiation

Sonic radiation is another form of energy that can attack molecular bonds and cause cracking of structures. There is a whole discipline of chemistry dealing with molecular changes, synthesis or decomposition, based on sonic energy input, which is called sonochemistry. In hydraulic fluids operating in a sonic energy environment, the first components to go are the polymers, such as the VII. With them goes the viscosity boost that the VII is contributing to the fluid. The solution is to use either very stable (to sonic energy) VII, or better, no VII at all. The additives also are subject to sonic shock, and many of the AW additives are known to crack up and precipitate out. In sonar systems used in submarines, a careless selection of hydraulic fluid can bring a whole set of problems due to fluid breakdown.

F. Fire Resistance

The best fire resistance in non-water-based hydraulic fluids is found in phosphate esters, fluorocarbons, and polyphenyl ethers. Water glycol systems (with over 35% water) have excellent fire-resistant characteristics, but the large amount of water tends to cause corrosion and fungus-related problems, and also their lubricity is low. Diesters and polyol esters are no better than aromatic mineral oils, although better than regular mineral oil (23). The PAOs have some mild fire-resistance characteristics as shown by some tests but not by others. Silanes can yield fire-resistant hydraulic fluids that are hard to burn but are not incombustible. Fluorocarbons or fluoroalcohols are closer to incombustible, but they have too high a specific gravity, high compression ratio, and problems with seal compatibility and additive insolubility. Some polyol esters, such as those with oleic acid (10), although not self-extinguishing, show good resistance to flammability. That is due to their low volatility and high ignition point (above 300°C). However, their unsaturation makes them prone to degradation, and on top of that they are vulnerable to hydrolysis and exhibit rather poor demulsibility.

The international standards for fire resistance are set forth in the ISO Standard 6743/H, 1982, Class L Classification, Family H (Hydraulic Systems). This family, which does not include automotive brake fluids or aircraft hydraulic fluids, classifies the fire-resistant fluids in four categories.

HFA: Solutions or emulsions containing more than 80% water. Service temperature +5 to +50°C.
HFB: Water-in-oil emulsions. Service temperature +5 to +60°C.
HFC: Water/polymer solutions or water/glycol containing less than 80% water. Service temperature –20 to +60°C.
HFD: Synthetic fluids containing no water. Service temperature –20 to +150°C.

That last category, HFD, is the main interest of this chapter. It includes phosphate esters, fluorocarbons, certain PAO formulations, silicate esters, certain polyol esters, silanes, etc. There are two major subdivisions of this category, the HFD-R for phosphate esters, and HFD-U for the other synthetic fluids.

Table 20.7 Relative Fire Resistance of Hydraulic Fluids

Product	Fire resistance
Mineral oil	Poor
PAOs	Fair
Polyol esters	Fair
Silicate esters	Fair
Polyglycols	Fair
Water/glycol	Excellent
Diesters	Fair
Silicones	Fair–good
Polyphenyl ethers	Good
Phosphate esters	Very good
Fluorocarbons	Excellent

Table 20.7 shows the generalized comparison of the fire resistance of a number of fire-resistant hydraulic fluids.

Table 20.8 presents several approximate physical properties of the most common fire-resistant fluids (24).

Table 20.9 compares various ignition characteristics of the most commonly used fire-resistant hydraulic fluids (19).

G. Toxicity and Environment

Paraffinic hydrocarbons and PAO are practically nontoxic and in limited amounts not threatening to the environment. Toxicity is poor in many phosphate esters. Actually some isomers in some aryl phosphates are neurotoxic and therefore strictly controlled by limits in specifications.

Biodegradability is a very desirable property when significant amounts of a hydraulic fluid leak into the environment. Low-viscosity PAO, especially the 2-cS grade, and less so the 4-cS grade, are reported to be biodegradable. Hydraulic fluids based on vegetable

Table 20.8 Approximate Physical Properties of Fire-Resistant Fluids

Property	Mineral oil	PAOs	Water/ Glycol	Phosphate ester (aryl)	Polyol ester
Specific gravity	0.87	0.81	1.07	1.13	0.91
Viscosity index	95	140	150	–30	185
Pour point, °C	–33	–59	–50	–21	–26
Operating range, °C	–5 to +75	–54 to +135	–30 to +30	5 to +80	–10 to +80
Autoignition temperature, °C	298	332	440	640	482
Vapor pressure	Low	Low	High	Low	Low
Fire resistance	Poor	Fair	Excellent	Very good	Good
Pump wear, ASTM D-2882, mg	22	10	100	25	15

Table 20.9 Comparison of Nonflammability of Fire-Resistant Fluids

Tests	Mineral oil	Polyol ester	Water/ glycol	Phosphate ester (aryl)
Autoignition temperature, °C	<350	482	435	640
High-pressure spray (Fed. 6052, MIL-F-7100)	Explosive ignition	—	No ignition	No ignition
Hot manifold (Fed. 6053, MIL-F-7100)	Instant ignition	Instant ignition	No ignition	No ignition
Pipe cleaner, number[a] (MIL-F-7100)	3	27	66	80

[a]Number of times before catching fire.

oils are considered to be very biodegradable. Rapeseed oil is now being used in many parts of Europe because of its good biodegradability.

IV. APPLICATIONS

The application of synthetic hydraulic fluids covers the full range of technological activities. For easier treatment, this is divided into five main categories of application: civil aviation, industry, marine, automotive, and military. The latter could, to a certain extent, be distributed to the other categories, but that would not do justice to the special nature of many of the military uses. In addition, it must be kept in mind that most of the progress in the synthetic hydraulic fluid area came from the military, which continues to spend considerable resources and time toward further advances.

A. Civil Aviation

1. Reasons for Use

Behind the extensive use of synthetic hydraulic fluids in aircraft are two factors. The first one is the need for a wide temperature operating range, a high-energy throughput, and compact systems, to satisfy the strict and ever-tightening requirements of modern aircraft. The second is the very real need for fire-resistant fluids for the vulnerable aircraft environment, especially the landing system, or wheel brakes, and units proximate to the engine heat flow. Thus the key properties that synthetic hydraulic fluids bring to the aircraft include:

Excellent low-temperature fluidity to operate in the coldest environment.
High-temperature stability—both thermal and oxidative—to allow prolonged full power operation without any fluid-related breakdowns.
Fire resistance for those systems that require it.
Compatibility with all materials that it may come in contact with, some of which are particular to aircraft systems.

2. Types of Fluids Used

A large number of synthetic hydraulic fluid types are used in aviation. Among them are PAOs, phosphate esters (both alkyl and aryl or mixed), silicate esters, silanes, fluorocarbons, fluorosilicones, fluoroglycols, polyphenyl ethers, etc. Many of those were developed for the military and bear military specification code numbers. More about them in the appropriate military applications section.

3. Phosphate Esters

The type of phosphate esters used in aircraft are mostly trialkyl phosphates, and they serve as the preferred fire-resistant hydraulic fluid. In aircraft—in distinction to other applications—excellent low-temperature fluidity for cold-weather operation is just as important a requirement as fire resistance. The alkyls—as opposed to aryls—can provide it. The finished formulation contains additives for improved oxidation stability, viscosity, and lubricity. Special additives are also frequently used to suppress valve erosion problems, reportedly due to complex electrochemical phenomena (25, 26). Contamination, especially with chlorides, can promote erosion of the metering edges, whereas proper filtration through activated clay helps to suppress the problem.

The seals, hoses, packings, and O-rings must be made from special elastomers, because natural rubber, Buna S, Buna N, and Neoprene are unsuitable. Among the appropriate types of rubber are butyl, silicone, fluoroelastomers, polytetrafluoroethylene (PTFE), and ethylene propylene diene monomer (EPDM). Because phosphate esters acting as solvents attack paints, such as conventional enamels and alkyd resins, epoxy resins and silicone enamels are recommended.

Airlines frequently use reprocessed phosphate ester fluid. Actually, reprocessing of a used synthetic hydraulic fluid, of any type, is normal practice in many applications.

4. PAO

Polyalphaolefin is a popular type of hydraulic fluid in aircraft. Properties such as excellent viscosity/fluidity, outstanding lubricity and AW characteristics (with the proper additives), good compatibility with conventional hydraulic seals and equipment, and a certain amount of fire-resistance improvement over mineral oil give PAO a wide range of applications. A finished PAO formulation usually contains a moderate amount of seal swell agents, such as a diester, to bring its polarity up to that of a mineral oil.

5. Polyphenyl Ethers

Polyphenyl ethers are used in some ultra-high-temperature hydraulic systems for advanced aircraft and spacecraft. These fluids are very stable thermally and oxidatively and they are also radiation resistant. This stability is due to the delocalization of π electrons with the substitution of aromatic groups for aliphatic. The meta linkage yields liquids, which allows for their use as hydraulic fluids. They also have good lubricity due to the formation of "sandwich compounds" on the metal surfaces. These form from aromatic ring systems and nonferrous metal ions, with d electrons of the metal interacting with π electrons of the aromatic compounds (22). Polyphenyl ethers swell rubber too much and, as with phosphate esters, they need special types of seal material.

6. Fluorosilicones

Fluorosilicone fluids are used in aircraft high-performance hydraulic systems. They possess excellent water resistance as well as resistance to chemicals. Thus, they can withstand the solvency of many solvents and they do not get washed out. Actuating drives is one such area of application. Fluorosilicones also have good load-carrying ability, which is helpful in the design of efficient systems.

7. Other Fluid Types

Chlorophenyl methyl silicone fluids are used in supersonic aircraft for high-temperature, low-flammability hydraulic fluids.

Silicate esters have been used in the Concorde Supersonic Transport, but such fluids are easily hydrolyzed, they tend to gel at high temperatures, and they are poor lubricants.

More relevant information on aircraft fluids is given at the military aircraft section.

B. Marine

Similar factors with those operating on aircraft are promoting the use of synthetic hydraulic fluids in marine applications. Fires can be devastating on a ship at sea. Shipboard fires are caused by electrical equipment or oil and fuel ignition. Oil escaping through a pinhole in a pipe of a hydraulic system at 207 bar, which is a normal operating pressure, can form a jet travelling over 10 m. Also, O-rings and flexible hoses can deteriorate with shock and noise and allow leaks. Therefore, fire-resistant fluids are used in many instances as hydraulic fluids on board ships. Two kinds of fire-resistant fluids are preferred: phosphate esters and water/glycol fluids. The phosphates are usually of the triaryl type.

Polyalphaolefin is extensively used in deck cranes of ships sailing between hot and temperate climates. In such cases, PAO retains its viscosity and the elastohydrodynamic lubrication factor remains satisfactory (27). A mineral oil thickened with VII would have sheared down, allowing high pump wear and increased clearances resulting in costly port delays. The alternative to PAO would have been extensive and expensive equipment modifications. Polyalphaolefin is also finding application in high-line transfer equipment, which allows transfer between ships. Piston pumps are usually the problem area in this job.

C. Industry

Synthetic hydraulic fluids are used extensively in industry. That includes nearly all sectors of industrial activity. Steel (28) and primary metals, machining and manufacturing, energy, chemical, and mining (29) are examples. Again, the main reasons for synthetic hydraulic fluids displacing mineral oils in industry are the same as with the previous categories: severe conditions, and safety concern.

1. Steel and Primary Metals

The steel industry, and generally the primary metals industry, which includes steel processing, aluminum and zinc die casting, metal forming processes, etc., must deal with fire hazards. This is so because molten or otherwise hot metal is being processed in close proximity to hydraulic control equipment. Because fluid leakage is always a distinct possibility, and the hot metal constitutes an available ignition source, fire is a high-probability event. Hence, fire-resistant hydraulic fluids are required, and that means synthetics or water-based fluids. Water-based fluids have excellent fire resistance but poor lubricity, pitting tendencies, and a restricted temperature range. In modern, high-pressure and high-temperature hydraulic systems, water is not a good choice. Phosphate ester hydraulic fluids, usually triarylphosphates, are successfully used on a large scale in industry. Examples are hydraulic systems for the handling of hot metal ingots or slub, die casting, including continuous casting machines, furnace controls, rolling mills, shears, and ladles.

Water/glycol fluids are also used despite their deficiencies, mostly in the zinc and aluminum die-casting industries or in less efficient hydraulic systems. These are formulations containing 40–50% water, 30–50% polyglycol thickeners, 0–20% propylene or ethylene glycol, and 1–2% additives. They must be constantly monitored to ensure that the water content does not fall below 35%, which would break their fire-resistance capabilities. Thus they are limited to temperatures of operation below 60°C.

Polyol ester-based fluids, many with phosphates in them, have been used as fire-resistant hydraulic fluids in some industrial applications. Although their fire resistance is no match to that of phosphate esters, polyol esters have some advantages. Among them are low specific gravity, high VI, easier recovery from water, and no requirement to change the seals when converting from mineral oil (24).

Polyol esters or other fire-resistant fluids are frequently used in the hydraulic systems of industrial robots, especially those used for welding. In such cases adequate filtration is of paramount importance because of the critical tolerances in servo valves and other intricate components.

When attempting to change over from a mineral to a phosphate ester hydraulic fluid, care must be taken to flush all mineral oil out. Mineral contamination above 3% could make the fluid flammable.

2. Mining

The mining industry, especially the below-ground section, is characterized by space scarcity, and also by heat, cold, water, dust, and dirt in the environment. Space constraints tend to minimize the size of the hydraulic fluid sump and the air space around it, and to increase operating pressures, temperatures, and loading (29). Besides, it makes servicing the hydraulic unit very difficult at a time when servicing is critical to deal with contamination factors at their worst. The biggest problem, however, is fire, which, in an underground mine, can mean many deaths. Therefore, most of the fluids used are fire-resistant hydraulic fluids. Phosphate esters are used extensively. Examples are continuous miners and associated equipment. In coal mining, they are used in hydrokinetic transmissions (fluid coupling) driving coal conveyors and in coal-face machinery such as power loaders. Water/glycol fluids are also used extensively in mining operations. Polyalphaolefin is used in many types of mining hydraulic equipment in areas where fire hazards are not great. Polyalphaolefin is chosen as superior to mineral oil under severe conditions and as requiring less down-time for servicing.

3. Manufacturing

Manufacturing covers a long list of activities where all kinds of machines, crafts, equipment, tools, and numerous items are fashioned. Production could also be included here, ranging from commodities to off-shore oil exploration platforms, etc. This sector also uses large amounts of synthetic hydraulic fluids.

Polyalphaolefin is a fluid type that is gaining favor in many areas, including sealed-for-life units, critical servovalves, machine tools, etc. Phosphate esters are also popular here in areas of fire hazard. Polyglycols, usually with water, are also used for their nonflammability. Perfluoropolyethers (PFPE) find use as hydraulic oil in vacuum pumps due to its extremely low vapor pressure and excellent chemical resistance.

4. Power Plants

Fire-resistant hydraulic fluids are used heavily in power plants. Specific examples are boiler control systems and hydraulic control circuits of steam turbines, including electrohydraulic control for throttle/governor mechanisms. Many large steam turbines are equipped with hydraulic circuits totally separate from the main bearing lubricant supply and they use phosphate esters. That ensures the shutdown of the turbine in the event of a fire and eliminates the danger from contact with superheated steam pipes if a leak or burst occurs.

The major manufacturer of steam turbine hydraulic systems has been recommending aryl phosphate esters for the last 35 years. Fluoroelastomer seals are used with those systems (30). Polyalphaolefin is also used, when a fire-resistant fluid is not required, to improve efficiency and lengthen service intervals. In nuclear power plants, in areas with significant radiation exposure, polyphenyl ether is most frequently used. Contrary to conventional hydraulic fluids, polyphenyl ethers are very radiation resistant and do not show large viscosity increases with time. This lengthens the useful life of the fluid and reduces down-time for service. In areas with no significant radiation exposure, such as the steam turbine hydraulic governor systems, triarylphosphates are used to advantage for their fire resistance (31).

5. Arctic Environment

Arctic conditions stress the importance of low temperature fluidity above everything else. This applies to both industrial and automotive applications. Only synthetic hydraulic fluids can meet this challenge. Polyalphaolefin is widely used, especially the lower-viscosity grades. Also, diesters of low viscosity are used in selected areas where water is not present.

6. Water Installations

Critical areas of water installations have to observe their own cardinal rule: Avoid pollution and contamination. Hydraulic systems in those areas tend to use polyglycol hydraulic fluids. Examples of such applications are sluices, dams, dredging boats, water-treatment plants, and swimming pools (32).

D. Automotive

Automobiles, trucks, and construction or off-highway equipment also use hydraulic systems in need of proper fluid. In most automobiles, there is no separate hydraulic system needing a synthetic fluid, except for the brakes. However, in mobile equipment, there are high-speed, high-pressure systems that use a large output pump. These are conditions where a synthetic fluid can offer improvements.

1. Mobile Equipment

Polyalphaolefin has been found to improve significantly both efficiency and durability of those hydraulic systems.

Phosphate esters have also been used where fire resistance is a requirement. In those systems, however, one has to watch for wear problems that arise if the acidity of the phosphate ester is allowed to exceed a neutralization number (NN) of 2.0. It appears that the metal surfaces catalyze the decomposition of the phosphate ester, in the presence of moist air, into acidic by-products. In cases of axial displacement pumps, the acidity has been found to reach NN 3.0 in only 300 h of operation and exceed NN 10.0 on 950 h. Acid pitting was then observed on the housing, and the system showed an inability to maintain pressure (33).

2. Brakes

The brakes of automobiles and other vehicles make up an area where no mineral oil is allowed.

The early brake fluids in the 1920s were a mixture based on castor oil, and the seals were made of leather. By the end of the 1930s, all U.S. automobiles had rubber seals in the brake hydraulic systems. The front wheel drive development caused a weight shift, and that increased considerably the braking torque requirements of the front units. Higher heat generation added to elevated brake line pressures, caused by vacuum-assist power boosters, have put a lot of stress on the parts of the system, including the seals and the fluid.

To understand the magnitude of these stresses, one must consider the following: The capacity of a modern brake system is very high. The thermal energy generated from one average deceleration within 61 m from 55 mph (89 kmh) to a complete stop is enough to boil off 450 g of water or to soften 16 cm^3 of steel. Temperature in excess of 650°C can be expected at the front brake pads and the brake fluid itself. Adjacent rubber seals may be heated to above 150°C. In a normal city traffic peak, temperatures of the system can exceed 200°C (34). Therefore, the choice of the brake fluid chemistry is truly critical for safety.

3. Brake Fluids

The brake fluid, which is a special hydraulic fluid, is purely synthetic. There are three main types of brake fluids as specified by the U.S. Department of Transportation (DOT) and essentially accepted worldwide (34): Type 3 (DOT-3), Type 4 (DOT-4), and Type 5 (DOT-5).

Table 20.10 summarizes the requirements and typical compositions of those three types of brake fluids (10, 19, 34).

The major danger for deterioration of the brake fluid is water pickup. The moisture drawn into the system is absorbed by the fluid. However, as the water content increases,

Table 20.10 Brake Fluids—Requirements and Composition

	DOT-3	DOT-4	DOT-5
A. Requirements			
Boiling point, °C, minimum	205	230	260
Wet boiling point, °C, minimum	140	155	180
Ignition point, °C, minimum	82	100	—
Viscosity, cSt, minimum			
100°C	1.5	1.5	1.5
50°C	4.2	4.2	
Low-temperature fluidity, cSt, maximum			
–40°C	1,500	1,300	
–55°C			900
pH	7.0–11.5	7.0–11.5	
B. Typical composition			
Base, by weight	Polyether, 10–20%	Boric ester of polyether, 30–40%	Silicone oil, 80–90%
	Glycol ether, 80–90%	Glycol ether, 60–70%	—
Rubber swell agent, by weight	—	—	Phosphate ester, 10–20%
Additives, by weight	1–2%	1–2%	0.1–0.2%

the boiling point of the brake fluid decreases. After a few years of operation, a drop in the boiling point of over 90°C is possible. Low boiling point facilitates boiling or vapor generation and at high temperatures (after a few stops) that can cause brake-fade problems. That means loss of brake pedal effect, resulting in a very dangerous situation.

DOT-3, based on glycols and polyglycols, is very vulnerable to moisture. As a matter of fact, the higher the boiling point of the polyglycol, the more hydroscopic it can be.

DOT-4, based on boric esters of polyglycol, has been formulated to give greater in-service stability to water pickup. The boric acid ester consumes the moisture and denies it a chance to hydrolyze the polyglycol.

DOT-5, based on silicone oil, has little fear of moisture, and it can operate at higher temperatures. It also exhibits excellent low temperature fluidity, as expected from the relative performances shown in Table 20.2. However, it is possible for unabsorbed water to collect at a low point, such as a steel cylinder or in the lines, and cause corrosion.

When using silicone fluids in automobiles, one has to remember that they can be a hindrance to adhesion and they must be completely eliminated when painting. Also, they must be kept away from electrical contacts because silicone, which has a high spreadability, is a good insulator.

The U.S. Postal Service and the military specify DOT-5 fluid in their vehicles, and large savings in maintenance costs are expected from it. Minor swelling problems of some elastomers by silicone fluids have brought about improved formulations that utilize additives to make the DOT-5 fluid behave closer to the glycol-based fluid.

Seals compatible with glycol fluids are EPDM and SBR.

Seals compatible with silicone fluids are fluoroelastomers.

E. Military

The military is credited with the main advances in hydraulic fluid technology due to their highly sophisticated requirements and willingness to finance new breakthroughs. Thus synthetic hydraulic fluids in the military constitute a higher section of the total hydraulic fluids volume than in any other industry.

1. Military Aircraft

The aircraft sector is dominated by synthetic fluids. There is a wide variety of types of synthetic hydraulic fluids for use in the military aircraft (35–37).

Polyalphaolefin is represented by the specification MIL-H-83282, and it is used in many applications with no imminent fire hazards. MIL-H-83282 (NATO designation H-536) has been used in Navy carrier aircraft (38) with excellent results. Polyalphaolefin formulations can typically contain an AW agent, such as Tricresyl phosphate (TCP), an antioxidant, such as a hindered phenol, and a rubber swell agent, such as an ester. They can also contain a VII to increase viscosity. However, their performance in modern aerospace hydraulic pumps appears to depend more on base-stock viscosity than on the kinematic viscosity of the VII-thickened fluid (39). Alkyl benzenes, of special structure, could be used as high-temperature hydraulic fluids in Air Force applications if certain production problems are solved (40). Silanes, both alkyl and aryl types, are also well represented. Silicate esters and disiloxanes are used for high-temperature applications or as coolants for packaged electronic systems in aircraft and missiles. Silicones, including chloro derivatives, are also used. Phosphonitriles, fluorocarbons, and fluoroglycols are used in certain niches where special properties are important for performance. Phosphate esters, both the trialkyl and triaryl type, dominate the fire-resistant segment of application. Truly nonflammable hydraulic fluids can be based on fluorinated hydrocarbons or esters (41).

Table 20.11 shows the military designations and chemistry of a representative number of military specifications. Some of them are of the fire-resistant category and mainly, but not exclusively, intended for aircraft (35).

2. Missiles and Rockets

Missile hydraulic systems are subjected to very severe temperature variations. In a bomb bay or on a missile launch platform at high altitude, temperatures might get down to –54°C. However, when the missile is deployed, the air friction on the missile skin can raise the temperature to 316°C.

Disiloxane can meet the temperature range requirement, but it has been shown to form, when in storage, gelatinous precipitates due to hydrolysis. This can clog hydraulic in-line filters causing pump cavitation and loss of hydraulic power. Disiloxane is also unstable in the presence of metals and corrodes steel.

The PAO-based fluids or silahydrocarbon fluids can overcome storage stability problems and perform at the required temperature range (38).

Silahydrocarbons are used extensively in most rocket hydraulic systems for rocket control. In addition to the above-mentioned wide temperature range, they have good hydrolytic stability, good thermal conductivity and thermal capacity, and good AW properties when fortified with AW additives. That makes them very dependable for their difficult task. However, silahydrocarbons are more expensive than PAO-based fluids.

3. Navy Ships

The ships of the Navy face the danger of shipboard fires, and they use extensively fire-resistant hydraulic fluids. These fluids can be phosphate esters types or water/glycol types (42, 43). Ships' requirements include very low pour points (below –30°C), low toxicity, lubricity durability, corrosion protection, material compatibility (pipes, couplings, seals, etc.), adequately high operating temperatures, and contamination control.

Table 20.11 Representative Military Specifications

Specification	Composition
MIL-H-83282	Polyalphaolefins (PAO)
MIL-H-83306	Trialkylphosphates plus VII
MIL-H-8446B	Silicate ester (canceled)
MIL-H-27601A	Silane
MIL-L-19457B	Triarylphosphates
MIL-H-19457 Type I	Triarylphosphates
MIL-L-9236B	Trimethylolpropane ester
MLO-54-408C	Tetradedecyl silane
MLO-56-280	Diphenyl di-*n*-dodecyl silane
MLO-56-578	Octadecyl trioctyl silane
MLO-54-540	Silicate ester
MLO-54-856	Silicate ester
MLO-59-287	Chlorophenylmethyl silicone
MLI-59-692	Bis(phenoxy phenoxy) benzene
MLO-63-25	Phenoxy base triphosphonitrile

Phosphate esters require a well-designed system to prevent water contamination. Water/glycol fluids have a low-temperature ceiling of operation and very low lubricity, but they allow for seawater contamination. The latter is important for equipment that interfaces with water or is located outside the hull (27, 43).

Fire resistance is such an important property for military applications that it generates a large number of technical publications and new products. A sample of the extensive work carried out on flammability of hydraulic fluids can be had by scanning the pertinent literature (4, 11, 35, 44–50).

F. Servicing and Maintenance

All hydraulic fluid systems require servicing and maintenance. In the case of synthetic hydraulic fluids, prevention of contamination and purification is even more critical. Good maintenance pays well in cost savings and trouble-free performance. Avoidance and exclusion of contaminants is the first basic rule. Constant monitoring of the fluid condition is the second. Good monitoring will be very helpful in maintaining peak performance of both fluid and hardware.

Particle counting and viscosity, acidity, infrared spectrum, and spectroscopic analysis for wear metals are recommended for most hydraulic fluids (51). For synthetic hydraulic fluids, such careful monitoring is even more important and cost-effective. Phosphate ester fluids filtered through activated clay or, better yet, activated alumina, are purified from acids created through hydrolysis. Attapulgus clay has been associated with production of deposits in turbine hydraulic systems (52–54) whereas alumina has not. Acidity, if left to increase, appears to allow foaming and promote corrosive attacks on the metals and other parts of the system. Worse yet, a degraded phosphate ester fluid accelerates further degradation (55). The filters should be changed when the acidity reaches NN 0.1 and certainly before NN 0.2 is exceeded (56).

When converting a hydraulic system from a mineral to a synthetic, such as a fire-resistant fluid, great care should be taken to flush the equipment well and to pay attention to compatibility with the seals and other parts of the system (24). Even a small amount of leftover mineral oil could greatly decrease the fire resistance of the new fluid (56). Also, cleaning with chlorine-containing solvents should be discouraged. Contamination of a phosphate ester fluid with chlorinated solvents or salts can cause severe problems with corrosion or electrochemical erosion, as mentioned previously. Similar problems can be introduced to many other types of synthetic hydraulic fluids through contamination.

REFERENCES

1. Wambach, W. E. (1983). Hydraulic systems and fluids, *Lub. Eng.*, 39, 483–486.
2. Papay, A. G., and C. S. Harstick (1975). Petroleum-based industrial hydraulic oils—Present and future developments, *Lub. Eng.*, 31, 6–15.
3. Snyder, C. E., Jr., L. J. Gschwender, K. Paciorek, R. Kratzer, and J. Nakahara (1986). Development of a shear stable viscosity index improver for use in hydrogenated polyalphaolefin-based fluids, *Lub. Eng.*, 42, 547–557.
4. Snyder, C. E., Jr., L. J. Gschwender, and W. B. Campbell (1982). Development and mechanical evaluation of nonflammable aerospace (–54°C to 135°C) hydraulic fluids, *Lub. Eng.*, 38, 41–51.
5. Anonymous (1987). Une autre conception performance des huiles hydrauliques, *Petrole Informations*, 98.

6. Papay, A. G. (1983). Oil-soluble friction reducers—Theory and application, *Lub. Eng.*, 39, 419–426.
7. Hydraulic fluids based on two centistoke synthetic hydrocarbons, U.S. Patent 4,537,696 (1985).
8. Kussi, S. (1986). Eigenshaften von Basisflüssigkeiten für synthetische Schmierstoffe, *Tribol. Schmierungstech.*, 33, 33–39.
9. Law, D. A., J. R. Lohuis, J. Y. Breau, and A. J. Harlow (1984). Development and performance advantages of industrial, automotive and aviation synthetic lubricants, *J. Synth. Lub.*, 1, 6–33.
10. Seki, H. (1989). Properties of hydraulic fluids and their application—Synthetic type fluids, *Junkatsu*, 34, 587–593.
11. Snyder, C. E., Jr., L. J. Gschwender, C. Tamborski, and G. J. Chen (1982). Synthesis and characterization of silahydrocarbons—A class of thermally stable wide-liquid-range functional fluids, *ASLE Trans.*, 25, 299–308.
12. Tamborski, C., G. J. Chen, D. R. Anderson, and C. E. Snyder, Jr. (1983). Synthesis and properties of silahydrocarbons, a class of thermally stable wide liquid range fluids, *Ind. Eng. Chem. Prod. Res. Dev.*, 22, 172–178.
13. Gupta, V. K., C. E. Snyder, Jr., L. J. Gschwender, and G. W. Fultz (1989). Thermal decomposition investigations of candidate high temperature base fluids I. Silahydrocarbons, *STLE Trans.*, 32, 276–280.
14. Snyder, C. E., Jr., C. Tamborski, L. J. Gschwender, and G. J. Chen (1982). Development of high-temperature (–40°C to 228°C) hydraulic fluids for advanced aerospace applications, *Lub. Eng.*, 38, 173–178.
15. Paige, H. L., C. E. Snyder, Jr., L. J. Gschwender, and G. J. Chen (1990). A systematic study of the oxidative stability of silahydrocarbons by pressure differential scanning calorimetry, *Lub. Eng.*, 46, 263–267.
16. Gupta, V. K., M. A. Stropki, T. J. Gehrke, L. J. Gschwender, and C. E. Snyder, Jr. (1990). Hydrolytic studies of some silicon-based high temperature fluids, *Lub. Eng.*, 46, 706–711.
17. Gschwender, L. J., C. E. Snyder, Jr., and A. A. Conte, Jr. (1985). Polyalphaolefins as candidate replacements for silicate ester dielectric coolants in military applications, *Lub. Eng.*, 41, 221–228.
18. Scott, R. N., L. O. Knollmueller, F. J. Milnes, T. A. Knowles, and D. F. Gavin (1980). Silicate cluster fluids, *Ind. Eng. Chem. Prod. Res. Dev.*, 19, 6–11.
19. Yagi, M. (1987). Synthetic lubricants—Application to industrial use, *Junkatsu*, 32, 121–125.
20. Conte, A. A. (1985). The action of organo-phosphate additives in polysiloxane fluids, *J. Synth. Lub.*, 2, 95–120.
21. Spikes, H. A. (1986). Applications review: Helicopters 1. Future helicopter transmission oils, *J. Synth. Lub.*, 3, 181–208.
22. Plagge, A. (1985). Gebrauchsiegenschaften synthetischer Schmierstoffe und Arbeitsflüsigkeiten, *Tribol. Schmierungstech.*, 32, 270–278.
23. Staley, C. (1979). Fire-resistant hydraulic fluids, Chemicals for Lubricants and Functional Fluids Symposium, London, November.
24. Wiggins, B. J. (1987). System conversions for fire-resistant hydraulic fluids, *Lub. Eng.*, 43, 467–472.
25. Phillips, W. D. (1988). The electrochemical erosion of servo valves by phosphate ester fire-resistant hydraulic fluids, *Lub. Eng.*, 44, 758–767.
26. Beck, T. R. (1983). Wear by generation of electrokinetic streaming current, *ASLE Trans.*, 26, 144–150.
27. Skinner, R. S. (1986). Synthetic lubricants—Why their extra cost can be justified, *Mar. Eng. Rev.*, August, 18–21.
28. Cichelli, A. E. (1983). Steel mill lubrication, *Lub. Eng.*, 39, 410–413.
29. Okon, L. W. (1983). Lubrication in the mining industries, *Lub. Eng.*, 39, 487–488.

30. Phillips, W. D. (1986). The use of triaryl phosphates as fire-resistant lubricants for steam turbines, *Lub. Eng.*, 43, 228–235.
31. International Guidelines for the Fire Protection of Nuclear Plants (1983). National Nuclear Risks Insurance Pools and Association Publication, September, p. 10.
32. Anonymous (1988). Polyglykole—Technische Anwendungen, *Erdöl Kohle*, 41, 229–230.
33. Perez, J. M., R. C. Hansen, and E. E. Klaus (1990). Comparative evaluation of several hydraulic fluids in operational equipment—A full scale pump stand test and the four ball wear tester, Part II. Phosphate esters, glycols, and mineral oils, *Lub. Eng.*, 46, 249–255.
34. Car, J. (1988). Elastomer materials for automotive hydraulic brake systems, *Lub. Eng.*, 44, 22–27.
35. Coordinating Research Council, Inc. (1986). *Flammability of Aircraft Hydraulic Fluids—A Bibliography*, CRC report no. 545, Atlanta, Georgia.
36. Snyder, C. E., Jr. (1982). Utilization of synthetic-based hydraulic fluids in aerospace applications, *Int. Jahrb. Tribol.*, 1, 409–418.
37. Snyder, C. E., Jr. (1979). Aerospace applications of synthetic hydraulic fluids, *Performance Testing of Hydraulic Fluids*, *Pap. Int. Symp*, London, England.
38. Gschwender, L. J., C. E. Snyder, Jr., D. R. Anderson, and G. W. Fultz (1984). Determination of storage stability of hydraulic fluids for use in missiles, *Lub. Eng.*, 40, 659–663.
39. Gschwender, L. J., C. E. Snyder, Jr., and S. K. Sharma (1988). Pump evaluation of hydrogenated polyalphaolefin candidates for a –54°C to 135°C fire-resistant Air Force aircraft hydraulic fluid, *Lub. Eng.*, 44, 324–329.
40. Gschwender, L. J., C. E. Snyder, Jr., and G. L. Driscoll (1990). Alkylbenzenes—Candidate high temperature hydraulic fluids, *Lub. Eng.*, 46, 377–381.
41. Snyder, C. E., Jr., and L. J. Gschwender (1984). Nonflammable hydraulic system development for aerospace, *J. Synth. Lub.*, 1, 188–200.
42. Eastaugh, P. R., M. R. O. Hargreaves, and H. J. Jones (1983). Fire hazards associated with warship hydraulic equipment, Institute of Mechanical Engineers Conference on Naval Engineering Present and Future, Bath, September.
43. Page, R. N. M. (1986). Selection of a fire-resistant fluid for hydraulic systems in Royal Navy ships, *Trans. Inst. Mar. Eng.* (Tech. Meet.), paper no. 10, 98, 9–14.
44. Gupta, V. K., L. J. Gschwender, C. E. Snyder, Jr., and M. Prazak (1990). Thermal stability characteristics of a non-flammable chlorotrifluoroethylene CTFE basestock fluid, *Lub. Eng.*, 46, 601–605.
45. Snyder, C. E., Jr., and L. J. Gschwender (1984). Non-flammable hydraulic fluid systems development for aerospace, *J. Synth. Lub.*, 1, 188–200.
46. Snyder, C. E., Jr., and L. J. Gschwender (1983). Fluoropolymers in fluid and lubricant applications, *Ind. Eng. Chem. Prod. Res. Dev.*, 22, 383–386.
47. Military Specification (1982). Hydraulic fluid, fire resistant, synthetic hydrocarbon base, aircraft, NATO Code no. H-537, MIL-H-83282, 10 February.
48. Military Specification (1982). Hydraulic fluid, rust inhibited, fire resistant, synthetic hydrocarbon base, MIL-H-46170B, 18 August.
49. Conte, A. A., and J. L. Hammond (1980). *Development of a High Temperature Silicone Fire-Resistant Hydraulic Fluid*, Report no. NADC 79248-60, Naval Air Development Center, Warminster, PA, 5 February.
50. Raymond, E. T. (1982). *Design Guide for Aircraft Hydraulic Systems and Components for Use with Chlorotrifluoroethylene Non-Flammable Fluids*, AFWAL-TR-2111, Air Force Aero Propulsion Laboratory, Wright-Patterson Air Force Base, OH, March.
51. Poley, J. (1990). Oil analysis for monitoring hydraulic oil systems, a step-stage approach, *Lub. Eng.*, 46, 41–47.
52. Phillips, W. D. (1983). The conditioning of phosphate ester fluids in turbine applications, *Lub. Eng.*, 39, 766–780.
53. Grupp, H. (1979). Aufban von schwer entflammbaren Hydraulikflüssigkeiten auf phosphor-

säure-esterbasis, Erfahrungen aus dem praktischen Einsatz in Kraftwerk, *Der Machinen Schaden*, 52, 73–77.
54. Tersiguel-Alcover, C. (1981). *La Filtration Des Esters Phosphates sur Alumine Activee*, EDF Report P 33/4200/81-24, June.
55. Shade, W. N. (1987). Field experience with degraded synthetic phosphate ester lubricants, *Lub. Eng.*, 43, 176–182.
56. Stark, R. (1985). Anwendugstechnische Richtlinie für schwerentflammbare Hydraulikflüssigheiten HSD, *Schmierungstechnik, Berlin*, 16, 285–286.

21
Metalworking

William L. Brown
Union Carbide Chemicals and Plastics Company Inc.
Tarrytown, New York

I. INTRODUCTION

Metalworking is a major industry in the United States today. Metalworking operations include rolling, forging, stamping, drawing, forming, cutting, and grinding. Practically all metal objects, from I-beams to screws, have undergone at least one metalworking operation.

The success of most metalworking operations is dependent on the use of good lubricants and coolants. There is a large variety of these metalworking fluids. They include fatty-acid-based soaps, formulated hydrocarbons, emulsified oils, and aqueous solutions. These products are used in operations ranging from the drawing of wire to the tapping of nut threads. In each case, the metalworking fluid has been adapted to satisfy the needs of a specific application.

This chapter reviews the uses of synthetic lubricants in various metalworking operations. It begins by defining metalworking and metalworking fluids. A brief review of the history of metalworking fluids is then presented, with particular attention being paid to the incorporation of synthetic lubricants.

The synthetic lubricants currently being used in metalworking fluids are defined. These products include polyalkylene glycols, esters, and synthetic hydrocarbons. The generic structures of these synthetic lubricants are illustrated and the physical properties that have led to their incorporation in metalworking fluids are described.

This chapter focuses on the use of polyalkylene glycols in metalworking fluids.

Polyalkylene glycols are of importance to the metalworking industry because of their unique solubility properties in water. The mechanism of how they work is described, and a number of applications are presented. The use of other synthetic lubricants in metalworking fluids is then reviewed.

The remainder of this chapter examines the future of synthetic lubricants in the metalworking industry. The effect of disposal regulations and the need for improved workpiece quality on the use of synthetic lubricants in metalworking fluids are discussed.

It should be pointed out here that the metalworking industry is very large and diverse. Metalworking fluid formulations are tailored to specific operations and locations. Formulations are affected by such factors as production speeds, inventory turnover, local water quality, ambient weather conditions, and effluent regulations. Therefore, this chapter will attempt only to give an overview of the uses of synthetic lubricants in the metalworking industry.

II. DEFINITIONS

Metalworking is the shaping of a metallic workpiece to conform to a desired set of geometric specifications. Metalworking can be divided into two basic categories, cutting and forming. In cutting operations, the blank is shaped by removing unwanted metal in the form of discrete chips. Cutting operations include turning, tapping, milling, broaching, and grinding. Metal-forming processes involve the plastic deformation of the workpiece into a desired shape. Drawing, hot and cold rolling, stamping, and forging are examples of metal-forming operations.

In both cutting and forming operations, the metalworking fluid plays a critical role. The two most important functions of the metalworking fluid are to provide adequate lubrication between the work piece and the tool or die and to remove the heat that is generated (1).

Lubrication can be defined as the reduction of friction between two moving surfaces. In metalworking operations, lubrication can be divided into two types, hydrodynamic and boundary or extreme pressure (EP). In hydrodynamic lubrication, the moving surfaces are always separated by a film of fluid or lubricant. The film thickness and the coefficient of friction are both a function of the parameter ZN/P, where Z is the lubricant's dynamic viscosity, N represents the relative velocity of the moving surfaces, and P is the applied load per unit area (2, 3). The faster the velocity or the more viscous the lubricant, the thicker the fluid film. The coefficient of friction will also increase because of a rise in viscous drag. Conversely, increasing the load or pressure will decrease the film thickness and the coefficient of friction as long as hydrodynamic conditions are maintained.

Boundary or EP lubrication is necessary when the pressures experienced are great enough to cause contact between the moving metal surfaces. The purpose of boundary or EP additives is to minimize the wear experienced when the surfaces rub against each other.

Boundary lubricant additives are polar compounds like fatty alcohols, acids, and esters, which absorb onto the metal surfaces forming thin, low-shear-strength films. These solid films help prevent metal–metal contact and thus reduce friction and wear (2, 4).

The EP lubricity additives used in metalworking operations are usually organic compounds that contain phosphorus, chlorine, or sulfur. During the metalworking process, these additives react with the metal surfaces, forming organic or organometallic films (2, 4). These films then act to reduce the force necessary to slide the surfaces past one another, while at the same time minimizing wear (5).

Cooling is the other critical function of a metalworking fluid. The transfer of heat away from the tool or die is affected by the specific heat and the heat of vaporization of the metalworking fluid. The specific heat is the amount of heat required to raise the temperature of 1 g of fluid 1°C. The larger this value, the more heat the fluid can absorb for an incremental temperature rise, resulting in more efficient cooling. The heat of vaporization is the amount of heat required to vaporize a gram of liquid. A fluid with a high heat of vaporization will also cool efficiently, because it absorbs large amounts of energy as it transforms from a liquid to a gas. Heat transfer in metalworking can also be affected by the fluid's boiling point, ambient temperature (5), viscosity, surface energy, and application method.

Good lubrication and cooling will prolong tool or die life, improve surface finishes, and permit higher production speeds. The cooling and lubricating properties of a metalworking fluid must be matched to the operation being performed. In high-cutting-speed operations like turning and grinding, cooling is of paramount importance. However, in lower-speed operations that involve heavy cuts or large deformations, the lubricating properties of the metal working fluid are critical (6).

In addition to providing adequate lubrication and cooling, metalworking fluids must protect the machine and the workpiece from corrosion and remove chips from the cutting zone. It is also important that the metalworking fluid be benign and nonirritating to the operator.

There are four major classes of metalworking fluids. These different types of straight oils, soluble oils, chemical solutions, and semichemical solutions, will now be defined.

Straight Oils. These products are derived primarily from petroleum fractions, although animal and vegetable oils are occasionally used (6). Synthetic lubricants can also be employed as straight oil base stocks. These base stocks, whether natural or synthetic, are usually compounded with various boundary and EP lubricity additives. Straight oil metalworking fluids contain no water and are sold "ready-to-use." They are excellent lubricants but have limited cooling capacity.

Soluble Oils. Soluble oils are actually oil-in-water emulsions, which take advantage of the lubricity of oils and the cooling properties of water. This is a very versatile class of metalworking fluids. In severe cutting or forming operations, they are usually formulated with chlorine- and sulfur-containing extreme pressure (EP) lubricity additives and diluted only one to five times with water (7). However, in light cutting or grinding operations, the soluble oil concentrate often contains only rust inhibitors and can be cut 20–100 times with water.

Chemical Solutions (Synthetics). Chemical metalworking fluids contain no petroleum oil and form true solutions when diluted with water. These metalworking fluids are often referred to as "synthetics." However, not all "synthetic" metalworking fluids contain synthetic lubricants as they are defined later in this chapter. Synthetic lubricants include polyalkylene glycols, various esters, and polyolefins. Therefore, to avoid confusion in this chapter, these oil-free, water-soluble products will be referred to as "chemical metalworking fluids" or "chemical solutions."

Chemical solutions were originally formulated for grinding and light duty cutting operations. However, the incorporation of synthetic lubricants and the development of improved water-soluble boundary and EP lubricity additives have greatly expanded the utility of this class of metalworking fluids.

Semichemical Solutions (Semisynthetics). Semichemical metalworking fluids are an attempt to get the best properties of both soluble oils and chemical solutions. The most common definition of a semichemical metalworking fluid is a product whose concentrate contains water-soluble additives, emulsifiers, and less than 20% petroleum oil (8).

III. METALWORKING FLUID BACKGROUND

The use of lubricants is known to date back to the times of the ancient Egyptians, who used fats to grease chariot wheels (9). However, the use of lubricants in metalworking operations is relatively recent. One of the earliest references to metalworking lubricants comes from the writings of Biringuccio in the early 1500s (10). He observed that it is important to use wax when drawing high-quality gold and silver wire. The widespread use of metalworking fluids, however, coincided with the industrial revolution which began in England in the late eighteenth century (2).

By the mid nineteenth century mineral oils were being widely used as metalworking lubricants (2). W. H. Northcott observed in 1868 that the use of oils as metalworking fluids greatly improved cutting speeds and tool life, reduced power consumption, and produced smoother cuts (11). Northcott also described how soda-water worked well as a metalworking fluid, despite some corrosion problems.

The development of high-speed cutting tools and the increased use of grinding began to show the limitations of oils as metalworking lubricants (2). Most of the energy used to cut or grind a metal workpiece is dissipated as heat (4, 8, 12). Thus, as cutting speeds increased, the cooling properties of the metalworking fluid became more important. While oils are excellent lubricants, their cooling capacity is limited. The use of oil-based metalworking fluids in higher-speed operations resulted in poor tool life and the generation of irritating smoke and fumes (13, 14).

In 1883, Taylor demonstrated the importance of water as a base for metalworking fluids (15). He showed that by using water as a metalworking fluid, cutting speeds could be increased by 30–40%. The primary reason for this increase in speeds is the excellent cooling properties of water relative to hydrocarbon oils. Water has a higher specific heat and heat of vaporization than hydrocarbon oils, accounting for its superior cooling ability (16).

By the end of the nineteenth century, the use of water to flood the cutting tool was quite common (12). However, as a metalworking fluid, water has two obvious drawbacks. First, water is a poor lubricant. Second, the use of water leads to the corrosion of both the tool and the workpiece (17).

The first attempts to solve these shortcomings occurred at the turn of the century. Sodium carbonate, or soda ash, was added to inhibit corrosion, and phosphates and soaps were included for lubricity enhancement (12, 14). However, over time the soda ash had a tendency to drop out of solution, depositing on machinery and thereby preventing smooth operation (12). Also, about the same time, mixtures of oils and water, loosely coupled by alkali, were being used to improve the performance of water as a metalworking fluid. Unfortunately, these systems did not adequately control corrosion and were inherently unstable (12).

Good-quality soluble oils were first developed in the 1920s (12, 14). They quickly became the dominant form of water-based metalworking fluid, although solutions containing sodium carbonate continued to be used in grinding operations through the 1950s (12, 14). Soluble oils effectively take advantage of the cooling properties of water while providing good lubricity and corrosion protection. The use of soluble oils instead of neat hydrocarbon-based metalworking fluids enabled manufacturers to increase their production rates without significant sacrifices in surface finish or tool or die life.

The use of soluble oils grew quickly at the expense of neat hydrocarbon metalworking fluids during the middle of this century. Besides being better coolants, soluble oils offer a number of other advantages. Soluble oils are cleaner than straight oil metalworking fluids. They do not fume or smoke significantly, and they pose a greatly reduced risk as a fire hazard (7). Also, it is easier to remove soluble oil residues from the machinery and the workpiece.

Another major factor that led to the growth of soluble oils is their versatility. The lubrication and cooling properties of soluble oil metalworking fluids can be adjusted by simply altering the dilution ratio or changing the type or amount of lubricity additives. In heavy-duty forming operations the soluble oil concentrate can be formulated with high concentrations of chlorine-, sulfur-, or phosphorus-containing extreme pressure (EP) lubricity additives and diluted only 1:1 or 1:2 with water prior to use (7). However, soluble oils are also used in high-speed, light-duty cutting and grinding operations after being diluted 20:1 to 100:1 with water. These light-duty products are often formulated only with rust inhibitors.

A further advantage of soluble oils relative to neat hydrocarbon-based metalworking fluids is cost. Although the price of a soluble oil concentrate is more expensive than an equivalent volume of straight oil metalworking fluid, the actual cost is significantly reduced due to dilution with water. Also, soluble oils are less prone to losses from dragout than straight oils, which results in lower fluid makeup costs (14).

Despite their obvious advantages, soluble oils did present some new problems. The most significant drawback associated with soluble oils is their susceptibility to bacterial attack. Microorganisms living in the water quickly learn to metabolize the emulsified organic compounds. They rapidly multiply, releasing unpleasant odors, and eventually splitting the emulsions by metabolizing the surfactants and lowering the fluid's pH (12). While the incorporation of biocides can reduce the magnitude of this deficiency, biological attack is still the biggest problem faced by the users of soluble oils.

Soluble oils presented several other problems. Soluble oils tend to emulsify tramp gear lubricants or hydraulic fluids, which can adversely affect the performance of the metalworking lubricant. Also, soluble-oil metalworking fluids are sensitive to water quality, especially pH and hardness (18).

In the early 1950s the users of grinding solutions containing soda ash began to find substitutes that provided good corrosion protection without forming troublesome deposits (12). These alternative corrosion inhibitors were based mostly on alkali nitrites, organic amines, or amine neutralized organic acids (12, 14, 19). These new solutions, referred to as chemical or synthetic metalworking fluids because they contained no petroleum oil, were quite successful. They provided good corrosion protection, excellent cooling, and adequate lubricity. They were also clean, very stable, and more resistant to biological attack than soluble-oil metalworking fluids (5).

Because of their initial success as grinding coolants, and the continuing search for cutting and forming lubricants that were more trouble-free than soluble oils, chemical metalworking fluids were tried in a number of more demanding cutting operations such as turning, milling, and tapping (14). Their performance in these operations was promising enough to encourage further development. The inclusion of water-soluble lubricity additives such as fatty acids, phosphate acid esters (20), and polyalkylene glycols significantly enhanced the lubricity of chemical metalworking fluids. Chemical metalworking fluids were soon being used in many cutting operations as well as applications involving light- and medium-duty drawing and forming (14, 19).

Chemical metalworking fluids offered a number of advantages over traditional soluble oils. These advantages, which still exist today, are summarized below.

Biological Stability. While biocides are still needed for optimum performance, chemical metalworking fluids are generally more resistant to biological attack than soluble oils (5). This increased biological stability results in less odor, better pH control, and longer sump life (5, 12). Longer fluid life means reduced raw material and disposal costs.

Solution Stability. Chemical metalworking fluids can be formulated to have good hard-water stability and be resistant to pH variations (18). They are also less prone to drag out than soluble oils, resulting in less coolant use and cleaner chips or swarf (12).

Cooling. Chemical metalworking fluids are significantly better coolants than soluble oils (5, 21). Better cooling means longer tool life and higher production speeds.

Cleanliness. The clarity of chemical metalworking fluids is very good, resulting in excellent workpiece visibility (5, 14). Not only are these products clear initially, but their ability to reject tramp gear and hydraulic oils helps them remain transparent (18). Chemical metalworking fluids also produce less oily residues on the machines and floor, thereby greatly reducing the risk of a fall (14).

However, chemical metalworking fluids were not a panacea. They presented the metalworking industry with the following problems:

Slideway Lubrication. Many chemical metalworking fluids, because of their good detergency, have a tendency to wash away slideway lubricants. This results in significant operational problems. Also, some chemical metalworking fluids can, over time, cause the machine's moving parts to stick. This sticking can be the result of water evaporating from some isolated coolant, leaving behind the more viscous corrosion inhibitors and lubricity additives. However, it can also be due to the deposition of additives that react with the calcium and magnesium ions in hard water to form insoluble residues (14). Gummy residues can also be formed when fluids containing amine borate corrosion inhibitors are used in hard water (22).

Over the years it has been found that slideway lubrication problems can be reduced by reformulating the concentrate with lower-viscosity components that have good hard-water stability and less detergency. Also, the use of detergent-resistant slideway lubricants and a more comprehensive machine lubrication program can help alleviate these problems (12).

Machine Lubrication. Upon switching from soluble oils to chemical metalworking fluids, some operators discover that their machinery is no longer being adequately lubricated. Because chemical metalworking lubricants by definition contain no hydrocarbons, they do not leave a residual oil film on the machine's moving parts (5). This problem can often be solved by following a routine machine lubrication program (12).

Dermatitis. Because of their good detergency and high pH, chemical metalworking fluids can wash away the skin's natural oils, resulting in dermatitis (8). The use of barrier creams or rubber gloves, along with the practice of good industrial hygiene, can help overcome this problem.

Paint and Seal Compatibility. Some of the components in chemical coolants can attack the seals and paints used in metalworking machinery. Resistant paints and seals are now available and can be found in much of the newer machinery (12).

Lubricity and Corrosion. While insufficient lubricity and corrosion protection limited the use of early chemical metalworking fluids, the development of improved water soluble additives has greatly expanded the utility of these products. They can now be employed in many of the applications that are currently using soluble oils.

The problems associated with chemical metalworking fluids, especially those pertaining to machine and slideway lubrication, led to the development of semichemical metalworking fluids. These products typically contain up to 20% hydrocarbon oil in their concentrate as well as surfactants, amines, and other water-soluble additives. Semichemical fluids were well established by the mid 1960s (5), and they are currently enjoying an increase in popularity. However, while they strive to incorporate the better machine lubrication of soluble oils, they also lose some of the bacterial resistance inherent in chemical metalworking fluids (12). In short, semichemical products, as their name suggests, are a compromise between soluble oils and chemical metalworking fluids.

By 1971, 33% of the metalworking fluids sold into cutting and grinding applications were soluble oils, chemical fluids, or semichemical products (23). However, these fluids are sold as concentrates, which, on average, are diluted about 20 times with water. Thus, approximately 90% of all metal removal operations were using water based metalworking fluids (24).

Both chemical and soluble-oil metalworking fluids have undergone significant formulation changes over the past 20 years due to toxicity concerns.

In the late 1960s the presence of nitrosamines in chemical metalworking fluids was discovered. Nitrosamines can be formed by reactions between nitrites and monoethanolamine or diethanolamine. Particular attention was given to the discovery of *N*-nitrosodiethanolamine (NDELA) in metalworking fluids. NDELA is a potent carcinogen to rats and hamsters (25), and it is known to absorb through human skin (26). Because of concerns about the exposure of machinists to NDELA, much work was done during the mid 1970s to find replacements for sodium nitrite (22). The best substitutes were found to be amine borates or the salts of organic acids and amines. By 1980 the use of nitrites in metalworking fluids had been essentially eliminated.

In the mid 1980s, oils containing polynuclear aromatics were identified as potential carcinogens. As a result, practically all oil-containing metalworking fluids were reformulated. These products now incorporate severely hydrotreated hydrocarbons as lubricant base stocks (27).

Toxicity continues to be a major issue affecting the use of the various types of metalworking fluids. The other factors currently shaping the metalworking fluid industry are waste minimization, disposal, environmental regulations, and workpiece quality.

III. SYNTHETIC LUBRICANTS

Synthetic lubricants are defined as products that are made from the controlled combination of discrete compounds. This distinguishes them from refined petroleum products. There are several major classes of synthetic lubricants that are used in metalworking operations. These lubricants are described below.

A. Polyalkylene Glycols

This diverse class of synthetic lubricants is made from the polymerization of alkylene oxides as shown in Fig. 21.1. The starter is usually an alkyl alcohol or a diol, but branched polymers are also made by alkoxylating triols or polyols. The oxide feed can be all ethylene oxide ($R' = H$), all propylene oxide ($R' = CH_3$), or mixtures of the two.

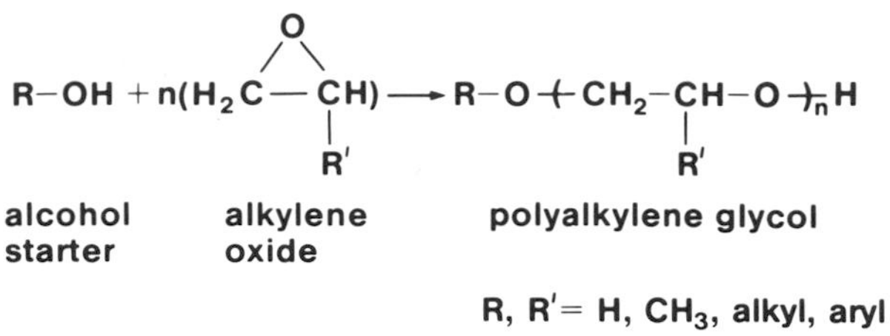

$$R{-}OH + n(H_2C{-}CH(R')\text{ [epoxide, O bridging]}) \longrightarrow R{-}O{-}\!\left(CH_2{-}CH(R'){-}O\right)_n{-}H$$

alcohol starter — alkylene oxide — polyalkylene glycol

R, R′ = H, CH_3, alkyl, aryl

Figure 21.1 Polyalkylene glycols.

Higher-molecular-weight monomers like butylene oxide can also be polymerized, but the commercial use of these products is very limited. The oxide feeds can be either mixed, resulting in a random polyalkylene glycol, or sequential, yielding a blocked structure. Blocked polyalkylene glycols are more surface-active than the random polymers, but they are also more likely to foam.

Polyalkylene glycols (PAGs) can be either soluble or insoluble in water, depending on the ratio of ethylene oxide to propylene oxide in the monomer feed. At room temperature, polyalkylene glycols made from a monomer feed consisting of more than approximately 20% ethylene oxide are water-soluble. Polyalkylene glycols that contain less than 20% polymerized ethylene oxide show good solubility in hydrocarbons.

Water-soluble polyalkylene glycols have been used extensively in chemical and semichemical metalworking fluids (24, 28–30). Oil-soluble polyalkylene glycols have been employed in both semichemical and straight-oil metalworking fluids as lubricity additives (31). They have also been used neat in applications where their excellent water washability, resistance to staining or sludge formation, or clean burn-off characteristics are important (32–35).

B. Esters

Esters are made from the condensation of organic acids and alcohols. This is a very diverse class of compounds, and the physical properties vary greatly depending on the acids and alcohols used. The most common classes of esters are monobasic acid esters, dibasic acid esters, polyol esters, and polyalkylene glycol esters (PAG esters). The generic structures of these compounds are shown in Fig. 21.2.

Monobasic acid esters and polyalkylene glycol esters are widely used in the metalworking industry. Monobasic acid esters such as methyl and butyl stearate and hexyl laurate have long been employed as lubricity additives in metalworking fluids (2, 32). In hydrocarbon-based products, these esters help reduce the interfacial tension between the oil and the metal surface, thereby increasing the ability of the fluid to penetrate between the workpiece and the tool or die (5). In water-based products, monobasic acid esters are most commonly used as additives in the hydrocarbon phase of soluble-oil or semichemical metalworking fluids, where they serve as boundary lubricants and corrosion inhibitors.

Polyalkylene glycol esters are used in chemical or semichemical metalworking fluids because of their solubility or good dispersity in water (36, 37). They are usually made from polyalkylene glycol diols, but monol and polyol started polyalkylene glycols can also be used. Polyalkylene glycol esters provide good lubricity and are synergistic with many corrosion inhibitors.

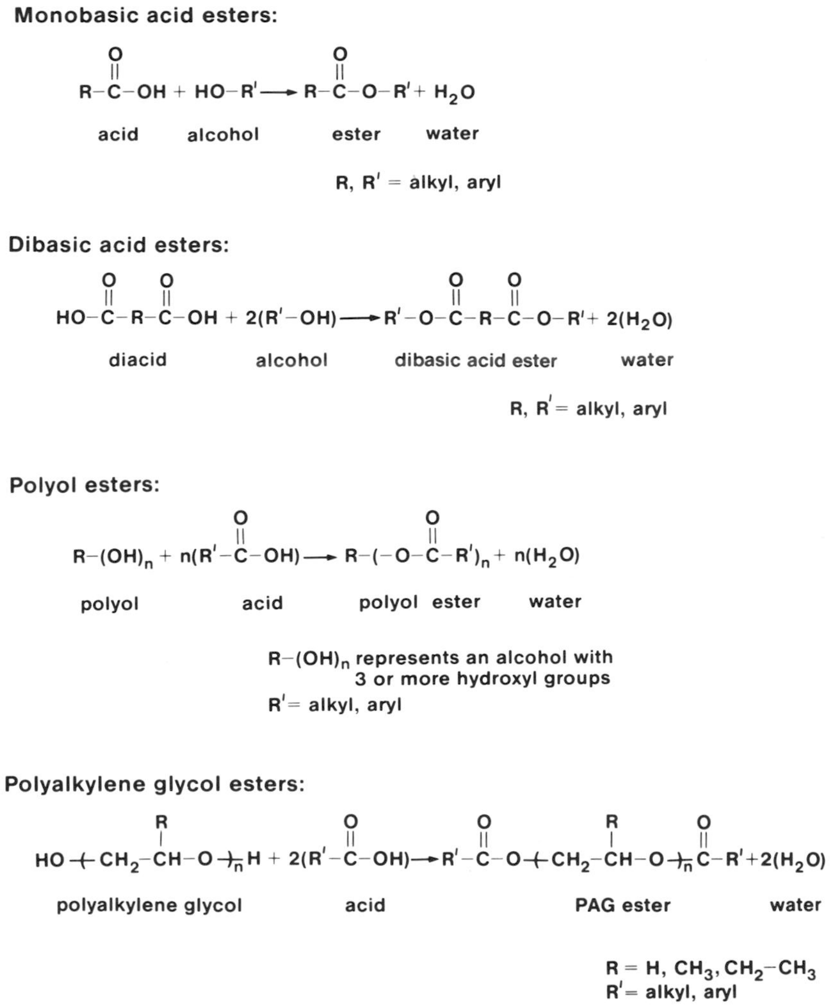

Monobasic acid esters:

$$R{-}\overset{O}{\overset{\|}{C}}{-}OH + HO{-}R' \longrightarrow R{-}\overset{O}{\overset{\|}{C}}{-}O{-}R' + H_2O$$

acid alcohol ester water

R, R′ = alkyl, aryl

Dibasic acid esters:

$$HO{-}\overset{O}{\overset{\|}{C}}{-}R{-}\overset{O}{\overset{\|}{C}}{-}OH + 2(R'{-}OH) \longrightarrow R'{-}O{-}\overset{O}{\overset{\|}{C}}{-}R{-}\overset{O}{\overset{\|}{C}}{-}O{-}R' + 2(H_2O)$$

diacid alcohol dibasic acid ester water

R, R′ = alkyl, aryl

Polyol esters:

$$R{-}(OH)_n + n(R'{-}\overset{O}{\overset{\|}{C}}{-}OH) \longrightarrow R{-}({-}O{-}\overset{O}{\overset{\|}{C}}{-}R')_n + n(H_2O)$$

polyol acid polyol ester water

$R{-}(OH)_n$ represents an alcohol with 3 or more hydroxyl groups
R′ = alkyl, aryl

Polyalkylene glycol esters:

$$HO{+}CH_2{-}\overset{R}{\overset{|}{C}H}{-}O{+}_nH + 2(R'{-}\overset{O}{\overset{\|}{C}}{-}OH) \longrightarrow R'{-}\overset{O}{\overset{\|}{C}}{-}O{+}CH_2{-}\overset{R}{\overset{|}{C}H}{-}O{+}_n\overset{O}{\overset{\|}{C}}{-}R' + 2(H_2O)$$

polyalkylene glycol acid PAG ester water

R = H, CH_3, $CH_2{-}CH_3$
R′ = alkyl, aryl

Figure 21.2 Structures of some esters that are used as synthetic lubricants.

C. Phosphate Esters

Phosphate esters are formed by reacting alcohols with phosphoric acid as shown in Fig. 21.3. Phosphate esters are used as synthetic lubricants in fire-resistant hydraulic fluids because of their resistance to burning (38). However, these synthetic lubricant base stocks have not found any significant use in metalworking applications.

Phosphate esters are widely used in metalworking fluids as EP lubricity additives (2, 5, 6, 32, 39, 40). Oil-soluble phosphate esters like tricresyl phosphate are well-known lubricity additives and are commonly used in straight-oil and soluble-oil metalworking fluids. Water-soluble phosphate esters can be made by reacting ethoxylated alcohols with phosphoric acid. These products are often used as lubricity additives in chemical metalworking fluids (6, 30).

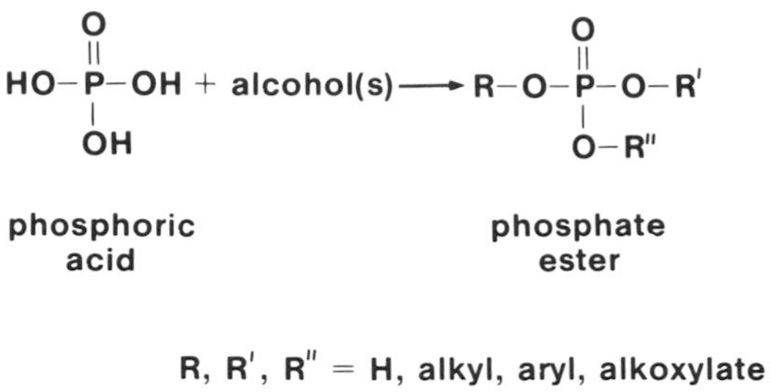

$$HO-\underset{\displaystyle OH}{\overset{\displaystyle O}{\overset{\|}{\underset{|}{P}}}}-OH + \text{alcohol(s)} \longrightarrow R-O-\underset{\displaystyle O-R''}{\overset{\displaystyle O}{\overset{\|}{\underset{|}{P}}}}-O-R'$$

phosphoric acid — phosphate ester

R, R′, R″ = H, alkyl, aryl, alkoxylate

Figure 21.3 Phosphate esters.

D. Synthetic Hydrocarbons

Synthetic hydrocarbons are made from the combination of olefins through their double bonds. The most common synthetic hydrocarbons are polyalphaolefins, dialkyl benzenes, and polyisobutylenes. The generic structures of these three types of synthetic hydrocarbons are shown in Fig. 21.4.

Of these three classes of synthetic hydrocarbons, only polyisobutylenes have found

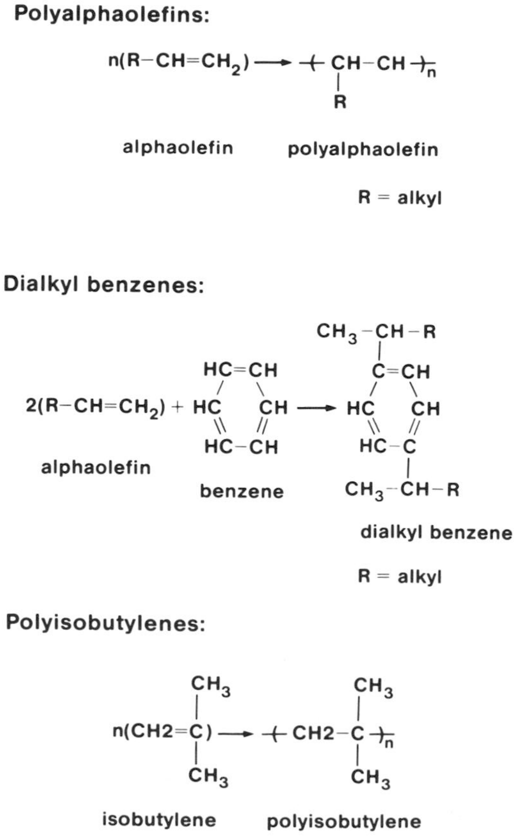

Polyalphaolefins:

$$n(R-CH=CH_2) \longrightarrow -\!\!\!\left[\underset{\displaystyle R}{\underset{|}{CH}}-CH\right]\!\!\!-_n$$

alphaolefin — polyalphaolefin

R = alkyl

Dialkyl benzenes:

$$2(R-CH=CH_2) + C_6H_6 \longrightarrow (CH_3-CH(R))_2C_6H_4$$

alphaolefin — benzene — dialkyl benzene

R = alkyl

Polyisobutylenes:

$$n(CH2=\underset{\displaystyle CH_3}{\overset{\displaystyle CH_3}{\overset{|}{\underset{|}{C}}}}) \longrightarrow -\!\!\!\left[CH2-\underset{\displaystyle CH_3}{\overset{\displaystyle CH_3}{\overset{|}{\underset{|}{C}}}}\right]\!\!\!-_n$$

isobutylene — polyisobutylene

Figure 21.4 Structures of some common synthetic hydrocarbons.

significant use in the metalworking industry. Their primary applications have been as nonstaining thickeners (2), or as base oils for high-temperature forming operations (32).

E. Other Synthetic Lubricants

Other synthetic lubricants and polymers have found use in metalworking operations. Polyacrylonitriles (41), styrene–maleic anhydride copolymers (42, 43), and polyacrylamides (44) have all been documented as providing good lubricity in water-based metalworking fluids. Polyvinylpyrrolidones, polyvinyl alcohols, and copolymers of acrylic acid or methacrylic acid and an acrylic ester have also been cited as lubricity additives for use in chemical metalworking fluids (45). However, the commercial use of these synthetic polymers is small. Also, a number of thermoplastic polymers have been used as solid lubricants in forming operations. A review of these polymers is presented by Shey (2). However, the use of solid lubricants is not covered in this chapter.

V. SYNTHETIC LUBRICANTS IN CHEMICAL METALWORKING FLUIDS

The most significant use of synthetic lubricants in the metalworking industry is in chemical solutions. The use of polyalkylene glycols in conjunction with water-soluble boundary or EP lubricity additives can significantly enhance the lubricity of chemical metalworking fluids and allow them to effectively compete with soluble oils in a variety of cutting and forming operations.

A. Polyalkylene Glycols in Chemical Metalworking Fluids

Polyalkylene glycols have a number of characteristics that make them ideal for use in chemical metalworking fluids. These properties include (24):

Water Solubility. Polyalkylene glycols can be synthesized that are soluble in water in all proportions. They are hydrolytically stable and are essentially unaffected by water quality or hardness.

Inverse Solubility. Polyalkylene glycols exhibit a property that is referred to as "inverse solubility." This property is characteristic of ethoxylated materials. As the temperature of an aqueous polyalkylene glycol solution increases, the solubility of the polyalkylene glycol in water decreases. Above a temperature known as the cloud point, the polyalkylene glycol will come out of solution, forming a hazy or cloudy dispersion. A 1% solution of a polyalkylene glycol below and above its cloud point is shown in Fig. 21.5.

The cloud point of a polyalkylene glycol is dependent on a number of factors. These include the polymer's ethylene oxide to propylene oxide ratio, starter, end groups, and molecular weight. It is also affected by the polyalkylene glycol concentration and the presence of other water-soluble compounds. In particular, the cloud point of polyalkylene glycols can be lowered by the presence of ionizing salts. The high-temperature insolubility of polyalkylene glycols can be used to greatly enhance the lubricating ability of chemical metalworking fluids.

Nonionic Behavior. Because polyalkylene glycols are nonionic, they can be used in combination with either cationic or anionic additives.

Low Reactivity. Polyalkylene glycols are noncorrosive to commonly used metals.

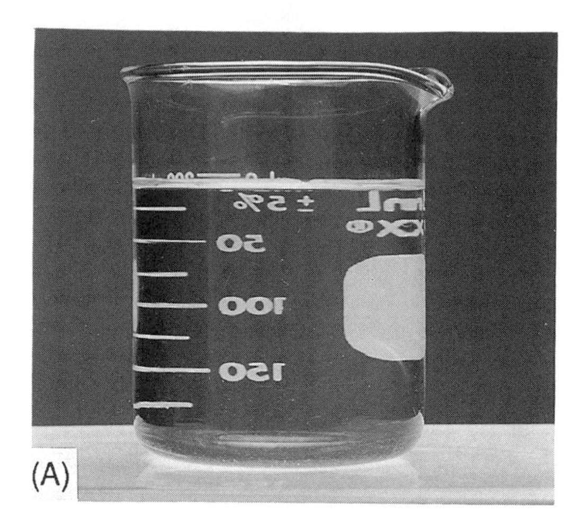

Figure 21.5 The inverse solubility (cloud point) of a 1% solution of polyalkylene glycol polymer in water: (A) PAG solution at ambient temperature, (B) insoluble PAG at elevated temperature.

Liquid Residues. Because polyalkylene glycols are stable and water-soluble, their residues are liquid and, if necessary, easily removed from machinery (24).

Resistant to Biological Attack. While polyalkylene glycols are biodegradable, their degradation rates are slow relative to those of fatty acids, phosphate esters, and many commonly used surfactants. This bioresistance leads to easier fluid maintenance and longer sump life and thus lower raw material and disposal costs.

Low Toxicity. Polyalkylene glycols exhibit low toxicity. However, as with all chemicals, the Material Safety Data Sheets should be studied before the product is used.

A typical formulation for a metalworking fluid employing a polyalkylene glycol is shown in Table 21.1. This type of formulation is usually diluted 10:1 to 30:1 with water for typical cutting operations. It can also be used at much higher concentrations. For severe drawing operations, dilutions of 1:1 or 2:1 with water might be used. For light grinding, dilutions up to 50:1 have been employed.

The purpose of the polyalkylene glycol is to provide hydrodynamic lubricity and

Table 21.1 Typical Chemical Metalworking Fluid Formulation

Component	Weight %
Polyalkylene glycol	10–15
Phosphate ester (or fatty acid)	5–10
Sulfurized fatty acid	0–5
Nonnitrite corrosion inhibitor	15
Triethanolamine	10–15
Biocide	[a]
Water	45–60

[a]Biocide added at concentrations recommended by the manufacturer.

inverse solubility. The fatty acid or phosphate ester provides enhanced wetting and boundary lubrication. The phosphate ester also has EP lubricating properties. In severe metalworking operations, a water-soluble sulfur-containing compound may be necessary for added EP lubricity. The nonnitrite inhibitor is to protect both the machine and the workpiece from corrosion. Triethanolamine serves as an inexpensive corrosion inhibitor by keeping the pH of the solution elevated, typically between 8.5 and 9.2. The ethanolamine also helps to solubilize some of the lubricity additives. Biological attack is retarded through the use of biocides. Other additives such as antifoams, dyes, and chelating agents can be added if needed.

The key to the effectiveness of polyalkylene glycols in chemical metalworking fluids is their inverse solubility. The following mechanism has been used to describe the role of the inverse solubility of polyalkylene glycols in metalworking fluids (2, 24, 30, 46). This mechanism, is illustrated in Fig. 21.6.

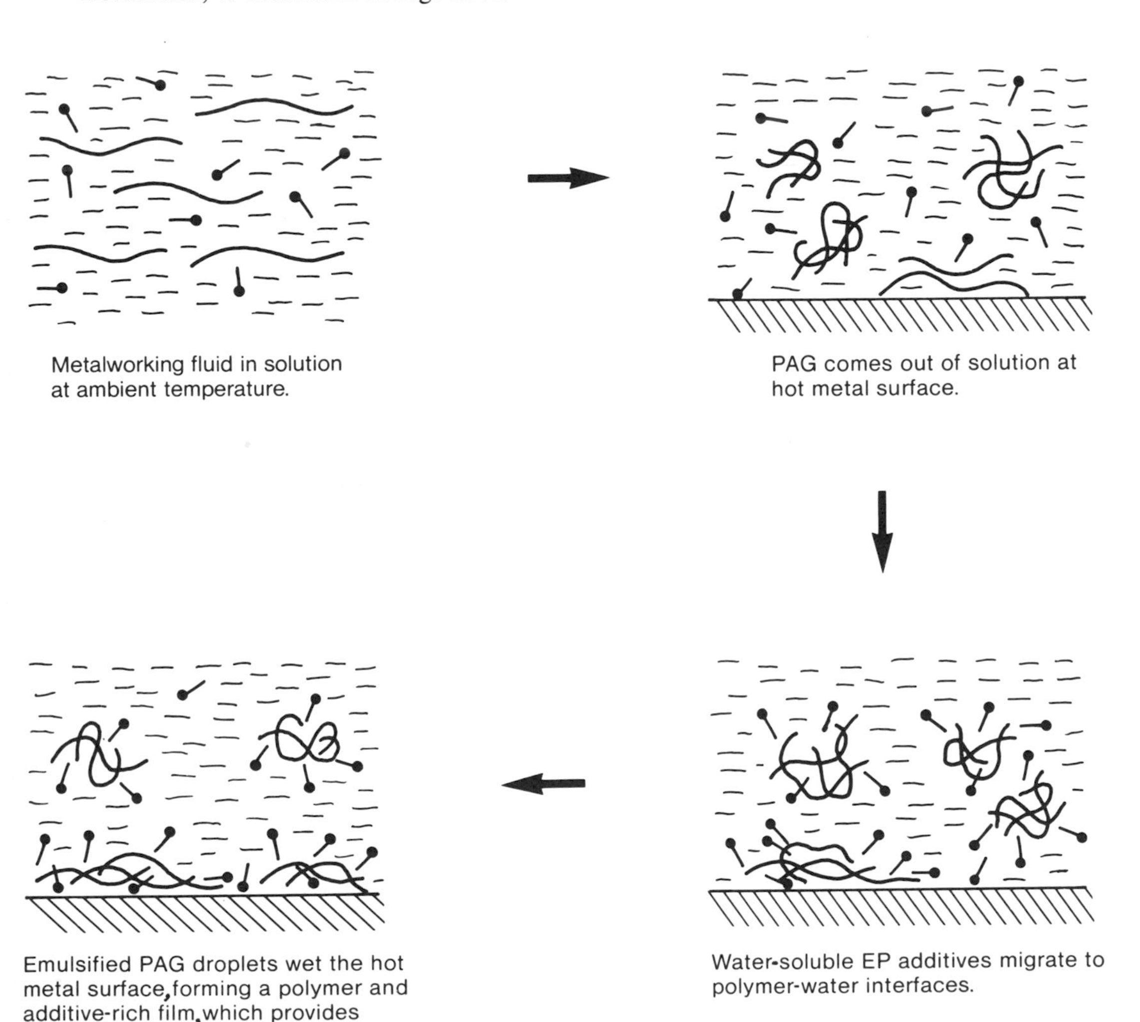

Metalworking fluid in solution at ambient temperature.
PAG comes out of solution at hot metal surface.
Water-soluble EP additives migrate to polymer-water interfaces.
Emulsified PAG droplets wet the hot metal surface, forming a polymer and additive-rich film, which provides excellent lubricity.

Figure 21.6 The inverse solubility (cloud point) mechanism of chemical metalworking fluids containing polyalkylene glycols.

Cloud-Point Mechanism of Chemical Metalworking Fluids that contain Polyalkylene Glycols.

1. At ambient temperatures, typical of those found in a machine sump, the metalworking fluid is completely soluble in water, forming a clear, transparent solution.
2. The metalworking fluid is brought into contact with the hot tool or die. The temperature of the fluid is elevated above the cloud point of the polyalkylene glycol, causing the polymer to come out of solution and form small, oil-like drops.
3. The lubricity additives in the metalworking fluid are surface-active and collect at the interface between the polyalkylene glycol and the water phases.
4. These polyalkylene glycol drops then preferentially wet the surfaces of the workpiece and the tool or die, much like the emulsified hydrocarbon droplets in a soluble-oil metalworking fluid. The result is the formation of a thin layer of concentrated polyalkylene glycol and lubricity additives. This polyalkylene glycol film has excellent lubricating properties. The ability of the polyalkylene glycol to concentrate the polar lubricity additives at the tool or die surface helps explain the synergy often noted between these products (6, 24, 30).
5. The spent chips and excess metalworking fluid fall back into the relatively cool machine sump, where the polyalkylene glycol goes back into solution. In contrast, the chips generated while using soluble oils are coated with a significant amount of hydrocarbon lubricant, which does not reemulsify. As a result, less lubricant is lost due to dragout when a chemical metalworking fluid is used.

Chemical metalworking fluids that do not contain polyalkylene glycols rely on their water-soluble additives coming in contact with the surface of the workpiece in order to achieve sufficient lubricity. The probability of this occurring is a function of the concentration of these lubricity additives. The addition of polyalkylene glycols, because of their inverse solubility, serves to concentrate the lubricity additives at the point of cut, or, in the case of a drawing or forming operations, at the hot surface of the die. Also, the polyalkylene glycol itself is an excellent hydrodynamic lubricant. Better lubricity can therefore be achieved at the same product dilution ratio when polyalkylene glycols are used in conjunction with water-soluble boundary or EP additives.

While the presence of polyalkylene glycols improves the efficiency of lubricant delivery to the hot surfaces of the workpiece and the tool or die, it is important to remember that the mechanism of these chemical metalworking fluids is temperature activated. At ambient temperatures the lubricity of chemical metalworking fluids that contain polyalkylene glycols is greatly reduced. Because they contain no oil, these fluids do not coat the relatively cool machine parts with a film of lubricant. This can lead to problems on machines that rely on the metalworking fluid for slideway or other mechanical lubrication. The incorporation of a small amount of emulsified oil to form a semichemical metalworking fluid is currently a popular solution to this problem. The adoption of a more rigorous machine lubrication program is also effective in some cases.

Since elevated temperatures are required in order for the polyalkylene glycol to come out of solution and coat the hot die surface, forming lubricants containing these polymers can have trouble during start-ups when the dies are cold. This problem is more severe when softer, nonferrous metals are being deformed. In these cases, unacceptable amounts of the softer workpiece material can transfer onto the die before operating temperatures can exceed the cloud point of the polyalkylene glycol. These cold start-up problems can sometimes be overcome by initially increasing the concentration of the metalworking fluid to the point where it provides sufficient lubricity at ambient temperatures. The incorporation of a polymeric thickener and the emulsification of small amounts of mineral oil are other possible solutions. Work is also being done to see if preheating the polyalkylene glycol containing metalworking fluid during the start-up can solve this problem.

B. Laboratory Studies on Polyalkylene Glycols in Chemical Metalworking Fluids

There are many lubricity tests used to evaluate metalworking fluids in the laboratory. The pin and V-block, four-ball, and Timken wear tests are all commonly used. However, much better correlations with actual metalworking operations are obtained when fluids are evaluated under cutting or deformation conditions.

Three studies undertaken to examine the role of polyalkylene glycols in chemical metalworking fluids are presented in this section. The first utilizes a plain strain compression test in which a lubricated aluminum coupon is plastically deformed. The remaining two studies used an instrumented lathe to evaluate different metalworking fluids while cutting a section of steel pipe. These studies provide strong evidence that supports the cloud-point mechanism of polyalkylene glycol-based chemical metalworking fluids that was described earlier in this chapter.

1. Plain Strain Compression Test: The Effect of PAG Cloud Point on the Performance of Chemical Metalworking Fluids

A plain strain compression press (47) was used to verify the cloud-point mechanism of chemical metalworking fluids containing polyalkylene glycols. The press consists of two rectangular male dies, their faces measuring 0.25 by 1.5 in. The dies were attached to a 10-ton hydraulic press. Each die was equipped with thermostatically controlled heating elements so that the temperature of the dies' surfaces could be controlled.

A 1-in-wide coupon made from a 0.16-cm-thick sheet of 5252 aluminum, H28 temper, was thoroughly cleaned. A bead of lubricant was then placed on each side of the coupon. The lubricated coupon was then placed between the male dies of the plain strain compression press as shown in Fig. 21.7 (30). The dies were then forced together under a predetermined pressure that was great enough to cause plastic deformation of the coupon. The thickness of the coupon after deformation was measured and the percent reduction was calculated using the equation

$$\% \text{ Reduction} = [1 - (\text{final thickness}/\text{initial thickness})] \times 100$$

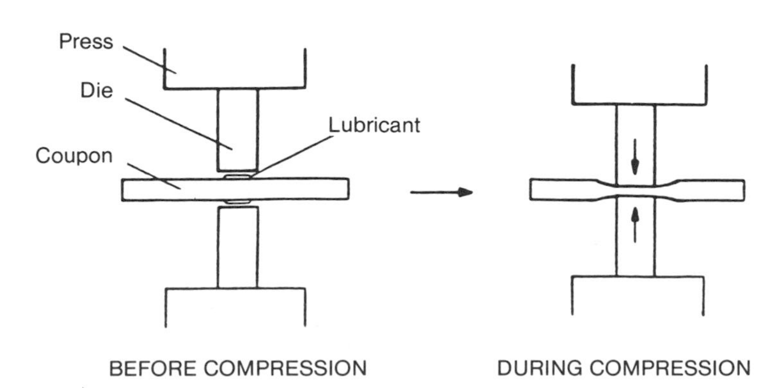

Figure 21.7 The deformation of an aluminum coupon in a plane strain compression press (30).

The greater the percent reduction, the better the lubricant.

Two chemical metalworking fluid concentrates were made up using the formulation shown in Table 21.2. The only difference between concentrates A and B was the polyalkylene glycol used. The polyalkylene glycol used in concentrate A had a 1% cloud point of 37°C, while the polymer in concentrate B had an inverse solubility temperature of 65°C. Concentrates A and B were tested for lubricity using the plain strain compression press, as were 5% solutions of each in water. Fluid C, a kerosene-based aluminum cold rolling fluid, was also tested.

In this study a lubricated aluminum coupon was placed between the dies. The dies were then brought together and held for 5 s at a pressure of 78,000 psi. The dies were then released and the percent reduction of the coupon determined. The lubricity of each fluid was tested at die temperatures ranging from 20 to 95°C. The results are plotted in Fig. 21.8 (30).

The effect of the cloud point of the polyalkylene glycol is evident from the results shown in Fig. 21.8. At all temperatures, concentrates A and B were excellent lubricants, giving percent reduction values of between 80 and 90%. This was to be expected, since the viscosities of concentrates A and B were quite high, 110 and 67 cSt (40°C), respectively. The slight improvement of lubricity with increasing temperature may have been due to the evaporation of water from the concentrate or from a mild cloud-point effect.

At low temperatures, below the cloud point of either polymer, the lubricity of the 5% aqueous solutions of concentrates A and B was quite low. The percent reductions achieved using these dilute solutions of concentrates A and B were 50 and 45%, respectively. Between 30 and 50°C, there was a sudden rise in the measured percent reductions of solution A, indicating a significant improvement in lubricity. This improvement in lubricity was centered around the cloud point of the product's polyalkylene glycol (37°C). Above 50°C, the 5% solution of fluid A provided lubricity equal to that of the concentrate.

The lubricity of the 5% solution of concentrate B showed the same behavior as the dilute concentrate A, only the increase in lubricity occurred between 60 and 75°C. As with the 5% solution of concentrate A, this improvement in lubricity bracketed the cloud point of the polyalkylene glycol used in concentrate B. Above 75°C, the dilute solution of fluid B provided the same lubricity as the concentrate.

The kerosene-based rolling lubricant showed poor lubricity at all temperatures, primarily because of its low viscosity of 4 cSt (40°C). What is important is that its lubricity decreased with increasing temperature. This is because the product's viscosity decreases as the temperature rises. While a more viscous hydrocarbon lubricant would provide much better lubricity, its performance would also be expected to decrease with increasing temperature due to a drop in viscosity.

Table 21.2 Metalworking Fluid Concentrate for Plane Strain Compression Study

Component	Weight %
Polyalkylene glycol	40
Fatty acid	10
Ethanolamine	20
Water	30

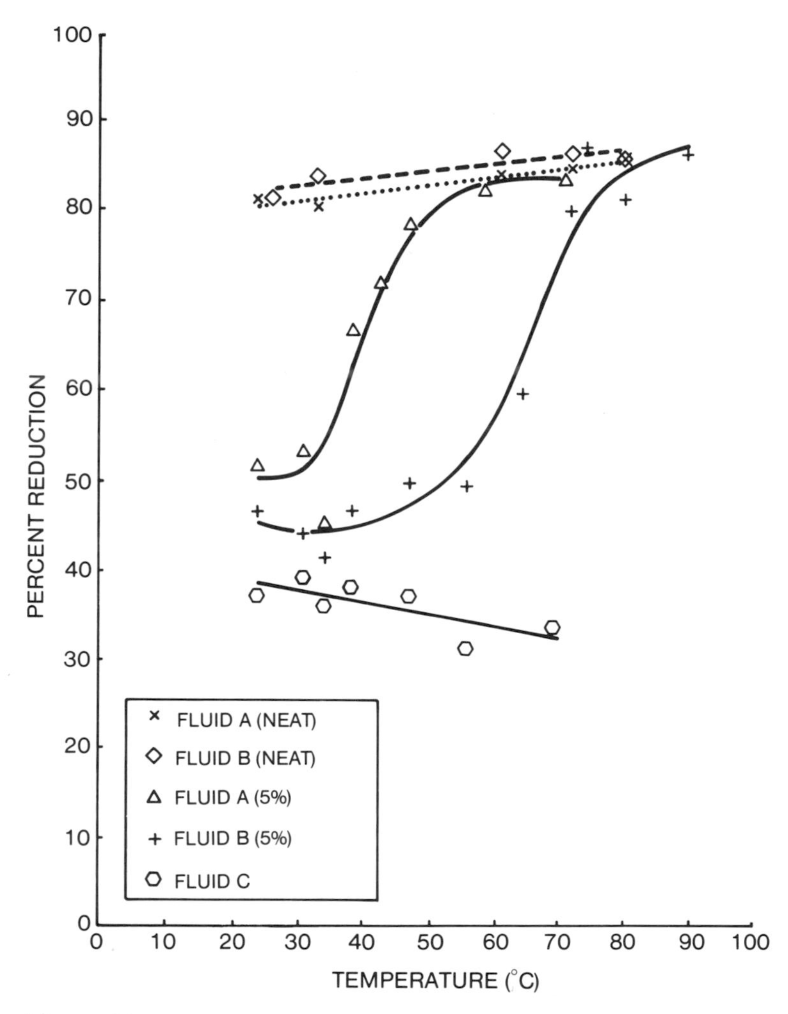

Figure 21.8 A demonstration of the cloud-point mechanism of chemical metalworking fluids containing polyalkylene glycols using a plane strain compression press (30).

The results of this test support the cloud-point mechanism of chemical metalworking fluids that contain polyalkylene glycols. It should be noted that this was a static test. When compared to a high-speed cutting operation, there was a relatively long time in this test for the polyalkylene glycol to come out of solution and form a lubricant film on the surfaces of the dies. However, while straight-oil lubricity is not achieved in many cutting and forming operations, the improvement in performance due to the use of polyalkylene glycols in chemical metalworking fluids is well documented (6, 24, 30, 46).

2. Lathe Test: The Effect of PAG Cloud Point on the Performance of Chemical Metalworking Fluids

A similar study was performed using a 16-in Loge & Shipley single-point turning lathe. The lathe used in this study was instrumented with a strain gauge and a thermocouple. The thermocouple was inserted in the cutting tool. The tool holder was mounted on the strain gauge. A piece of 1026 steel pipe was end-cut while the metalworking fluid under evaluation was directed onto the rake face of the tool. The tool temperature and the vertical cutting force were monitored throughout the duration of the test. Both cutting force (2, 48, 49) and tool temperature (2, 5, 13, 50) have been shown to correlate with tool life. The lower the tool temperature and the cutting force, the more effective the metalworking fluid.

To show the effect of the inverse solubility of the polyalkylene glycol in a chemical metalworking fluid, five concentrates were made up differing only in the cloud point of the polyalkylene glycol used (51). The concentrates tested are listed in Table 21.3. Each concentrate was diluted 20:1 with water and then tested under cutting conditions 1–5, with condition 5 being the most severe.

The results of this test are shown in Table 21.4. A good correlation exists between decreasing cloud point and enhanced performance. Fluids A and B, with cloud points of 37 and 52°C, respectively, cut effectively under all five conditions. Fluid C, which contains a polyalkylene glycol with a cloud point of approximately 95°C, failed under condition 5. Fluids D and E, whose cloud points were both greater than 100°C, failed at condition 4. Fluid F, which contained no polyalkylene glycol, failed at condition 3. The results of this work demonstrate the cloud-point mechanism of polyalkylene glycol containing chemical metalworking fluids in an actual metal-removal operation.

Table 21.3 Metalworking Fluid Concentrates for Cloud-Point Study Using a Lathe[a]

Sample	%TEA	%H20	%FA	%PAG	PAG MW	PAG viscosity (cSt, 40°C)	PAG cloud point 1% solution, °C)
A	18	32	10	40	2700	400	37
B	18	32	10	40	2700	400	52
C	18	32	10	40	2700	400	95
D	18	32	10	40	2500	280	>100
E	18	32	10	40	2000	[b]	>>100
F	18	72	10	0	—	—	—

[a]TEA, triethanolamine; FA, fatty acid; MW, number-average molecular weight.
[b]Product is a solid at room temperature due to high ethylene oxide content.

Table 21.4 Lathe Test Results for Cloud-Point Study

Sample	Condition 1		Condition 2		Condition 3		Condition 4		Condition 5	
	T	F	T	F	T	F	T	F	T	F
A	94	490	100	540	104	550	114	610	122	680
B	92	500	100	540	104	570	116	640	126	710
C	100	500	102	540	108	540	108	640	Failure	
D	96	500	102	540	108	570	Failure			
E	96	510	106	540	102	580	Failure			
F	100	520	186	548	Failure					

[a]T, Cutting tool temperature in degrees Celsius; F, vertical cutting force in pounds-force; failure indicates either severe vibration or tool breakage.

3. Lathe Test: The Synergy between PAGs and Fatty Acids or Phosphate Esters

The instrumented lathe was used to demonstrate the synergy that exists between polyalkylene glycols and polar, water-soluble lubricity additives (30). Five concentrates were made up so that the sum of the polyalkylene glycol and the lubricity additive was equal to 40% by weight. The percentage of polymer ranged from 40 to 0 while the percentage of lubricity additive went from 0 to 40. The concentrates, shown in Table 21.5, were then diluted 20:1 with water and evaluated on the lathe.

Two different studies were run. In the first, the lubricity additive was a fatty acid. The results from this study are shown in Fig. 21.9. The minimum cutting force was achieved with polyalkylene glycol levels of between 20 and 30%. Maximum cooling, represented by the minimum tool temperature, occurred with between 10 and 20% polymer in the concentrate. These curves show that combinations of the polyalkylene glycol and fatty acid perform better than either additive by itself.

In the second study, a water-soluble phosphate acid ester was employed as the EP lubricity additive. The data from this study are shown in Fig. 21.10. In this case, minimum cutting force was achieved with polyalkylene glycol levels of between 10 and 20% in the concentrate. The minimum tool temperature occurred between 0 and 30% polymer. Again, combinations of polyalkylene glycol and lubricity additive performed better than either additive used individually. This synergy is further evidence supporting the cloud-point mechanism of chemical metalworking fluids containing polyalkylene glycols.

C. Modified PAGs in Chemical Metalworking Fluids

Polyalkylene glycols provide good lubricating properties in a large number of metalworking operations. They are also very stable in hard water, are low foaming, and are resistant to biological attack. When used in combination with additives such as fatty acids or phosphate esters, the lubricating and cooling properties of the metalworking fluid are enhanced. Unfortunately, both fatty acids and phosphate esters are sensitive to hard water, prone to foaming, and susceptible to biodegradation. In an attempt to achieve the benefits of these combinations without the drawbacks, two types of modified polyalkylene glycols have been commercialized. These modified polyalkylene glycols are either esterified or grafted with organic acids.

Polyalkylene glycol esters are made from the condensation reaction between organic acids and the terminal hydroxyl groups of the alkoxylated polymer. These products, like

Table 21.5 Metalworking Fluid Concentrates used in the Lathe Synergy Study

% PAG	% Lubricity additive	% TEA	% Water
40	0	40	20
30	10	40	20
20	20	40	20
10	30	40	20
0	40	40	20

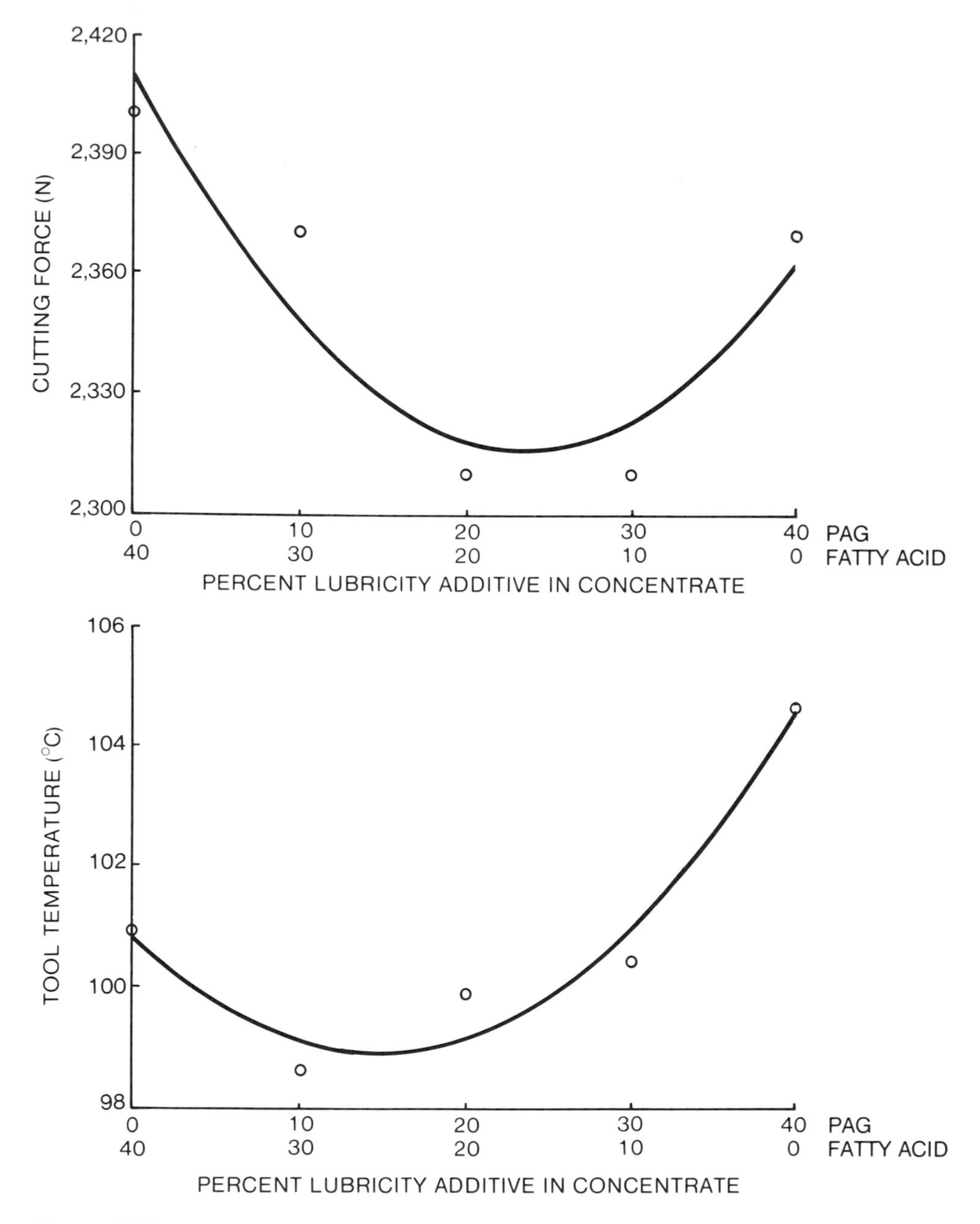

Figure 21.9 A demonstration of the synergy between polyalkylene glycols and fatty acids in chemical metalworking fluids using an instrumented lathe (30).

polyalkylene glycols, exhibit inverse solubility in water and thus behave in an analogous manner. Polyalkylene glycol esters exhibit good boundary lubricating properties (36, 52), yet have better hard-water stability and are less likely to foam than blends of fatty acids and unmodified polyalkylene glycols (22).

A second method of modifying polyalkylene glycols that has gained commercial

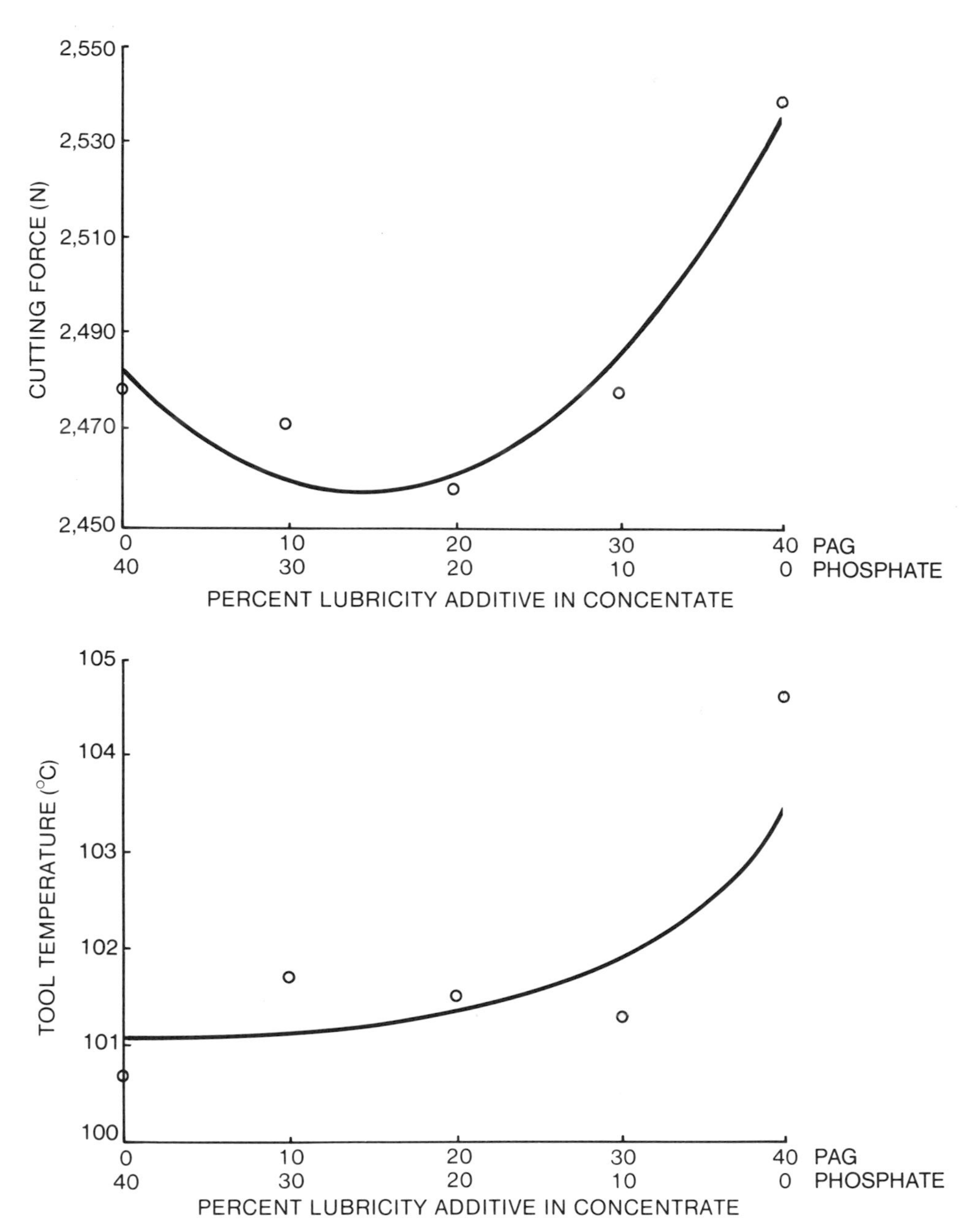

Figure 21.10 A demonstration of the synergy between polyalkylene glycols and phosphate esters in chemical metalworking fluids using an instrumented lathe (30).

acceptance is the addition of organic acid functionalities through grafting technology. The result is an anionic polyalkylene glycol that has organic acid groups randomly attached to the polymer's backbone with hydrolytically stable covalent bonds. These modified polyalkylene glycol polymers exhibit inverse solubility and excellent hydrodynamic and boundary lubricity. Because the acid groups are randomly attached to the polymer, the foaming tendencies of this type of product are significantly lower than those of blends of fatty acids and polyalkylene glycols. Since the polyalkylene glycol itself is very water-soluble at room temperature, the grafted polymer has excellent hard-water stability. Its resistance to biological attack is also very good (53).

D. Applications of PAGs in Chemical Metalworking Fluids

Polyalkylene glycols have been used in chemical metalworking fluids for the past 40 years. Because of their good lubricity and synergy with other water-soluble lubricity additives, they have helped expand the applications of chemical metalworking fluids.

Chemical metalworking fluids containing polyalkylene glycols have all of the advantages characteristic of this product class. These advantages include excellent cooling, good biological and hard-water stability, transparency, and cleanliness. The presence of the polyalkylene glycols in chemical metalworking fluids enables the formulation of products with enhanced lubricity properties, allowing them to compete directly with heavy-duty soluble oils in many applications.

While polyalkylene glycols or modified versions have been used in chemical metalworking fluids for many years (19, 24, 37), performance data comparing these products to other types of metalworking fluids are relatively scarce. Articles comparing different classes of metalworking fluids to each other, like chemical solutions to soluble oils or soluble oils to straight oils, are readily found in the literature. However, the specific formulation information needed to examine the effects of synthetic lubricants in these products is rarely included. The rest of this section reviews some of the documented work that demonstrates the effect of polyalkylene glycols, both normal and modified, in chemical metalworking fluids. The use of polyalkylene glycol-based metalworking fluids in grinding, tapping, hobbing, rolling, drawing, and forming operations is described.

1. Grinding

Levesque and McCabe published a paper in 1984 describing their experience with an acid grafted polyalkylene glycol in three different grinding operations (53).

In the first case a mineral seal oil fortified with approximately 5% of a fatty acid was used to grind hardened steel balls. Odor, general housekeeping problems, and a concern about the flammability of the oil caused them to switch to a water-based metalworking fluid.

A chemical metalworking fluid based on a mixture of caprylic acid and a polyalkylene glycol was selected for the ball-grinding operation. Initially the product worked very well. During use, however, the effectiveness of this fluid gradually decreased. The addition of concentrate was necessary to return the coolant to acceptable performance levels. Analytical testing of the coolant during use showed that the drop in performance was due to the selective depletion or biodegradation of the caprylic acid. Rather than continually monitor their fluid for acid concentration, the authors switched to a dilute solution of an amine-neutralized, acid-grafted polyalkylene glycol. This modified polyalkylene glycol performed well.

The second case study involved a centerless roller grinding operation. A heavily fortified soluble oil was used in a 7,000-gal central sump. The problem with the soluble oil was that it covered the regulatory wheels with a layer of oil and metal fines, which prevented the achievement of the desired tolerances. The authors decided to switch to a chemical metalworking fluid because of cleanliness of this type of product.

The soluble oil was replaced with a chemical metalworking fluid containing a fatty acid ester, a boric acid corrosion inhibitor, and a biocide. Initially the product worked well, but within a month the surface finishes degraded to an unacceptable level. The depletion of the fatty ester was determined to be the problem.

The authors then switched this system to the amine-neutralized, acid-grafted polyalkylene glycol fortified with a carboxylic acid-based corrosion inhibitor. At the time the paper was written the roller grinding operation had been using this chemical metalworking fluids for 16 months with essentially no problems due depletion or biodegradation.

In the third study conducted by the authors, the same chemical metalworking fluid based on the acid grafted polyalkylene glycol was used in a double-disk surface grinding operation. The life of this coolant system was more than 8 months, compared with only 2–3 months with the previously used soluble oil.

2. Tapping

In 1984 Nash and Colakovic published a study on the affect of a polyalkylene glycol ester and three other lubricity additives on the performance of chemical metalworking fluids during the tapping of high-silicone-content aluminum blanks (52).

The fluids were evaluated using a tapping torque test machine (54). The taps employed were HSS, 3-flute, 10–15 mm. The tap surface speed was 0.508 m/s. The nut blanks were made from high silicone A380.1 aluminum with a 16–25 mms non-heat-treated surface finish.

The chemical metalworking fluids were diluted to 10% of their original concentration with 140 ppm hardness water. The torque required to tap the nut blanks was recorded and compared to that achieved when using a straight-oil metalworking fluid. This control oil was a neat cutting fluid fortified with chlorine- and sulfur-containing lubricity additives. The percent tapping efficiency was then calculated using the equation

$$\% \text{ Efficiency} = (\text{control oil torque/test fluid torque}) \times 100$$

The higher the percent tapping efficiency, the better the metalworking fluid.

A series of chemical metalworking fluids was made up to determine the effects of four water-soluble lubricity additives on tapping efficiency. The additives tested were a polyalkylene glycol ester that exhibits inverse solubility, an alkyl acid phosphate, a sulfurized oleic acid, and a chlorinated oleic acid. Each fluid also contained a carboxylate salt for corrosion protection and triethanolamine to solubilize the lubricity additives and provide reserve alkalinity. The physical characteristics of the components used to make up the test concentrates are shown in Table 21.6.

Table 21.6 Physical Characteristics of Fluid Components

Component	Kinematic Viscosity (cSt) 40°C	Kinematic Viscosity (cSt) 100°C	Acid Number	Miscellaneous Properties
Water	—	—	—	140 ppm Hardness
Carboxylate salt	300	—	170	140 Base number
TEA	222	14	—	
PAG ester	2,750	230	15	1% Cloud point = 77°C
Alkyl acid phosphate	95	18	330	11% Phosphorus
Sulfurized oleic acid	190	20	185	8.5% Sulfur
Chlorinated oleic acid	150	22	140	30% Chlorine

The compositions of the different fluids tested and their average percent tapping efficiency are shown in Table 21.7. It can be seen from these data that the largest positive effects of a single component come from the addition of the polyalkylene glycol ester or the alkyl acid phosphate. Furthermore, the combination of the PAG ester and phosphate was synergistic, providing lower tapping torques and thus better efficiency than the sulfurized, chlorinated reference oil. The addition of the sulfurized oleic acid resulted in a small improvement over the base formulation and also showed synergy when combined with the polyalkylene glycol ester. The use of the chlorinated oleic acid had no significant affect on tapping efficiency.

3. Hobbing

Katsuki et al. described work that they had done to evaluate the performance of water-based metalworking fluids in a gear-hobbing operation (55). In this study the durability of the hob was evaluated using a fly-tool cutting test on a milling machine. This fly-tool cutting evaluation was set up to correlate closely with an actual gear-hobbing operation.

The metalworking fluid concentrates were diluted with water. During the cutting operation, the face of the hob was flooded with the diluted metalworking fluid. Grooves were cut into the workpiece to correspond to the manufacture of 14.7 gears. The corner and center wear of the hob were then evaluated. The smaller the wear scars were, the more effective the metalworking fluid.

Four water-soluble polyalkylene glycols of different molecular weights were evaluated and compared to a chlorinated fatty oil that was considered an excellent gear-cutting lubricant. The polyalkylene glycols were all diluted 10 times with water and then compared to the oil-based standard. The polyalkylene glycols evaluated are characterized in Table 21.8. The results are shown in Table 21.9.

It can be seen from the data in Table 21.9 that the center wear of the chlorinated fatty oil control is lower than that of any of the polyalkylene glycols. The differences are more extreme at the intermediate cutting speeds of 86 and 117 m/min. At these intermediate speeds, increasing the molecular weight of the polyalkylene glycol tends to decrease the center wear. However, at cutting speeds of 62 and 159 m/min, there was essentially no difference seen among the performances of the different molecular weight polymers.

The corner wear experienced when using the polyalkylene glycols was equivalent to

Table 21.7 Test Fluid Concentrate Compositions and Aluminum Tapping Torque Efficiency

	Solution Concentrates															
Component	Base	1	2	3	4	5	6	7	8	9	10	11	12	13	14	15
Water	84	74	80	70	80	70	76	66	80	70	76	66	76	66	72	62
Carboxylate salt	8	8	8	8	8	8	8	8	8	8	8	8	8	8	8	8
TEA	8	8	8	8	8	8	8	8	8	8	8	8	8	8	8	8
PAG ester	—	10	—	10	—	10	—	10	—	10	—	10	—	10	—	10
Phosphate	—	—	4	4	—	—	4	4	—	—	4	4	—	—	4	4
Sulfurized acid	—	—	—	—	4	4	4	4	—	—	—	—	4	4	4	4
Chlorinated acid	—	—	—	—	—	—	—	—	4	4	4	4	4	4	4	4
Percent efficiency, mean value	79	91	90	101	84	97	92	101	80	81	97	99	88	91	84	91

Table 21.8 Polyalkylene Glycols Evaluated in Hobbing Study

Sample	MW	Viscosity (cSt at 30°C)	VI	Diluted Viscosity (cSt at 40°C)
PAG-1	1,675	205	203	1.29
PAG-2	2,430	276	210	1.41
PAG-3	4,750	1,590	290	2.16
PAG-4	11,800	27,500	427	5.26

that of the chlorinated fatty oil at a cutting speed of 62 m/min. As cutting speed increased, however, the corner wear seen when using the oil-based control rose much more quickly than with the polyalkylene glycols. There was no significant affect of polyalkylene glycol molecular weight on corner wear.

A second study was performed to determine the affect of polyalkylene glycol concentration on hobbing performance. Polymer PAG-4 was tested over a wide range of dilutions and compared to water and the chlorinated fatty oil control at a cutting speed of 159 m/min. The results are shown in Table 21.10.

From this study it can be seen that even at very low concentrations the presence of the polyalkylene glycol yields very low corner wear relative to water. At concentrations equal to or greater than 0.62%, the use of the PAG-4 solution also results in lower corner wear than the chlorinated fatty oil control and equivalent center wear. It is interesting to note that while corner wear decreased with increasing polyalkylene glycol concentration, center wear increased.

The authors of this study found that while the use of the polyalkylene glycols provided superior performance as measured by lower wear, the corrosion protection provided by these solutions was insufficient. It was found that the addition of rust preventatives could significantly improve the corrosion protection provided by the polyalkylene glycol. However, the addition of these corrosion inhibitors did cause hob wear to increase at higher cutting speeds.

4. Machining and Rolling

A number of chemical metalworking fluid formulations based on polyalkylene glycols and mixtures of these polymers with polyvinylpyrrolidone, polyvinyl alcohol, polyacrylates, and polymethacrylates were patented by Marx (45). Several of these formulations and their applications are shown in Table 21.11. The formulations in this table were used successfully in a number of machining and sheet-rolling operations.

Table 21.9 Results from Hobbing Evaluation

Cutting speed	Center wear (mm)					Corner wear (mm)				
(m/min)	Control	PAG-1	PAG-2	PAG-3	PAG-4	Control	PAG-1	PAG-2	PAG-3	PAG-4
62	0.07	0.11	0.12	0.11	0.12	0.10	0.10	0.10	0.10	0.10
86	0.10	0.28	0.22	0.16	0.18	0.17	0.11	0.11	0.11	0.11
117	0.14	0.48	0.50	0.37	0.29	0.27	0.12	0.12	0.12	0.15
159	0.18	0.21	0.20	0.20	0.20	0.50	0.16	0.16	0.16	0.15

Table 21.10 Effect of Polyalkylene Dilution Ratio on Hobbing Performance

PAG-4 concentration (%)	Center wear (mm)			Corner wear (mm)		
	PAG-4	Water	Control oil	PAG-4	Water	Control oil
0.10	0.11	0.11	0.18	0.60	1.02	0.50
0.20	0.11			0.48		
0.33	0.12			0.60		
0.62	0.16			0.22		
1.25	0.17			0.21		
2.50	0.18			0.20		
5.00	0.16			0.11		
10.00	0.20			0.11		

Marx also described how the replacement of soluble oils with chemical metalworking fluids containing the aforementioned water-soluble polymers enabled cutting speeds to be increased in a number of sawing, planing, and drilling applications. These metalworking fluids were also successful at replacing straight oils in deep hole drilling operations. This resulted in significant cost savings.

Marx then relates how aqueous solutions containing 15–20% polyalkylene glycol or mixtures of polyalkylene glycols and some of the higher-molecular-weight polymers mentioned previously can be used for deep-drawing steel and stainless steel parts. These solutions can also be used for the drawing of wire and tubing. An advantage of these polymers in drawing operations is that their residues can be easily washed off with water.

Table 21.11 Formulations and Applications of Chemical Metalworking Fluids Containing Polyalkylene Glycols

	Concentrate formulation (%)			
Component	1	2	3	4
PAG	20	15	15	20
Polyvinylpyrrolidone	—	5	5	—
Polyvinyl alcohol	—	—	—	4
Amine phosphate	—	—	—	—
CI[a]	—	—	46	46
TEA	6	6	—	—
Water	74	74	24	30
Dilution ratio	20:1	20:1	20:1	10:1
Application	Steel sheet rolling, brass and copper sheet formation	Rolling of thin foils (Cu, Sn, Au, or brass)	Steel tapping, rod and channel formation	Steel planing, milling, and cutting

[a]CI is a corrosion inhibitor package containing 16 parts of benzoic acid, 9 parts of TEA (triethanolamine), 15 parts of triethanolamine phosphate, and 6 parts of morpholine.

5. Blanking and Drawing

The early use of polyalkylene glycol solutions in blanking and drawing operations was documented by Sweatt and Langer in 1951 (37). The properties of the polyalkylene glycol used in these chemical metalworking fluids are shown in Table 21.12.

A 25% solution of this polyalkylene glycol in water with a small amount of corrosion inhibitor was used to replace a petroleum oil-based lubricant in a blanking and a drawing operation. The two applications are summarized in Table 21.13. While the good lubricity of the polyalkylene glycols is evident from the increased tool life, their cleanliness and water washability was also important.

6. Cold Forming

A chemical metalworking fluid formulation for use in cold forging operations was patented by Felton (46). The formulation is shown in Table 21.14. This metalworking fluid was used neat to successfully form $\frac{3}{4}$-in hexagonal nut blanks from a $\frac{3}{4}$-in rod of AISI 1038 steel. The nuts were formed at a rate of 2 blanks each second. This operation took five separate steps. The die used in each step was flooded with lubricant between hits. The trial was run without problems for 5 h, indicating good lubrication.

The finished blanks were bright and shiny when the polyalkylene glycol-based metalworking fluid was used, as opposed to the dull, scorched appearance that was achieved when employing a straight-oil forming lubricant. Also, the smoke generated during this operation was greatly reduced when the forming lubricant was switched from the oil to the polyalkylene glycol-based product.

Table 21.12 Polyalkylene Glycol Characteristics

Property	Value
Molecular weight	1,600
Viscosity	
40°C (cSt)	132
100°C (cSt)	26
Specific gravity (20/20°C)	1.05
Water solubility (25°C)	Complete
1% Cloud point (°C)[a]	50

[a]Inverse solubility temperature.

Table 21.13 Applications of Aqueous Polyalkylene Glycol Solutions

Operation	Workplace material	Metalworking fluid	Number of pieces per die refinishing
Blanking and pressing	Annealed spring steel	Oil	35,000– 50,000
		PAG solution[a]	100,000–120,000
Drawing shells	Nickel-plated steel	Oil	25,000– 30,000
		PAG solution[a]	>65,000

[a]Solutions contained 25% polyalkylene glycol.

Table 21.14 Cold-Forming Lubricant Formulation

Component	Volume %	Component Description
Polyalkylene glycol	32	Molecular weight = 2,200 40 wt% Ethylene oxide (EO) 60 wt% Propylene oxide (PO)
Sulfurized fatty acid	5	14 wt% Sulfur
Chlorinated fatty acid	5	35 wt% Chlorine
Glycerine	3	
Potassium nitrite	2	
Potassium hydroxide	1.75	
Silicone defoamer	0.10	
Water	51.15	

7. Drawing and Forming

The use of a polyalkylene glycol-based chemical metalworking fluid in a number of drawing and forming operations was described by Brown (56). The formulation used is shown in Table 21.15. This formulation was used at various dilutions with water depending upon the severity of the operation. Up to 5% of a sulfurized fatty acid was added for additional extreme pressure lubrication when needed.

The operations in which this formulation was used are summarized in Table 21.16. The first three operations are described in more detail in the following sections.

Trailer Hitches. Trailer hitches were being drawn from HRPO AKDQ C 10-08 (0.141–0.151 in thick) drawing-quality steel using an 800-ton press. Ten hitches were drawn each minute. Using a straight-oil metalworking lubricant that was fortified with a chlorinated paraffin, the dies had to be refinished every 15,000–20,000 parts. While the lubricity provided by the chlorinated oil was good, it was very difficult to completely remove the lubricant from the formed pieces. Also, the operators complained of skin irritation while using the chlorinated product.

The straight oil was therefore replaced by the polyalkylene glycol-based metalworking fluid shown in Table 21.10. This formulation was fortified with 5% of the sulfurized fatty acid and was used without further dilution with water. Die life remained unchanged at 15,000–20,000 hitches per refinishing. However, the polyalkylene glycol-based lubri-

Table 21.15 Polyalkylene Glycol-Based Drawing and Forming Lubricant

Component	Weight %
Polyalkylene glycol	15%
Phosphate acid ester	6%
Corrosion inhibitor (nonnitrite)	15%
Triethanolamine	10%
Sulfurized fatty acid	0–5%
Water	49–54%

Table 21.16 Applications of Polyalkylene Glycol-Based Forming Lubricant

Application	Process description	Dilution ratio, H_2O:Lube	Previous product	Observations
Trailer hitches	Severe draw	Neat	Chlorinated paraffin	Nonstaining, nonirritating, good die life
Water heater tops and bottoms	Blanking and one 2 in. draw	20:1	R&O sol oil[a] (dil. 20:1)	Water washable, good rust protection
Water-heater jacket tops	Cone spinning 40% elongation	20:1	Chlorinated sol oil (dil. 10–20:1)	Good cooling and lubrication
Oven liners	5-Stage operation (four 1/2-in. draw, one shear/punch)	2:1	Chlorinated sol oil (dil. 2:1)	Water washable, no staining
Lawnmower bodies	Blanking operation and one 4-in. draw	2:1	Chlorinated sol oil (dil. 2:1)	Excellent rust protection, good die life

[a]Sol oil, soluble oil metalworking fluid; dil., dilution.

cant provided excellent corrosion protection while the formed parts were stored, and could be easily removed with aqueous cleaning systems prior to subsequent operations. More importantly, the operators found the product to be benign.

Water-Heater Tops and Bottoms. A soluble-oil metalworking fluid formulated with rust inhibitors and diluted 20 times with water was being used to form water heater tops and bottoms. The operation involved a 4-in draw of 9- to 10-gauge mild cold-rolled steel. The pieces were produced at a rate of 10 per minute. The problem with the soluble oil was that it was very difficult to completely remove from the formed pieces. The presence of residual oil caused defects in the enamel coating, which in turn led to the premature corrosion of the water heater.

The chemical metalworking fluid shown in Table 21.15 with no sulfurized fatty acid was diluted 20:1 with water and used to replace the soluble oil. The polyalkylene glycol-based metalworking fluid provided equivalent lubricity when compared to the soluble oil. More importantly, this chemical metalworking fluid was easily washed off of the formed parts, virtually eliminating defects in the enamel coatings.

Cone Spinning of Water-Heater Jacket Tops. Tops for water-heater jackets were made in a cone-spinning operation. A 16-in-diameter disk made from mild cold-rolled steel was blanked in a separate step. During the cone-spinning operation, this disk was spun at 1,500–1,700 rpm and made to undergo a 40% elongation. The wheel of the cone-spinning machine was flooded with a 20:1 dilution of a chlorinated soluble oil. This metalworking fluid provided insufficient cooling and lubricity, which resulted in the discoloration of the workpiece.

The chlorinated soluble oil was replaced with the chemical metalworking fluid shown in Table 21.15 fortified with the sulfurized fatty acid. This product was used as a 5%

solution in water. The lubricity and cooling properties of this product were excellent. Also, this chemical metalworking fluid provided the spun workpiece with corrosion protection during storage for more than 60 days.

To summarize, polyalkylene glycol-fortified chemical metalworking fluids are effective coolants and lubricants in a wide range of metal removal and deformation operations. They have replaced both soluble oils and straight-oil metalworking fluids. The polyalkylene glycol-containing products have all of the beneficial properties of chemical metalworking fluids including excellent cooling, ease of maintenance, and cleanliness. In addition, they exhibit enhanced lubricity, which enables them to compete with heavy-duty soluble oils and, in some cases, straight-oil metalworking fluids. The properties of polyalkylene-based chemical metalworking fluids make them especially desirable in applications where workpiece staining or water solubility are important.

VI. SYNTHETIC LUBRICANTS IN SEMICHEMICAL METALWORKING FLUIDS

Semichemical metalworking fluids are, as their name implies, a hybrid of soluble oils and chemical solutions. Their main advantage is that they are cleaner and better coolants than soluble oils but still contain emulsified hydrocarbons, which provide good corrosion protection and lubricity to both the tool or die and the machinery.

A typical formulation for a semichemical metalworking fluid concentrate is shown in Table 21.17. This concentrate will then be diluted 20:1 to 40:1 with water for most cutting or grinding operations. For more severe metalworking operations, higher concentrations are employed.

Synthetic lubricants in semichemical metalworking fluids are used primarily as water-soluble lubricity additives. Canter et al. (36) described the use of polyalkylene glycol esters in semichemical metalworking fluid formulations. A number of esters were made up by reacting polyalkylene glycols with either one or two equivalents of a fatty methyl ester. The polyalkylene glycols used had ethylene oxide to propylene oxide ratios of 5:1, 3:1, and 1:1. The fatty groups evaluated were pelargonate, laurate, oleate, and stearate. Many of these compounds displayed inverse solubility properties in water.

The polyalkylene glycol esters were incorporated into the base formulation shown in

Table 21.17 Typical Semichemical Concentrate Formulation

Component	Weight %
Mineral oil	5–20
Emulsifiers	5–20
Couplers	0–5
Corrosion inhibitors	5–10
EP additives	0–10
Water-soluble lubricity additives	0–20
Biocides	[a]
Water	40–70

[a]As recommended by the manufacturer.

Table 21.18. The oil to emulsifier base ratios were adjusted to enable the formation of a stable microemulsion for each polyalkylene glycol ester tested. The reference semichemical fluid formulation is also shown in Table 21.18.

All of the semichemical metalworking fluid formulations containing polyalkylene glycol esters provided significantly better lubricity on a pin and V-block wear test at a 10:1 dilution with water than the reference fluid at the same concentration. Two of the formulations containing dioleate esters as well as the product made with the dipelargonate ester were also low-foaming and provided better corrosion protection than the reference fluid.

Another growing use for polyalkylene glycols in semichemical metalworking fluids is as coupling agents to help solubilize the corrosion inhibitor packages (34).

In theory, synthetic hydrocarbons like polyalphaolefins could be used in place of the emulsified mineral oil in a semichemical formulation. The cost of this substitution usually outweighs the benefits in most applications. However, there are a small number of cases where such formulations are marketed for use in operations where no mineral oils are allowed yet an emulsified lubricant phase is still desired.

VII. SYNTHETIC LUBRICANTS IN SOLUBLE-OIL METALWORKING FLUIDS

Soluble-oil metalworking fluids still make up a majority of the water-based lubricants used today. They are well accepted, and the presence of a hydrocarbon oil provides many operators and machinists with a significant degree of comfort. A typical soluble-oil concentrate formulation is shown in Table 21.19.

The use of synthetic lubricants in soluble-oil metalworking fluids is very small. Water-soluble synthetic lubricants are not used in soluble-oil formulations. Synthetic hydrocarbons or esters could be emulsified along with or in place of the mineral oil, but the cost of such substitutions usually outweighs the benefits. However, as with semichemical metalworking fluids, there are specialty applications where synthetic hydrocarbons are emulsified to make "synthetic" soluble oils. This is usually done to satisfy customers who require oil-free metalworking fluids but still need an emulsified product. Such formulations can also be used in applications where the workpiece is susceptible to staining.

Another application for synthetic lubricants is in rolling oils for steel. These products

Table 21.18 Semichemical Test and Reference Formulations

Component	Test formulation (% component)	Reference fluid (% component)
Naphthenic oil	3–13	8
Emulsifier base	9–20	12
PAG ester	10	—
Amine-borate	6	6
Triazine	2	2
Propylene glycol methyl ether	1.5	1.5
Water	60.5	70.5

Table 21.19 Typical Soluble-Oil Concentrate Formulation

Component	Weight %
Mineral oil	70–85
Emulsifiers	10–20
Coupling agents	1–5
Corrosion inhibitors	5–10
EP additives	0–10
Biocide	[a]
Water	0–5

[a]As recommended by the manufacturer

are usually soluble oils containing mineral oils and tallow fats. These natural fats are being replaced in some cases by synthetic esters made from the reaction between pentaerythritol or trimethylolpropane and C-12 to C-18 fatty acids (34).

VIII. SYNTHETIC LUBRICANTS IN STRAIGHT-OIL METALWORKING FLUIDS

Straight-oil metalworking lubricants continue to be the products of choice in a large number of metalworking operations, particularly those involving drawing and forming or low-speed, high-severity metal removal. A typical straight-oil formulation is shown in Table 21.20.

Straight-oil metalworking fluids can be formulated with synthetic lubricants instead of mineral oils. The predominant synthetic lubricants used in straight-oil metalworking fluids are polyalkylene glycols, polyisobutylenes, and esters. The major advantages of these products are their low staining and clean burn-off characteristics. Several applications of straight-oil metalworking fluids based on synthetic lubricants are described next.

The use of neat polyalkylene glycols in two drawing operations is described by Sweatt and Langer (37). Their work is summarized in Table 21.21. In both operations die life was significantly increased by switching from an oil-based product to a polyalkylene glycol. When using an oil-soluble polyalkylene glycol to draw 85-15 brass, they also found that there was a significant reduction in tarnish. More importantly, it was possible

Table 21.20 Typical Formulation of a Straight-Oil Metalworking Fluid

Component	Weight %
Mineral oil	75–100
Corrosion inhibitors	0–5
EP additives	5–20
Boundary lubricity additives	0–10
Antioxidants	0–2

Table 21.21 Applications of Neat Polyalkylene Glycols as Forming Lubricants

Operation	Workpiece material	Metalworking lubricant	Number of pieces per die refinishing
Drawing	85-15 Brass	Oil	5,000
		PAG (oil-soluble) (27 cSt, 40°C)	17,000
Drawing	Sheet iron	Oil	50
		PAG (H_2O-soluble) (130 cSt, 40°C)	250

to solder the brass pieces after drawing without first having to remove the lubricant. The use of a water-soluble polyalkylene glycol to draw sheet iron not only increased die life by a factor of five but also enabled the parts to be thoroughly cleaned using a water wash.

Similar observations were noted by J. M. Russ, Jr. (1951, private communication, Union Carbide Corp.), while making 3-in bubble caps. The caps were made from 0.064-in-thick copper sheet. The operation involved a 2.5-in draw of the copper blanks. The original lubricant was lard oil. The use of the lard oil led to sticking between the die and the workpiece and discoloration of the finished caps. The lard oil was replaced by an 80-cSt (40°C) water-soluble polyalkylene glycol. The use of this synthetic lubricant eliminated workpiece sticking and discoloration and also improved the surface finish of the drawn caps.

The addition of oil-soluble polyalkylene glycols to aluminum sheet and foil rolling lubricants has also been explored. Whetzel et al. (31) described how the addition of polyalkylene glycols to light mineral oils resulted in an increase in fluid performance. The percent reduction in thickness achieved when rolling 1100 aluminum alloy under a rolling load of 5,500 lb per inch of strip width increased 10–15% when 4% polyalkylene glycol was added to the mineral oil base.

Another application for oil-soluble polyalkylene glycols is in vanishing oils. The polyalkylene glycols are dissolved in low-molecular-weight hydrocarbons having flash points of less than 140°F. The vanishing oil is then applied to the workpiece, where the volatile carrier evaporates, leaving a thin, uniform, polyalkylene glycol film. This film provides excellent lubricity and is easy to remove from workpiece. In many cases the workpiece does not have to be cleaned.

Polyalkylene glycols, polyisobutylenes, and alkyl benzenes are all finding use in wire drawing compounds as carriers of dispersed solid lubricants. The major advantage of these synthetic lubricants in wire drawing operations is their clean burn-off characteristics during annealing (32, 34).

Polyisobutylenes are also used as mineral oil thickeners in a wide variety of metalworking applications. Their high molecular weight, low staining characteristics, and tendency to volatilize completely at high temperatures without leaving varnishes make them well suited for this use (32).

There is little indication that straight oils containing synthetic lubricants are being used in cutting applications. The benefits associated with synthetic lubricants do not make up for the added cost in this segment of the metalworking industry. However, it is possible that polyalphaolefins could be used as cutting-oil base stocks in some specialty operations (34).

IX. MARKET OUTLOOK

A. Market Size

The total annual consumption of metalworking fluids in the United States is currently estimated to be between 90 and 130 million gallons. This includes straight oils and water-dilutable concentrates (57–59). Approximately 60–70% of these products are used in metal removal operations, while the remaining metalworking fluids and concentrates are employed in forming applications (58, 59).

It is estimated that 40% of the metalworking fluids sold are straight oils, while soluble oils, chemical, and semichemical fluid concentrates make up the remaining 60% (59). The water-dilutable metalworking fluids can be subdivided further. Approximately 60% of these products are soluble oils, while the remaining 40% are divided evenly between chemical and semichemical metalworking fluid concentrates (59). Taking the dilution of the soluble-oil, chemical, and semichemical concentrates into account, the total annual consumption of metalworking fluids in the United States has been estimateded to be 3.2 billion gallons (60).

It is very difficult to determine the number of pounds of synthetic lubricants that go into each of the four segments of metalworking fluids. There are two major reasons for this. First, formulators are hesitant to give out information regarding the amount of synthetic lubricants they use, in order to protect their formulation strategies. Second, the producers of synthetic lubricants do not know what percent of their products sold to formulators goes into metalworking fluids. This is because the formulators of metalworking fluids are also likely to compound and sell other products like hydraulic fluids, gear lubricants, quenchants, and compressor lubricants. All of these products can be formulated with synthetic lubricants, and it is therefore very difficult for the polymer suppliers to know in what applications their products are being used.

B. Future of Synthetic Lubricants in Metalworking Fluids

There are three major factors that will influence the shape of the metalworking fluid market over the next decade. The first involves waste minimization, disposal, and environmental impact. The second factor is workpiece quality. The third is metalworking fluid toxicity.

1. Waste Minimization and Disposal

The most important factor influencing the metalworking fluid market today is waste minimization and disposal. Waste disposal regulations are regional and may vary considerably between different municipalities. The appropriate choice of metalworking fluid may be greatly affected by these local regulations. The advantages and disadvantages of the four classes of metalworking fluids with respect to waste minimization and disposal are summarized next.

Straight-oil metalworking fluids are relatively easy to maintain. With the absence of water, and assuming the product does not contain large amounts of fatty compounds, bacterial activity is minimal when compared to water-based products. Once the useful life of the straight-oil metalworking fluid is over, the used product can be burned for fuel value or recycled (61). However, the disposal of straight-oil products can be made significantly more difficult by the presence of chlorinated paraffin lubricity additives. Also, stricter air quality standards are in some cases making it more difficult to use oil-based metalworking fluids because of mist formation, smoke generation, and the evolution of volatile hydrocarbons.

Of the water-based metalworking fluids, soluble oils are the most difficult to maintain. They are very susceptible to attack from microorganisms. They also tend to emulsify tramp oils and can be sensitive to water quality. Because they are two-phase systems, they are not always amenable to comnnonly used fluid maintenance techniques such as ion exchange, ultrafiltration, and centrifugation.

From a disposal point of view, soluble oils are relatively easy to treat. Soluble oils can be split rather easily into an oil phase, which can be incinerated or reclaimed, and a water phase. Often this water phase can be sent directly to the local POTW (publicly owned treatment works). However, as water regulations become increasingly strict, the number of cases where this aqueous phase does not meet disposal regulations is growing. In these cases, secondary treatment is required. Soluble oils that contain chlorinated paraffins are becoming more difficult to get rid of. It is becoming harder to incinerate chlorine-containing compounds, and local POTWs are starting to closely regulate the amount of chlorinated organic products that they will accept in a waste stream.

Chemical fluids are the easiest metalworking fluids to maintain. They are much more resistant to biological attack than soluble oils. Because they form true solutions in water, they are amenable to a wide variety of treatment techniques. They can be ion exchanged to keep the water hardness under control, centrifuged to remove metal fines and tramp oils, and filtered to remove solids and emulsified oils. Large central systems containing chemical metalworking fluids have been maintained for periods of several years.

Because chemical metalworking fluids are easier to maintain than soluble oils, they generally last longer. This significantly reduces the volume of waste metalworking fluid that is generated. However, because all of the components are water-soluble, removing the organic components from the water is difficult. Whether or not this is a problem depends on the local regulations. Often a POTW will accept a spent chemical metalworking fluid if it is found to be compatible with their treatment system. Sometimes a surcharge is levied.

Semichemical metalworking fluids fall somewhere between soluble oils and chemical products. They are usually easier to maintain than soluble oils. However, for disposal purposes, semichemical products often take on the worst characteristics of soluble oil and chemical metalworking fluids.

2. Workpiece Quality

Product quality is becoming extremely important. As result, tolerance of workpiece corrosion, staining, and coating defects is decreasing. Two major causes of staining and coating defects are the presence of corrosive additives in the metalworking fluid and the incomplete removal of the lubricant prior to the coating process.

Chlorinated hydrocarbons are one of the most commonly used EP lubricity additives in straight-oil and soluble-oil metalworking fluids. However, during storage, the residual chlorine can cause significant staining of the metal workpieces. Because of their staining tendencies and the fact that their disposal is becoming increasingly difficult, much work is underway to develop replacements. This work is providing significant opportunities for the use of synthetic lubricants in metalworking fluids.

The complete removal of residual metalworking fluids from the workpiece can be difficult when either straight-oil or soluble-oil products are used. Residual lubricant can prevent the adherence of coatings, like paints or enamels, causing unacceptable defects.

These problems are becoming more widespread as the use of vapor degreasers and solvent cleaning processes is coming under pressure for various environmental reasons. The need for water-washable metalworking fluids should result in the increased use of chemical and semichemical products. It will also favor the use of neat polyalkylene glycols in straight-oil applications because solvent cleaning operations can be omitted due to the polymer's water solubility. The complete burn-off characteristics of polyalkylene glycols, polyisobutylenes, and alkyl benzenes will become more important because they sometimes enable the elimination of a cleaning operation prior to various high-temperature operations.

As the need for improved workpiece quality grows, the production of defective parts will become unacceptable. The resistance to the higher cost of synthetic lubricants should therefore decrease as the price of seconds due to inadequate cleaning increases.

3. Toxicity

Health concerns will continue to have a strong effect on this industry due to the high exposure of the operator to the metalworking fluids. Chlorinated paraffins have been coming under pressure, but concerns seem to be decreasing as more toxicity testing is done on the products currently being used. It is difficult to foresee the next toxicity issue, but it too will have a major impact on future formulations and market segmentation.

X. CONCLUSIONS

Synthetic lubricants will continue to play a major role as water-soluble lubricity additives in chemical and semichemical metalworking fluids. The synthetic lubricants most commonly used in these metalworking fluids are polyalkylene glycols and their ester or acid derivatives. Metalworking fluids based on these products are excellent coolants and lubricants. They are in general easy to maintain, low in toxicity and environmental impact, nonstaining, and easy to remove from the finished workpiece. As environmental issues become more important, the use of this class of synthetic lubricants in metalworking fluids is likely to increase.

In straight-oil metalworking fluids, the use of synthetic lubricants is basically limited to specialty applications. Synthetic esters, polyisobutylenes, and polyalkylene glycols are all used in applications where clean burn-off or nonstaining characteristics are important.

In conclusion, the use of synthetic lubricants in metalworking fluids will grow. Increasingly strict environmental regulations affecting the workplace and air and water quality will favor the use of water-based products and therefore polyalkylene glycol lubricants. Environmental and disposal-related concerns will also reduce the use of solvent cleaning systems and chlorinated lubricity additives. Both of these factors should also favor the use of synthetic lubricants. The emphasis on product quality and the increasing cost of seconds will also increase the consumption of synthetic lubricants. As the costs associated with waste disposal, fluid maintenance, and workpiece quality all rise, the performance advantages of synthetic lubricants will outweight their higher initial costs and lead to their increased application in metalworking lubricants.

ACKNOWLEDGMENTS

I gratefully acknowledge the contributions and guidance that I received from Dr. Paul L. Matlock and Dr. Nye A. Clinton. I especially want to thank my wife Kathy for her support, advice, and technical input. I also want to thank my sons Robby and Tommy for their patience during the writing of this chapter.

REFERENCES

1. Shaw, M. C. (1957). *Metal Cutting Principles*, 3rd ed., M.I.T. Press, Cambridge, MA.
2. Schey, J. A. (1983). *Tribology in Metalworking*, American Society for Metals, Metals Park, OH.
3. Ellis, E. G. (1968). *Fundamentals of Lubrication*, Scientific Publications (G.B.) Ltd., Broseley, Shropshire.
4. Booser, R. E. (1988). *Handbook of Lubrication*, volume II, CRC Press, Boca Raton, FL.
5. Springborn, R. K. (1967). *Cutting and Grinding Fluids: Selection and Application*, American Society of Tool and Manufacturing Engineers, Dearborn, MI.
6. Kajdas, C. (1989). Additives for metalworking lubricants—A review, *Lub. Sci.*, **1**,385–409.
7. Robin, M. (1978). Forming with water base lubricants. *Manufacturing Eng.*, **81**,53–55.
8. Barber, S. J., and W. H. Millett (1974). Water: "New" metalworking solution?, *American Machinist*, **118**,95–100.
9. Evans, E. A. (1963). *Lubricating and Allied Oils*, Chapman & Hall Ltd., London.
10. Biringuccio, V. (1959). *The Pirotechnia of Vannoccio Biringuccio (1540)*, transl. by C. L. Smith and M. T. Gnudi, Basic Books, Inc., New York.
11. Northcott, W. H., (1868). *A Treatise on Lathes and Turning*, Longman, Green & Co., London.
12. Edwards, J., and E. Jones (1977). Synthetic cutting fluids, *Tribol. Int.*, 10,29–31.
13. Kelly, R. (1982). Synthetic can-drawing fluids for D & I operations, *Lub. Eng.*, **38**,675–680.
14. Morton, I. S. (1971). Water base cutting fluids still a ?, *Ind. Lub. Tribol.*, **23**,57–62.
15. Taylor, F. W. (1907). On the art of cutting metals. *Trans. ASME*, **28**,31–58.
16. Beaton, J., J. M. Tims, and R. Tourret (1964–1965). Function of metal-cutting fluids and their mode of action, *Proc. Inst. Mech. Eng.*, **170**,193–214.
17. Sluhan, C. A. (1963). Cutting fluids, *Am. Soc. Tool Manuf. Eng.*, **62**, paper no. 399.
18. Thornhill, F. H. (1971). Synthetic cutting oils, *Ind. Lub. Tribol.*, **23**,70–72.
19. Langer, T. W., and F. M. Blake (1953). Inhibited polyoxyalkylene glycol fluids, U.S. Patent no. 2,624,708.
20. Mould, R. W., H. B. Silver, and R. J. Syrett (1977). Investigations of the activity of cutting oil additives, part V—The EP activity of some water-based fluids. *Lub. Eng.*, **33**,291–298.
21. Sluhan, C. A. (1960). Some considerations in the selection and use of water soluble cutting and grinding fluids. *Lub. Eng.*, **16**,110–118.
22. Hunz, R. P. (1984). Water-based metalworking lubricants, *Lub. Eng.*, **40**,549–553.
23. Manufacturing Research Institute of American Machinist, Data Sheet MRI-12.
24. Mueller, E. R., and W. H. Martin (1975). Polyalkylene glycol lubricants: Uniquely water soluble, *Lub. Eng.*, **31**,348–356.
25. Jarvholm, B., P. A. Zingmark, and B. G. Osterdahl (1991). High concentration of n-nitrosodiethanolamine in a diluted commercial cutting fluid, *Am. J. Ind. Med.*, **19**,237–239.
26. Notice of formulators of metalworking fluids—potential risk from nitrosamines (1984). *EPA Chemical Advisory*, TS-799, September.
27. Ladov, E. N. (1986). Evaluating and communicating the carcinogenic hazards of petroleum derived lubricant base oils and products. *Lub. Eng.*, **42**,272–277.
28. Klamann, D. (1984). *Lubricants and Related Properties*, Verlag Chemie, Deerfield Beach, FL.
29. Mullins, R. M., Miller, P. R., and Bucko, R. J. (1984). *A Comparison of Matrix and Nonmatrix Tapping Torque Test Procedures in the Evaluation of Experimental Cutting Fluids*, ASLE preprint No. 84-AM-3C-1.

30. Brown, W. L. (1988). The role of polyalkylene glycols in synthetic metalworking fluids, *Lub. Eng.*, **44**,168–171.
31. Whetzel, J. C., Jr., F. Chapel, and S. Rodman (1964). Metal working lubricant, U.S. Patent no. 3,124, 531.
32. Singh, R. V. (1982). Synthetic metal working lubricants, *3rd Int. Kolloq.—Schmierstoffe in der Metallbearbeitung*, **1**,32.1–32.7.
33. Przybylinski, J. L. (1981). Diethanol disulfide as an extreme pressure and anti-wear additive in water soluble metalworking fluids, U.S. Patent no. 4,250,046.
34. Williamson, E. I. (1986). Commercial developments in synthetic lubricants—A European overview, part 2, *J. Synth. Lub.*, **3**,45–53.
35. Russ, J. M., Jr. (1947). *"UCON" Synthetic Lubricants and Hydraulic Fluids,* ASTM technical paper no. 77:3–11, ASTM, Philadelphia.
36. Canter, N. M., J. J. Chaloupka, and G. J. Fischesser (1988). The use of ethylene oxide/propylene oxide (EO/PO) esters as additives in semisynthetic metalworking formulations, *Lub. Eng.*, **44**,257–261.
37. Sweatt, C. H., and T. W. Langer (1951). Some industrial experiences with synthetic lubricants, *Mech. Eng.*, **73**,469–476.
38. Hobson, P. D. (1955). *Industrial Lubrication Practice*, The Industrial Press, New York.
39. Beiswanger, J. P. G., W. Katzenstein, and F. Krupin (1964). Phosphate ester acids as load-carrying additives and rust inhibitors for metalworking fluids, *ASLE Trans.*, **7**,398–405.
40. Smith, G. F., and M. K. Budd (1976). Lubricants for cold working of aluminum, U.S. Patent no. 3,966,619.
41. Mateeva, S., and I. Glavchev (1980). Some operational characteristics of a hydrolysed polyacrylonitrile-based cutting fluid, *Tribol. Int.*, **13**,69–71.
42. Stram, M. A. (1972). Lubricant compositions, U.S. Patent no. 3,657,123.
43. Grower, H. D., B. G. Grower, and D. Young (1971). Aqueous lubricating compositions containing salts of styrene-maleic anhydride copolymers and an inorganic boron compound, U.S. Patent no. 3,629,112.
44. Janatka, V., and E. P. Kirwan (1971). Lubricant-coolant, U.S. Patent no. 3,563,859.
45. Marx, J. (1976). Synthetic lubricant for machining and chipless deformation of metals, U.S. Patent no. 3,980,571.
46. Felton, G. F., Jr. (1976). Low smoking lubricating composition for cold heading operations, U.S. Patent no. 3,983,044.
47. Guminski, R. D., and J. J. Willis (1960). Development of cold-rolling lubricants for aluminum alloys, *J. Inst. Metals,* **88**,481–492.
48. Richter, J. P. (1977). Cutting oil performance—A significant new machining test, ASLE preprint no. 77-LC-2C-3.
49. DeChiffre, L. (1980). Lubrication in cutting—Critical review and experiments with restricted contact tools, *ASLE Trans.*, **24**,340–344.
50. Ham, I. (1968). Fundamentals of tool wear, *Am. Soc. Tool Manufacturing Eng.*, MR68-617.
51. Brown, W. L. (1987). Notebook No. 12811, Union Carbide Chemicals and Plastics Company Inc.
52. Nash, J. C., and N. Colakovic (1985). Effect of synthetic additives on the performance of aluminum tapping fluids, *Lub. Eng.*, **41**,721–724.
53. Levesque, A., and M. McCabe (1984). Improved synthetic coolants using a modified polyalkylene glycol, *Lub. Eng.*, **40**,664–666.
54. Faville, W. A., and R. M. Voitik (1978). The Falex tapping torque test machine, *Lub. Eng.*, **34**,193–197.
55. Katsuke, A., T. Ueno, H. Matsuoka, and M. Kohara (1985). Research on soluble cutting fluids for gear cutting—The influence of dilution ratio and the effect of synthetic fluids, *Jpn. Soc. Mech. Eng.*, **28**,735–743.

56. Brown, W. L. (1990). The use of polyalkylene glycols in metal forming and drawing lubricants, presented at the STLE Annual Meeting, Denver, CO, unpublished.
57. Steigerwald, J. C. (1989). *Report on the Volume of Lubricants Manufactured by the Independent Lubricant Manufacturers in 1989*, Independent Lubricant Manufacturers Assoc. (ILMA), Alexandria, VA.
58. National Petroleum Refiners Association (1989). *1989 Report on U.S. Lubricating Oil Sales*, National Petroleum Refiners Assoc. Washington, DC.
59. The Lubrizol Corp. (1991). *Metalworking Fluid Trends*, 491 404–7.
60. HWB fluid market seen ready for sharp growth (1985). *Chem. Marketing Rep.*, 5/20, 27.
61. Childers, J. C. (1989). Metalworking fluids—A geographical industry analysis, *Metalworking Topics*, **1**(2),1–4.

22
Automotive Trends

John M. Collins
Ethyl Corporation
Southfield, Michigan

I. HISTORY OF AUTOMOTIVE FLUIDS

Lubricants in today's automotive applications have been evolving for many centuries. When machinery came into use with the Industrial Revolution in the late 1700s, people soon realized that a grease or oil, derived from animal or vegetable sources, made the equipment run better, last longer, use less energy (i.e., break fewer leather driving belts), and make less noise. Compared to this early machinery, the early automobile was a precision instrument and required the better types of fluids in order to survive and serve its user.

Henry Ford, Gottlieb Daimler, Rudolf Diesel, and the other pioneers in the development of the automotive vehicle and its components all used lubricants in their equipment. While it is true that these people were mainly concerned with the mechanics of their inventions, they drew upon the experience accumulated since the invention of the wheel when specifying which lubricant to use and recommend.

In the intervening years, many varieties of automotive fluids have come into being and are still required today.

Lubricants for the engine.
Lubricants for gears and transmissions.
Greases for bearings and slow-moving parts.
Hydraulic fluids to transmit power in braking, steering, and power equipment systems.

The lubricants for engines, gears, transmissions and greases, and the hydraulic fluids came to be derived from crude oil. These fluids in their natural state no longer provide the protection that today's equipment demands. The moving parts of today's vehicles operate at much higher temperatures than before, resulting in the oil breaking down. Additives are used to enhance the oil's ability to lubricate, to protect critical areas of the engine and transmission train, to give better resistance to breakdown, and to prevent the sludge that is formed from interfering with the operation of the equipment.

In severe duties, it has become necessary to consider using better base stocks than mineral oils. Synthetic materials like polyalphaolefins (PAO) and esters are blended into mineral oils to make a part-synthetic oil, or are used directly to make a fully synthetic lubricant. With appropriate additive treatment, superior lubricants become possible.

As the need for better lubricants arose, the original equipment manufacturers (OEMs) established tests and specifications to define the viscosity and quality of oil required in their equipment. The Society of Automotive Engineers (SAE) established a schedule of viscosity grades, for engine oils and gear oils that has gained general acceptance worldwide. In North America, the oil industry, through the American Petroleum Institute (API), established quality standards for engine and gear oils that are accepted in that part of the world. Similar oil quality standards have been developed in Japan and in Europe.

Hydraulically assisted systems have replaced many of the older human-powered, mechanically levered control systems. These hydraulic systems are found in such items as braking systems, steering systems, power winches, cranes, actuators, and hydraulic driving motors for large tracked mobile equipment like bulldozers. The hydraulic fluids used must flow easily at all temperatures and be capable of resisting the shearing actions inherent in these systems.

II. CURRENT STATUS OF AUTOMOTIVE FLUIDS

In order to have a meaningful look at the future of automotive fluids and the likelihood of synthetic-based fluids being used, it is necessary to have an understanding of the current situation. In a word, the situation is in a state of transition, with the specifications for engine oils, transmission oils, and axle oils all being revised simultaneously. This portion of the chapter reviews the changes that are in progress on an application-by-application basis.

A. Engine Oils

Most of today's engine oils are derived from crude oil. Synthetic engine oils, usually based on polyalphaolefins (PAO) derived from the processing of ethylene, have been a niche market favored by vehicle enthusiasts. Recently, more suppliers of synthetic engine oils have appeared and are promoting the superior performance of their products in protecting the customers' engines under all conditions. Current engine oil qualities and those attained with synthetic base stocks are discussed in Chapter 15 on automotive crankcase applications.

1. Gasoline-Fueled Engines

The American Petroleum Institute (API) uses the designation "SG" to denote top quality engine oil for use in gasoline-fueled engines. Recently the API SG quality was redefined to include the requirements of the North American OEMs effective January 1, 1992. In reality, many marketers of passenger car motor oil (PCMO) already offer a product that meets these specifications in order to fulfill the OEM requirement that any marketer wishing to be considered for factory-fill business must also sell the same product on the open market.

The Motor Vehicle Manufacturers' Association (MVMA) in conjunction with the Japanese Automobile Manufacturers Association (JAMA) established the International Lubricant Standards Association (ILSAC) to address what it saw as shortcomings in the API standards, compliance, and timeliness. The recently issued ILSAC specification GF-1 for engine oils is intended to cover the needs of American and Japanese cars sold in North America. It does not include the European engine oil specifications.

The MVMA established its North American Lubrication Standards and Approval System (NALSAS) to control approvals of oils meeting the standards and to monitor compliance. The MVMA and API are working to reconcile differences between their two systems of oil quality control so that the vehicle owner will have a full understanding of oil quality needs.

Conventional mineral oil formulations have been adequate for API SG quality engine oils. The OEM specifications will become more stringent as manufacturers redesign their engines to meet the upcoming changes in emission and fuel economy regulations. The ILSAC specification should be seen as only the first of many stepwise improvements in quality requirements. Restriction of oil volatility, to the Comité des Constructeurs D'Automobiles du Marché Commun standard CCMC G-4 used in Europe, will force many formulators to move to hydrocracked/hydrotreated base stocks or even part-synthetic blends, while a move to the CCMC G-5 standard with a SAE 5W–30 viscosity oil will necessitate the use of fully synthetic formulations.

The OEMs, in their owners' manuals, specify the use of engine oils meeting their quality specifications. Customer compliance with North American OEM engine oil recommendations has been less than the OEMs would like to see. Acceptance of the fuel-efficient SAE 5W–30 viscosity that the OEMs used in their regulatory compliance testing has been only 25% of that intended (1). This nonacceptance of the SAE 5W–30 viscosity grade can be attributed to two primary causes:

Lack of customer confidence in that oil to provide proper engine lubrication.
Pricing of the oil higher than the SAE 10W–30 grade.

Owners of performance cars generally follow or exceed the OEM recommendations, with many of them utilizing synthetic lubricants. The engines of the new generation use less oil, have tighter clearances, and in many cases run hotter. One vehicle maker has issued an engine oil specification for a performance vehicle that will require the use of a synthetic base stock to meet targets for volatility control and resistance to oil breakdown. This OEM has recognized that a synthetic oil provides superior performance when conditions are severe.

European automakers have their own lubricant requirements in addition to industry standards. Their CCMC G-4 specification can generally be met by lubricants based on mineral oils, although hydrocracked oils are sometimes incorporated. The CCMC G-5 specification is viewed by the Europeans as the oil quality needed for future engines as regulations tighten and customers demand better performance and durability. It has physical and performance requirements that exceed the abilities of mineral oils alone and cause oil manufacturers to incorporate severely hydrogenated mineral oils in their formulations. Such oils of SAE 5W–30, 5W–40, or 5W–50 viscosity use polyalphaolefin

(PAO) base stock to meet the performance targets of low viscosity and low oil loss through evaporation.

The Japanese OEMs generally request oils meeting API SG quality plus extra tests that are specific to individual companies. Oils meeting all requirements of an OEM are identified and sold as "Genuine Oils" and are used by the OEM at its service locations.

It should be recognized that meeting any one of the three groups of engine oil specifications does not guarantee meeting either of the others.

2. Diesel-Fueled Engines

In North America, diesel engines are used in the large, commercial trucks that haul commodities between cities and distribute them at their destinations, and in most interurban and city buses. Some of the larger pickup trucks and a few imported models of cars also use diesel engines. By contrast, in Europe this type of engine is very popular in passenger cars as well as in truck and bus fleets.

Diesel engines have different requirements for their engine oils from those required for gasoline engines. Failure to use the proper oils will lead to early engine failures, particularly in the intercity trucks, many of which log 125,000 miles (200,000 km) per year or more. The American Petroleum Institute (API) uses the designation CF to denote top quality diesel engine oil. The designation is further subdivided into CF, CF-2, and CF-4 quality to distinguish oils for older engines, two-stroke-cycle engines, and four-stroke-cycle engines, respectively. Most engine oils of CF-4 quality meet the test requirements of both diesel and gasoline engines. These "universal" engine oils, while simplifying the marketing of the oil, cause concern to the diesel engine manufacturers, many of whom feel that their needs are "compromised" to some degree.

With diesel engines being redesigned in the 1990s to reduce their emissions, engine tolerances are being reduced and internal heat flows altered, in addition to changes to the fuel combustion processes. These new close-tolerance heavy-duty engines must remain in emission compliance for 290,000 miles (464,000 km). To do this, wear must be virtually eliminated. Each manufacturer is concentrating on the needs of its engines, and selection of an engine oil to satisfy those engines is of paramount importance. The oil recommended must make minimal contribution to emissions and at the same time provide excellent wear protection for the engine parts. The engine makers also wish to continue to improve the fuel consumption figures of their engines, as fuel cost is the largest single operating expense for a trucker.

Two-stroke-cycle diesel engines are subject to the same emission regulations, but because of their design, it will be more difficult to meet those regulations. Many of these engines are used to power urban buses and, in that service, come under stringent regulations, which were delayed from 1991 to 1993. The API quality CF-2 is the current industry standard for two-stroke-cycle diesel engines in North America, although individual engine manufacturers require additional tests specific to the needs of their engines. As in the case of four-stroke-cycle engines, absolute minimization of wear will be a necessity. Users of two-stroke-cycle engines have generally favored engine oils of single viscosity grades such as SAE 30 or SAE 40. These are always mineral-oil-based.

European diesel engine makers require lubricants meeting the CCMC specification D-4 and the SHPD (super high performance diesel) standard. The Japanese diesel engine makers use the API CC and CD quality specifications with extra tests specific to the individual OEM. These specifications, both European and Japanese, are discussed in Chapter 15 with automotive crankcase applications.

The "uniqueness" of specifications, based on manufacturers' spheres of influence, is diminishing as European and Japanese manufacturers seek acceptance of their units in North America. This includes meeting the emission standards and satisfying the customers' desires for durability. Lubricant manufacturers seeking to claim that they meet all these specifications are forced to look to the best base stock and additive package combination. Polyalphaolefins alone, or in combination with the best mineral oils, will help formulators meet all these specifications simultaneously.

In a word, there are major changes occurring in the diesel engine field and the lubricants that will be needed.

3. Other Engines

While gasoline- and diesel-fueled engines predominate throughout the world, much work is being done to adapt engines to run on mixtures of methanol with gasoline or diesel fuel. The engines oils used with the totally hydrocarbon fuels are being shown to be inadequate for the methanol fuels. Reformulation and testing of new and different oils is ongoing and may indicate a need for synthetic base stocks. The need for minimal wear and minimal contribution to emissions is just as strong for these engines as with gasoline- and diesel-fueled engines.

Propane- and natural gas-fueled engines are used in many countries. These engines, utilized in stop-and-go inner city driving, run more cleanly and seem to be adequately served by mineral oil-based engine oils. Unless their typical service becomes more rigorous with much higher temperatures, synthetic engine oils will not likely be used.

B. Transmission and Differential Gear Oils

In basic terms, a transmission is a device that allows the power output of the engine to be delivered to the wheels under various combinations of engine speed and vehicle speed. A transmission oil functions as lubricant, coolant, and as a hydraulic fluid. The differential divides the power to the rear wheels on some vehicles. Its gear oil functions primarily as a lubricant.

1. Manual Transmissions

The lubricant used in manual transmissions, or gearboxes as they are often known, must prevent gear wear and remove excess heat. In this application, the oil is subjected to very severe shearing forces, which may have the effect of degrading the viscosity and quality of the oil. European and Japanese vehicles call for API GL-4 gear oil quality or a manufacturer's specification. In North America, transmission oils conform to API GL-4 gear oil quality or to an engine oil quality standard. Most of these specifications are currently being met by mineral oil-based formulations.

Some passenger cars and light-duty trucks use engine oil of the same quality and viscosity as used in the vehicle's engine, usually API SG/CE and SAE 5W–30 or 10W–30. This oil is changed approximately every 30,000 miles (50,000 km). Some of the manufacturers of heavy truck transmissions favor diesel engine oil, in this case SAE 40 or 50 viscosity, with changeouts at 50,000 miles (80,000 km). Others call for gear oil with API GL-4 quality and SAE 80W–90 or 85W–140 viscosity.

Eaton Corporation has established a specification for a fully synthetic transmission oil of SAE 50 viscosity that passes a lengthy field test. Owners of Eaton transmissions who follow this oil specification can use oil drain intervals of 250,000 miles (400,000 km) and will have a warranty of 750,000 miles (1,200,000 km). This use of a synthetic oil is transforming the heavy truck transmission and axle (see following section on differential gears) market as their competitors move to meet this new challenge.

2. Automatic Transmissions

The automatic transmission fluid (ATF) performs as a hydraulic fluid in the torque converter, as a lubricant to prevent wear throughout the unit, and as a coolant that carries heat to a small heat exchanger. Most manufacturers of automatic transmissions recommend that the ATF be changed out at some frequency from 20,000 to 40,000 miles (32,000–64,000 km), particularly if the vehicle has been used for trailer towing. Owners frequently ignore these recommendations and operate as if the fluid were immortal, replacing it only when having the transmission repaired. It is an expensive lesson in vehicle maintenance.

The major automatic transmission fluid specifications have been written by the original equipment manufacturers (OEMs), with the best known ones being:

General Motors: Dexron® IIE, Dexron® IID (obsolete January 1, 1993), Allison Transmission Division C-4.
Ford Motor Company: Mercon®, Ford CJ and F (old requirements).
Chrysler Corporation: Mopar® ATF Plus.
Mercedes Benz: Sheets 236.1, 236.6, 236.7, and 236.9.
Japanese OEMs have individual "Genuine" ATFs.

The ATFs required by these companies are similar enough to one another that it is possible for lubricant manufacturers to have an ATF formulation that meets more than one of them.

The introduction of electronically controlled transmissions, with the promise of more developments to come, has created a need for a smaller viscosity change in the ATF across the range of operating temperatures. The new Dexron® IIE specification, with its requirement for a much lower viscosity at very cold temperatures, will provide improved operability at those temperatures. Such an ATF will allow the transmission to operate more consistently in spite of temperature changes.

While most automatic transmission fluids have been made from mineral oils, meeting this Dexron® IIE specification has forced many lubricant blenders to use very select mineral oils or to incorporate hydrocracked/hydrotreated oils in their new formulations in order to attain the desired viscometrics. These base stocks are also very useful in engine oils of SAE 5W-30 viscosity, and are not readily available in the volumes required to meet all needs. In January 1993, when General Motors withdraws its approvals on the older Dexron® IID formulations, many blenders will have to choose where and how they utilize such stocks. Some blenders will undoubtedly incorporate PAO-based synthetic stocks in their formulations.

3. Other Transmissions

The completely variable transmission or constantly varying transmission (CVT) uses a double pulley and belt arrangement to change engine speed to wheel speed through an infinite range of values. It is a relatively new design of transmission, and its current use is limited to smaller engines and vehicles. The lubricant provides wear protection and cooling for the unit.

While an automatic transmission fluid is used by some makers as the lubricant, the needs of this unit have not been fully explored and another fluid, perhaps totally different from any in use today, may be indicated.

4. Differential Gears

The lubricant used in differentials, or axles as they are often known, must prevent wear of the hypoid gears. In this application, the oil is subjected to very severe shearing forces, which may have the effect of degrading the viscosity and quality of the oil. European and Japanese vehicles usually call for API GL-5 gear oil quality or a manufacturer's specification. In North America, differential oils conform to API GL-5 gear oil quality. All of these specifications are currently being met by mineral oil-based formulations.

Eaton Corporation has established a specification for a fully synthetic differential gear oil that meets the U.S. military specification MIL-L-2105D, has SAE 75W, 80W–90, or 85W–140 viscosity, and has passed a lengthy field test. Owners of Eaton differentials who follow this oil specification can use oil drain intervals of 250,000 miles (400,000 km) and will have a warranty of 750,000 miles (1,200,000 km). As mentioned in the section on transmission oils, Eaton's use of synthetic oils is transforming the heavy truck transmission and axle market as its competitors move to meet this new challenge.

C. Hydraulic System Oils

1. Mobile Hydraulic Systems

Light mineral oils with small amounts of additives are used as hydraulic fluids in various types of hydraulic systems. The oil transmits pressure throughout the system, and the pressure causes movement of selected mechanical components. These oils usually meet the specifications of the manufacturers of the hydraulic pumps including Vickers, Hägglunds-Denison, and Cincinnati-Milicron.

Currently these specifications reflect the concerns of the pump makers regarding wear in their components. While it is a very valid concern, it is not the only challenge facing the hydraulic fluid. With very few exceptions, the testing done to satisfy these specifications does not encompass changes to or degradation of the fluid in other parts of the overall system.

Mobile equipment and the on-board hydraulic systems must operate in extremes of temperature, both cold and hot, and require fluids that flow easily and do not degrade to sludge and varnish. Should a leak or rupture occur, the fluid, being under high pressure, will spray in all directions. Control of flammability can be very important in many cases. In some jurisdictions, OEMs are being asked/required to use oils that, if spilled, will not harm the environment.

Those who design and install these systems do not set specifications that reflect the needs of the system but seem to go along with what is recommended for the hydraulic pump. The shortcoming of this approach has been shown when builders of earth-moving equipment asked for biodegradable fluids and found that none had been approved by the hydraulic pump manufacturers.

2. Braking Systems

Brake fluids have evolved into very durable and specialized hydraulic fluids, ones that are essentially factory-filled-for-life. Most approved fluids are based on silicones to endure the high temperatures that can be generated in stopping the vehicle. They are not replaced unless something else goes wrong with the brake system, such as seal leakage or line rupture.

III. PRESSURES AND DRIVING FORCES ON THE AUTOMOTIVE INDUSTRY AND LIKELY RESPONSES

A. Regulations

1. Environmental

Before any vehicle can be offered for sale in North America, Europe, or Japan, the manufacturer must prove that it meets the current emission standards and will continue to do so for many years. Proposals have been put forward in the United States to markedly reduce the currently allowed passenger car emissions and to require compliance for twice as long (100,000 miles/160,000 km). Another proposal will set emission standards based on testing at 20°F (–7°C), in addition to standards at 68–80°F (20–27°C).

The longer duration coupled with the reduced allowable emissions means that the engine must not suffer *any* wear if it is to be in compliance at 100,000 miles. The colder test temperature, besides affecting the engine combustion characteristics, means that the viscosity of the oil and its effect on engine starting and warmup, and in the energy-consuming drag in other components, becomes more critical.

Probable responses by the OEMs include:

Use smaller vehicles and engines to reduce emissions per mile, and work them harder to retain performance. Better lubricants will be needed.

Redesign engines to reduce the emissions that remain.

Design engines to run on fuels other than gasoline and diesel fuel (i.e., methanol, propane, natural gas).

Sell nonfuel vehicles like electric cars.

Renew their efforts to convince the consumer that engine wear will not be a problem when using SAE 5W–30 viscosity engine oil that meets OEM specifications.

Demand a distinct improvement in oil quality and durability to ensure that the engine does not wear out of specification.

Demand lubricants with reduced volatility to minimize the contribution of these fluids to emissions. These last two points may require a move to synthetic or part-synthetic engine oils.

Utilize low-viscosity transmission and gear oils to minimize the engine load contribution from those components.

2. Energy Conservation

Current U.S. regulations mandate fuel economy standards for cars and light trucks, and provide penalties when the manufacturer's fleet fails to meet the standard. In 1990, the corporate average fuel economy (CAFE) standard of 27.5 miles/gal (mpg) or 8.5 liters/100 km was not met by cars of the domestic automakers and some luxury import cars. Their penalty is $5.00 times the number of cars sold, for every 0.1 mpg their fleet average was below 27.5 mpg. For 1 million cars and a 0.2 mpg shortfall, the total penalty would be $10,000,000. While several proposals to increase the CAFE standard are being discussed, it would appear that a value of 30–32 mpg (7.3–7.8 liters/100 km) is a possibility by 1996.

The proposal to do emission testing at 20°F (–7°C) will result in data that will allow calculation of fuel economy values at that lower temperature, values that will be distinctly worse than in the standard test at 68–80°F (20–27°C). In addition to the possibility that penalties may be set for noncompliance with future cold-temperature CAFE standards, the OEM faces the usual competitive pressures over who has the better fuel economy figures.

Probable responses by the OEMs include:

Promote the sales of smaller cars and restrain, through pricing, the sales of larger, heavier cars.
Downsize cars, eliminating most six-passenger cars.
Substitute lighter materials.
Improve vehicle streamlining, particularly for operation at higher speeds where fuel consumption goes up markedly.
Promote fuel economy benefits of SAE 5W–30 engine oils in the 20°F tests. As mentioned previously, they will have to renew their efforts to convince the consumer that engine wear will not be a problem when using this oil. This is a potential field for synthetic and part-synthetic engine oils.
Try hybrid engines with duel fuels for CAFE credits.
Use low-viscosity oils in gears, manual transmissions, and automatic transmissions to minimize drag in the 20°F test. Synthetic fluids, with their excellent low-temperature flow properties, are a distinct possibility in these applications.

3. Safety

Safety-related regulations are intended to prevent the vehicle from failing in a manner that would create a hazard to people and property, and to protect the occupants when things go wrong through mechanical or human failings. In many cases, compliance with the regulations results in an increase in the vehicle weight, and a negative impact on the vehicle's fuel economy and emissions ratings.

Probable responses by the OEMs include:

Incorporate airbags and antilock brakes on more cars; 100% of the North American new car sales may have them by 1996.
Apply strong, lightweight materials in vehicles.
Redesign vehicles and components to accept new saftey concepts such as the Mercedes-Benz corner crash.
Improve fuel economy rating, as previously discussed, to overcome the impact of weight increases related to safety.
Require fire-resistant or nonflammable fluids for hydraulic systems located close to hot surfaces.

B. Competitive Forces

The supplying of vehicles has become a worldwide industry, with automakers on opposite sides of the globe competing in many markets. These companies compete for the money that a buyer is willing to invest in a vehicle. Customers are given the choice of many very different makes and models, styles, and technologies. Each company utilizes any advantages available to it and attempts to minimize or overcome the disadvantages that impede it.

At the point where the auto company meets the customer, the sales decision is made on the basis of money and customer satisfaction. The customer determines the *worth* of the transaction.

1. Cost Control

Profit is the necessary result for any business that intends to continue to exist. In the simplest terms, it is what is left after you have sold your products and paid your bills. The selling price of a car is governed by market forces, leaving the automaker to concentrate on controlling costs in the manufacturing and distribution systems. He is always looking for ways to make even the smallest component less expensively or to eliminate it altogether.

Probable responses by the OEMs include:

Manufacture more components and vehicles in parts of the world with lower labor costs.

Utilize sealed units. If these units contain a fluid, it will have to be of superior quality in order to provide as long a service life as possible. Synthetic fluids are superior candidates for these applications.

Extend in-depth quality control to more and more aspects of their business, and to their suppliers.

2. Customer Satisfaction

In the eyes of the car salesman, the customer wants Cadillac satisfaction at a Chevrolet price, or, if you wish, a Rolls Royce at the price of a Citroen 2CV. In a word, the customer wants the best satisfaction for the money. Satisfaction is found in such concepts as ease of operation, performance, handling, safety, fuel economy, minimal maintenance requirement, reliability, and a warranty to fix those things that do go wrong.

Maintenance is a topic that none of the customers relish. They see it as an inconvenience that takes the vehicle out of service and, in North America, leaves them stranded in a community that frequently has no good public transit system to use as an alternative. They are not really convinced that their car needs all the maintenance that is specified in the owner's manual and are quite willing to skip some maintenance intervals. They frequently have a poor view of the reputation of the service people for honesty, workmanship, speed of service, and cost control. They would welcome the opportunity of never doing any vehicle service. At the same time, this vehicle is expected never to break down or at least to have a warranty to cover the costs incurred for any breakdown while they own it.

Probable responses by the OEMs include:

Use smaller engines with higher output derived from multivalve cylinders, turbo- or superchargers, and electronics to handle downsizing, emissions, and CAFE changes without losing performance satisfaction.

Educate customers to the necessity of using top-quality engine oils for these vehicles to avoid damage to their investment. The OEM engine oil specifications will likely necessitate synthetic or part-synthetic engine oils.

Redesign transmissions and gears, and computer-link them to the engine for operation and handling attributes.

Establish quick lubes at dealerships for fast oil changes. This could improve the likelihood that a lubricant meeting the OEM recommendations is used.

Improve servicing at dealerships for greater customer satisfaction. The OEM servicing is a weak link in repeat car sales.

Minimize servicing needs where possible by the use of filled-for-life fluid applications. Synthetic lubricants with their superior resistance to degradation and their all-temperature viscometrics will find application in this type of service.

Extend warranties. There are liabilities with attendant costs. How can customer abuses be controlled?

IV. POSSIBLE USE OF SYNTHETIC FLUIDS

Synthetic fluids for use as engine oils, gear oils, transmission fluids, or in any other automotive application will have to earn a place in a market dominated by mineral oils. Being more expensive, they will have to provide worthwhile benefits where mineral oils are inadequate. These benefits can be a net reduction in vehicle costs to the OEM, a more attractive product to the new car buyer, reduced maintenance, and the ability to keep the vehicle performing like new for a longer period of time.

The best choice may be a part-synthetic lubricant where the synthetic constituent provides fluid properties not attainable with an all-mineral formulation. Control of oil volatility is an example. Such a lubricant will have a cost that is intermediate between mineral oil and synthetic fluid costs and can be expected to have properties and performance that are intermediate as well.

A. Engine Oils

Synthetic engine oils are well known in race-car circles where their superb thermal/oxidative resistance helps the oil to survive and lubricate at extremely hot temperatures. The low volatility of these oils helps the engine retain the oil quantity necessary for proper lubrication, avoiding oil starvation in critical zones in the engine. This application will continue, with polyalphaolefin-based synthetics expected to continue their dominance.

Many passenger cars, such as Chevrolet Corvettes, are sold with performance-quality engines. The owners of these cars, while not necessarily racing them, do subject the engine and its oil to periods of high output and temperature. Some of these cars are equipped with turbochargers, where the proximity of the lubricated rotating parts to the hot exhaust gases causes the thermal/oxidative stability of the engine oil to be severely stressed. Synthetic engine oils are used by some owners, and this trend is expected to increase markedly in the near future. It will be driven by market forces.

General Motors' engine oil specification 9985847 covers factory-fill engine oil in 1992 Corvettes. It contains performance limits that cannot be met by other than a fully synthetic engine oil. GM is expected to recommend that the car owner use an oil meeting this specification when doing oil changes. Acknowledgment by the world's largest car maker that its top line performance car benefits from a synthetic engine oil will cause other performance car owners to reassess their needs.

Top-quality synthetic engine oils are becoming more readily available in North America. As of May 1991, six well-known marketers of engine oils and a number of much smaller performance-oriented niche marketers were offering fully synthetic SAE 5W–30, 5W–40, or 5W–50 oils. Other major oil marketers can be expected to follow with their brands.

Marketplace competition is expected to result in a lowering of costs for an engine oil change using synthetic oils. As of May 1991, this change can be made for $30, versus $22 when a top grade of mineral oil is used. It will not likely drop to the same price as mineral oil but will be low enough to attract a certain segment of the marketplace.

Partly synthetic passenger car engine oils are not common in North America. No marketing category distinguishes them, with the result that the oil marketers who do use some synthetic base stock are unable to claim a product differentiation from the more common all-mineral-oil formulations. As these marketers begin to sell oils meeting the ILSAC GF-1 specification, some of them will have to use part-synthetic formulations. It is expected that they will claim product differentiation based on better performance than the all-mineral oils in an effort to recover the additional costs.

Passenger car engine oils in Europe will continue to utilize the CCMC G-4 and G-5 standards with part-synthetic oils used in both categories, and fully synthetic SAE 5W–30, 5W–40, and 5W–50 oils in the G-5 line. No tightening of these standards is seen before the year 2000. The shift from mineral oils to part-synthetic and fully synthetic oils will keep pace with the flow of new cars into the market.

With the current restructuring of the CCMC, the European requirements will eventually be united with those of ILSAC. At that time a new ILSAC specification incorporating the quality standards used in Europe will ensure the recognition of part-synthetic oils in North America. Fully synthetic SAE 5W–30, 5W–40, and 5W–50 oils will either be covered in this new specification or have a separate one. It is estimated that by 1995, as much as 75% of the North American passenger car engine oil sold will meet GF-1 or higher qualities (2).

The makers and users of four-stroke cycle diesel engines have historically favored oils of SAE 15W–40 viscosity. Their efficacy at protecting engines has been proven over many years and in countless numbers of trucks. With substantial changes required to enable the new generation of diesel engines to meet the 1994 particulate emission standards, OEMs are looking at what they can accomplish with changes in engine lubricants. Their needs are:

Minimal contribution to particulate emissions.
Little or no wear during the regulated period to ensure staying within the emission regulations.
Continue or improve on current oil change intervals.
Reduce fuel consumption if possible.

Synthetic and part-synthetic engine oils using polyalphaolefins are being seriously examined for this service. Some of them are SAE 10W–30 viscosity. The features of interest are:

Excellent low-temperature flow properties. Much of the metal-to-metal wear occurs on cold start-ups when the engine oil cannot be pumped fast enough to prevent wear at the points of metal-to-metal contact.
Retention of high-temperature viscosity due to a lack of sheardown protects the engine at other times.
Resistance to thermal and oxidative degradation keeps engine parts clean and functioning as designed.
Low volatility minimizes oil loss and the lubricant contribution to emissions.
Use of SAE 10W–30 oil in place of SAE 15W–40 oil may reduce fuel consumption.

At present there are a few such oils in the marketplace, but their acceptance has been very limited and their impact on the particulate emission problem is undetermined at this time. It is expected that the demands being made on diesel engines for trucks, buses, and some cars will result in the introduction of part-synthetic and fully synthetic engine oils for these vehicles.

The makers and users of two-stroke-cycle diesel engines have generally used single-viscosity-grade oils. These oils lack the lighter lube oil fractions of the multigrade oils and should have lesser volatility. It may be possible to reduce their contribution to emissions even further by moving to part-synthetic or fully synthetic formulations. Improvements in wear control should also be possible, but no information on test work to confirm this has been made public to date.

B. Transmission and Differential Gear Oils

Transmission oils, whether used in manual or automatic units, and gear oils are candidates for factory-fill-for-life applications. If the unit is well designed and precisely assembled with no way for external materials to get in or oil to leak out, these units should be capable of lasting for the life of the vehicle. At the same time, it is expected that the OEMs will move to computer integration of transmission operation with engine operation and other parameters. This will involve greater use of electronically controlled automatic transmissions.

The fluid used, whether an engine oil, gear oil, automatic transmission fluid, or a totally new specification, will have to provide superior performance for the life of the vehicle. Significant requirements will be:

Retention of initial viscometrics. No sheardown, either temporary or permanent; no thickening due to oxidation.
No wear of the transmission parts or gears.
No fluid degradation leading to sludge, varnish, or other deposits.

It is expected that polyalphaolefin base stocks with optimized additive packages will be instrumental in fulfilling these requirements. The excellent high- and low-temperature viscometrics achieved without the use of viscosity index improvers, the thermal and oxidative stability, the low volatility, and the ability to incorporate very effective additive packages for control of wear will all contribute to their acceptance.

Part-synthetic formulations for these applications are expected to be less likely to succeed. They will be unable to match the performance of the fully synthetic oils and will require more frequent changeouts. Their lesser cost will not make up for the increased costs of the vehicle being out of service more frequently.

The OEMs will have to find a way to economically justify this factory-fill-for-life concept. The present logic dictates that the delivered cost of the vehicle is of paramount importance. Features attractive to the customer are classified as options and billed separately. Designers will have to go back to basic principles. As much attention must be paid to how a lubricant contributes to the performance of an item as is currently paid to the selection of metallurgy and construction techniques.

It is possible that the use of a synthetic lubricant in this type of service might allow some of the following actions:

Reducing size of or totally eliminating transmission coolers. This would result in a reduction in vehicle weight, freeing up of some space in the crowded engine compartment, and, in the case of total elimination, a reduction in installation costs.
Sealing transmissions and differentials to exclude air, moisture, and dirt. The dipstick used to check the fluid level in transmissions could then be eliminated, reducing construction costs and freeing a little space in the engine compartment.
Marketing the feature that during the life of the vehicle less oil will be required, recycled, or destroyed.

C. Hydraulic System Oils

The hydraulic pump manufacturers are evaluating synthetic fluids in some very specific applications. It is expected that while they will not abandon the mineral oil-based formulations that are adequate in many applications, they will create specifications for superior performance, specifications that will require synthetic base stocks. Part-synthetic formulations are seen as less likely, as the user who wants a distinctly better fluid will be unlikely to go halfway.

The pump manufacturers are now being asked to work with the system designer to ensure that their pumps can live with the fluid after it has been in service for a long time. Because dirt, debris, sludge, and varnish will impact the performance and wear of the pump, the hydraulic OEMs have an interest. While a filter will help keep the fluid cleaner, it represents a pressure drop in the hydraulic system and, when coated with sludge and debris, a likely and unnecessary loss in system performance.

Hydraulic fluids based on synthetic base stocks will have excellent flow properties at a wide variety of operating temperatures, resulting in minimal pressure drops at filters and other restrictions. Their thermal/oxidative stability will allow operation at elevated temperatures without the fluid breaking down into sludge or varnish. This reduction in solids will reduce the load on the filters and help extend pump life. The ability to blend synthetic base stocks together and achieve high-viscosity-index fluids that do not shear down in these applications will allow reconsideration of design bases for both the pumps and the system as a whole.

V. CONCLUSIONS

The 1990s are seeing an intense period of change in the field of automotive lubricants. Specifications for gasoline and diesel engine oils, manual and automatic transmission fluids, and axle oils are all in a state of flux, with further changes expected. The demands of the vehicle makers for greater performance and protection of their equipment are pushing the mineral-oil-based formulations to their limits.

Mineral oil-based lubricants will still have many applications. In many situations they will be blended with hydrocracked or synthetic base stocks to provide some improvement in lubricant quality. These blends will be identified as an improvement but will still be inadequate to meet the most stringent specifications.

The OEMs, having seen the attributes of part-synthetic and fully synthetic lubricants, will be looking for the best in their fill-for-life applications. Besides giving superior performance, these long-life lubricants will reduce the quantity of all lubricants requiring recycling or disposal during the life of a vehicle.

Synthetic automotive lubricants and synthetic base stocks have come of age.

GLOSSARY OF TERMS

API, American Petroleum Institute. Consensus group within the oil industry responsible, among other duties, for defining lubricant quality standards, and for enforcing compliance with those standards where their symbol is used.

CAFE, corporate average fuel economy. The average fuel economy of the fleet of cars sold by a manufacturer. It is calculated separately for domestic production and imports, and each result is compared to a standard. The penalty is $5.00 times the number of cars sold, for every 0.1 mile per gallon (mpg) their fleet average was below the standard. For 1 million cars and a 0.2 mpg shortfall, the penalty would total $10,000,000.

CEC, Conseil Européen de Coordination. In English, it is the European Coordinating Council and is the governing body for European standards.

CCMC, Comité des Constructeurs D'Automobiles du Marché Commun. European motor vehicle builders association.

Crankcase. The sump at the bottom of the engine, which covers the engine crankshaft and in which the oil sits when it is not circulating through the engine.

HDD, heavy-duty diesel. An abbreviation for diesel engines themselves and also for the oil used in those engines.

ILSAC, International Lubricant Standards Association. Association of MVMA and JAMA to develop unified lubricant standards. Specification GF-1 was issued in 1991 for use in North America. CCMC has declined to participate at this time.

JAMA, Japan Automobile Manufacturers' Association. Speaks for Japanese OEMs worldwide.

MVMA, Motor Vehicle Manufacturers' Association. A North American OEM association. As part of ILSAC it participates in promoting the ILSAC engine oil standards.

NALSAS, North American Lubricant Standards and Approval System. This body has been established to monitor and control compliance with the engine oil requirements of the North American OEMs.

OEM, original equipment manufacturer. The company that makes the vehicle or unit. Plural form is OEMs.

PCMO, passenger car motor oil. An abbreviation for engine oil used in gasoline-fueled engines.

SAE, Society of Automotive Engineers. Consensus body responsible, among other duties, for establishing the need for and the details of automotive and aerospace standards. Lubricant viscosity standards are their responsibility.

REFERENCES

1. Collins, J. M. (1989). Ethyl Corporation, Industrial Chemicals Division, Southfield, MI.
2. Lubrizol Petroleum Chemicals Company. *1991 Trends Rev.*, Wickliffe, OH.

23
Industrial Trends

Raymond B. Dawson
Ethyl Corporation
Baton Rouge, Louisiana

Charles Platteau
Ethyl S. A.
Brussels, Belgium

I. INTRODUCTION

The industrial sector of the market for functional fluids and lubricants is vast. In many divisions of the market, it is taken broadly to include all sectors excluding those associated with transportation and the military. To endeavor to look at each element of the industrial market and to predict trends in the utilization of synthetic fluids would be an enormous undertaking beyond the scope of a single chapter. Fortunately, many of the driving forces shaping the needs of individual elements are common to the bulk of the industrial sector. Therefore, we first concentrate on the development of general market needs and the implications for the development of synthetic fluids. In the succeeding sections of the chapter, these needs are exemplified in relation to selected sectors of the market. We have tried not to be too speculative and have limited our focus to those trends expected to develop over the next decade or so.

In this chapter, we differentiate between synthetic fluids and water-based fluids. Within the metalworking industry water-based fluids are often referred to as synthetic fluids, but it is more consistent with the scope of this book to limit synthetic fluids to those that use a synthesized fluid as the principal base stock. We also exclude very-high-viscosity-index hydrocracked oils from our definition of synthetic. We believe that these fluids are more appropriately described as very highly refined mineral oils, although some may argue that they are synthetic on the basis that their production process involves chemical synthesis.

Currently the use of synthetic fluids within the industrial market is relatively small. In volume terms, about 6% of the U.S. industrial market for oils and functional fluids is based on synthetic fluids (1). In Europe about 5% of industrial functional fluids are either synthetic or semisynthetic. The share of the market held by synthetic fluids will increase, but only gradually, and only as the use of a synthetic fluid can be justified in terms of improved efficiency, improved quality, or if concerns over safety or the environment require their use. Synthetic fluids generally will increase their presence in the market through gradual application in niche areas. However, it is possible that in some areas legislative action will, either directly or indirectly, force a wholesale change in fluid type from a mineral oil to a synthetic. This already has been seen in the case of lubricants for use in R134a compressors and could develop in some areas of white oil application.

II. MARKET NEEDS

A. Cost-Performance Balance

Cost-performance is key to the development of synthetic fluid based products. In general, the synthetic fluid will be more expensive than its mineral-oil-based competitor. Therefore, the use of a synthetic fluid in place of a mineral oil may only be justified if its performance offsets the higher price. In an increasing number of current and developing applications this can be demonstrated. The properties of synthetic fluids that allow superior performance are shared by many different fluid types, the differences between fluids being questions of degree or questions of relative combinations of different properties.

Many industrial processes and associated equipment are being developed to operate at higher temperatures. Higher temperatures can allow faster processing, and hence gains in productivity and greater efficiency can be realized. Alternatively, products of superior quality may be produced. These higher temperatures are leading to a demand for less volatile and more thermally and oxidatively stable fluids. Many synthetic fluids show improved thermal and oxidative stability over mineral oils. They also tend to show, for a given viscosity, lower volatility and higher viscosity index. For moderate increases in continuous operating temperature up to about 180°C, polyalphaolefins or dibasic acid esters may be suitable. Beyond 180°C, up to about 215°C, polyol esters or polyalkylene glycols offer improved stability. At yet higher temperatures polyphenylene oxides or perfluorinated polyethers could be required. All of these fluids are capable of intermittent service at higher temperatures than those given. Research into high-temperature lubricants is an active area, often first stimulated by military or transportation sectors of the market rather than the industrial sector. Research into lubricants for use in low-heat-rejection engines, in particular, may lead to new classes of fluids that will find eventual application in the industrial sector (2, 3).

In cold conditions, low-temperature fluidity becomes an important requirement. Specially refined mineral oils are capable of operation down to less than –55°C, but there are indications that these fluids are becoming less available. Synthetic fluids often exhibit good low-temperature characteristics. Polyalphaolefins, depending on viscosity, have pour points down to less than –60°C. Silicone fluids also exhibit excellent low-temperature fluidity, as do certain perfluorinated polyethers. For specialized applications under very cold conditions such low-temperature properties may remove the necessity for auxiliary heating systems or reduce the power needed from a starting system. More precise hydraulic control is also potentially available. In the medium term, the market for low-temperature lubricants and fluids is not expected to expand greatly, but in the longer term the need to unlock the natural resources in the subarctic and possibly antarctic regions will expand this niche area.

In more general areas of application, a synthetic fluid may allow longer periods between oil drains and associated reductions in down time more than pay for the higher cost. In the extreme, synthetic fluids may allow the development of "fill-for-life" fluid systems that will simplify equipment design and construction by, for example, removing the need for oil drain plugs. Such simplifications can be vital elements in potential cost reduction programs to maintain and improve competitive position. Simplification of design and reduced capital costs can also be realized through the use of low-flammability fluids such as silicones, polyol esters, and phosphate esters.

Of course, the cost of the fluid is fundamental to the cost-performance equation, and the final choice between different synthetic fluids may well be determined by the relative differences between the prices of these fluids. Relative pricing can be expected to vary somewhat as a function of supply/demand relationships, which will impact not only on the pricing policies of synthetic fluid suppliers but also on the costs of raw materials and production. However, for the majority of established fluids, relative prices are likely to stay broadly as they are today. The prices of developmental synthetic fluids can be expected to fall dramatically if they manage to establish themselves and economies of scale are brought into play. It also needs to be borne in mind that the price of base fluid is often a minor component of the overall cost of a finished functional fluid. There will be costs of additives, blending and packaging costs, and costs of distribution. There are also the overheads associated with research and development and product promotion. For relatively small-volume, niche applications these costs can be very high. As volumes increase and as research and development (R&D) expenses decrease we can expect to see a trend to lower finished fluid prices. It is also increasingly recognized that the total cost of using a particular fluid is more than just the initial fluid price and its potential life—of increasing importance will be the cost of fluid disposal.

B. Environmental Acceptability

Concern for the environment is becoming an ever-larger consideration in the selection of functional fluids. In many cases a particular synthetic fluid will be more environmentally acceptable than its lower-cost mineral-oil-based competitor and at the same time be technically superior to vegetable-based oils or water-based fluids. New legislation and associated regulations can force a market to switch to new fluid types, even when the technical performance of the new fluid is inferior to the old fluid and its price much higher. This certainly has been the case in the transformer oil market and for the mining industry, where legislation has driven polychlorinated biphenyl-based fluids out of the market. We expect to see this trend increase. Environmental acceptability manifests itself in a number of ways, and in the following sections we briefly review the major factors.

1. Biodegradability

Biodegradability is generally regarded as a favorable fluid characteristic from an environmental standpoint. The rationale is that such a fluid will readily break down in the environment to nontoxic, simple compounds, ultimately carbon dioxide and water. Accumulation of fluid in the environment is limited, and long-term problems associated with that accumulation are avoided.

It is necessary to use the term "biodegradable" with care. Nearly all fluids are, to some extent, biodegradable, with the possible exception of largely inorganic fluids such as silicones. Certainly mineral oils, given sufficient time, do biodegrade, and many texts will describe mineral oils as being biodegradable. The important factor is the rate of biodegradation under ambient conditions. Required biodegradability must therefore be defined in terms of a test protocol selected to reflect adequately the conditions under which a fluid may be released into the environment. In Europe, the common protocol used to determine relative biodegradability is the CEC-L-33-T82 test, which was originally designed as a test to evaluate the biodegradability of two-stroke oils designed for use in marine outboard engines. This test is now being used to evaluate fluids designed for very different applications. It is unlikely that the CEC-L-33-T82 test will be accepted as the protocol for determining biodegradability in governmental regulations. For example, European Economic Community (EEC) regulations currently being developed to categorize environmentally hazardous materials are likely to use OECD (Organization for Economic Cooperation and Development) or related protocols. In the future we are likely to see a more sophisticated approach, with the development of test protocols reflecting more accurately real-world conditions.

Niche markets for biodegradable fluids already are becoming established. These are for use in applications where loss of fluid to the environment might be anticipated—for example, hydraulic fluids used in mobile equipment or bar oils for chain saws. Many of these fluids are based on vegetable oils, most often rapeseed oil. The biodegradable synthetic fluid must then compete with the cheaper vegetable-oil-based product and will only gain significant market share where the superior technical performance of the synthetic oil leads to a cost-performance benefit.

The classes of synthetic fluid that show a much higher rate of biodegradation compared to mineral oils are the esters and polyglycols. These classes of fluid are already well established in certain application areas. Their major competition will come from vegetable oils, notably rapeseed oil. We anticipate active development of these classes of synthetic fluid to maximize biodegradability and at the same time achieve levels of performance, notably thermooxidative and hydrolytic stability, well in excess of that achievable with the use of vegetable-oil-based products.

2. Waste Disposal/Fluid Recycling

Concern for the environment will continue to focus attention on how industry disposes of its waste. This is a trend that we can expect to accelerate through the 1990s as the disposal of waste becomes a major problem in many industrial areas. This will result in a much higher level of fluid management than we see today. There will be increased focus on reducing the overall level of waste, and the disposal of waste will be tightly controlled. The clean-up and recycling of waste fluids will become increasingly attractive, and fluid suppliers or specialized companies will provide industry with disposal, clean-up, and recycling services. This trend will favor increased use of highly stable synthetic fluids on several counts.

First, the synthetic will offer longer service life, and hence the overall level of waste material is reduced and the frequency of recycling will decrease. Second, highly stable synthetic fluids may lend themselves more readily to clean-up because the level of additivation may be less in a synthetic fluid and the degree of fluid degradation during clean-up processing may be less. The price differential between a recycled synthetic fluid and a recycled mineral oil may thus be less than that between virgin fluids. All of these factors may shift the cost performance balance in favor of the more stable synthetic fluids.

It should be noted that the characteristics of a readily recyclable fluid are not those of the relatively inexpensive biodegradable vegetable oils, a fact that could favor synthetic fluids designed to offer good recyclability together with ready biodegradability.

3. *Fluid Toxicity*

Synthetic fluids can be produced to very high levels of purity with minimal contamination. Many fluids are essentially composed of a very few or even a single molecular species. The toxicological characteristics of the fluid therefore can be carefully assessed and quality standards can be maintained over time. Mineral oils are a mix of many thousands of different molecules and different batches of nominally the same oil can be different in terms of detailed composition as the crude source changes and refining conditions are changed. Even with highly refined white mineral oils there are concerns over potential toxicological hazards (4), and we may see replacement of these fluids in certain applications. How much of this potential replacement might fall to synthetic fluids rather than vegetable oils is a question of cost performance.

4. *Ozone Depletion*

Since 1974, the world community has become increasingly concerned with reported depletion of the ozone layer, particularly over the southern polar region (5). As a result of this the Montreal protocol (6), subsequently modified at London (7), has set down a timetable for a reduction, and eventual elimination, of the use of fluids and gases whose molecules contain chlorine. By the year 2000, it is targeted that chlorofluorocarbons (CFCs) will be eliminated as aerosol propellants, in urethane foam production, as solvents, and as refrigerants. These changes will affect other functional fluids currently used in the same environment as the CFCs. In particular, refrigerator compressors working with R134a in place of R12 will require new lubricants.

5. *Global Warming*

Parallel with the rising concern about atmospheric ozone depletion, we see growing apprehension over the problem of global warming. It is probable that the mid 1990s will see increased regulations to control the emissions of gases that contribute to global warming. Such regulations will impact on industry's selection of functional fluids and on their use and disposal.

C. Quality

The drive for improved product quality will continue as companies seek a competitive edge for their products. There is good reason to believe that this will result in increased use of synthetic fluids, which, because of their own properties, will allow processes to operate more consistently and in such a manner as to produce higher-quality end-products. Often such improvements in quality will be accompanied by improvements in overall productivity. Examples can be found in the textile industry, where higher quality and more efficient production in developed countries can offset lower labor costs in the developing world, and in the development of new heat-transfer fluids designed to meet particular requirements in the control of complex chemistries required for the production of pharmaceuticals and certain specialty chemicals (8).

D. Safety

In response to concerns for the environment, there will continue to be improvements in overall safety of industrial processes and operations. The major area for increasing the safety factor has been and will continue to be in the reduction of fire hazards. Low-flammability fluids already have major shares of the fluid market in particularly sensitive situations such as metal casting, hot metal rolling, and in the mining industry. Many halogen-containing fluids currently are being displaced from these markets because of environmental concerns. This provides opportunities for alternate fluids. In any situation where there is the possibility of a mineral oil leaking onto surfaces at temperatures in excess of the autoignition temperature of the oil (260–371°C), there is a real risk of fire. In the electric power-generating industry, for example, there is increased risk of fire because steam temperatures have risen as a result of turbine manufacturers seeking to improve operating efficiency. This is resulting in a trend to replace mineral-oil-based turbine lubricants by lubricants based on fire-retardant fluids (9).

Other safety issues include reducing the amounts of oil-based mists in the atmosphere of certain work environments. The degradation of fluids to airborne smokes and mists is a related problem. Both problems have been of particular concern in the textile industry. We anticipate that synthetic-based fluids will continue to find niche application areas to help improve the work environment. In the chemical industry, specialized synthetic fluids can help minimize the risks involved should the fluid come into contact with particularly reactive chemicals. For example, perfluorinated polyethers are finding application as heat-transfer fluids in vessels containing hot oleum (R. J. De Pasquale, Ausimont U.S.A. Inc., personal communication, April, 1991).

III. SELECTED MARKET-SECTOR TRENDS

A. Hydraulic Fluids

Developments in the market for hydraulic fluids are being driven by three main drivers. First, there is the environmental acceptability of fluids; second, there is the development of smaller, high-pressure hydraulic systems; and third, there is the issue of safety in the workplace. All of these factors favor replacement of mineral-oil-based hydraulic fluids by alternate fluid systems afforded by vegetable oils, water-based fluids, and synthetic fluids.

Environmental concerns cover two areas. The first addresses the general question of waste fluid disposal. This will favor synthetic fluids with their potential for longer service life before requiring replacement. We are likely to see increased service to industrial operations in the area of "fluid management." This will cover not only recommendations for fluid use but also the management of fluid disposal. The development of facilities to recycle fluids may well further favor the use of more stable synthetic fluids in that they can be more suitable for recycling, and the relatively high costs of initial fill may be offset by longer life and potential recycling.

The second environmental concern addresses the fate of fluids once they enter the natural environment. For mobile hydraulic equipment in particular there is concern over potential, noncontrollable leaks of fluid into the environment. This has resulted in a developing market for readily biodegradable fluids. This market will continue to develop. Most biodegradable fluids developed to date have been based on rapeseed oils rather than on the more expensive synthetic fluids. Vegetable oils, however, are inherently rather poor with respect to oxidative and hydrolytic stability, and parts of this market may require a higher level of stability but still a relatively high rate of biodegradation. This will require the use of more expensive synthetic fluids, probably of the polyol ester type. How much of the market will be satisfied by vegetable oils and how much will require a higher level of performance will depend on experience, the development of new specifications, and the relative cost-performance characteristics of the finished fluids.

The development of smaller, high-pressure hydraulic systems will promote the use of more stable fluids able to withstand higher temperatures and will provide greater protection to pumps. Synthetic-based fluids will show the best technical performance and will penetrate this market sector. The potential reduction in fluid volume within such a high-pressure hydraulic system will also reduce the differential in cost between filling the system with a less expensive fluid and a more expensive fluid.

For many years, the temperature characteristics of certain synthetic fluids have been put to good advantage, particularly with respect to low temperature operability. The lower dependence of viscosity on temperature for many synthetic fluids over conventional mineral oils can give improved "feel" to manually controlled hydraulic systems as found on much digging and construction equipment. Operators of such equipment sense a very real difference in system response when starting from cold as opposed to warmer operation. Given that such mobile systems may also require the use of relatively biodegradable fluids, we see a further factor favoring the use of polyol ester-based fluids over vegetable oil-based systems, because vegetable oils have a relatively high pour point. The use of polyalphaolefins in such systems has also been discussed (10).

Low-flammability or nonflammable hydraulic fluids find particular application in situations where fire hazard presents an unacceptable risk. The mining and steel industries are the major consumers of such fluids. Over the last 10 years, we already have seen the replacement of polychlorinated biphenyls (PCBs) as hydraulic fluids in the mining industry due to the ecological and toxicological hazards associated with this class of chemical. In some sectors of the market, other chlorinated materials such as chlorinated benzyltoluene found favor as PCB replacements, but these materials have also come under increasing environmental pressure and are themselves being phased out (11, 12). The major low-flammability synthetic fluids used in the market today are phosphate esters and long-chain polyol esters or mixtures of these fluids with mineral oils. These fluids compete with each other and with the well-established classes of water-containing fluids. Polyalkylene glycols, long used in water–glycol fire-resistant fluids, have also been promoted as possible fire-resistant fluids in their own right. The key factors in determining which fluid is preferred in a given application are the degree of fire resistance required, the performance of the fluid as a hydraulic fluid, materials compatibility, price, and environmental acceptability. The degree of fire resistance required often is regulated by government authorities; hence, markets show significant geographic differences. The trend has been for polyol esters to win market share where their level of fire resistance is acceptable. This has occurred particularly in the steel industry in the United States. The principal reasons are lower cost and improved materials compatibility. This trend is expected to continue, but the fact remains that phosphate esters are inherently superior to polyol esters in terms of fire resistance, and hence they are likely to maintain their position in those market sectors requiring a higher level of fire resistance. The market for fire-resistant fluids is significantly driven by regulatory considerations, and today's trends could be either reinforced or reversed by regulatory changes. Such changes might occur as the trend to create larger, harmonized economic markets between countries develops.

We may also see the development of niche markets for new fire-resistant synthetic fluids. Such fluids are unlikely to contain chlorine or bromine because of environmental considerations, which focuses attention on phosphorus-containing fluids where a relatively high level of fire resistance is required. Phosphazenes are a class of fluids that have some promise in this area, but their high cost is likely to limit their potential.

B. Turbine Oils

Heavy-duty gas turbine designs are driven by the desire for higher efficiency, which means a trend toward higher gas temperatures. These developments are being made possible through the development and use of new heat-resistant alloys and ceramics. As a result, the lubricant is exposed to increasingly high temperatures, particularly in the center bearings and exhaust end bearings. This will lead to increased use of synthetic lubricants with improved thermooxidative stability. Polyalphaolefins and polyol ester fluids are likely to find areas of application. Gas turbines derived from aircraft turbines already use the same synthetic ester-based lubricants as the jet engine, and this is likely to continue for the foreseeable future.

The development of steam turbines has been driven by the same desire for improved efficiency. As steam temperatures have risen, the need for synthetic fluids has developed. Most notably, the fire risk of mineral oils leaking onto steam lines at temperatures well in excess of the autoignition point of the oil has led to the use of phosphate esters in the electrohydraulic turbine governor system. However, it is also recognized that many fires have their origin in the lubrication system. For example, a German report (13) has indicated that 43% of fires in power stations originated in the lubrication system, and 49% from the hydraulic system; in the remainder, the source could not be identified. Thus, we expect to see some advance in the use of fire-resistant fluids in the lubrication system, particularly in those environments where the results of a fire could be particularly devastating. Phosphate esters are the obvious candidates, and their use has been reviewed (9). It also has been argued that mineral-oil-based turbine lubricants rely to some extent on regular topping up of the system with fresh oil to maintain performance. A reduction in top-up rate coupled with generally higher temperatures could significantly reduce the life of the mineral-oil-based lubricant and require the use of a synthetic fluid (14). Polyalphaolefin fluids or blends of polyalphaolefin with mineral oil would be favored for use. Longer-term innovations to increase turbine efficiency include the use of ammonia/water mixtures, which may also require a new approach to lubrication.

C. Industrial Gear Oils

The dominant trends in the lubrication of industrial gears are toward higher temperatures, longer life, and lower friction. These trends are the result of the designers' targets of lower cost, simplified designs, and improved efficiency. More power through smaller gearboxes results in higher lubricant temperatures. The longer life of a lubricant reduces maintenance costs and, in the limit of a "fill-for-life" potential, there are considerable savings to be achieved from removing drain plugs and filler caps and simplifying design and manufacture. All of these factors are leading to the gradual introduction of synthetic-based oils. In addition, the synthetic lubricant can offer significant advantages where very low temperatures are likely to be encountered. The leading candidate for use as the base stock for synthetic gear oils is polyalphaolefin. Polyalphaolefin-based gear lubricants when combined with the correct additive pack show much improved thermooxidative stability over mineral-oil-based lubricants. They also show a potential for reducing energy losses in the gear box. Frictional losses are reduced, which manifests itself as lower oil temperature in service. The lower oil temperature also helps increase service life. Polyglycols are reported (14) to show particular advantages in worm gears where relative motion of teeth is entirely sliding and frictional losses can be considerable. The use of the polyglycol reduces friction and lowers operating temperature.

D. Compressor Oils

The compressor market can be split into three sectors. These are air compressors, compressors for other gases, and refrigeration compressors. Industrial air compressors make up the largest sector and have seen the successful introduction of synthetic-based lubricants. The reason for this is that the air compressor exposes the lubricant to air at high temperatures, which results in the ready oxidation of mineral-based oils. Frequent changing of a mineral-oil-based lubricant is therefore necessary for continued operation of the compressor. The replacement of the mineral oil by a synthetic oil can greatly increase the time between lubricant changes, and the greater expense of the synthetic oil is more than offset by the reduction in down-time and by the extended oil life. For example, in a rotary-screw air compressor the time between oil changes can be extended from 1,000–2,000 h of operation to between 4,000 and 8,000 h by replacing a mineral-oil-based lubricant with a polyalphaolefin-based lubricant. In general, synthetic lubricants based on polyalphaolefins or esters have enjoyed the most success. Polyalphaolefins have tended to find greater acceptance for use in rotary-screw compressors, and esters have tended to be favored in reciprocating compressors. The reason for this is based on a perceived greater degree of lubricity from the ester-based lubricant, which is beneficial in the prevention of wear between the piston and cylinder. However, the real-world situation is more complex than a simple comparison between the fundamental properties of ester types versus polyalphaolefins, because in the finished air compressor lubricant there will also be chemical additives and associated carrier oils, and in the case of a largely polyalphaolefin-based lubricant there will almost certainly be between 5 and 15% of an ester component to ensure good seal swell characteristics as well as to provide some improved lubricity and additive solubility. We expect the share of this market held by synthetic-based lubricants to increase. The desire for longer drain intervals will continue to drive the market in the direction of the synthetic lubricant. Longer drain intervals will be desirable not only from an efficiency standpoint but also because of the increasing concerns with used oil disposal. In applications where biodegradability is a factor, we expect polyol ester-based lubricants to gain over polyalphaolefin-based lubricants. This might be expected in the "mobile compressor" sector of the market, where spills and leakage into the environment are more likely.

Changes in the market for refrigeration compressor lubricants are driven by mandated changes in the refrigerant fluid used. This is most evident in the compressors used for automotive air conditioning and in the compressors found in domestic refrigerators and freezers. These compressor types traditionally have used the chlorofluorocarbon R12 as the refrigerant fluid with mineral-oil- or alkyl-benzene-based lubricants. Recent international agreements (6, 7) have mandated the phased replacement of chlorofluorocarbons with materials of lower ozone depletion potential. The R12 in automotive air-conditioning units and domestic refrigeration units is most likely to be replaced by the hydrofluorocarbon (HFC) R134a. R134a is a much more polar molecule than R12. For that reason, it differs from R12 in that it is not miscible with hydrocarbon-based lubricants such as mineral oils, alkyl benzenes, or polyalphaolefins. Miscibility is an important requirement in the designs of these systems. For the last few years a considerable amount of work has been devoted to developing new lubricants for use with R134a. Early work focused on polyalkylene glycol-based lubricants, with more recent developments being made in the area of ester-based lubricants. Other chemistries investigated have included derivatives of unsaturated polyalphaolefin oligomers and fluorinated molecules, but to date these have not been successful for technical and economic reasons. The first-generation lubricant for use with R134a in automotive air-conditioning units likely will be polyalkylene glycol based. For domestic refrigerators, it is more probable that the lubricant of choice will be ester based. The 1990s will see the development of this significant new market for synthetic compressor oils, amounting to some 5,000 tons of lubricant for automotive air conditioners and some 16,000 tons of lubricant for R134a compressors in domestic refrigerators and freezers each year. Only if R134a does not prove to be the refrigerant of choice is this scenario going to change. That would not seem to be very likely, but there are some voices raised against R134a on the basis of its potential contribution to global warming, and the favored refrigerant might then be R152. If this, or some other so far unrecognized negative factor, becomes an issue, then this whole area will become open again.

Another result of the phase-out of CFCs could be an increase in the use of ammonia-based refrigeration systems. This also could result in increased use of synthetic lubricants, probably based on polyalphaolefin, although hydrotreated, highly paraffinic-based lubricants are also likely to share a part of this market sector. Lubricants based on these fluids show greater thermal and chemical stability with ammonia than do conventional oils. This results in reduced sludge and varnish formation, which allows extended drain interval (15). The polyalphaolefin-based lubricant has the advantage of superior low-temperature fluidity. In large industrial ammonia compressors polyalkylene glycols have found application as lubricants. Ester-based lubricants are not suited for use with ammonia because their use results in excessive sludge formation.

For propane and other hydrocarbon refrigerants polyglycol-based lubricants have proved suitable. These fluids have the advantage that they do not dissolve the hydrocarbon gas and thus do not suffer from dilution and consequent loss of viscosity as do mineral oils (14).

In gas compressors, the nature of the gas determines the type of lubricant to be used. Certain polyglycols, as mentioned above, have the advantage that they do not dissolve hydrocarbons. This makes them particularly suitable for the lubrication of natural gas compressors. In addition, these polyglycols are toxicologically acceptable for use in low-density polyethylene compressors. The use of the polyglycol-based lubricant results in extended drain interval, reduced wear, and improved efficiency. However, the low-density polyethylene market is mature and is expected to decline as a market for synthetic lubricants. New production is focused on low-density linear polyethylene, which is a low-pressure process that does not require the performance level of polyglycol lubricants. Oxygen compressors are either designed to run dry or, in order to minimize the risk of explosion, a low-flammability lubricant, often a phosphate ester, is used.

E. Paper-Mill Oils

Synthetic lubricants are beginning to establish niche positions in the pulp and paper industry. The key to their use is improvement in the plants overall productivity. The industry typically will follow original equipment manufacturers' recommendations for the lubrication of new or revamped equipment. The trend is to run hotter and for newer machines to have smaller lubricant reservoirs. Synthetic lubricants offer greater stability than the currently used mineral oils, and their advantages are being recognized by OEMs, component manufacturers, and plant lubrication engineers. Polyalphaolefin-based lubricants are likely to take a significant position in coming years, particularly in drying and calendering operations.

Synthetic greases also are developing niche application areas. For example, they are used in roller bearings in such areas as the wet-end press section where the combination of high-loading, high-temperature, and acidic conditions impose particularly severe operating conditions. A slow but steady growth of niche applications such as these is anticipated for a variety of synthetic and part-synthetic lubricants. Where savings justify it, even lubricants as expensive as perfluorinated polyether greases will find their niche.

F. Industrial Greases

The need for greater appreciation of the lubrication of bearings was highlighted at a recent NLGI meeting (16). It was estimated that 43% of premature rolling bearing failures are caused by improper lubrication. In many cases it was indicated that this was due to the use of mineral-oil grease being used in applications where their limit of performance is being exceeded. Synthetic industrial greases are thus finding increasing niche application areas as they offer increased service life under conditions of higher load and higher temperature. The benefit of using such greases is the potential for a significant reduction in maintenance and down-time. For the same reason there is a trend toward the use of sealed-for-life bearings. This trend also will develop the use of synthetic lubricants. Specialized synthetic lubricants also find their niches in potentially hostile chemical environments where advantage arises from superior stability.

G. Metalworking Oils

Metalworking consumes a very large volume of fluids, many of which are based on water solutions or emulsions. Indeed, as mentioned in our introduction, the term "synthetic fluid" is often used in this area to describe water-based systems. It is beyond the scope of this chapter to discuss these systems in detail. Suffice it to say that in this industry the use of synthetic organic fluids has largely been as components in aqueous systems. Polyalkylene glycols find use, in aqueous solution, in cutting oils and quenching oils. Long-chain polyol esters are finding increased use in steel-rolling oil formulations. The development of their use has been and will continue to be driven by the desire for greater efficiency, higher quality, and ever-increasing environmental pressures. These pressures already have had considerable impact in limiting the use of chlorinated materials, nitrites, and phenols in finished emulsions. The need to reduce smoke formation in the workshop will require the use of increasingly stable and less volatile materials, which will continue to develop the use of synthetic organic fluids in formulations.

H. Food-Contacting Oils

Considerable quantities of oils are used in processes that bring them directly or indirectly into contact with food. The vast majority of these are white oils composed of mineral hydrocarbons refined to sufficient degree to be regarded as food grade. In the United States, fluids coming into contact with food require appropriate Food and Drug Administration (FDA) approval, but different regulations apply in different parts of the world.

White mineral oils have come under increasing scrutiny regarding their suitability for use where food contact is involved, and the results of a recent Shell study (17) have attracted much attention, particularly in Europe (4). If, as a result of these concerns, white mineral oils are banned from significant sectors of this market, then there will be considerable opportunities for alternative fluids. For many applications it is likely that vegetable oils could find a ready market, but in certain processes it could prove necessary to use synthetic fluids because of a requirement for superior stability.

A prime example of this could be in the production of polystyrene. At the moment white mineral oil is used in the production of polystyrene as an internal lubricant and extender. Up to about 6% can be used, depending on the polystyrene grade. In use, polystyrene goods come into contact with foods and thus the mineral oil has to be of suitable food quality. In Europe, it is estimated that about 50,000 tons of white mineral oil are consumed each year in this application. Vegetable oils are unlikely to be suitable for this application because the process temperatures are too high, and hence there exists the possible development of a significant new market for suitably approved synthetic fluids.

It is impossible to predict how the situation may develop, but this is a striking example of how toxicological concerns could result in new regulations that would have a significant impact on the fluids market.

I. Heat-Transfer Fluids

The heat-transfer fluid market already has a substantial sector of synthetic fluids. For the world market, close to 20% of the annual volume of some 20 million gallons is synthetic fluid. In dollar terms, however, this $100 million market is split such that somewhat more than half its value is derived from the synthetic fluids (8, 18). Synthetics offer improved thermal stability and also improved low-temperature characteristics over mineral-oil-based fluids. Even when the heat-transfer function is only concerned with high temperatures, better low-temperature fluidity can prevent system shutdowns and the need to steam-trace lines in cold ambient conditions. The types of synthetic fluids used and a comparison of their properties, temperature range of operation, and design parameters have recently been reviewed (19, 20). This is a relatively mature market, and general growth is expected to be at a few percent a year, with synthetics gradually increasing their share of the total market. New opportunities are presenting themselves in relatively lower-temperature systems associated with the process control of drug and specialty chemical production, and this sector of the market is predicted to show the fastest growth (8, 21).

J. Textile Fluids

The textile industry in developed countries continues to come under tremendous pressure from lower-labor-cost developing countries. To compete, the textile industry is responding through advanced technology to reduce the labor component of production, to increase production efficiency, and to produce superior quality goods. In addition, the industry is facing considerable environmental pressure in terms of the workplace and the disposal of waste. Both of these factors are having some impact on the fluids used in process oil formulations. However, this is an industry that is extremely cost conscious, and synthetic fluids will penetrate the mineral oil share of the market slowly and only to the extent that a clear cost/performance benefit is realized.

In the production of texturized polyester and nylon yarns, greater efficiency is continuously sought through higher spinning and texturizing speeds and minimizing down-time of equipment needed for cleaning of hot plates and replacing of texturizing disks. The primary spin finishes applied after spinning and prior to texturizing today are largely synthetic based. Typically, these synthetic-based finishes are applied as an aqueous emulsion. Greater speed generally demands higher plate temperatures, and this requires the lubricant in the spin-finish formulation to have higher thermal stability or at least to decompose in a "clean" manner, leaving no deposits on the hot plate. Deposits on the hot plate impact on the heat transfer from plate to yarn and cause variability in quality in terms of texture and uniform dyeability. Variability in dyeability causes havoc in the production of first-quality fabrics and cannot be tolerated. Deposits on the hot plate also can lead to increased filament breakage, which reduces efficiency. Filament breakage becomes more of a problem as finer filament yarns are produced. Finer filament yarns are increasing their share of the market because of the generally superior aesthetics of fabrics and garments produced from such yarns. A higher smoke point is desirable to reduce contamination of the workplace, and there is growing concern over the fate of used spin-finish lubricants in the environment. Synthetic lubricants, often esters, are used in these spin-finish formulations because their superior thermal stability and higher smoke points make their use cost-effective versus mineral or vegetable oils.

Many other process oils used in the textile industry, such as coning oils, are largely mineral oil based. This is because the temperatures met in downstream processing are much lower than those found on texturizing hot plates and the cost-effectiveness of synthetic fluids is more difficult to demonstrate. Only if regulatory action demands it will we see a major switch away from these mineral-oil-based systems. However, there will continue to be particular situations where a synthetic-based oil will solve particular problems and their use will be justified. In particular, the higher smoke points of certain synthetic fluids over mineral oils is an attractive feature that will become more important. Synthetic fluids are to be found in ring spinning and twisting. The benefits are seen largely in increased efficiency through less down-time and lower energy consumption. In particular, they can be used to replace grease on solid rings, and they have the added benefit of cleaning petroleum-fouled rings and maintaining clean rings almost indefinitely.

Specialty yarns are a small but established market for synthetic fluids. The higher prices of such yarns mean that a more expensive fluid can be tolerated as long as its superior performance justifies its use. Synthetic fluids, silicones in particular, find application in the lubrication of spandex yarns. As the demand for such yarns increases, we can expect to see a corresponding increase in the quantity of synthetic fluid consumed. Silicones also find application in the lubrication of polyester staple yarns used as the stuffing in pillows, quilts, and duvets. This is because the particular frictional characteristics of silicone oils have proven themselves most suitable in maintaining the bulking characteristic of the staple. Such applications will continue to be developed as niche opportunities for the synthetic fluids.

K. Drilling Mud Fluids

Drilling muds were traditionally water based, and today there is still much development effort put into improving their qualities and performance. Indeed, one of the objectives of such development activity is the replacement of oil-based muds, which earlier displaced water-based muds in severe drilling environments such as those found in the North Sea. The more robust oil-based muds allowed for much more efficient drilling, reducing drilling time by 30% or more. Unfortunately, oil-based muds have proven to be unacceptable environmentally, particularly in the sea. In the North Sea, this has resulted in ever-tightening regulations governing oil-based muds and the quantity of mud allowed to be discharged along with cuttings into the sea (22). This has provided the stimulus for the development of muds based on synthetic fluids that will prove environmentally acceptable. Perhaps the first example of such a fluid is the ester around which Baroid developed its "Petrofree" mud. This is thought to be an isobutyl rape oil ester (23). A second fluid introduced into this market is described as a polyether and has been used by Akers. The principal characteristic of these fluids is a high degree of biodegradability, which, it is argued, will render them more environmentally acceptable in this area of application. Baroid supports its claims by describing biodegradation tests in seawater under both aerobic and anaerobic conditions (24). This market is still at a developing stage, and the final key for success will depend on the regulations imposed by the government authorities. However, we believe that the fundamental drive for environmentally improved muds will represent a significant opportunity for synthetic fluids tailored to provide the right balance of properties. The potential market is several tens of thousands of tons a year, which is very sizeable by the standards of today's markets for synthetic fluids.

L. Electrorheological Fluids

When considering future industrial fluid trends, it is appropriate to review the potential for the development of systems utilizing electrorheological fluids. Electrorheological fluids have been known for more than 50 years (25), but the last decade in particular has seen a resurgence in research and development activity in the creation of such fluids and their potential applications. The near-instantaneous response of an electrorheological fluid to an applied electric field is potentially a very attractive property in designing systems that will respond to and take advantage of the high speed with which a computer can detect, manipulate, and respond to signals. Added to this, such fluids offer the prospect of simplified design and fewer moving parts, which leads to greater reliability at lower cost. The development of anhydrous systems (26–28) gives the promise of extending the temperature range over which such fluids can operate.

It is clear that some of the projections made in the mid to late 1980s for commercial application of this technology and its subsequent growth have been too optimistic. At this writing, there are no commercial applications for these fluids. However, further breakthroughs in the development of electrorheological fluids, or the design of systems depending on them, could create a substantial market. The areas of application of any significant volume are likely to come first in the automotive market rather than in the industrial market, but if these develop, industrial applications will follow.

If electrorheological fluids do become commercially significant, they represent a substantial potential market for synthetic fluids. Certainly, electrorheological fluids can be produced using many different fluids, including mineral and vegetable oils, as the continuous phase (29), but we believe that it is likely that a potentially commercial fluid will require careful control of the properties of the continuous phase and dispersed phase. The fact that a synthetic fluid can be made to very precise quality-control limits will favor synthetic fluids over mineral-oil-based or vegetable-oil-based fluids.

REFERENCES

1. Como, D. J. (1990). *Eng. Dig.*, November, 10.
2. Sutor, P., E. A. Bardasz, and W. Bryzik (1990). SAE Technical Paper 900687, International Congress & Exposition, Detroit, MI, February 26–March 2.

3. Marolewski, T. A., R. J. Slone, and A. K. Jung (1990). SAE Technical Paper 900689, International Congress & Exposition, Detroit, MI, February 26–March 2.
4. Watts, P. (1989). *BIBRA Bull.*, 28, 59–65.
5. Molina, M. J., and F. S. Rowland (1974). *Nature*, 249, 810–812.
6. United Nations Environment Program (1987). Montreal Protocol on Substances That Deplete the Ozone Layer, Final Act, United Nations.
7. London Meeting of the Parties to the Montreal Protocol (June 1990).
8. Parkinson, G. (1989). *Chem. Eng.*, February, 39.
9. Electric Power Research Institute (1989). EPRI NP-6542, Project 2969-2, Final Report, September.
10. Blackwell, J. W., and D. K. Walters (1987). Developments in Hydraulics for the Construction Industry, presented at Public Works Congress 1987, The British Fluid Power Association, Birmingham, England, April.
11. *Chemistry in Britain* (1990). May, p. 404.
12. Phillips, W. D. (1990). Synthetic Fire Resistant Fluids, presented at the College of Petroleum Studies Synthetic Lubricants Course SP 5, St. Catherine College, Oxford, December.
13. Kaspar, K. (1977). *Der Maschinenschaden*, 50(3), 87–92.
14. Hatton, D. R. (1990). Trends in Industrial Synthetic Lubricants, presented at the College of Petroleum Studies, Synthetic Lubricants Course SP5, St. Catherine College, Oxford, December.
15. Short, G. D. (1990). *Jrn. Soc. of Tribololism and Lubrication Eng.* 46(4), 239–247.
16. Wunsch, F. (1990). Synthetic Fluid Based Lubricating Greases, presented at the 1990 Annual Meeting of the National Lubricating Grease Institute, Denver, CO, October 28–31.
17. Clark, D. G. (1987). Shell Oil Company, External Report SBER.87.010, June.
18. Business Communications Company, Inc. (1989). *Industrial Fluids: Opportunities and Markets*, Business Communications Company, Inc., report C-091, December, p. 111.
19. Green, R. L., A. H. Larsen, and A. C. Pauls (1989). *Chem. Eng.*, February, 90–98.
20. Seifert, W. F. (1989). *Chem. Eng.*, February, 99–104.
21. Gruver, M. E., and R. Pike (1988). *Chem. Eng.*, December, 149–152.
22. Scruton, M. (1990). *Petrol. Rev.*, October, 502–505.
23. Muller, H., C. P. Herold, S. von Tapavicza, D. J. Grimes, J. M. Braun, and S. P. T. Smith, European Patent Application 0 374 671, June, 1990.
24. Peresich, R. (1990). *Offshore*, September, 32–33, 55.
25. Winslow, W. M. (1949). *J. Appl. Phys.*, 20(12), 1137–1140.
26. Block, H., and J. P. Kelly, U.S. Patent 4,687,589. (1987).
27. Filisko, F. E., and W. E. Armstrong, U.S. Patent 4,744,914. (1988).
28. Goosens, J., G. Oppermann, and W. Grape, U.S. Patent 4,702,855. (1987).
29. Block, H., and J. P. Kelly (1988). *J. Phys. D: Appl. Phys.*, 21, 1661–1677.

24
Aerospace

Carl E. Snyder, Jr.
and Lois J. Gschwender
Air Force Wright Laboratory
Wright-Patterson Air Force Base, Ohio

Synthetic fluids and lubricants have been used in aerospace equipment for many years. Aerospace applications are very demanding on fluids and lubricants. The major reason that aerospace applications are so demanding is that there is a large concern about the weight associated with aerospace systems. Since significant costs are incurred with flying every pound of an aerospace system, all elements of every system are the smallest, lightest available. This results in minumum volumes of fluids and lubricants used, the smallest heat exchangers possible, smaller reservoirs, smaller pumps and actuators, etc. The result is that fluids and lubricants in aerospace applications are required to withstand extremely severe levels of stress because small volumes are used and must operate at high temperatures generated in the application, as well as at the extremely low temperatures in which aerospace equipment is required to operate. In general, synthetic fluids and lubricants are required for aerospace applications due to the wide temperature range over which they must operate. In comparison, nonaerospace applications are generally not as concerned about the amount of fluid used or the overall weight of the system and do not put as much demand on fluids and lubricants. However, as nonaerospace applications become more sophisticated and the synthetic fluids and lubricants become less exotic and more readily available at lower costs, synthetics will be more widely used. Different classes of synthetics have been used for different aerospace application areas.

To better define the scope of aerospace applications, it must be recognized that the largest-volume applications occur in aircraft equipment. Although the excellent performance characteristics of synthetic fluids and lubricants have also resulted in their use in spacecraft, missiles, and satellites, these volumes are significantly smaller and therefore not as well known or defined. The arrangement of the fluid classes discussed in this chapter is based on the larger-volume applications of the class of synthetics as a primary method of grouping. Lower-volume applications will be mentioned as appropriate as they are being discussed. Most of the classes of synthetic lubricants covered in this chapter have been covered in detail in earlier chapters of this book. Therefore, references are cited only when a specific class of synthetic lubricants is not covered elsewhere in this book or when the information is extremely important.

The two major areas of application of synthetic fluids and lubricants are gas turbine engine lubricants and hydraulic fluids. Applications involving synthetics that are of lower volume are greases, coolants, and inertial guidance damping fluids. This method of grouping is not meant to indicate that a critical, low-volume application where a synthetic lubricant is the only choice is not equally, if not more, important.

I. LIQUID LUBRICANTS

The largest-volume application area of liquid lubricants in aerospace is gas turbine engine oils. The most widely used class of synthetic lubricants for gas turbine engine oils is the esters. Selection of esters was driven by their wide usable temperature range and their excellent thermooxidative stability in the presence of metals. The environment in which they are required to operate includes extremely low temperatures (down to –55°C), at which their viscosity must be low enough to permit the engines to start, as well as high bulk fluid temperatures (up to 200°C), at which they must provide lubrication for the main shaft bearings in the engines. The ester-based lubricants used for the engine lubrication application are described in military specifications MIL-L-7808 (1) and MIL-L-23699 (2). Those specifications currently describe ester-based turbine engine lubricants that operate from –54 to 175°C and from –40 to 175°C, respectively. Materials conforming to those specifications are adequate to meet the lubrication requirements for most current aerospace gas turbine engine. However, in an attempt to improve the fuel efficiency of turbine engines, higher operational temperatures are predicted for near-term advanced engines. It is anticipated that those requirements can be met by modifications of the upper-temperature operational requirements of the current specifications, and the properties of the ester based lubricants that meet them, to 200°C. This will require a careful balance of ester base stocks and improved additives to achieve the balance of viscosity–temperature properties and excellent thermooxidative stability, as well as other requirements for a gas turbine engine lubricant (3). Further advanced engine concepts will require that different classes of synthetic lubricants be utilized, as will be discussed. In addition to turbine engine lubrication, the esters are used in aerospace applications as low-temperature greases such as MIL-G-23827 (4), gear oils such as DOD-L-85734 (5), and to some degree as instrument lubricants such as MIL-L-6085 (6).

When aircraft engine operational temperatures exceeded the limits of ester-based lubricants, another class of synthetic lubricants with significantly higher high-temperature stability was utilized. That class of synthetic lubricants is the polyphenylethers (7). The liquid lubricant described in military specification MIL-L-87100 (8) has an upper operational temperature of 300°C. In addition, MIL-L-87100 has excellent fire resistance as demonstrated by a flash point in excess of 450°C and autogenous ignition temperature of 450°C. The major deficiency of this class of liquid lubricants is that they have extremely poor low-temperature operational capability, as they have pour points of +5°C and higher, limiting their lower use temperature to +15°C. In addition, the current formulation described by the specification of these fluids has relatively poor lubricity characteristics compared to other classes of liquid lubricants. Those limitations, coupled with their high cost ($1,000+ per gallon), have limited their use to applications where no other liquid lubricants would function. As more efficient gas turbine engines operating at higher temperatures are developed, the polyphenylethers, either as MIL-L-87100 or an advanced version of the specification, will find increased applications.

When the capabilities of the polyphenylethers are exceeded or when liquid lubricants are required to be capable of operating not only at the elevated temperatures where polyphenylethers operate but also at the more typically required low temperatures of –40°C and below, it is anticipated that the liquid lubricant of choice will be based on a perfluoropolyalkylether (PFPAE) (9). Commercial versions of this class of synthetic lubricants are currently available with the potential for providing a liquid lubricant capable of operating over a –54 to 300°C temperature range. Research and development programs are currently underway to increase the upper temperature to at least 345°C. The major deficiency of this class of synthetic lubricants is the lack of suitable additive technology. The chemical behavior of the PFPAE fluids is so different from other nonperfluorinated lubricants that the additives used to enhance the properties of other lubricants are not even soluble in PFPAE fluids. There are very limited examples of additives that are soluble in PFPAE fluids, and those were all specifically synthesized to be soluble in PFPAE fluids (10–12). While this class of fluids has very attractive and impressive properties as unformulated fluids, their true potential cannot be realized until a supporting technology base of performance-improving additives has been developed. The types of additives required for PFPAE fluids to have properties appropriate for their use as liquid lubricants in aerospace applications are (a) metal deactivator/stability additive, (b) rust inhibitor, and (c) lubricity additive. The PFPAE synthetics are used in oxidatively stable greases as described in military specification MIL-G-27617 (13). Other potential aerospace applications for formulated PFPAE fluids include long-life lubricants for space, instrument lubricants, and high-temperature nonflammable hydraulic fluids.

II. HYDRAULIC FLUIDS

Synthetic-based hydraulic fluids are widely used in aerospace. The nonsynthetic hydraulic fluid that the synthetics replaced in both commercial and military aircraft is described in specification MIL-H-5606 (14). Synthetic hydraulic fluids were developed to replace MIL-H-5606 in order to provide increased fire safety. MIL-H-5606 is a naphthenic mineral-oil-based hydraulic fluid that has proved to be an adequate aerospace hydraulic fluid from an operational aspect. However, the high flammability hazard associated with its use is well known (15). The commercial aircraft industry recognized this hazard first and, in conjunction with the fluid industry, developed a fire-resistant hydraulic system around the phosphate ester class of synthetics. It was necessary to develop an entirely new hydraulic system because the phosphate esters are not compatible with the same seals, paints, wiring insulation, etc. that are used in aircraft using a hydrocarbon-based hydraulic system. In addition, hydraulic system components had to be modified to provide optimum performance with the new phosphate-ester-based hydraulic fluids. The phosphate ester hydraulic fluids are described in AS1241A (16).

The military community did not follow the commercial industry in the switch from MIL-H-5606 to phophate esters. This decision was driven primarily by the noncompatibility of the phosphate esters with the aircraft systems and ground-services equipment originally designed to use the hydrocarbon-based MIL-H-5606. In fact, mixtures of MIL-H-5606 and AS1241 hydraulic fluids resulted in gel formation, requiring excessive maintenance to correct the problem. In addition, the aggressive solvency of the phosphate esters toward seals, paints, and wiring insulation used in aircraft with hydrocarbon-oil-based hydraulic systems prevented their consideration as a retrofit option. The military conversion from MIL-H-5606 to a fire-resistant synthetic-based hydraulic fluid required that another new class of synthetic fluids be developed, that is, synthetic hydrocarbon fluids based on polyalphaolefins. The synthetic hydraulic fluids based on polyalphaolefins are described in military specification MIL-H-83282 (17). MIL-H-83282 was developed to replace MIL-H-5606 as a no-retrofit, drain-and-fill replacement. This required total compatibility with the materials used in MIL-H-5606 systems and with the MIL-H-5606 system designs. Most military aircraft were converted to MIL-H-83282 by 1985. The only aircrafts for which the conversion was not approved were those for which acceptable operation at –54°C would be compromised by the higher viscosity of MIL-H-83282 at lower temperatures. MIL-H-83282 is described as a –40 to 204°C hydraulic fluid, compared to –54 to 135°C for MIL-H-5606. A recently completed development program has provided a polyalphaolefin-based fire-resistant hydraulic fluid with –54°C viscosity equivalent to MIL-H-5606 (18,19). The improved fire-resistant properties of MIL-H-83282 over MIL-5606, which have resulted in significant reductions in hydraulic fluid fire damage, include (a) higher flash and fire points, (b) higher autogenous ignition temperature, (c) lower flame propagation rate, and (d) improved resistance to gunfire ignition (15). The conversion of aircraft from MIL-H-5606 to MIL-H-83282 was accomplished by both drain-and-fill and attrition methods, both of which were equally successful and without problem.

Other quite important, but smaller-volume, applications of polyalphaolefins (PAO) are greases such as MIL-G-81322 (20), instrument lubricants such as MIL-L-85812 (21), and liquid coolants (22). The PAO-based greases provide excellent usable temperature range and good reliability with low maintainability requirements. Instrument lubricants based on PAO have successfully replaced the difficult-to-obtain paraffinic-based mineral-oil instrument lubricants previously used. The PAO-based coolants meeting the properties defined in MIL-C-87252 (23) are in the process of replacing another class of synthetic fluids, the orthosilicate esters, as dielectric and liquid coolants in military electronic systems (22). The polyalphaolefins have excellent properties as lubricants and hydraulic fluids, and their compatibility with mineral oils and systems designed to use mineral-oil-based lubricants and fluids makes them excellent candidates for use in newly emerging aerospace systems; they are also outstanding candidates as replacements for mineral-oil-based products when they either become difficult to obtain or can no longer provide adequate performance.

Both phosphate ester and polyalphaolefin hydraulic fluids have been excellent hydraulic fluids, which, due to their fire-resistant properties, have significantly reduced the hydraulic fluid fire hazards in both commercial and military aircraft. However, they are not nonflammable, but are capable of ignition if sufficient energy (temperature, flame, etc.) is available. On current and future aircraft, high-fire-hazard areas where hydraulic fluids are used exist in brake systems, in which brake temperatures can approach 1600°C on an aborted takeoff, and around engine nacelles where the temperatures are in excess of 800°C. Both of these conditions exceed the autogenous ignition temperatures and flash and fire points of both phosphate-ester- and polyalphaolefin-based hydraulic fluids. As the costs of our aircraft and other aerospace systems continue to increase, it becomes even more important to minimize the possibility of losing these aircraft to hydraulic fluid fires. The development and validation of a completely nonflammable hydraulic fluid and compatible seals have recently been completed (24–26). The synthetic hydraulic fluid is based on chlorotrifluoroethylene oligomers (CTFE) and is described in military specification MIL-H-53119 (27). The CTFE-based hydraulic fluid is not compatible with hydraulic systems designed for use with other hydraulic fluids and therefore requires that hydraulic systems be designed around its unique properties. MIL-H-53119 is specified for use from –54 to 175°C and is compatible with a number of elastomeric seals. One of the major disadvantages of MIL-H-53119 is that it has significantly higher density, which results in a serious penalty for use in aerospace applications. In order to overcome this penalty, higher pressure hydraulic components were developed and systems were designed and validated. At higher pressures, 55.2 MPa (8000 psi), the penalties associated with the higher density are minimized due to the extremely small volumes of hydraulic fluid required. If the weight penalty were not important, MIL-H-53119 could be used at lower pressures and could provide nonflammable hydraulic systems for a variety of application areas. Another important application for higher-molecular-weight versions of CTFE as well as polymers of bromotrifluoroethylene (BTFE) is as high-density flotation/damping fluids for inertial guidance systems.

III. OTHER

The only class of synthetic fluids that has been developed for quite some time that has not been discussed in this chapter is the silicones. The silicone class of synthetics has some very interesting properties that would make it seem to be a serious candidate for a wide number of aerospace applications. The most important of those is the extremely good viscosity–temperature properties the silicone fluids possess, especially the polydimethylsiloxanes. However, the silicones also possess two less desirable properties that make them less desirable for the two major volume applications in aerospace, that is, gas turbine engine lubricants and hydraulic fluids. The more significant deficiency is their inability to provide lubrication for steel-on-steel rubbing surfaces. Lubricity additives are generally not effective in silicones. This deficiency has limited their use as both liquid lubricants and hydraulic fluids. In addition, another deficiency that limits their use as hydraulic fluids is their low bulk modulus, or high propensity for compressibility. This requires compensation in hydraulic system design in the form of larger actuators than would be required for less-compressible fluids. The larger actuator would compensate for the "sponginess" of the fluid and would provide satisfactory service, but the weight of the hydraulic system would be significantly increased, which is unacceptable for aerospace applications. However, silicones have been used in a variety of greases (28) that are widely used in aerospace applications.

Another member of the silicon-containing class of synthetic fluids is the silicate ester class (7). This class of synthetic fluids has had two areas of application: wide-temperature-range hydraulic fluids and coolants. The original application of the silicate esters as a hydraulic fluid was as described in military specification MIL-H-8446 (29). This specification, which has been canceled due to lack of current systems requiring the fluid, described a hydraulic fluid for use over the temperature range of –54 to 204°C. The silicate esters were the most acceptable class of hydraulic fluids for that requirement. Their major deficiency was their propensity to hydrolyze with moisture that got into the hydraulic system. The resulting hydrolysis products were an alcohol, which degraded the fire resistance of the fluid, and a gelatinous precipitate that clogged system filters and the small orifices that exist in hydraulic systems, resulting in high levels of maintenance. Similar hydrolysis problems were experienced with the silicate-ester-based coolants, described in military specification MIL-C-47220 (30). This problem with hydrolysis, which resulted in a high level of maintenance, has led to the recent substitution of the polyalphaolefin-based coolant MIL-C-87252 for MIL-C-47220 in many military aerospace applications.

IV. DEVELOPMENTAL SYNTHETIC FLUIDS AND LUBRICANTS

The synthetic fluids and lubricants discussed so far in this chapter have found significant application in the aerospace industry or else have had significant production capability and potential applications identified. In this section, classes of newly emerging synthetic lubricants and fluids are discussed, with the properties which make them so promising.

The first class of newly emerging synthetics is the silahydrocarbon, or tetraalkylsilane, class. While this class of synthetics has been known for quite some time, its potential application in the aerospace industry had not been significantly advanced until recently (31). The largest-volume application for the silahydrocarbons is as wide-temperature-range, high-temperature, fire-resistant hydraulic fluids. Their excellent viscosity–temperature properties make them excellent candidates, as they can be used down to –54°C while still maintaining adequate viscosity at elevated temperatures to provide adequate film thickness for lubrication. Their excellent stability at temperatures up to 370°C permits their extended use at elevated temperatures. Since these fluids contain aliphatic carbon–hydrogen bonds, oxygen must be excluded at these elevated temperatures. Another very important aerospace application is liquid space lubricants (32,33). Their excellent viscosity–temperature characteristics permit the selection of extremely high molecular weight (1,000–1,500 amu) silahydrocarbon fluids to be used. These fluids have extremely low volatility, which makes them excellent for long-life, noncontaminating liquid lubricants for space.

Another class of synthetic fluids and lubricants that are still in the stages of development are the *n*-alkyl benzenes (34). These fluids have excellent thermal stability and very good viscosity–temperature properties. One of the advantages these fluids have over the polyalphaolefin and silahydrocarbon classes for use at high temperature is their improved solubility for performance-improving additives. The benzene ring appears to provide significant solubility enhancement for the typically polar performance-improving additives that are essential to provide the required performance. Their most promising aerospace application is as a wide-temperature-range, high-temperature hydraulic fluid. The major obstacle in reducing them to the application is lowering the cost of production, while maintaining the wide liquid-range stability.

REFERENCES

1. MIL-L-7808J Military Specification (1982). Lubricating oil, aircraft turbine engine, synthetic base, NATO code number 0-148, 11 May.
2. MIL-L-23699D Military Specification (1990). Lubricating oil, aircraft turbine engine, synthetic base, NATO code number 0-156, 9 October.
3. Gschwender, L. J., C. E. Snyder, Jr., and G. A. Beane, IV (1987). Military aircraft 4-cSt gas turbine engine oil development, *Lub. Eng.*, 43(8), 654–659.

4. MIL-G-23827B Military Specification (1983). Grease, aircraft and instrument, gear and actuator screw, NATO code number G-354, 20 June.
5. DOD-L-85734 (1985). Lubricating oil, helicopter transmission system, synthetic base, 21 February.
6. MIL-L-6085C Military Specification (1991). Lubricating oil: Instrument, aircraft, low volatility, 5 February.
7. Gunderson, R. C., and A. W. Hart, eds. (1962). *Synthetic Lubricants*, Reinhold Publishing, New York.
8. MIL-L-87100 Military Specification (1976). Lubricating oil, aircraft turbine engine, polyphenyl ether base, 12 November.
9. Snyder, C. E., Jr., and L. J. Gschwender (1983). Fluoropolymers in fluids and lubricant applications, *I and EC Proc. R and D*, 22,383–386.
10. Tamborski, C., and C. E. Snyder, Jr. (1977). Perfluoroalkylether substituted aryl phosphines and their synthesis, U.S. Patent 4,011,267, March 8.
11. Tamborski, C., and C. E. Snyder, Jr. (1984). Perfluoroalkylether substituted phenyl phosphines, U.S. Patent 4,454,349, June 12.
12. Sharma, S. K., L. J. Gschwender, and C. E. Snyder, Jr. (1990). Development of a soluble lubricity additive for perfluoropolyalkylether fluids, *J. Synth. Lub.*, 7, 15–23.
13. MIL-G-27617D Military Specification (1984). Grease, aircraft and instrument, fuel and oxidizer resistant, 14 November.
14. MIL-H-5606E Military Specification (1978). Hydraulic fluid, petroleum base; Aircraft missile and ordnance, NATO code number H-515, 26 January.
15. Snyder, C. E., Jr., A. A. Krawetz, and T. Tovrog (1981). Determination of the flammability characteristics of aerospace hydraulic fluids, *Lub. Eng.*, 37(12), 705–714.
16. AS 1241A (1983). Fire Resistant Phosphate Ester Hydraulic Fluid for Aircraft, Society of Automotive Engineers, Warrendale, PA, March.
17. MIL-H-83282 Military Specification (1986). Hydraulic fluid, fire resistant, synthetic hydrocarbon base, aircraft, metric, NATO code number H-537, 25 March.
18. Gschwender, L. J., Snyder, C. E., Jr., and Fultz, G. W. (1986). Development of a −54°C to 135°C synthetic hydrocarbon-based, fire-resistant hydraulic fluid, *Lub. Eng., 42,* 485–490.
19. Gschwender, L. J., Snyder, C. E., Jr., and Sharma, S. K. (1988). Pump evaluation of hydrogenated polyalphaolefin candidates for a −54°C to 135°C fire-resistant Air Force aircraft hydraulic fluid, *Lub. Eng. 44,* 324–329.
20. MIL-G-81322D Military Specification (1982). Grease, aircraft, general purpose, wide temperature range, 2 August.
21. MIL-L-85812 Military Specification (Issue Pending). Lubricating oil, instrument, ball bearing, synthetic hydrocarbon.
22. Gschwender, L. J., C. E. Snyder, Jr., and A. A. Conte, Jr. (1985). Polyalphaolefins as candidate replacements for silicate ester dielectric coolants in military applications, *Lub. Eng.*, 41, 221–228.
23. MIL-C-87252 Military Specification (1988). Coolant fluid, hydrolytically stable, dielectric, 2 November.
24. Snyder, C. E., Jr., and L. J. Gschwender (1980). Development of a nonflammable hydraulic fluid for aerospace applications over a –54°C to 135°C temperature range, *Lub. Eng.*, 36, 458–465.
25. Snyder, C. E., L. J. Gschwender, and W. B. Campbell (1982). Development and mechanical evaluation of nonflammable aerospace –54°C to 135°C hydraulic fluids, *Lub. Eng.*, 38, 41–51.
26. Gschwender, L. J., C. E. Snyder, Jr., and S. K. Sharma (1992). Development of a –54°C to 175°C high temperature nonflammable hydraulic fluid MIL-H-53119 for Air Force systems, *Lub. Eng.*, in review.
27. MIL-H-53119 Military Specification (1991). U.S. Army, Hydraulic fluid, nonflammable, chlorotrifluoroethylene base, 1 March.

28. MIL-G-25013E Military Specification (1983). Grease, aircraft, ball and roller bearing, NATO code number G-372, 20 June.
29. MIL-H-8446B Military Specification (1959). Hydraulic fluid, nonpetroleum base, aircraft, 12 March.
30. MIL-C-47220B Military Specification (1982). USAF, Coolant fluid, dielectric, 29 December.
31. Snyder, C. E., L. J. Gschwender, C. Tamborski, G. Chen, and D. R. Anderson (1982). Synthesis and characterization of silahydrocarbons—A class of thermally stable wide liquid range functional fluids, *ASLE Trans.*, 25, 299–308.
32. Paciorek, K. J. L., J. G. Shih, R. H. Kratzer, B. B. Randolph, and C. E. Snyder, Jr. (1990). Polysilahydrocarbon synthetic fluids I. Synthesis and characterization of trisilahydrocarbons, *I and EC Prod R and D,* 29, 1855–1858.
33. Snyder, C. E., Jr., L. J. Gschwender, B. B. Randolph, K. J. L. Paciorek, J. G. Shih, and G. J. Chen (1992). Research and development of low volatility long life silahydrocarbon based liquid lubricants for space, *Lub. Eng., 48,* 325–328.
34. Gschwender, L. J., C. E. Snyder, Jr., and G. Driscoll (1990). Alkyl benzenes—Candidate high-temperature hydraulic fluids, *Lub. Eng.*, 46, 377–381.

25 Environmental Impact

Brad F. Droy*
Ethyl Corporation
Baton Rouge, Louisiana

I. INTRODUCTION

As the decade of the 1990s unfolds, an ever-increasing demand placed on the chemical industry is the development of materials that are environmentally friendly. For the most part, we as a society (including the chemical industry) are uneducated as to the complexity and definition of the concept of "environmental friendliness." As an industry, we realize that we must meet the demands of a population that is becoming increasingly educated on environmental issues. To become environmental stewards (and to be perceived as such by society) could greatly benefit the industrial sector. This will not be an easy task to accomplish. To this end, it seems prudent first to outline and define the complexity of the problem before one attempts to describe what the environmental impact of a class of materials would be on the environment.

To properly address the environmental impact of a chemical, one should analyze the entire life-cycle of the material. Such a "cradle-to-grave" analysis (Fig. 25.1) is currently being used to develop ecolabeling schemes in Europe and elsewhere (1). In this approach, the impact of each life-cycle stage is addressed and applied to the overall effect a particular product will have on the environment. The four major quantifiable end points in the life-cycle analysis are energy output, wastes, reuse, and recycle capabilities. Although extremely complex, the life-cycle analysis is the only correct way to address and objectively quantify the complete environmental impact a material will have. The life-cycle analysis is unique in that it focuses on the entire life history of a product, not just on

---

**Current affiliation:* Woodward-Clyde Consultants, Baton Rouge, Louisiana

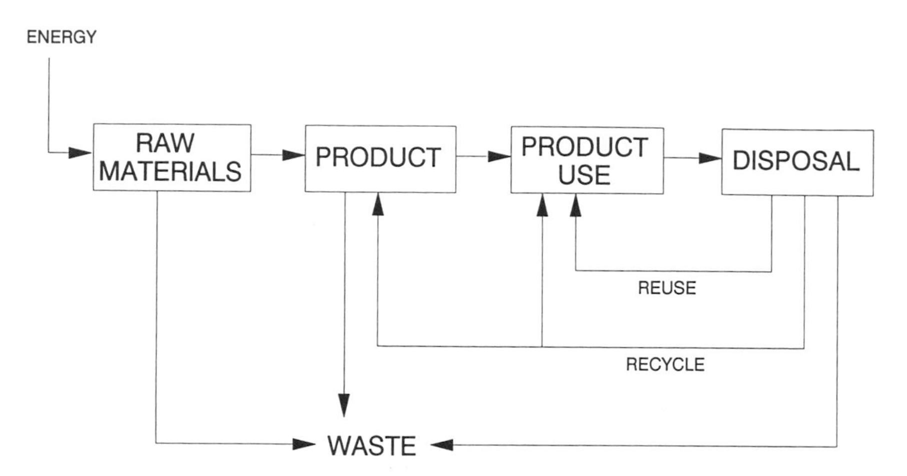

Figure 25.1 Schematic representation of a "cradle-to-grave" analysis. Adapted from ref. 1.

one specific aspect such as plant emissions. Although in its infancy stages, the life-cycle analysis will undoubtedly be important to chemical industry research and development practices in the future. For instance, substitution of alternative raw materials or use of material and product manufacturing practices that utilize less energy input and produce less waste could lead to the production of a more marketable, environmentally friendly material.

Once a life-cycle analysis is completed, a clear picture of a material's overall environmental impact is produced. This assessment may include information on the energy and resources needed to produce the material, what portions may be reusable, what the recycle potential is, and, very importantly, the risk of the material and waste products of manufacture to health and environment. Therefore, the life-cycle analysis will help industry produce a material in the most environmentally sound manner.

What society will demand and/or tolerate with respect to the environment, however, is unpredictable. A case in point is the Alar scare. Here, primarily due to misinformation and subsequent panic, the public demanded discontinuation of use of Alar although the scientific evidence has shown the material was not harmful to people. On the other hand, millions of people fill their cars with gasoline each day and expose themselves to fumes that are known to be hazardous, at least to laboratory animals. In addition, it is a well-known fact that the earth has a limited supply of oil reserves that eventually will be exhausted. Yet the public continues to use gasoline (primarily due to convenience) as if it is nonhazardous and inexhaustible. These examples illustrate the complexity of the economic, political, and public perception issues that add to the already complex quantifiable aspects of defining the environmental impact of chemicals on health and environment. A treatise, therefore, on the political, economical, and psychological aspects of environmental friendliness is well beyond the scope of this chapter. However, these concepts should be addressed in any holistic analysis of environmental impact.

Two major areas within the cradle-to-grave matrix that have received the most attention are raw materials use and depletion and toxicity of waste products. A third area, recycling, is becoming increasingly popular. Resource conservation is extremely important to the overall quantitative assessment of the environmental impact of chemicals. The quantitation of the energy requirements associated with materials acquisition, use, disposal, and recycle would be a tremendous undertaking for one chemical, not to mention the enormous task it would be for an entire class of substances. Even to just consider waste analysis within the confines of the cradle-to-grave approach, one would need to define and quantitate the toxicity of the material, exposure potential to the material, toxicity and exposure aspects of various waste materials, processes and by-products, and finally, risk assessment and characterization. Clearly, science and industry have a long way to go in providing clear answers to the myriad of environmental issues facing us now and in the future.

Therefore, because of the lack of vital information in numerous categories of the life-cycle analysis, this chapter will attempt to address only a very small portion of the nature of the environmental impact of synthetic lubricants. Even within this small microcosm of the total picture, the issue of a clear definition of environmental impact is complex. Nonetheless, it is at least a start toward providing a definition of the potential effect of anthropogenic chemicals on the environment.

II. GENERAL TOXICITY AND ENVIRONMENTAL FATE CONSIDERATIONS: ENVIRONMENTAL LABELING

Environmental labeling schemes (i.e., ecolabeling) are being developed and instituted by governments worldwide to help consumers become increasingly aware of environmental issues and to aid them in selecting products that are environmentally friendly. There is no doubt that future market trends will be significantly impacted by consumer demands for environmentally sound products. In the Introduction, I have provided a general overview of the enormous complexity of the cradle-to-grave analysis that must be instituted to provide a quantitative answer to the question of environmentally friendly. In this section, I provide some general toxicity and environmental fate considerations that are used in developing ecolabeling schemes, give examples of some early attempts at ecolabeling, and finally, describe an aquatic toxicology and environmental fate ecolabeling decision tree.

A. Toxicity

Although mammalian toxicology is integral to an understanding of the biological relevance (i.e., human or environmental health significance) of industrial chemicals, much of the early work in environmental labeling has been related to ecotoxicology and aquatic toxicology. Therefore, most of the discussion on general toxicity and environmental fate considerations will be focused on these subdisciplines of toxicology.

The toxic potential of a particular chemical substance is evaluated by addressing the intrinsic toxicity of the material itself *and* understanding the magnitude of exposure to an individual, population, community, or ecosystem. Exposure to chemicals is determined by the complex interrelationship of several factors, including emission rate, mobility of the chemical substance in an ecosystem, degradation, accumulation potential, climatic conditions, and population density. Only when these two parameters, exposure and toxicity, are evenly evaluated can one fully understand the toxic impact a substance may have on the environment. In other words, an extremely toxic chemical may not pose a threat to the environment if one eliminates or minimizes the exposure to the material. On the other hand, a material that is fairly nontoxic may become an environmental concern if populations are exposed to large quantities.

A third factor that usually receives little or no attention, but is equally important to the overall toxicity assessment process, is sensitivity or susceptibility of the target species or population to a xenobiotic. For the most part, species for toxicology testing have been selected for environmental impact studies because of their sensitivity to pollutants, availability, and the specific niche they occupy in an ecosystem. A myriad of test organisms have been evaluated and are available for use in bioassays. Besides the typical mammalian species such as rats, mice, guinea pigs, and dogs, other species such as sheepshead and fathead minnows, earthworms, mallard ducks, rainbow trout, bluegill sunfish, insects, benthic invertebrates, mollusks, and bacteria are used in bioassays. New test methods using alternate species are being developed at a fast pace to keep up with the growing list of new regulations on chemical manufacture and use.

As a science, toxicology is fundamentally related to exposure. Such exposure- (i.e., dose-) response relationships are typically used to define the toxicity of a given material to a certain group. Toxicity evaluation end points include lethality and effects on reproduction and growth. The most commonly determined endpoint is the LC_{50} (lethal concentration 50), which is defined as the concentration of material (in air or water) needed to kill 50% of a population within a specified time period. With direct oral or dermal doses this parameter is referred to as the LD_{50} or lethal dose 50. This one parameter is the most important number used by regulatory bodies worldwide in assessing the toxicity of a particular material.

B. Environmental Fate

Secondary to the acute toxicity of chemicals, in regard to ecotoxicity, is bioaccumulation potential. Bioaccumulation is the concentration (i.e., accumulation) of chemicals from water or food into living organisms. Bioaccumulation is determined by the degree of uptake, distribution, metabolism, and elimination of a chemical in an organism. For instance, a material that is rapidly taken up and widely distributed, but metabolized and eliminated, will not likely bioaccumulate. Similarly, a material that is not taken up cannot bioaccumulate. However, some highly lipid-soluble materials will be taken up readily and stored in lipid (i.e., fat). A classic example is the polychlorinated biphenyls (PCBs), which have caused great alarm in areas such as the Great Lakes ecosystem because of their presence, persistence, and accumulation in fish.

In aquatic species, most of the bioaccumulation occurs via direct uptake from water rather than from food. The ratio expressed by the concentration in an organism divided by the concentration in water is called the bioconcentration factor or BCF. Like the LC_{50}, the BCF is generated and used for environmental hazard assessment purposes. Often, though, an expensive and difficult bioconcentration test can be replaced by an estimation of the relative lipophilicity of the material and subsequent prediction of the BCF. For instance, it is a well-documented fact that, for the most part, a linear relationship exists between lipophilicity and bioconcentration in aquatic species. An estimation of lipophilicity can be made by determining the octanol/water partition coefficient (log P). This physicochemical parameter (log P) is determined by measuring the ratio of a material as it partitions into or out of octanol or water. The logarithm of this ratio is the log P. Today, with the help of computer programs that analyze structure–activity relationships, this number can be calculated without actual laboratory experimentation. In general, the higher the log P, the greater is the likelihood that a material will bioconcentrate.

However, materials with a log P of < 1 or > 6 or 7 are not expected to bioconcentrate

(2). Until just recently, regulatory bodies had generally categorized all materials with log P > 3 as bioaccumulators. It is now recognized that materials with high log P will not bioconcentrate for several well-documented reasons (3). These include the likely adsorption and partitioning of highly lipophilic materials onto the organic fraction of sediment, lack of absorption of hydrophobic chemicals across fish gills due to low water solubility, and molecular volume considerations. Figure 25.2 illustrates a schematic representation of the linear and inverse relationships between log P and BCF. This principle can be extremely important from a marketing and regulatory perspective. For instance, materials with high log P values (i.e., > 7), including synthetic lubricants such as polyalphaolefins, normally considered as bioaccumulators, are now considered innocuous from a bioaccumulation perspective. This no doubt increases the marketability and lessens the regulatory pressure on these products.

Besides toxicity and bioaccumulation, a third major criterion used in developing ecolabeling schemes is persistence or the rate and extent of degradation. Within the matrix of ecotoxicity assessment, degradation may override harmful characteristics such as toxicity and bioaccumulation. For instance, a chemical that does not persist may not be an ecological threat, even though the material is toxic and can potentially bioaccumulate. Degradation can occur by abiotic and biotic means including hydrolysis, photolysis, and biodegradation. Therefore, before conducting tests for biodegradation, one should assess the hydro- and photolytic capabilities of a substance. However, from an ecolabeling perspective, it seems that most of the interest in degradation is in the area of biodegradation.

Determining the biodegradation capabilities of a material is very difficult. This is because most of the test methods for biodegradation tests have been developed for use with water-soluble materials. However, many of the materials, including synthetic lubricants, produced by industry, and in which biodegradation assessment is in demand, are hydrophobic. Biodegradation test method development for water-insoluble substances is in progress to deal with this problem.

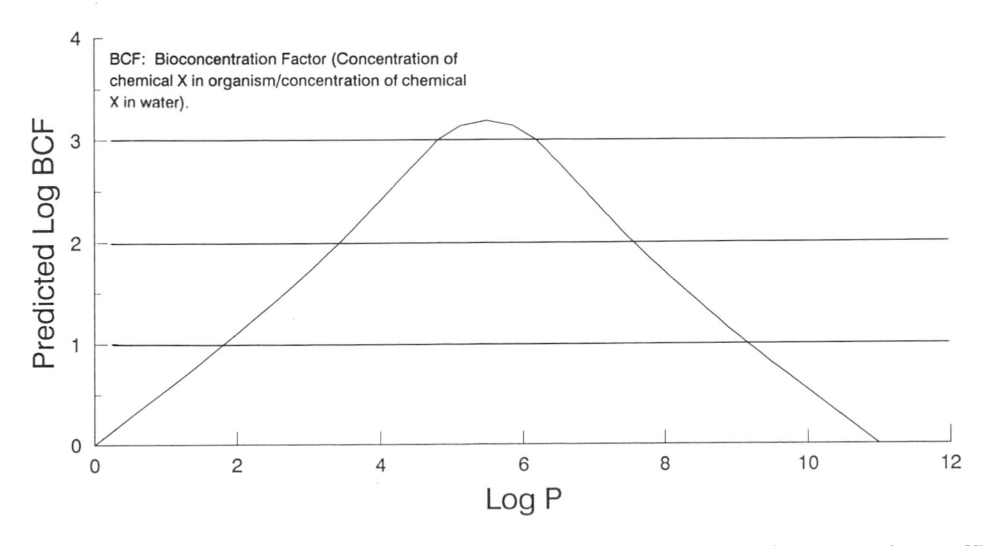

Figure 25.2 Schematic representation of the relationship between octanol–water partion coefficient and bioconcentration in aquatic species. Note the inverse relationship between these two parameters as log P increases past 6.

Biodegradation can be defined or described by several acceptable means. Materials can be described as being readily, inherently, or relatively biodegradable under either aerobic or anaerobic conditions. In general, tests for ready biodegradability are the most stringent. These tests usually measure oxygen consumption or carbon dioxide production is a closed aqueous system over a period of 28 days. A material that biodegrades in a ready biodegradability test would be considered to likely degrade in a rapid and complete manner in the environment. Tests for inherent biodegradability are less stringent than ready biodegradation tests. In these tests, conditions are made more favorable for degradation by increasing exposure times and ratio of test compound mass to bacteria. However, a material that degrades in this test may not be considered to be biodegradable in the environment. Recently, a test for relative biodegradability has been developed. This test, the CEC L-33-T-82 test, was designed for water-insoluble materials and compares the biodegradation of a component versus the degradation of reference mineral oils. This test, though not as yet accepted for regulatory purposes or complete in its development, is gaining acceptance in Europe.

C. Ecolabeling Schemes

An example of an early pollution classification (ecolabeling) scheme that is still effectively used today to regulate hazardous materials shipped overseas is MARPOL (Marine Pollution Act). MARPOL, the full effect of which was exerted in 1987, primarily uses acute aquatic toxicity and bioaccumulation potential in determining pollution categories for hazardous chemical shipment. In the MARPOL scheme, chemicals are classified into four pollution categories based on the magnitude of acute toxicity (LC_{50}) and bioaccumulation potential. For instance, a material that is toxic to aquatic life (i.e., $LC_{50} < 10$ ppm) and that bioconcentrates will be classed into Pollution Category A, the most severe class. The shipping and disposal of such a material will be stringently scrutinized and restricted with such a pollution classification. However, biodegradation is not used significantly in the MARPOL scheme.

In 1990, the European Economic Community (EEC) developed criteria for the designation and application of a specific ecolabel (or avoidance thereof). Under the seventh amendment to Directive 67/548 EEC, which was passed to regulate dangerous substances in 1967, this amendment is better known as the "Dangerous to the Environment" legislation. Under this law, materials that are shown to be toxic to aquatic organisms and bioaccumulate will be "stamped" with a dead tree/dead fish symbol. Persistence (biodegradation), although important to this legislation in regard to assessing the potential for long-term adverse effects, like MARPOL, is not directly related to the assignment of the ecolabel.

Based on review of current and proposed regulations worldwide, I have developed a decision tree for environmental acceptance in regard to aquatic toxicity and environmental fate (Fig. 25.3). The decision tree illustrates that acute aquatic toxicity is the primary criterion for environmental acceptance designation. For the most part as described previously, this is universally accepted. Therefore, LC_{50} values below 10 ppm are unacceptable in all cases. Next in significance is ready biodegradability. Clearly, a material that has low toxicity to aquatic life ($LC_{50} > 100$ ppm) and that is readily biodegradable would be environmentally acceptable. If a material is not readily biodegradable and a question exists as to potential persistence, then one must evaluate bioconcentration possibilities—hence, the introduction of the log P and fish bioconcentra-

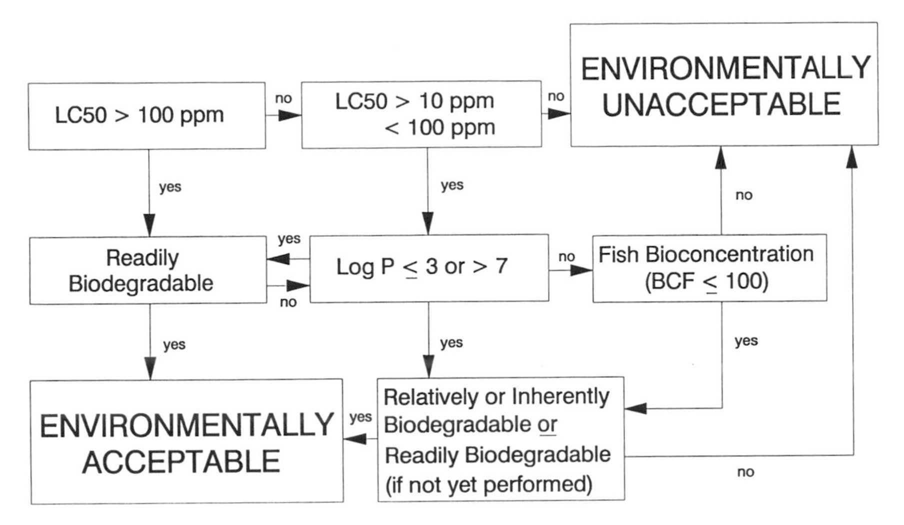

Figure 25.3 Flow chart illustrating a decision tree for environmental acceptance/unacceptance in regard to aquatic toxicity and environmental fate.

tion test criteria. If a chemical possesses a log P between 3 and 7, the material would be considered a suspect bioconcentrator, and then a definitive fish bioconcentration test could be run. A "passing" grade for this test is a BCF $\leq$ 100. If the material that is not readily biodegradable (or is slightly toxic with an $LC_{50} > 10$ or < 100 ppm) passes the log P and/or fish bioconcentration test, it must subsequently pass a test for "relative" (i.e., CEC L-33-T-82) or "inherent" biodegradability in order to be deemed environmentally acceptable.

This proposal is similar to a recently proposed U.S. hazard communication standard process for determination of aquatic hazard. However, the present proposal possesses an international flavor by encompassing worldwide regulations and takes into consideration some of the marketing pressure that the chemical industry will face in the future in regard to ecotoxicity labeling issues. As stated previously, most labeling standards today do not stress biodegradation of products. The present proposal requires that a material be at least partially biodegradable to be considered environmentally acceptable. Scientifically, I agree with the rationale that biodegradation may not be a significant concern if a material does not bioaccumulate. In fact, the present proposal shows that one must address bioconcentration if a material is not readily biodegradable. However, the increasing emphasis placed on biodegradation from a regulatory and marketing viewpoint, especially in Europe, indicates that more emphasis should be placed on biodegradation than shown in other ecolabeling proposals. A case in point is the German Blue Angel label for chainsaw lubricants. Under this labeling scheme, biodegradation is more important than aquatic toxicity or bioconcentration as a criterion for label designation. A chainsaw lubricant can carry the Blue Angel label only if all major components meet OECD ready biodegradability criteria and all minor components are inherently biodegradable (nonpersistent). Perhaps the German scheme is a forerunner of things to come.

Hopefully, in the previous two sections, readers have gained an appreciation for the complexity of the issue at hand and have obtained some insight on some of the general toxicology considerations that go into the hazard assessment process and subsequent impact analysis of a chemical on the enviroment. In the last section of this chapter, I present some specific toxicological information on some of the synthetic lubricants discussed in this volume. Although in general there is a paucity of toxicology information on synthetic lubricants, we can at least make a start down the long road of environmental impact analysis.

III. TOXICITY OF SYNTHETIC LUBRICANTS

The following toxicity information on various synthetic lubricants was obtained through review of several databases and textbooks, and from information obtained from other authors of this book.

A. Polyalphaolefins

Polyalphaolefins (PAOs) are not expected to cause acute or chronic toxicity to humans if exposure occurs by the dermal, oral, or inhalation routes. In addition, PAO are not expected to cause skin or eye irritation. This toxicological evaluation is based on the very limited toxicity data available on PAO (Ethyl Corporation). Currently, the only data available on PAO are tests of acute toxicity including oral LD_{50}, skin and eye irritation tests, inhalation LC_{50}, and tests for comedogenicity (chloracne formation potential).

Although little information (one acute study that was obtainable; Ethyl Corporation) exists on the aquatic toxicity of PAOs, they are not expected to be toxic to aquatic organisms. Furthermore, no literature information was found in regard to bioaccumulation or biodegradation of PAOs. However, although PAOs are highly lipid-soluble with calculated log P values of > 10 (Ethyl Corporation data), they are not expected to bioaccumulate for reasons explained earlier in this chapter. For instance, it seems that for hydrocarbons in general, toxicity increases as alkyl chain length, lipid solubility, and log P values increase and as water solubility decreases. The water solubility component seems to exert a cutoff point beyond which compounds are not sufficiently soluble to be toxic. The physicochemical characteristics of PAOs (i.e., low water solubility, high log P) seem to place them beyond this cutoff point in water solubility, which renders them inert in aquatic systems.

Low-viscosity PAOs (i.e., 2 and 4 cSt) can be classified as relatively biodegradable in aqueous systems as defined by the CEC L-33-T-82 test (Ethyl Corporation data). However, this does not mean that PAOs are biodegradable. As discussed earlier, the concept of biodegradation is very difficult to define. Biodegradability may be difficult to show conclusively due to the extreme low water solubility of PAOs and their subsequent lack of bioavailability for degradation by bacteria. In water, PAOs may adsorb to organic particulates or form colloidal suspensions, thereby decreasing the available surface area for bacterial biodegradation. The inability of bacteria to interact with the hydrocarbon substrate would produce very little apparent biodegradation as defined by currently accepted protocols such as the closed bottle test for ready biodegradability.

Therefore, in general, PAOs, due to the property of low aqueous solubility, should be relatively inert in aquatic systems in regard to acute toxicity, biodegradability, and bioaccumulation.

B. Silahydrocarbons

In general, silahydrocarbons are not expected to be acutely toxic via the oral, dermal or inhalation routes of exposure. Oral and dermal LD_{50} values for octyl/decyl silahydrocarbons have been reported to be >5.0 g/kg and > 2.0 g/kg, respectively (4). In the same study, inhalation LC_{50} of this silahydrocarbon was reported to be > 4.8 mg/liter (at saturation). Silahydrocarbons have very low vapor pressures (i.e., < 1 mm Hg); therefore, inhalation should not be a biologically significant route of exposure. No skin irritation or sensitization is expected to occur upon exposure. However, silahydrocarbons may be slightly irritating to the eyes (4). No chronic effects are expected following long-term exposure to these substances.

Very little information is available in regard to ecotoxicity and environmental fate of silahydrocarbons. One reference to a Swiss study on a C6/8 silahydrocarbon indicated this material to be virtually nonbiodegradable. However, this claim could not be confirmed. Because of the extreme low water solubility of silahydrocarbons, they are not expected to adversely impact aquatic environments.

C. Chlorotrifluoroethylene

Although chlorinated hydrocarbons include some toxicologic agents such as carbon tetrachloride and vinyl chloride that are classical to the field of toxicology as potent liver and kidney toxins and carcinogens, chlorotrifluoroethylene does not seem to be included in that class. As these materials are used, chlorotrifluoroethylene does not seem to pose much of a hazard to health and environment. Information supplied by Halocarbon Products Corporation (Hackensack, NJ) indicates the oils and greases these materials are components of, are fairly innocuous. For instance, these mixtures have very low vapor pressures (1 mm Hg or less). Therefore, they should not be much of an inhalation hazard. A representative oil containing chlorotrifluoroethylene caused no deaths in rats dosed at 2.5 g/kg/day orally for 21 consecutive days. Toxicology studies conducted on these oils showed a potential for liver toxicity in rodents, but not in primates. These oils are not skin irritants, nor are they mutagenic.

These materials, as they are used, are very water insoluble. Therefore, based on the low acute toxicity and the low potential for bioavailability to aquatic organisms, these materials are not expected to pose a significant threat to aquatic biota. However, as with any highly lipophilic material, the potential for bioaccumulation should not be overlooked or disregarded until definitive testing has taken place.

D. Polyalkenylglycols

In general, glycols, due to their low vapor pressures, pose little inhalation hazard (5). Low-molecular-weight glycols such as ethylene glycol are known to be toxic to humans, producing lethality with as little as 1.4 ml/kg oral dose. However, high lethality seems to be associated with this specific low-molecular-weight glycol. For instance, propylene glycol has acute oral LD_{50} values of 9–32 ml/kg in laboratory animals. It seems, therefore, that by analogy, polyalkenylglycols should not pose much of a threat to mammalian species.

Information supplied by Union Carbide indicates their polyalkenylglycol fluids and lubricants to be resistant to rapid biooxidation. These materials appear to biodegrade slowly and are nontoxic to fish and, perhaps, other aquatic organisms as well. For instance, aquatic toxicity data supplied by Union Carbide indicates fathead minnow 96-h LC_{50} values to be 8,100 ppm for one fluid and > 10,000 ppm for another. Therefore, there materials do not seem to have the potential for much harm to the environment.

E. Silicones

Silicones, due to their chemical inertness (6), are not expected to cause any significant adverse effects to health or environment. Halogenated silicones such as methylchlorosilanes are corrosive.

F. Phosphazene Fluids

Very little toxicological information was found on phosphazene fluids. In two studies conducted by E. R. Kinkead et al. (1989) of the Harry G. Armstrong Aerospace Medical Research Laboratory at Wright-Patterson Air Force Base in Dayton, Ohio, cyclotriphosphazene (CAS number 291–37-2) hydraulic fluids were found to be nontoxic both by inhalation and through repeated dermal exposure following 21-day exposures in rats and rabbits, respectively. These results back up the general idea that phosphazenes are generally nontoxic.

G. Polybutenes

Information supplied by BP Chemicals Ltd. (UK) provides an excellent toxicity overview for polybutenes (CAS number 9003-29-6). As stated in their technical data sheets, polybutenes are materials of very low biological activity. In fact, these materials have been accepted for use for many years as components of cosmetics, surgical adhesives, and pharmaceutical preparations. Polybutenes have been shown to cause minor irritation to the eyes of rabbits with only slight conjunctival involvement and no corneal effects. This material is not a sensitizer but will cause some minor skin irritation upon exposure. Polybutenes are not toxic by dermal absorption, as indicated by dermal LD_{50} values in excess of 25 g/kg.

Polybutenes are nontoxic orally in both acute and chronic exposure situations. Oral LD_{50} values are reported to be greater than 15.4–34.1 g/kg in rats. Furthermore, no treatment-related effects were seen in dogs fed 1,000 mg/kg/day or in rats fed with up to 20,000 ppm (2%) in the diet for 2 years. The low vapor pressure and high viscosity of polybutenes make the inhalation route of exposure unlikely.

Environmental data on polybutenes show these materials to be nonhazardous to aquatic life. BP Chemicals report rainbow trout TLm_{96} (i.e., LC_{50}) and 48-h *Daphnia* EC_{50} (effective concentration 50) data of > 1000 ppm and no effect from exposure to 100% water-soluble fraction, respectively, indicating these materials to be nontoxic to fish and invertebrates. Furthermore, the International Maritime Organization (IMO) rates polybutenes as nonhazardous to aquatic species and as nonbioaccumulators. However, like most highly water insoluble materials, polybutenes are not readily biodegraded, as indicated by showing less than 3% biodegradation in a 28-day ready biodegradation test (60% is needed for "passing").

H. Phosphate Esters

For the sake of space, the discussion on phosphate esters will be on tributyl phosphate (CAS number 126-78-8). According to information supplied by FMC Corporation (Philadelphia), although several million pounds of tricresyl phosphate (a potent neurotoxin) are used as additives in synthetic fluids, most is being replaced by "synthetic" trisubstituted derivatives of *ortho*-phosphoric acid, which includes tributyl phosphate.

However, tributyl phosphate possesses some significant toxicity in its own right. For instance, sites of toxicity include the central nervous system, skin, and lungs (7). This toxicity review indicates that tributyl phosphate is not acutely toxic orally, as indicated by an oral LD_{50} of 3 g/kg in rats. However, the material may induce anesthesia, paralysis, and death under high exposure conditions, which are not likely to occur. The most relevant toxicity to be concerned with is the likely irritation of the skin, lungs, and eyes upon contact with tributyl phosphate. Tributyl phosphate is not mutagenic.

A literature search in the AQUIRE data base shows aquatic LC_{50} reports from 5 to 9 ppm in rainbow trout (96 h) to 330 ppm in *Daphnia magna* (24 h). This indicates a potential for significant toxicity to aquatic organisms. However, based on the structure of this material, I would not expect tributyl phosphate to persist in aqueous systems.

I. Alkylated Aromatics

Very little information was available on the alkylated aromatics that are used in synthetic lubricants. However, as they are employed in lubricants, they are not expected to be of much harm to health and environment.

IV. CONCLUSION

In general, synthetic lubricants do not seem to be of much concern toxicologically. However, to assess the full environmental impact of these materials, a quantitative "cradle-to-grave" assessment is necessary.

ACKNOWLEDGMENTS

I would like to thank Anne Lemmon for the literature searches she performed and Virginia Yarbrough for the help with graphics. Also, I wish to thank all of the authors for the vital information they supplied to me.

REFERENCES

1. Proctor and Gamble Company (1991). Resource and environmental profile analysis of oleochemical and petrochemical alcohols. Report prepared by Franklin Associates, Ltd.
2. McKim, J., Schmieder, P., and Veith, G. (1985). Absorption dynamics of organic chemical transport across trout gills as related to octanol-water partition coefficient. *Toxicol. Appl. Pharmacol.,* 77, 1–10.
3. Leuterman, A. J. J., and B. F. Droy, (1991). *N*-Octanol/water partition/coefficient: Test requirement for North Sea applications. *Oil Gas J.* (in press).
4. Maltie, D. R. (1989). Harry G. Armstrong Medical Research Laboratory, Toxicology Branch, Wright-Patterson Air Force Base, Dayton, OH, presentation at the SAE meeting, Dallas, TX, October 2.
5. Andrews, L. S., and R. Snyder (1986). Toxic effects of solvents and vapors. In *Casaret and Doull's Toxicology, The Basic Science of Poisons,* eds. C. D. Klaassen, M. O. Amdur, and J. Doull, Macmillan, New York, pp. 636–668.
6. *Kirk-Othmer Encyclopedia of Chemical Technology* (1982). vol. 20, p. 954, John Wiley and Sons, Canada.
7. American Conference of Governmental Industrial Hygienists (1986). *Documentation of the Threshold Limit Values and Biological Exposures Indices,* 5th ed., American Conference of Governmental Industrial Hygienists, p. 591.

26
Commercial Developments

E. Ian Williamson
The College of Petroleum and Energy Studies
Oxford, England

I. INTRODUCTION

Synthetic lubricants have been known and used for almost 50 years—1991 was in fact the fiftieth anniversary of the first jet aircraft flights.

Solid development has been taking place ever since: the first polyalkylene glycols in 1945, diesters in 1951, phosphate esters in 1953, polyol esters in 1963, and so on. However, in the 1980s a much wider general acceptance of synthetic lubricants has taken place, largely because of technological developments raising the level of performance requirements in equipment and engines.

Essentially, in the 1980s, we have been in a market-development phase, where both the oil industry and the equipment manufacturers have been unsure as to which synthetic fluid would most closely meet their needs; hence there has been a period of strong interproduct competition in the industry, that is to say, synthetic versus synthetic. This in turn has led to synthetic/synthetic blends, and to semisynthetics (or synthetic/mineral oil blends), as well as increasing competition from very-high-viscosity-index (VHVI) hydrocracked or isomerized mineral oil base stocks.

Around 60 end uses for synthetics have now been commercialized for automotive, industrial, functional/process fluid, and aviation lubricant applications, including military requirements. The commercial development of synthetic lubricants and fluids is still continuing, and will probably achieve more breakthroughs in terms of end uses in the 1990s, as issues such as volatility, biodegradability, toxicity, and disposability come to have a significant effect on the lubricants industry.

The cost differentials between synthetics and mineral oil base stocks are the major impediment to the future growth of synthetics. This factor, above all, remains the limiting factor, so that after all these years of development, synthetics still account for only about 2% of the 30 million tons of lubricants and fluids consumed annually in the world [excluding the former USSR, China, and the remaining centrally planned economies (CPEs)].

II. TRENDS IN THE LUBRICANTS BUSINESS AND BASE OILS PROFITABILITY

A small number of international oil companies—primarily Shell, Exxon, Mobil, Texaco, and BP—dominated the global lubricants business until the end of the 1960s. In the 1970s the oil business was rocked by the two oil crises and the rise of state oil companies, of which there are now over 100, worldwide. In response, the majors retrenched and reorganized and are now again emerging as both technical and marketing leaders. In many countries they are now being invited back as partners by the state oil companies.

The following are the key trends that are affecting the lubricants business in the 1980s and 1990s:

1. Continued high level of competitive activity.
2. Increased commitment by the majors to both the base oils and finished lubricants businesses.
3. Concentration of lubricant businesses into managed profit centers—mainly by the majors—leading to higher efficiency and some global product launches.
4. Rationalization of blending plants and improvement of plant efficiency and quality control, such as BS5750/ISO 9000.
5. Reduction in the number of independents in the market—smaller blenders being acquired, plus product line exchanges.
6. Stronger emphasis on marketing and distribution: advertising, packaging, brand differentiation, and use of effective sales channels.
7. Continued interest by chemical companies in the lubricants business, as suppliers of additives and synthetic base oils.
8. Demands by automotive, aviation, and industrial equipment manufacturers for enhanced performance levels, leading to the wider use of synthetics and production of unconventional base oils (UBOs) by the majors.
9. Environmental issues, from toxicity of additives, through disposal, to biodegradability.

The lubricants market has always been extremely competitive, and is now even more so. The independents and sector product specialists have generally sought business at the expense of the majors, either on price or performance grounds. They need high levels of technology at their fingertips, plus ever-increasing marketing expertise, to continue to do so, as the majors are increasing their investment in product development, marketing, and production methods.

The lubricants business was originally targeted by the majors as a profitable outlet for lubricant base stocks, at prices approximately twice those of crude oil. The key problem in lubricants profitability has always been the majors' continued willingness to sell base stocks on the wholesale market—that is, to their competitors—because they could neither control the overall market, nor consume their own base oil production completely. In addition, the additive companies have been either separate companies, joint ventures, or otherwise separate profit centers and have, similarly, been willing to provide their research and development technology to independents.

The results are obvious—the independents, with their lower overheads, have always been able to undercut the majors' prices, so that profitability on the average *finished* lubricant has always been under pressure. Contrary to general belief, the majors do not usually subsidize the transfer prices for base stocks within their own companies, but try to charge realistic, arm's-length prices. These are comparatively easy to set (and hard to challenge) when a market in base oils demonstrably exists. These transfer price rules are occasionally broken, if a significant tender is being bid and the volume is considered vital to the interests of the major.

The profitability of base stock manufacture is fairly easy to maintain. As happened in both 1974 and 1980, feedstocks can be diverted to refinery cracking units, until "shortages" force the prices up. This is a self-adjusting mechanism, which supply managers are quite familiar with, linked to visible markets, such as Rotterdam. Lubricant feedstocks provide attractive yields of gasoline and middle distillates and are premium stocks for use in catalytic cracking units.

Another problem affecting profitability is the high servicing costs of the business, in terms of skilled manpower. Although the original equipment manufacturers (or OEM) representatives can be justified by the majors and the additive companies, at the other end of the marketing chain, the lubrication survey done by field engineers is an expensive exercise tying up a specialist for 2 or more weeks on a single customer report.

Nevertheless, despite all these problems, the majors and large independents remain heavily committed to the business and are concentrating their efforts. Shell now has separate business streams for lubricants, a key factor in their international launch of Helix. Mobil has been able to coordinate the international promotion of Mobil 1, and Esso has launched Ultra internationally.

Blending plants have always suffered from low productivity and high inventories, as a result of random customer ordering and the wide range of products on offer—500 or more per plant. The availability of computers and automated blending allowed the companies to install the latest technology and close down old blending plants. In the United Kingdom, Shell (Stanlow), Esso (Purfleet), Mobil (Birkenhead), and Texaco (Manchester) all have new plants built in the late 1980s.

These same ideas are being adopted by the independents, but they are also now acquiring other independents, in order to increase volumes, prior to rationalization. Century Oils (now part of Fuchs) has installed the latest blending technology, and Hunting Lubricants acquired six or seven companies, including Filtrate, ROC, and H G Allcards, in the last 2 years, before themselves being bought by Carless Lubricants, which in turn was bought by Kuwait Petroleum. Oleofiat formed a joint company with Fuchs of Germany, whereby its motor oil brands (Valvoline and Oleoblitz) went to Oleofiat and Fuchs Lubrificante took over Oleofiat's Industrial Division. Cepsa, one of the three major lubricants marketers in Spain, has now been taken over by ELF. Nynas of Sweden, already owned 50% by PdVSa, has had its remaining 50% acquired by Neste Oy. Thus, two state-owned companies now control this former independent.

Brand differentiation and better marketing techniques are driving the lubricants business worldwide. Packaging to "high-tech" levels and widespread advertising have improved and increased brand awareness. Television advertising is now commonplace, as is support for large- and small-scale motor sport, with Grand Prix car racing at the one end of the scale and motorcycle events at the other.

The chemical companies remain interested in the lubricants business, especially the additives companies, but few of the synthetic base oil producers have moved as far as direct marketing, except Huels with Anderol, Hatco in the United States, and Union Carbide.

Demands for a lower NOACK value by VW and CCMC (in G5) are boosting manufacture of unconventional mineral base oils. Chevron, Esso, Repsol, and Total are new investors in this field. This is giving some majors a special "edge" on quality and cost competitiveness, which they are exploiting to the full in the market place. The U.S. oil market is now lagging behind Europe, in terms of product quality specifications (especially volatility) and formulations. Low prices and market confusion between 5W–30, 10W–30, 10W–40, and 15W–40 are not helping.

Overall these changes are bringing improvements both to the consumer and to company profitability. They will make the lubricants business one of the most dynamic sectors of the oil business in the 1990s. It is felt that the slimming down, modernizing, and restructuring processes will continue. Those independent lubricant companies that survive will need to find ever-better ways of doing business. They will need to cross frontiers, and they will have to find those particular niches, like Castrol, Croda, Houghton Quaker Chemical, Les Emulgateurs and Oemeta, where they can do something better than anyone else. The majors will survive because of their technology innovation, international base, and strong OEM recommendation positions; the additive companies will survive and the national oil companies will probably go on expanding, but there will not be room for everyone!

III. SYNTHETIC BASE STOCK SUPPLY—WHICH COMPANIES ARE INVOLVED?

There are four types of companies in the lubricants business worldwide, as shown below (these figures exclude the former centrally planned economies and Eastern Europe):

Blending and marketing companies—1700 companies.
Mineral oil base stock producers—65.
Synthetic base stock producers—55.
Additive companies—15.

The independent marketing and blending companies are too numerous to list. There are about 400 in Western Europe and over 800 in the United States. The largest is Castrol, which has worldwide operations. Idemitsu-Kosan is the largest in Japan; Fuchs is now number one in Europe if Castrol is counted as an international.

Mineral base oil supplier numbers are increasing slightly, as national oil companies invest and the business becomes more internationalized; some are also synthetic base stock producers. Overall, there are now about 100 potential suppliers of mineral and synthetic bases. All the synthetic base oil companies in Western Europe, the United States, and Japan are shown in Tables 26.1–26.3.

On the supply side we are fortunately now not plagued with a major increase in the number of new market entrants. There even seem to be some "accidental" or structural withdrawals, with the closure of some polybutene (PB) plants in the United States and in Antwerp, or the sale of the Atochem polyalkylene glycol (PAG) plant at Chocques to ICI in 1988. The polyalphaolefin supply situation in Europe has now eased, with the new 13,000-ton Mobil plant at Gravenchon now on-stream, together with a 36,000-ton plant

Table 26.1 Synthetic Base Stock Suppliers in Western Europe—1991

Base Stock	Supplier	Location
Polyalphaolefins	Ethyl Corporation	Feluy
	Exxon Paramins	Le Havre
	Mobil Chemicals	Gravenchon
	Neste Oy	Berrengen
Alkylbenzenes	Chevron Chemical	Gonfreville
	Exxon Paramins	Le Havre
	Wintershall	Ibbenburen
Polyol and dibasic esters[a]	Akzo Chemie	Duren
	BASF	Ludwigshafen
	BP Chemicals	Hull
	FMC (Ciba-Geigy)	Manchester
	DS Industries	Copenhagen
	Henkel	Dusseldorf
	Hoechst	Gendorf
	Huels/Anderol	Gelsenkirchen
	ICI	Baleycourt/Wilton
	Oleofina	Oelegem
	Quaker Chemical	4 plants in Europe
	Societe NYCO	Conflan Sainte Honorine
	Unichema	Gouda/Bromborough
Polybutenes	Amoco Fina	Antwerp[b]
	BASF	Ludwigshafen
	BP Chemicals	Grangemouth
	BP Chimie	Lavera
	Exxon Chemicals	Cologne
	Lubrizol	Le Havre
Polyalkylene glycols	BASF	Ludwigshafen
	Bayer	Leverkusen
	Berol Nobel	Stenungsund
	BP Chemicals	Hythe
	Dow Chemical	Terneuzen
	Harcros Chemicals	Manchester
	Hoechst	Gendorf/Frankfurt
	Huels	Gelsenkirchen
	ICI	Wilton/Choques
	Montedipe (Auschem)	Priolo
	Shell Chemicals	Pernis/Carrington
Phosphate esters	Bayer	Leverkusen
	Ciba-Geigy	Manchester
	Coalite	Bolsover

[a]All chain lengths.
[b]Currently closed.

Table 26.2 Major Synthetic Base Stock Suppliers in North America—1991

Base Stock	Supplier	Location
Polyalphaolefins	Chevron Chemical	Cedar Bayou, LA
	Ethyl Corporation	Deer Park, TX
	Mobil Chemical	Beaumont, TX
	Uniroyal	Ontario, Canada
Alkylbenzenes	Vista Chemical (RWE)	Baltimore, MD
Esters for lubricants and fluids (all chain lengths)	Akzo Chemicals	Gallipolis Ferry, WV
	Emery (Henkel)	Cincinnati, OH
	Hatco Corporation	Fords, NJ
	Hercules (Aqualon)	Louisiana, MO
	Houghton Group	Allentown, PA
	Huels America	Chestertown, MD
	Inolex	Philadelphia, PA
	Mobil Chemical	Edison, NJ
	Quaker Chemicals	Conshohocken, PA
	Quaker Chemicals	Detroit, MI
	Quaker Chemicals	Pomona, CA
Polybutenes	Amoco Chemicals	Texas City, TX
	Amoco Chemicals	Whiting, IN
	Exxon Chemicals	Bayway, NJ
	Lubrizol	Houston, TX
Polyalkylene glycols	BASF Corporation	Washington, NJ
	BASF Corporation	Wyandotte, MI
	Dow Chemical	Freeport, TX
	Ethox	Greenville, SC
	Mazer (PPG Group)	Gurnee, IL
	Texaco Chemicals	Austin, TX
	Olin Corporation	Brandenburg, KY
	Union Carbide	South Charleston, WV
Phosphate esters (industrial)	Akzo Chemie	Gallipolis Ferry, WV
	FMC	Nitro, WV
	Monsanto	Bridgeport, NJ
	Monsanto	St. Louis, MO

for Ethyl at Feluy, and Neste Oy on-stream in Berrengen, Belgium. European capacity for PAO will stand at 80,000 tons by 1992, compared to 135,000 tons in the United States and Canada (Ethyl, Mobil, Chevron, and Uniroyal). Ethyl Corporation, having taken over Emery's PAO plant, is now the largest producer, with 120,000 tons capacity.

In the areas of diesters, polyol esters, and polyglycols, plant capacities are meaningless for these batch or reactor processes, which usually produce a multiplicity of products, or grades, of esters or PAGs. Production volumes can usually be readily and rapidly

Table 26.3 Synthetic Base Stock Suppliers in Japan—1991

Base Stock	Supplier
Polyalphaolefins	Lucant (Idemitsu Kosan) plan 1994
Alkyl benzenes	Idemitsu Kosan Matsumura Oil Institute Nichiyu Chemical Nisseki Resin Chemical
Esters	Daihachi Chemical Henkel-Hakasui Kokura Gosei Sanken Kako Shin-Nippon Rika
Polybutenes	Idemitsu Petrochemical Nichiyu Chemical Nisseki Resin Chemical
Polyalkylene glycols	Asahi Denka Nippon Oil and Fat Sanyo Chemical
Phosphate esters	Ajinimoto Daihachi Chemical Kashima Kogyo

increased by running extra shifts. No shortages of supply are predicted for these synthetics, or for their raw materials, in the foreseeable future.

As regards fire-resistant phosphate esters, the market has been under pressure both from water glycol and from long-chain polyol esters from Quaker Chemical. In the United States mineral oil/phosphate ester mixes are also used. Bayer has now effectively withdrawn from phosphate ester lubricants, to concentrate on plasticizers. They were the main supplier to the German mining industry. In the United Kingdom, Albright and Wilson gave up about 5 years ago and sold out to Ciba-Geigy, which then closed down the Albright and Wilson plant at Rainham in Essex. In 1992 Ciba-Geigy sold its industrial fluids business to FMC of the United States. Coalite has just expanded its small phosphate ester plant in the United Kingdom.

In polybutenes, BP is the largest producer in Europe, with 115,000 tons of capacity at Grangemouth and Lavera, while Amoco is the biggest U.S. supplier, with 185,000 tons of capacity. Exxon Chemicals has 40,000 tons of capacity in Europe and 90,000 tons in the United States, making it number two worldwide. Altogether there were 570,000 tons of capacity (60% United States, 35% Europe, 5% Japan) in 1991.

Finally, returning to the polyglycols, which are felt to be a good growth area, Bayer is reported to have come into this business in the last 2 or 3 years, and BP has also been marketing aggressively, having taken over Union Carbide's production facilities in Europe in the early 1980s. The old Lankro/Diamond Shamrock plant in Manchester is now owned by Harcros Chemicals and is the only nonintegrated PAG producer in Western Europe.

There have been no supply problems in 1990–1991 for the major synthetic bases. The key issue has been price, and base stock marketers have now seen significant easing of the upward price pressures on ethylene and propylene, which were affecting their costs in the late 1980s.

IV. HOW LARGE IS THE LUBRICANTS MARKET?

So far it has been not been possible to obtain any reliable data on the lubricants position of Eastern Europe, including the former USSR. Figures quoted publicly indicate that the former USSR consumes more lubricants than the United States and Canada, which together use 8.5 million tons per annum. This is simply not credible, given the very low car and truck population, the lack of roads (the former USSR has less mileage of asphalt/tarmac roads than the United Kingdom!), and the state of the industry in the East. Broad indications for the existing or former CPE areas are Eastern Europe, 1–2 million tons, USSR, 5 million tons; and China, 2 million tons.

Excluding Eastern Europe for now, Table 26.4 gives rounded off figures for lubricants production and consumption in 1990, by region. Since additives will account on average for 7–8% of all lubricants (auto 13%, industrial 5%) and plant operating rates rarely exceed 85% for sustained periods, probably these plants are able to produce 30–31 million tons of lubricants, assuming full blending has taken place.

Additional conventional base oil capacity is most likely to be built in Asia/Pacific, in the Middle East, and in South America.

As regards the industrialized areas, there is major reluctance to close down profitable base oil production units, and indeed a desire to enhance the capability of existing plants, by investing in unconventional base oil production. It may be noted that BP, while closing the fuels sections of both the Llandarcy and Dunkirk refineries, kept the lubricants units open. Nynas has expanded output at Nynashamn with a new hydrogen unit. KPC at Rotterdam has completed a major revamp.

More lubricant base oil capacity of the unconventional type has come on-stream using hydrocracker residue streams, as does BP at Lavera, which was the pioneer of this technology. Union-Fuchs' new plant (10,000 tons) came on-stream in 1987. Energoinvest at Modrica in Yugoslavia was rebuilt following a fire. Total at Vlissingen plans to extract a base oil stream from the hydrocracker. Repsol has a similar plan.

Shell at Petit Couronne in France is working its existing slack wax isomerization unit hard and reportedly has been buying in extra slack wax and also cutting more selectively to improve product quality. Esso at Fawley, UK, has built a wax isomerization plant. All

Table 26.4 Finished Lubricant Demand and Base Stock Supply by Region—1990 Estimates (Thousands of Metric Tons/Percentages)

	Plants	Capacity	Percentage	Demand	Percentage
United States/Canada	40	14,000	42	9,500	35
Western Europe	30	7,500	22	5,700	21
Asia Pacific	27	6,000	18	6,800	25
Central/South America	16	3,500	10	2,600	10
Middle East	7	1,500	5	1,000	4
Africa	8	900	3	1,500	5
Total	128	33,400	100	27,100	100

in all, unconventional base oil (UBO) capacity could be around 200,000 metric tons in Western Europe by 1993. In the United States, Chevron at Richmond, California has unveiled new UBO technology known as Iso Cat-bewaxing in 1992. Other countries, such as Venezuela, are also considering new UBO plants.

V. DEVELOPMENTS IN THE SYNTHETIC LUBRICANTS BUSINESS

The synthetics business is moving steadily forward, in terms of both market penetration and sales volumes. The process, which began in 1980, when so many chemical companies became interested in the subject, as a result of their perception that specialty chemicals were in fashion and that the price gap between mineral oils and chemicals would narrow, has now been challenged, since the opposite occurred in 1986–1991, when crude oil prices fell and mineral base oils followed, to around $350 or $400 per ton at the wholesale level.

It was an unfortunate coincidence that basic chemical "shortages" occurred in 1987–1988, pushing ethylene and some derivative prices to disturbing levels and bulk polyalphaolefins to nearly $2000 per ton. Although these prices have now fallen by 25%, it caused some loss of confidence in polyalphaolefins, and the threat of substitution from VHVI mineral oils increased from then on.

As a result of these pricing swings, now moving somewhat in the opposite direction, with the 1990–1992 downturn in the chemical industry, the price ratio of PAO to mineral oil base stocks has varied between 5.7 and 3.5 to 1, within a 2–7 year period. It has been a good lesson, and highlights the need for the lubricants industry to understand the driving forces in synthetics production and pricing. Similarly, the chemical industry needs to understand the economics of the base oil industry, which has remained linked to crude pricing for 30 years. Only the VHVI oils have been able to break this pattern.

On the automotive lubricants front, sales of semisynthetic engine oils are rising steadily across Europe, but particularly in the German-speaking countries, France, and Italy. Sales of fully synthetic lubricants (except Mobil 1 and Castrol RS/Syntron) are believed to be flattening, as some full-synthetic products are de-emphasized by companies on the grounds of consumer resistance to the high costs. (An average lubricant cost for a 4–5 liter change of Mobil 1 in the UK costs $45, versus $15 for Castrol GTX.) The increased sales of semisynthetics are now leading to significant increases in demand for both PAO and diesters from lubricants blenders. With the introduction of 5W–30 grades in the German-speaking countries, these volumes will continue to rise. In the United States sales of 5W–30 and 10W–30 grade products have been slowly increasing at the expense of 10W–40 and higher viscosities. New products have been launched by Amoco, Castrol, Kendall, Pennzoil, and Quaker-State. Due to government imposition of CAFE (corporate average fuel economy) on the car companies, the fuel efficiency of U.S. cars has increased by 50% in a decade, so low-viscosity engine oils are very much in favor, as they assist in meeting the CAFE targets.

In the industrial markets we are seeing steady growth for PAO, PB, PAGs, and long-chain polyol esters, particularly in compressors, bearings/gears, and circulation systems and for fire-resistant hydraulics and metalworking applications. Long-chain polyol esters have become the dominant technology in steel-rolling oils, but are now also coming into use for biodegradable outdoor lubricants, as used in forestry, waterways, and construction. On the down side, the position of phosphate esters in fire-resistant fluids of the HFD type is weak, as is the long-term position of polyglycols in brake fluids, with the introduction of DOT 4, now widely used in Europe but not yet in the United States.

Overall the message from the markets is that the technical understanding of synthetics' unique properties has been increasing for some years. Some applications have been lost by one synthetic, but gained by another. Where demonstrable advantages can be assessed, for example in high-performance compressors, there is a solid acceptance at all levels, from OEMs, lubricant companies, and customers—this is the key to success. If any of the three elements of consensus are missing, there is only disappointment and frustration.

Despite significant efforts over a decade, however, the market penetration of synthetics has been patchy, especially outside Europe, which received a major boost in the 1980s from the introduction of the NOACK volatility test and is now receiving a second from biodegradability, or toxicity problems, generally summarized as environmental isues.

Overall, synthetics and semisynthetics still have less than 2% of the lubricants market. What is a realistic target?—probably 5% within a decade would be a major achievement, with an ultimate upper limit of 10% (i.e., 3 million tons). This would fundamentally only come about with major changes in engine oil make-up or performance requirements. It would come more quickly if the U.S. market moved to higher-quality specifications, especially on volatility for motor oils.

VI. END USE MARKETS FOR SYNTHETIC LUBRICANTS

Table 26.5 highlights the overall market sectors of current and future interest for synthetic lubricants. All in all, within these sectors, some 60 end uses for synthetics have been identified, as shown in the Competition Matrix in Table 26.6, which shows where synthetics are competing with each other, as well as with both conventional mineral and VHVI oils.

In order to arrive at some conclusions, in the text that follows, these sectors will be grouped into six major areas:

Automotive lubricants
Industrial lubricants
Functional fluids
Metalworking fluids
Process fluids
Aviation lubricants
Military uses

A. Automotive Lubricants

1. Engine Oils for Automobiles and Trucks

The main feature of the markets in 1987–1991 has been the continued efforts by most companies to upgrade their top-tier oils to 10W–30 or 10W–40 in Europe, while the bulk of the demand remains at 15W–40; there is a confused mixture within multigrades of 10W–30 or -40 and 15W–40 in the United States in roughly equal proportions, with sales of 5W–30 at about 15% of the average level of the other three grades. Semisynthetic 5W oils are now being introduced in the German-speaking countries in Europe.

Much as been written and discussed concerning the specific options that the blenders have to produce high-performance 5W–30 or 10W–30 and 10W–40 oils, but they boil down to polyalphaolefins, esters, or VHVIs, as a partial replacement for the light mineral oils in the formulation. The requirement in Europe is to pass the NOACK volatility test

Table 26.5 Markets of Future Interest for Synthetic Lubricants and Fluids

Ground transportation markets
Crankcase oils
Two stroke oils
Transmission fluids
Brake fluids
Auto air-conditioning
Greases
Industrial markets
Gas turbines for generators and mechanical drives
Circulation, gear, bearing, and chain lubricants
Air, gas, and refrigeration compressors
Hydraulic and fire-resistant fluids
Food-grade machinery lubricants
Greases and wire rope dressings
Functional fluids
Heat-transfer and solar fluids
Electrical and insulating oils
Compensators, governors, and control oils
Metalworking fluids
Cutting oils
Rolling oils
Drawing, stamping, and forming fluids
Quenchants
Process products
Cable compounds, oils, and gels
Oil-based drilling fluids
Fluxes and release agents
Textile finishes
Aviation and military lubricants
Gas turbines
Aircraft piston engine oils
Hydraulic fluids
Instrument oils
Greases

(now at 13% weight loss with VW and CCMC G5), but increasingly, oxidation stability is also becoming regarded as significant, and the VHVI grades are excellent in this regard. Of the major oil marketers, Shell is using VHVI, and so is BP, which calls its product LHC (Lavera hydrocracked component). Mobil is mainly promoting the fully synthetic Mobil 1, and Castrol is supporting Syntron X in the United Kingdom (named Castrol RS in Germany), which is another full synthetic. Esso, with Ultra and EX2, is also using polyalphaolefins as a synthetic component but may switch to VHVI from their new U.K. plant.

Table 26.6 Synthetic Lubricants and Fluids—The Competition Matrix in 1991—I

Lubricant	Conventional mineral oil	PAO	PAG	Polybutene	Diester	SC polyol ester	LC polyol ester	Phosphate ester	Alkyl benzene	VHVI
Automotive lubricants										
Engine oils (four-stroke)	•	•			•	•				•
Two-stroke oils	•			•	•		•			
Transmissions (automatic)	•	•			•					•
Gear oils (multigrade)	•	•		•	•	•				•
Shock absorbers	•	•								
Auto air conditioning			•			•				
Brake fluids			•							
Greases	•	•				•	•			
Industrial lubricants										
Gas turbines—industrial(I)/aircraft type (A)	•(I)	•(I)			•(A)	•(A)				
Gears, circulation, bearings	•	•	•	•	•	•				
Industrial gearboxes (filled-for-life compounds)			•							
Air compressors	•	•	•		•	•				•
LDPE compressors (HP only)	•[a]	•	•	•						
Oxygen compressors								•		
Ethylene vinyl acetate compressors				•						
Natural gas compressors (pipelines/reinjection)			•							
Chemical gas compressors (butadiene and VCM)			•							
High-purity gas compressors (helium, nitrogen)			•							
Refrigeration compressors	•	•	•[b]			•			•	
General hydraulic fluids	•	•	•				•			•

Fire-resistant hydraulic fluids			•				•	•	
Greases	•	•	•	•	•	•			
Food-grade oils and greases	•[a]	•	•				•		
Wire rope lubricants	•			•					
Chainsaw oils (engine)					•		•		
Industrial chain lubricants (hot conditions)			•	•					
Functional fluids									
Heat-transfer fluids	•	•	•			•			•
Solar fluids		•							
Electrical fluids/transformers	•	•		•		•			•
Offshore compensators			•			•	•	•	
Electrohydraulic controls/governors in utilities							•	•	
Electrostatic precipitators							•		
Metalworking fluids									
Cutting oils (soluble)	•		•				•		
Steel rolling oils							•		
Steel stamping (hot and cold)			•	•					
Steel pressing	•			•					
Steel continuous casting	•			•			•		
Aluminum stamping/forging			•						
Aluminum can stock rolling							•		
Aluminum can drawing		•	•	•			•		
Aluminum foil rolling			•	•					
Copper rolling				•					
Copper tube drawing				•					
Wire drawing			•	•			•		
Quenchants	•		•						
Process fluids									
Cable compounds/oils		•		•					•
Circuit board fluxes			•						
Fibre optics		•							

Table 26.6 *(continued)*

Lubricant	Conventional mineral oil	PAO	PAG	Polybutene	Diester	SC polyol ester	LC polyol ester	Phosphate ester	Alkyl benzene	VHVI
Process fluids *(continued)*										
Drilling fluids (oil-based)	•	•								
Mold release agents (plastics/rubber)			•							
Textile lubricants and finishes (polyester/polyamide)	•[a]	•	•							
Aviation lubricants										
Gas turbines					•	•				
Piston engines	•	•								•
Hydraulics (civil aircraft)								•		
Greases	•	•			•	•				
Military uses										
Gas turbines for aircraft					•	•				
Gas turbines for tanks						•				
Avionics (Mil-C-87252)		•								
Grease (air) (Mil-G-81322)		•								
Hydraulics (ground) (Mil-H-46170)		•								
Hydraulics (air) (Mil-H-83282)		•								
Instruments (Mil-L-6085A)		•				•				
Arctic engine oil (Mil-L-46167)		•			•				•	

[a]White oils are used, usually meeting FDA/BP/DIN specifications.
[b]Compressors using ammonia as refrigerant.

Table 26.7 gives typical formulations now in use in Western Europe, where the emphasis is on 5W/X synthetics and 10W/X top-tier oils, formulated from PAO, esters, and VHVI mineral oils.

A number of less well-known marketers, such as Beverol and the French independents (Motul, Hafa, Unil, etc.), are known to be using fully synthetic PAO-based oils. All in all, 1987–1991 has been a "vintage" period for PAO, and the total market appears to have exceeded 45,000 tons per year of PAO demand, for the first time, in Western Europe (including industrial sales).

As the total market expands, it appears that semisynthetics are slowly winning out over full synthetics, and in this area PAO is still seen as the major component, but competition from diesters is increasing. In the United States the combination of lack of a uniformly accepted volatility test, very low oil pricing, and widespread acceptance of "cheap oil, changed often" at "Jiffy" type 10-minute oil change outlets, has allowed moderate quality oils to dominate the market. In 1989–1991, however, the independents in particular are fighting back with new product launches such as Conoco XTREME (5W–30), Synquest (Quaker State, 5W–50), and Amoco's Ultimate (PAO/ester).

At present GM is against the NOACK test, and Ford uses ASTM but may switch to NOACK. There is an MVMA simulated distillation test (ASTM D-2887) that specifies 20% volatility. Ford thinks a 17% requirement will come in soon. There is also a GM high-temperature/high-shear test for the Chevrolet Corvette, with a 9% oil volatility requirement. The net result is that PAO sales into the U.S. branded automotive lubricant market of 3,300,000 tons totaled only 15,000 tons in 1990s, while in Europe about 35,000 tons was used.

2. Motorcycles and Outboard Engines

In general motorcycle and competition two-strokes use, PAO, esters, and polybutenes are vying with each other for different sectors of the market. A number of companies are making strong efforts on the competition scene, notably Castrol, Rock Oil, Silkolene, P J Harvey (United States), and Putoline (Holland). Additionally Bel-Ray, the U.S. market leader, and Bardahl are marketing synthetics to the general motorcycle market. Petrofina remains a strong long-term marketer of general two-stroke oils based on polybutenes.

Motorcycle lubricants are seen very much as a specialist niche for synthetics, since the bulk of the lubricants are sold through specialist shops. In other words, the change in the form of retailing, coupled with the idea of motor bikes as sporting, or "fun" machines,

Table 26.7 Typical Formulations Now Found in West European Fully Synthetic and Semisynthetic Retail Engine Lubricants

SAE ratings	Mineral oil, %	Type of synthetic base
Full synthetics		
5W–50 to 15W–50	< 10	PAO + diester or polyol ester (20%)
	< 10	Diester (90%) + PAO/additives (10%)
Semisynthetics		
5W–30	20–70	PAO or VHVI
10W–30	20–70	PAO + diester or VHVI
10W–40	50–80	PAO or VHVI
10W–50	50–70	Polyol ester or VHVI

rather than transport, has changed the demand for lubricants and the perception of lubricants. The majors have responded, and Shell, for instance, is offering VHVI-based motorcycle lubricants, which are sold at most of its service stations as well as motorcycle outlets; the brand name is Quattro. Castrol has also launched synthetic motorcycle lubricants, under the names Castrol 545 and 747.

Modern outboard two-stroke engines have evolved into sophisticated units. These are often multicylinder, with advanced oil injection systems, electronic ignition, and engine capacities of over 3 liters. The Evinrude Euro 4 engine test has been used to study lubricity. The connecting rod plain bearings in the engine are particularly sensitive to lubrication quality. Test work has shown that normal mineral-based lubricants of Boating Industry Association (BIA) TC-W quality pass at 50 : 1 but fail at 100 : 1 due to inadequate connecting rod bearing lubrication. In contrast, synthetic-based lubricants have been demonstrated to extend the fuel oil ratio to 100 : 1 and, in some cases, even to 150 : 1.

Outboard motor boat engine oils have also been moving steadily from mineral oil bases to diesters in Europe, to meet environmental concerns over water pollution. The biodegradability test for this, modified, is CEC-L-33-TT-82, and comparative tests of biodegradability are given in Table 26.8.

3. Transmission and Gearboxes

In the automotive manual transmission and axle oils sector, SAE 75W–90 is only slowly finding favor. Mobil now even has this on general sale in smaller packs in Europe. For heavy duty the mainly polybutene-based SAE 80W–140 grades are still selling well. The use of SAE 75W–90 or 80W–140 gear oils is reputed to save between 4 and 10% on fuel consumption. The use of these grades, when based entirely on mineral oils, has reportedly led to rear axle wear problems. This problem has been eliminated by the incorporation of about 20–25% esters into the formulation of 75W–90, and a patented PB from Lubrizol into the 80W–140 formulation (the Lubrizol product is considered here to be a semisynthetic, rather than an additive package). Mobil's 75W–90 is fully synthetic, as is Nyco's and others in Europe. Shell and Esso sell similar products in the United States.

According to a synthetics paper from Castrol (1), lubricant oil temperatures at high speeds can be 30°C cooler with a synthetic. In the United States Eaton Axles has issued new Road-Ranger specifications for full synthetics. After the first oil change (3,000–5,000 miles) the fluids can be run for up to 250,000 miles without changing, giving real cost benefits to users.

Demand for SAE 75W–90 rear axle oils is expecting to grow. The attitude of the

Table 26.8 Biodegradability of Synthetic- and Mineral-Based Oils [Method" CEC-L-33-T-82 (modified)]

Base fluid type	Viscosity (100°C), cSt	Percentage biodegraded
Mineral oil 500 solvent neutral	10.4	24
Mineral oil 150 solvent neutral	5.2	17
Polyalphaolefin	7.8	0
Polyalphaolefin	3.6	9
Ester 1	12.7	12
Ester 2	13.5	85
Ester 3	17.2	87

original equipment manufacturers (OEMs) is critical, as the majority of vehicles now have filled-for-life transmissions. The market for 75W–90 is already strong in Scandinavia, with its severe winters. In the past axle oils have been frozen solid, when engines were still able to start.

The annual demand in the United States for axle and gearbox fluids is around 15,000 tons, for both cars and trucks. At present, sales of synthetics are less than 1,000 tons, but this figure is expected to rise to 3,000–4,000 tons by the mid 1990s, mainly 75W–90 for trucks, as well as cars.

4. Automatic Transmissions

Technology steadily moves ahead in this market, with the introduction of the "thinking" electronic transmission, rather than one based on hydraulics. The GM Dexron specifications have dominated this area for many years. Dexron II E is a filled-for-life specification, which could be formulated with PAOs. So far there is not much evidence of synthetics succeeding in this important outlet, which appears, on the surface, to be "made" for high-viscosity-index synthetics.

5. Brake Fluids

Brake fluids represent one of the major existing outlets for polyalkylene glycols. There are two main brake fluid formulations in use in Western Europe, both based on specifications issued by the U.S. Department of Transportation (DOT). They are DOT 3 and DOT 4. The formulations differ markedly in polyalkylene glycol content, as shown in Table 26.9.

Brake fluids have been continually improved, especially in terms of boiling point, starting from the SAE J 7OR3 specification of the 1960s. However, brake fluid has largely become a filled-for-life fluid and is rarely changed. It is now being widely recognized that brake fluids deteriorate in service after 2–3 years, due to water absorption. In general, the higher the boiling point of the brake fluid, the more hygroscopic the glycols are. Boiling points are still trending slightly upward. Typical boiling points in use in Europe are now 220–250°C; 220°C is the minimum requirement to meet DOT 4.

The DOT 4 fluids are formulated to give greater "in service" stability, especially as regards water absorption. The formulation of DOT 4 blocks the ether linkage and hydroxyl group, which are vulnerable to moisture pickup.

There has been a major swing to DOT 4 fluids in the last 5 years in Europe, but not in the United States. Naturally this has been reflected in lower sales of PAGs to the formulators. This trend will of course continue to work its way through the car "fleet" over the next 5–10 years. At the peak, in the mid 1980s, PAG sales were about 6,000–9,000 tons for consumption in Europe. This figure will decline to about

Table 26.9 Typical West European Brake Fluid Formulations

Raw material	DOT 3 (percentage)	DOT 4 (percentage)
Monoethylene/monopropylene glycol	3–10	None
Mixed glycol ethers	70–80	Balance
Polyalkylene glycols	20–25	5
Mixed glycol ether borates	None	15–45
Additive package	2	2
	100	100

3,000 tons by the early 1990s. In the United States there is currently a trend toward lower-molecular-weight polyglycols.

6. Automotive Air-Conditioning

This market, of course, is much larger in the United States than anywhere else in the world. As a result of the chlorofluorocarbon (CFC) controversy, new refrigerants are being introduced, notably R134A. The compatible synthetics seem to be either polyglycols or polyol esters, with PAGs currently dominant.

Sales in the United States of synthetics for this application will total 2,500 tons in 1991, with 1,000 tons for GM, about the same for Ford and Chrysler combined, and 500 tons for other manufacturers including the Japanese.

7. Automotive Greases

PAO- and ester-based greases are being used for automotive purposes, but the bulk of the market remains with conventional products such as lithium-based greases. However, in Western Europe, concern expressed for the environment, with particular emphasis on buses and trucks with automatic grease applicators, is leading to the adoption of polyol-ester-based greases, using long-chain (i.e, C18 oleic acid) technology. At present this market is in its infancy, having really begun in 1990, but it is expected to grow.

B. The Industrial Markets

In the industrial lubricants field, although the market is dominated by large oil companies, with wide industrial product ranges of sometimes 300–500 grades, it is also practicable to sell directly to some OEMs and end users, as has been done by synthetics suppliers to the refrigeration and air compressor markets.

Synthetics have proved strong contenders in six main areas:

Gas turbines.
Circulation, gear, and bearing oils.
Industrial chain lubricants.
Compressors of all types.
Hydraulic and fire-resistant fluids.
Other industrial products, including greases, food-grade lubricants, and wire rope dressings.

1. Gas Turbines

Gas turbines fall into two main types:

1. The direct use of aero-derived turbines for electricity generation and mechanical drive applications (compressors and pumps).
2. Industrial gas turbines, pioneered by General Electric (GE), but also now developed by ABB, GEC, and Siemens. These are also used for electricity and cogeneration but are now on a much larger scale than the aero engines. For example, while the Rolls-Royce RB211 aero-engine generates 25 MW, the latest industrial GE Frame 9 F produces 210 MW of electricity.

Almost all aero-derived engines operating on oil rigs in the North Sea, or on Russian/Ukranian gas pipelines for pumping, use aero-engine oils, for example, diesters for Rolls-Royce Avons and polyol esters for RB 211s. On the other hand, the large industrial gas turbines have tended to use high-performance mineral turbine oils. Now, however, there are examples of some using synthetics, based on PAO/ester mixes. In the 1990s the market outlook for heavy-duty industrial gas turbines is exceptionally good, as natural-gas-fired combined-cycle power stations of high efficiency and moderate capital cost are built in large numbers, in response to environmental pressures.

Since privatization, in the United Kingdom alone, around 15–25 new power stations of 300–900 MW each have been proposed, mainly using combinations of at least two industrial gas turbines and one steam turbine. Since 1960, General Electric, the market leader, has shipped over 5,000 units of industrial gas turbines.

2. *Circulation, Gear, Bearing, and Chain Lubricants*

This is a very wide field. Every industrial plant or factory contains gearboxes or bearings. Circulation systems are usually confined to large machinery units, such as the bearing lubrication systems encountered in paper and steel mills. Large chains are mainly used in conveyor systems, but the difficult applications are mainly those involving severe heat from ovens etc. Although the PAGs enjoy a broad industrial market, increased usage of synthetic alternatives is expected in future. The PAO and PAO/ester blends are the most likely competitors. It is believed that sales of synthetics into these general gear areas of the industrial market have increased significantly over the past few years, as a result of major sales efforts, especially by Mobil, and total sales in Western Europe could now be around 10,000 tons per annum of all synthetics in this sector.

Circulation Systems. Some circulation systems, such as those feeding lubricant to gearboxes operating paper mill drying cylinders, run at high bulk oil temperatures. Particularly in Scandinavia and the United States, it has been found that PAG lubricants enable the mills to be run faster and at a higher temperature, without deterioration of the lubricant through oxidation. Higher paper output has been achieved. Large paper mill manufacturers are now recommending synthetics for start-up. Polypropylene glycol is the most commonly used synthetic base for this application, but PAOs are also being introduced. In low-temperature conditions, found in cold storage applications and occasionally in steel works and opencast mining, alkyl benzenes and PAOs are being used as circulation lubricants.

Gears. Various methods are used to apply gear lubricants. Application may be by drip feed, splash, spray, or lubricant bath. Large industrial gear sets, using spur or helical gears operating under moderate loads, are usually lubricated by circulating systems. The heavier-viscosity turbine oils are generally suitable for this application. In enclosed systems the temperatures reached may be high enough to necessitate the use of premium oils, with high oxidation resistance. These oils should also possess satisfactory antifoam properties and, if water is present, good antirust and good demulsibility properties. Where higher loadings are encountered in industrial gears, lead, sulfur, and phosphorus compounds have been commonly used, to improve the load-carrying capacity of the lubricant. In the last decade more sophisticated compounds have been developed to replace lead, for health reasons.

Synthetic lubricants of the PAG type are finding wide acceptance as industrial gear lubricants. They are particularly suitable for worm gears, because of their lower coefficient of friction. Their ability to lubricate at high bulk oil temperatures, or under heavy loading conditions, has expanded the number of applications. Polyalphaolefins are also now being used, incorporating a special extreme pressure (EP) addititive package, for oil temperatures in the 50–180°C range. In the United Kingdom David Brown, Highfield, J. H. Fenner, and Moss Gears have approved PAO-based products.

Thickened oil compounds made from PAG bases have now been used successfully for some time in packed-for-life gearboxes. David Brown has been able to mass-manufacture small gearboxes without drain plugs, which can therefore be mounted in any position. A further advantage of this synthetic compound is that the gearboxes can be shipped for use in any climate, without the necessity to specify different lubricants. In the United States, U.S. Electric Motors uses a special PAG-based compound (ISO 220) in filled-for-life gearboxes.

Another useful application for PAGs is worm gears on externally operating coal pulverizers, where previously there was a need to heat the cylinder oil lubricant by steam coils before start-up; using PAG lubricants, these do not now need such elaborate preheating.

Bearings. The high pressures encountered on large calender machines producing rubber or plastics can result in bulk oil temperatures of over 200°C. Up to about 150°C mineral oils are still used, but for higher temperatures the PAGs have been accepted as the solution. Again, PAOs have also been introduced for this application, and even some PAO/ester blends in the United States.

Although the PAGs enjoy a broad industrial market, increased usage of synthetic alternatives is expected in future. Polyaolefins and PAO/ester blends are the most likely competitors. It is believed that sales of synthetics into these sectors of the industrial market have increased significantly over the past few years, as a result of major sales efforts, especially by Mobil, and total sales in Western Europe could now be around 10,000 tons per annum.

3. Industrial Chain Lubricants

Industrial chains are often working under severe conditions of heat, in textile works (stenter chains), in car factories, or in pottery and glass kilns. Such high-temperature conveyor bearings have always been a difficult lubrication problem. Often molybdenum disulfide-carrying products have been used. The PAOs, PBs, and trimellitates are now proving successful for temperatures up to 200°C. Some manufacturers in the United States also market ester-based products for this application, for use up to 280°C. One major supplier estimates a total of 1,000 tons of sales of synthetics into "hot" applications in the United States, such as ovens, glassworks, stenters, and conveyors.

4. Compressors of All Types

This is a complex market, because of the wide range of gases and compressor types in service. Described next are the main gases, types of compressor, the competing synthetic products being sold into the end use, or those uses where only one synthetic has proved satisfactory. Also see Table 26.10. The text highlights the situation and rationale for a particular lubricant usage in a particular compressor.

Refrigeration Compressors. Refrigeration compressor oils must lubricate moving parts effectively, while satisfying requirements not demanded of other lubricants. These oils are expected to give many years of trouble-free service, often without replacement or loss make-up, or other maintenance attention. Lubrication, of course, is the primary function of a refrigerator compressor oil; it must, in addition, have other properties, because

It mixes intimately with the refrigerant used in the system.
It is carried over in small amounts into the refrigerant lines.
It is in direct contact with the motor windings in hermetic units.
It is exposed to both temperature extremes: high temperatures at the compressor discharge valve, and very low temperatures at the expansion valve.

Table 26.10 Main Compressor Classifications and Synthetic Lubricants Being Used—1991

Compressor classification	Types of synthetics used
Air (reciprocating, vane, and screw)	PAOs and PAO/ester blends Diesters and polyol esters (some)
Chemical gases (butadiene and vinyl chloride monomer)	PAGs
Ethylene vinyl acetate	Polybutenes
Liquefied natural gas (on board ship)	PAGs
Low-density polyethylene (LDPE)	PAGs PBs and PB/white oil blends PAO/white oil blends
Natural gas compression and transmission	PAGs
Oxygen	Phosphate esters
Refrigeration	Mainly alkylbenzenes, some PAOs for CFCs PAGs (ammonia compressors) Esters (for R134A/R125B refrigerants)

Traditionally, these compressors have been lubricated with wax-free naphthenic oils, because of the need for low pour points. Wax separation in paraffinic oils at low temperatures can cause clogging of the expansion valve and lines. In recent years there have been strong commercial efforts made in this market with alkylbenzenes. The alkylbenzenes have good miscibilty with fluorocarbon refrigerants such as R22 and R502, and have lower critical solution temperatures, down to between –60 and –100°C. Typical low-temperature applications are the production of antibodies, nitrogen peroxide, liquid oxygen, and soluble coffee. The alkylbenzenes can be used in either reciprocating or rotary compressors, including the modern screw type. Blends of alkylbenzene and naphthenic oils are also being marketed, primarily aimed at balanced lubrication properties.

Mobil has been making strong efforts in this field with PAO-based oils. It is expected that penetration of alkylbenzenes and PAOs into this market will increase. Currently sales in Europe are believed to be about 3,000 tons, either to the initial fill market or for special low-temperature industrial applications. The bulk of the market is probably going to remain with naphthenics in the foreseeable future, but a number of manufacturers are now using alkylbenzenes for initial fill.

Some of the newer refrigerants, such as R134A/R125B introduced as replacements for CFCs, are causing difficulties for PAOs, which are not compatible. It appears so far that esters offer the best solution.

Air and Gas Compressors. This sector of the industrial market covers a wide range of compressor types and the compression of air, oxygen, helium, natural gas, liquified petroleum gas (LPG), ethylene (to make low density polyethylene, LDPE), and ethylene vinyl acetate.

In these compressors, the main problems of lubrication center around the need for high oxidation stability of the lubricant, in order to resist degradation and deposits at increasing discharge temperatures and, on the other hand, the problems of cross-leakage between oil and air (or gas). In the case of oxygen compressors this could lead to an explosion, and in the case of ethylene compressors used to make low-density polyethylene, the need is for a lubricant meeting food and drug regulations, since LDPE products are often in contact with food.

The compressor market is changing technically, with a general trend to higher discharge temperatures on reciprocating compressors and, in addition, a move away from reciprocating, toward vane and screw types. For reciprocating compressors, following investigations into explosions caused by deposits (and subsequent blockage) forming in the discharge pipes, new lubricant specifications, calling for low-carbon-forming oils, have now been widely accepted. The Neurop (POT) oxidation test, a safety rather than a performance test, is one of these. "Heart cut" mineral oils with high viscosity index (VI) and molecules of the same size are being used to satisfy these tests. The mineral oils are satisfactory for discharge temperatures up to 220°C. Synthetics are also being used, including PAGs, PAOs, diesters, and polyol esters. The PAO/ester blends have proved the most widely accepted products in the U.S. market, with a high share of the total business. Ingersoll-Rand has their own formulas based on esters. Additionally, PAGs are being used in Ingersoll-Rand and Sullair compressors.

In oil-flooded screw-type compressors, the oil circulates in the compressor and acts as both lubricant and coolant, coming out with the air stream, through the air receiver and filter pack, to return to the compressor. Residence time is important. Encapsulation of the compressor to reduce noise adds to the operating temperature. Oil changes are now made annually (4,000 h), instead of quarterly (1,000 h). The lubrication of screw compressors is a growth area for synthetics, and diesters are now being widely used, as well as PAO/ester blends. However, VHVI hydrorefined mineral oils are also proving satisfactory.

Looking at the specialized gases, the oxygen compressor manufacturers are seeking ways of reducing the risk of explosions. Some oxygen compressors of the large, separate-cylinder-feed, reciprocating type are tending to use polytetrafluoroethylene (PTFE) coatings and operate "dry," or oil free. Some are using phosphate esters.

Compressors for reinjection of natural gas, or pumping on gas transmission lines, with the pumps usually being driven by gas turbines, sometimes use polyol ester lubricants, if operating on a common oil system; however, many are separate, and use PAGs. Shipboard compressors, maintaining gas pressure on liquified natural gas (LNG)/LPG ships, are concerned with gas dilution of the lubricant and subsequent bearing failure from too low an oil viscosity. The PAGs are being used for these compressors, because of the lower solubility of the gas in the synthetic lubricant.

Ethylene compressors, making LDPE, also represent a significant market for synthetics. These large compressors, chiefly Burckhardt, Ingersoll-Rand, and Nuovo-Pigone, are now using PBs, PAOs and PAGs (sometimes blended with white oils) instead of the previously used white oils and "O" wax. The synthetics are less susceptible to ethylene dilution, and also pass the Food and Drug Administration (FDA) requirements. Apart from air compressors, the largest single end use for synthetics identified has been the LDPE compressors, where, for example, PAG sales worldwide are estimated at 15,000 tons, including 5,000 tons in the United States, 6,000 tons in Western Europe, 2,000 for Asia, and 2,000 elsewhere, including 1,500 tons for the former USSR.

The total demand for compressor lubricants in Western Europe is around 50,000 tons. As an order-of-magnitude estimate, it is believed that total synthetics sales are currently in the range of 15,000 tons. There is now competition from VHVI mineral oils. At least a similar volume is used in the United States, where VHVI competition is not so strong. A volume of 4,000 tons of PAGs is known to be used in chemical, natural gas, and helium compressors. The United States of course has a large highly developed network of natural gas pipelines and gas production.

5. Hydraulic and Fire-Resistant Fluids

Hydraulics represent the largest single end use in the industrial market. Synthetic fluids, however, have only really been successfully sold into the fire-resistant fluid sector. This is the main sales outlet for phosphate esters and a major market for PAGs. Competition comes also from invert emulsions (water-in-oil), particularly in the United Kingdom, and from long-chain polyol esters.

Table 26.11 gives a comparison of the various competing fluids, in terms of performance. In choosing a fluid, one of the key factors is the sealing system. For example, it is not usually practicable to switch from phosphate ester to water/glycol, but it is possible to switch from phosphate ester to, for example, a long-chain polyol ester, such as Quintolubric from Quaker Chemical.

Phosphate esters have obtained a significant market and are particularly strongly entrenched in the steel industry. Phosphate ester sales to the mining industry are for specialized applications, such as continuous miners and hydrokinetic transmissions (fluid couplings), driving coal conveyors, but in Europe, not in the United States. Steam turbine governor control mechanisms are a small but important market, the aim being to ensure shutdown of the turbine in the event of fire. The major problem with the phosphate esters is their toxicity. If carefully handled and used in well-maintained systems they are satisfactory, but if leakage is uncontrolled, they can be hazardous to both operatives and the environment. As a result, phosphate esters have been losing their market share in recent years, to water/glycol-based fluids for new systems and to long-chain polyol esters for existing systems, or for applications where mineral oils are being replaced by fire-resistant (FR) fluids for insurance purposes.

Water/glycol fluids usually contain 35–50% water, 0–30% propylene or ethylene glycol, 30–60% PAG thickeners, and a balanced additive package. Their main markets are die casting (where they have taken over from phosphate esters), some machine tools, the steel industry, underground mining (for hydraulic motors and hydrostatic transmissions), and the automotive industry (for welding and induction hardening equipment).

Water-in-oil emulsions are still widely used by steel plants in the United States and by British Coal, for hydraulic motors and hydrostatic transmissions. However, in the rest of the European Economic Community (EEC) these emulsions are not considered to have sufficient fire resistance, and in France and Germany water/glycols are used in preference.

Table 26.11 Industrial and Fire-Resistant Fluids—Comparison of Properties and Performance Capabilities

Characteristic	Mineral oils	Phosphate ester	Water/glycol solutions	Water-in-oil emulsions	Oil-in-water emulsions	Ester products + additives
Specific gravity	0.85	1.3	0.90	0.93	1.0	0.90
Fire resistance	None	Excellent	Excellent	Fair	Good	Good
Maximum operating temperature	80°C	90°C	65°C	65°C	65°C	100°C
Open fire point	170–250°C	320°C	None	None	None	340°C
Auto ignition temperature	330°C	500–600°C	600°C	400°C	—	460°C
Lubrication properties	Good	Good	Fair	Good	Poor	Excellent

The steel industry and the automotive industry have used these water-in-oil emulsions, but in general they are losing favor to fluids with better fire resistance. The 90/10 (or 95/5) fluids are basically 5–10% mineral oil blends in water. Naturally they are very low cost, and are widely used for nondynamic hydraulic oil uses such as hydraulic rams and pit-props. Considerable work is being carried out to improve these fluids with additives and possibly the incorporation of synthetics. In the United States sales of 95/5 or HWB (high-water-based) fluids were about 3,500 tons (as sold) in 1990.

Long-chain polyol-ester-based fluids have been achieving significant sales in the mining, steel, and car industries, because they combine nontoxicity, ease of interchangeability with phosphate esters, and a reasonable level of fire resistance. As well as in mining and steel plants, new markets developed include electrostatic precipatators and paper-mill hydraulic systems, where there is a fire risk. Food processing plants with a fire risk, such as gas- or LPG-heated chicken farms or dairies, are also now using polyol esters in their equipment.

Summary. In general, water/glycol (HFC) fluids and the more recently developed long-chain polyol esters are increasing, while water-in-oil emulsions and phosphate esters are declining. On a longer view, the water/oil fluids (or HFA fluids), based either on mineral oil or mineral oil/synthetic blends, and both including improved additive packages, could show worthwhile growth.

Annual demand for water/glycol fluids is around 25,000 tons and phosphate esters about 4,500 tons in Western Europe. Polyol ester sales are believed to have reached 2,000 tons per annum. As yet there does not appear to be any evidence of other synthetics being used. The 1990 figures for Western Europe are given in Table 26.12. In the United States, water/glycol sales as FR fluids were 26,000 tons in 1990, with phosphate esters at 4,000 tons and long-chain polyol esters 6,000 tons.

6. *Other Industrial Products Including Greases, Wire Rope Dressings, and Food-Grade Lubricants*

Higher-performance greases, based on synthetics of all types—PAO, PAG, PB, and polyol esters (all chain lengths)—are being widely promoted. Hatco in the United States has a polyol ester grease (HATCO 3000) for use on stenter frames, and other applications with operating temperatures of up to 280°C.

In Europe the biodegradability issue is bringing new greases to the market, and here long-chain polyol esters are being offered for outdoor usage, on off-road equipment and buses. Similarly, for forestry chainsaw oils, long-chain esters are used together with diester-based two-stroke oils for the chainsaw engine.

Polybutene-based products are available for "high tackiness" applications on over-

Table 26.12 Estimated Usage of Hydraulic Fluids (All Types)—Western Europe—1990 (Metric Tons)

Product	ISO classification	Tonnage	Percent
Mineral oils	—	400,000	87
Oil-in-water	HFA	15,000	3
Water-in-oil	HFB	12,000	3
Water-polymer (i.e., polyglycols/glycols)	HFC	26,000	6
Full synthetic fluids (phosphate ester, polyol ester)	HFD	6,500	1
Totals		459,500	100

head conveyors in car plants, where drip-off would be unacceptable. Polybutene-based products are also now in use in wire rope lubricant formulations, traditionally bitumen/asphalt based.

Lubricants based on PAGs, PAO, and PB can be formulated to pass FDA tests. Union Carbide offers a range of fully formulated PAG-based extreme pressure lubricants for food machinery, where accidental contact can occur. All components can be identified in the FDA Regulations 21CFR 178.3570(a). In addition to providing nontoxicity, better lubricity, higher VI, and oxidative and thermal stability, the lower pour points and viscosities have led to energy saving of up to 8%, compared to white oils or mineral oils, in food machinery gearboxes. As discussed later under aluminum rolling, lubricants for machinery rolling foil is a growing outlet for PB-formulated lubricants. The PAO producers are also attempting to enter into the "food-grade" markets with FDA grades.

C. Functional Fluids

1. Definition

Functional fluids may be defined as those industrial products that form an integral part of a particular piece of equipment, without which the equipment will not operate or may suffer damage: heat-transfer fluids, electrical insulating fluids, and fluids for offshore motion compensators are good examples. It is important to differentiate functional fluids from process applications, where the product is actually consumed or converted in the manufacture of the final product, for example, rubber process oils or textile finishes.

Some products are very difficult to classify. For example, drilling fluids—functional or process?—or hydraulic fluids—lubricant or functional fluid? We have tried, in such cases, to follow industry practices.

2. Heat Transfer and Solar Fluids

This is a complex area, and highly specialized. The largest volumes tend to be in major circulation systems at chemical or process plants. This business is usually awarded at the initial-fill stage, with sales thereafter only as top-up or infrequent change. It is important to offer technical support at the design and commissioning stage of these large "one-off" systems.

The fluid used is dependent on the heat-transfer temperature required. About 45% of the overall heat-transfer fluid market is met with mineral oils. The total worldwide market is 50,000 tons per annum, and synthetics represent about 9,000 tons (20% of the total) per annum. Thus liquids represent 65% of the market. The balance are high-temperature vapor-phase systems (25%) and intermediate-temperature systems (10%).

The main synthetic contenders are polyglycols, silicones, and specialized hydrocarbons. Mineral oils have a tendency to foul, and also have a relatively narrow operating range (up to 300°C) and poor thermal stability. PAGs also have some problems of thermal stability and narrow operating range (165–250°C). However, they are in wide use in smaller systems, such as plastic injection molding machines, mobile heat exchangers (for composites), or small chemical reactors. Silicones have high thermal stability and wide operating range, but poor heat-transfer properties. The specialized hydrocarbons, such as hydrogenated polyphenyls (–10 to 345°C), alkyl aromatics (50–300°C), or polyphenyls (75–400°C) are offered to provide the right balance of properties. Monsanto and Dow Chemical are leaders, with the widest range of specialist products on offer. BP, Bayer, Huels, ICI, Texaco, Nippon Steel, and Wibarco (with alkyl aromatics) are suppliers with limited ranges of products. These well-established aromatic compounds, phenyls and diphenyl oxide type materials, are coming under increasing government pressure for control and restriction in the United States. Uniroyal and Ethyl Corporation have been actively promoting PAOs in this market sector, as potential replacements.

Solar fluids for roof-type water heaters are another heat-transfer market, which, although already in wide use in countries like the United States, Italy, Cyprus and Israel, is likely to grow, in an energy-conscious and "green" world. Low-viscosity PAOs (2 cS), with metal deactivators, are finding outlets as solar fluids.

3. Electrical and Insulating Oils

Naphthenic base oils have traditionally been used in transformers and capacitors, for those applications where fire risk was not being considered, that is to say, outdoors. The naphthenics were believed to have higher oxidation stability than paraffinics and, with their lower wax contents, the pour points were suitable, on average, for transformers operating outdoors down to –35°C (pour points of –45°C). In fact, special naphthenics have been refined with pour points down to –60°C, but for applications with potential operating temperatures below –50°C, synthetics have generally been used.

Transformers operating indoors have tended in the past to use fire-resistant insulating fluids, such as polychlorinated biphenyls (PCBs). However, under EC regulations, many end users are now prohibited from using PCBs. The EC regulations are not as strict as those that have led to the almost total replacement of PCBs in all applications in Sweden, the country in which their environmental persistence was first recorded in the late 1960s. The total market for PCBs in Western Europe is estimated to have been about 2,000 tons, of which some 800 tons were in use as dielectric fluids. A further 900 tons was used by the German mining industry, in conjunction with phosphate esters, for hydrokinetic transmission and other fire-resistant hydraulic uses, underground.

The excellent range of properties exhibited by PCBs as dielectrics is illustrated by the number of products that have been introduced to cover the same field adequately. In general, silicone-based products have been found satisfactory for those transformer applications less dependent on electricity-related properties, where the main function is as a heat transfer medium; high-voltage capacitor uses have required a range of other products. Silicones and other specialty-type products have probably gained the biggest share of the PCB replacement market. They are particularly suitable for new equipment, which can be designed around the dielectric. Dow Corning has been marketing a polymethylsiloxane product, and silicone-based products are also offered by Bayer and Rhone Poulenc for transformers.

Other developments in this field have come from GEC in the United Kingdom, whose Micanite and Insulator Co subsidiary has completed work on natural fatty-acid-based esters, which it is marketing under the name of Midel. GEC sees the main market as being for retrofilling existing equipment, where silicone-based products are less suitable. Another ester product has been developed by Rhone Poulenc, in cooperation with the Swedish cable company, Asea Kabel. The fluid is a nonchlorinated ester, benzyl neocaprate. Further possibilities for PCB replacement exist with the polyalphaolefins. Uniroyal Chemicals has introduced a range of high-molecular-weight PAO fluids, which it claims are less costly than silicones and which have been approved by the fire authorities.

The demand for insulating oils of all types, including mineral oils, is estimated at 150,000 tons in Western Europe, covering both the initial-fill and top-up and replacement markets. The expected demand for synthetics of all types is thought to be in the 2,000 ton range.

4. Specialized Functional Fluids

Some highly specialized, but low-volume, areas can be grouped together for ease of reference. These include offshore motion compensators, electrohydraulic controls and governor mechanisms in power stations, and electrostatic precipitators.

Offshore motion compensators are hydraulic mechanisms on floating drilling rigs and drill ships that allow the vessels to move, while the drill string remains attached to the sea bed and the well riser. In general, fire-resistant fluids are used, as "offshore" is designated a hazardous area and hence double protected. Phosphate esters, PAGs, and long-chain polyol esters have been successfully sold into this market.

Similarly, governor control mechanisms in power stations need to effectively shut down the turbines, even if the station is on fire. This is a well-established, long-term market outlet for phosphate esters.

Electrohydraulic controls in power plants, with large (6 tons/2,000 gallons) reservoirs, have proved to be a new outlet for long-chain polyol esters in the 1980s. Similarly, electrostatic precipitators, fitted to reduce the dust or solids content of fluids or gases (such as stack gases in power stations), are beginning to use long-chain polyol esters, obviously again for fire-hazard applications.

D. Metalworking Fluids

Metalworking can be defined as changing the physical dimensions of a piece of metal, referred to as the workpiece. Metalworking operations can be split into two categories, cutting and forming. Cutting is the removal of metal from the workpiece in the form of chips. Cutting operations involve turning, milling, tapping, drilling, and grinding. In forming processes, the workpiece is subjected to enough pressure to cause the metal to flow into a predetermined configuration. Forming operations include hot and cold rolling, drawing, stamping, and coining. In both cutting and forming operations, friction from metal-to-metal contact causes wear and heat generation. Additional heat is generated as metal is stretched and deformed. Excessive heat buildup will result in tool wear and poor surface finish. The purpose of metalworking fluids is to minimize friction by providing good lubricity and to remove heat from the tool and workpiece.

The metalworking business in Western Europe is very large, at about 360,000 tons of total demand. This is broken down into about 200,000 tons of cutting oils, divided approximately 50: 50 into neat oils and soluble oils (as supplied), together with 85,000 tons of rolling oils and the balance in stamping and drawing fluids, quenchants and protectives. The U.S. market totaled about 600,000 tons in 1990, broken down as follows: 53% metal removal fluids, 31% forming fluids, 9% protectives, and 7% metal treatment products.

Pricing is highly competitive, and synthetics have had a mixed reception in recent years in general-purpose cutting oils. However, one major U.S. synthetic base supplier reported a 40% increase in sales between 1985 and 1990 in the cutting fluids sector. On the other hand, in the more specialized markets of steel and aluminium rolling, drawing, stamping, and quenching fluids, synthetics have scored notable successes in the 1980s.

1. Soluble Cutting Oils

In the cutting oil field change is constant, and the current situation is that the main favored products is semisynthetic: that is, it retains a mineral oil content, but contains an array of other products, mainly based around a CI (corrosion inhibitor) pack. Some of the semisynthetic cutting oils contain up to 20 components, and it is very difficult to persuade formulators to confirm what is being used in the final product. In the 1970s polyglycols were introduced as components for cutting fluids, but suffered technical drawbacks from paint stripping and the removal of lubricants from machine slideways. As a result, synthetics received a "poor press" in the cutting oil business. The move to semisynthetics in the 1980s was an attempt to rebalance the situation, by again adding mineral oils to the formulation. Nevertheless, it appears that cutting oils are still a target for the synthetics suppliers, with primarily PAGs trying and gaining success especially for very hard or exotic metals. The PAGs have also been used by formulators as the lubricity base for water-soluble cutting and grinding fluids. They work by taking advantage of the phenomenon of inverse solubility, which means that a material becomes less soluble in water as the solution temperature increases, as it does at the metalworking workface, between tool and piece. The PAG comes out of solution and coats/protects the metal surfaces at the critical time.

In addition to components, Union Carbide also offers fully formulated extreme-pressure metalworking fluids, as well as bases to which can be added fatty acids or phosphate esters. The PAGs are also used as coupling agents in CI packs, which represent about 2% of the final fluid.

Another recent development has been the introduction of long-chain polyol esters as "neosynthetics." These are used at high levels (30–50%) in the emulsion concentrates, to replace the EP additive, sulfur or chlorine. Such formulations are used for machining hard alloys, like silicone aluminum, and for deep hole or gun drilling applications.

Overall, however, volumes of synthetics sold into this sector remain small, and the bulk of the market remains either with traditional mineral oil emulsions or the multicomponent "semisynthetics," which are mainly complex chemicals, rather than the synthetic base fluids under review here.

2. Neat Cutting Oils

There appears to be no evidence that any synthetic bases are being used to manufacture neat or straight cutting oils. This is a very cost sensitive market, so the bulk of demand will remain with mineral oils. PBs or oil-soluble PAGs could be candidates for special formulations.

3. Rolling Oils (Steel, Aluminum, and Copper)

In steel rolling oils the move towards long-chain polyol esters is continuing, and it is expected that they will have a totally dominant share of the steel market in Europe early in the 1990s. Fundamentally these products are based on trimetholyl propane (TMP) or pentaerythritol esters. Earlier formulations were based on emulsifiable oils (70% mineral oil, 25% tallow fats, 5% emulsifiers). The TMP esters were originally developed by Quaker Chemical, but this technology has now been adopted by other specialist suppliers in both Europe and the United States. However, while these esters now have an estimated 90% of the European market, development in the United States has been much slower, and only about 35% of the steel market is currently using long-chain esters.

In cold rolling of aluminum foil, where all products used must be nontoxic, because the foil is used for food wrapping, the main fluid used for the rolling process is a nontoxic gas oil, mainly produced by solvents companies. Also in wide use are isoparaffinic solvents such as Exxon's Norpar. A single large mill can use 3,000 tons of solvents per annum. Rolling is at 200–300 m per minute at 675°C. The base solvent has a viscosity of 1.8 cSt at 40°C and a boiling range of 200–250°C. Load-bearing additives are incorporated, which enable increased reductions to be taken, before breakdown of the lubricants. It is known that polybutene is in use as an additive. because of its ability to depolymerize and evaporate. Similarly there is evidence in the United States that PAGs are used as an additive/rolling aid at a ratio of 5% of the total solvent.

Additionally, during 1988–1991, in foil rolling plants in Europe, PB-formulated machinery lubricants have been increasing their share of the business, they ensure compatibility and nontoxicity in the event of cross-leakage to the foil.

For copper rolling, the annual demand for fluids is about 15,000 tons in Western Europe. Emulsifiable mineral oil/fatty acid blends are used, to avoid staining problems due to water, and metal passivators are also added. Fatty acid esters are replacing straight fatty acids. Polybutenes have been used for hot rolling of copper, but it is not known if the products are yet established.

4. Stamping, Pressing, and Forming

In hot stamping, the primary products used are mixtures of graphite and water, from companies such as Acheson Colloids. The concern of the hot-stamping companies is that if they use any kind of mineral product or chemical product, there will be a "fire-flash" before the product actually serves its purpose, as a carrier of the graphite to the point of contact. Nevertheless, there is still interest in the possibility of using polyglycols or polybutenes for these applications. More technical work is needed before progress will be made, as this is a very conservative industry.

In cold-pressing lubricants for sheet metal, it is believed that the traditional special semifluid pastes are still the dominant products, coming from organizations such as the Houghton Group. BP confirms that it is selling polybutenes for stainless steel pressing, again presumably because of the clean burn-off and nonstaining nature of the PB. The PAGs and PBs are also being used in automatic cold stamping of components, such as spark-plug bodies. The clean burn-off of the products during annealing is a major advantage. Several suppliers have been trying synthetics in recent years for these applications.

In the stamping and forging of aluminum in the United States, the industry has recently been replacing chlorinated paraffins, for environmental reasons. Removal of stamping fluid is of course needed, and here PAGs, which can be water-rinsed, are finding an outlet in preference to products needing washing off with 1,1,1-trichloroethane.

5. Wire and Tube Drawing

In the drawing of wire, the products that were used for many years were based on soaps and fats. Clearly these are now being replaced by more modern and environmentally acceptable products. It has been reported that some 2,500 tons of polyglycols are being sold into the wire-drawing sector in Europe, and this usage for PAGs has also been confirmed in the United States. BP Chemicals also advises that polybutenes are being used in stainless steel wire drawing and also in copper tube drawing. On the other hand, in the production of shaped aluminum extrusions (e.g., for windows or patio doors) it would appear that water–graphite mixtures are still the norm. These are swabbed onto the ram area of the extruder, and not used as a die lubricant.

6. Aluminum Can Stock and Can Drawing

These are specialized markets, where a number of companies, like Mobil, have strong positions. Additionally, specialized formulators like NALCO and FERO in the United States compete, and some aluminum companies make up their own formulations on site.

Nontoxicity and FDA approval are, of course, mandatory requirements. It is known that PAGs are in use in small amounts, as are PBs. Similarly, long-chain polyol esters, emulsified at the consuming plants, are in use for aluminium can stock drawing in the United States.

7. Quenching Fluids

In the quenching fluids sector, attempts by producers of synthetic quenchants to push up their market share continue without abatement. Union Carbide was the first producer in this sector, with its Quenchant A, and subsequently ICI and others have moved into the market. Polymer quenchants are much better than mineral oils in environmental terms, because of the elimination of fire hazards and the need for less protection equipment. They also improve working conditions, due to the elimination of smoke and fumes. Additionally, because they are diluted with water to an average of 15% (range 10–20%), there are lower initial costs and reduced "drag-out." The main products competing in this market are PAGs, polyvinyl alcohols (PVA), polyvinyl pyrrolidone (PVP), and polyacrylates. The two leaders at present are PAGs and polyacrylates, there having been a number of problems with PVA and PVP.

Union Carbide's product range has been extended to Quenchants A, B, E, HT, and RL. These are sold directly to large users such as GM or Boeing, or to specialized commercial heat-treating companies.

In 1985 total U.S. polymer sales were 18,000 tons, in a quenchant market of 73,000 tons (as sold). By 1990, polymer quenchant sales in the United States had reached 33% of the total (25,000 tons out of 75,000 tons), the balance being mineral oils. About half the total polymers are PAG-based, giving sales of 12,000 tons of PAG-based quenchant fluids (as sold) in 1990.

Metalworking Fluids—A Summary

An immense amount of concern has been seen in the metalworking fluids business in recent years, as numerous previously accepted formulations, and additives such as chlorine, have come under threat from environmental regulations. Disposal and biodegradability are other major issues. Thus the industry now has to concern itself with "cradle-to-grave" problems, within which performance on the job is only one criteria.

In Sweden, rapeseed-based natural cutting oils were introduced as early as the mid 1980s by AB Karlshamn. In the 1990s we may expect to see synthetic ester-based cutting oils in use, notwithstanding the initial costs, due to their ready biodegradability. Similarly, oil-soluble PAGs will expand their market share for special applications, such as slideway lubricants for machine tools. High-molecular-weight polybutenes are another contender, being completely nontoxic. Table 26.13 summarizes the products identified in use in Western Europe in 1990.

E. Process Fluids

1. Cable Compounds, Impregnants, and Optical Fiber Gels

The term "cable compounds" needs further definition and emphasis. This term is concerned with those compounds used for the impregnation of paper-insulated electric power cables and for filling interstitial spaces in plastic-insulated telecommunication cables. It includes blends of oils (which may be mineral oil or synthetic), and formulations of such blends with additives to achieve particular characteristics, such as higher vicosity and a nonmigratory performance during service.

Table 26.13 Types of Synthetic Base Stocks Used in Metalworking Products in Western Europe—1990

Category	Types of synthetic used	Comments
Soluble cutting oils	PAGs Long-chain polyol esters	Growing for special applications, such as screw cutting, hard alloys, or deep hole drilling
Neat cutting oils	—	Almost none used
Rolling oils (steel, aluminum, copper)	Long-chain polyol esters PBs	Dominant in steel Small—for aluminum only
Quenching fluids	PAGs	Significant, but competing with polyacrylates, PVP, PVA, etc.
Drawing, warm forging and stamping fluids, wire drawing	PAGs, PBs	Wire drawing is large, other uses small
Aluminum can stock and can drawing	PAGs, PBs Long-chain polyol esters	FDA approval needed for all products

Even in this limited and specialized field there is a wide diversity of compound types. The primary differentiation is between power cables and telephone cables, as follows.

Power cables include:

"Self-contained" hollow-core cable.
"Pipe-type" cable.
"Solid" cable (mass impregnated—nondraining or viscous liquid).

Telephone cables may be "fully filled" plastic-insulated cables, or optical-fiber cables.

It is important to distinguish between the terms "impregnating," as applied to power cables, and "filling," as applied to telephone cables. In a power cable, the compound and the paper together form a composite dielectric, which has greatly superior properties of insulation than could be achieved with either component separately. The compound forms an integral part of the cable insulation, and in the manufacture of the cable every effort is made to ensure that no air or vacuous spaces remain in the dielectric, following the process of impregnation.

In a telephone cable it is the extruded plastic around the conductors that forms the primary insulation. The "filling" compound has a secondary insulating function, but is primarily introduced to prevent the penetration of water into and along the length of the cable, in the event of rupture of the cable oversheath. Pockets of air can be tolerated within the interstitial spaces of the cable core, providing a continuous channel does not exist along a significant length of the cable. Nationally or internationally agreed quality control tests are applied to manufactured cable lengths, to ensure compliance with specification requirements.

Cables of the hollow fiber type require oils of the lowest possible viscosity, compatible with the need for an acceptably high flash or fire point, to minimize the possibility of fire hazard. Naphthenic mineral oils having a carefully controlled degree of aromaticity are still used for this application, but currently alkyl benzenes are more often specified.

Pipe-type cables utilize a fairly low-viscosity oil to transmit the hydraulic pressure within the pipe, but a considerably higher viscosity compound to impregnate the cable cores. Polybutenes have replaced mineral oils to a very great extent in this design.

Solid-type cables were traditionally impregnated with blends of viscous oils, to which refined natural resin (colophonium) had been added. This addition resulted in a very greatly increased viscosity at ambient temperature, without unduly increasing the viscosity at cable impregnating temperatures (commonly between 115 and 135°C). In this way problems resulting from compound drainage in service were reduced, while processing times during manufacture could be kept down, in the interests of economics. Polybutenes of approximately 900–1200 molecular weight have been widely used in the past two decades for the same purpose; their disadvantage of high viscosity has been compensated for by improved cable manufacturing techniques. However, since the 1960s, by far the greater proportion of solid cables have been impregnated with MIND (mass–impregnated nondraining) compounds. These can be based either on mineral oils or on polybutenes, to which are added suitable waxes, which result in a high melting point.

In the telecommunications markets, metallic conductor cables use compounds made from either mineral oils/waxes, polybutenes/waxes, mineral oils, or block copolymers. For optical-fiber cables, the tube or slot filling formulations are based on low-viscosity mineral oils/polybutenes or high-molecular-weight PAOs, gelled with micronized silica. For filling interstitial spaces, wax/mineral oil or wax/polybutene blends are used.

Cable compounds for power and telephone usage are usually formulated by specialists such as Dussek-Campbell (Castrol), BP Chemicals, Exxon Chemicals, Witco, or Schumann. The market for traditional compounds is around 20,000 tons per annum in Western Europe, and that for optical-fiber gels around 5,000 tons worldwide in 1990. As regards impregnants and cable oils, the sales of PBs in Western Europe now total about 10,000–15,000 tons into these electrical and cable applications, and those of alkyl benzenes around 7,000 tons. Thus these sectors are of great importance to synthetic base stock producers. They are the largest single outlet identified for alkyl benzenes.

2. Oil-Based Drilling Fluids

Due to high temperatures and pressures, there are significant advantages to the use of oil-based drilling muds, including good lubricity coefficients, easier drilling through salt, potash or gypsum, reduced drill-point torque and drag, plus corrosion protection of the drill pipe. In general, faster drilling, longer bit life, and higher working temperatures are possible. Although oil-based muds use less additives and are reusable, there have been concerns about harm to marine life if muds are disposed of at sea. Therefore the shrimp test or Krangen-Krangen test is used in the North Sea to assess the effects of mud oils on sea life.

The range of possible viscosities used on drilling fluids is quite wide, and varies from light mineral process oils through nontoxic gas oils to odorless kerosine. There may just be a chance that completely nontoxic synthetic fluids could find a niche in this market, despite their higher costs. In the future this is likely also to depend on their biodegradability, thus bringing esters into contention.

At its height the North Sea was using about 50,000 tons of oil bases for drilling muds annually. This figure halved in the period after the 1986 crude oil price collapse, but has now stabilized and is growing again.

3. Circuit Board Fluxes

A new end use for synthetics has been developed in printed circuit board fluxes, where clean burn-off is an essential requirement. The PAGs dominate this market. Potential sales are around, 5,000 tons per annum worldwide, with of course strong demand on the Pacific Rim.

4. Mold Release Agents

This is another hard-to-define market. Mineral oil emulsions have been used as concrete mold release agents for many years. The end use identified here is for mold release in the plastics and rubber industries, where clean lift-off and noncompatibility with the rubber or plastic are essential characteristics. The PAGs are the most popular product on a price/performance basis, with silicones also competing in this sector, although at much higher prices.

5. Textile Fiber Lubricants

Synthetic fibers are polymers of very high molecular weight. The nature of polymers is such that they have very little inherent lubricity. In every stage of their manufacture, from polymerization to production of finished consumer articles, it is necessary to provide lubrication to aid the control of friction. This lubrication is provided as a "spin finish" for primary fiber production. Depending on the end use, this finish may be removed and secondary textile mill finishes provided for subsequent processing and even for point-of-sale handling.

The demands of the textile industry for processing aids to control the fiber friction at different stages of processing are infinite. This has led to a supporting industry providing the necessary expertise, often integrating chemical manufacture, textile evaluation, and blending facilities. For simplicity the requirements can be classified into two main groups, spin finishes and textile mill finishes, as follows.

Spin finishes:

Provide the correct balance of fiber-to-fiber and fiber-to-guide friction.
Provide static control during polymer spinning and subsequent processing.
Seek to provide a product that has optimal thermal stability and volatility characteristics.
Give rapid fiber surface wet out.
Have good anticorrosive properties.
Have good stability at high and low temperatures.
Have minimal effect on the fastness of dyes and pigments.
Are easy to remove in washing-off baths.

Textile mill finishes:

Provide correct frictional and static control as with spin finishes.
Provide effective control for a full range of natural and synthetic fibers and blends.
Provide efficient performance in a range of textile operations.
Are compatible with other processing aids both within the system and with spin finishes.
Can be easily removed and have no adverse effect on the dyeing or dye fastness ratings.
Do not leave deposits that are difficult to remove, or interfere with machine performance.
Are stable over a wide range of temperatures and have an extended shelf life.

Originally, white mineral oils were widely used for this application, however, PAGs have now taken over from white oils almost completely, due to their controllability (via molecular weight) and water solubility. The PAGs are used mainly in nylon and polyester finishes. Probably 90% of PAG consumption is in this area, rather than in wool or natural fibers. There are no figures available on actual PAG consumption within the many finish formulations.

Worldwide, including the former CPE countries, about 175,000 tons of spin and textile mill finishes was estimated to have been used in 1990. In the United States around 40,000 tons was used. Western Europe consumed around 30,000 tons and the Pacific Rim about 25,000 tons in 1990. A large number of major companies (i.e., Akzo, BASF, CEC, Croda, Henkel, Hoechst, ICI, Montedison, Shell, and Unichema) are involved in this business, together with small specialist formulators.

6. Food-Grade Process Oils

For many years white mineral oils have been used for direct food contact applications, a high-volume example being dough-knife lubricants in bakeries. A large bakery can use 50–100 tons per annum of white oil; similarly, for accidental or occasional food contact applications, white oils have been used as mentioned earlier, for LPDE compressors.

In 1990 in the United Kingdom there was a temporary ban imposed (subsequently rescinded) on the use of white oils in direct food applications, as a result of tests that found the white oils were retained in animal tissues. This issue is not yet resolved. Concern also revolves around naphthenic compounds in some hydrogenated white oils and their interaction with bread bleaches, to form polychlorobiphenyls.

If the white oil ban is subsequently imposed, then a new market for synthetics may open up. However, low-cost competition from natural vegetable oil esters, such as rapeseed oil, is likely to secure the bulk of the direct food application market, rather than synthetics.

As an indicator of volumes, worldwide white oil consumption in 1990, for all applications (including cosmetics, personal care products, agricultural spray oils, pharmaceuticals, paper making, or fiber batching), totaled around 800,000–900,000 tons. The direct food applications include bread making, rice polishing, "glossing" of fruit and bread, and, it is believed, dedusting sprays used in grain conveyors in the United States, where white oils are currently used.

F. Aviation Markets

1. Civil

The bulk of civilian aviation lubricant demand is for gas turbines in jet aircraft, followed by aviation piston engine oils. Development costs for jet lubricants are extremely high, as are the costs of flight tests. Effectively, only Castrol, Esso, Mobil, Nyco, and Shell remain in the aviation gas turbine business, even BP having withdrawn in 1980, as others did in the 1970s.

The first generation of gas turbine lubricants, known as Type I oils, were of 3 cSt viscosity, mainly based on diesters, principally azealates and sebacates. One other earlier type of lubricant (the Type I, 7.5-cSt oil) was thickened with a polyalkylene glycol; this product was used primarily in turboprops, to combat the heavier loadings. The civil turboprop aircraft that used Type I oils are now mainly phased out, except for a few Viscounts, plus of course HS748s and Fokker F27s, used on regional routes; in Western Europe most aircraft now use Type II or Type III oils, which are all based on polyol esters and are of 5 cSt viscosity. Table 26.14 shows the various lubricant specifications. Diester-based type I oils are still in use, at a level of about 1,500 tons per annum in

Table 26.14 Categories of Aviation Gas Turbine Lubricants, Together with Main Accepted Specifications (United States, United Kingdom, and France)

Type	Viscosity (100°C), cSt	Base	Specifications
I	3	Diester[a]	Mil-L-7808 C to F (USA)[b] Air 3513 (F)[b]
I	7.5	Diester	DEngRD 2487 Issue 4 Amend 1 (UK) Air 3517 (F)
II	3	Polyol ester	Mil-L-7808J (USA)[b] Air 3514A (F)
II	5	Polyol ester	Mil-L-23699C/D (USA)
III	4	Polyol ester	Mil-L-7808(?)—1990 USAF
III	5	Polyol ester	DEngRD 2497 (UK) XAS-2354 (USA)

[a]Mixed diester/polyol esters sometimes used in U.S. engines.
[b]Original specification (C to H) now superseded by Mil-L-7808J and higher serials.

Europe, for both civil aircraft as mentioned and for aircraft retained by the military (such as the VC 10 transport). Additionally, a good proportion of diesters is used in jet engines for generating, gas pumping, or industrial use, such as the Rolls-Royce Avon. Almost no diesters are in use now in the United States.

Esso and Mobil are the market leaders in jet lubricants, so over 70% of the gas turbine lubricants used in Western Europe are actually imported from the United States in finished form. Exxon has its major blending plant at Bayway, NJ, and Shell, Nyco, and Castrol blend in Europe. The leading base ester suppliers in the United States are Hatco and Mobil, followed by Emery (now Henkel), while in Europe, Unichema, Ciba-Geigy, and Nyco are the main producers of esters approved for aviation usage. Nyco is a major supplier to the French Air Force, and also a significant exporter of synthetics to the former USSR.

In the piston engine aviation market, which is of course dominated by engine lubricants for small, light aircraft, high-quality single-grade mineral oils have been the preferred choice for many years. However, Shell has now introduced a PAO/mineral oil multigrade (Aeroshell Oil W 15W–50), and it is understood that this is enjoying increasing popularity, with sales of over 1,500 tons per anum (that is, 30% of the total demand of 5,000 tons) in the United States in 1990.

In the field of aircraft hydraulics, civilian aircraft are using specially developed types of fire-resistant phosphate esters in their hydraulic systems. Monsanto and Chevron of the United States are the two suppliers to this market sector on a global basis. The requirement is for phosphate esters capable of operating at very low temperatures. Monsanto makes alkyl-aryl esters "in-house" at Bridgeport, NJ, and possibly in St. Louis, MO. Chevron offers mixed triaryl/trialkyl esters. Total demand for phosphate esters is believed to be about 3,000–4,000 tons. Military aircraft tend not to use fire-resistant hydraulic fluids, but more recently have been switching from naphthenic mineral oils to PAO/ester blends, following losses in action in Vietnam, through fires from hydraulic leakages.

The overall demand for "aviation" hydraulic fluids worldwide is estimated at about 15,000 tons, including the former CPE countries. About half this demand is mineral oil, and it should also not be overlooked that "aviation quality" or "superclean" and military specification fluids are demanded for ground equipment, and even in earthmoving equipment, by nonmilitary organizations.

Aviation greases are of course made to the same stringent standards as other products for the aircraft industry, and have to withstand the same extremes of temperature. A number of companies are offering greases using PAOs or esters as bases, and these are proving to be increasingly successful in this demanding market.

Despite its intrinsic appeal, the aviation business does not utilize large volumes of lubricants. The lubricants in gas turbines are rarely changed, but are simply topped up at the end of long flights, so that there is a continual renewal of the charge. Additionally, one company in the United Kingdom has Rolls-Royce approval to recondition gas turbine lubricants. A small but steady business exists in recycling and reconstituting used gas turbine lubricants and returning them to their owners, all over the world.

It is estimated that the total consumption for both military and civilian aviation is in the order of 70,000 tons per annum, broken down as follows: 45% for gas turbine oils, 25% for piston engines, 25% for hydraulic fluids, and the balance of 5% for greases, compounds and miscellaneous uses.

2. Military Uses

Practically all military lubricant applications are covered by specifications, issued by the United States (Mil-L etc.), the United Kingdom (D Eng), France (AIR), or the former USSR, with NATO codes also available as a cross-reference.

Obviously, many synthetic products, such as gas turbine lubricants, are common to both civilian and military uses. However, there are differences in specifications and usage, in that, for example, the U.S. Air Force prefers to use a 3-cSt polyol ester, while most civilian aircraft and the U.S. Navy prefer 5 cSt. There are now signs of change here, and a new 4-cSt polyol ester specification is being introduced soon. So far, Castrol (with Castrol 4000) is the only approved supplier. The U.S. Army has a tank, the M1 Abrams, which is the only one in the world with a gas turbine engine. This has a requirement for a polyol ester lubricant meeting the Mil-L-23699D specification.

A further longstanding but small U.S. Army requirement has been for an Arctic engine oil (specification Mil-L-46167). A product meeting this specification was in great demand by oil companies in the Alaskan oilfields, during their development, but since then sales have dropped sharply. The predominant formulation used was a dialkyl benzene-based lubricant, supplied by Conoco (now sold by Vista Chemical); however, the specification can also be met with a 70% PAO/30% ester formulation.

The PAO-based formulations have also been in use for several years in military hydraulic fluids for aircraft (Mil-H-83282) and ground equipment (Mil-H-46170). Sales of these fluids were around 9,000–10,000 tons in 1991 worldwide. The 83282 specification replaces the naphthenic mineral oil specification (Mil-H-5606) and is known to be used in F14, F16, and F/A18 fighters of the U.S. Air Force, Navy, and Marines. The formulation is a mixed 65% PAO/35% diester, but it is believed that a polyol ester may be introduced soon. The 46170 specification formulation includes rust inhibitors, for tanks and other ground-equipment hydraulics, used when in storage.

Additionally, PAO and esters have been introduced for instrument applications, where the recognized specification is Mil-L-6085A. Avionics uses include dielectric heat-transfer fluids, used in closed systems for radars and ECM systems. A U.S. Air Force specification Mil-C-87252 calls for a 2-cSt PAO for these applications.

A wide range of military greases for aviation has been specified, including those with NATO codes such as G382 based on mineral oils (–40 to 120°C operating range), or silicones and perfluorosilicones (G372 and G398). As far as synthetics are concerned, G354 is a diester-based grease (–75 to 120°C operating), G395 is a nonsoap PAO-based grease (–55 to 175°C opeating), and G363 is a complex ester-based grease that is hydrocarbon resistant.

Finally, mention should be made of the former USSR, which surprisingly has used large volumes of naphthenic mineral oils for aircraft gas turbine lubrication for many years. However, the former USSR also had a polyol ester specification, B-3V for a 5 kinematic viscosity (KV) (100°C) lubricant, and VNII NP-50-1-4-U for a 3.2-KV diester-based lubricant, as well as IPM-10 for a 3-KV diester/synthetic hydrocarbon-blended product.

VII. CONCLUSIONS

The synthetic lubricants and fluids business remains tough, competitive, and slow going, despite having penetrated nearly 60 end uses over the years.

However, synthetic lubricants is no longer a pioneering industry, since the 1980s has been the decade when many end-user customers and OEMs learned to accept synthetics as inherently better in performance than conventional mineral oils. As synthetics know-how, experience, and confidence have grown, market-driven companies such as Castrol, Mobil, Quaker Chemical, Shell, and Union Carbide have become increasingly supportive of the benefits and marketing opportunities of synthetic or semi-synthetic products. This has led to a wider range of new uses and products becoming available.

As far as the military and the OEMs are concerned, more specifications are being issued, like MIL-H-46170, VW500 or Road Ranger from Eaton, which can only be met with synthetics, or unconventional base oils—hence opening up wider opportunities.

Environmental, health, and safety issues are also now playing a major role in boosting the usage of synthetics. Some examples are the growth in demand for esters and rapeseed/ester blends in biodegradable lubricants, the substitution of PAGs for chlorinated paraffins, in the forging and stamping of aluminum, or the increasing use of long-chain polyol esters in plants such as paper converters, where there may be a fire risk. Ease of disposal will also grow as an issue in the lubricants business, as it is in the plastics business today, which may help some synthetics.

The dynamism of the synthetics business is reflected in the many sectors that have been opened up in the last decade or so. Examples are two-stroke engines, auto air-conditioners, aircraft piston engines, military hydraulics and instruments, extensions in the usage of fire-resistant fluids, offshore compensators, solar fluids, electrostatic precipitators, fiber-optic gels, circuit board fluxes, and metalworking applications—especially rolling, drawing, stamping, and pressing operations.

Pricing relationships between the synthetics and between mineral oils, unconventional base oils, and synthetics are now more widely understood, together with product performance advantages or limitations. This is now leading to more careful and accurate formulation choices, especially in those areas where confusion reigned in the early 1980s, such as motor oils. Polyalphaolefins, esters, and their new competition from VHVI base oils are now the preferred blending stocks for advanced formulations, including 5W/10W-based oils with low volatility.

However, despite all these efforts, synlubes remain (except perhaps in European motor oils), a conglomeration of essentially niche markets. As a result, the overall demand for synthetics is not going to grow at the rapid rates once forecast in the early 1980s, unless there is, in particular, a more rapid swing to 5W and the introduction of a NOACK type low volatility test for U.S. automobiles and truck engine oils. Considering that most of the market there is already on at least 10W oils, this is of significant importance for synlubes volumes, as it has been in Europe. The road ahead for synthetic oils remains interesting, but not spectacular. Success will come from hard work and the dedication of efforts to meeting real customers needs and solving technical problems.

Structurally, the suppliers of synthetic base oils are primarily chemical companies, while the bulk of lubricants are sold by oil companies. To succeed, chemical company know-how about end uses and performance needs to match that of the oil companies. The lubricants business has in general been dominated by engineers, not chemists, and the blending of these two disciplines, or their partnership, is the key to success.

REFERENCE

1. Coffin, P. S., C. M. Lindsay, A. J. Mills, H. Lindenkamp, and J. Fuhrman (1990). The application of synthetic fluids to automotive lubricant development: Trends today and tomorrow, *J. Synth. Lubr.*, 7(2), 123–143.

Index